B I O L O G Y *Concepts and Applications*

The biological perspective can sharpen your ability to perceive relationships between seemingly unrelated aspects of the world of life.

Think about a red fox in the shadows cast by a snow-dusted spruce tree. Both organisms face the challenges of cold, dry winters. Spruce trees could not survive without the protective covering of thick-walled, packed-together cells at the surface of their needlelike leaves; and the fox could not survive without its insulative fur coat. Liquid water is not freely available in winter, but the trees manage to conserve it; the only openings through which precious water can escape are not at the leaf surface but rather are sunken below it. The fox forages actively in winter and gets water from its food. Although food supplies dwindle in winter, the fox stays alive by not being finicky about what it eats— small live mammals, large dead ones, arthropods, even plant material are fair game, so to speak. The fox also uses the spruce forest to good advantage for cover and for escape routes.

Unlike the leaves of deciduous species, spruce leaves must function as food factories for more than one season. This means they are more vulnerable to damage from air pollution. And when forests are cleared for agriculture, why do you suppose the fox raids the chickenhouse?

BIOLOGY
Concepts and Applications

CECIE STARR
Belmont, California

General Advisors and Contributors

JOHN ALCOCK
Arizona State University

ROBERT COLWELL
University of Connecticut

EUGENE KOZLOFF
University of Washington

WILLIAM PARSON
University of Washington

SAMUEL SWEET
University of California, Santa Barbara

Wadsworth Publishing Company
Belmont, California
A Division of Wadsworth, Inc.

Biology Editor: Jack C. Carey

Editorial Assistants: Kathryn Shea, Olivia Wirthman

Art Director and Designer: Stephen Rapley

Production Coordinator: Robin Lockwood, Bookman Productions

Production Services Manager: Jerry Holloway

Manufacturing: Randy Hurst

Editorial Production: Carolyn McGovern, copy editor; Martha Ghent, Gloria Joyce, Mimi McCarty, Beverly McMillan, Ed Serdziak

Permissions and Photo Research: Marion Hansen

Marketing: Joy Westberg

Artists: Lewis Calver, Joan Carol, Raychel Ciemma, Robert Demarest, Ron Ervin, Enid Hatton, Darwen Hennings, Vally Hennings, Joel Ito, Illustrious, Inc., Robin Jensen, Keith Kasnot, Julie Leech, Laszlo Meszoly, Leonard Morgan, Victor Royer

Production artists: Susan Breitbard, Renee Deprey, Natalie Hill, Carole Lawson, Judith Levinson, Joan Olson, Jill Turney

Cover design: Stephen Rapley

Compositor: Monotype Company, Inc.

Color Processing: H&S Graphics Inc.

Printing: R. R. Donnelley & Sons Company

Printed in the United States of America
2 3 4 5 6 7 8 9 10 — 95 94 93 92 91

Library of Congress Cataloging-in-Publication Data

Starr, Cecie.
   Biology: concepts and applications / Cecie Starr.
       p.   cm.
   Includes bibliographical references and index.
   ISBN 0-534-13368-1
   1. Biology.   I. Title.
QH307.2.S73   1991
574—dc20   90-19733
                                               CIP

PREFACE

ON CONCEPTS AND APPLICATIONS

The biological perspective is one of the most useful and rewarding tools for interpreting the world around us. Students can use it to understand the past and predict possible futures for ourselves and all other kinds of organisms on this planet; they can use it to pick their way logically through the environmental, medical, and social landmines that are an inevitable part of our lives today.

By definition, having "perspective" means being able to view things in terms of their relative importance or relationships to one another. (We have no perspective, for instance, when we can't see the forest for the trees.) The problem is, biology as it exists today does not automatically lend itself to such wonderful views.

In just the past few decades, biological research expanded explosively in many new directions. Molecular biology, cell biology, genetics, evolution, anatomy and physiology, and ecology have all been ferociously stirred up by numerous discoveries, and the dust has yet to settle. Introductory textbooks often reflect the intellectual uncertainties by giving nearly equal weight to too many details (too many trees). Are some long-standing models and theories outmoded or incorrect? Are models and theories from the cutting edge of research reliable enough to include in a basic text? If textbook authors do not sort through and evaluate the relative importance of such details, how can students be expected to understand what biology is about?

Besides this, how much weight should be given to social, medical, and other human-related issues that might snap students to attention? Pepper the chapters indiscriminately with too many examples of applications, and students might lose sight of the thread of basic biological information. Be too stingy with applications, and the chapters might put students to sleep.

Ideally, authors should (1) consult with research specialists to identify major trends in each field, (2) distill the basic concepts dominating each field, and (3) carefully select applications to show students how those concepts can be used to shed light on their daily lives. *A balance must be struck between accurate core material and applications.*

This book reflects the ideas of many instructors and specialist reviewers who understand the need for this balance. Some of them worked previously with me on one or more editions of *Biology: The Unity and Diversity of Life*, which has been class-tested by more than a million students. As you can see from the list of reviewers on page ix, they include dedicated teachers, our guardians of reading level and depth of coverage; and well-known researchers, our guardians of accuracy. There simply is no way to describe the thoughtful effort each gave to helping create this new book, *Biology: Concepts and Applications*. I can only salute their commitment to quality in education.

HOW THIS BOOK EVOLVED

For several years, many instructors have been asking for a concepts-oriented book that is concise yet balanced in its coverage. Before the writing began, Wadsworth and I consulted with instructors at more than fifty colleges to determine which topics they would like to see included or left out. The consensus among them proved startling, and it strongly influenced our thinking.

Our consultations also revealed broad support for the following objectives:

1. Identify the key concepts and select topics that reflect the research trends in all major fields.

2. Present those concepts and trends in light of two major themes in biology—evolution and energy flow.

3. Use interesting, informative applications to stimulate student interest.

4. Write in a clear, engaging style, without being patronizing or teleological.

5. Don't overload students with too many technical terms.

6. Present the material accurately but not at a high level.

7. Create easy-to-follow line illustrations and select photographs that enhance understanding of concepts.

8. Include enough human biology to enhance understanding of our evolution, behavior, and ecology as well as our body structure and functioning.

9. Include enough comparative biology to convey a sense of the unity and diversity among living things.

10. Include enough examples of problem solving and experiments to provide familiarity with a scientific approach to interpreting the world.

Meeting all those objectives was quite a challenge; yet I had fun writing the book and believe that students will enjoy reading it. The material is streamlined but not insulting of their capacity to learn. Sentences are short, simple, and to the point; examples range from the entertaining to the sobering.

THE ILLUSTRATION PROGRAM

I confess to an eighteen-year obsession with writing and creating illustrations simultaneously, to the everlasting pain of my publisher. It takes almost as much time to research, develop, and integrate art with the text as to write the manuscript itself. The obsession extends to specifying how illustrations and text will be arranged on each book page. With *Biology: Concepts and Applications*, I wanted the page layouts to make maximum pedagogical sense yet still look clean and inviting. At the same time, approximately 250 paintings, 270 line drawings, and 600 photographs had to be shoehorned into a book that would be less than 600 pages long.

Chapters on inherently challenging topics typically start out with an overview illustration to orient students before getting into details. Chapters 5 and 6 (photosynthesis and aerobic respiration) are good examples of this. Similarly, overview illustrations in Chapters 25, 26, 28, and 29 work together to remind students of the functional links among the digestive, circulatory, respiratory, and urinary systems of vertebrates.

To help students comprehend what, exactly, they are looking at, icons (pictorial representations) positioned next to the main art show where the pathways or structures being depicted are located in a cell, multicelled body, or some other system. Zoom-sequence illustrations that move from the macroscopic to the microscopic serve a similar purpose.

Careful use of color makes hard-to-visualize topics more tangible and helps students track the information being presented. See, for example, the illustrations of biochemistry, DNA structure, and protein synthesis (in Chapters 2 through 6, 11, and 12). Notice how proteins are color-coded green, carbohydrates pink, lipids yellow and gold, DNA blue, and RNA orange. Hard-to-interpret micrographs come with simple explanatory diagrams. Full-color anatomical paintings help give students a sense of the splendid internal complexity of organisms, including ourselves.

In many illustrations, visual and written summaries are combined to make concepts easier to grasp. Wherever possible, information is broken down into a series of steps, which are far less threatening than one large, complicated diagram. Students find this approach useful, particularly with regard to the illustrations on mitosis, meiosis, and protein synthesis. It works just as effectively for such complex topics as antibody-mediated immunity, reflex pathways, and neural functioning.

The art program for the book was not developed in a vacuum, of course. It depended on the writings and models developed by respected researchers; it depended on working closely with outstanding photographers and artists. The names of these individuals appear in the Credits and Acknowledgments section at the book's end.

STUDY AIDS

Each chapter begins with a list of key concepts. These are not hastily scribbled marketing gimmicks; the choices truly reflect the conceptual thrust of the book. Also, summary statements of concepts within the text itself help keep readers on track; these are identified by bright blue lines above and below.

Several end-of-chapter study aids further reinforce key concepts. Each chapter concludes with a *summary* in list form, *review questions*, a *self-quiz*, a list of *selected key terms*, and *recommended readings*. Each question and key term has a page number tying it to the relevant text page. *Endpapers* at the back of the book provide a handy index to all applications.

Besides the concepts orientation and the pedagogically effective art program just described, there are numerous *genetics problems* to give students a better grasp of the principles of inheritance. The *glossary* includes all of the text's boldface definitions. It has pronunciation guides and origins of words, when such information will make seemingly formidable words less so. The *index* is comprehensive, because students may find a door to the text more quickly through finer divisions of topics.

The first of four appendices has *metric-English conversion charts*. The second is a brief *classification scheme* that students can use for reference purposes; the third, *detailed solutions* to the genetics problems; and the fourth, *answers* to self-quizzes.

SUPPLEMENTS

Nineteen supplements enhance the effectiveness of this book. *Full-color transparencies* and *35mm slides* of many illustrations from the book are all labeled with large, boldface type. A *Test Items* booklet has 2,500 questions by several outstanding test writers. The questions also are available in electronic form on IBM, Apple IIe, and Macintosh.

An *Instructor's Resource Manual* has the following selections for each chapter: a chapter outline; chapter objectives; a list of the boldface and italic terms; a detailed lecture outline, correlated with the transparencies; suggestions for lecture presentations; suggestions for classroom and laboratory demonstrations; suggested discussion questions; research paper topics; and annotations for filmstrips and videos.

Lecture outlines as presented in the *Instructor's Resource Manual* are also available on a data disk for those who wish to modify the material. In addition, an *Instructor's Manual for Videodisks* correlates the textbook with available videodisks.

A *Study Guide* has learning aids organized by chapter section. This allows students to focus on smaller amounts of material and skip over unassigned sections. It has a detailed summary, a list of key terms, learning objectives, and a self-quiz for each section in the chapter. The *chapter objectives* are available on disk as part of the testing file for those who wish to modify or select portions of the material. An *electronics study guide* consists of multiple-choice questions different from those in the test-item booklet. No matter which answer the students pick, they get specific feedback on why their answer is correct or incorrect. A 100-page *Answer Booklet* has detailed answers to all of the book's end-of-chapter review questions.

A version of *Stella II,* a software tool for developing critical thinking skills, is available to users of the book. Stella II helps students understand the hows and whys of biological processes, ranging from aerobic respiration to ecosystem dynamics and evolution. It provides a set of simple building blocks that can be used to piece together the relationships inherent in a biological system. Students control the learning process as they incorporate assumptions into their model; then, through simulation and animation, they discover the dynamic implications of their assumptions. Instructors can also lead the class in the construction or simulation of a model,

providing students with immediate feedback based on their assumptions.

A set of about 400 *flashcards* with 800 items is available. For each chapter, one card lists basic concepts, another lists learning objectives, and additional cards cover all boldface terms, with pronunciation guides.

Four anthologies are available. *Contemporary Readings in Biology* has articles on additional applications of interest to students. *Science and the Human Spirit: Contexts for Writing and Learning* helps students learn to write effectively about biology. *Ethical Issues in the New Reproductive Technologies* has thought-provoking discussions of some major issues of our time. *The Game of Science* gives students a realistic view of what science is and what scientists do.

A *Laboratory Manual* contains thirty-three experiments and exercises. Many are divided into distinct parts that can be assigned individually, depending on the time available. All have a similar format, with objectives, discussion (introduction, background, and relevance), list of materials for each part of an experiment, procedural steps, pre-lab questions, and post-lab questions. An *Instructor's Manual* accompanies the laboratory manual. It covers quantities, procedures for preparing reagents, time requirements for each portion of the exercise, hints to make the lab a success, and vendors of materials with item numbers.

A COMMUNITY EFFORT

Jack Carey, Wadsworth's biology editor, does things in a big way. Over the years, he has asked well over 1,500 instructors and research specialists to share their experiences and understandings with us through reviews. And reviews. And reviews. It is safe to say that not many authors have had the benefit of so much firsthand knowledge and so many different ways of looking at the same topic. For example, although no fewer than thirty specialists helped me create a chart that correlates earth and life history, no one was completely sure about how to represent the changing range of diversity over time, and finally a noted authority on the subject, John Sepkoski, Jr., took time out to noodle a sketch. The final version appears on page 198. Just as many reviewers checked out other original illustrations, including the carbon cycle on page 534.

After all the detailed reviews, correspondence, and phone calls, many of us have nearly become an extended family, united in a shared commitment to education. The book benefits from the writings of John Alcock and Robert Colwell on animal behavior and ecology, respectively. Eugene Kozloff did not even get ruffled when he got requests at odd hours for, say, a *really* good example of comparative embryology or *really* accurate references on nautiloid shell structure through the ages. George Cox graciously endured pesterings on fine points. Richard Falk, an experienced microscopist, kindly calculated magnifications for all micrographs. Richard (Greg) Rose and John Jackson meticulously monitored and advised me on manuscript development. Through detailed reviews of all draft manuscripts, Virginia Buckner, Gina Erickson, Virginia Latta, Monica Macklin, Carroll Rawn, John Rickett, Gerald Summers, and Sandra Winicur helped monitor content and level. Ronald Hodgson, Brian Myres, Ivan Palmblad, Barbara Pleasants, and Michael Smiles made valuable suggestions when we launched the project.

I commend Robin Lockwood for enduring the excruciating production demands with intelligence and grace; and Hal Lockwood, Richard Lynch, Renee Deprey, and Judith Levinson for doing the same. I commend Jerry Holloway for being so unflappably competent; Stephen Rapley for coming up with such a beautiful book design and cover; Marion Hansen for tracking down the very best photographs; Kathy Shea for being so efficient and invariably pleasant; and Randy Hurst for riding herd on manufacturers.

Thanks go to Carolyn McGovern for her meticulous copy editing; to Martha Ghent, Gloria Joyce, Mimi McCarty, Beverly McMillan, Ed Serdziak for their fine editorial work; and to our captive student readers, Kristen Carey and Ryan Carey. More thanks go to Lewis Calver, Joan Carol, Raychel Ciemma, Robert Demarest, Ron Ervin, Enid Hatton, Darwin and Vally Hennings, Joel Ito, Keith Kasnot, Robin Jensen, Julie Leech, Laszlo Meszoly, Leonard Morgan, and Victor Royer for their magnificent artwork; and to production artists Susan Breitbard, Natalie Hill, Carole Lawson, Joan Olson, and Jill Turney.

But most of all I thank you, Jack, for surviving the strength of my convictions.

CECIE STARR
October 1990

REVIEWERS

Peter Armstrong, *University of California, Davis*
Aimée Bakken, *University of Washington*
Jack Bradbury, *University of California, San Diego*
B. R. Brinkley, *University of Alabama, Birmingham*
Mark Brinson, *East Carolina University*
Virginia Buckner, *Johnson County Community College*
Harold Burton, *State University of New York, Buffalo*
Clyde Calvin, *Portland State University*
A. Kent Christensen, *University of Michigan Medical School*
Robert Colwell, *University of Connecticut*
George Cox, *San Diego State University*
Jerry Coyne, *University of Chicago*
David Gale Davis, *University of Alabama*
Fred Delcomyn, *University of Illinois, Urbana*
Katherine Denniston, *Towson State University*
Patrick Doyle, *Middle Tennessee State University*
Gordon Edlin, *University of Hawaii, Manoa*
Gina Erickson, *Highline Community College*
Paul Ewald, *Amherst College*
Richard Falk, *University of California, Davis*
David Flesch, *Mansfield University*
Joseph Fondacaro, *Marion Merrell Dow, Inc.*
Christine Foyer, *I.N.R.A. Laboratories du Metabolisme, Versailles, France*
Jeffrey Froehlich, *University of New Mexico*
Fred Funk, *Northern Arizona University*
Saul Genuth, *Case Western Reserve University*
Jeffrey Gidday, *Washington University School of Medicine*
W. M. Hess, *Brigham Young University*
Walter Hewitson, *Bridgewater State College*
Ronald Hodgson, *Central Michigan University*
John Jackson, *North Hennepin Community College*
Leonard Johnson, *University of Tennessee College of Medicine*
Patricia Jones, *Stanford University*
Florence Juillerat, *Indiana University–Purdue University, Indianapolis*
Bryce Kendrick, *University of Waterloo*
John Kimball, *Tufts University*
Eugene Kozloff, *University of Washington*
Howard Kutchai, *University of Virginia Medical School*
William Lassiter, *University of North Carolina, Chapel Hill*
Virginia Latta, *Jefferson State Junior College*
George Lefevre, *California State University, Northridge*
Robert Little, *Medical College of Georgia*
Monica Macklin, *Northeastern State University*
Michael Madigan, *Southern Illinois University, Carbondale*
Lynn Margulis, *University of Massachusetts, Amherst*
Eleanor Marr, *Dutchess Community College*
Joyce Maxwell, *California State University, Northridge*

J. R. McClintic, *California State University, Fresno*
F. M. Anne McNabb, *Virginia Polytechnic Institute and State University*
Thomas Mertens, *Ball State University*
Elizabeth Moore-Landecker, *Glassboro State College*
Mary Ellen Morbeck, *University of Arizona*
William Morrison, *Shippensburg University*
David Morton, *Frostburg State College*
Richard Murphy, *University of Virginia*
Brian Myres, *Cypress College*
Arnold Nemerofsky, *State University of New York, New Paltz*
David Norris, *University of Colorado*
Ansa Ojanatra, *University of Turku, Finland*
Merle Olson, *University of Texas Health Science Center*
Ivan Palmblad, *Utah State University*
William Parson, *University of Washington*
Thomas Parsons, *University of Toronto*
Jan Pechenik, *Tufts University*
Barbara Pleasants, *Iowa State University*
Frank Powell, *University of California, San Diego*
Carroll Rawn, *Seton Hall University*
John Rickett, *University of Arkansas, Little Rock*
Richard G. Rose, *West Valley College*
Cleon Ross, *Colorado State University*
Thomas Rost, *University of California, Davis*
Frank Salisbury, *Utah State University*
William Schlesinger, *Duke University*
Jurgen Schnermann, *University of Michigan School of Medicine*
Prem Sehgal, *East Carolina University*
Roger Sharp, *University of Nebraska, Omaha*
David Shoemaker, *Emory University School of Medicine*
Marilyn Shopper, *Johnson County Community College*
Roger Sloboda, *Dartmouth College*
Michael Smiles, *State University of New York, Farmingdale*
Ralph Smith, *University of California, Berkeley*
Janet Stein-Taylor, *University of Illinois, Chicago*
Gerald Summers, *University of Missouri, Columbia*
Samuel Sweet, *University of California, Santa Barbara*
Ian Tizard, *Texas A&M University*
James Trammell, Jr., *Arapahoe Community College*
Heinz Valtin, *Dartmouth Medical School*
Margaret Warner, *Indiana University, Krannert Institute of Cardiology*
Mark Weiss, *Wayne State University*
George Welkie, *Utah State University*
Mark Wheelis, *University of California, Davis*
Brian Whipp, *University of California, Los Angeles, Center for the Health Sciences*
Sandra Winicur, *Indiana University, South Bend*

CONTENTS IN BRIEF

INTRODUCTION

1   Methods and Organizing Concepts in Biology   2

UNIT I / THE CELLULAR BASIS OF LIFE

2   Chemical Foundations for Cells   16
3   Cell Structure and Function   37
4   Ground Rules of Metabolism   58
5   Energy-Acquiring Pathways   68
6   Energy-Releasing Pathways   80

UNIT II / PRINCIPLES OF INHERITANCE

7   Cell Division and Mitosis   94
8   Meiosis   104
9   Observable Patterns of Inheritance   117
10 Chromosome Variations and Human Genetics   128
11 DNA Structure and Function   143
12 From DNA to Proteins   150
13 Recombinant DNA and Genetic Engineering   163

UNIT III / PRINCIPLES OF EVOLUTION

14 Microevolution   172
15 Macroevolution   189
16 Human Evolution: A Case Study   212

UNIT IV / EVOLUTION AND DIVERSITY

17 Viruses, Monerans, and Protistans   222
18 Fungi and Plants   239
19 Animals   258

UNIT V / PLANT STRUCTURE AND FUNCTION

20 Plant Tissues   292
21 Plant Nutrition and Transport   306
22 Plant Reproduction and Development   316

UNIT VI / ANIMAL STRUCTURE AND FUNCTION

23 Animal Tissues, Organ Systems, and Homeostasis   336
24 Protection, Support, and Movement   346
25 Digestion and Human Nutrition   360
26 Circulation   374
27 Immunity   390
28 Respiration   406
29 Water-Solute Balance   418
30 Neural Control and the Senses   426
31 Endocrine Control   449
32 Reproduction and Development   462

UNIT VII / ECOLOGY AND BEHAVIOR

33 Population Ecology   492
34 Community Interactions   506
35 Ecosystems   523
36 The Biosphere   542
37 Human Impact on the Biosphere   569
38 Animal Behavior   586

Appendix I   Units of Measure
Appendix II   A Classification System
Appendix III Answers to Genetics Problems
Appendix IV   Answers to Self-Quizzes
Credits and Acknowledgments
Glossary of Biological Terms
Index

DETAILED TABLE OF CONTENTS

INTRODUCTION

1 **METHODS AND ORGANIZING CONCEPTS IN BIOLOGY**   **2**

KEY CONCEPTS   2
Shared Characteristics of Life   3
DNA and Biological Organization   3
Interdependency Among Organisms   4
Metabolism   5
Homeostasis   5
Reproduction   6
Mutation and Adapting to Change   6
Life's Diversity   7
Five Kingdoms, Millions of Species   7
An Evolutionary View of Diversity   7
The Nature of Biological Inquiry   10
On Scientific Methods   10
COMMENTARY: *Testing the Hypothesis Through Experiments*   11
Limitations on Science   13
Summary   13

UNIT I / THE CELLULAR BASIS OF LIFE

2 **CHEMICAL FOUNDATIONS FOR CELLS**   **16**

KEY CONCEPTS   16
Organization of Matter   17
The Structure of Atoms   17
Isotopes   18
Bonds Between Atoms   18
COMMENTARY: *Dating Fossils, Tracking Chemicals, and Saving Lives—Some Uses of Radioisotopes*   19
The Nature of Chemical Bonds   21
Ionic Bonding   22
Covalent Bonding   22
Hydrogen Bonding   23
Properties of Water   23
Acids, Bases, and Salts   25
Acids and Bases   25
The pH Scale   25
Dissolved Salts   26
Buffers   26
Carbon Compounds   27
Families of Small Organic Compounds   27
Functional Groups   27
Condensation and Hydrolysis   27
Carbohydrates   28
Lipids   29
Proteins   31
Nucleotides and Nucleic Acids   33
Summary   34

3 **CELL STRUCTURE AND FUNCTION**   **37**

KEY CONCEPTS   37
The Nature of Cells   37
Basic Cell Features   37
Cell Size and Microscopy   38
Cell Membranes   40
Membrane Structure and Function   40
Diffusion   41
Osmosis   42
Available Routes Across Membranes   42
Endocytosis and Exocytosis   44

Prokaryotic Cells—The Bacteria   45
Eukaryotic Cells   45
    Function of Organelles   45
    Typical Components of Eukaryotic Cells   46
    The Nucleus   46
    Cytomembrane System   50
    Mitochondria   52
    Specialized Plant Organelles   52
    The Cytoskeleton   52
    Cell Surface Specializations   55
Summary   55

4   GROUND RULES OF METABOLISM   **58**

    KEY CONCEPTS   58
Energy and Life   58
The Nature of Metabolism   60
    Energy Changes in Metabolic Reactions   60
    Metabolic Pathways   60
Enzymes   61
    Enzyme Structure and Function   62
    Effects of Temperature and pH on Enzymes   63
    Control of Enzyme Activity   63
Cofactors   64
ATP: The Main Energy Carrier   64
    Structure and Function of ATP   64
    The ATP/ADP Cycle   65
Electron Transport Systems   65
Summary   66

5   ENERGY-ACQUIRING PATHWAYS   **68**

    KEY CONCEPTS   68
From Sunlight to Cellular Work: Preview of the Main Pathways   69
Photosynthesis   70
    Simplified Picture of Photosynthesis   70
    Chloroplast Structure and Function   70
Light-Dependent Reactions   71
    Light Absorption   71
    ATP and NADPH Formation   73
Light-Independent Reactions   76
    Calvin-Benson Cycle   76
    C4 Plants   77
Summary   78

6   ENERGY-RELEASING PATHWAYS   **80**

    KEY CONCEPTS   80
ATP-Producing Pathways   80
Aerobic Respiration   80
    Overview of the Reactions   80
    Glycolysis   81
    Krebs Cycle   85
    Electron Transport Phosphorylation   86
Anaerobic Routes   87
    Lactate Fermentation   88
    Alcoholic Fermentation   88
Alternative Energy Sources in the Human Body   88
    COMMENTARY: *Perspective on Life*   90
Summary   90

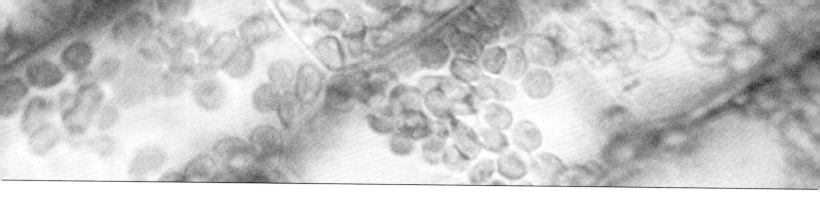

UNIT II / PRINCIPLES OF INHERITANCE

7 CELL DIVISION AND MITOSIS 94

KEY CONCEPTS 94
Dividing Cells: The Bridge Between Generations 94
Overview of Division Mechanisms 95
Some Key Points About Chromosome Structure 95
Mitosis, Meiosis, and the Chromosome Number 96
Mitosis and the Cell Cycle 97
Stages of Mitosis 97
Prophase 100
Metaphase 100
Anaphase 100
Telophase 100
Cytokinesis 101
Summary 102

8 MEIOSIS 104

KEY CONCEPTS 104
On Asexual and Sexual Reproduction 104
Overview of Meiosis 105
Think ''Homologues'' 105
Overview of the Two Divisions 106
Stages of Meiosis 106
Prophase I Activities 106
Separating the Homologues 109
Separating the Sister Chromatids 110
Meiosis and the Life Cycles 110
Gamete Formation 110
More Gene Shufflings at Fertilization 112
Meiosis Compared with Mitosis 113
Summary 113

9 OBSERVABLE PATTERNS OF INHERITANCE 117

KEY CONCEPTS 117
Mendel's Insights into Patterns of Inheritance 117
Mendel's Experimental Approach 118
Some Terms Used in Genetics 118
The Concept of Segregation 119
Testcrosses 121
The Concept of Independent Assortment 122
Variations on Mendel's Themes 122
Dominance Relations 122
Interactions Between Different Gene Pairs 123
Multiple Effects of Single Genes 124
Environmental Effects on Phenotype 124
COMMENTARY: Sickle-Cell Anemia 125
Continuous Variation in Traits 125
Summary 126

10 CHROMOSOME VARIATIONS AND HUMAN GENETICS 128

KEY CONCEPTS 128
Return of the Pea Plant 128
Autosomes and Sex Chromosomes 129
Linkage and Crossing Over 131
Chromosome Variations in Humans 132
Autosomal Recessive Inheritance 132
Autosomal Dominant Inheritance 134
X-Linked Recessive Inheritance 134
Changes in Chromosome Structure 135
Changes in Chromosome Number 137
COMMENTARY: Prospects and Problems in Human Genetics 138
Summary 141

11 DNA STRUCTURE AND FUNCTION 143

KEY CONCEPTS 143
Discovery of DNA Function 143
DNA Structure 145
Components of DNA 145
Patterns of Base Pairing 146

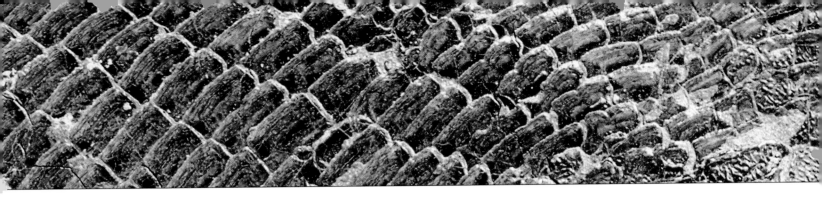

DNA Replication	147
Assembly of Nucleotide Strands	147
Replication Enzymes	147
Organization of DNA in Chromosomes	148
Summary	148

12   FROM DNA TO PROTEINS   **150**

KEY CONCEPTS	150
Protein Synthesis	150
Transcription	151
Classes of RNA	152
Translation	152
Mutation and Protein Synthesis	155
Controls Over Gene Activity	157
Gene Control in Prokaryotes	158
Gene Control in Eukaryotes	158
COMMENTARY: *Cancer—When Gene Controls Break Down*	160
Summary	161

13   RECOMBINANT DNA AND GENETIC ENGINEERING   **163**

KEY CONCEPTS	163
Recombinant DNA Technology	163
Plasmids	163
Producing Restriction Fragments	164
DNA Amplification	165
COMMENTARY: *RFLPs and Genetic Fingerprinting*	166
Expressing a Cloned Gene	166
Mapping the Human Genome	167
Genetic Engineering: Risks and Prospects	167
Genetically Engineered Bacteria	167
Genetically Engineered Plants	168
Genetically Engineered Animals	168
COMMENTARY: *Human Gene Therapy*	169
Summary	169

UNIT III / PRINCIPLES OF EVOLUTION

14   MICROEVOLUTION   **172**

KEY CONCEPTS	172
Emergence of Evolutionary Thought	172
Microevolutionary Processes	177
Variation in Populations	177
Mutation	179
Genetic Drift	180
Gene Flow	180
Natural Selection	180
Evidence of Natural Selection	181
COMMENTARY: *Sickle-Cell Anemia—Lesser of Two Evils?*	182
Stabilizing Selection	182
Directional Selection	184
Disruptive Selection	184
Sexual Selection	184
Speciation	185
Defining the Species	185
Divergence and Isolation	185
Summary	187

15   MACROEVOLUTION   **189**

KEY CONCEPTS	189
Evidence of Macroevolution	189
The Fossil Record	189
Dating Fossils	190
Comparative Morphology	191
Comparative Biochemistry	193
Macroevolution and Earth History	194
Origin of Life	194
Drifting Continents and Changing Seas	197
Extinctions and Adaptive Radiations	201
The Archean and Proterozoic Eras	202
The Paleozoic Era	202
The Mesozoic Era	204
The Cenozoic Era	204
COMMENTARY: *The Dinosaurs—A Tale of Global Impacts, Radiations, and Extinctions*	205
Organizing the Evidence—Phylogenetic Classification	208
Summary	210

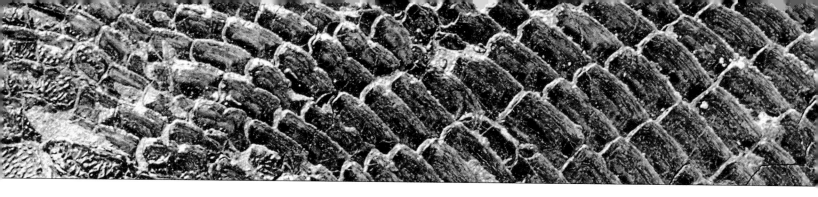

16 HUMAN EVOLUTION: A CASE STUDY   **212**

KEY CONCEPTS   212
The Mammalian Heritage   212
The Primates   213
　Trends in Primate Evolution   213
　Primate Origins   215
The Hominids   216
　Australopiths   217
　Stone Tools and Early *Homo*   218
　Homo erectus   218
　Homo sapiens   218
Summary   219

17 VIRUSES, MONERANS, AND PROTISTANS   **222**

KEY CONCEPTS   222
Viruses   222
Monerans   225
　Characteristics of Bacteria   225
　Major Groups of Bacteria   226
　About the "Simple" Bacteria   229
The Rise of Eukaryotic Cells   229
Protistans   229
　COMMENTARY: *Speculations on the Origin and*
　Evolution of Eukaryotes   230
　Slime Molds   232
　Euglenids   232
　Chrysophytes   233
　Dinoflagellates   234
　Flagellated and Amoeboid Protozoans   234
　Sporozoans   235
　Ciliated Protozoans   236
Summary   237

18 FUNGI AND PLANTS   **239**

KEY CONCEPTS   239
Part I. Kingdom of Fungi   240
　General Characteristics of Fungi   240
　Major Groups of Fungi   240
　COMMENTARY: *A Few Fungi We Would Rather*
　Do Without   242
　Lichens and Other Symbionts   244
Part II. Kingdom of Plants   245
　General Characteristics of Plants   245
　Evolutionary Trends Among Plants   245
　Red, Brown, and Green Algae   246
　Bryophytes   248
　Lycophytes, Horsetails, and Ferns   249
　Existing Seed Plants   250
　Gymnosperms   252
　Angiosperms—The Flowering Plants   254
Summary   256

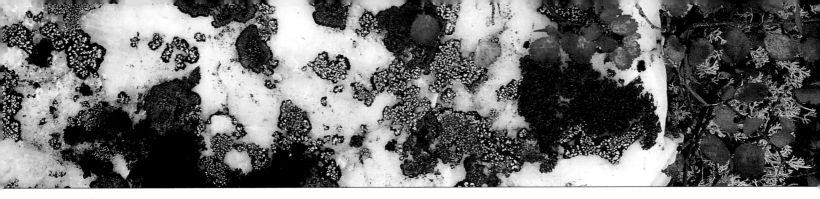

| 19 | ANIMALS | 258 |

KEY CONCEPTS	258
Overview of the Animal Kingdom	259
General Characteristics of Animals	259
Body Plans	259
Sponges	261
Cnidarians	263
Flatworms	265
Turbellarians	265
Flukes	265
Tapeworms	266
Roundworms	266
Rotifers	266
COMMENTARY: *A Rogues' Gallery of Parasitic Worms*	267
Two Main Evolutionary Roads	268
Mollusks	269
Gastropods	270
Bivalves	271
Cephalopods	271
Annelids	272
Arthropods	273
Arthropod Adaptations	273
Chelicerates	274
Crustaceans	274
Insects and Their Relatives	276
Echinoderms	278
Chordates	280
Invertebrate Chordates	281
Evolutionary Trends Among the Vertebrates	281
Fishes	283
Amphibians	285
Reptiles	285
Birds	286
Mammals	288
Summary	289

UNIT V / PLANT STRUCTURE AND FUNCTION

| 20 | PLANT TISSUES | 292 |

KEY CONCEPTS	292
The Plant Body: An Overview	293
Shoot and Root Systems	293
Plant Tissues	293
How Plant Tissues Arise: The Meristems	294
Monocots and Dicots Compared	295
Shoot System	295
Arrangement of Vascular Bundles	295
Arrangement of Leaves and Buds	297
Leaf Structure	297
COMMENTARY: *Uses and Abuses of Leaves*	299
Root System	299
Taproot and Fibrous Root Systems	299
Root Structure	300
Woody Plants	303
Herbaceous and Woody Plants Compared	303
Tissue Formation During Secondary Growth	303
Early and Late Wood	303
Summary	304

| 21 | PLANT NUTRITION AND TRANSPORT | 306 |

KEY CONCEPTS	306
Nutritional Requirements	307
Water Absorption	309
Water Transport and Conservation	310
Transpiration	310
Control of Water Loss	310
Mineral Uptake and Accumulation	312
Transport Through the Phloem	313
Storage and Transport Forms of Organic Compounds	313
Translocation	313
Pressure Flow Theory	314
Summary	314

22 PLANT REPRODUCTION AND DEVELOPMENT 316

KEY CONCEPTS	316
Reproductive Modes	316
Gamete Formation in Flowers	317
Floral Structure	317
Microspores to Pollen Grains	317
Megaspores to Eggs	318
Pollination and Fertilization	318
Pollination	318
Fertilization and Endosperm Formation	318
COMMENTARY: *Coevolution of Flowering Plants and Their Pollinators*	320
Embryonic Development	323
Seed Formation	323
Fruit Formation and Seed Dispersal	323
Patterns of Growth and Development	326
Seed Germination and Early Growth	326
Effects of Plant Hormones	327
Plant Tropisms	328
Biological Clocks and Their Effects	329
COMMENTARY: *From Embryo to the Mature Oak*	331
Summary	331

UNIT VI / ANIMAL STRUCTURE AND FUNCTION

23 ANIMAL TISSUES, ORGAN SYSTEMS, AND HOMEOSTASIS 336

KEY CONCEPTS	336
Overview of Animal Tissues	337
Epithelial Tissue	337
Connective Tissue	337
Muscle Tissue	339
Nervous Tissue	341
Major Organ Systems	341
Homeostasis and Systems Control	342
The Internal Environment	342
Mechanisms of Homeostasis	342
Summary	344

24 PROTECTION, SUPPORT, AND MOVEMENT 346

KEY CONCEPTS	346
Integumentary System	346
Functions of Skin	346
Structure of Skin	346
Skeletal System	348
Functions of Bones	348
Characteristics of Bone	349
Skeletal Structure	350
COMMENTARY: *On Runner's Knee*	352
Joints	352
Muscular System	353
Comparison of Muscle Tissues	353
Skeletal Muscle Contraction	353
Control of Contraction	355
Skeletal-Muscular Interactions	356
COMMENTARY: *Athletes and Anabolic Steroids*	357
Summary	358

25 DIGESTION AND HUMAN NUTRITION   **360**

KEY CONCEPTS   360
Digestive System   361
    Human Digestive System: An Overview   362
    Into the Mouth, Down the Tube   363
    The Stomach   363
    The Small Intestine   364
    The Large Intestine   366
Human Nutritional Requirements   366
    Energy Needs and Body Weight   366
    Carbohydrates   367
    Lipids   367
    COMMENTARY: *Extreme Eating Disorders—Anorexia Nervosa and Bulimia*   368
    Proteins   368
    Vitamins and Minerals   370
Nutrition and Metabolism   371
Summary   372

26 CIRCULATION   **374**

KEY CONCEPTS   374
Circulatory System: An Overview   374
Characteristics of Blood   376
    Functions of Blood   376
    Blood Volume and Composition   376
Human Circulatory System   378
    Blood Circulation Routes   378
    The Human Heart   379
    COMMENTARY: *On Cardiovascular Disorders*   381
    Blood Pressure in the Vascular System   384
    Hemostasis   385
    Blood Typing   386
Lymphatic System   387
    Lymph Vascular System   387
    Lymphoid Organs   388
Summary   388

27 IMMUNITY   **390**

KEY CONCEPTS   390
Nonspecific Defense Responses   391
    Barriers to Invasion   391
    Phagocytes   391
    Complement System   391
    Inflammation   391
Specific Defense Responses: The Immune System   393
    Overview of the Defenders   393
    Recognition of Self and Nonself   393
    Primary Immune Response   394
    Antibody Diversity and the Clonal Selection Theory   396
    COMMENTARY: *Cancer and the Immune System*   396
    Secondary Immune Response   398
Immunization   398
Abnormal or Deficient Immune Responses   399
    Allergies   399
    Autoimmune Disorders   399
    Deficient Immune Responses   399
    COMMENTARY: *AIDS—The Immune System Compromised*   400
    Case Study: The Silent, Unseen Struggles   402
Summary   403

28 RESPIRATION   **406**

KEY CONCEPTS   406
Respiratory Systems   406
    Specialized Respiratory Surfaces   406
    Human Respiratory System   410
Air Pressure Changes in the Lungs   412
Gas Exchange and Transport   412
    Gas Exchange in Alveoli   412
    Gas Transport Between Lungs and Tissues   412
Controls Over Respiration   413
    Matching Air Flow to Blood Flow   413
    COMMENTARY: *When the Lungs Break Down*   414
    Hypoxia   416
Summary   416

29 SOLUTE-WATER BALANCE   **418**

KEY CONCEPTS   418
Control of Extracellular Fluid   419
   Water Gains and Losses   419
   Solute Gains and Losses   419
   Urinary System of Mammals   420
   Nephron Structure   420
Urine Formation   421
   Filtration of Blood   421
   Reabsorption of Water and Solutes   422
   COMMENTARY: *On Fish, Frogs, and Kangaroo Rats*   423
   Secretion   424
   Acid-Base Balance   424
   Kidney Failure   424
Summary   424

30 NEURAL CONTROL AND THE SENSES   **426**

KEY CONCEPTS   426
What Nervous Systems Do   426
Neurons   427
   Structure and Function of Neurons   427
   Neural Messages   428
   Synapses   431
Paths of Information Flow   433
Vertebrate Nervous Systems   434
   Peripheral Nervous System   434
   Central Nervous System   436
The Human Brain   437
   The Cerebral Hemispheres   437
   Memory   438
   Emotional States   438
   States of Consciousness   438
   COMMENTARY: *Drug Action on Integration and Control*   439
Sensory Input   441
   Classes of Receptors   441
   Somatic Senses and Pain Perception   441
   Taste and Smell   442
   Hearing   442
   Vision   444
Summary   447

31 ENDOCRINE CONTROL   **449**

KEY CONCEPTS   449
"The Endocrine System"   449
   Discovery of Hormones   449
   Types of Signaling Molecules   450
The Hypothalamus-Pituitary Connection   450
   Posterior Lobe Secretions   451
   Anterior Lobe Secretions   451
Selected Examples of Hormonal Control   454
   Adrenal Glands   455
   Thyroid Gland   455
   Parathyroid Glands   456
   Pancreatic Islets   456
   Pineal Gland   458
Signaling Mechanisms   458
   Steroid Hormone Action   458
   Nonsteroid Hormone Action   458
Summary   460

32 REPRODUCTION AND DEVELOPMENT   **462**

KEY CONCEPTS   462
The Beginning: Reproductive Modes   463
Basic Patterns of Development   464
   Stages in Development   464
   Mechanisms of Development   465
Human Reproductive System   467
   Male Reproductive Organs   467
   Female Reproductive Organs   470
   Menstrual Cycle   470
   Sexual Union   473
From Fertilization to Birth   475
   Fertilization   475
   Implantation   475
   Membranes Around the Embryo   476
   The Placenta   477
   Embryonic and Fetal Development   477
   Birth and Lactation   479
   COMMENTARY: *Mother as Protector, Provider, Potential Threat*   480
Postnatal Development, Aging, and Death   482
Control of Human Fertility   483
   Some Ethical Considerations   483
   COMMENTARY: *Sexually Transmitted Diseases*   485
Summary   488

UNIT VII / ECOLOGY AND BEHAVIOR

33 POPULATION ECOLOGY — 492

KEY CONCEPTS	492
Ecology Defined	493
Population Dynamics	494
Population Size and Patterns of Growth	494
Checks on Population Growth	496
Survivorship Curves	497
Human Population Growth	498
How We Began Sidestepping Controls	499
Present and Future Growth	500
Controlling Population Growth	500
Questions About Zero Population Growth	503
Summary	504

34 COMMUNITY INTERACTIONS — 506

KEY CONCEPTS	506
Characteristics of Communities	506
The Concepts of Niche and Habitat	506
Types of Species Interactions	508
Mutually Beneficial Interactions	509
Competitive Interactions	509
Categories of Competition	509
Competitive Exclusion	509
Consumer-Victim Interactions	510
"Predator" Versus "Parasite"	510
Dynamics of Predator-Prey Interactions	510
Prey Defenses	513
Parasitic Interactions	514
Community Organization, Development, and Diversity	515
Resource Partitioning	515
COMMENTARY: *Hello Lake Victoria, Goodbye Cichlids*	516
Effects of Predation on Competition	516
Species Introductions	516
Succession	518
Patterns of Species Diversity	520
Summary	521

35 ECOSYSTEMS — 523

KEY CONCEPTS	523
Characteristics of Ecosystems	523
Structure of Ecosystems	525
Trophic Levels	525
Food Webs	527
Energy Flow Through Ecosystems	528
Primary Productivity	528
Major Pathways of Energy Flow	528
Ecological Pyramids	529
Biogeochemical Cycles	531
Hydrologic Cycle	533
Carbon Cycle	534
COMMENTARY: *Greenhouse Gases and a Global Warming Trend*	536
Nitrogen Cycle	538
Transfer of Harmful Compounds Through Ecosystems	539
Summary	540

36 THE BIOSPHERE — 542

KEY CONCEPTS	542
Characteristics of the Biosphere	543
Biosphere Defined	543
Global Patterns of Climate	543
The World's Biomes	547
Deserts	550
Dry Shrublands and Woodlands	551
Grasslands	552
Forests	554
Tundra	557
The Water Provinces	558
Lake Ecosystems	558
Marine Ecosystems	560
COMMENTARY: *El Niño and Oscillations in the World's Climates*	566
Summary	567

37 HUMAN IMPACT ON THE BIOSPHERE   **569**

    KEY CONCEPTS   569
Environmental Effects of Human Population Growth   569
Changes in the Atmosphere   570
    Local Air Pollution   570
    Acid Deposition   571
    Damage to the Ozone Layer   573
Changes in the Hydrosphere   574
    Consequences of Large-Scale Irrigation   574
    Maintaining Water Quality   574
Changes on Land   576
    Solid Wastes   576
    Conversion of Marginal Lands for Agriculture   576
    Deforestation   576
    COMMENTARY: *Tropical Forests—Disappearing Biomes?*   578
    Desertification   580
A Question of Energy Inputs   580
    Fossil Fuels   581
    Nuclear Energy   582
    COMMENTARY: *Biological Principles and the Human Imperative*   583
Summary   584

38 ANIMAL BEHAVIOR   **586**

    KEY CONCEPTS   586
Genes, Hormones, and Behavior   587
    Genetic Basis of Behavior   587
    Hormonal Effects on Behavior   587
Instinct and Learning   587
    Instinctive Behavior   587
    Learned Behavior   588
    Imprinting   588
The Adaptive Value of Behavior   589
    Adaptive Feeding Behavior   590
    Anti-Predator Behavior   590
    Adaptive Reproductive Behavior   590
Social Behavior   592
    Social Communication   592
    Costs and Benefits of Social Life   594
    Social Life and Self-Sacrifice   596
    Evolution of Altruism   597
    Human Social Behavior   598
Summary   598

Appendix I   Units of Measure

Appendix II   A Classification System

Appendix III   Answers to Genetics Problems

Appendix IV   Answers to Self-Quizzes

Credits and Acknowledgments

Glossary of Biological Terms

Index

BIOLOGY *Concepts and Applications*

FACING PAGE: *A spectacular view of the sun—the ultimate driving force for life on earth.*

INTRODUCTION

1 METHODS AND ORGANIZING CONCEPTS IN BIOLOGY

KEY CONCEPTS

1. All organisms are alike in these respects: Their structure, organization, and interactions arise from the properties of matter and energy. They obtain and use energy and materials from their environment, and they make controlled responses to changing conditions. They grow and reproduce, with DNA in each organism providing the instructions for traits that are passed on from one generation to the next.

2. Organisms show great diversity in their structure, function, and behavior. The diversity is an outcome of natural selection and other evolutionary processes.

3. The theories of science are based on systematic observations, hypotheses, predictions, and relentless testing. The external world, not internal conviction, is the testing ground for scientific theories.

Buried somewhere in that mass of nerve tissue just above and behind your eyes are memories of first encounters with the living world. Still in residence are sensations of discovering your hands and feet, your family, the change of seasons, the smell of rain-drenched earth and grass. In that brain are traces of early introductions to a great disorganized parade of insects, flowers, frogs, and furred things—mostly living, sometimes dead. There, too, are memories of questions—"*What is life?*" and, inevitably, "*What is death?*" There are memories of answers, some satisfying, others less so.

Observing, asking questions, accumulating answers—in this manner you have acquired a store of knowledge about the world of life. Experience and education have been refining your questions, and no doubt the answers are more difficult to come by. What *is* life? What defines the living state? The answers you get may vary, depending, for example, on whether they come from a physician, a pastor, or a parent of a severely injured girl who is being maintained by mechanical life support systems because her brain no longer functions.

Yet, despite the changing questions, certain characteristics of life remain the same. Leaves still unfurl during the spring rains. Animals are born, they grow, reproduce, and die even as new individuals of their kind replace them. The world of life has not changed in these respects; it is just that your perceptions about them have deepened.

It is scarcely appropriate, then, for a book to claim that it is your introduction to biology—"the study of life"—when you have been studying life ever since awareness of the world began penetrating your brain. The subject is the same familiar world you have already thought about for many years. That is why this book claims only to be biology *revisited*, in ways that may help carry your thoughts about life to deeper, more organized levels of understanding.

Let us return to the question, What is life? The question has no simple answer, for it addresses a story that has been unfolding in countless directions for several billion years! To biologists, "life" is an outcome of its ancient molecular beginnings and its degree of organization. "Life" is a way of capturing and using energy and materials. "Life" is a commitment to programs of growth and development; it is a capacity for reproduction. "Life" is **adaptive**—meaning that organisms adjust to changing environmental conditions, both in the short term and through successive generations. Clearly, these definitions can only hint at the meaning of life. Deeper insight comes with wide-ranging study of life's characteristics.

Throughout this book you will encounter examples of living things—how they are constructed, how they function, where they live, what they do. The examples provide evidence for certain concepts that, when taken together, will give you a sense of what "life" is. The concepts are summarized in this chapter to give perspective on things to come. You may also find it useful to refer to this summary later on, as a way of reinforcing your grasp of details.

SHARED CHARACTERISTICS OF LIFE

DNA and Biological Organization

Picture a croaking frog squatting on a rock. Without even thinking about it, you know that the frog is alive and the rock is not. At a much deeper level, however, the difference between them blurs. They and all other things are composed of the same particles (protons, electrons, and neutrons). The particles have become organized into atoms, according to the same physical laws. At the heart of those laws is something called **energy**—a capacity to make things happen, to do work. Energetic interactions bind atom to atom in predictable

patterns, giving rise to what we call molecules. Energetic interactions among molecules hold a rock together—and they hold a frog together.

It takes a special type of molecule called deoxyribonucleic acid, or **DNA**, to set living things apart from the nonliving world. DNA molecules contain the instructions for assembling each new organism from a few kinds of "lifeless" molecules. By analogy, think of what you can do with a little effort and a pile of just two kinds of ceramic tiles. When you glue the tiles together according to certain blueprints, different patterns of organization emerge (Figure 1.1). Similarly, life emerges from lifeless matter with a DNA "blueprint," some raw materials, and energy.

The structure and organization of nonliving *and* living things arise from the fundamental properties of matter and energy.

The structure and organization *unique* to living things starts with instructions contained in DNA molecules.

Look carefully at Figure 1.2, which outlines the levels of organization in nature. The quality of "life" actually emerges at the level of cells. A **cell** is the basic living unit. This means it has the capacity to maintain itself as an independent unit and to reproduce, given appropriate sources of energy and raw materials. Amoebas and many other single-celled organisms lead independent lives. A **multicelled organism** has specialized cells ar-

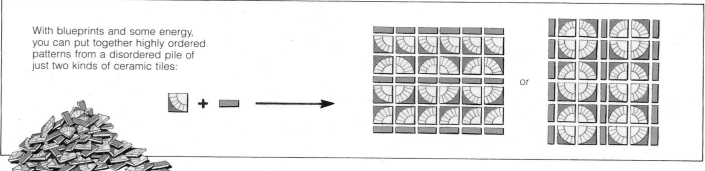

With blueprints and some energy, you can put together highly ordered patterns from a disordered pile of just two kinds of ceramic tiles:

or

Figure 1.1 Emergence of organized patterns from disorganized beginnings. Two ceramic tile patterns are shown here. (Can you visualize other possible patterns using the same two kinds of tiles?) Similarly, the organization characteristic of life emerges from simple pools of building blocks, given energy sources and specific DNA "blueprints."

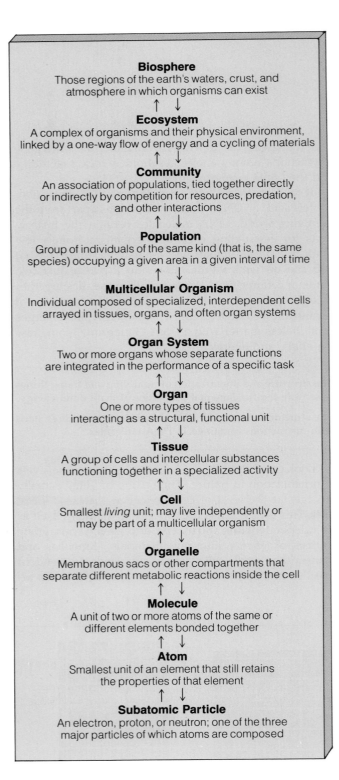

Biosphere
Those regions of the earth's waters, crust, and atmosphere in which organisms can exist

↑ ↓

Ecosystem
A complex of organisms and their physical environment, linked by a one-way flow of energy and a cycling of materials

↑ ↓

Community
An association of populations, tied together directly or indirectly by competition for resources, predation, and other interactions

↑ ↓

Population
Group of individuals of the same kind (that is, the same species) occupying a given area in a given interval of time

↑ ↓

Multicellular Organism
Individual composed of specialized, interdependent cells arrayed in tissues, organs, and often organ systems

↑ ↓

Organ System
Two or more organs whose separate functions are integrated in the performance of a specific task

↑ ↓

Organ
One or more types of tissues interacting as a structural, functional unit

↑ ↓

Tissue
A group of cells and intercellular substances functioning together in a specialized activity

↑ ↓

Cell
Smallest *living* unit; may live independently or may be part of a multicellular organism

↑ ↓

Organelle
Membranous sacs or other compartments that separate different metabolic reactions inside the cell

↑ ↓

Molecule
A unit of two or more atoms of the same or different elements bonded together

↑ ↓

Atom
Smallest unit of an element that still retains the properties of that element

↑ ↓

Subatomic Particle
An electron, proton, or neutron; one of the three major particles of which atoms are composed

Figure 1.2 Levels of organization in nature.

ranged into tissues, organs, and often organ systems. Although each cell depends on the integrated activities of other cells in the body, it still has the capacity for independent existence. (Such cells can be removed from the body and kept alive under controlled conditions.)

At a more inclusive level of organization we find the **population:** a group of single-celled or multicelled organisms of the same kind—that is, of the same *species*—occupying a given area. A flock of penguins is an example. The populations of whales, seals, fishes, and all other organisms living in the same area as the penguins make up a **community.** The next level includes the community *and* its physical and chemical environment; this is the **ecosystem.** The most inclusive level of organization is the **biosphere:** those regions of the earth's waters, crust, and atmosphere in which organisms live.

The increasingly complex levels of biological organization start with energetic interactions at the molecular level and extend through the biosphere.

Interdependency Among Organisms

With few exceptions, a flow of energy from the sun maintains the great pattern of organization in nature. Some single-celled and multicelled organisms (plants, mostly) trap and convert sunlight energy to chemical energy that can be used to build sugars, starch, and other biological molecules from simple raw materials in the environment. This process is called **photosynthesis.** In another process, called **aerobic respiration,** cells release energy stored in molecules—energy that can be used to drive cellular activities.

Think of plants as being food "producers" for the world of life. Animals are "consumers." Directly or indirectly, they feed on energy stored in plant parts. Bacteria and fungi are "decomposers." When they feed on tissues or remains of other organisms, they break down complex molecules to simple raw materials—which can be recycled back to producers (Figure 1.3).

Energy flows to, within, and from single cells and multicelled organisms; it flows within and among populations, communities, and ecosystems. As you will see, interactions among organisms even influence the cycling of carbon and other substances on a global scale, and they influence the earth's energy "budget." Understand the extent of these interactions and you will gain insight into the greenhouse effect, acid rain, and many other modern-day problems.

All organisms are part of webs of organization in nature, in that they depend directly or indirectly on one another for energy and raw materials.

Metabolism

Photosynthesis and aerobic respiration are examples of metabolic activity, which occurs only in living things. **Metabolism** refers to the cell's capacity to (1) extract and transform energy from its surroundings and (2) use energy in ways that ensure its maintenance, growth, and reproduction. In essence, metabolism means "energy transfers." In photosynthesis, for example, energy is transferred from the sun to a molecule called ATP, then ATP transfers energy to molecules that the cell uses as building blocks or tucks away as energy reserves.

Living things show metabolic activity: their cells acquire and use energy to stockpile, tear down, build, and eliminate materials in ways that promote survival and reproduction.

Homeostasis

It is often said that organisms alone "respond" to the environment. Yet a rock also responds to the environment, as when it yields to gravity and tumbles downhill or changes shape slowly under the battering of wind, rain, or tides. The real difference is this: Organisms show *controlled* responses to change.

Life happens to be maintained within narrow limits. Your body, for example, can withstand only so much heat or cold. Your body also must rid itself of harmful substances. Certain foods must be available to it, in certain amounts. Yet temperatures shift, harmful substances may be encountered, and food is sometimes plentiful and sometimes scarce.

You and all other organisms have built-in means of adjusting to change. These internal adjustments help maintain operating conditions within some tolerable

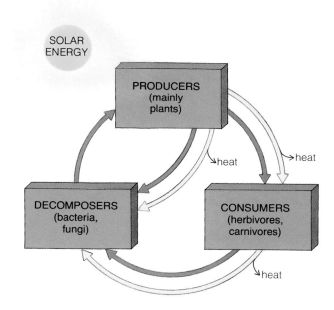

Figure 1.3 Energy flow and the cycling of materials in the biosphere. Here, grasses of the African savanna are producers that provide energy directly for zebras (herbivores) and indirectly for lions and vultures (carnivores). The wastes and remains of all these organisms are energy sources for decomposers, which cycle nutrients back to the producers.

range. The capacity for maintaining the "internal environment" is called **homeostasis.** Single cells and multi-celled organisms both have homeostatic controls. For example, your cells rapidly take up energy-rich sugar when their metabolic activity increases. Birds fluff their feathers and so retain body heat when the outside temperature drops.

Homeostasis implies constancy, a sort of perpetual bouncing back to a limited set of operating conditions. In some respects, constancy is indeed vital. Your red blood cells will not function unless they are bathed in water that contains fairly exact amounts of dissolved components. Your body works so that this fluid is always much the same. Yet organisms also adjust to certain irreversible changes. We might call this *dynamic* homeostasis, for the adjustments actually shift the body's form and function over time.

A simple example will do here. In humans, irreversible chemical changes trigger *puberty*, the age at which sexual reproductive structures mature and become functional. At puberty, the body steps up its secretions of hormones (including testosterone in males and estrogen in females). The increased secretions call for new events such as the menstrual cycle. This cycle includes a rhythmic buildup of substances that prepare the female body for pregnancy, then disposal of substances when pregnancy does not occur. Homeostasis is still operating, but developmental events now demand new kinds of adjustments in the body.

All organisms respond to changing conditions through use of homeostatic controls, which help maintain favorable operating conditions for the body.

Some homeostatic controls keep conditions within some tolerable range throughout the life cycle.

Some homeostatic controls help the body adjust to changes in the internal state as the life cycle unfolds.

Reproduction

We humans tend to think that we enter the world rather abruptly and are destined to leave it the same way. Yet we and all other organisms are more than this. We are part of an immense, ongoing journey that began billions of years ago. Even before birth, the sperm and egg destined to form a new human being are developing according to instructions passed on through countless generations. Each new human body proceeds through many stages of development that prepare it, ultimately, for *reproduction*. With reproduction, the journey of life continues.

What comes to mind when you think of a moth? Do you think simply of a winged insect? What about the

a　　　　　　　　　　　　　　　　b

Figure 1.4 "The insect"—a continuum of developmental stages, with new adaptive properties emerging at each stage. Shown here: the development of a giant moth, from egg (**a**) to larval stage (**b**), to pupal form (**c**), to emergence of the reproductive form of the adult moth (**d,e**).

tiny egg deposited on a branch by a female moth (Figure 1.4)? The egg contains all the instructions necessary to become an adult moth. By those instructions, the egg develops into a caterpillar—a larval form adapted for rapid feeding and growth. The caterpillar eats and increases in size until an internal alarm clock goes off. Then the body enters a pupal stage of wholesale remodeling. Some cells die, and other cells multiply and become organized in different patterns that lead to the adult moth. The adult has organs in which egg or sperm develop. And its wings are brightly colored and move at a frequency that can attract a potential mate. In short, the adult is adapted for reproduction.

None of these stages is "the insect." The insect is a series of organized stages—and the instructions for each stage were written long before the formation of that moth egg.

Each organism arises through *reproduction* (the production of offspring by one or more parents).

Each organism is part of a reproductive continuum that extends back through countless generations.

Mutation and Adapting to Change

The word **inheritance** refers to the transmission, from parents to offspring, of structural and functional patterns characteristic of each species. In living cells, hereditary instructions are encoded in molecules of DNA. Hereditary instructions assure that offspring will resemble their parents, but they also permit *variations* on the basic

c   d   e

plan. (For example, some humans are born with six fingers on each hand instead of five.) The variations arise through **mutations**, which are changes in the structure or number of DNA molecules.

Most mutations are harmful, for the separate bits of information in DNA are part of a coordinated whole. For example, a single mutation in a tiny bit of human DNA may lead to hemophilia or some other genetic disorder.

Yet some mutations may prove to be harmless, even beneficial, under prevailing conditions. One type of mutation in light-colored moths leads to dark-colored offspring. When a dark moth rests on a soot-covered tree, bird predators do not see it. Where most trees are soot-covered, as in industrial regions, light moths are more likely to be seen and eaten—so the dark form has a better chance of surviving and reproducing. Under such conditions, the mutant form is more adaptive.

In all organisms, DNA is the molecule of inheritance: its instructions for reproducing traits are passed on from parents to offspring.

Mutations introduce variations in heritable traits.

Although most mutations are harmful, some give rise to variations in form, function, or behavior that turn out to be adaptive under prevailing conditions.

LIFE'S DIVERSITY

Five Kingdoms, Millions of Species

Until now, we have focused on life's *unity*—on characteristics shared by all organisms. Superimposed on this shared heritage is immense *diversity*. Many millions of different kinds of organisms, or **species**, inhabit the earth. Many millions more existed in the past and became

extinct. Early attempts to make sense of life's diversity led to a classification scheme in which each species was assigned a two-part name. First came the **genus** name (plural, genera), which was given to all the species having perceived similarities to one another. Then came the name of a species within that genus. For instance, *Ursus maritimus* is the scientific name of the polar bear; *Ursus americanus* is the name for the black bear.

Life's diversity is further classified by using more inclusive groupings. For example, similar genera are placed in the same *family*, similar families into the same *order*, then similar orders into the same *class*. Similar classes are placed into a *division* or *phylum* (plural, phyla), which is assigned to a *kingdom*. Today, most biologists recognize the following kingdoms:

Monera	*Bacteria; single cells of relatively little internal complexity. Producers or decomposers.*
Protista	*Protistans; single cells of considerable internal complexity. Producers or consumers.*
Fungi	*Fungi; mostly multicelled. Decomposers.*
Plantae	*Plants; mostly multicelled. Mostly producers.*
Animalia	*Animals; multicelled. Consumers.*

Figure 1.5 shows a few representatives of the five kingdoms of life.

An Evolutionary View of Diversity

Organisms are so much alike in so many ways, what could possibly account for their diversity? In biology, a key explanation is called *evolution by means of natural selection.*

Ever since life originated, each organism has required a steady supply of energy and materials. Yet think about the times when you have encountered shortages (for example, of water, gasoline, electricity, or lettuce). In

Organisms In All Five Kingdoms Show These Characteristics:

1. Complex structural organization based on instructions contained in DNA molecules.

2. Directly or indirectly, dependence on other organisms for energy and material resources.

3. Metabolic activity by the single or multiple cells composing their body.

4. Use of homeostatic controls that maintain favorable operating conditions in the body despite changing conditions in the environment.

5. Reproductive capacity, by which the instructions for heritable traits are passed from parents to offspring.

6. Diversity in their form, in the functions of their various body parts, and in their behavior. Such traits are adaptations to changing conditions in the environment.

7. The capacity to evolve, based ultimately on variations in traits that arise through mutations in their DNA.

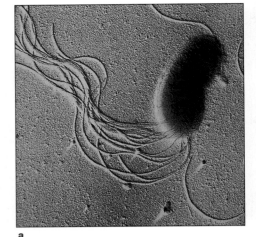
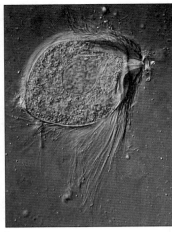

a   b

d

Figure 1.5 Representatives of the five kingdoms of life.

Kingdom Monera. A bacterium (**a**), seen with the aid of a microscope. The single-celled bacteria making up this kingdom live nearly everywhere, including in or on other organisms. The ones in your gut and on your skin outnumber the cells making up your body. Fortunately, animal cells are much larger than bacterial cells are, so you are only a few percent "bacterial" by weight.

Kingdom Protista. A parasitic trichonomad (**b**), discovered in a termite gut. This kingdom of single-celled organisms has poorly defined boundaries; many lineages seem to have evolutionary connections with plants, fungi, and animals.

Kingdom Fungi. A stinkhorn fungus (**c**). Many fungi are major decomposers. Even a single elm tree can shed 400 pounds of leaves in one season. Without decomposers, communities would gradually be buried in their own garbage.

Kingdom Plantae. A grove of California coast redwoods (**d**). Like nearly all members of the plant kingdom, redwoods produce their own food through photosynthesis. Flower of a plant called a composite (**e**). Its intricate pattern guides bees to the flower's nectar. The bees get food and the plant gets pollinated. Many organisms are locked in such mutually beneficial interactions.

Kingdom Animalia. Male bighorn sheep (**f**) competing for females. Like all animals, they cannot produce their own food and so must eat other organisms. Like most animals, they move about far more than organisms of other kingdoms (most of which do not move about at all).

the past, as today, resources were not always abundant. *So the organisms in each population must have been demanding a share of limited resources.*

Imagine, next, that variant forms of traits arose through DNA mutations and proved to be adaptive. ("Traits" are aspects of an organism's body, its functioning or behavior.) Because individuals with the variant traits had some advantage, they tended to survive and reproduce with greater frequency. Through successive generations, the variations became the more common forms. In other words, **evolution** occurred; the character of the population changed over time.

Long ago, the naturalist Charles Darwin used pigeons as an example of how selection can occur within a population. Domesticated pigeons show great variation in their traits (Figure 1.6). Darwin pointed out that

c

e

f

pigeon breeders who wish to promote a certain trait, such as a black tail, will select individual pigeons having the most black in their tail feathers. By permitting only those birds to mate, they will foster the trait and eliminate others from their captive population. Thus Darwin used *artificial* selection as a model for *natural* selection.

In later chapters, we will be looking at the mechanisms and consequences of natural selection. For now, keep in mind these key points:

1. Members of a population vary in form, function, and behavior, and much of this variation is heritable.

2. Some varieties of heritable traits are more adaptive than others; they improve chances of surviving and reproducing.

3. Because members with adaptive traits are more likely to reproduce, their offspring tend to make up more of the reproductive base of new generations. This tendency, called differential reproduction, means that adaptive traits increase in frequency in the population.

4. Natural selection is the name given to this process of differential reproduction among the variant members of a population. It is an evolutionary process; it can change the character of a population over time.

5. As a result of natural selection and other evolutionary processes, variations have accumulated in different lines of organisms. Life's diversity is the sum total of these variations.

Figure 1.6 A few examples of the more than 300 varieties of domesticated pigeons. Such forms have been derived, by selective breeding, from the wild rock dove (**a**).

THE NATURE OF BIOLOGICAL INQUIRY

On Scientific Methods

Evolution is a major phenomenon of nature—that is, a **principle**—and natural selection is one explanation of how it occurs. Awareness of evolution developed over centuries, as naturalists and travelers observed and collected specimens of living and extinct organisms, then asked questions about the remarkable diversity those specimens represented. It became clear that almost all organisms alive today are very different from organisms of the remote past. Eventually the evidence was overwhelming. The differences among living and extinct species had to be a consequence of evolution—of changes in lines of descent that have accumulated since life began. But *how* did those changes occur? As you will see in later chapters, it took a great deal of scientific inquiry to come up with plausible answers.

If we were to idealize the route from a question about such a major aspect of the natural world to a fundamental explanation for it, we might end up with a list like this:

1. Ask a question (or identify a problem).

2. Make one or more **hypotheses,** or educated guesses, about what the answer (or solution) might be. This means using the process of **induction:** sorting through clues, hunches, and observations, then combining bits of information and logic to produce a general statement (the hypothesis).

3. *Predict* what the consequences might be if a hypothesis is valid. This process of reasoning from a general statement to predicting consequences is called **deduction** (and sometimes the "if-then" process).

4. Devise ways to *test* those deductions by making observations, developing models, or performing experiments.

5. Repeat the tests as often as necessary to determine whether results will be consistent and as predicted.

6. Report objectively on the tests and on conclusions drawn from them.

7. Examine alternative hypotheses in the same manner.

This route represents what might be called a scientific approach to interpreting the natural world. Hypotheses are proposed, then deductions are made and tested (see *Commentary*). There is more than one method of doing this. Insights arise from accident and intuition as well as from methodical search. Some individuals adhere to existing procedures, others make up new procedures as they go. No matter what the individual method, how-

Testing the Hypothesis Through Experiments

William H. Leonard, Clemson University

How is it that scientists probe so skillfully into the monument of life and discover so much about its foundations? What is it about their manner of thinking that yields such precise results? Simply put, scientific inquiry routinely depends on systematic observation and test.

Observations can be made directly, through systems of vision, hearing, taste, olfaction, and touch. They can be made indirectly, through use of special equipment (such as a microscope) that extends the range of perception. With practice, we can become skilled at *making systematic observations*. This means focusing one or more senses on a particular object or event in the environment, and screening out the "background noise" of information that probably has no bearing on our focus.

Hypothesizing means putting together a tentative explanation to account for an observation. When a hypothesis is scientific, it is *testable* through experiments. Experiments are devised to test whether predictions that can be derived from the hypothesis are correct. Thus the hypothesis must be constructed so that it provides a framework for stating the results of an experiment. Its content must be more specific than a problem statement, and often it is worded in the negative. Why is this so? Scientists tend to accept tentatively a plausible idea until it is shown to be false. It is difficult to prove experimentally that a hypothesis is true, because its validity would have to be demonstrated for all possible cases and under all possible conditions. Scientists therefore continue to test hypotheses by devising experiments that might show them to be false. If they succeed, then the hypothesis must be modified or discarded. That is why hypotheses are expressed in the negative.

For example, "DDT concentrations of 0.0001 percent by weight in the food of laboratory rats will not have harmful effects on the maintenance of the rat population over five years." If experiments reveal harmful effects at that dosage, then the hypothesis is not correct, and support is given to the idea that DDT is harmful.

Testing the hypothesis through experiments is at the heart of scientific inquiry. The goal is to control all variables except the one under study. Variables are events or conditions subject to change. For example, variables that are common to many biological experiments are the amount of light, temperature, and moisture. Others are concentrations of substances and numbers of organisms (population density) in a defined space. There are three general categories of variables:

independent variables   *the condition or event under study*

dependent variables   *conditions or events that could change because of the presence of, or change in, an independent variable*

controlled variables   *conditions that could affect the outcome of an experiment but that do not, because they are held constant*

An experimenter observes or manipulates one independent variable at a time, to identify any effects it has on dependent variables. If more than one independent variable were studied simultaneously, it would not be clear which one was responsible for the observed experimental results.

In one classic experimental design, a population of organisms is divided into two groups. The experimental group is the one subjected to the independent variable; the control group is not. All other variables are held the same in both groups. Thus, any differences that show up in test results for the two groups can be attributed to the independent variable. The illustration on the next page is an example of the use of experimental and control groups. This experiment has been used to test the hypothesis that laboratory rats ingesting DDT with normal food will lose weight, show less resistance to disease, and have a lower reproductive rate than rats not ingesting DDT. Notice that the rats were randomly assorted into either the experimental group or the control group. *Randomization* ensures that both groups are representative (or equivalent) samples of the original population.

Collecting and organizing test results is a necessary process in biological experiments. Data tables or graphs are used to organize and display information for analysis.

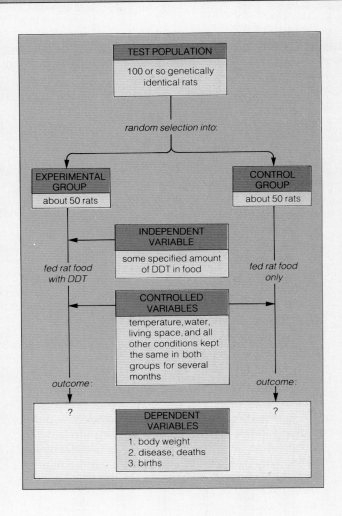

An example of a classic experimental design in biology. The experiment is designed to test the hypothesis that DDT ingested with food will not have harmful effects on laboratory rats over a period of time. With all other variables held constant, test results should refute or support the hypothesis.

Graphs are especially useful in illustrating trends or patterns. Data analysis is less mechanical and more conceptual than collecting and organizing the information. Often, statistical tests are used to determine if differences between experimental data and control data are *significant* or are likely due only to chance. If it can be argued that the differences are due to chance only, then it can also be argued that the independent variable had no effect. For example, say that at the end of an experiment on DDT effects, the average adult rat weight was 187.4 grams in the DDT-fed group, and 206.7 grams in the control group. Is this difference significant enough to suggest that there was an actual effect? The use of mathematical tools characteristic of statistical analysis could help in finding an answer.

Generalizing from test results requires careful and objective analysis of the data gathered. Usually, the hypothesis under test is accepted or rejected on the basis of conclusions drawn. A statement is written about what new insights (if any) have been gained into the original problem. Apparent trends are noted when the same data appear in test results gathered over a period of time. Often, further questions and hypotheses are posed in an attempt to guide additional studies of the problem.

ever, the bottom line in science is this: *Hypotheses must be testable—and other scientists must be able to duplicate or verify the tests.*

No scientist can put forward a hypothesis and demand that we believe it, no questions asked. Even if a test seems to verify it, the hypothesis is valid *only in terms of that test*—and further testing may show it to be invalid. That is why you won't (or shouldn't) hear a scientist say, "There is no other explanation!" More likely you will hear, "Observations and tests made so far have not disproved the hypothesis."

In science, a **theory** is a related set of hypotheses which, taken together, form an explanation about some aspect of the natural world. In modern science, the term theory is not used lightly. It is bestowed only on explanations that can be relied upon with a very high

degree of confidence. Even a theory is not beyond scrutiny, of course, for new observations and test results may lead to its refinement or replacement. Far from being a disaster, this activity stimulates the development of even more adequate explanations.

Obviously, individual scientists would rather have their names associated with useful explanations than with useless ones. But they must be objective and keep asking themselves: "Can my ideas be disproved through observations and tests?" This does not mean all scientists are objective all of the time or even most of the time; no one can lay claim to that. Rather, scientists are expected to put aside their individual pride or biases by testing their ideas, even in ways that might prove them wrong. Even if one scientist doesn't, or won't, *others will*—for science proceeds as a community that is both cooperative and competitive. Ideas are shared, with the understanding that to expose errors is just as important as to applaud insights.

Limitations on Science

The call for objectivity strengthens the theories that emerge from scientific studies. Yet it also puts limits on the kinds of studies that can be carried out. Beyond the realm of what can be analyzed with the methods and technology available, certain events remain unexplained. Why do we exist, for what purpose? Why does any one of us have to die at a particular moment and not another?

Answers to such questions are *subjective*. They come from within, as an outcome of all those experiences and mental connections shaping the consciousness of each individual. Because individuals differ so enormously in this regard, subjective answers do not readily lend themselves to scientific analysis.

This is not to say that subjective answers are without value. No human society can function without a shared commitment to standards for making judgments, however subjective those judgments might be. Moral, aesthetic, economic, and philosophical standards vary from one society to the next. But all guide their members in deciding what is important and good, and what is not. All attempt to give meaning to what we observe and what we do.

Every so often, scientists stir up controversy when they explain part of the world that was previously considered beyond natural explanation, or belonging to the *supernatural*. This is sometimes true when moral codes are interwoven with religious narratives, which grew out of observations by ancestors. Questioning some long-standing view of the world may be misinterpreted as questioning morality, even though the two are not remotely synonymous.

For example, centuries ago Nicolaus Copernicus studied the movements of planets and stated that the earth circles the sun. Today the statement seems obvious. Back then, it was heresy. The prevailing belief was that the Creator had made the earth (and, by extension, humankind) the immovable center of the universe! Not long afterward a respected professor, Galileo Galilei, studied the Copernican model of the solar system. He thought it was a good one and said so. He was forced to retract his statement publicly, on his knees, and to put the earth back as the fixed center of things. (Word has it that when he stood up he muttered, "But it moves nevertheless.")

Today, as then, society has its sets of standards. Today, as then, those standards may be called into question when a new, natural explanation runs counter to a supernatural belief. When this happens it doesn't mean that scientists as a group are less moral, less lawful, less sensitive, or less caring than any other group. It means only that their individual and collective work has been guided by one additional standard: *The external world, not internal conviction, must be the testing ground for scientific beliefs.*

Systematic observations, hypotheses, predictions, tests—in all these ways, science differs from systems of belief that are based on faith, force, authority, or simple consensus.

SUMMARY

1. All organisms are alike in the following characteristics:
   a. Their structure, organization, and interactions arise from the basic properties of matter and energy.
   b. They rely on metabolic and homeostatic processes.
   c. They have the capacity for growth, development, and reproduction.
   d. Their heritable instructions are encoded in DNA.

2. There are millions of different kinds of organisms. Attempts to find meaning in life's diversity led to classification schemes based on perceived similarities and differences in physical traits. Each distinct kind of organism was called a species, distinct species resembling one another more than they resembled other species were grouped into the same genus, and so on with increasingly inclusive groupings into family, order, class, phylum, and kingdom.

3. Diversity among organisms arises through mutations that introduce changes in the DNA. These changes lead to heritable variation in the form, functioning, or behavior of individual offspring.

4. Individuals in a population vary in their heritable traits, and the variations influence their ability to survive and reproduce. Under prevailing conditions, certain varieties of a given trait may be more adaptive than others. They will be "selected" and others eliminated through successive generations. The changing frequencies of different traits changes the character of the population over time; it evolves. These points are central to the principle of evolution by natural selection.

5. There are many scientific methods of gathering information. Scientists typically use these processes:
   a. Hypothesizing about the meaning of some aspect of the world (for example, proposing that the tremendous diversity among organisms, both living and extinct, came about by evolution).
   b. Predicting what the consequences will be if a hypothesis is valid (the "if-then" process).

c. Testing the hypothesis in ways that can be duplicated or verified by others, as through systematic observation and through experimentation.

6. A principle is a major phenomenon of nature, such as evolution. A theory is a related set of hypotheses that together form an explanation of some phenomenon. An example is the theory of how the process of natural selection brings about evolution.

7. Scientific theories are based on systematic observations, hypothesizing, predictions, and tests. The external world, not internal conviction, is the testing ground for those theories.

Review Questions

1. For this and subsequent chapters, make a list of the **boldface** terms that occur in the text. Write a definition next to each, then check it against the one in the text. (You will be using these terms later on.)

2. Why is it difficult to give a simple definition of life? (For this and subsequent chapters, *italic numbers* following review questions indicate the pages on which the answers may be found.) *2*

3. What does *adaptive* mean? Give some examples of environmental conditions to which plants and animals must be adapted in order to stay alive. *2*

4. Study Figure 1.2. Then, on your own, arrange and define the levels of biological organization. What concept ties this organization to the history of life, from the time of origin to the present? *4*

5. In what fundamental ways are all organisms alike? *4*

6. What is metabolic activity? *5*

7. What is DNA? What is a mutation? Why are most mutations likely to be harmful? *7*

8. Outline the one-way flow of energy and the cycling of materials through the biosphere. *5*

9. What does evolution mean? *8*

10. Witnesses in a court of law are asked to "swear to tell the truth, the whole truth, and nothing but the truth." What are some of the problems inherent in the question? Can you think of a better alternative? *10*

11. Design a test to support or refute the following hypothesis: The body fat in rabbits appears yellow in certain mutant individuals—but only when those mutants also eat leafy plants containing a yellow pigment molecule called xanthophyll. *11*

Self-Quiz   (Answers in Appendix IV)

1. The complex patterns of structural organization characteristic of life are based on instructions contained in _____.

2. Directly or indirectly, all living organisms depend on one another for _____.

3. _____ is the ability of organisms to extract and transform energy from the environment and use it during maintenance, growth, and reproduction.

4. _____ means maintaining the body's internal operating conditions within a tolerable range even when environmental conditions change.

5. Diverse structural, functional, and behavioral traits are considered to be _____ to changing conditions in the environment.

6. The capacity to evolve is based on variations in traits, which originally arise through _____.

7. Organisms show _____, the ability to transmit instructions for heritable traits from parents to offspring.

8. Living and nonliving things _____.
    a. have exactly the same kinds of atoms and molecules
    b. share the same basic properties of matter and energy
    c. have the same energy content
    d. all of the above

9. Homeostatic controls _____.
    a. allow organisms to share raw materials
    b. activate the reproduction of organisms
    c. allow organisms to respond to changing conditions
    d. produce important mutations in organisms

10. That each of us has great, great, great, great, grandmothers and grandfathers is an example of a unique property of life known as _____.
    a. metabolism
    b. homeostasis
    c. reproduction
    d. organization

11. A scientific approach to explaining various aspects of the natural world includes all of the following except, _____.
    a. hypothesis
    b. testing
    c. faith and simple consensus
    d. systematic observations

12. A related set of hypotheses that collectively explain some aspect of the natural world is a scientific _____.
    a. prediction
    b. test
    c. theory
    d. authority
    e. observation

Selected Key Terms

adaptive *2*	evolution *8*	mutation *7*
aerobic respiration *4*	Fungi *7*	photosynthesis *4*
Animalia *7*	genus *7*	Plantae *7*
biosphere *4*	homeostasis *6*	population *4*
cell *3*	hypothesis *10*	principle *10*
community *4*	inheritance *6*	Protista *7*
DNA *3*	metabolism *5*	reproduction *6*
ecosystem *4*	Monera *7*	species *7*
energy *3*	multicelled organism *3*	theory *12*

FACING PAGE: *Living cells of a green alga* (Elodea), *with their chemical factories called chloroplasts.*

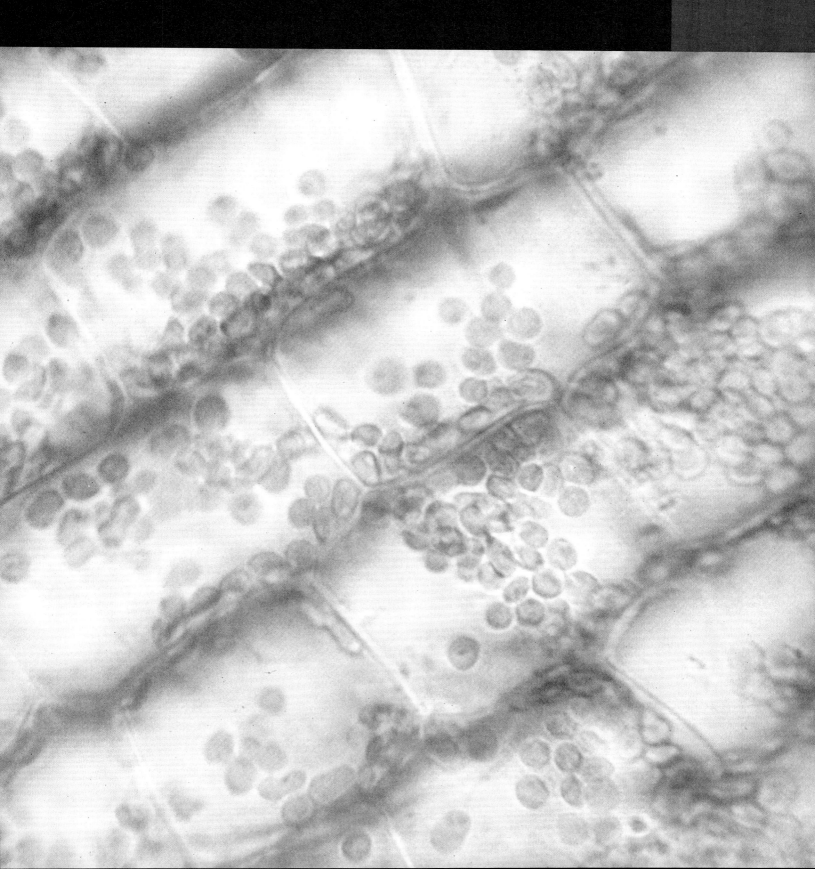

2 CHEMICAL FOUNDATIONS FOR CELLS

KEY CONCEPTS

1. Atoms give up, acquire, or share their electrons with other atoms in specific ways, and these interactions are the basis for the structural organization and activities of all living things.

2. Chemical bonds are unions between the electron structures of different atoms. The most common bonds in biological molecules are ionic bonds, hydrogen bonds, and covalent bonds.

3. Life depends on the properties of water, including its temperature-stabilizing effects, cohesiveness, and capacity to dissolve many substances.

4. Even though cells continuously produce and use hydrogen ions (H$^+$) during chemical reactions, they have the means to maintain the H$^+$ concentration within narrow limits.

As you read this page, thousands of chemical reactions are proceeding inside you in ways that keep your body running smoothly. Whether it is daytime or the middle of the night, streams of sunlight are reaching half of the earth's surface; and countless plants are converting energy contained in the sun's rays into forms that can be used for assembling the carbohydrates and other building blocks of roots, stems, and leaves. Your very life depends on breathing in the oxygen released into the atmosphere during chemical reactions such as these, the first of which occurred over 2.5 billion years ago.

In the past two centuries—a mere blip of evolutionary time—we have managed to discover what chemical substances are made of, how they can be transformed into different substances, and what it takes to accomplish the transformations. Some of the products of these discoveries are synthetic fabrics, fertilizers, vaccines, antibiotics, and the plastic components of refrigerators, computers, television sets, jet planes, and cars.

Much of our chemical "magic" brings us great benefits *and* monumental problems. For example, fertilizer applications and other agricultural practices maintain food supplies for the 5.3 billion people on earth; without them, much of the human population would simply starve to death. But weeds don't understand that the fertilizers are for crop plants, and plant-eating insects don't understand that the crops are for us, not them. Each year they gobble up or ruin about 45 percent of what we grow. In 1945 we began to battle them with

Figure 2.1 Cropduster with its rain of pesticides.

synthetic pesticides, which kill weeds as well as insects, worms, rodents, and other animals that threaten not only our food supplies but also our health and peace of mind, our pets, and ornamental plants. In 1988 alone, Americans managed to spread more than a billion pounds of pesticides through homes, gardens, offices, industries, and farmlands (Figure 2.1).

Among the insect killers are the carbamates, organophosphates (including malathion), and halogenated compounds (including chlordane). Most are neurotoxins; they block vital communication signals between nerve cells. Some remain active for days, others for weeks or years. They kill many of nature's insect eaters, including dragonflies and birds, as well as the targeted pests. Over time, the pests build up resistance to the pesticides, for reasons that will become apparent in later chapters. And only 10 percent of the insecticides being sold today have been assessed for potential health hazards.

Consider that *we* inhale pesticides, ingest them with food, or absorb them through skin. After entering the body, many pesticides can cause headaches, rashes, asthma, and bronchitis in susceptible people, and can increase their vulnerability to chronic infections. Susceptibility depends on genetic makeup, overall health, nutritional habits, and concurrent exposure to other toxic substances. Exposure to certain pesticides can trigger hives, joint pain, and other moderate allergic reactions in about 11 million Americans; they can trigger life-threatening immune reactions in another 5 million.

Maintaining our food supplies, industries, and health depends on chemistry—and so does our chance of reducing harmful side effects of its application. You owe it to yourself and others to gain greater understanding of chemical substances and their interactions. By demystifying the "magic" of chemistry, you will be better equipped to assess the benefits and risks of its application to the world of life.

ORGANIZATION OF MATTER

"Matter" is anything that occupies space and has mass. It includes the solids, liquids, and gases around you and within your body. All forms of matter are made of one or more fundamental substances called elements. About ninety-two elements occur naturally on earth. It takes only four kinds—hydrogen, carbon, nitrogen, and oxygen—to make up most of your body (Figure 2.2).

The Structure of Atoms

Look at some water in a glass, then imagine your eyes probing ever deeper into its underlying structure. First you would discover molecules composed of hydrogen and oxygen. By definition, a **molecule** is a unit of two or more atoms (of the same or different elements) bonded together. In turn, each kind of **atom** is the smallest unit of matter that is unique to a particular element. Below the level of atoms, you always find the same types of particles. The particles are protons, neutrons, and electrons, the universal building blocks of atoms.

Protons and neutrons make up the atom's core region, or nucleus. Protons have a positive charge (p^+); neutrons are electrically neutral. Electrons, which have

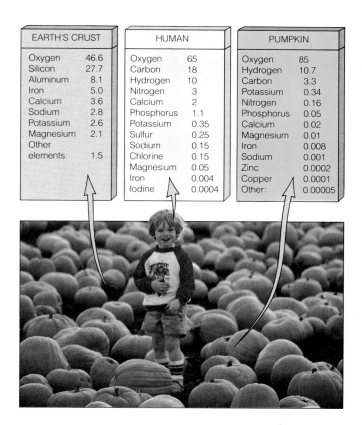

Figure 2.2 Comparison of the proportions of different elements in the earth's crust, the human body, and a pumpkin as percentages of the total weight of each.

Table 2.1	Atomic Number and Mass Number of Elements Common in Living Things		
Element	Symbol	Atomic Number	Most Common Mass Number
Hydrogen	H	1	1
Carbon	C	6	12
Nitrogen	N	7	14
Oxygen	O	8	16
Sodium	Na	11	23
Magnesium	Mg	12	24
Phosphorus	P	15	31
Sulfur	S	16	32
Chlorine	Cl	17	35
Potassium	K	19	39
Calcium	Ca	20	40
Iron	Fe	26	56
Iodine	I	53	127

Figure 2.3 Chemical bookkeeping.

Symbols for elements are used in writing *formulas*, which identify the composition of compounds. (For example, water has the formula H_2O. The subscript indicates two hydrogen atoms are present for every oxygen atom.) Symbols and formulas are used in *chemical equations*: representations of reactions among atoms and molecules.

In written chemical reactions, an arrow means "yields." Substances entering a reaction (*reactants*) are to the left of the arrow. Products of the reaction are to the right. For example, the overall process of photosynthesis is often written this way:

$$6CO_2 \quad + \quad 6H_2O \quad \longrightarrow \quad C_6H_{12}O_6 \quad + \quad 6O_2$$

6 carbons	12 hydrogens	6 carbons	12 oxygens
12 oxygens	6 oxygens	12 hydrogens	
		6 oxygens	

Notice there are as many atoms of each element to the right of the arrow as there are to the left (even though they are combined in different forms). Atoms taking part in chemical reactions may be rearranged but they are never destroyed. The *law of conservation of mass* states that the total mass of all materials entering a reaction equals the total mass of all the products.

When thinking about cellular reactions, keep in mind that no atoms are lost, so the equations you use to represent them must be balanced in this manner.

a negative charge (e^-), are attracted to the nucleus. (They move rapidly around it and actually occupy most of the atom's volume.) An atom has just as many protons as electrons, so it has no *net* charge, overall.

The number of protons in the nucleus, called the **atomic number**, is different for each element. Only the hydrogen atom has one proton; its atomic number is 1. Only the carbon atom has six protons; its atomic number is 6. Table 2.1 lists other examples. The total number of protons *and* neutrons in the nucleus is the **mass number**. For example, the mass number of a carbon atom with six protons and six neutrons is 12. (The relative masses of atoms are also called atomic weights.)

As you will see, knowing an atom's atomic number and mass number tells us something about how it will interact (if at all) with other atoms. Those values give us an idea of whether that atom can give up, acquire, or share electrons with other atoms. *Such electron activity is the basis for the organization of materials and the flow of energy through the living world.*

Isotopes

Although all atoms of an element have the same number of protons, they may vary slightly in how many neutrons they have. Atoms having the same atomic number but a different mass number are **isotopes**. Thus "a carbon atom" might be carbon 12 (containing six protons, six neutrons), carbon 13 (six protons, seven neutrons), or carbon 14 (six protons, eight neutrons). These can be written as ^{12}C, ^{13}C, and ^{14}C. All isotopes of an element have the same number of electrons, so they interact with other atoms in the same way. Accordingly, cells can use any isotope of carbon (or of some other element) for a given metabolic reaction.

You have probably heard of radioactive isotopes, or radioisotopes, which are unstable and tend to break apart (decay) into more stable atoms. The *Commentary* describes some of the ways radioisotopes are used in research, in medicine, and in establishing the age of fossil-containing rocks.

All atoms of an element have the same number of electrons and protons, but they can vary slightly in the number of neutrons. The variant forms are isotopes.

BONDS BETWEEN ATOMS

Let's turn now to the nature of reactions among atoms. In case you are not familiar with such reactions, take a moment to review Figure 2.3, which summarizes a few conventions used in describing them.

Dating Fossils, Tracking Chemicals, and Saving Lives—Some Uses of Radioisotopes

In the winter of 1896, the physicist Henri Becquerel tucked a heavily wrapped rock of uranium into a desk drawer, on top of an unexposed photographic plate. A few days later, he opened the drawer and discovered a faint image of the rock on the plate—apparently caused by energy emitted from the rock. One of his coworkers, Marie Curie, gave the name "radioactivity" to the phenomenon.

As we now know, radioisotopes are unstable atoms, with too many protons or neutrons in the nucleus. The instability causes them to capture or emit electrons or some other particle. This spontaneous process, called radioactive decay, continues until the original isotope has changed to a new, stable isotope, one that is not radioactive.

Radioactive Dating

Each type of radioisotope has a characteristic number of protons and neutrons, and it decays spontaneously at a characteristic rate into a different isotope. The *half-life* is the time it takes for half the nuclei in any given amount of a radioactive element to decay into another element. The half-life cannot be modified by temperature, pressure, chemical reactions, or any other environmental factor. That is why radioactive dating is such a reliable method of determining the age of rock layers in the earth—hence the age of fossils they may contain. To determine the age of a particular rock, we can compare the amount of one of its radioisotopes with the amount of the decay product for that isotope.

For example, ^{40}potassium has a half-life of 1.3 billion years and decays to ^{40}argon, a stable isotope. The age of anything that contains ^{40}potassium can be determined by measuring the ratio of ^{40}argon to ^{40}potassium. In this way, researchers have dated fossils that are millions, even billions, of years old (Figure a). Radioactive dating with ^{238}uranium, which has a half-life of 4.5 billion years, indicates that the earth formed 4.6 billion years ago. The Table lists the useful ranges of the main radioisotopes used in dating methods.

Main Radioisotopes Used in Dating			
Radioisotope (unstable)	Stable Product	Half-Life (years)	Useful Range (years)
^{87}rubidium ⟶	^{87}strontium	49 billion	100 million
^{232}thorium ⟶	^{208}lead	14 billion	200 million
^{238}uranium ⟶	^{206}lead	4.5 billion	100 million
^{40}potassium ⟶	^{40}argon	1.3 billion	100 million
^{235}uranium ⟶	^{207}lead	704 million	100,000
^{14}carbon ⟶	^{14}nitrogen	5,730	0–60,000

a A well-preserved fossilized frond from a tree fern (left), one of many species that lived between 320 million and 250 million years ago. A fossilized sycamore leaf that dropped 50 million years ago (above).

Tracking Chemicals

Emissions from radioisotopes can be detected by a scintillation counter and other devices. This means that isotopes can be used as *tracers*. They can be used to identify the pathways or destination of a substance that has been introduced into a cell, the human body, an ecosystem, or some other "system."

For example, because all isotopes of an element have the same number of electrons, they all interact with other atoms in the same way. Accordingly, cells can use isotopes of carbon for a given metabolic reaction. Carbon happens to be a key building block for photosynthesis. By putting plant cells in a medium enriched in ^{14}carbon, researchers identified the steps by which plants take up carbon and incorporate it into newly forming carbohydrates. Tracers also are helping us increase crop production by providing insights into how plants use synthetic fertilizers and naturally occurring nutrients.

What about medical applications? As one example, the thyroid is the only gland in the body to take up iodine. A tiny amount of the radioisotope ^{123}iodine can be injected into a patient's bloodstream, then the thyroid can be scanned with a scintillation counter. This is called a radioisotope scan. Figure b shows examples of what these scans may reveal.

Saving Lives

In nuclear medicine, radioisotopes are used to diagnose and treat diseases. Patients with irregular heart beats use pacemakers, which are powered by the energy emitted from ^{238}plutonium. (This otherwise dangerous radioisotope is sealed in a case to prevent its emissions from damaging body tissues.) With PET (positron-emission tomography), radioisotopes provide diagnostic information about abnormalities in the metabolic functions of specific tissues. The radioisotopes are attached to glucose or some other biological molecule, then they are injected into a patient, who is moved into a PET scanner. When cells in certain tissues absorb the glucose, the radioisotopes give off energy that can be used to produce a vivid image of the variations in metabolic activity (see Figure c).

Finally, some cancer treatments make use of the fact that radioisotopes can damage or destroy living cells. In radiation therapy, localized cancers are deliberately bombarded with ^{226}radium or ^{60}cobalt.

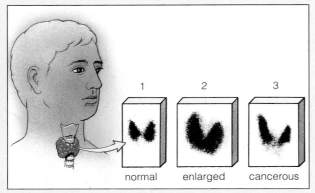

1   2   3

normal   enlarged   cancerous

b   Scans of human thyroid glands after ^{123}iodine was injected into the bloodstream. The thyroid normally takes up iodine (including radioisotopes) and uses it in hormone production. (1) Uptake by a normal gland. (2) Enlarged gland of a patient with a thyroid disorder. (3) Cancerous thyroid gland.

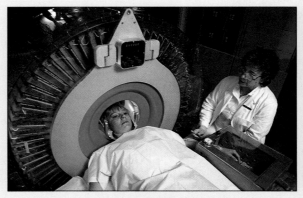
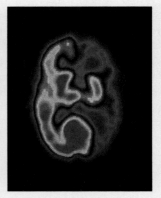

c   Patient being moved into a PET scanner. The inset shows a vivid image of a brain scan of a child with a severe neurological disorder; the different colors signify differences in metabolic activity in one half of the brain; the other half shows no activity.

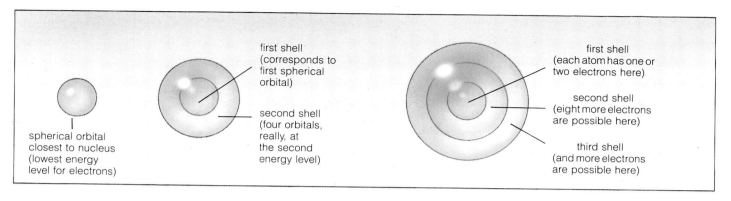

first shell
(corresponds to
first spherical
orbital)

second shell
(four orbitals,
really, at
the second
energy level)

spherical orbital
closest to nucleus
(lowest energy
level for electrons)

first shell
(each atom has one or
two electrons here)

second shell
(eight more electrons
are possible here)

third shell
(and more electrons
are possible here)

Figure 2.4 How electrons are arranged in atoms. In every atom, one or at most two electrons occupy a ball-shaped volume of space (an orbital) close to the nucleus, which at this scale would be an invisible speck at the ball's center. This orbital is at the lowest energy level.

At the next (higher) energy level, there can be as many as eight more electrons (two in each of four orbitals), but the orbital shapes get tricky. For our purposes, we can simply visualize the orbitals at each energy level as being somewhere in a "shell" of the sort sketched here.

The Nature of Chemical Bonds

A chemical bond is a union between the electron structures of atoms. Usually an atom gives up, gains, or shares one or more electrons with another atom. Some atoms do this rather easily; others do not. The differences in bonding behavior arise through differences in the number and arrangement of electrons in the atoms of each kind of element.

Picture three narcissistic actresses arriving at the Academy Awards ceremony wearing the same bright red designer dress. Each has a compulsion to be in the limelight but dreads being photographed next to the others. Two might maneuver themselves *near* the center of attention while scooting away from each other to some extent, but by unspoken agreement, all three never, ever stay in the same place at the same time.

Electrons behave roughly the same way. They are attracted to the protons of a nucleus but repelled by other electrons that may be present. They spend as much time as possible near the nucleus and as far away from each other by moving in different *orbitals*, which are regions of space around the nucleus in which electrons are likely to be at any instant. Each orbital has enough room for two electrons, at most (Figure 2.4).

A simple although not quite accurate way to think about electron orbitals is to imagine them occupying the region inside *shells* around the nucleus. The shell closest to the nucleus has one orbital; it can hold no more than two electrons. The next shell can have as many as eight electrons, two in each of four orbitals. Successive shells can have still more electrons, as Table 2.2 shows.

Hydrogen, the simplest atom, has one electron in its first (and only) shell. Sodium, with eleven electrons, has a lone electron in its outermost shell (Figure 2.5).

Table 2.2	Electron Distribution for a Few Elements					
			Electron Distribution		
Element	Chemical Symbol	Atomic Number	First Shell	Second Shell	Third Shell
Hydrogen	H	1	1	—	—
Helium	He	2	2	—	—
Carbon	C	6	2	4	—
Nitrogen	N	7	2	5	—
Oxygen	O	8	2	6	—
Neon	Ne	10	2	8	—
Sodium	Na	11	2	8	1
Magnesium	Mg	12	2	8	2
Phosphorus	P	15	2	8	5
Sulfur	S	16	2	8	6
Chlorine	Cl	17	2	8	7

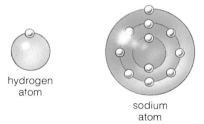

hydrogen atom

sodium atom

Figure 2.5 Distribution of electrons (yellow dots) for hydrogen and sodium atoms. Each atom has a lone electron (and room for more) in its outermost shell. Atoms having such vacancies tend to enter into reactions with other atoms.

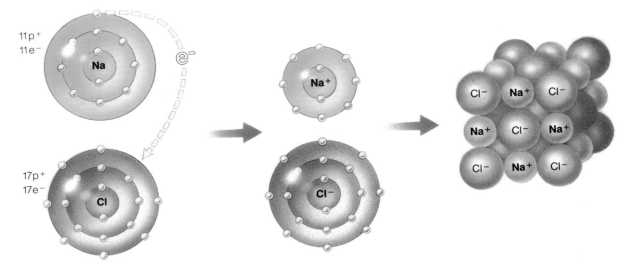

a A sodium atom has the same number of protons and electrons; it has no net charge. The same is true for a chlorine atom. But sodium tends to give up the lone electron in its outer shell to chlorine.

b The resulting sodium ion still has 11 protons but only 10 electrons; the extra proton gives it a positive charge. The chloride ion has 18 electrons but only 17 protons; the extra electron gives it a negative charge.

c The mutual attraction between the oppositely charged ions may hold them together as an ionic bond. Such bonds occur between NaCl units in table salt.

Figure 2.6 Ionic bonding in sodium chloride (NaCl).

In other words, both kinds of atoms have a "vacancy" in an orbital in their outermost shell—*and such atoms tend to react with other atoms.* As you can see from Table 2.2, atoms that tend to enter into reactions include not only hydrogen but also carbon, nitrogen, and oxygen—the main building blocks of organisms.

Electrons of an atom are arranged in regions of space around the nucleus (orbitals), but there can be no more than two electrons in each orbital.

Atoms with unfilled orbitals in their outermost shell tend to react with other atoms.

Ionic Bonding

Sometimes the balance between the protons and electrons of an atom is disturbed so much that one or more electrons are knocked out of the atom, pulled away from it, or added to it. When an atom loses or gains one or more electrons, it becomes positively or negatively charged; and in this state it is an **ion.**

When an atom loses or gains electrons, another atom of the right kind must be nearby to accept or donate the electrons. Since one loses and one gains electrons, *both* become ionized. Depending on the surroundings, the two ions can go their separate ways or stay together through the mutual attraction of their opposite charges. An association of two oppositely charged ions is an **ionic**

bond. Table salt, or NaCl, has ions of sodium (Na^+) and chloride (Cl^-) linked together this way (Figure 2.6).

An ion is an atom (or a compound) that has gained or lost one or more electrons, and so has acquired an overall positive or negative charge.

In an ionic bond, a positive and a negative ion are linked by the mutual attraction of opposite charges.

Covalent Bonding

Often, an attraction between two atoms is not quite enough for one to pull electrons completely away from the other. The atoms end up sharing electrons in a **covalent bond.** We can use a line to represent a single covalent bond between two atoms, as in H—H. If we want to focus on the number of electrons being shared, we can use a dot to represent each one:

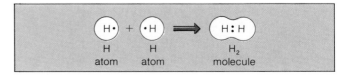

In a double covalent bond, two atoms share two pairs of electrons. An example is the O_2 molecule, or O=O. In a triple covalent bond (such as N≡N), two atoms share three pairs of electrons.

Covalent bonds may be nonpolar or polar. In a *nonpolar* covalent bond, both atoms exert the same pull on shared electrons. The word nonpolar implies no difference between the two "ends" (or poles) of the bond. An example is the H—H molecule; its hydrogen atoms (with one proton each) are equally attractive to the shared electrons.

In a *polar* covalent bond, atoms of different elements do not exert the same pull on electrons (because they have different numbers of protons). The more attractive atom ends up with a slight negative charge that is balanced out by the other atom, which ends up with a slight positive charge. In other words, the bond between them has no *net* charge, but it still shows a slight polarity.

A water molecule (H—O—H) has two polar covalent bonds. Its electrons are less attracted to the hydrogens than to the oxygen, which has more protons in its nucleus.

In a covalent bond, atoms share electrons.

If electrons are shared equally, the bond is nonpolar. If they are not shared equally, the bond is polar (slightly positive at one end and slightly negative at the other).

Hydrogen Bonding

In a **hydrogen bond,** an atom of a molecule interacts weakly with a hydrogen atom in the vicinity that is already taking part in a polar covalent bond. (The hydrogen, which has a slight positive charge, is attracted to the slight negative charge of the other atom.) Hydrogen bonds can form between two different molecules, as shown here:

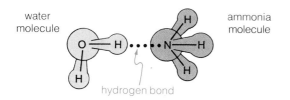

As you will see, hydrogen bonds between water molecules impart some structure to liquid water.

Hydrogen bonds also can form between two different parts of the same molecule, as they do repeatedly in DNA. Although such bonds are individually weak, they collectively help stabilize the structure of DNA and other large biological molecules.

In a hydrogen bond, a hydrogen atom of a molecule interacts weakly with another atom already taking part in a polar covalent bond.

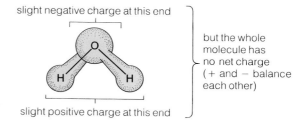

Figure 2.7 Polarity of a water molecule.

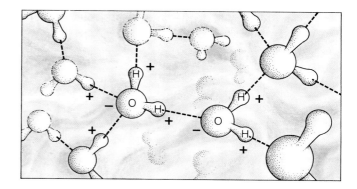

Figure 2.8 Hydrogen bonds between water molecules in liquid water.

Properties of Water

A water molecule carries no net charge, but it shows some polarity as a result of its electron arrangements and bond angles. The whole molecule is slightly negative at the oxygen "end" and slightly positive at the "end" where the two hydrogens are positioned (Figure 2.7). Because of their polarity, water molecules hydrogen-bond with one another. Figure 2.8 shows how their hydrogen atoms form weak hydrogen bonds with oxygen atoms of neighboring molecules in liquid water.

How do other substances behave in water? Polar substances are **hydrophilic** (water-loving); they are attracted to one end or the other of a water molecule and may form weak hydrogen bonds with it. Nonpolar substances are **hydrophobic** (water-dreading); they tend to be repelled by water. Shake a bottle containing water and salad oil (a hydrophobic substance), then put it on a counter. The oil and water hold little attraction for each other. Gradually, hydrogen bonds reunite the water molecules (they replace ones that were broken when you shook the bottle). As they do, the oil molecules are pushed aside and forced to cluster in droplets or in a film at the water's surface.

Hydrogen bonds and hydrophobic interactions underlie three properties of water: its internal cohesion, temperature-stabilizing effects, and capacity to dissolve many substances.

Cohesion. A substance with cohesive properties resists rupturing when stretched (placed under too much tension). Hydrogen bonds impart cohesion to water. For example, where air and water meet at a pond surface, hydrogen bonds exert a constant inward pull on the uppermost water molecules. They impart a high surface tension that resists penetration by leaves and small insects (Figure 2.9). Cohesion also helps pull up water through narrow cellular pipelines in the roots, stems, and leaves of plants.

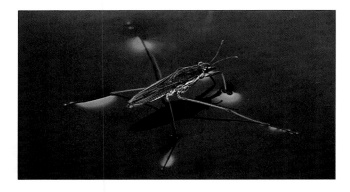

Figure 2.9 Water strider "walking" on water.

Temperature Stabilization. Hydrogen bonds contribute to water's temperature-stabilizing effects. The temperature of a substance is a measure of how fast its molecules are moving. When water is heated, its molecules cannot move faster until hydrogen bonds among them are broken. Compared to other substances, water can absorb considerable heat before its temperature increases greatly.

When water becomes heated past a certain point, molecules at its surface absorb enough heat energy to break hydrogen bonds and escape. Some water has changed from the liquid to the gaseous state, a process called *evaporation*. Energy required for evaporation is drawn mostly from the surrounding liquid, so the process lowers the surface temperature of water. After aerobic exercise, you cool off by perspiration and evaporative water loss.

Water also resists freezing. Water is liquid at room temperature because hydrogen bonds between its molecules constantly break and form again, permitting some freedom of movement. Only when the temperature drops below 0°C do the molecules become locked in the bonding pattern of ice (Figure 2.10).

Solvent Properties. Because of the polar nature of its molecules, water is an excellent solvent for ions and polar molecules. A solvent is any fluid in which one or more substances can be dissolved. The dissolved substances themselves are called **solutes**.

What does "dissolved" actually mean? Consider what happens when you pour some table salt into water.

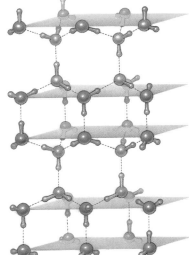

Figure 2.10 Hydrogen bonding pattern between the water molecules in ice. Dashed lines signify hydrogen bonds.

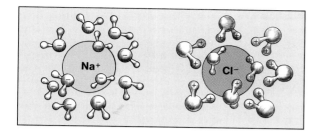

Figure 2.11 Spheres of hydration around charged ions.

The salt crystals separate into Na⁺ and Cl⁻ ions. Water molecules cluster around each positively charged ion with their "negative" ends pointing toward it. They also cluster around each negatively charged ion with their "positive" ends pointing toward it, as Figure 2.11 shows. These "spheres of hydration" shield charged ions and keep them from interacting, so the ions can remain dispersed in water. *A charged substance is "dissolved" in water when spheres of hydration form around its individual ions or molecules.* This happens to solutes in cells, in the sap of maple trees, in your blood, and in the body fluids of all other organisms.

Cell structure and function depend on three properties of water: its internal cohesion, temperature-stabilizing effects, and capacity to dissolve many substances.

ACIDS, BASES, AND SALTS

Acids and Bases

Some substances release **hydrogen ions** (H⁺) when they dissolve in water. We call this type of substance an **acid**. Hydrochloric acid (HCl), for example, separates into H⁺ and Cl⁻. Other substances release ions that can *combine with* hydrogen ions. This type of substance is a **base**. Sodium hydroxide (NaOH) is a base. It separates into Na⁺ and a hydroxide ion (OH⁻), which can combine with H⁺ to form water.

The pH Scale

The H⁺ concentrations inside and outside a cell affect its activities. The **pH scale** is used to measure the concentration of hydrogen ions in different solutions. The part of the scale we are concerned with ranges from 0 (most acidic) to 14 (most basic). The midpoint, 7, represents a neutral solution in which H⁺ and OH⁻ concentrations are the same. A change in one unit in pH means a tenfold change in H⁺ (Figure 2.12).

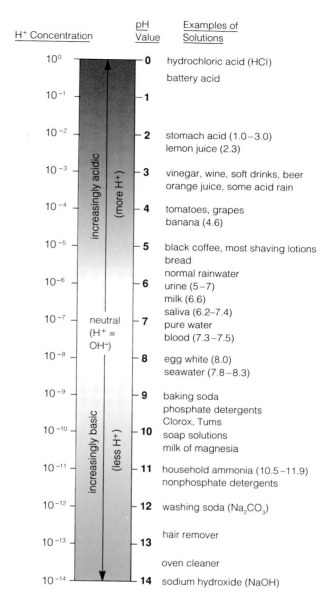

Figure 2.12 The pH scale, in which a liter of fluid is assigned a number according to the number of hydrogen ions present. The most useful part of the scale ranges from 0 (most acidic) to 14 (most basic), with 7 representing neutrality.

A change of only 1 on the pH scale means a tenfold change in hydrogen ion concentration. For example, the gastric juice in your stomach is ten times more acidic than vinegar, and vinegar is ten times more acidic than tomatoes.

Pure water has a pH of 7. A solution with a pH of less than 7 has more H^+ than OH^- ions. The opposite is true of any solution with a pH of more than 7. *The greater the H^+ concentration, the lower the pH value.*

Living cells are sensitive to pH, and their interior usually is kept close to neutrality. However, the environmental pH may be much greater or less than this. Cells of sphagnum mosses grow in peat bogs, where the pH is 3.2 to 4.6 (acidic). Roundworms thrive in places where the pH is 3.4. Fluids bathing most cells of your body range between 7.35 and 7.45. Industrial wastes affect the pH of rain (Figure 2.13).

Figure 2.13 Sulfur dioxide emissions from a coal-burning power plant; special camera filters revealed these otherwise invisible emissions. Together with other airborne pollutants, sulfur dioxides dissolve in atmospheric water to form acidic solutions. They are a major component of acid rain.

Dissolved Salts

We have mentioned salts in passing but have not yet defined them. A **salt** is an ionic compound, formed when an acid reacts with a base. Sodium chloride can form this way:

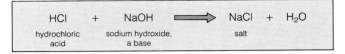

Many salts dissolve into ions that play vital roles in cells. For example, ions of potassium (K^+) and sodium take part in the "messages" traveling through the nervous system. Calcium ions (Ca^{++}) take part in cell movements, cell division, nerve function, muscle contraction, and blood clotting. Ions in your body fluids affect the movement of water and dissolved substances into and out of your cells.

Buffers

Cells continually produce and use hydrogen ions, yet pH does not shift drastically inside them. Why? In most cells, buffers and other mechanisms keep the internal pH fairly constant. **Buffers** are molecules that combine with hydrogen ions, release them, or both in response to changes in pH. When a metabolic reaction produces an excess of H^+, buffers accept the excess. When other

Figure 2.14 Carbon compounds. There is a Tinkertoy quality to carbon compounds, in that a single carbon atom can be the start of truly diverse molecules assembled from "straight-stick" covalent bonds. Consider the *hydrocarbons*, which consist only of hydrogen and carbon. Methane (CH_4) is the simplest hydrocarbon. If you were to strip one hydrogen from methane, the result would be a *methyl group:*

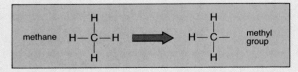

Now imagine that two methane molecules are each stripped of a hydrogen atom and bonded together. If the resulting structure were to lose a hydrogen atom, you would end up with an *ethyl group:*

You could go on building a continuous chain, with all the carbon atoms arranged in a line:

To such linear carbon chains, you could add branches:

You might even have chains coiled back on themselves into rings, represented in any of these ways:

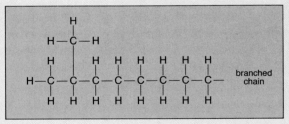

reactions deplete H^+, buffers can dole out reserves. Carbonic acid is one of the body's major buffers. It separates into a bicarbonate ion and H^+ in water:

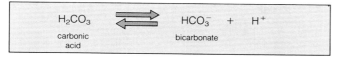

$$H_2CO_3 \rightleftharpoons HCO_3^- + H^+$$

carbonic acid   bicarbonate

In turn, bicarbonate can combine with hydrogen ions to form carbonic acid.

Buffer molecules combine with or release hydrogen ions in response to changes in cellular pH.

CARBON COMPOUNDS

By far, oxygen, hydrogen, and carbon are the most abundant elements in your body; they account for ninety-three percent of its weight (Figure 2.2). Much of the oxygen and hydrogen is linked together as water molecules. But those two elements also are linked in significant amounts to carbon—the most important structural element in the body.

Carbon can form as many as four covalent bonds with other atoms (Figure 2.14). In cells, carbon atoms are linked into chains or rings that serve as backbones for strandlike, sheetlike, and chunky molecules—some of which have thousands or millions of atoms. The carbon-based molecules assembled in cells are known as **organic compounds.** In contrast, water, carbon dioxide, and other simple, "inorganic" compounds have no carbon chains or rings.

Families of Small Organic Compounds

Organic compounds having no more than about twenty carbon atoms are grouped into four main families: the simple sugars, fatty acids, amino acids, and nucleotides. Usually these compounds are dissolved in the cellular fluid. They are used as energy sources or building blocks for larger molecules, including polysaccharides, lipids, proteins, and nucleic acids.

Simple sugars, fatty acids, amino acids, and nucleotides serve as energy sources and building blocks for the large molecules present in cells.

Functional Groups

The structure and behavior of organic compounds depend on the properties of their **functional groups,** which are atoms covalently bonded to the carbon backbone.

For example, fats and waxes have methyl groups in which hydrogen atoms form nonpolar covalent bonds with carbon atoms (Figure 2.15). Water cannot form hydrogen bonds with nonpolar substances; that is why fats and waxes do not dissolve in water. Sugars and other alcohols have hydroxyl groups (—OH) attached to the backbone. Water can form hydrogen bonds at these groups; that is why sugars dissolve in water.

Condensation and Hydrolysis

Small organic compounds don't just get together on their own to form larger ones. Their union depends on the action of **enzymes,** a special class of proteins that speed up reactions between specific substances. In **condensation,** enzymes speed up the covalent linkage of small molecules in a reaction that may also yield water. One molecule is stripped of an H atom and another is stripped of an —OH group, then the two molecules are

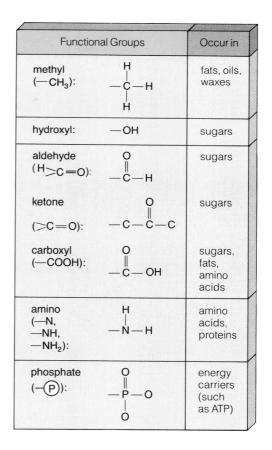

Figure 2.15 Examples of functional groups that confer distinctive properties upon carbon compounds.

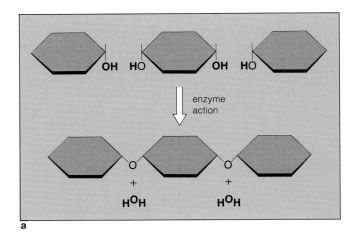
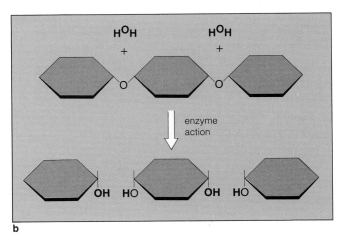

Figure 2.16 (a) Condensation of three subunits into a larger molecule. Water can be formed during the reaction. (b) Hydrolysis of a molecule into three subunits.

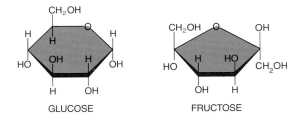

GLUCOSE   FRUCTOSE

Figure 2.17 Glucose and fructose, two monosaccharides.

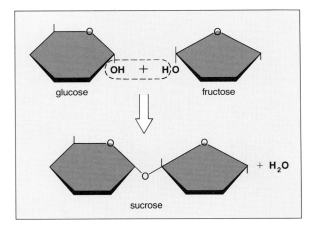

glucose   fructose

sucrose   + H₂O

Figure 2.18 Condensation of two monosaccharides into a disaccharide.

joined. At the same time, the H and —OH combine to form a water molecule (Figure 2.16).

Cells often obtain building blocks for large organic compounds by breaking apart other molecules. **Hydrolysis,** a common breakdown process, is like condensation in reverse. Covalent bonds are split, and an H atom and an —OH group derived from water become attached to the fragments (Figure 2.16).

Carbohydrates

A **carbohydrate** is a simple sugar or a large molecule composed of sugar units. Carbohydrates are the most abundant biological molecules. All cells use them as structural materials, transportable packets of quick energy, and storage forms of energy.

A **monosaccharide,** or one sugar unit, is the simplest carbohydrate. ("Saccharide" comes from a Greek word meaning sugar.) Sugars are soluble in water, most are sweet-tasting, and the most common have five or six carbon atoms. Ribose and deoxyribose (which occur in RNA and DNA, respectively) are in this category. So is glucose (Figure 2.17), a sugar you will encounter repeatedly in this book. Glucose is the main energy source for most organisms and the precursor, or "parent" molecule, of many other organic compounds.

An **oligosaccharide** has two or more sugar units. The kinds with two sugars are known as *disaccharides.* Sucrose (with a glucose and a fructose unit) is the form in which carbohydrates are transported in leafy plants (Figure 2.18). We make table sugar by extracting and crystallizing sucrose from plants such as sugar cane. Lactose (a glucose and a galactose unit) is present in milk. Oligosaccharides with three or more sugar units are usually attached as short side chains to proteins and other molecules. Some side chains have roles in cell membrane function and in immunity.

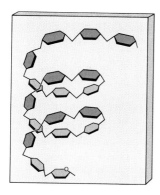

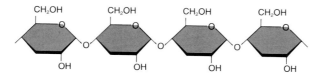

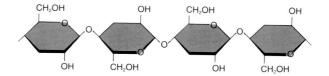

Figure 2.19 Oxygen bridges between the glucose subunits of amylose, a form of starch. The boxed inset depicts the coiling of an amylose molecule, which is stabilized by hydrogen bonds.

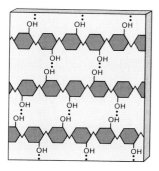

Figure 2.20 Structure of cellulose, which is composed of glucose subunits. Neighboring cellulose molecules link together at—OH groups to form a fine strand. Such strands may be twisted together and then coiled to form cellulose threads.

A **polysaccharide** is a straight or branched chain of hundreds or thousands of sugar units, of the same or different kinds. The most common ones—glycogen, starch, and cellulose—have glucose units. Glycogen is a glucose-storage form in animals. Starch is a glucose-storage form in plants. Starch molecules can be quickly hydrolyzed when cells require free glucose. Cellulose, a structural material in plant cell walls, is tough, fibrous, and insoluble. It can withstand so much weight and stress, it has been likened to steel rods in reinforced concrete. The different properties of starch and cellulose arise from differences in bonding alignments between their glucose units (Figures 2.19 and 2.20).

Lipids

Lipids are compounds of mostly carbon and hydrogen that show little tendency to dissolve in water, but they do dissolve in nonpolar solvents (such as ether). Some lipids serve as energy reserves; others are components of membranes and other cell structures. Here we will focus on two types: lipids with or without fatty acid components.

Lipids with Fatty Acids. A **fatty acid** has a long, unbranched carbon backbone with a —COOH group at the end (Figure 2.21). Three common lipids (the glycerides, phospholipids, and waxes) have fatty acids, stretched out like tails.

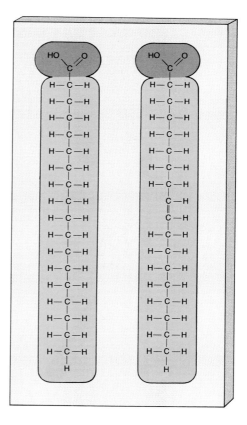

Figure 2.21 Structural formulas for stearic acid (left) and oleic acid (right). The stearic acid is saturated. The oleic acid, with its double bond in the carbon backbone, is unsaturated.

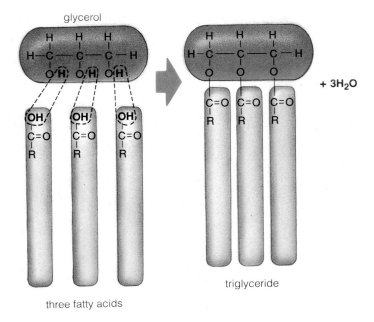

glycerol

+ 3H₂O

three fatty acids

triglyceride

Figure 2.22 Condensation of fatty acids into a triglyceride. Here, the R signifies the "rest" of the carbon chain in each fatty acid molecule.

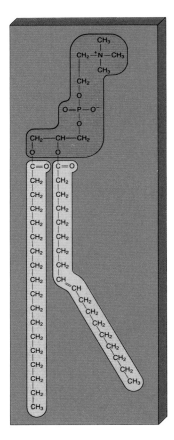

Figure 2.23 Structural formula of a typical phospholipid found in animal cell membranes. The hydrophilic head is shown in orange, and the hydrophobic tails in gold.

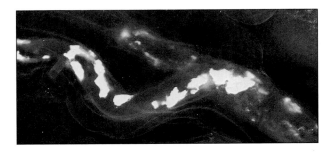

Figure 2.24 Plaques in blood vessels that service the heart. Such plaques form when fats and cholesterol become deposited in the blood vessel wall, then calcium salts and a fibrous net form over the abnormal mass. Blood clots may form here and narrow or block the arteries, leading to a heart attack.

Glycerides are the body's most abundant lipids and its richest source of energy. A **glyceride** molecule has one, two, or three fatty acid tails attached to a backbone of glycerol (Figure 2.22). The terms monoglyceride, diglyceride, and triglyceride refer to whether one, two, or three fatty acid tails are present. Such molecules form the substances we call fats and oils.

Saturated fats, including butter and lard, tend to be solids at room temperature. "Saturated" means all the carbon atoms in the fatty acid tails are joined by single C—C bonds and as many hydrogen atoms as possible are linked to them. In saturated fats, the tails snuggle together in parallel array.

Unsaturated fats, or oils, tend to be liquid at room temperature. In this case, one or more double bonds occur between the carbon atoms in the fatty acid tails (Figure 2.21). Oils are liquid because the double bonds create kinks that disrupt packing between tails.

Some amounts of unsaturated fats are important in nutrition. Immature rats that were placed on a fat-free diet grew abnormally, their hair fell off, their skin turned scaly, and they died young. The conditions never developed in other rats that were fed small amounts of linoleic acid, an unsaturated fat.

A **phospholipid** has a glycerol backbone. Attached to the backbone are two fatty acid tails, a phosphate group, and a small hydrophilic group (Figure 2.23). As you will see in later chapters, phospholipids are the main components of cell membranes.

Waxes also have long-chain fatty acids. Some wax secretions form coatings that help protect, lubricate, and maintain the pliability of skin and hair. Others help make feathers water-repellant. In many plants, waxes and a lipid (cutin) form the cuticle, a covering on the surface of aboveground parts that helps prevent water loss. The cuticle gives many leaves and fruits, such as apples, a shiny appearance.

Lipids with No Fatty Acids. Among the lipids that have no fatty acid tails, we find the steroids. All **steroids** have the same backbone of four carbon rings, but they vary in the number, position, and type of functional groups attached to it:

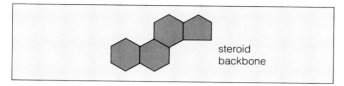
steroid
backbone

You have probably heard of cholesterol. This steroid is a key component of animal cell membranes. It is used in the synthesis of vitamin D, which functions in the development of bones and teeth. But excess cholesterol also plays a role in atherosclerosis, a disorder in which lipids become deposited in the walls of arteries. The deposits build up, arteries narrow, and blood flow may be obstructed (page 381 and Figure 2.24).

Many hormones, including sex hormones, are steroids. Hormones help regulate the body's growth, development, and reproduction, as well as its everyday functioning. Bodybuilders and athletes sometimes use certain hormonelike steroids to increase their muscle mass. Unfortunately, use of those steroids can lead to pronounced behavioral disorders, liver damage, and other abnormalities.

Proteins

Proteins are large molecules composed of amino acids. An **amino acid** is a small organic molecule having an amino group, an acid group, and one or more atoms called its R group:

amino
group

H_3N^+ — C — H

COO⁻ acid group

R R group

Proteins are assembled from only twenty or so different kinds of amino acids (Figure 2.25 shows some of them). Yet proteins are the most diverse of all biological molecules. They include enzymes, which make specific reactions proceed faster than they would on their own. They also include molecules concerned with cell movements, storage, and transport. Many hormones are proteins; so are antibodies, substances that help defend the body against disease-causing agents. Proteins also are structural materials, the stuff of bone and cartilage, hoof and claw.

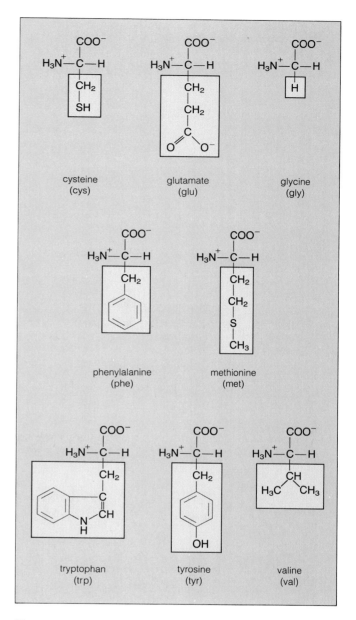

Figure 2.25 Structural formulas for eight of the twenty common amino acids. The R groups are within the light green boxes.

Figure 2.26 Condensation of a polypeptide chain from four amino acids.

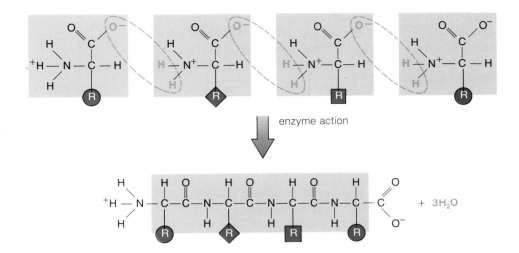

enzyme action

+ 3H₂O

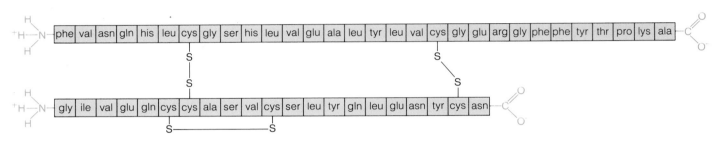

Figure 2.27 Linear sequence of amino acids in cattle insulin, as determined by Frederick Sanger in 1953. This protein is composed of two polypeptide chains, linked by disulfide bridges (—S—S—).

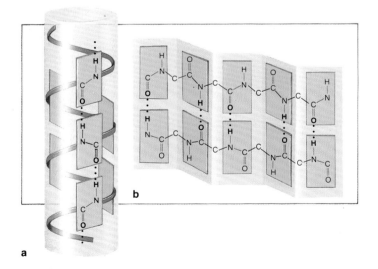

b

a

Figure 2.28 Hydrogen bonds (dotted lines) in a polypeptide chain. Such bonds can give rise to a coiled chain (**a**) or sheetlike array of chains (**b**).

Protein Structure. Protein formation begins when amino acids are strung together, one after the other. A covalent bond (called a peptide bond) forms between the amino group of one amino acid and the carboxyl group of another (Figure 2.26). Three or more linked amino acids are called a **polypeptide chain.**

Different sequences of amino acids occur in different proteins. For example, the two chains making up the protein insulin in cattle always have the sequences shown in Figure 2.27. The specific sequence of amino acids in a polypeptide chain is the protein's *primary* structure.

The sequence of amino acids influences what shape a protein can assume, what its function will be, and how it will interact with other substances. It does so in two major ways. First, oxygen and other atoms of different amino acids in the sequence take part in hydrogen bonds that hold the chain in a coiled or extended pattern (Figure 2.28). The coiled or extended pattern, based on regular hydrogen bonding, is the protein's *secondary* structure.

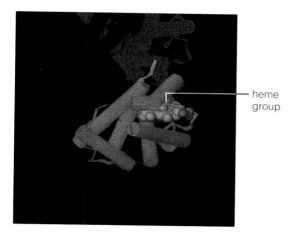

Figure 2.29 Model of the three-dimensional structure of one of the four polypeptide chains in a hemoglobin molecule. Red cylinders represent parts of the chain that are helically coiled. Irregular loops that disrupt the regular coiling are shown as ribbons. The amino acid proline causes such disruptions at four different points in the primary sequence of amino acids. When oxygen is transported in blood, it binds with an iron-containing component (heme group) of the molecule.

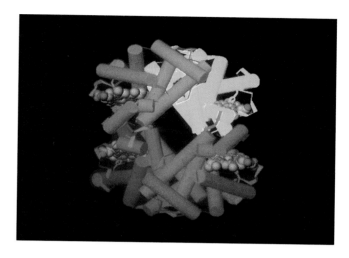

Figure 2.30 Quaternary structure of the protein hemoglobin in human blood. Hemoglobin is a red pigment circulating in animal blood and carrying vital oxygen to tissues. The hemoglobin molecule consists of four polypeptide chains, held together by numerous bonds that are individually weak but collectively strong. (Compare Figure 2.29.)

Second, different R groups in the sequence interact and determine the way the chain can bend and twist into its three-dimensional shape. The shape resulting from R-group interactions is the protein's *tertiary* structure. Figure 2.29 is an example.

Some proteins have *quaternary* structure, meaning they incorporate two or more polypeptide chains. The resulting protein can be globular, fiberlike, or some combination of the two. Hemoglobin is an example (Figure 2.30).

The amino acid sequence of a polypeptide chain dictates the final three-dimensional structure of a protein—and this structure dictates how the protein will interact with other cell substances.

Protein Denaturation. Interactions holding a protein (or some other molecule) in its normal, three-dimensional shape can be disrupted. This type of disruption, called **denaturation,** can occur after exposure to high temperatures (typically above 60°C) or to chemical agents. The polypeptide chain unwinds or changes shape and the protein is no longer functional.

For example, the "white" of an uncooked chicken egg is a concentrated solution of the protein albumin.

When you cook an egg, heat does not affect the strong covalent bonds of albumin's primary structure—but it destroys the weaker bonds responsible for its secondary and tertiary structure. Although denaturation can be reversed for some kinds of proteins when normal conditions are restored, albumin isn't one of them. There is no way to uncook a cooked egg.

Nucleotides and Nucleic Acids

The small organic compounds called nucleotides are central to life. Each **nucleotide** contains three components: a five-carbon sugar (ribose or deoxyribose), a nitrogen-containing base (either a single-ringed structure called a pyrimidine or a double-ringed purine), and a phosphate group. For example, one nucleotide has the three components hooked together in this way:

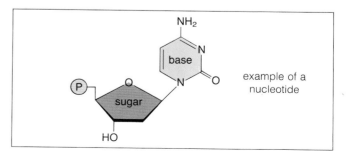

Three kinds of nucleotides or nucleotide-based molecules are the adenosine phosphates, the nucleotide coenzymes, and the nucleic acids.

Adenosine phosphates are relatively small molecules that function as chemical messengers within and between cells, and as energy carriers. Cyclic adenosine monophosphate (cAMP) is a chemical messenger. Adenosine triphosphate (ATP) is a nucleotide that serves as an energy carrier.

Nucleotide coenzymes transport the hydrogen atoms and electrons necessary in metabolism. Two coenzymes, abbreviated NAD^+ and FAD, are modified forms of the vitamins niacin and riboflavin, respectively.

Nucleic acids are large single- or double-stranded chains of nucleotide units. A phosphate bridge connects the sugars of each nucleotide, and the bases stick out to the side (Figure 2.31a). The sequence in which the four kinds of bases follow one another varies among nucleic acids.

Deoxyribonucleic acid (DNA) and the ribonucleic acids (RNAs) are built according to the plan just outlined. DNA is usually a double-stranded molecule that twists helically about its own axis (Figure 2.31b). The bases of one strand are connected by hydrogen bonds to bases of the other strand. You will be reading more about these molecules in chapters to come. For now, it is enough to know this: (1) Genetic instructions are encoded in the sequence of bases in DNA, and (2) RNA molecules function in the processes by which genetic instructions are used in building proteins.

SUMMARY

1. Atoms are composed of protons, neutrons, and electrons. The number and arrangement of electrons dictate how atoms can form molecules, what the properties of molecules will be, and how (or whether) molecules will interact.

2. An isolated atom has a net electric charge of zero. An ion is an atom or compound that has gained or lost one or more electrons; it has acquired an overall positive or negative charge.

3. All atoms of an element have the same number of protons but they can vary in the number of neutrons. Variant forms of atoms of an element are isotopes. Radioisotopes are used as tracers in biological systems, and they can help determine the age of fossils (remains of once-living organisms).

4. Atoms of hydrogen, carbon, nitrogen, and oxygen (the main structural elements of cells) have vacancies in their outermost shell. They tend to form bonds with other atoms, thereby gaining enough electrons to fill the vacancies.

5. In ionic bonds, a positive and a negative ion remain together by the mutual attraction of their opposite charges. In covalent bonds, atoms share one or more electrons. In hydrogen bonds, a polar covalently bonded atom is weakly attracted to a hydrogen atom that is already bonded to something else.

6. Acids release hydrogen ions (H^+) in water. Bases release ions that can combine with H^+. At pH 7, the H^+ and OH^- concentrations are equal. Buffers and other mechanisms maintain cellular pH at about 7.

7. The properties of water influence the organization and behavior of cells and their environment. These properties include its cohesiveness and its capacity to dissolve other substances.

8. Simple sugars, fatty acids, amino acids, and nucleotides serve as building blocks for the cell's large molecules (mainly polysaccharides, lipids, proteins, and nucleic acids). Some also serve as energy sources. Table 2.3 summarizes the main carbon compounds in living things.

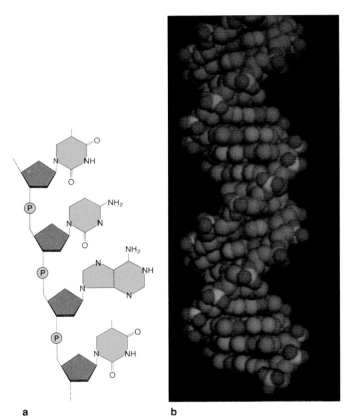

a   **b**

Figure 2.31 (**a**) Examples of bonds between nucleotides in a nucleic acid molecule. (**b**) Model of a segment of DNA, a molecule that is central to maintaining and reproducing the cell. The nucleotide bases are shown in blue.

Table 2.3 Summary of the Main Carbon Compounds in Living Things

Category	Main Subcategories	Some Examples and Their Functions	
CARBOHYDRATES *contain an aldehyde or a ketone group, one or more hydroxyl groups*	**Monosaccharides**	Glucose	Structural roles, energy source
	Oligosaccharides	Sucrose (a disaccharide)	Form of sugar transported in plants
	Polysaccharides	Starch Cellulose	Food storage Structural roles
LIPIDS *are largely hydrocarbon, generally do not dissolve in water but dissolve in nonpolar substances*	**Lipids with fatty acids:** *Glycerides*: one, two, or three fatty acid tails attached to glycerol backbone	Fats (e.g., butter) Oils (e.g., corn oil)	Forms in which energy is stored
	Phospholipids: phosphate group and (often) two fatty acids attached to glycerol backbone	Phosphatidylcholine	Key component of cell membranes
	Waxes: long-chain fatty acid tails attached to alcohol	Earwax Waxes in cutin	Protective barrier to inner ear Water retention by plants
	Lipids with no fatty acids: *Steroids*: four carbon rings; the number, position, and type of functional groups vary	Cholesterol	Component of animal cell membranes; can be rearranged into other steroids (e.g., bile acids, sex hormones)
PROTEINS *are polypeptides (up to several thousand amino acids, covalently linked)*	**Fibrous proteins:** Individual polypeptide chains, often linked into tough, water-insoluble molecules	Keratin Collagen	Structural element of hair, nails Structural element of bones
	Globular proteins: One or more polypeptide chains folded and linked into globular shapes; many roles in cell activities	Polymerases Hemoglobin Insulin Antibodies	Enzyme function Oxygen transport Hormonal effect on metabolism Tissue defense
NUCLEOTIDES *are units (or chains) having a five-carbon sugar, phosphate, and a nitrogen-containing base*	**Adenosine phosphates**	ATP	Energy carrier
	Nucleotide coenzymes **Nucleic acids** Chains of thousands to millions of nucleotides	NAD^+, $NADP^+$ DNA, RNAs	Transport of protons (H^+) and electrons from one reaction site to another Storage, transmission, translation of genetic information

Review Questions

1. Define element, atom, molecule, and compound. What are the six main elements (and their symbols) in most organisms? *17-18*

2. Explain the differences among an atom, an ion, and an isotope. *17, 22*

3. Explain the differences among covalent, ionic, and hydrogen bonds. *22-23*

4. What is the difference between a hydrophilic and a hydrophobic interaction? Is a film of oil on water an outcome of bonding between the molecules making up the oil? *23*

5. Define an acid, a base, and a salt. On a pH scale from 0 to 14, what is the acid range? Why are buffers important in living cells? *25-26*

6. What type of bond is associated with the temperature-stabilizing, cohesive, and solvent properties of water? Is that bond also important in hydrophobic interactions? *23-24*

7. What are the four main families of small organic molecules used in cells for the assembly of carbohydrates, lipids, proteins, and nucleic acids (the large biological molecules)? *27*

8. Identify which of the following is the carbohydrate, fatty acid, amino acid, and polypeptide: *28-32*

   a. $^+NH_3$—CHR—COO$^-$
   b. $C_6H_{12}O_6$
   c. (glycine)$_{20}$
   d. $CH_3(CH_2)_{16}COOH$

9. Is this statement true or false? Enzymes are proteins, but not all proteins are enzymes. *31*

10. Describe the four levels of protein structure. How do the side groups of a protein molecule influence its interactions with other substances? Give an example of what happens when the bonds holding a protein together are disrupted. *33*

11. Distinguish between the following: *28-34*
   a. monosaccharide, polysaccharide
   b. peptide, polypeptide
   c. glycerol, fatty acid
   d. nucleotide, nucleic acid

Self-Quiz   *(Answers in Appendix IV)*

1. Atoms are constructed of protons, neutrons, and _____.

2. Ions are produced when atoms gain or lose _____.

3. The chemical properties of an element are determined by the number and arrangement of its _____.

4. _____ can be used to trace particular substances in the metabolism of biological systems to determine the age of fossil-bearing rocks.

5. Electrons are shared in a(n) _____ bond.
   a. covalent
   b. ionic
   c. hydrogen
   d. double ionic

6. Polar substances are _____; nonpolar substances are _____.
   a. hydrophilic; also hydrophilic
   b. hydrophilic; hydrophobic
   c. hydrophobic; also hydrophobic
   d. hydrophobic; hydrophilic

7. _____ combine with hydrogen ions; _____ release them in response to changes in cellular pH.
   a. acids; bases
   b. bases; acids
   c. buffers; buffers
   d. a and c

8. The atoms of certain elements are more abundant in cell structure than others. They are _____.
   a. carbon, nitrogen, calcium, and oxygen
   b. carbon, oxygen, hydrogen, and nitrogen
   c. carbon, oxygen, hydrogen, and magnesium
   d. carbon, nitrogen, iron, and oxygen

9. Match each type of molecule with the correct description.
   a. carbohydrate   _____ long chain of amino acids
   b. phospholipid   _____ energy carrier
   c. protein   _____ glycerol, fatty acids, and a
   d. DNA   phosphate
   e. ATP   _____ long chain of nucleotides
                     _____ one or more sugar monomers

Selected Key Terms

acid *25*	fatty acid *29*	nucleic acid *34*
amino acid *31*	functional group *27*	nucleotide *33*
atom *17*	hydrogen bond *23*	pH scale *25*
atomic number *18*	hydrogen ion (H$^+$) *25*	phospholipid *30*
base *25*	hydrolysis *28*	polypeptide chain *32*
buffer *26*	hydrophilic *23*	protein *31*
carbohydrate *28*	hydrophobic *23*	proton *17*
condensation *27*	ion *22*	radioisotope *19*
covalent bond *22*	ionic bond *22*	salt *26*
denaturation *33*	lipid *29*	solute *24*
electron *17*	mass number *18*	
enzyme *27*	molecule *17*	

Readings

Hegstrom, R., and D. Kondepudi. January 1990. "The Handedness of the Universe." *Scientific American* 262 (1):108–115.

Lehninger, A. 1982. *Principles of Biochemistry.* New York: Worth. Classic reference book in the field.

Miller, G. T. 1991. *Chemistry: A Contemporary Approach.* Third edition. Belmont, California: Wadsworth. Simple introduction to basic chemical concepts and their application to everyday life.

Scientific American. "The Molecules of Life." October 1985. This entire issue is devoted to articles on current insights into DNA, proteins, and other biological molecules. Excellent illustrations.

CELL STRUCTURE AND FUNCTION 3

KEY CONCEPTS

1. Cells are the smallest units having the complex organization, metabolic and reproductive behavior, and other characteristics of life. All cells have a plasma membrane surrounding an inner region of cytoplasm. Eukaryotic cells have a nucleus and other organelles (membrane-bound compartments) within the cytoplasm; prokaryotic cells (bacteria) do not.

2. Cell membranes afford control over the exchange of substances between a cell and its environment. They also have mechanisms for cell-to-cell recognition and for receiving outside information that can change cell activities. Lipids form the basic bilayer structure of membranes. Proteins associated with the bilayer carry out most known membrane functions.

3. Water, oxygen, carbon dioxide, and many other small molecules having no net charge can diffuse across the lipid bilayer of membranes. Ions and most large molecules cannot do this; they must be actively or passively transported across by membrane proteins.

THE NATURE OF CELLS

Basic Cell Features

As small as it may be, a cell is a *living* thing engaged in the risky business of survival. Think about just one of the challenges being met by the single, free-living cells called diatoms (Figure 3.1). Some diatoms live in estuaries, where seawater mixes with currents from rivers and streams. Seawater is much "saltier" than fresh water (many more ions are dissolved in seawater). Even though ion concentrations in estuaries change with the tides, they cannot be allowed to change as much inside the cell body, otherwise the diatom would die.

So think of the cell for what it is: a tiny, organized bit of life in a world that is, by comparison, disordered and sometimes harsh. *No matter what goes on outside, the cell must bring in certain substances, keep out others, and conduct its internal activities with great precision.*

Cells do show great diversity in structure and function, as you might gather by comparing a diatom with one of your liver cells. However, they are alike in a few

Figure 3.1 Diatoms—each a free-living, single cell. They live in "glass houses" made of silica from their own secretions, and they represent some of the diversity possible on the basic cellular plan.

10 μm

basic respects. All cells have a plasma membrane surrounding an inner region of cytoplasm. In bacterial cells, the DNA is concentrated in part of the cytoplasm; in all other cells, it is concentrated in the nucleus:

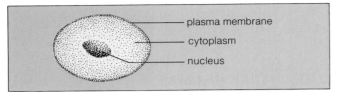

plasma membrane
cytoplasm
nucleus

The basic cellular regions and their functions can be defined in this way:

1. Plasma membrane. This outermost membrane of a cell separates internal metabolic events from the environment and allows them to proceed in organized, controlled ways. It does not totally isolate the interior; many substances move across the membrane. It also has receptors for external molecules that can alter cell activities.

2. Nucleus. This membrane-bound compartment contains hereditary instructions (DNA) and other molecules that function in how the instructions are read, modified, and dispersed.

3. Cytoplasm. The cytoplasm is everything enclosed by the plasma membrane, except for the nucleus. In all but bacterial cells, it has compartments in which specific metabolic reactions occur. It includes particles and filaments bathed in a semifluid substance. The filaments form a "skeleton" that imparts shape and permits movement of cell structures.

Cell Size and Microscopy

Generally, cells and their structures are too small to be observed with the unaided eye. For example, a string of about 2,500 red blood cells would only be as long as your thumbnail is wide! Yet our awareness of the cellular world dates back to the seventeenth century (Figure 3.2). Over time, observations led microscopists to believe that (1) all organisms are made of cells, (2) the cell is the basic *living* unit of organization, and (3) all cells arise from preexisting cells. These three hypotheses, called the **cell theory,** are still valid.

Today we use light microscopes to explore cellular details down to about 0.2 micrometer and electron microscopes for details even smaller (Figures 3.2 and 3.3). To give you a sense of the sizes involved, one of your red blood cells is about 6 to 8 micrometers across.

a Robert Hooke's compound microscope and his drawing of dead cork cells.

Figure 3.2 The microscope—gateway to the cell.

Emergence of the Cell Theory. Early in the seventeenth century, Galileo Galilei arranged two glass lenses in a cylinder. With this instrument he looked at an insect and came to describe the stunning geometric patterns of its tiny eyes. Thus Galileo was the first to record a biological observation made through a microscope.

(a) At mid-century, Robert Hooke looked at a thin slice of cork through his microscope and saw tiny, empty compartments. He gave them the name *cellulae* (meaning small rooms)—and this was the origin of the biological term "cell." Hooke actually was looking at walls of dead cells, which is what cork is made of, although he did not think of them as being dead because he did not know cells could be alive.

Given the simplicity of their instruments, it is amazing that the pioneers in microscopy saw as much as they did. Antony van Leeuwenhoek had great skill in lens construction and possibly the keenest vision. He even observed a bacterium—a type of cell so small it would not be seen again for another two centuries!

In the 1800s, improvements in lens design brought cells into sharper focus. Theodor Schwann and Matthias Schleiden concluded that all animal and plant tissues are composed of cells. Each cell, they said, develops as an independent unit even though its life is influenced by the whole plant or animal.

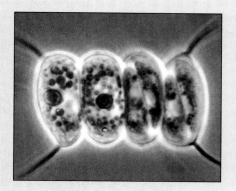

b Light micrograph (phase-contrast).

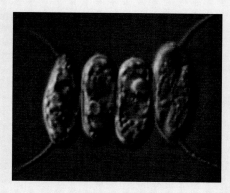

⊢ 10μm ⊣

As on other micrographs in this book, the short bar provides a reference for size. Each micrometer (μm) is only 1/1,000,000 of a meter.

c Light micrograph (Nomarski process).

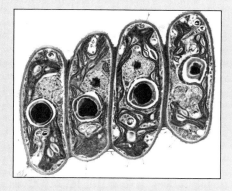

d Transmission electron micrograph, thin section.

e Scanning electron micrograph.

Rudolf Virchow studied cell reproduction and figured out where cells come from. These observations were distilled into the cell theory: *All organisms are composed of one or more cells, the cell is the basic living unit of organization for all organisms, and all cells arise from preexisting cells.*

Microscopy Today. The micrographs in **b** through **e** compare the type of detail revealed by modern microscopes. (A *micrograph* is simply a photograph of an image formed with a microscope.) The specimen is a green alga, and the magnification is the same in all cases.

The **compound light microscope** has glass lenses that bend incoming light rays to form an enlarged image of a specimen. If you wish to observe a *living* cell, it must be small or thin enough for light to pass through. Also, cell parts must differ in color and density from their surroundings—but most are nearly colorless and optically uniform in density. Cells are stained (exposed to dyes that react with some parts but not others), but staining can alter the parts and kill the cell. Dead cells begin to break down at once, so they are pickled or preserved before staining.

No matter how good a glass lens system is, when the diameter of the object being viewed is magnified by 2,000 times or more, cell parts appear larger but are not clearer. The properties of wavelengths of visible light limit the resolution of smaller details.

Transmission electron microscopes have great magnifying power because they use magnetic lenses to focus electrons (which respond to magnetic force). Electrons are particles but they also behave like waves—and their wavelengths are much shorter than those of visible light. Details of a cell's internal structure show up best with transmission electron microscopes.

With a *scanning electron microscope,* a narrow electron beam is directed back and forth across a specimen's surface, which has been coated with a thin metal layer. Electron energy triggers the emission of electrons in the metal. The emission patterns can be used to form an image. Scanning electron microscopy provides a three-dimensional view of surface features.

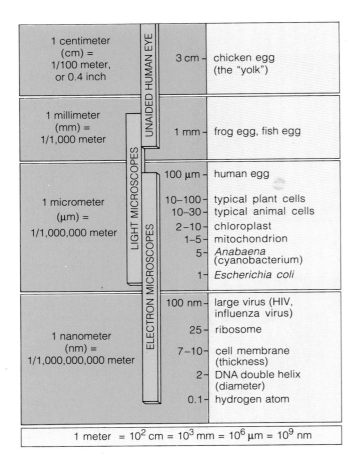

1 centimeter (cm) = 1/100 meter, or 0.4 inch		3 cm –	chicken egg (the "yolk")
1 millimeter (mm) = 1/1,000 meter		1 mm –	frog egg, fish egg
1 micrometer (μm) = 1/1,000,000 meter		100 μm –	human egg
		10–100 –	typical plant cells
		10–30 –	typical animal cells
		2–10 –	chloroplast
		1–5 –	mitochondrion
		5 –	*Anabaena* (cyanobacterium)
		1 –	*Escherichia coli*
1 nanometer (nm) = 1/1,000,000,000 meter		100 nm –	large virus (HIV, influenza virus)
		25 –	ribosome
		7–10 –	cell membrane (thickness)
		2 –	DNA double helix (diameter)
		0.1 –	hydrogen atom

UNAIDED HUMAN EYE / LIGHT MICROSCOPES / ELECTRON MICROSCOPES

1 meter = 10^2 cm = 10^3 mm = 10^6 μm = 10^9 nm

Figure 3.3 Units of measure used in microscopy. The micrometer is used in describing whole cells or large cell structures. The nanometer is used in describing cell ultrastructures and large organic molecules.

CELL MEMBRANES

Membrane Structure and Function

Before we look closely at the cell, let's think about its membranes, since these are so essential for its organization and activities. The main components of cell membranes are phospholipids and proteins. A **phospholipid**, recall, has a hydrophilic (water-loving) head and two fatty acid tails, which are hydrophobic (water-dreading):

polar (hydrophilic) head

nonpolar (hydrophobic) tails

What happens when many phospholipid molecules are immersed in water? Hydrophobic interactions may force them to cluster together into two layers, with all the fatty acid tails sandwiched between the hydrophilic heads. This arrangement, called a **lipid bilayer,** is the structural basis of all cell membranes:

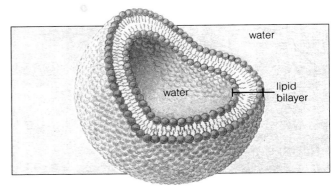

water

water

lipid bilayer

Figure 3.4 shows the **fluid mosaic model** of membrane structure. The bilayer is the "fluid" part because its lipid molecules show quite a bit of movement. They move sideways and their tails flex back and forth, so neighboring lipids cannot become packed into a solid layer. Lipids with short tails and unsaturated (kinked) tails also disrupt the packing. The membrane is said to have a "mosaic" quality because it is an intricate composite of lipids and proteins. The proteins carry out most membrane functions.

Some membrane proteins are open or gated channels for water-soluble substances. Others are like pumps that transport substances across the lipid bilayer with the help of energy from ATP. Still others transfer electrons. Certain proteins help cells recognize each other and stick together in a tissue. Receptor proteins are binding sites for hormones and other substances that trigger alterations in metabolism or cell behavior. For example, the binding of the hormone somatotropin turns on enzymes that crank up the machinery for cell growth and division.

The features common to all cell membranes can be summarized this way:

1. Cell membranes are composed largely of lipids (especially phospholipids) and proteins.

2. In membranes, lipid molecules have their hydrophilic heads at the two outer faces of a bilayer and their fatty acid tails sandwiched in between.

3. The lipid bilayer is the basic *structure* of cell membranes and serves as a hydrophobic barrier between two fluids. (The *plasma membrane* separates the fluids that are inside and outside the cell. *Internal cell membranes* separate different fluids within the space of the cytoplasm.)

4. Proteins embedded in the bilayer or positioned at its surfaces carry out most membrane *functions*.

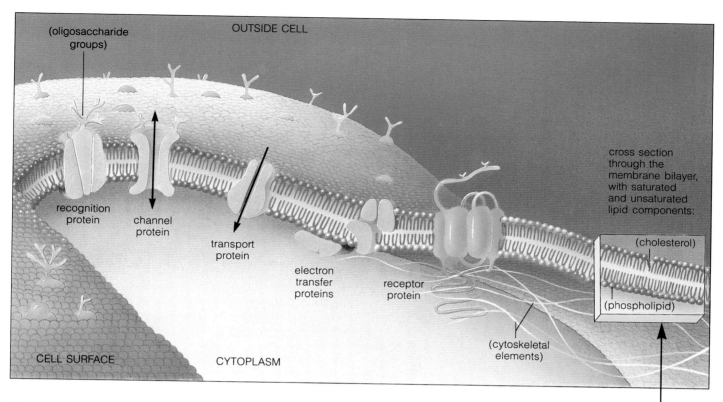

Figure 3.4 Increasingly enlarged views of the plasma membrane, based on the fluid mosaic model of membrane structure. The micrograph shows the plasma membrane of a pancreatic cell, thin section.

Diffusion

Why do substances travel one way or another across cell membranes? In the absence of other forces, molecules (or ions) of a given type move down their **concentration gradient,** which means they tend to move to a region where they are less concentrated. They are driven to do so because they constantly collide with one another, millions of times a second. Random collisions do send the molecules back and forth, but the *net* movement is outward from the region of greater concentration.

The random movement of like molecules down a concentration gradient is called **diffusion.** The molecules of each substance diffuse *independently* of other molecules that may be present. When you put a few drops of food coloring into one end of a pan filled with water, the dye molecules diffuse in one direction—to the region where they are less concentrated. And the water molecules move in the opposite direction, to the region where *they* are less concentrated (Figure 3.5).

Diffusion is the random movement of like molecules or ions down their concentration gradient.

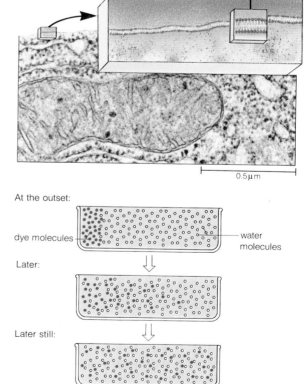

Figure 3.5 Diagram of the diffusion of dye molecules in one direction and water molecules in the opposite direction in a pan of water.

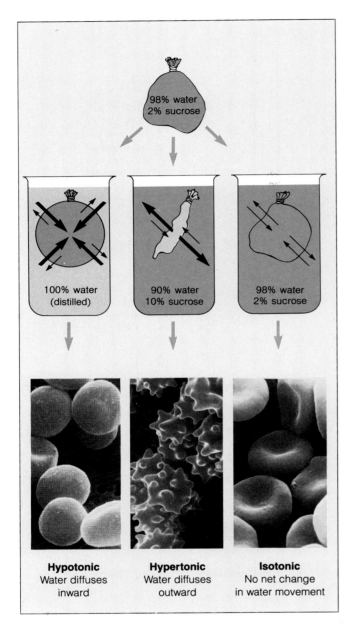

Figure 3.6 Effects of osmosis in different environments. The sketches show why it is important for cells to be matched to solute levels in their environment. (In each sketched container, arrow width represents the relative amount of water movement.)

The micrographs correspond to the sketches. They show the kinds of shapes that might be seen in red blood cells placed in hypotonic solutions (influx of water into the cell), hypertonic solutions (outward flow of water from the cell), and isotonic solutions (internal and external solute concentrations are matched).

Red blood cells have no special mechanisms for actively taking in or expelling water molecules. Hence they swell or shrivel up if solute levels in their environment change.

As you will see, the movement of different substances between two regions may be influenced by more than concentration gradients. It may be influenced also by gradients in pressure, temperature, and electric charge. In addition, mechanisms built into the cell membrane have profound influence on movements across it.

Osmosis

We use the term **osmosis** for the movement of water across membranes in response to solute concentration gradients, a pressure gradient, or both. (Solutes are any substances dissolved in water.) Osmosis only occurs at a "differentially permeable" membrane, meaning some substances but not others can cross it.

Osmotic movements are affected by the relative concentrations of all the solutes present in the two fluids bathing both sides of the membrane. For cells, this means the fluid inside and the fluid outside the plasma membrane. When solute concentrations are equal in both fluids, or *isotonic*, there is no net osmotic movement of water in either direction. When the solute concentrations are not equal, one fluid is *hypotonic* (has less solutes) and the other is *hypertonic* (has more solutes). Water molecules tend to move from a hypotonic fluid to a hypertonic one.

If cells did not have mechanisms for adjusting to differences in solute concentrations, they would shrivel or burst, as Figures 3.6 and 3.7 illustrate.

Osmosis is the movement of water across a differentially permeable membrane in response to solute concentration gradients, a pressure gradient, or both.

Available Routes Across Membranes

Many small, electrically neutral molecules move readily across the lipid bilayer of cell membranes by simple diffusion. Water, oxygen, carbon dioxide, and ethanol are among them. (Remember, even though a water molecule shows polarity, it carries no *net* charge.)

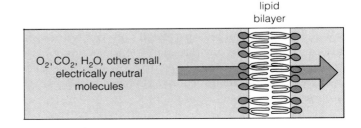

In contrast, glucose and other large, electrically neutral molecules almost never diffuse across the lipid bilayer.

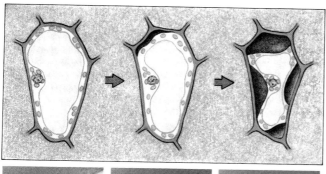

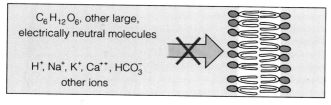

a b c

Figure 3.7 Effects of osmosis on plant cells. Most land plants grow in hypotonic soil, so water tends to move into their cells by osmosis. The cells have fairly rigid walls outside the plasma membrane; when water is absorbed, pressure increases against the wall. *Turgor pressure* refers to the internal pressure on a cell wall resulting from the inward osmotic movement of water.

If turgor pressure becomes high enough to counter the effects of cytoplasmic solutes, water also will be squeezed back out. When the *outward* flow equals the *inward* flow, cell walls cannot collapse and the soft plant parts stay erect. When soil solutes reach high concentrations, however, wilting occurs.

(**a**) At the start of this experiment, 10 grams of salt (NaCl) in 60 milliliters of water are added to a pot containing tomato plants. (**b**) The plant starts collapsing after about 5 minutes. (**c**) Wilting is severe in less than 30 minutes. The corresponding sketches show progressive plasmolysis (shrinking of cytoplasm away from the cell walls).

Neither do positively or negatively charged ions, no matter how small they are:

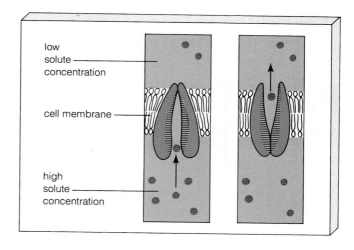

$C_6H_{12}O_6$, other large, electrically neutral molecules

H^+, Na^+, K^+, Ca^{++}, HCO_3^- other ions

Membrane proteins passively or actively transport these substances in one direction or the other.

In **passive transport,** solutes do not move across the lipid bilayer; they cross the membrane through the interior of proteins embedded in the bilayer. The proteins allow specific solutes to move *only* in the direction that diffusion would take them, so the movement is called "facilitated diffusion" (Figure 3.8).

In **active transport,** a transport protein receives an energy boost, usually from ATP, that allows it to move solutes with or against their concentration gradients across the membrane (Figure 3.9). One active transport system, the sodium-potassium pump, helps maintain high levels of potassium and low levels of sodium inside the cell.

By these mechanisms, cells and membrane-bound compartments within them are supplied with raw materials and rid of wastes. The mechanisms also help maintain pH and fluid volume within a functional range.

low solute concentration

cell membrane

high solute concentration

Figure 3.8 Possible mechanism of facilitated diffusion. The sketch shows a channel protein as if it were sliced through its midsection. Water-soluble molecules bind with the hydrophilic groups present on the interior surface of the protein (the shaded areas). The binding changes the protein shape in a way that allows a bound molecule to cross the membrane.

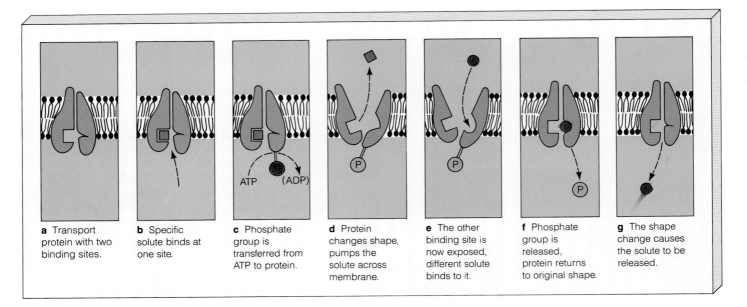

a Transport protein with two binding sites.

b Specific solute binds at one site.

c Phosphate group is transferred from ATP to protein.

d Protein changes shape, pumps the solute across membrane.

e The other binding site is now exposed, different solute binds to it.

f Phosphate group is released, protein returns to original shape.

g The shape change causes the solute to be released.

Figure 3.9 Simplified picture of an active transport system in animal cell membranes. Transport of one kind of solute across the membrane is coupled with transport of another kind in the opposite direction. The transport protein receives an energy boost from ATP and so undergoes changes in its shape that are necessary for the transport process.

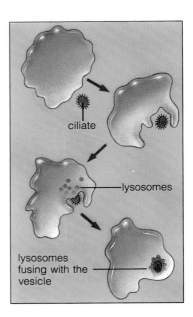

Figure 3.11 Endocytosis in an amoeba, a single-celled organism. The amoeba engulfs ciliates and other single cells. (Lysosomes are bags of digestive enzymes.)

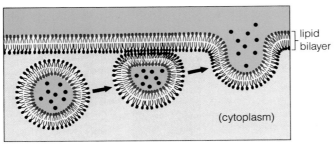

a Exocytosis

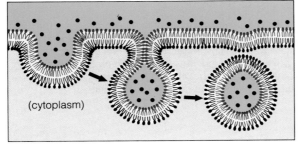

b Endocytosis

Figure 3.10 (**a**) Fusion of a vesicle with the plasma membrane during exocytosis. (**b**) Formation of a vesicle during endocytosis.

Endocytosis and Exocytosis

Cells construct many small sacs, or **vesicles,** out of membranes. Different vesicles transport or store substances within the cytoplasm. In **exocytosis,** for example, vesicles form inside the cytoplasm, then move to the plasma membrane and fuse with it, so that their contents are moved to the outside (Figure 3.10a). In **endocytosis,** a small patch of plasma membrane encloses particles at or near the cell surface, then it sinks in and pinches off, forming a vesicle that moves into the cytoplasm (Figures 3.10b and 3.11).

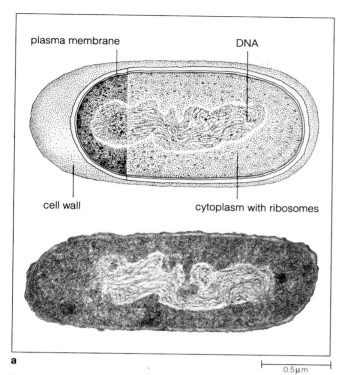

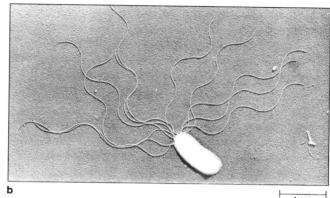

Figure 3.12 Bacterial body plans. (**a**) Sketch and transmission electron micrograph of *Escherichia coli.* (**b**) Surface view of *Pseudomonas marginalis*, which is equipped with bacterial flagella. Like other types of flagella, these structures are used in propelling the cell through its environment.

PROKARYOTIC CELLS— THE BACTERIA

Let's now turn to the characteristics of specific cell types, starting with bacteria. Bacteria are the smallest cells and, in structural terms, the simplest to think about. Most have a rigid or semirigid **cell wall** around the plasma membrane. The wall, formed by secretions from the bacterium, supports the cell and imparts shape to it (Figure 3.12). Beneath the wall, a plasma membrane controls the movement of substances into and out of the cytoplasm. Bacterial cells have a small volume of cytoplasm, with many ribosomes dispersed through it.

A **ribosome** has two subunits, each composed of RNA and protein molecules. In all cells, not just bacteria, ribosomes are workbenches for making proteins; new polypeptide chains are constructed at the ribosomal surface (page 154).

Bacterial cells are said to be **prokaryotic** because they do not have a nucleus. (The DNA simply occupies an irregularly shaped region of cytoplasm.) The word prokaryotic means "before the nucleus," and it implies that some forms of bacteria existed on earth before cells with a nucleus evolved.

EUKARYOTIC CELLS

Function of Organelles

All cells except bacteria are **eukaryotic.** They contain many **organelles,** which are membranous sacs and other compartmented portions of the cytoplasm. The most conspicuous organelle is the nucleus (eukaryotic means "true nucleus").

No chemical equipment in the world can match the eukaryotic cell for the sheer number of reactions that proceed in so small a space. Many of the reactions are incompatible. For example, a starch molecule can be put together by some reactions and pulled apart by others— but a cell would gain nothing if the reactions occurred at the same time on the same molecule. Reactions proceed smoothly in cells, largely for these reasons:

Organelles physically separate chemical reactions (many of which are incompatible) in the space of the cytoplasm.

Organelles separate different reactions in time, as when molecules are produced in one organelle, then used later in other reaction sequences.

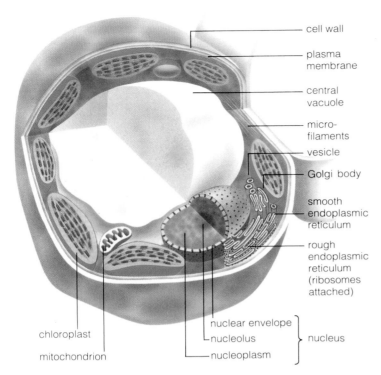

cell wall

plasma
membrane

central
vacuole

micro-
filaments

vesicle

Golgi body

smooth
endoplasmic
reticulum

rough
endoplasmic
reticulum
(ribosomes
attached)

chloroplast

mitochondrion

nuclear envelope
nucleolus
nucleoplasm
} nucleus

Figure 3.13 Generalized sketch of a plant cell, showing the types of organelles that may be present.

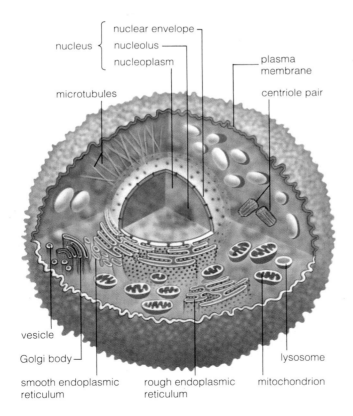

nuclear envelope
nucleolus
nucleoplasm
} nucleus

microtubules

plasma
membrane

centriole pair

vesicle

Golgi body

lysosome

smooth endoplasmic
reticulum

rough endoplasmic
reticulum

mitochondrion

Figure 3.14 Generalized sketch of an animal cell, showing the types of organelles that may be present.

Typical Components of Eukaryotic Cells

In general, eukaryotic cells contain the following organelles, each with specific functions:

nucleus	*physical isolation and organization of DNA*
endoplasmic reticulum	*modification of polypeptide chains into mature proteins; lipid synthesis*
Golgi bodies	*further modification, sorting, and shipping of proteins and lipids for secretion or for use in cell*
lysosomes	*intracellular digestion*
transport vesicles	*transport of a variety of materials to and from organelles and plasma membrane*
mitochondria	*ATP formation*

Besides the organelles just listed, eukaryotic cells have many thousands of *ribosomes*, either "free" in the cytoplasm or attached to certain membranes. Cells have a *cytoskeleton*, an internal network of protein filaments. The cytoskeleton is responsible for the cell's shape, internal organization, movement of structures and organelles, and often movement through the environment.

Only photosynthetic eukaryotic cells have *chloroplasts*, organelles of food production and storage. Often, fungal and plant cells have fluid-filled sacs called *central vacuoles*. For many protistans, fungi, and plants, a *cell wall* surrounds the plasma membrane.

Figures 3.13 through 3.16 show where organelles and structures might be located in a typical plant or animal cell. Keep in mind that calling them "typical" is like calling a squid or cactus a "typical" animal or plant; mind-boggling variation exists on the basic plan.

The Nucleus

Carbohydrates, lipids, proteins, and nucleic acids are the main building blocks of cell architecture. It takes the special class of proteins called enzymes to build and use those molecules as a cell grows, maintains itself, and reproduces. Thus, *cell structure and function begin with proteins—and instructions for building the proteins themselves are contained in DNA.*

A membrane-bound compartment, the **nucleus,** isolates the DNA in eukaryotic cells. A nucleus serves two functions: it helps control access to the DNA and simplifies DNA packaging for cell division. Every nucleus has the components listed in Table 3.1.

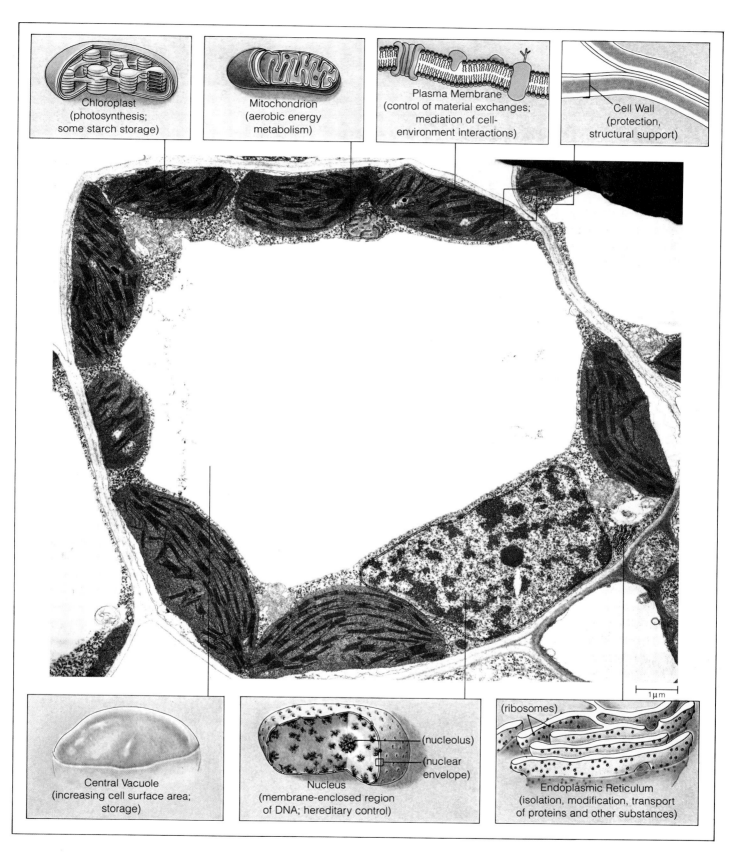

Figure 3.15 Transmission electron micrograph of a plant cell from a blade of Timothy grass, cross-section.

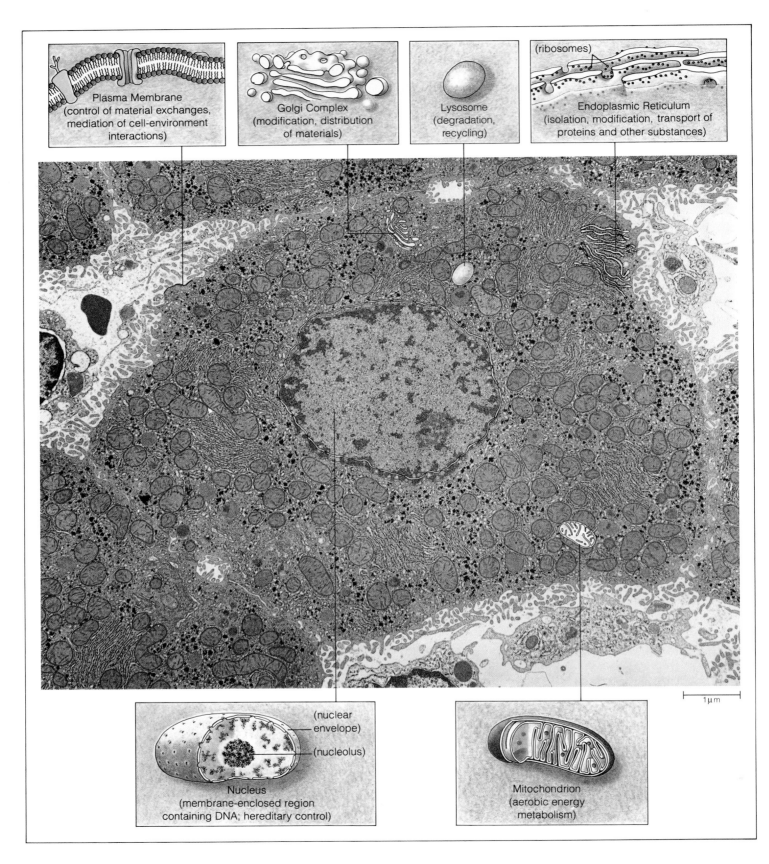

Plasma Membrane
(control of material exchanges, mediation of cell-environment interactions)

Golgi Complex
(modification, distribution of materials)

Lysosome
(degradation, recycling)

(ribosomes)

Endoplasmic Reticulum
(isolation, modification, transport of proteins and other substances)

1μm

(nuclear envelope)

(nucleolus)

Nucleus
(membrane-enclosed region containing DNA; hereditary control)

Mitochondrion
(aerobic energy metabolism)

Figure 3.16 Transmission electron micrograph of an animal cell from a rat liver, cross-section. Liver cells have an unusually large number of mitochondria.

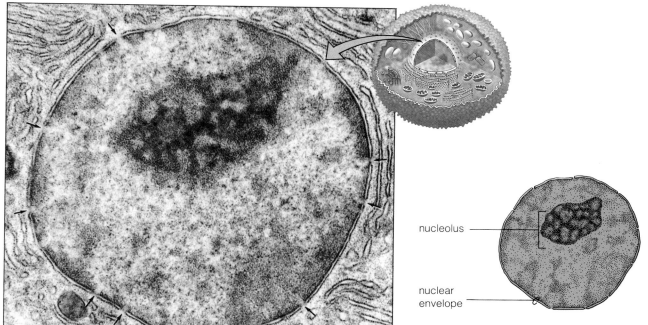

nucleolus

nuclear
envelope

Figure 3.17 Transmission electron micrograph of the nucleus from a pancreatic cell, cross-section. Arrows point to pores in the nuclear envelope.

Nuclear Envelope. The outermost part of the nucleus is a membrane called the **nuclear envelope.** The envelope is a double-membrane structure; it consists of two lipid bilayers. Ribosomes dot the membrane's outer surface, and pores extend across it at regular intervals (Figures 3.16 and 3.17). The nuclear envelope seems to be the boundary for controlled exchanges between the nucleus and cytoplasm, with its pores being passageways.

Nucleolus. As cells grow, two or more dense masses of irregular size and shape develop in the nucleus. Each mass is a **nucleolus** (plural, nucleoli), a region where subunits of ribosomes are assembled before shipment out of the nucleus. Figure 3.17 shows a nucleolus in a nondividing cell.

Chromosomes. Eukaryotic DNA is threadlike, with a great number of proteins attached to it like beads on a chain. Some of the proteins are enzymes; many form a scaffold for organizing the DNA during cell division.

Before a cell divides, its DNA molecules are duplicated (both new cells get all the necessary DNA instructions this way). Then the duplicated molecules fold and twist into condensed structures, proteins and all. Early microscopists could see only the condensed structures, and they called them **chromosomes** (colored bodies). Today, we call DNA and its proteins a chromosome regardless of whether it is in threadlike or condensed form.

Table 3.1	Components of the Nucleus
Component	Definition
Nuclear envelope	Double-membraned, pore-riddled boundary between the nuclear interior and the cytoplasm
Nucleolus	Dense cluster of the types of RNA and proteins used to assemble ribosomal subunits
Nucleoplasm	Fluid portion of the nucleus
Chromosomes	DNA molecules and numerous proteins attached to them

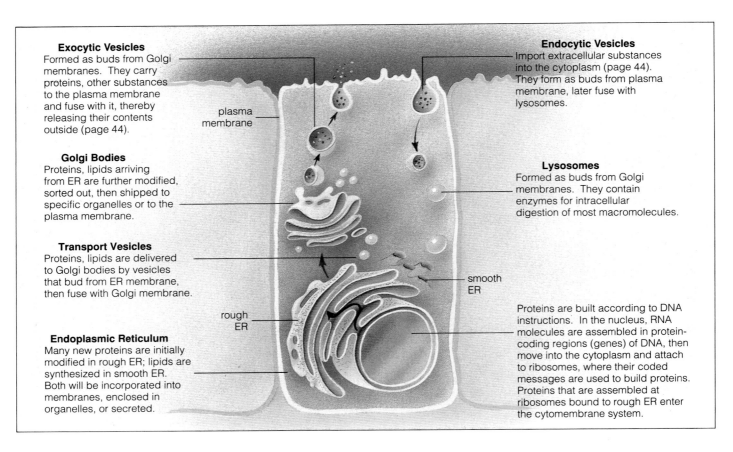

Exocytic Vesicles
Formed as buds from Golgi membranes. They carry proteins, other substances to the plasma membrane and fuse with it, thereby releasing their contents outside (page 44).

Golgi Bodies
Proteins, lipids arriving from ER are further modified, sorted out, then shipped to specific organelles or to the plasma membrane.

Transport Vesicles
Proteins, lipids are delivered to Golgi bodies by vesicles that bud from ER membrane, then fuse with Golgi membrane.

Endoplasmic Reticulum
Many new proteins are initially modified in rough ER; lipids are synthesized in smooth ER. Both will be incorporated into membranes, enclosed in organelles, or secreted.

plasma membrane

rough ER

Endocytic Vesicles
Import extracellular substances into the cytoplasm (page 44). They form as buds from plasma membrane, later fuse with lysosomes.

Lysosomes
Formed as buds from Golgi membranes. They contain enzymes for intracellular digestion of most macromolecules.

smooth ER

Proteins are built according to DNA instructions. In the nucleus, RNA molecules are assembled in protein-coding regions (genes) of DNA, then move into the cytoplasm and attach to ribosomes, where their coded messages are used to build proteins. Proteins that are assembled at ribosomes bound to rough ER enter the cytomembrane system.

Figure 3.18 Cytomembrane system. Endoplasmic reticulum, transport vesicles, Golgi bodies, and endocytic vesicles are components of the secretory pathway of this system (dark arrows).

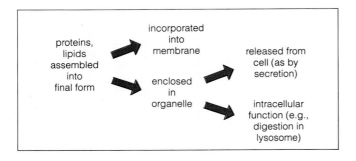

Figure 3.19 Outline of the flow of proteins and lipids through the cytomembrane system.

Cytomembrane System

The polypeptide chains of proteins are assembled in the cytoplasm. What happens to the newly formed chains? Many are stockpiled with other materials dissolved in the cytoplasm; others enter the cytomembrane system. The **cytomembrane system** includes the *endoplasmic reticulum, Golgi bodies, lysosomes,* and a variety of *vesicles.* As Figures 3.18 and 3.19 indicate, proteins and lipids take on their final form and are distributed by way of this system.

Endoplasmic Reticulum and Ribosomes. The membrane of **endoplasmic reticulum,** or ER, has rough and smooth regions, owing largely to the presence or absence of ribosomes on the surface of the ER that faces the cytoplasm.

Rough ER has many ribosomes attached, and it often is arranged as stacked, flattened sacs (Figure 3.20). Polypeptide chains have polysaccharide groups attached to them as they pass through rough ER. These chains are destined for secretion outside the cell or for delivery to several organelles. Rough ER is abundant in secretory cells, such as cells of the pancreas that produce and secrete digestive enzymes.

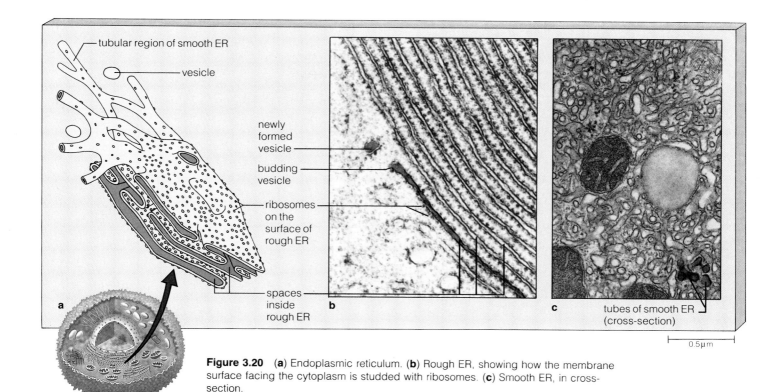

- tubular region of smooth ER
- vesicle
- newly formed vesicle
- budding vesicle
- ribosomes on the surface of rough ER
- spaces inside rough ER

a

b

c

tubes of smooth ER (cross-section)

0.5 μm

Figure 3.20 (**a**) Endoplasmic reticulum. (**b**) Rough ER, showing how the membrane surface facing the cytoplasm is studded with ribosomes. (**c**) Smooth ER, in cross-section.

Smooth ER is free of ribosomes and curves through the cytoplasm like interconnecting pipes. It is not involved in protein synthesis; it is a specific region of ER from which vesicles carrying newly synthesized proteins and lipids are budded. Smooth ER is highly developed in seeds and in animal cells that secrete steroid hormones. Drugs and some harmful by-products of metabolism are inactivated in the smooth ER of liver cells.

Golgi Bodies. In a **Golgi body,** many proteins and lipids undergo final processing, then are sorted out and packaged for specific destinations. A Golgi body outwardly resembles a stack of pancakes, usually curled at the edges (Figure 3.21). The topmost pancakes bulge at the edges, then the bulges break away as vesicles. Some secretory cells concentrate and store products in these vesicles until the cell receives signals to release them. Figure 3.19 shows where the Golgi bodies fit in the secretory pathway.

Lysosomes. Some vesicles budding from Golgi bodies become **lysosomes,** the main organelles of digestion inside the cell. Different enzymes in these membrane bags can break down polysaccharides, nucleic acids, proteins, and some lipids. Lysosomes fuse with the vesicles carrying a variety of substances or worn-out cell parts to be degraded. Lysosomes also can destroy bacteria and foreign particles.

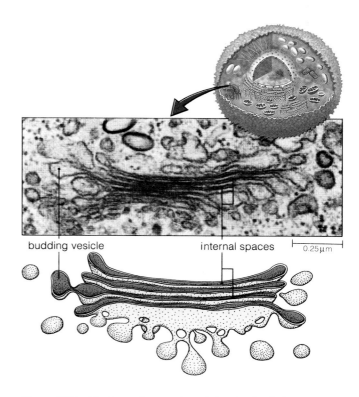

budding vesicle   internal spaces   0.25 μm

Figure 3.21 Electron micrograph and sketch of a Golgi body.

Mitochondria

In the **mitochondrion** (plural, mitochondria), much of the energy stored in carbohydrates is released and used to form molecules such as ATP—which provides energy to drive a variety of cellular reactions. Mitochondria can extract far more energy from carbohydrates than can be done by any other means, and they do so with the help of oxygen. When you breathe in, you are taking in oxygen for your mitochondria.

There may be a dozen to a thousand mitochondria in eukaryotic cells. Each has an outer membrane facing the cytoplasm and an inner membrane, usually with many deep, inward folds (Figure 3.22). The double-membrane system creates two compartments that are used in ATP formation.

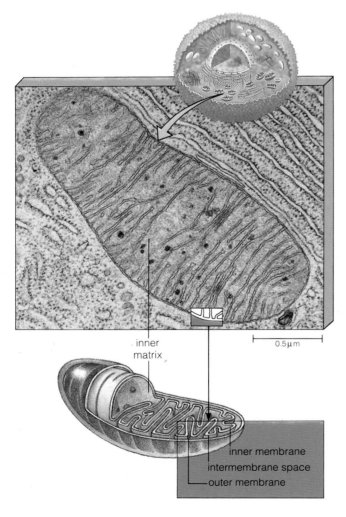

inner
matrix

0.5μm

inner membrane
intermembrane space
outer membrane

Figure 3.22 Micrograph and generalized sketch of a mitochondrion.

Specialized Plant Organelles

Chloroplasts. The **chloroplast** functions in photosynthesis. Like a mitochondrion, a chloroplast has a double-membrane system (Figure 3.23). Parts of the inner membrane are often organized as stacked disks; here, pigments, enzymes, and other molecules trap sunlight energy and take part in ATP formation. Sugars and other products of photosynthesis form in the semifluid substance that surrounds the stacks.

Chloroplasts often are oval or disk shaped and may be green, yellow-green, or golden-brown. The color depends on the kinds and amounts of light-absorbing pigment molecules in their membranes. Chlorophyll, a green pigment, is an example.

Central Vacuoles. Mature, living plant cells often have a fluid-filled **central vacuole** (Figure 3.15). This organelle usually occupies fifty to ninety percent of the cell interior, so there is only a narrow zone of cytoplasm between the vacuole and the plasma membrane.

A central vacuole can store amino acids, sugars, ions, and toxic wastes. It also increases cell size and surface area. During growth, plant cell walls enlarge under the force of water pressure that builds up inside the vacuole. The cell itself enlarges under this force, and its increased surface area enhances the rate at which it can absorb minerals.

The Cytoskeleton

Each cell type has a characteristic shape and internal organization made possible by its own tiny **cytoskeleton.** This interconnected system of bundled fibers, slender threads, and lattices extends from the nucleus all the way to the plasma membrane (Figure 3.24). Its main components are *microtubules, microfilaments,* and *intermediate filaments* assembled from protein subunits:

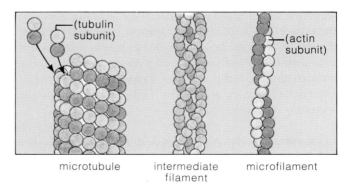

(tubulin subunit)

(actin subunit)

microtubule   intermediate filament   microfilament

Some parts of the cytoskeleton are *transient;* they appear and disappear at different times in the life of a cell. Thus microtubules assemble when it is time to form

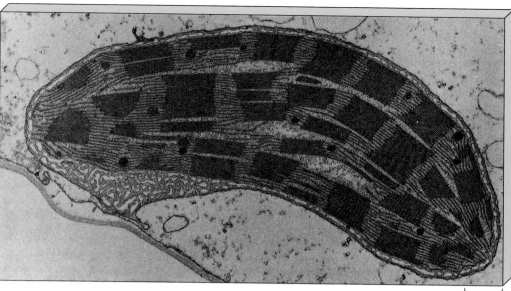

Figure 3.23 Micrograph and generalized sketch of a chloroplast. A semifluid substance (the stroma) surrounds an elaborate system of membrane compartments. Commonly, many of the compartments are organized as stacks of flattened disks (grana, singular, granum). Like mitochondria, chloroplasts have a double-membrane envelope.

0.5μm

chloroplast
envelope

outer membrane ⎯

intermembrane space ⎯

inner membrane ⎯

granum

stroma

a "spindle" for moving chromosomes during cell division, then disassemble when the task is done. But many parts of the cytoskeleton are *permanent*. For example, permanent filaments in skeletal muscle cells are the basis of contraction. Flagella and cilia, to be described next, are also permanent.

Flagella and Cilia. Microtubule-containing structures propel many free-living eukaryotic cells through their surroundings. These structures are called **flagella** (singular, flagellum). Certain protistans, including the ones that cause African sleeping sickness, have one or more flagella. So do sperm cells.

Cilia (singular, cilium), another type of microtubular structure, typically are arrayed at the cell surface (Figure 3.25). Many free-living cells use cilia for stirring up their surroundings. For example, airborne bacteria and other particles are drummed out of your lungs when many thousands of cells lining the air tubes beat their cilia in coordination.

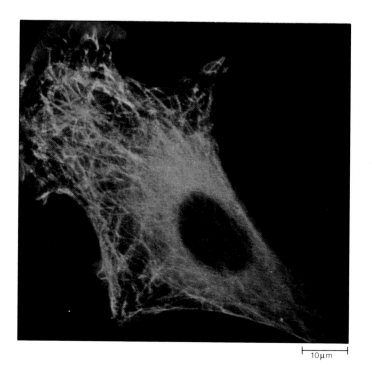

Figure 3.24 Cytoskeleton of a fibroblast, a type of cell that gives rise to certain connective tissues in animals. It has been made visible by "fluorescence microscopy." (Molecules that bind only to specific proteins were labeled with fluorescent dyes, then injected into the cell. The glow from the bound molecules marked the location of three different proteins.) In this composite of three images, actin (blue) and vinculin (red) are associated with microfilaments. Tubulin (green) is associated with microtubules.

10μm

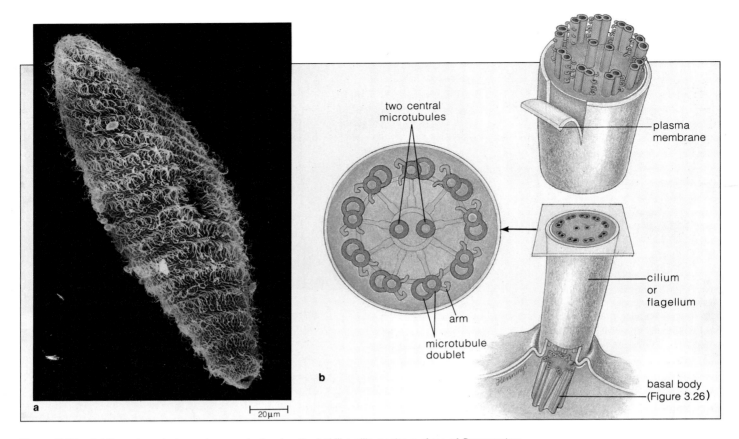

Figure 3.25 (**a**) Scanning electron micrograph showing the hairlike cilia on the surface of *Paramecium*. (**b**) The 9+2 array of microtubules in a cilium or flagellum.

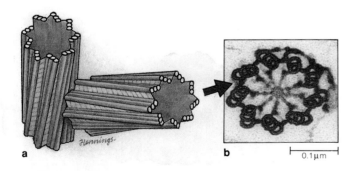

Figure 3.26 (**a**) A pair of centrioles, which occur near the nucleus of many cells. Centrioles apparently help organize the cytoskeleton; in many species they are patterns for basal bodies, which give rise to the microtubular core of cilia and flagella. (**b**) Electron micrograph of a basal body, thin section.

Flagella are longer and less numerous than cilia, but both have the same organization (Figure 3.25). Nine pairs of microtubules ring two central microtubules, in what is called a *9+2 array*.

What Organizes the Cytoskeleton? Microtubules play key roles in the cell's internal organization, movement, and division. But what organizes the microtubules themselves? That seems to depend on the number, type, and location of small masses of proteins and other substances in the cytoplasm. These masses are called "microtubule organizing centers" (or MTOCs for short).

An MTOC near the nucleus in most animal cells includes a pair of **centrioles.** Each is a short cylinder of triplet microtubules (Figure 3.26). This centriole pair may govern the plane of cell division. During development, each cell must divide at a prescribed angle relative to the other cells. Successive division planes influence the shape of the embryo and, eventually, the adult form. Centrioles also serve as patterns for assembling **basal bodies,** which in turn organize the growth of microtubules in flagella or cilia.

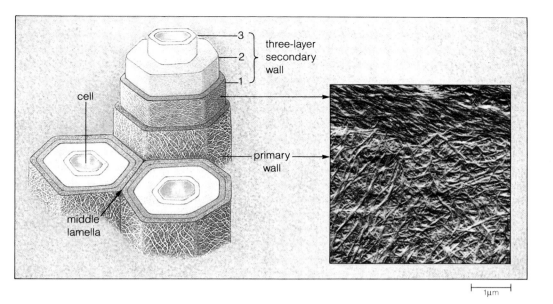

Figure 3.27 Primary and secondary walls of plant cells. Strands in the micrograph are mostly cellulose. Adjacent cells are cemented together at primary walls. The cementing material (middle lamella) contains pectin compounds.

During growth, cellulose strands are bundled together and added to the developing primary wall. Afterward, many plant cells also deposit materials that form an inner, rigid secondary wall, often of several layers. Cutin, suberin, and waxes commonly are embedded in cell walls at the plant's surface; they play a protective role and help reduce water loss.

Cell Surface Specializations

Cell Walls.   For many cells, surface deposits outside the plasma membrane form coats, capsules, sheaths, and walls. **Cell walls** occur among bacteria, protistans, fungi, and plants. (Animal cells do not produce walls, although some secrete products to the surface layer of tissues in which they occur.) Figure 3.27 shows the structure of plant cell walls.

Most cell walls have carbohydrate frameworks. They generally provide support and resist mechanical pressure, as when they keep plant cells from stretching too much when they expand with incoming water during growth. Even the most solid-looking walls have microscopic spaces that make them porous, so water and solutes can move to and from the plasma membrane.

The Extracellular Matrix.   A meshwork called the **extracellular matrix** holds animal cells and tissues together. The matrix influences how cells in a tissue will divide and what their shape will be; its composition influences cell metabolism. Its components include collagen and other fibrous proteins, proteins with oligosaccharides attached, and specialized polysaccharides that form a jellylike "ground substance." Nutrients, hormones, and other molecules readily diffuse from cell to cell through the ground substance.

Cell Junctions.   In multicelled organisms, each living cell interacts with its physical surroundings *and* with its cellular neighbors at **cell junctions.** At tissue surfaces, cells link tightly together and so keep the interior of the organism (or organ) from being indiscriminately exposed to the outside world. In tissues, cells of the same type recognize one another and stick together. Finally, in tissues where cells must act in coordinated fashion (as in heart muscle), the cells share channels to exchange signals, nutrients, or both.

We will consider the extracellular matrix and cell junctions in later chapters. For now, the point to remember is this: A multicelled organism depends on coordinated linkage and communication between cells.

SUMMARY

1. All living things are made of one or more cells, each cell is the basic living unit, and a new cell arises only from cells that already exist. These are the three points of the cell theory.

2. All cells have cytoplasm and a plasma membrane, which acts as a boundary for exchanges between the cell's interior and its surroundings. Eukaryotic cells have a nucleus, an organelle in which the DNA resides. Prokaryotic cells (bacteria) do not have a nucleus.

3. Eukaryotic cells have membranous compartments called organelles. Table 3.2 lists the main organelles and the structures common to prokaryotes or eukaryotes.

4. Cell membranes consist mostly of phospholipids and proteins. A lipid bilayer gives the membrane its structure and its impermeability to water-soluble substances. Proteins in the bilayer or at one of its surfaces carry out most membrane functions.

5. Membrane functions include: control of substances moving into and out of cells, compartmentalization of the cytoplasm into specialized organelles, signal reception, and cell-to-cell recognition.

Table 3.2 Summary of Typical Components of Prokaryotic and Eukaryotic Cells

Cell Component	Function	Prokaryotic	Eukaryotic			
		Moneran	Protistan	Fungus	Plant	Animal
Cell wall	Protection, structural support	✔*	✔*	✔	✔	none
Plasma membrane	Regulation of substances moving into and out of cell	✔	✔	✔	✔	✔
Nucleus	Physical isolation and organization of DNA	none	✔	✔	✔	✔
DNA	Encoding of hereditary information	✔	✔	✔	✔	✔
RNA	Transcription, translation of DNA messages into specific proteins	✔	✔	✔	✔	✔
Nucleolus	Assembly of ribosomal subunits	none	✔	✔	✔	✔
Ribosome	Protein synthesis	✔	✔	✔	✔	✔
Endoplasmic reticulum	Modification of many proteins into mature form; lipid synthesis	none	✔	✔	✔	✔
Golgi body	Final modification of proteins, lipids; sorting and packaging them for shipment inside cell or for export	none	✔	✔	✔	✔
Lysosome	Intracellular digestion	none	✔	✔*	✔	✔
Mitochondrion	ATP formation	**	✔	✔	✔	✔
Photosynthetic pigment	Light-energy conversion	✔*	✔*	none	✔	none
Chloroplast	Photosynthesis, some starch storage	none	✔*	none	✔	none
Central vacuole	Increasing cell surface area, storage	none	none	✔*	✔	none
Cytoskeleton	Cell shape, internal organization, basis of cellular motion	none	✔*	✔*	✔*	✔
Complex flagellum, cilium	Movement	none	✔*	✔*	✔*	✔

*Known to occur in at least some groups.
**Aerobic reactions do occur in many groups, but mitchondria are not involved.

6. Substances cross a membrane by the following mechanisms:

a. Diffusion (natural, unassisted movement of solutes down a concentration gradient)

b. Osmosis (movement of water across a membrane in response to solute concentration gradients, a pressure gradient, or both)

c. Facilitated diffusion (movement of a substance through a membrane protein, in the direction that diffusion would take it)

d. Active transport (movement of a substance, with or against a concentration gradient, through a transport protein that must receive an energy boost to operate)

Review Questions

1. State the three principles of the cell theory. 38

2. Most cells share three structural features: a nucleus, cytoplasm, and a plasma membrane. Describe the functions of each. 38

3. Describe the fluid mosaic model of plasma membranes. 40

4. List the structural features of all cell membranes. 40

5. Describe some functions of membrane proteins. 43

6. Diffusion accounts for the greatest volume of substances moving into and out of cells. How does diffusion work? 41

7. What is osmosis, and what causes its occurrence? 42

8. Explain the difference between active and passive transport mechanisms. 43

9. Suppose you want to observe details of the surface of an insect's compound eye. Would you benefit most from a compound light microscope, transmission electron microscope, or scanning electron microscope? *39*

10. Eukaryotic cells generally contain these organelles: nucleus, endoplasmic reticulum, Golgi bodies, lysosomes, microbodies, and mitochondria. Describe the function of each. *46*

11. Describe the structure and function of chloroplasts. *52*

12. What is a cytoskeleton? How do you suppose it might aid in cell functioning? *52*

13. Cell walls occur among which organisms: bacteria, protistans, plants, fungi, or animals? Are cell walls solid or porous? *45,56*

14. With a sheet of paper, cover the Table 3.2 column entitled Function. Then name the primary functions of the cell structures listed in this table. *56*

Self-Quiz *(Answers in Appendix IV)*

1. All living things are composed of one or more _____.

2. The _____ is the boundary where materials are exchanged between the inside and outside of a cell.

3. Unlike eukaryotic cells, prokaryotic cells _____.
   a. do not have a plasma membrane
   b. have RNA, not DNA
   c. do not have a nucleus
   d. all of the above

4. Organelles _____.
   a. are membranous
   b. are present in all cells
   c. separate chemical reactions in time and space
   d. all of the above

5. Plant cells but not animal cells have a _____.
   a. mitochondrion
   b. nucleolus
   c. ribosome
   d. cell wall

6. Eukaryotic DNA is isolated and organized in the _____.
   a. nucleolus
   b. nucleus
   c. lysosome
   d. Golgi body

7. Cell membranes consist largely of a _____.
   a. carbohydrate bilayer and proteins
   b. protein bilayer and phospholipids
   c. phospholipid bilayer and proteins
   d. nucleic acid bilayer and proteins

8. Membrane functions are accomplished principally by _____.
   a. proteins in or on the bilayer
   b. phospholipids in or on the bilayer
   c. nucleic acids in or on the bilayer
   d. hormones in or on the bilayer

9. _____ is the movement of a solute down its concentration gradient.
   a. osmosis
   b. active transport
   c. diffusion
   d. facilitated diffusion

10. In response to solute concentration gradients or pressure gradients, water moves across a membrane by _____.
    a. osmosis
    b. active transport
    c. diffusion
    d. facilitated diffusion

11. The passive movement of a substance *through channel proteins* as it follows its concentration gradient across a cell membrane is called _____.
    a. osmosis
    b. active transport
    c. diffusion
    d. facilitated diffusion

12. The energy-assisted movement of a substance *through transport proteins* that span the cell membrane is called _____.
    a. osmosis
    b. active transport
    c. diffusion
    d. facilitated diffusion

13. Match each type of organelle with its correct function.
    a. mitochondrion   _____ protein synthesis
    b. chloroplast   _____ movement
    c. ribosome   _____ intracellular digestion
    d. ER   _____ sort and package lipids and
    e. nucleolus   proteins for shipment
    f. Golgi body   _____ photosynthesis
    g. lysosome   _____ protein modification, lipid
    h. flagellum, cilium   synthesis
       _____ ATP formation
       _____ ribosome subunit assembly

Selected Key Terms

active transport *43*	flagellum *53*
cell wall *45*	Golgi body *51*
central vacuole *52*	lipid bilayer *40*
centriole *54*	lysosome *51*
chloroplast *52*	microtubule *52*
chromosome *49*	mitochondrion *52*
cilium *53*	nuclear envelope *49*
concentration gradient *41*	nucleolus *49*
cytomembrane system *50*	nucleus *38*
cytoplasm *38*	organelle *45*
cytoskeleton *52*	osmosis *42*
diffusion *41*	passive transport *43*
endocytosis *44*	plasma membrane *38*
endoplasmic reticulum (ER) *50*	prokaryotic cell *45*
eukaryotic cell *45*	ribosome *45*
exocytosis *44*	vesicle *44*

Readings

Bretscher, M. October 1985. "The Molecules of the Cell Membrane." *Scientific American* 253(4):100–108.

deDuve, C. 1985. *A Guided Tour of the Living Cell.* New York: Freeman. Beautifully illustrated introduction to the cell; two short volumes.

Shih, G., and R. Kessel. 1982. *Living Images: Biological Structures Revealed by Scanning Electron Microscopy.* Boston: Science Books International. Paperback.

4 GROUND RULES OF METABOLISM

KEY CONCEPTS

1. Cells have the capacity to trap and use energy for building, stockpiling, breaking apart, and eliminating substances in ways that contribute to survival and reproduction. These activities are called metabolism.

2. The complex organization characteristic of life is maintained by a steady input of energy. To stay alive, cells must replace the energy that they inevitably lose during each metabolic reaction. Directly or indirectly, the sun is the source of energy replacement for nearly all organisms.

3. The concentrations of substances in cells are maintained, increased, or decreased through the coordination of different metabolic pathways.

When you look at a living cell through a microscope, you are watching a form pulsing with activity. Through its movements, the cell is identifying and taking in raw materials suspended in the water droplet on the slide. Even as you watch it, the cell is using energy and materials as it builds and maintains its membranes, its stores of chemical compounds, its DNA, its pools of enzymes. It is alive, it is growing, it may divide in two. Multiply this activity by *65 trillion cells* and you have an inkling of what goes on in your own body as you sit quietly, observing that single cell!

With this chapter we turn to the dynamics of cellular life—that is, to metabolism. **Metabolism** refers to all those chemical reactions by which cells acquire and use energy for synthesizing, accumulating, breaking apart, and eliminating substances in ways that contribute to survival and reproduction.

ENERGY AND LIFE

Energy is a capacity to make things happen, to cause change, to do work. You use energy to put a hard wax finish on a car. (Buffing a paste wax breaks attractions between molecules, and the hard, glossy finish results from new molecular positions.) A cell also uses energy to make things happen, as when it makes large molecules from smaller, simpler ones.

Figure 4.1 All events large and small, from the birth of stars to the death of a microorganism, are governed by laws of energy. Shown here, eruptions on the sun's surface and, to the right, *Volvox*—each sphere a colony of microscopically small single cells able to capture sunlight energy that indirectly drives their life processes.

You (or the cell) cannot create your own energy from scratch; you must borrow it from someplace else. That, basically, is the message of the **first law of thermodynamics:**

The total amount of energy in the universe remains constant. More energy cannot be created, and existing energy cannot be destroyed. It can only undergo conversion from one form to another.

Consider what this law means. The universe has only so much energy, distributed in a variety of forms. One form can be converted to another, as when corn plants absorb sunlight energy and convert it to the chemical energy of starch. By eating corn, you can extract and convert its energy to other forms, such as mechanical energy for your movements. With each conversion, a little energy escapes to the surroundings as heat. (Your body steadily gives off about as much heat as a 100-watt light bulb because of ongoing conversions in your cells.) However, none of the energy vanishes; it just ends up someplace else (Figure 4.2).

This brings us to the *quality* of energy available. Energy concentrated in a starch molecule is high quality, since it lends itself to conversions. Heat energy spread out in the atmosphere is low quality since, for all practical purposes, it can't be gathered up and converted to other forms.

It happens that the amount of low-quality energy in the universe is increasing. The reason is that no energy conversion can ever be 100 percent efficient—some of the energy goes off as heat. That is the point of the **second law of thermodynamics:**

The spontaneous direction of energy flow is from high-quality to low-quality forms. With each conversion, some energy is randomly dispersed in a form (usually heat) that is not as readily available to do work.

Without energy to maintain it, any organized system tends to become disorganized over time. **Entropy** is a measure of the degree of its disorder. Think about the Egyptian pyramids—originally organized, presently crumbling, and many thousands of years from now, dust. The *ultimate* destination of the pyramids and everything else in the universe is a state of maximum entropy, because there won't be any high-quality energy available to make things happen, to do work. Billions of years from now, everything probably will be at the same temperature—and energy conversions as we know them will never happen again.

Can it be that life is one glorious pocket of resistance to the rather depressing flow toward maximum entropy? After all, every time a new organism grows, energy becomes more concentrated and organized, not less so! Yet a simple example will show that the second law does indeed apply to life on earth.

The primary source of energy for nearly all organisms is the sun—which is steadily losing energy. Plants intercept some sunlight energy, then they lose energy to other organisms that feed, directly or indirectly, on plants. At each energy transfer along the way, more heat is added to the universal pool. Overall, then, energy is still flowing in one direction. *The world of life maintains a high degree of organization only because it is being resupplied with energy lost from someplace else.*

There is a steady flow of sunlight energy into the interconnected web of life, and this compensates for the steady flow of energy leaving it.

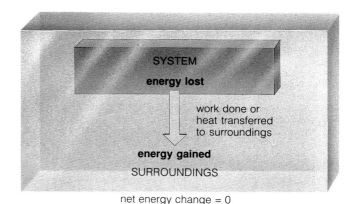

net energy change = 0

Figure 4.2 The nature of energy. According to the first law of thermodynamics, the total energy content of any system and its surroundings remains constant. "System" means all matter within a specific region, such as a plant, a DNA molecule, or a galaxy. The "surroundings" can be some specified region in contact with the system or can be as vast as the entire universe.

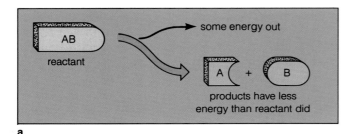

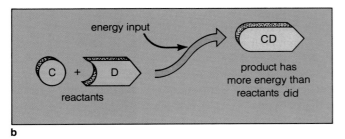

a

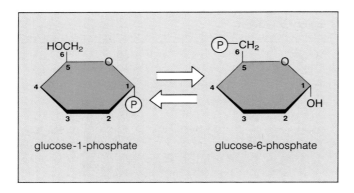

b

Figure 4.3 Energy changes in metabolic reactions. (**a**) In exergonic reactions, products have less energy than the reactants did. Some energy released by the reactions is harnessed to do cellular work. The breakdown of glucose into carbon-containing bits is an example (page 28).

(**b**) In endergonic reactions, products have more energy than the reactants did. An energy input drives these reactions. An example is photosynthesis, in which sunlight energy drives the linkage of carbon dioxide and water into sugars and other compounds of higher energy content.

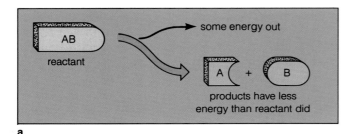

glucose-1-phosphate   glucose-6-phosphate

Figure 4.4 A reversible reaction. In all cells, phosphate can become attached to a glucose molecule in ways that prime the glucose to enter metabolic reactions. With high concentrations of glucose-1-phosphate, the reaction tends to run in the forward direction; with high concentrations of glucose-6-phosphate, it runs in reverse. (The "1" and "6" identify the particular carbon atom of the glucose ring to which phosphate is attached.)

THE NATURE OF METABOLISM

Energy Changes in Metabolic Reactions

The next two chapters describe the reactions of photosynthesis and aerobic respiration, the main pathways of energy flow in the world of life. Those reactions will make more sense if you keep the following three concepts in mind.

First, in the cellular world, substances present at the end of a reaction (the products) may have less or more energy than the starting substances (the reactants) had. Reactions showing a net *loss* in energy are exergonic (meaning "energy out"). Reactions showing net *gain* in energy are endergonic ("energy in"). They occur only when extra energy (as from sunlight) is fed into them, as Figure 4.3 indicates.

Second, most reactions are *reversible*. This means that when the concentration of product is high enough, then some number of product molecules will revert to reactants. Reversible reactions are indicated by arrows running in the "forward" and "reverse" directions (Figure 4.4).

Third, unless other events in the cell keep it from doing so, a reaction that is reversible approaches *dynamic equilibrium*. Then, a forward and reverse reaction proceed at equal rates (Figure 4.5).

Metabolic Pathways

In the cellular world, building up or tearing down substances requires control over the directions in which different reactions proceed. Why? Cells use only so many molecules of different substances at a given time, and they have only so much space to hold any excess. If cells were to produce more than they could use, put into storage, or secrete, the excess might cause problems.

For example, in the genetic disorder phenylketonuria (PKU), a series of reactions is blocked and a substance (phenylalanine) reaches high concentrations in the body. The excess enters into reactions that produce phenylketones, and the accumulation of those substances leads to severe mental retardation.

Normally, cells maintain, increase, and decrease the concentrations of substances by coordinating a variety of metabolic pathways. A **metabolic pathway** is an orderly series of reactions, the steps of which are quickened with the help of specific enzymes. Most sequences are linear; some are circular (Figure 4.6). Branches often link different pathways, with products of one pathway serving as reactants for others.

Overall, the main metabolic pathways are degradative or biosynthetic. In **degradative pathways,** carbohydrates, lipids, and proteins are broken down in step-

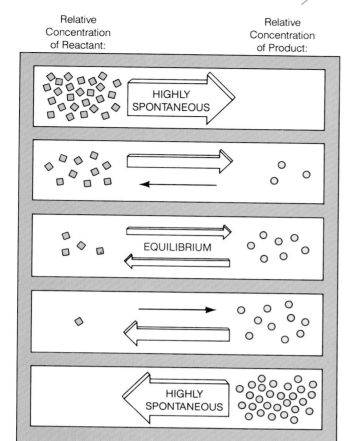

Relative Concentration of Reactant:

Relative Concentration of Product:

HIGHLY SPONTANEOUS

EQUILIBRIUM

HIGHLY SPONTANEOUS

Figure 4.5 Chemical equilibrium. With high concentrations of reactant molecules, reactions generally proceed most strongly in the forward direction. With high concentrations of product molecules, they proceed most strongly in reverse. At chemical equilibrium, the *rates* of the forward and reverse reactions are equal.

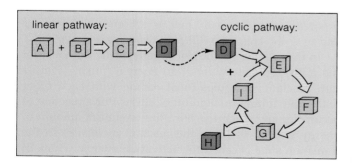

linear pathway:

cyclic pathway:

Figure 4.6 Linear and cyclic metabolic pathways.

wise reactions that lead to products of lower energy. Some of the energy released is used to do cellular work, some is stored, and some is lost as heat. In **biosynthetic pathways,** small molecules are assembled into lipids, proteins, and other large biological molecules.

The participants in metabolic pathways can be defined in the following ways:

reactants	*substances able to enter into a reaction; also called substrates*
enzymes	*proteins that catalyze (speed up) reactions*
cofactors	*small molecules and metal ions that help enzymes or that carry atoms or electrons stripped from the substrate to another site*
energy carriers	*mainly ATP, which readily donates energy to diverse reactions*
end products	*the substances present at the conclusion of a metabolic pathway*

Let's take a quick look at some of these substances and at their roles in metabolism.

ENZYMES

A cup of sugar left undisturbed for twenty years changes very little. But when you eat sugar, it rapidly undergoes chemical change. Enzymes synthesized by some of your cells account for the difference in the rate of change. **Enzymes** are proteins with enormous catalytic power, meaning they greatly enhance the rate at which specific reactions approach equilibrium. They do not make anything happen that would not eventually happen on its own. They just make it happen faster—often more than a million times faster.

Each type of enzyme is highly selective about its **substrates,** the reactants it will deal with. For example, thrombin (an enzyme involved in blood clotting) helps break a peptide bond *only* when the bond is between these two amino acids in this order:

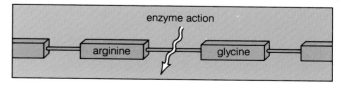

enzyme action

arginine   glycine

Enzyme molecules are not permanently altered or used up in reactions involving their substrates; they can be used over and over again.

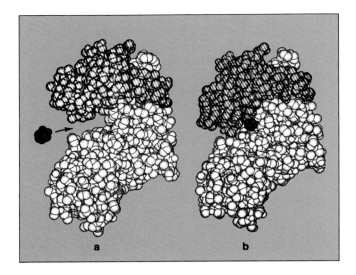

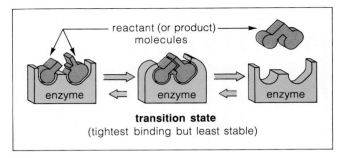

reactant (or product) molecules

enzyme   enzyme   enzyme

transition state
(tightest binding but least stable)

Figure 4.8 Induced-fit model of enzyme-substrate interactions. Only when the substrate is bound in place is the enzyme's active site complementary to it. The most precise fit occurs during a transition state that precedes the reaction. An enzyme-substrate complex is short-lived, partly because only weak bonds hold it together.

Figure 4.7 Model of the induced fit between an enzyme (hexokinase) and its bound substrate (a glucose molecule, shown here in red).

(**a**) The cleft into which the glucose is heading is the enzyme's active site. (**b**) In this enzyme-substrate complex, notice how the enzyme shape is altered temporarily: the upper and lower parts now close in around the substrate.

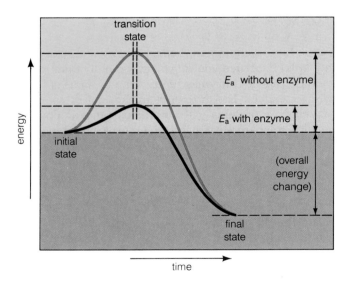

Figure 4.9 Energy hill diagram showing the effect of enzyme action. An enzyme greatly enhances the rate at which a reaction proceeds because it lowers the required activation energy (E_a). In other words, not as much collision energy is needed to boost the reactant molecules to the crest of the energy hill (the transition state).

Enzyme Structure and Function

At least one surface region of an enzyme molecule is folded in the shape of a crevice, called an **active site,** where a specific reaction is catalyzed (Figure 4.7).

As long ago as 1890, Emil Fischer thought the shape of some region of the enzyme's surface matches that of its substrate, like a lock precisely matching its key. Today we know the match is not so rigid. In Daniel Koshland's **induced-fit model,** an active site making contact with its substrate almost but not quite matches it (Figure 4.8). This means the bonding between them is not as strong as it could be. Even so, the interaction is enough to induce structural changes in the active site and to distort the bound substrate. When substrates fit most precisely into the active site, they are in an activated condition called the "transition state." Now the reaction proceeds spontaneously, just as a boulder pushed up and over the crest of a hill rolls down on its own.

In the cellular world, reactants typically do not enter the transition state without an energy "push." Simply put, reactant molecules must collide with each other with some minimum amount of energy for a reaction to occur. For any given reaction, the minimum amount of energy needed to bring the reactant molecules to the transition state is called the **activation energy.** Think of that amount as the "energy hill" over which the molecules must be pushed (Figure 4.9).

An enzyme increases the rate of a given reaction by *lowering* the required activation energy. How? One way is that weak but extensive bonding at the active site puts the enzyme's substrates in positions that promote reaction. (In contrast, reactants colliding on their own do so from random directions, so mutually attractive chemical groups may not make contact and reaction may not occur.)

An enzyme enhances the rate of a given reaction by lowering the activation energy required for it.

Effects of Temperature and pH on Enzymes

Each type of enzyme functions best within a certain temperature range (Figures 4.10 and 4.11). When the temperature becomes too high, reaction rates decrease sharply. The increased thermal energy disrupts weak bonds holding the enzyme in its three-dimensional shape, and denaturation occurs (page 33). Because the active site becomes altered, substrates cannot bind to it.

Brief exposure to high temperatures can destroy enzymes and adversely affect metabolism. This happens during extremely high fevers. When body temperature reaches 44°C (112°F), death generally follows.

Most enzymes function best in neutral surroundings, around pH 7. Higher or lower pH values generally disrupt the enzyme's three-dimensional shape and its function (Figure 4.12). Pepsin is one of the exceptions; this enzyme functions best in the extremely acidic fluid of the stomach.

Enzymes function only within limited ranges of temperature and pH.

Control of Enzyme Activity

Through controls over enzymes of different pathways, cells direct the flow of nutrients, wastes, and other substances in suitable ways. Certain controls govern the number of enzyme molecules available. Some accelerate or slow down the production of enzymes; others stimulate or inhibit the activity of enzymes already formed.

For example, **inhibitors** can bind with enzymes and interfere with their function. One type inhibits trypsin, a protein-digesting enzyme. Cells in the pancreas produce and secrete trypsin into the small intestine. Prior to secretion, trypsin is kept isolated in vesicles (in an inactive form). The inhibitor shuts down any molecules that do escape packaging. Without this safeguard, trypsin could be unleashed against the proteins of tissues and blood vessels in the pancreas. In *acute pancreatitis*, this enzyme and others are activated prematurely, sometimes with fatal results.

Some enzymes are governed by **allosteric control**. In addition to the active site, "allosteric" enzymes have control sites where specific substances can bind and

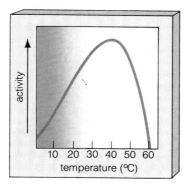

Figure 4.10 Effect of increases in temperature on enzyme activity.

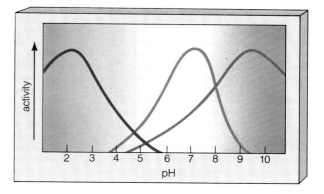

Figure 4.11 Visible effects of environmental temperature on enzyme activity. In Siamese cats, the fur on the ears and paws contains more dark-brown pigment (melanin) than the rest of the body. A heat-sensitive enzyme controlling melanin production is less active in warmer body regions, and this results in lighter fur in those regions.

Figure 4.12 Effect of pH on enzyme activity. The brown line charts the activity of an enzyme that is fully functional in neutral solutions. The red line charts the activity of one that is functional in basic solutions; the purple line, in acidic solutions.

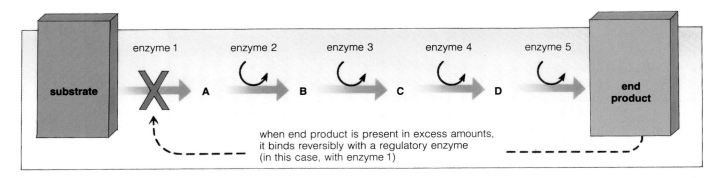

Figure 4.13 Example of feedback inhibition of a metabolic pathway. Here, the end product binds to the first enzyme in the pathway leading to its formation. When the product concentration drops, fewer molecules are around to inhibit the regulatory enzyme in the pathway, so production can rise again. Feedback inhibition allows concentrations of substances to be adjusted quickly to the cell's requirements.

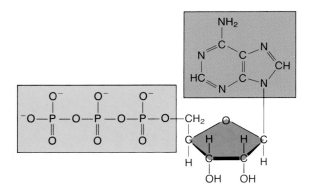

Figure 4.14 Structural formula for adenosine triphosphate, or ATP. The triphosphate group is shaded in gold, the sugar ribose in pink, and the adenine portion in blue.

COFACTORS

Some enzymes speed up the transfer of electrons, atoms, or functional groups. Many must be assisted by non-protein components called **cofactors,** either to help catalyze the reaction or to serve fleetingly as the transfer agents.

Cofactors include some large organic molecules that function as *coenzymes.* Two examples are **NAD⁺** (nicotinamide adenine dinucleotide) and **NADP⁺**. They deliver protons (H^+) and electrons stripped from a substrate to other reaction sites. As you will see, NAD^+ has roles in carbohydrate breakdown. $NADP^+$ takes part in photosynthesis.

Some *metal ions* also serve as cofactors. They include ferrous iron (Fe^{++}), which is a component of cytochrome molecules. The **cytochromes** are electron-transfer proteins found in cell membranes, such as the membranes of chloroplasts and mitochondria.

alter enzyme activity. For example, when a cell produces tryptophan molecules faster than it uses them, the unused molecules bind to and inhibit an allosteric enzyme. The enzyme happens to be required for tryptophan synthesis. So control of tryptophan production is a form of **feedback inhibition:** the output of the process works in a way that inhibits further output. Figure 4.13 illustrates this control mechanism.

Enzymes act only on specific substrates, and controls over their activity are central to the directed flow of substances into, through, and out of the cell.

ATP: THE MAIN ENERGY CARRIER

Structure and Function of ATP

Sunlight, recall, is the primary energy source for the web of life. Before the sun's energy can be used in cell activities, it must be transformed into the chemical energy of **ATP** (adenosine triphosphate). Also, cells cannot directly use the chemical energy of carbohydrates or other large organic molecules; they must first transform it to the energy of ATP. As Figure 4.14 shows, ATP is composed of adenine (a nitrogen-containing compound), ribose (a five-carbon sugar), and a triphosphate (three linked phosphate groups).

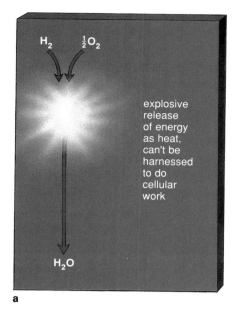
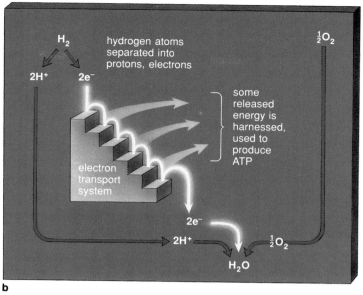

Figure 4.15 (**a**) When hydrogen and oxygen are made to react (say, by an electric spark), energy is released as heat. (**b**) In cells, the same type of reaction is made to occur in many small steps that allow much of the released energy to be harnessed in usable form. These "steps" are electron transfers, often between molecules that operate together as an electron transport system.

ATP provides energy for biosynthesis, active transport across cell membranes, and molecular displacements (such as those underlying muscle contraction). In fact, ATP transfers energy to or from nearly *all* metabolic pathways.

ATP directly or indirectly transfers energy to or from almost all metabolic pathways.

The ATP/ADP Cycle

In the **ATP/ADP cycle,** an energy input drives the linkage of ADP (adenosine diphosphate) and a phosphate group (or inorganic phosphate) into ATP; then the ATP donates a phosphate group elsewhere and becomes ADP:

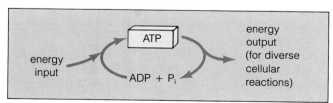

Adding phosphate to a molecule is called **phosphorylation.** What is important about it? When a molecule becomes phosphorylated by ATP, its store of energy generally increases *and it becomes primed to enter a specific reaction.*

With the ATP/ADP cycle, cells have a renewable means of conserving energy and transferring it to specific reactions. The ATP turnover is breathtaking. Even if you were bedridden for twenty-four hours, your cells would turn over approximately 40 kilograms (88 pounds) of ATP molecules simply for routine maintenance!

ELECTRON TRANSPORT SYSTEMS

ATP is produced during both photosynthesis and aerobic respiration, and its production depends on **electron transport systems.** Such systems consist of enzymes and cofactors, bound in a cell membrane, that transfer electrons in a highly organized sequence. One molecule "donates" electrons, and the next in line "accepts" them. Each time a donor gives up electrons, it is said to be "oxidized." Each time an acceptor acquires electrons, it is "reduced." *Oxidation-reduction reaction* merely means an electron transfer.

Electron transfers can occur after an atom (or a molecule) absorbs enough energy to boost one or more of its electrons to a shell farther from its nucleus (page 73). An excited electron quickly returns to the lowest energy level available to it—and it gives off energy when it does this. *Electron transport systems "intercept" excited electrons and make use of the energy they release.*

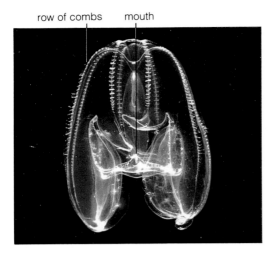

row of combs   mouth

Figure 4.16 Visible effects of electron transfers. This comb jelly (a relative of jellyfishes) is giving a good display of *bioluminescence*, or luminescent flashes in body tissues. Such flashes also occur in several groups of bacteria, fungi, fishes, and insects, including fireflies. Substances called luciferins give up electrons. The electrons return to a lower energy level—and light is emitted when they do. Special enzymes (luciferases) catalyze the reactions.

By analogy, think of an electron transport system as a staircase (Figure 4.15). Electrons "raised" (excited) to the top of the staircase have the most energy. They drop down the staircase, one step at a time (they are transferred from one electron carrier to another). With each drop, some energy being released can be harnessed to do work—for example, to make hydrogen ions move in ways that establish pH and electric gradients across membranes. Such gradients, as you will discover, are central in ATP formation.

Electron transfers may seem rather abstract. Figure 4.16 might help make the idea more concrete; it shows a striking effect of electron transfers in the world of life.

SUMMARY

1. Cells acquire and use energy to synthesize, accumulate, break down, and rid themselves of substances in controlled ways. These activities, which sustain cell growth, maintenance, and reproduction, are called metabolism.

2. Cellular use of energy conforms to two laws of thermodynamics. According to the first law, energy can be converted from one form to another but the total amount in the universe never changes. According to the second law, with each energy conversion, some energy is dispersed in a form that is not as readily available to do work.

3. Cells lose energy during metabolic reactions, but in nearly all cases, they replace it with energy derived directly or indirectly from the sun.

4. A metabolic pathway is a stepwise sequence of reactions, involving these substances:
   a. Reactants or substrates: the substances that enter a reaction
   b. Enzymes: proteins that serve as catalysts (they speed up reactions). Enzymes do not change the expected outcome of a reaction. They only change the rate at which the reaction proceeds.
   c. Cofactors: coenzymes (including NAD^+) and metal ions that help catalyze reactions or carry functional groups stripped from substrates.
   d. Energy carriers: mainly ATP, which readily donates energy to other molecules; most metabolic pathways run directly or indirectly on ATP energy.
   e. End products: the substances formed at the end of a metabolic pathway

5. During many electron transfers (oxidation-reduction reactions), energy is released that can be used to do work—for example, to make ATP.

Review Questions

1. State the first and second laws of thermodynamics. Which law deals with the *quality* of available energy, and which deals with the *quantity*? Give some examples of high-quality energy. *59*

2. Does the living state violate the second law of thermodynamics? In other words, how does the world of living things maintain a high degree of organization, even though there is a universal trend toward disorganization? *59*

3. Describe an enzyme and its role in metabolic reactions. *61*

4. Define substrate and active site. Why is binding at an active site a readily reversible event? *61, 62*

5. The high temperatures associated with severe fevers can impair cell functioning. Explain why. *63*

6. What are the three molecular components of ATP? What is the function of ATP, and why is phosphorylation of a molecule by ATP so important? *64, 65*

7. What is an oxidation-reduction reaction? What is its function in cells? *65*

Self-Quiz *(Answers in Appendix IV)*

1. When energy is lost during metabolic activities, it is replaced with more energy that is supplied directly or indirectly by the __Sun__.

2. Two laws of __THERmodyNAmics__ govern how cells acquire, convert, and transfer energy during metabolic reactions.

3. During each metabolic reaction, some energy is lost, usually as __Heat__.

4. Cells power their activities by:
   a. deploying mechanisms that create energy
   b. converting forms of energy that already exist
   c. recycling all the heat energy lost during reactions
   d. b and c

5. The ultimate source of energy for nearly all organisms on earth is _____.
   a. food
   b. water
   c. the sun
   d. ATP

6. Which of the following is *not* an aspect of metabolism?
   a. breaking down large molecules into simpler ones
   b. constructing large molecules from simpler ones
   c. acquiring energy
   d. using energy
   e. all of the above are aspects

7. Metabolic pathways _____.
   a. are stepwise series of chemical reactions
   b. are quickened with the aid of specific enzymes
   c. tear down and assemble molecules
   d. all of the above

8. The main energy carriers in cells are _____.
   a. NAD$^+$ molecules
   b. cofactors
   c. ATP molecules
   d. enzyme molecules

9. Enzymes _____.
   a. enhance reaction rates
   b. function within limited ranges of temperature and pH
   c. act only on specific substrates
   d. all of the above

10. Match each substance in a metabolic pathway with its correct description.
    a. reactant   __c__ a coenzyme or metal ion
    b. enzyme   __e__ the substance formed at the end
    c. cofactor   of a metabolic pathway
    d. energy carrier __d__ mainly ATP
    e. end product   __A__ a substance that enters a reaction
                       __b__ a protein that catalyzes a reaction

Selected Key Terms

activation energy *62*
active site *62*
allosteric control *63*
ATP *64*
ATP/ADP cycle *65*
biosynthetic pathway *61*
cofactor *64*
cytochrome *64*
degradative pathway *60*
electron transport system *65*
end product *61*
energy *58*

entropy *59*
enzyme *61*
feedback inhibition *64*
first law of thermodynamics *59*
metabolic pathway *60*
metabolism *58*
NAD$^+$ *64*
NADP$^+$ *64*
oxidation-reduction reaction *65*
phosphorylation *65*
second law of thermodynamics *59*
substrate *61*

Readings

Atkins, P. 1984. *The Second Law.* New York: Freeman.

Doolittle, R. 1985. "Proteins." *Scientific American* 253(4):88–99.

Fenn, J. 1982. *Engines, Energy, and Entropy.* New York: Freeman. Deceptively simple introduction to thermodynamics; good analogies. Paperback.

Fersht, A. 1985. *Enzyme Structure and Mechanism.* Second edition. New York: Freeman.

5 ENERGY-ACQUIRING PATHWAYS

KEY CONCEPTS

1. Organic compounds, with their carbon backbones, are the key building blocks and energy stores for life. Plants and other photosynthetic autotrophs assemble their own organic compounds. They use carbon dioxide from the air as a source for the carbon, and they trap sunlight energy that drives the synthesis reactions.

2. Photosynthesis is the main biosynthetic pathway by which carbon and energy enter the web of life. Animals and other heterotrophs must obtain their carbon and energy from organic compounds already built by the autotrophs.

3. Photosynthesis involves two sets of reactions. In the light-dependent reactions, pigment molecules give up electrons when they absorb light energy. Movement of the electrons through transport systems leads to the formation of ATP and NADPH (or ATP alone). In the light-independent reactions, the ATP provides energy to drive the joining of carbon and oxygen (from carbon dioxide) with hydrogens and electrons (from NADPH). The reactions produce sugar phosphates, which are used to form sucrose, starch, and other end products of photosynthesis.

Just before dawn in the Midwest the air is dry and motionless. The heat that has scorched the land for weeks still rises from the earth and hangs in the air of a new day. There are no clouds in sight. There is no promise of rain. For hundreds of miles, crops stretch out, withered or dead. All the marvels of modern agriculture can't save them now. In the absence of one vital resource—water—life in each cell of those many thousands of plants has ceased.

In Los Angeles, a student reading the morning newspaper complains that the Midwest drought will probably cause a hike in food prices. In Washington, D.C., economists calculate the crop failures in terms of decreased tonnage available for domestic consumption and export; government officials brood about what it means to the nation's balance of payments. In Ethiopia, a child with bloated belly and spindly legs waits passively for death. Even if food donations were to reach her now, it would be too late. Deprived of food resources too long, some cells of her body will never grow normally again.

You are about to explore ways in which cells trap and use energy. The cellular pathways may seem at first to be far removed from the world of your interests. *Yet the food molecules on which you and nearly all other organisms depend cannot be built or used without those pathways and the raw materials required for their operation.* We will return repeatedly to this point in later chapters, when we address topics such as human nutrition, human population growth and the environmental limits on agriculture, genetic engineering of new crop plants, and the effects of pollution on food production.

Figure 5.1 Links between photosynthesis and aerobic respiration—the main energy-acquiring and energy-releasing pathways in the world of life.

FROM SUNLIGHT TO CELLULAR WORK: PREVIEW OF THE MAIN PATHWAYS

For all living things, cell structure begins with organic compounds (which have a carbon backbone), and cell activities are driven by chemical energy stored in many of those compounds. The questions become these:

1. Where does the carbon come from in the first place?

2. Where does the energy come from to drive the linkage of carbon and other atoms into organic compounds?

3. How does the energy inherent in those compounds become available to do cellular work?

The answers vary, depending on whether you are talking about autotrophic or heterotrophic organisms.

Autotrophs obtain carbon and energy from the physical environment; they are "self-nourishing" (which is what autotroph means). Their carbon source is carbon dioxide (CO_2), a gaseous substance all around us in the air and dissolved in water. Only the *photosynthetic* autotrophs can get energy from sunlight. Plants, some protistans, and some bacteria fall in this category. A few kinds of bacteria are *chemosynthetic* autotrophs; they get energy by stripping electrons from sulfur or some other inorganic substance.

Heterotrophs are not self-nourishing; they feed on autotrophs, each other, and organic wastes. They must get carbon and energy from organic compounds *already built* by autotrophs. Animals, fungi, many protistans, and most bacteria are heterotrophs.

It follows, from the above, that carbon and energy enter the web of life primarily by **photosynthesis.** And energy stored in organic compounds as a result of photosynthesis can be released by one of several pathways that begin with breakdown reactions known as **glycolysis.** Of all the degradative pathways, **aerobic respiration** releases the most energy. Figure 5.1 shows the links between photosynthesis and aerobic respiration, the two main topics of this chapter and the next.

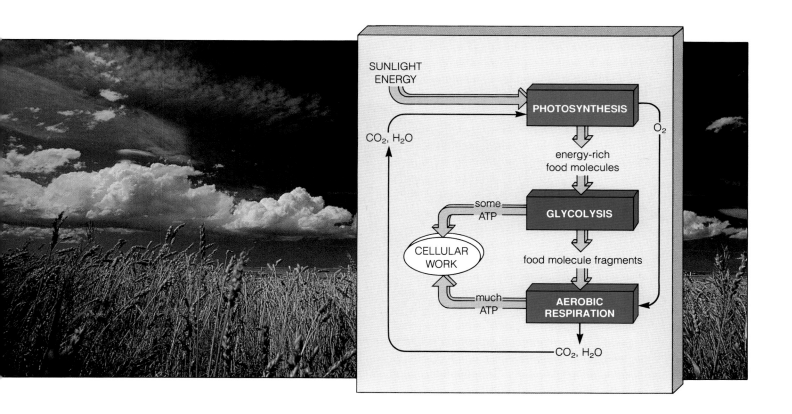

Figure 5.2 Functional zones of a chloroplast from the leaf of a sunflower plant (*Helianthus*). The light-dependent reactions of photosynthesis occur at thylakoid membranes, and they lead to ATP and NADPH formation. The light-independent reactions occur in the stroma. They lead to production of sugars and other carbon-containing molecules. (**a**) Section through a sunflower leaf, showing chloroplast-containing cells. (**b**) Chloroplast in cross-section. (**c**) Two of the grana. (**d**) Where photosynthetic reactions occur.

PHOTOSYNTHESIS

Simplified Picture of Photosynthesis

Photosynthesis consists of two sets of reactions. In the **light-dependent reactions,** sunlight energy is absorbed and converted to chemical energy, which is transferred to ATP and NADPH. (Here you may wish to refer to pages 64 and 73.) In the **light-independent reactions,** sugars and other compounds are assembled with the help of ATP and NADPH. Photosynthesis is typically summarized this way:

$$12H_2O + 6CO_2 \xrightarrow{\text{sunlight}} 6O_2 + C_6H_{12}O_6 + 6H_2O$$

Here, glucose ($C_6H_{12}O_6$) is shown as an end product (to keep the chemical bookkeeping simple). The reactions don't really stop there, however. Newly formed glucose and other simple sugars are linked at once into sucrose, starch, and other carbohydrates—the true end products of photosynthesis.

Chloroplast Structure and Function

In plants, photosynthesis proceeds in organelles called chloroplasts. Each chloroplast contains a **thylakoid membrane,** folded to form flattened channels and stacked disks known as **grana** (singular, granum). The interconnected space inside the disks and channels serves as a compartment for hydrogen ions that are used in ATP

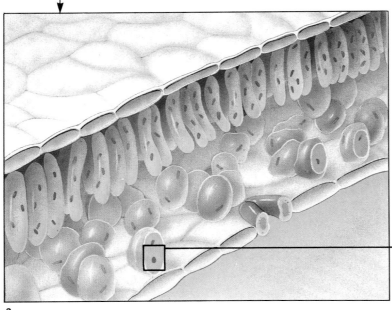

a

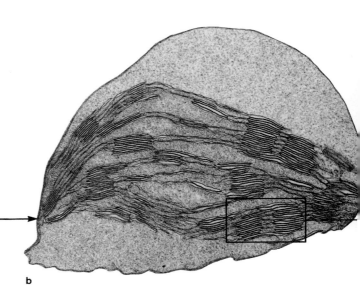

b

formation. Surrounding the disks and channels is the **stroma,** the zone inside the chloroplast where the actual products of photosynthesis are formed (Figure 5.2).

If you could line up 2,000 chloroplasts, one after another, the lineup would be no wider than a dime. Imagine the many millions of chloroplasts in just one lettuce leaf, each a tiny factory for producing sugars and starch—and you get an idea of the magnitude of metabolic events required to feed you and all other organisms on earth.

LIGHT-DEPENDENT REACTIONS

Three events unfold in the first stage of photosynthesis. Light energy is absorbed, electron and hydrogen transfers lead to ATP and NADPH formation, and electrons are replaced in the molecules that give up electrons at the start of the reactions.

Light Absorption

Light-Trapping Pigments. The thylakoid membrane contains **pigments,** which are photon-absorbing molecules. A photon is a packet of light energy. Some photons have more energy than others, and the differences correspond to different wavelengths of light. Organisms use only a small range of wavelengths for photosynthesis, vision, and other light-requiring processes. Most of those wavelengths are the ones we see as colors of light (Figures 5.3 and 5.4).

Each type of pigment absorbs certain wavelengths and reflects the rest. Leaves appear green because they have **chlorophyll** pigments that absorb blue and red wavelengths but reflect green. **Carotenoid** pigments absorb violet and blue wavelengths but reflect yellow, orange, and red. Leaves also contain carotenoids and other pigments, but their presence is usually masked by the more abundant chlorophylls. Collectively, photosynthetic pigments can absorb most of the wavelength

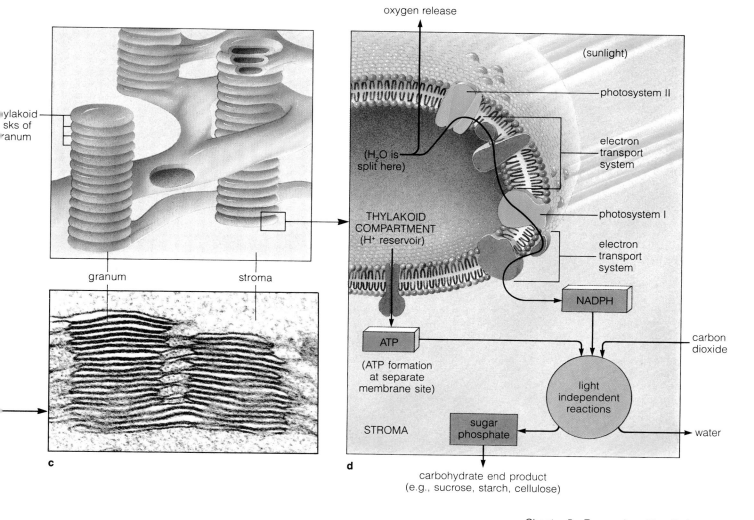

Figure 5.3 (**a**) Where wavelengths of visible light occur in the electromagnetic spectrum. Photosynthesis, vision, and other light-requiring processes typically use wavelengths ranging from about 400 to 750 nanometers. Shorter wavelengths (such as ultraviolet and x-rays) are so energetic they break bonds in organic compounds, so they can destroy cells. Longer wavelengths (such as infrared) are not energetic enough to drive the formation of NADPH during photosynthesis. NADPH is necessary for the biosynthetic programs of large, multicelled plants.

(**b**) Wavelengths absorbed by some photosynthetic pigments. Peaks in the ranges of absorption correspond to the measured amount of energy absorbed and used in photosynthesis. Colors used here correspond to the colors reflected by each pigment type. (Thus chlorophylls absorb blue and red wavelengths best and reflect wavelengths in between.) Together, different photosynthetic pigments can absorb most of the energy available in the spectrum of visible light.

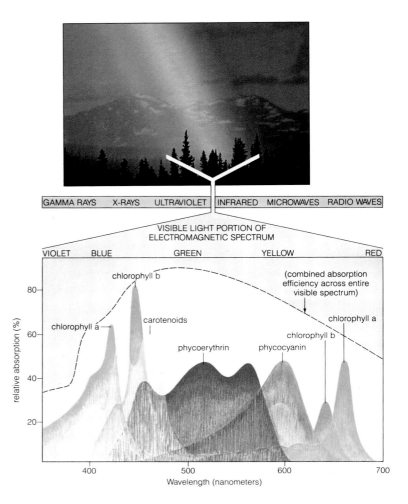

GAMMA RAYS	X-RAYS	ULTRAVIOLET	INFRARED	MICROWAVES	RADIO WAVES

VISIBLE LIGHT PORTION OF
ELECTROMAGNETIC SPECTRUM

VIOLET   BLUE   GREEN   YELLOW   RED

chlorophyll b

(combined absorption efficiency across entire visible spectrum)

chlorophyll a

chlorophyll a

carotenoids

chlorophyll b

phycoerythrin   phycocyanin

relative absorption (%)

80

60

40

20

400   500   600   700

Wavelength (nanometers)

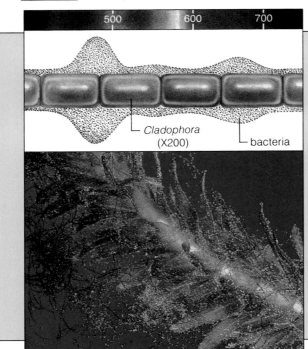

500   600   700

Cladophora
(X200)   bacteria

Figure 5.4 T. Englemann's 1882 experiment, which revealed the most effective wavelengths for photosynthesis by *Cladophora*, a filamentous green alga.

Oxygen is a by-product of photosynthesis, as the photograph (left) of the aquatic plant *Elodea* indicates. (The bubbles of oxygen sometimes are visible on sunny days.) Oxygen also is used by many organisms in the energy-releasing pathway called aerobic respiration. Englemann suspected that aerobic (oxygen-using) bacteria living in the same places as the alga would congregate in areas where oxygen was being produced.

Englemann used a crystal prism to cast a tiny spectrum of colors on a microscope slide. Then he positioned an algal filament to run parallel with the spectrum (as shown above). The bacteria did indeed cluster next to the filament where the most oxygen was being released. And those regions corresponded to colors (wavelengths) being absorbed most effectively—in this case, violet and red.

maple in summer

maple in autumn

Figure 5.5 Changes in leaf color in autumn. Chloroplasts of mature leaves contain chlorophylls, carotenoids (including the yellow carotenes and xanthophylls), and other pigments, each of which absorbs certain wavelengths of light. The intense green color of chloroplasts usually masks the presence of other pigments. In autumn, however, the gradual reduction in daylength and other factors trigger the breakdown of chlorophyll, and additional colors show through.

Also in autumn, water-soluble anthocyanins accumulate in the central vacuoles of leaf cells. These pigments appear red if plant fluids are slightly acidic, blue if basic (alkaline), and colors in between at intermediate levels of acidity.

energy available in the spectrum of visible light as Figure 5.3 indicates.

Photosystems. The pigments of chloroplasts are organized as **photosystems,** or clusters of 200 to 300 molecules. Nearly all those molecules simply "harvest" sunlight. When they absorb photon energy, one of their electrons gets boosted to a higher level:

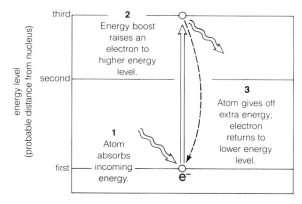

The electron returns almost at once to a lower level, and extra energy is released when it does. The energy hops from one molecule to another until it reaches a few special chlorophyll molecules that act like an energy trap. Those chlorophylls alone give up the electrons used in photosynthesis. The trapped energy drives the transfer of one of their electrons to an acceptor molecule—the next participant in the light-dependent reactions.

ATP and NADPH Formation

Electrons expelled from a chlorophyll molecule pass through one or two **electron transport systems.** Each system is a series of molecules bound in the thylakoid membrane (Figure 5.2d). Each accepts and then donates electrons to the next molecule in line. Energy released at certain transfers drives the attachment of phosphate to ADP, forming an ATP molecule. We call this *photophosphorylation,* because the pathway depends on an earlier input of light energy. As you will now see, such pathways are either cyclic or noncyclic.

Cyclic Pathway. A special chlorophyll complex (P700) occurs in a type of pigment cluster called **photosystem I.** In the simplest ATP-producing pathway, electrons "travel in a circle" from P700, through a transport system, then back to P700:

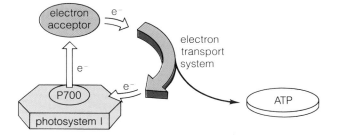

The cyclic pathway is probably the oldest means of ATP production. Early photosynthetic autotrophs were no larger than existing bacteria, so their body-building programs could scarcely have been enormous. They could have used ATP alone to build organic compounds even though such reactions are rather inefficient. (ATP carries energy only to sites where organic compounds are built; the electrons and hydrogen atoms required must be obtained by other means.) However, energy from the cyclic pathway alone would not have sustained the evolution of larger photosynthesizers, including leafy plants.

Noncyclic Pathway. Today, leafy plants rely mostly on a noncyclic pathway to produce ATP. In this pathway, the electrons do not return to the photosystem that gave them up. Rather, after moving through two transport systems, they end up in NADPH. And hydrogen and electrons carried by NADPH can be used *directly* in the synthesis of organic compounds!

As Figures 5.6 and 5.7 show, the noncyclic pathway begins at **photosystem II,** with its own special chloro-

phyll complex (P680). When P680 absorbs light energy, it gives up an electron to an acceptor molecule. From there, the electron moves through a transport system—and then to chlorophyll P700 of photosystem I.

The excited electron has not yet returned to the lowest available energy level. When the P700 absorbs light energy, electrons are boosted even higher and passed to a second transport system. Transport systems, recall, are like steps on an energy staircase—and this boost places electrons at the top of a higher staircase. There is enough energy left at the bottom of this staircase to attach two electrons and a hydrogen ion (H^+) to $NADP^+$, the result being NADPH.

In the noncyclic pathway, electrons flow in one direction to NADPH. In the meantime, the P680 molecule that gives up electrons in the first place is getting replacements—from water. Inside the thylakoid compartment, water molecules are being split into oxygen, "naked" protons (that is, hydrogen ions), and electrons. Photon energy indirectly drives this reaction sequence, which is called **photolysis** (Figure 5.6).

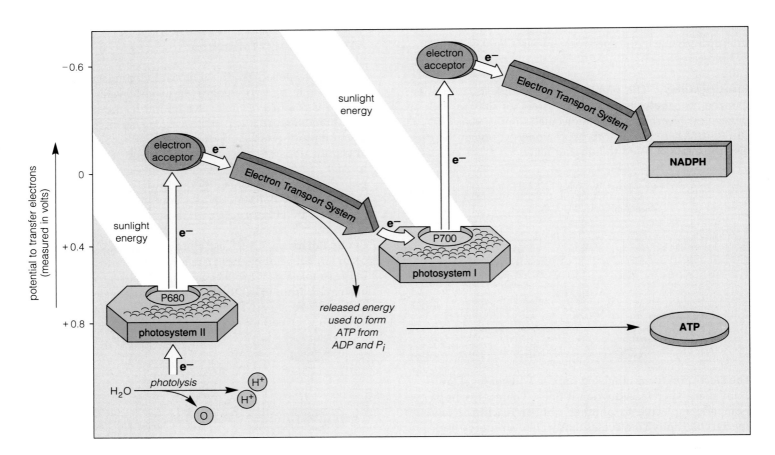

Figure 5.6 Noncyclic photophosphorylation, which yields NADPH as well as ATP. Electrons derived from the splitting of water molecules (photolysis) travel through two photosystems, which work together in boosting the electrons to an energy level high enough to lead to NADPH formation. Figure 5.7 provides a closer look at the mechanism by which ATP forms.

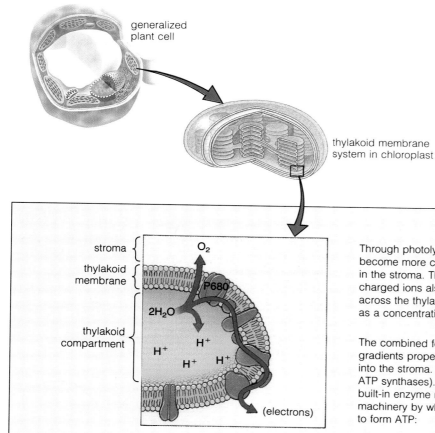

generalized
plant cell

thylakoid membrane
system in chloroplast

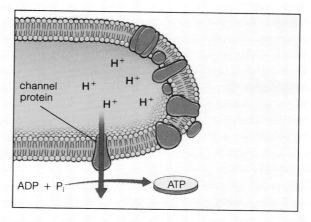

stroma
thylakoid
membrane

thylakoid
compartment

O₂

P680

2H₂O

H⁺

H⁺

H⁺   H⁺

(electrons)

Through photolysis and electron transport, hydrogen ions become more concentrated in the thylakoid compartment than in the stroma. The lopsided distribution of those positively charged ions also creates a difference in electric charge across the thylakoid membrane. An electric gradient as well as a concentration gradient has been established.

The combined force of the concentration and electric gradients propels hydrogen ions out of the compartment and into the stroma. The ions flow through channel proteins (called ATP synthases). The proteins span the membrane and have built-in enzyme machinery. The ion flow drives the enzyme machinery by which ADP combines with inorganic phosphate to form ATP:

Figure 5.7 A closer look at ATP formation in chloroplasts. At the start of the noncyclic pathway of photosynthesis, water molecules are split into oxygen, "naked" protons (or hydrogen ions), and electrons. (The reaction sequence is called photolysis.) The oxygen is released as an end product and the electrons are sent through transport systems. The hydrogen ions accumulate inside the thylakoid compartment of the chloroplast, as sketched above.

Hydrogen ions also accumulate in the compartment when electron transport systems are operating. (This is true of both the cyclic and noncyclic pathway.) When certain molecules of the transport system accept electrons, they also pick up hydrogen ions from the stroma and release them inside the compartment:

channel
protein

H⁺

H⁺

H⁺

H⁺   H⁺

ADP + Pᵢ   ATP

The idea that concentration and electric gradients across a membrane drive ATP formation is known as the *chemiosmotic theory*.

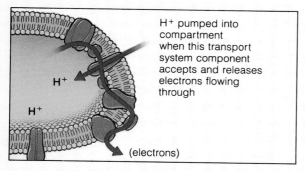

H⁺ pumped into
compartment
when this transport
system component
accepts and releases
electrons flowing
through

H⁺

H⁺

(electrons)

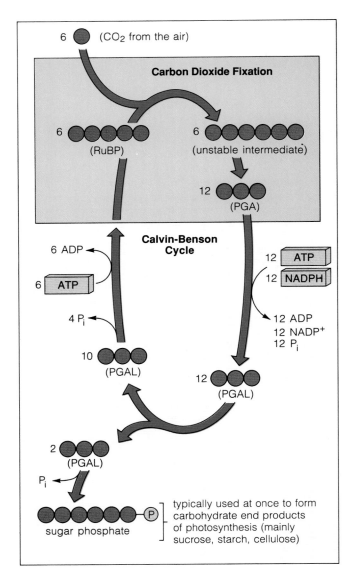

6 ● (CO₂ from the air)

Carbon Dioxide Fixation

6 ●●●●● (RuBP)

6 ●●●●● (unstable intermediate)

12 ●●● (PGA)

Calvin-Benson Cycle

6 ADP ←

6 ATP

12 ATP

12 NADPH

4 Pᵢ ←

12 ADP
12 NADP⁺
12 Pᵢ

10 ●●● (PGAL)

12 ●●● (PGAL)

2 ●●● (PGAL)

Pᵢ ←

●●●●●●—(P)   typically used at once to form carbohydrate end products of photosynthesis (mainly sucrose, starch, cellulose)

sugar phosphate

Figure 5.8 Summary of the light-independent reactions of photosynthesis. The carbon atoms of the different molecules are depicted in red. All of the intermediates have one or two phosphate groups; for simplicity we show only the one on the "end product" (sugar phosphate). The phosphate groups that have been detached from molecules during the reactions are designated Pᵢ, meaning "inorganic phosphate."

Oxygen atoms split from water molecules are by-products of the noncyclic pathway. Oxygen has been accumulating ever since this pathway emerged, over 3.5 billion years ago. It profoundly changed the earth's atmosphere. And it made possible aerobic respiration, the most efficient pathway for releasing energy from organic compounds. If the noncyclic pathway had not emerged, you and all other animals would not be around today, breathing the oxygen that helps keep your cells alive.

With cyclic photophosphorylation, ATP alone forms.

With noncyclic photophosphorylation, ATP and NADPH form.

Oxygen, a by-product of the noncyclic pathway, changed the earth's atmosphere and made aerobic respiration possible.

During the light-dependent reactions, a total of eighteen ATP molecules form for every six oxygen molecules that are released. All of those ATP molecules will be used during the light-independent reactions by which sugar phosphates are assembled.

LIGHT-INDEPENDENT REACTIONS

ATP provides energy and NADPH provides hydrogen atoms and electrons for the "synthesis" part of photosynthesis. The air around photosynthetic cells provides the carbon and oxygen (in the form of carbon dioxide). The reactions take place in the stroma of chloroplasts. They are called "light-independent" because they do not depend directly on sunlight.

Calvin-Benson Cycle

The heart of the light-independent reactions is the **Calvin-Benson cycle.** This cyclic pathway was named after its discoverers, Melvin Calvin and Andrew Benson. As Figure 5.8 shows, the reactions begin with **carbon dioxide fixation.** The carbon atom in a carbon dioxide molecule becomes affixed to **RuBP** (ribulose bisphosphate), a molecule having a backbone of five carbon atoms. The result is an unstable six-carbon intermediate that splits at once into two molecules of **PGA** (phosphoglycerate).

Each PGA receives a phosphate group from ATP. The resulting intermediate receives H⁺ and electrons from NADPH to form **PGAL** (phosphoglyceraldehyde). It takes six carbon dioxide molecules to produce twelve PGAL. Most of the PGAL becomes rearranged into new RuBP molecules—which can be used to fix more carbon.

But two PGAL are joined together to form a "sugar phosphate." Such sugars have phosphate groups attached that prime them for further reaction.

The Calvin-Benson cycle yields enough RuBP to replace the ones used in carbon dioxide fixation. The ADP, NADP+, and phosphate leftovers are sent back to the light-dependent reaction sites, where they are converted once more to NADPH and ATP. The sugar phosphate formed in the cycle can serve as a building block for the plant's main carbohydrates, including sucrose, starch, and cellulose. Synthesis of those compounds by different metabolic pathways marks the conclusion of the light-independent reactions.

During the Calvin-Benson cycle, carbon is "captured" from carbon dioxide, a sugar phosphate forms in reactions requiring ATP and NADPH, and RuBP (needed to capture the carbon) is regenerated.

Figure 5.9 summarizes the main reactants, intermediates, and products of both the light-dependent and light-independent reactions of photosynthesis.

C4 Plants

Temperature and water availability can affect carbon dioxide fixation in leafy plants. Leaves have a waxy covering that retards moisture loss, and water escapes mainly through tiny passages (stomata) across the leaf's surface. On hot, dry days the passages are closed and water is conserved. But carbon dioxide can't enter the leaf and oxygen builds up inside. The stage is set for a wasteful process called "photorespiration." In this process, oxygen instead of carbon dioxide becomes attached to the RuBP used in the Calvin-Benson cycle, with different results:

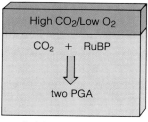

Calvin-Benson cycle predominates

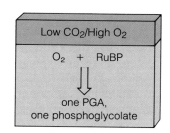

photorespiration predominates

Formation of sugar phosphates depends on PGA. When photorespiration wins out, less PGA forms and the plant's capacity for growth suffers.

However, crabgrass, sugarcane, corn, and other **C4 plants** can continue to build carbohydrates even when

the ratio of carbon dioxide to oxygen is unfavorable. They fix carbon not once but *twice*. Carbon fixation in certain cells produces oxaloacetate, a four-carbon compound. Hence the name "C4" plants. (Three-carbon PGA is produced in "C3" plants.) The four-carbon compound is transferred to different cells, where the carbon dioxide is released and picked up by the Calvin-Benson cycle (Figure 5.10).

Kentucky bluegrass, wheat, rice, and other C3 species have the advantage in regions where temperatures drop below 25°C; they are less sensitive to cold. But C4 species are more abundant in regions with the highest temperatures during the growing season. For example, eighty percent of all native species in Florida are C4 plants—compared to zero percent in Manitoba, Canada.

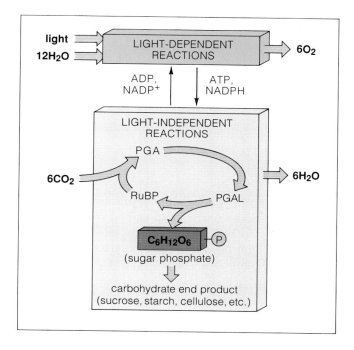

Figure 5.9 Summary of the main reactants, intermediates, and products of photosynthesis corresponding to the equation:

$$12H_2O + 6CO_2 \xrightarrow{\text{sunlight}} 6O_2 + C_6H_{12}O_6 + 6H_2O$$

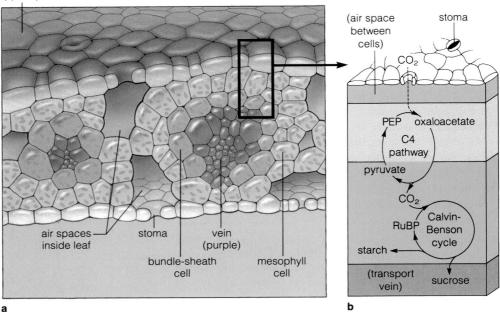

upper epidermis of leaf

(air space between cells)
stoma
CO₂
PEP   oxaloacetate
C4 pathway
pyruvate
CO₂
RuBP   Calvin-Benson cycle
starch
(transport vein)   sucrose

air spaces inside leaf
stoma
vein (purple)
bundle-sheath cell
mesophyll cell

a   b

Figure 5.10 C4 pathway. (**a**) This is the internal structure of a leaf from corn (*Zea mays*), a typical C4 plant. Notice how the photosynthetic bundle-sheath cells (dark green) surround the veins and are in turn surrounded by photosynthetic mesophyll cells (lighter green). (**b**) C4 plants have a carbon-fixing system that *precedes* the Calvin-Benson cycle.

SUMMARY

1. Plants and other photosynthetic autotrophs use sunlight (as an energy source) and carbon dioxide (as the carbon source) for building organic compounds. Animals and other heterotrophs must obtain carbon and energy from organic compounds already built by autotrophs.

2. Photosynthesis is the main biosynthetic pathway by which carbon and energy enter the web of life. It consists of two sets of reactions:

   a. The light-dependent reactions take place at the thylakoid membrane system of chloroplasts, and they produce ATP and NADPH (or ATP alone).

   b. The light-independent reactions take place in the stroma around the membrane system. They produce sugar phosphates that are used in building the end products of photosynthesis (e.g., sucrose and starch).

3. These are the key events of the light-dependent reactions:

   a. Photosystems are clusters of photosynthetic pigments in the thylakoid membrane. Light absorption causes the transfer of electrons from a photosystem to an acceptor molecule, which donates them to a transport system in the membrane.

   b. Operation of electron transport systems causes H^+ to accumulate inside the thylakoid membrane system. This produces concentration and electric gradients that drive the formation of ATP.

   c. In the cyclic pathway, electrons travel in a circle, back to the photosystem that originally gave them up. This pathway yields ATP only.

   d. The noncyclic pathway also yields ATP, but the electrons from a photosystem travel through two transport systems and end up in NADPH. Electrons derived from the splitting of water molecules replace the ones that the photosystem gives up.

4. These are the key events of the light-independent reactions:

   a. The ATP produced in the light-dependent reactions provides energy and NADPH provides hydrogen atoms and electrons for the "synthesis" part of photosynthesis.

   b. Sugar phosphates form by operation of the Calvin-Benson cycle. The cycle begins when carbon dioxide from the air is affixed to RuBP, making an unstable intermediate that splits into two PGA.

   c. Each PGA receives a phosphate group from ATP. The resulting molecule receives H^+ and electrons from NADPH to form PGAL. Two of every twelve PGAL are used to produce a sugar phosphate; the rest are rearranged to regenerate RuBP for the cycle.

Review Questions

1. Define the difference between autotrophs and heterotrophs, and give examples of each. In what category do photosynthesizers fall? 69

2. Summarize the photosynthesis reactions in words, then as an equation. Distinguish between the light-dependent and the light-independent stage of these reactions. 69, 70

3. Oxygen is a product of photolysis. What *is* photolysis, and does it occur during the first or second stage of photosynthesis? 74

4. Describe where the light-dependent reactions occur in the chloroplast, and name the molecules formed there. Do the same for the light-independent reactions. 70

5. A thylakoid compartment is a reservoir for which of the following substances: glucose, photosynthetic pigments, hydrogen ions, fatty acids? 71

6. Sketch the reaction steps of noncyclic photophosphorylation, showing where the excited electrons eventually end up. Do the same for the cyclic pathway. Which pathway has the greater energy yield? 74–76

7. Which of the following substances are *not* required for the light-independent reactions: ATP, NADPH, RuBP, carotenoids, free oxygen, carbon dioxide, enzymes? 76

8. Suppose a plant carrying out photosynthesis were exposed to carbon dioxide molecules that contain radioactively labeled carbon atoms ($^{14}CO_2$). In which of the following compounds will the labeled carbon first appear: NADPH, PGAL, pyruvate, PGA? 76

9. How many CO_2 molecules must enter the Calvin-Benson cycle to produce one sugar phosphate molecule? Why? 76

Self-Quiz *(Answers in Appendix IV)*

1. Molecules with backbones of _____ serve as the main building blocks of all organisms.

2. Photosynthetic autotrophs use _____ from the air as their carbon source and _____ as their energy source.

3. Carbon and energy enter the web of life mainly through _____.

4. _____ must obtain their carbon and energy from compounds constructed by autotrophs.

5. In plant cells, light-*dependent* reactions occur _____.
    a. in the cytoplasm
    b. at the plasma membrane
    c. in the stroma
    d. in the grana

6. In plant cells, light-*independent* reactions occur _____.
    a. in the cytoplasm
    b. at the plasma membrane
    c. in the stroma
    d. in the grana

7. In the light-dependent reactions, _____.
    a. carbon dioxide is incorporated into carbohydrates
    b. ATP and NADPH are formed
    c. carbon dioxide accepts electrons
    d. sugar phosphates are formed

8. When light is absorbed by a photosystem, _____.
    a. sugar phosphates are produced
    b. electrons are transferred to an acceptor molecule
    c. RuBP accepts electrons
    d. the light-dependent reactions are initiated
    e. both b and d are correct

9. The Calvin-Benson cycle begins when _____.
    a. light is available
    b. light is not available
    c. carbon dioxide is attached to RuBP
    d. electrons leave a photosystem

10. In the light-independent reactions, ATP furnishes phosphate groups to _____.
    a. RuBP
    b. NADP$^+$
    c. PGA
    d. PGAL

11. Match each event in photosynthesis with its correct description.

a. cyclic pathway	_____ uses RuBP; produces PGA
b. noncyclic pathway	_____ uses ATP and NADPH
c. carbon dioxide fixation	_____ forms NADPH
d. formation of PGAL	_____ produces ATP and NADPH
e. Transfer of H$^+$ and electrons to NADP$^+$	_____ produces ATP only

Selected Key Terms

autotroph *69*	PGA *76*
C4 plant *77*	PGAL *76*
Calvin-Benson cycle *76*	photolysis *74*
carbon dioxide fixation *76*	photophosphorylation *76*
chemiosmotic theory *75*	photosynthesis *69*
chlorophyll *71*	photosystem *73*
electron transport system *73*	RuBP *76*
granum *70*	stroma *71*
heterotroph *69*	sugar phosphate *77*
light-dependent reaction *70*	thylakoid membrane *70*
light-independent reaction *70*	

Readings

Hinkle, P., and R. McCarty. March 1978. "How Cells Make ATP." *Scientific American* 238(23):104–123. How the chemiosmotic theory explains ATP formation in both chloroplasts and mitochondria.

Miller, K. 1982. "Three-Dimensional Structure of a Photosynthetic Membrane." *Nature* 300:5887.

Moore, P. 1981. "The Varied Ways Plants Tap the Sun." *New Scientist* 12:394–397. Clear, simple introduction to the C4 plants.

Youvan, D., and B. Marrs. 1987. "Molecular Mechanisms of Photosynthesis." *Scientific American* 256:42–50.

6 ENERGY-RELEASING PATHWAYS

KEY CONCEPTS

1. Plants produce and use ATP during the reactions of photosynthesis. They also produce ATP by degrading organic compounds such as glucose, and this ATP serves as the main energy carrier for nearly all other reactions in plant cells. Bacteria, protistans, fungi, and animals also produce ATP by degradative reactions. The most common degradative pathway is aerobic respiration.

2. Aerobic respiration proceeds through three stages. First, glucose is partially degraded to pyruvate with a net energy yield of two ATP. Second, the pyruvate is completely degraded to carbon dioxide and water. Electrons liberated during the breakdown reactions are delivered to a transport system. Third, energy is released as electrons are transferred through the system, and it drives the formation of as many as thirty-six or more ATP for every glucose molecule. At the end of the system, free oxygen combines with the electrons and with H$^+$ to form water.

3. Over evolutionary time, photosynthesis and aerobic respiration have become linked on a global scale. The oxygen by-products of photosynthesis serve as final electron acceptors for the aerobic pathway. And the carbon dioxide and water released in aerobic respiration are raw materials used in building organic compounds during photosynthesis:

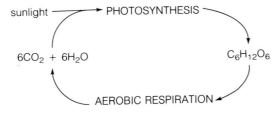

It is one of the quirks of the human mind that plants just aren't thought about very often as *living* organisms. Even vegetarians who become distressed at the thought of eating the flesh of an animal can relish the flesh of a pineapple. Perhaps it is understandable. Lacking autotrophic equipment of our own, we have to depend on something to produce energy for us; and if we carried a concern for the sanctity of life too far, we'd all starve to death.

And yet, at the biochemical level, there is an undeniable unity among organisms. Animals, plants, fungi, protistans, and most bacteria extract energy from organic compounds in much the same way you do. In fact, there is remarkable similarity in the flow of energy through all organisms—hence through the biosphere. We will return to this idea in the *Commentary* at the chapter's end.

ATP-PRODUCING PATHWAYS

The prime energy carrier for all forms of life is adenosine triphosphate, or **ATP** (page 64). By donating a phosphate group to many kinds of reactants and other molecules, ATP primes them to enter a reaction. And it does this for nearly every kind of reaction in cells.

Plants produce and use ATP during photosynthesis. But plants and all other organisms also produce ATP through degradative pathways that release energy from carbohydrates, lipids, or proteins. The main degradative pathways are **aerobic respiration** (which requires free oxygen for its operation) and **fermentation** (which does not). Here we focus on how ATP forms when a common carbohydrate—glucose—is degraded.

AEROBIC RESPIRATION

Overview of the Reactions

Of all degradative pathways, aerobic respiration produces the most ATP for each glucose molecule being dismantled. Whereas fermentation has a net yield of two ATP, the aerobic route yields *thirty-six* ATP or more. If you were a microscopic bacterium, you would not require much ATP. Being large, complex, and highly active, you depend absolutely on the high ATP yield of aerobic respiration (Figure 6.1).

The aerobic route is often summarized in the following way when glucose is the starting material:

$$C_6H_{12}O_6 + 6O_2 \implies 6CO_2 + 6H_2O$$

glucose   carbon
                           dioxide

Keep in mind that the aerobic route is actually three stages of reactions. In the first stage, **glycolysis,** glucose is partially degraded to pyruvate. In the second stage, which includes the **Krebs cycle,** pyruvate is completely degraded to carbon dioxide and water. Neither stage produces much ATP. However, while the glucose molecule is being broken apart, protons (H^+) and electrons stripped from it are delivered to a transport system. That system is used in the third stage of reactions, **electron transport phosphorylation,** which yields many ATP. Oxygen withdraws the electrons from the transport system (Figure 6.2).

Glycolysis

"Glycolysis" refers to the partial breakdown of glucose (or some other carbohydrate) into two molecules of pyruvate. Glycolysis takes place in the cytoplasm. En-

Figure 6.1 Typical net energy yield of aerobic respiration, the only pathway that can deliver enough ATP for strenuous activity, provided plenty of oxygen and glucose are available.

$$C_6H_{12}O_6 + 6O_2 \implies 6CO_2 + 6H_2O$$

glucose   carbon
                           dioxide

typical energy yield = 36 ATP

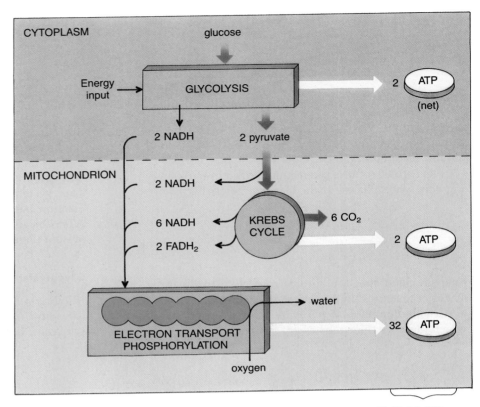

Gorillas and most other organisms (including the plant upon which the g rilla is dining) use the aerobic pathway.

Figure 6.2 Overview of aerobic respiration, the main energy-releasing pathway, with glucose as the starting material. In glycolysis, glucose is partially degraded to pyruvate. (Glycolysis is also the initial set of reactions in other pathways, including fermentation.) In the Krebs cycle, pyruvate is completely degraded to carbon dioxide. Coenzymes (NADH and FADH$_2$) accept protons and electrons being stripped from intermediates of the reactions and deliver them to an electron transport chain. Oxygen accepts the electrons from the transport system. From start (glycolysis) to finish, the aerobic pathway typically has a net energy yield of thirty-six ATP.

zymes speed the reactions, with intermediate molecules produced at one step serving as substrates for the next enzyme in the sequence (page 61).

Glucose has a backbone of six carbon atoms to which hydrogen atoms are attached (Figure 2.17). Here we show the backbone in simplified fashion:

The first steps of glycolysis are *energy-requiring;* they do not proceed without an energy input. Enzymes transfer two phosphate groups (from two ATP) to the glucose backbone. The backbone splits apart, and two molecules of **PGAL** (phosphoglyceraldehyde) form. Formation of PGAL marks the start of the *energy-releasing* steps of glycolysis (Figure 6.3).

Enzymes remove protons and electrons from the two PGAL and transfer them to NAD$^+$, forming two NADH.

Recall that NAD$^+$ is a "reusable" coenzyme. It picks up protons and electrons stripped from a substrate, delivers them to another reaction site, and so becomes NAD$^+$ again.

Next, each substrate molecule gives up a phosphate group to ADP. Thus two ATP have formed by what is called a **substrate-level phosphorylation.** (Unlike electron transport phosphorylation, this mechanism does not require oxygen or a transport system.) The same thing happens again, for a total of four ATP. But remember, two ATP were invested at the start of glycolysis, so the *net* yield from one glucose is two ATP.

Glycolysis ends with the formation of two pyruvate molecules, each having a three-carbon backbone.

***

Glycolysis produces two pyruvate, two NADH, and two ATP (net) for each glucose molecule degraded.

***

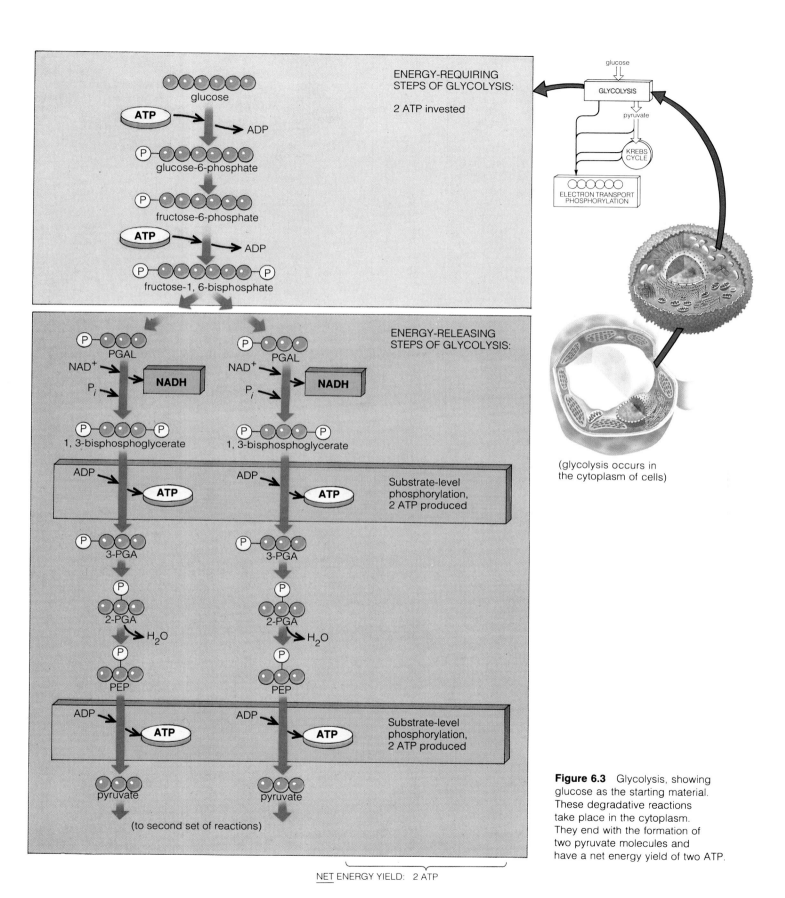

ENERGY-REQUIRING STEPS OF GLYCOLYSIS:

2 ATP invested

glucose

glucose-6-phosphate

fructose-6-phosphate

fructose-1, 6-bisphosphate

ENERGY-RELEASING STEPS OF GLYCOLYSIS:

PGAL

NADH

1, 3-bisphosphoglycerate

Substrate-level phosphorylation, 2 ATP produced

3-PGA

2-PGA

PEP

Substrate-level phosphorylation, 2 ATP produced

pyruvate

(to second set of reactions)

NET ENERGY YIELD: 2 ATP

(glycolysis occurs in the cytoplasm of cells)

glucose

GLYCOLYSIS

pyruvate

KREBS CYCLE

ELECTRON TRANSPORT PHOSPHORYLATION

Figure 6.3 Glycolysis, showing glucose as the starting material. These degradative reactions take place in the cytoplasm. They end with the formation of two pyruvate molecules and have a net energy yield of two ATP.

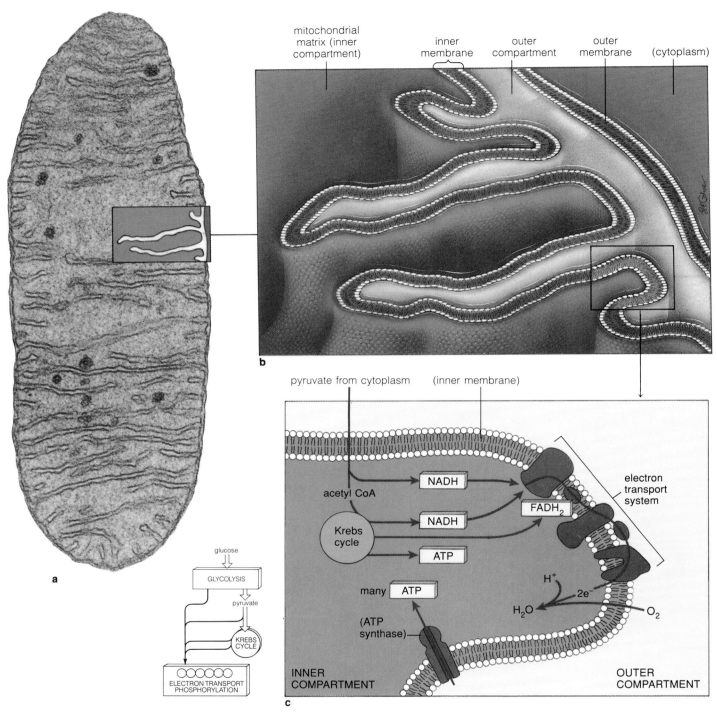

mitochondrial matrix (inner compartment) — **inner membrane** — **outer compartment** — **outer membrane** — **(cytoplasm)**

a

b

glucose

GLYCOLYSIS

pyruvate

KREBS CYCLE

ELECTRON TRANSPORT PHOSPHORYLATION

pyruvate from cytoplasm — (inner membrane)

acetyl CoA

NADH

Krebs cycle

NADH

FADH₂

ATP

many ATP

(ATP synthase)

electron transport system

H⁺

2e⁻

H₂O

O₂

INNER COMPARTMENT

OUTER COMPARTMENT

c

Figure 6.4 Functional zones of the mitochondrion. (**a**) Mitochondrion from a bat pancreatic cell, thin section.

(**b**) The inner mitochrondrial membrane divides the inside of this organelle into two compartments.

(**c**) The inner compartment is the site of the second stage of aerobic respiration, including Krebs cycle activities. The coenzymes NAD⁺ and FAD pick up H⁺ and electrons (forming NADH and FADH₂). They transfer their cargo to transport systems embedded in the inner membrane. Operation of these systems during the third stage sets up H⁺ concentration and electric gradients across the membrane. The gradients are coupled to ATP formation by enzyme systems (ATP synthases).

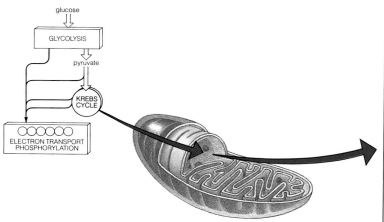

Krebs Cycle

In the second stage of the aerobic pathway, pyruvate enters the mitochondrion, an organelle of the sort shown in Figure 6.4. There the pyruvate is broken down completely, with its carbon and oxygen atoms ending up in carbon dioxide and water. The carbon dioxide exhaled by your lungs comes from the breakdown reactions.

Take a look at Figure 6.5. It shows that the second stage begins with conversion of each pyruvate to acetyl-CoA. The conversion product becomes attached to oxaloacetate, the point of entry into the Krebs cycle. (The cycle was named after Hans Krebs, who began working out its details in the 1930s.)

The Krebs cycle itself serves three functions. *First,* some reaction steps liberate H^+ and electrons for transfer to the coenzymes NAD^+ and FAD. *Second,* one reaction step serves to produce ATP (by substrate-level phosphorylation). *Third,* many reaction steps are concerned with juggling the intermediate molecules back into the form of oxaloacetate. Cells have only so much oxaloacetate, and the cycle would shut down if this compound were not regenerated.

The second stage adds only two more ATP to the small yield from glycolysis. *But it also loads many more coenzymes with H^+ and electrons that can be used in the third stage of the aerobic pathway:*

Glycolysis:		2 NADH
Pyruvate conversion preceding Krebs cycle:		2 NADH
Krebs cycle:	2 FADH$_2$	6 NADH
Total electron carriers sent to third stage of aerobic pathway:	2 FADH$_2$ +	10 NADH

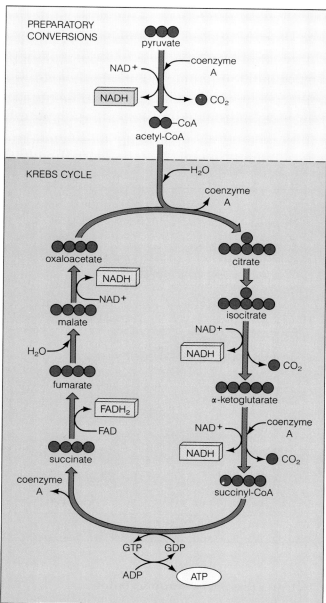

Figure 6.5 The Krebs cycle and the preparatory reactions preceding it. For *each* pyruvate molecule, 3 CO$_2$, 1 ATP, 4 NADH, and 1 FADH$_2$ are formed. But remember the steps shown occur *twice* for each glucose molecule broken down.

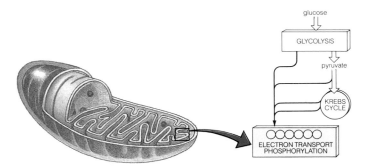

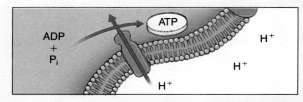

Figure 6.6 Electron transport phosphorylation. The reactions occur at transport systems and enzyme systems (ATP synthases) embedded in the inner mitochondrial membrane. The transport system consists of enzymes and other proteins (including cytochrome molecules) that operate one after the other, as shown in Figure 6.4c.

The membrane itself creates two compartments. The reactions begin in the inner compartment, when NADH and $FADH_2$ give up H^+ and electrons to the transport system. Electrons are accepted and passed through the system, but the H^+ is left behind—in the outer compartment:

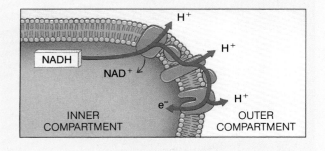

Soon there is a higher concentration of H^+ ions in the outer compartment than in the inner one. In other words, concentration and electric gradients now exist across the membrane. The H^+ ions follow the gradients and move back into the inner compartment. They do this by flowing through the ATP synthases that span the membrane. Energy associated with the flow drives the coupling of ADP and inorganic phosphate into ATP:

Do these events sound familiar? They should: ATP forms in much the same way in chloroplasts (Figure 5.7). The idea that concentration and electric gradients across a membrane drive ATP formation is called the chemiosmotic theory.

Electron Transport Phosphorylation

In the third stage of the aerobic pathway, coenzymes give up H^+ and electrons to transport systems. These systems consist of enzymes and other proteins embedded in the inner membrane of the mitochondrion (Figures 6.4 and 6.6).

The components of the transport system are arranged in series. Electrons are passed down the line, so to speak, and with each transfer they lose some energy. As Figure 6.4 shows, operation of the transport system drives the coupling of ADP and inorganic phosphate into ATP. At the end of the line, free oxygen combines with the electrons and with hydrogen ions to form water. By serving as the final electron acceptor, oxygen pulls electrons through the transport system.

Operation of the transport system commonly leads to the formation of thirty-two ATP. This brings the total net yield of the aerobic pathway to thirty-six ATP for each glucose molecule metabolized.

In aerobic respiration, the pyruvate from glycolysis is completely degraded to carbon dioxide and water.

NAD^+ and FAD accept protons (H^+) and electrons stripped from substrates of the reactions and deliver them to an electron transport system. Oxygen is the final acceptor of those electrons.

From glycolysis (in the cytoplasm) to the final reactions (in the mitochondrion), this pathway commonly yields thirty-six ATP for every glucose molecule.

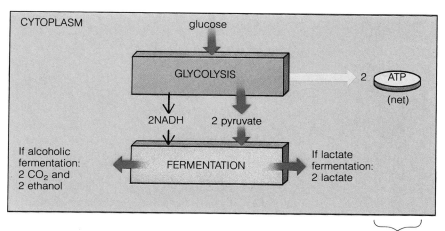

Energy Yield from Either Fermentation Route: **2ATP**

Figure 6.7 Overview of fermentation, a type of degradative pathway in which an intermediate or product of the reactions themselves serves as the final electron acceptor. The photograph shows the dustlike coating on grapes that contains yeasts, single-celled organisms that use a fermentation pathway.

ANAEROBIC ROUTES

In terms of metabolism, "aerobic" refers to a degradative pathway in which oxygen is the final acceptor of electrons stripped from glucose. When you do "aerobic" exercises, your skeletal muscle cells are being given a workout and require more oxygen and glucose than usual.

Aerobic respiration proceeds in muscle cells as long as plenty of oxygen and glucose are available. What happens when those cells contract too vigorously and exceed the body's capacity to deliver oxygen to them? They switch to an anaerobic pathway when oxygen levels are low. In an **anaerobic pathway,** something besides oxygen is the final electron acceptor. However, muscle cells cannot use this alternative for long. The anaerobic pathway has such a low energy yield that the cells quickly exhaust their glycogen stores. (Glycogen is the main storage form for glucose in animal cells.) Once the stores are depleted, muscles fatigue quickly and lose their ability to contract.

Many microbes rely exclusively on anaerobic pathways. Some are indifferent to the presence or absence of oxygen; they include bacteria that are used in the production of yogurt. Other microbes are "strict anaerobes" that die on exposure to oxygen. Among them are the bacteria that cause botulism, tetanus, and other diseases.

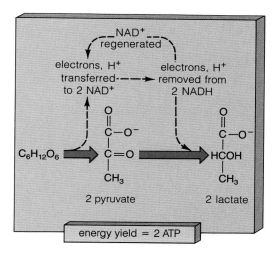

Figure 6.8 Lactate fermentation. The net ATP yield is from glycolysis, the first stage of the route.

Let's now look briefly at two anaerobic routes. Both are **fermentation pathways,** in which the final electron acceptor is actually an intermediate or product of the glucose molecule being degraded (Figures 6.7 through 6.9). An "outside" electron acceptor is not needed at all for the reactions.

Lactate Fermentation

In **lactate fermentation,** pyruvate from glycolysis is converted to lactate, a three-carbon compound. Figure 6.8 shows the reaction sequence. Sometimes lactate is called "lactic acid." However, it is more accurate to refer to the ionized form of the compound (lactate), which is far more common in cells.

One group of bacteria produces lactate exclusively as the fermentation product; milk or cream turned sour is a sign of their activity. Muscle cells can use lactate fermentation as a pathway for a short time under anaerobic conditions, as described earlier. When you breathe hard after exercising, oxygen levels in your blood rise and the lactate is converted to carbon dioxide and water.

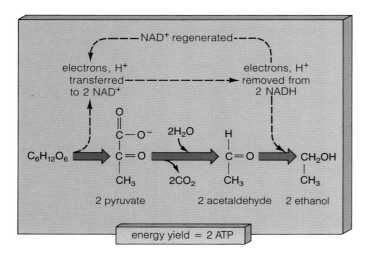

Figure 6.9 Alcoholic fermentation. The net ATP yield is from glycolysis, the first stage of the route.

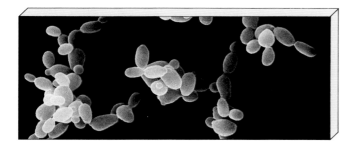

Figure 6.10 Budding cells of the yeast that makes bread dough rise.

Alcoholic Fermentation

In **alcoholic fermentation,** pyruvate from glycolysis is broken down to acetaldehyde. The acetaldehyde accepts electrons from NADH and so become ethanol. Figure 6.9 shows the reaction sequence.

Yeasts are single-celled fungi that rely on alcoholic fermentation. Figure 6.10 shows cells of the species that makes yeast dough "rise." The dough expands because carbon dioxide, the gaseous product of alcoholic fermentation, is accumulating within it. The alcohol in beer, distilled spirits, and wine results from the fermentation activities of similar types of yeast.

Keep in mind that glucose is not completely degraded by a fermentation route, so considerable energy still remains in the products. No more ATP is produced, beyond the two molecules from glycolysis; the final steps serve only to regenerate NAD^+.

A fermentation pathway has a net yield of two ATP (from glycolysis). NAD^+ is regenerated during the reactions.

Also keep in mind that the low energy yield of fermentation and other anaerobic pathways is quite enough for some microbes. It can even help carry some otherwise "aerobic" cells through times of stress. But it is not enough to sustain the activities of large, multicelled organisms.

ALTERNATIVE ENERGY SOURCES IN THE HUMAN BODY

Our cells require a steady supply of carbohydrates, lipids, and proteins for energy and raw materials. When we eat more carbohydrates than our cells are calling for, the excess can be stored as glycogen, most notably in liver or muscle cells, and as fat. Excess fats are tucked away as glistening droplets in the cells of adipose tissue. Between meals, when demands for raw materials and energy exceed dietary intake, the body can draw upon a variety of stored organic compounds.

Of the foods we eat, carbohydrates—glucose especially—are the main source of energy. Our cells tap their stores of complex carbohydrates (glycogen, primarily) to release glucose subunits. They degrade lipids next, then the body's proteins as a last resort. Figure 6.11 shows the points of entry into the aerobic pathway for complex carbohydrates, fats, and proteins. Chapter 25 describes the digestion of these compounds into their simpler components. Here we simply outline how fats and proteins can be used as alternative energy sources.

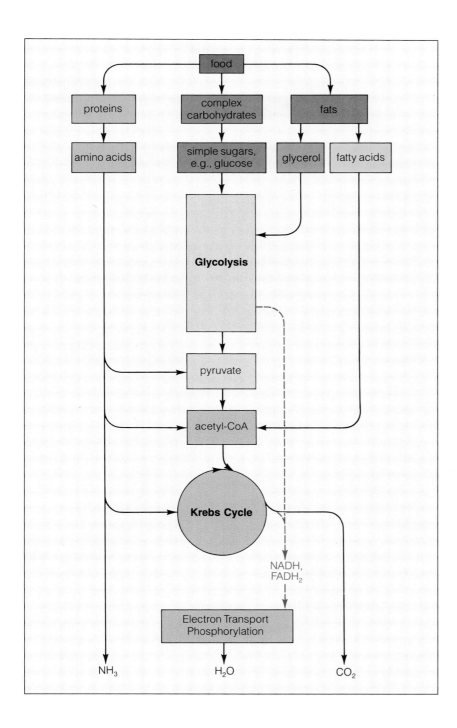

Figure 6.11 Points of entry into the aerobic pathway for complex carbohydrates, fats, and proteins after they have been reduced to their simpler components in the human digestive system.

A fat molecule consists of a glycerol head and one, two, or three fatty acid tails. Once it has been digested into its component parts, the glycerol can be converted to PGAL and inserted into the glycolytic reactions (Figure 6.3). The long carbon backbone of each fatty acid can be split many times into two-carbon fragments, these being easily converted into acetyl-CoA and picked up for the Krebs cycle (Figure 6.5). Whereas each glucose molecule has six carbon atoms, a fatty acid has many

more—and its degradation produces many more ATP molecules.

Following protein digestion, the amino acid subunits can have their amino group (NH_3) removed. The remnants can be fed into the Krebs cycle, where the hydrogens can be removed from the remaining carbon atoms and transferred to coenzymes. The amino groups are converted into ammonia, a waste product, and removed from the body by the kidneys.

SUMMARY

1. ATP, the main energy carrier in cells, can be produced by photosynthesis. It also can be produced by aerobic respiration or fermentation, these being degradative pathways by which chemical energy is released from glucose (or some other organic compound).

2. Glycolysis, the partial breakdown of a glucose molecule, is the first stage of the main degradative pathways. It takes place in the cytoplasm and requires an initial energy input from ATP. During glycolysis, two ATP (net), two NADH, and two pyruvate molecules are produced for each glucose molecule.

3. In aerobic respiration, oxygen is the final acceptor of electrons stripped from glucose. The pathway proceeds from glycolysis, through the Krebs cycle, and through electron transport phosphorylation. Its net energy yield is commonly thirty-six ATP per glucose molecule.

4. In the second stage of aerobic respiration, pyruvate from glycolysis is converted to a form that can enter the Krebs cycle (a cyclic metabolic pathway). The conversion reactions and the two turns of the cycle itself produce eight NADH, two $FADH_2$, and two ATP. In this stage, the glucose molecule is degraded completely to carbon dioxide and water.

5. The coenzymes NADH and $FADH_2$ (formed during glycolysis and the Krebs cycle) deliver electrons to a transport system embedded in the inner membrane of mitochondria. The third stage of the aerobic pathway (electron transport phosphorylation) depends on electron transfers through the system *and* on phosphorylation reactions that take place at channel proteins spanning the membrane.

6. Operation of the transport system sets up H^+ concentration and electric gradients across the membrane. H^+ moves down the gradients, through channel proteins. Energy associated with the flow drives the coupling of ADP and inorganic phosphate to form ATP. Oxygen accepts electrons from the transport system and combines with H^+ to form water.

7. Many microbes are not metabolically equipped to use oxygen as a final electron acceptor. They rely instead on alcoholic fermentation, lactate fermentation, or other anaerobic pathways. Also, some cells that normally use the aerobic pathway (such as skeletal muscle cells) can switch to anaerobic routes when oxygen supplies are low.

8. Compared with aerobic respiration, the fermentation routes have a small net yield (two ATP, from glycolysis), because glucose is not completely degraded. Following glycolysis, the remaining reactions serve to regenerate NAD^+.

Perspective on Life

In this unit, you have read about pathways by which cells trap, store, and then release energy to drive their activities. Over evolutionary time, the main pathways—photosynthesis and aerobic respiration—have become interconnected on a grand scale.

When life began, free oxygen was absent from the earth's atmosphere. Most likely, early single-celled organisms produced ATP by pathways similar to glycolysis. And given the anaerobic conditions, fermentation routes must have predominated.

Photosynthetic organisms emerged more than 3 billion years ago, and they turned out to be a profound force in evolution. Oxygen, a by-product of their activities, began to accumulate in the atmosphere. Some photosynthesizers were opportunistic about the increasing oxygen levels. Perhaps through mutations in their metabolic machinery, they gained the capacity to use oxygen as an electron acceptor for degradative reactions—and in time some cells abandoned photosynthesis entirely. Among those cells were the forerunners of animals and other organisms able to survive with aerobic machinery alone.

With aerobic respiration, life became self-sustaining, for its final products—carbon dioxide and water—are precisely the materials used to build organic compounds in photosynthesis! Thus the flow of carbon, hydrogen, and oxygen through the energy pathways of living organisms came full circle:

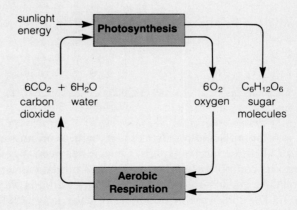

Perhaps one of the most difficult connections you are asked to perceive is the link between yourself—a living, intelligent being—and such remote-sounding things as energy, metabolic pathways, and the cycling of carbon, hydrogen, and oxygen. Is this really the stuff of humanity?

Think back, for a moment, to the description of a water molecule. A pair of hydrogen atoms competing with an oxygen atom for a share of the electrons joining them doesn't exactly seem close to our daily lives. But from that simple competition, the polarity of the water molecule arises. As a result of the polarity, hydrogen bonds form between water molecules. And that is a beginning for the organization of lifeless matter that leads, ultimately, to the organization of matter in all living things.

For now you can imagine other kinds of molecules interspersed in water. Many are nonpolar and resist interaction with the water molecules. Others are polar and respond by dissolving in it. And the lipids among them (with water-soluble *and* water-insoluble regions) spontaneously assemble into a two-layered film. Such lipid bilayers are the basis for all cell membranes, hence all cells. The cell has been, from the beginning, the fundamental *living* unit.

With the boundary afforded by a cell membrane, chemical reactions can be contained and controlled. The essence of life *is* chemical control. This "control" is not some mysterious force. It is a chemical responsiveness to energy changes and to the kinds of molecules present in the environment. It operates by "telling" a class of protein molecules—enzymes—when and what to build, and when and what to tear down.

And it is not some mysterious force that creates the proteins themselves. DNA, the slender double strand of heredity, has the chemical structure—*the chemical message*—that allows molecule faithfully to reproduce molecule, one generation after the next. Those DNA strands tell many billions of cells in your body how countless molecules must be built and torn apart for their stored energy.

So yes, carbon, hydrogen, oxygen, and other organic molecules represent the stuff of you, and us, and all of life. But it takes more than molecules to complete the picture. You are alive because of the way molecules are organized and maintained by a constant flow of energy. It takes outside energy from sources such as the sun to drive their formation. Once molecules are assembled into cells, it takes outside energy derived from food, water, and air to sustain their organization. Plants, animals, fungi, protistans, and bacteria are part of a web of energy use and materials cycling that ties together all levels of biological organization. Should energy fail to reach any part of any level, life there will dwindle and cease.

For energy flows through time in only one direction—from forms rich in energy to forms having less usable stores of it. Only as long as sunlight flows into the web of life—and only as long as there are molecules to recombine, rearrange, and recycle—does life have the potential to continue in all its rich diversity.

In short, life is no more *and no less* than a marvelously complex system of prolonging order. Sustained by energy transfusions, it continues because of a capacity for self-reproduction—the handing down of hereditary instructions. With those instructions, energy and materials are organized, generation after generation. Even with the death of the individual, life is prolonged. With death, molecules are released and can be recycled once more, providing raw materials for new generations. In this flow of matter and energy through time, each birth is affirmation of our ongoing capacity for organization, each death a renewal.

Review Questions

1. ATP can be produced when carbohydrates are degraded. Phosphorylations and electron transfers are needed to do this. Describe the roles of these reactions in the main energy-releasing pathways. *82*

2. Which energy-releasing pathways occur in the cytoplasm? In the mitochondrion of eukaryotes? *83, 84*

3. Is the following statement true? Your muscle cells cannot function at all unless they are supplied with oxygen. *87*

4. Glycolysis is the first stage of all the main pathways by which glucose is degraded. Can you define those pathways in terms of the final electron acceptor for their reactions? If you include the two ATP molecules formed during glycolysis, what is the *net* energy yield from one glucose molecule for each pathway? *82, 85*

5. In anaerobic routes of glucose breakdown, further conversions of pyruvate do not yield any more usable energy. What, then, is the advantage of the conversions? *88*

6. Describe the functions of the Krebs cycle. Describe the functions of electron transport phosphorylation. *85, 86*

Self-Quiz (Answers in Appendix IV)

1. Plants as well as bacteria, protistans, fungi, and animals can produce ATP by degrading _____.

2. Glucose can be degraded by way of two anaerobic pathways, called _____ and _____, as well as by aerobic respiration.

3. In the first stage of aerobic respiration, glucose is partially broken down to _____, which in the second stage is broken down completely to _____ and _____.

4. ATP is best described as _____.
   a. a high-energy phosphate compound
   b. a primary source of chemical energy
   c. being produced by plants, animals, bacteria, protistans, and fungi
   d. all of the above

5. Which of the following is *not* a product of glycolysis?
   a. two NADH
   b. two pyruvate
   c. two H_2O
   d. two ATP

6. Glycolysis occurs in which part of the cell?
   a. nucleus
   b. mitochondrion
   c. plasma membrane
   d. cytoplasm

7. The final acceptor of the electrons stripped from glucose during aerobic respiration is _____.
   a. water
   b. hydrogen
   c. oxygen
   d. NADH

8. Electron transport systems for the aerobic reactions are located in the _____.
   a. cytoplasm
   b. inner mitochondrial membrane
   c. outer mitochondrial membrane
   d. stroma

9. The flow of _____ through channel proteins in the inner mitochondrial membrane provides the energy to couple ADP and inorganic phosphate to form ATP.
   a. electrons
   b. hydrogen ions
   c. NADH
   d. $FADH_2$

10. Match each type of metabolic reaction with its function:
   _____ glycolysis
   _____ fermentation
   _____ Krebs cycle
   _____ electron
      transport
      phosphorylation

   a. produces ATP, NADH, and CO_2
   b. degrades glucose into two pyruvate
   c. regenerates NAD^+
   d. flow of H^+ through channel proteins that drives ATP formation

Selected Key Terms

aerobic respiration *80*
alcoholic fermentation *88*
anaerobic pathway *87*
ATP *80*
electron transport
   phosphorylation *81*
$FADH_2$ *82*
glycolysis *81*
Krebs cycle *81*
lactate fermentation *88*
mitochondrion *84*
NADH *82*
substrate-level
   phosphorylation *82*

Readings

Becker, W. 1986. *The World of the Cell.* Menlo Park, California: Benjamin/Cummings. Chapters 7 and 8 are a good place to start for further readings on anaerobic and aerobic metabolism.

Brock, T., B. Smith, and M. Madigan. 1988. *Biology of Microorganisms.* Fifth edition. Englewood Cliffs, New Jersey: Prentice-Hall. Clear descriptions of the energy-releasing pathways of microbes.

Lehninger A. 1982. *Principles of Biochemistry.* New York: Worth. Clear, accessible introduction to metabolic pathways.

FACING PAGE: *Human sperm, one of which will penetrate this mature egg and so set the stage for the development of a new individual in the image of its parents.*

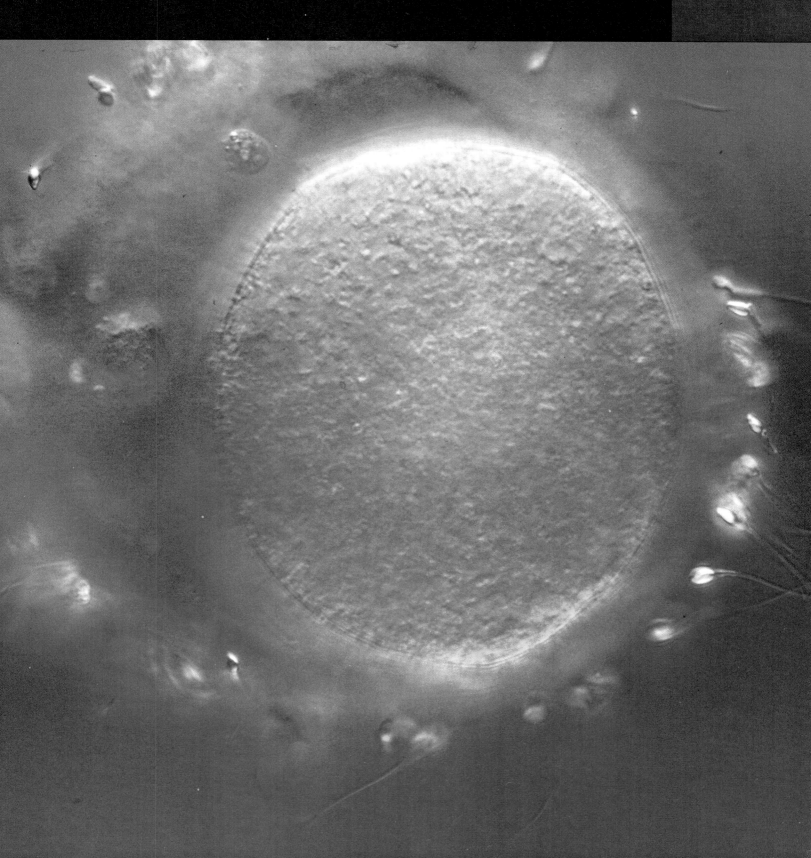

7 CELL DIVISION AND MITOSIS

KEY CONCEPTS

1. Each cell of a new generation will not grow or function properly unless it receives a duplicate of all the parental DNA and a portion of the cytoplasm from the parental cell. In eukaryotes, DNA is parceled out to daughter cells by mitosis or meiosis, both of which are nuclear division mechanisms.

2. A chromosome is a DNA molecule with certain proteins attached to it, and the cells of each species have a characteristic number of them. "Diploid" cells contain two of each type of chromosome characteristic of the species.

3. Mitosis maintains the number of chromosomes from one cell generation to the next. Thus each daughter cell formed by the division of the nucleus and cytoplasm of a diploid parental cell will be diploid also.

4. Mitosis proceeds through four continuous stages: prophase, metaphase, anaphase, and telophase. Actual cytoplasmic division (cytokinesis) occurs toward the end of the nuclear division or at some point afterward.

5. Mitosis is the basis of asexual reproduction of single-celled eukaryotes as well as the growth of multicelled eukaryotes. Meiosis occurs only in germ cells, a cell lineage set aside for reproduction.

DIVIDING CELLS: THE BRIDGE BETWEEN GENERATIONS

Have you ever stopped to think about how your body developed into its current splendid self? Starting with the fertilized egg in your mother's body, a single cell divided in two, then the two into four, and so on until billions of cells were growing, developing in specialized ways, and dividing at different times to produce your various body parts (Figure 7.1). Nerve cells in your brain ceased dividing before you were born; yet ongoing cell divisions are still replacing the lining of your small intestine every five days. Even now, in a cell lineage set aside for reproduction, divisions are probably proceeding that will give rise to sperm or eggs—the reproductive bridge to the next generation.

In biology, **reproduction** means producing a new generation of cells or multicelled individuals. Reproduction begins with the division of single cells. And the ground rule for cell division is this:

Each cell of a new generation must receive a duplicate of all the parental DNA and enough cytoplasmic machinery to start up its own operation.

DNA, recall, contains the genetic instructions for making proteins. Some proteins serve as structural

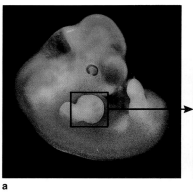 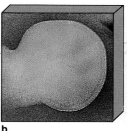  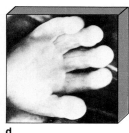

Figure 7.1 Development of the human hand by way of cell divisions and other processes. Individual cells resulting from the mitotic cell divisions are clearly visible in (**e**). The hand is turned palm upward in (**d**).

a   b   c   d

materials; many serve as enzymes during the synthesis of carbohydrates, lipids, and other building blocks of the cell. Unless new cells receive the necessary DNA instructions, they will not grow or function properly.

Also, the cytoplasm of the parental cell already has operating machinery—enzymes, organelles, and so on. When a daughter cell inherits what looks merely like a blob of cytoplasm, it really is getting "start-up" machinery for its operation, until it has time to use its inherited DNA for growing and developing on its own.

Overview of Division Mechanisms

In multicelled plants and animals, cell division typically begins with mitosis or meiosis and ends with cytokinesis. **Mitosis** and **meiosis** are *nuclear* division mechanisms— the means by which DNA is sorted out and distributed into new nuclei for the forthcoming daughter cells. The actual splitting of a parental cell into two daughter cells occurs by way of **cytokinesis,** or *cytoplasmic* division.

Mitosis is the basis for bodily growth (through repeated cell divisions). It also is the basis for asexual reproduction in many plants, animals, and other organisms. Meiosis occurs only in germ cells, a cell lineage set aside for sexual reproduction. By definition, sexual reproduction is a process that begins with meiosis,

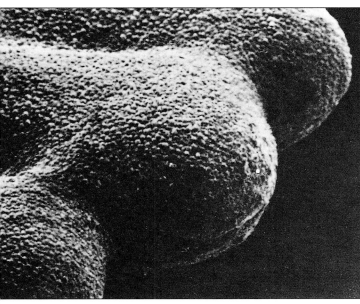

e

proceeds through the formation of gametes (sex cells, such as sperm and eggs), and ends at fertilization. At fertilization, a sperm nucleus and egg nucleus fuse together in the zygote, the first cell of the new individual.

This chapter focuses on mitosis, and the chapter to follow, on meiosis. As you will see, the two division mechanisms have much in common but they differ in their end result. Both are limited to eukaryotes; the prokaryotes (bacteria) use a different cell division mechanism, as described on page 226.

Some Key Points About Chromosome Structure

Before you track the distribution of DNA into daughter cells, reflect for a moment on its structural organization. Many proteins are attached to eukaryotic DNA, and they generally are equal in mass to the DNA itself. Together, the DNA and proteins form a structure called the **chromosome.**

Between divisions, a chromosome is stretched out in threadlike form. It is still threadlike when it is duplicated prior to cell division, and the two threads remain attached for awhile as **sister chromatids** of the chromosome. Each "thread," of course, is a DNA double helix with its associated proteins, which we show here as a simplified version of a current model:

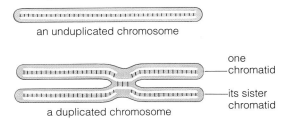

an unduplicated chromosome

one chromatid

its sister chromatid

a duplicated chromosome

Notice the small region of the chromosome where the DNA does not appear to be completely duplicated. This is the **centromere,** a region having attachment sites for microtubules that will help move the chromosome during nuclear division:

centromere

Keep in mind that this model is highly simplified. The DNA double helix in each chromatid is actually two

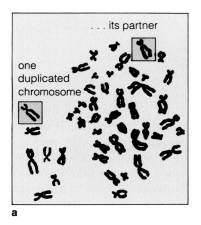

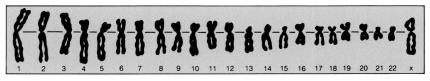

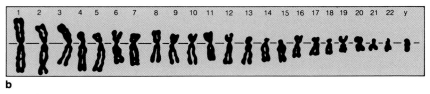

Figure 7.2 (**a**) Photograph of the 46 chromosomes from a diploid cell of a human male. All are in the duplicated state. (**b**) By cutting apart and arranging the chromosomes according to length and shape, we see that there are two sets of 23 chromosomes, with all the chromosomes in one set having a partner, or homologue, in the other. The partners don't pair at all during mitosis, but they pair up with each other during meiosis.

molecular strands twisted together repeatedly like a spiral staircase (that's what the "double helix" means), and it is much longer than can be shown here.

Mitosis, Meiosis, and the Chromosome Number

All individuals of the same species have the same number of chromosomes in the cells making up their body. (These are called somatic cells, to distinguish them from the germ cells.) For example, there are 46 chromosomes in your somatic cells, 48 in a gorilla's, and 14 in a garden pea's. *With mitosis, the chromosome number is maintained, division after division:*

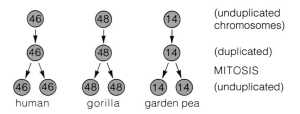

To get an initial sense of the difference between mitosis and meiosis, take a look at Figure 7.2, which shows all the chromosomes in a human somatic cell. Notice every two chromosomes have the same length and shape; their hereditary instructions also deal with the same traits. The two corresponding chromosomes are **homologous chromosomes.** Homologues don't pair at all during mitosis, but they pair with each other during meiosis. (You may have noticed that the X and Y chro-

mosomes don't look alike, but they pair during meiosis so we still call them homologues.)

During meiosis, the chromosome number is reduced by half for forthcoming gametes. And not just any half—each gamete ends up with one of each pair of homologous chromosomes. (It doesn't matter which of the two it gets.) Reducing the chromosome number requires two nuclear divisions:

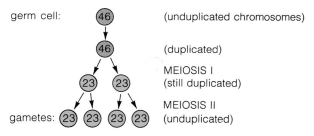

We say that gametes end up with a **haploid** number of chromosomes, meaning "half" the parental number. Later, when a sperm nucleus and an egg nucleus fuse at fertilization, the **diploid** number is restored. "Diploid" means having two chromosomes of each type in the somatic cells of sexually reproducing species.

Mitosis is a type of nuclear division that *maintains* the parental number of chromosomes for forthcoming cells. It is the basis for bodily growth and, in some cases, asexual reproduction of eukaryotes.

Meiosis is a type of nuclear division that *reduces* the parental chromosome number by half—to the haploid number. It occurs only in germ cells used in sexual reproduction.

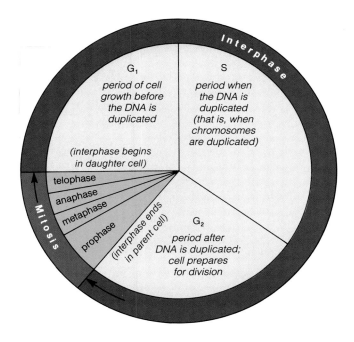

Figure 7.3 Eukaryotic cell cycle. This drawing has been generalized; the length of different stages varies greatly from one type of cell to the next.

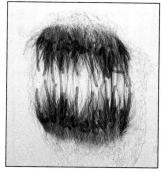

a Prophase

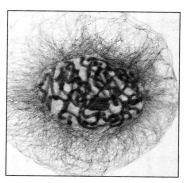

c Anaphase

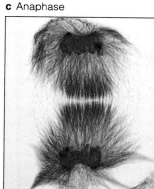

b Metaphase

d Telophase

MITOSIS AND THE CELL CYCLE

Mitosis is a very small part of the **cell cycle,** a recurring sequence of events that extends from the time of a cell's formation until its own division is completed. Normally, a cell destined to enter mitosis spends about ninety percent of the cell cycle in **interphase,** when it increases its mass, approximately doubles the number of its cytoplasmic components, and finally duplicates its DNA (Figure 7.3).

The cell cycle varies in its duration, but it is fairly consistent for all cells of a given type. Your brain cells are arrested at interphase and never will divide again. In contrast, the cells of a newly forming sea urchin may double in number every two hours.

STAGES OF MITOSIS

When a cell makes the transition from interphase to mitosis, it stops constructing new cell parts. Profound changes now proceed one after the other, through four stages. The sequential stages of mitosis are called **prophase, metaphase, anaphase,** and **telophase.**

Before considering details of mitosis, take a look at Figure 7.4, which indicates the extent of chromosome movements through the different stages. Chromosomes do not move about on their own. They are moved by a

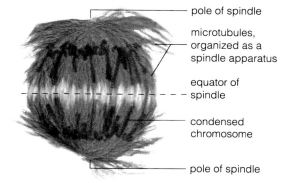

Figure 7.4 Appearance of the spindle apparatus during the four stages of mitosis in a cell from a plant seed. Microtubules of the spindle are stained red, and the chromosomes are stained blue. At prophase, microtubules of the cytoskeleton are outside the nuclear envelope, but they are becoming organized to form a spindle. By metaphase, the nuclear envelope has broken down and the spindle microtubules have penetrated the nuclear region.

spindle apparatus composed of microtubules. Remember that microtubules are components of the cytoskeleton. When a nucleus is about to divide, microtubules disassemble and then reassemble into a spindle just outside the nucleus (Figure 7.5). The spindle establishes two poles toward which the chromosomes will be moved during division.

prophase

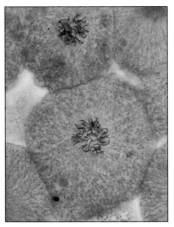

INTERPHASE	MITOSIS

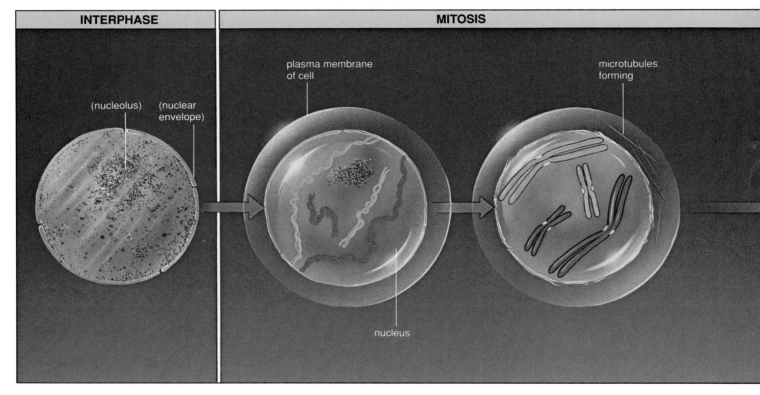

plasma membrane of cell

microtubules forming

(nucleolus) (nuclear envelope)

nucleus

Nucleus at Interphase

The DNA is duplicated, then the cell prepares for division.

Early Prophase

The DNA and associated proteins start condensing into the threadlike chromosome form. (Chromosomes are already duplicated.) Two chromosomes derived from the male parent are in green; their homologues from the female are pink.

Late Prophase

Chromosomes continue to condense. Microtubules start to assemble outside the nucleus; they will form the spindle. Centrioles (if present) are moved by the microtubules toward opposite poles. The nuclear envelope starts breaking up.

Figure 7.5 Mitosis: the nuclear division mechanism that maintains the parental chromosome number in daughter cells. Shown here, a diploid animal cell (with pairs of homologous chromosomes, derived from two parents).

For the sake of clarity, only two pairs of homologous chromosomes are shown in the diagram and the spindle apparatus is simplified. With rare exceptions, the picture is more involved than this, as indicated by the micrographs of mitosis in a whitefish cell.

metaphase

anaphase

telophase

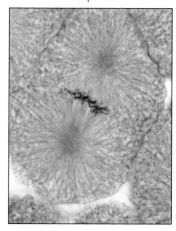
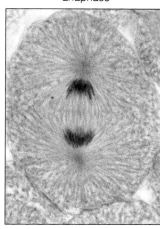
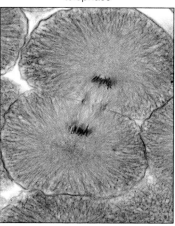

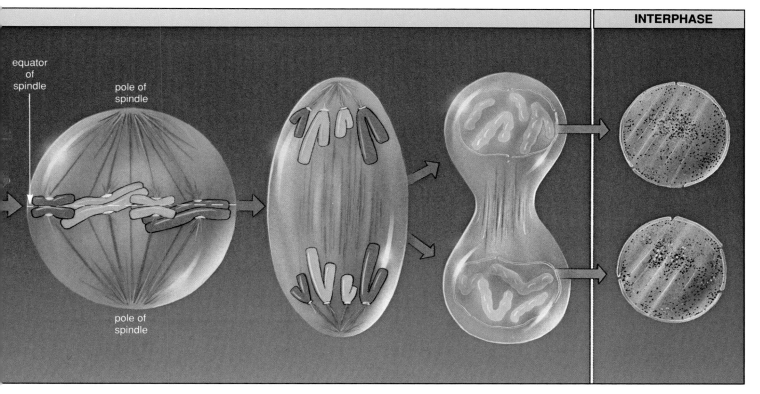

Metaphase
Sister chromatids of each chromosome are attached to the spindle. All chromosomes are now lined up at the spindle equator.

Anaphase
Sister chromatids of each chromosome will now be separated from each other and moved to opposite poles.

Telophase
Chromosomes decondense. New nuclear membranes start forming. Most often, cytokinesis occurs before the end of telophase.

Interphase
Two daughter nuclei are formed, each with a diploid number of chromosomes (the same as the parental nucleus).

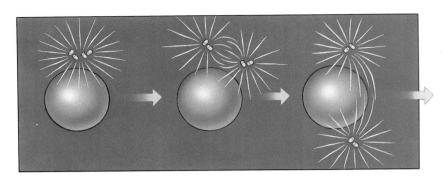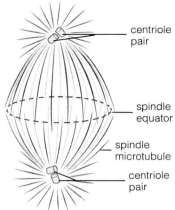

Figure 7.6 Formation of one type of mitotic spindle in eukaryotic cells having centrioles. Microtubules form two "half-spindles" that meet at the spindle equator.

In many multicelled organisms, a microtubule organizing center near the nucleus dictates where the two spindle poles will occur (page 97). Sometimes this center includes a pair of centrioles that were duplicated earlier (at the end of interphase). A centriole pair is moved to each pole of the spindle during its formation. *Where* the centrioles end up seems to influence the orientation of the cytoskeleton in each daughter cell. In turn, that orientation influences the direction of cell division—hence the shape of the developing embryo.

Prophase

Prophase, the first stage of mitosis, is evident when chromosomes become visible in the light microscope as threadlike forms. ("Mitosis" comes from the Greek *mitos,* meaning thread.) Each chromosome was duplicated earlier, during interphase, so it already consists of two sister chromatids joined at the centromere. In late prophase, each chromosome becomes condensed into thicker, rodlike forms (Figure 7.5).

Metaphase

Late in prophase, the nuclear envelope breaks up and the spindle is free to penetrate the nuclear area (Figure 7.6). At early metaphase, the chromosomes seem to go into a frenzy. This happens when spindle microtubules make their first pass at harnessing the chromosomes. With the first random contacts, each chromosome spins around its long axis, and it is yanked back and forth until its sister chromatids are firmly oriented toward opposite poles. The yanking continues until the chromosomes lie halfway between the poles, at the spindle equator (Figure 7.5). Thus two events dominate metaphase: orientation of sister chromatids toward opposite poles, and alignment of all chromosomes halfway between the poles.

Anaphase

Two events characterize anaphase: sister chromatids of each chromosome are separated from each other, and those former partners are moved to opposite poles. Once they do separate, they are no longer referred to as chromatids. Each is now an independent chromosome:

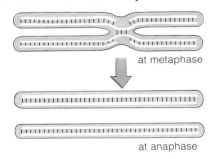

Telophase

Telophase begins once the separated chromosomes arrive at opposite spindle poles. Now the chromosomes decondense into the threadlike form. Small vesicles (the fragments of the old nuclear envelope) fuse together to form patches of membrane alongside the chromosomes. Patch joins with patch, and eventually a new, continuous nuclear envelope separates the hereditary material from the cytoplasm. Once the nucleus is completed, telophase is completed—and so is mitosis.

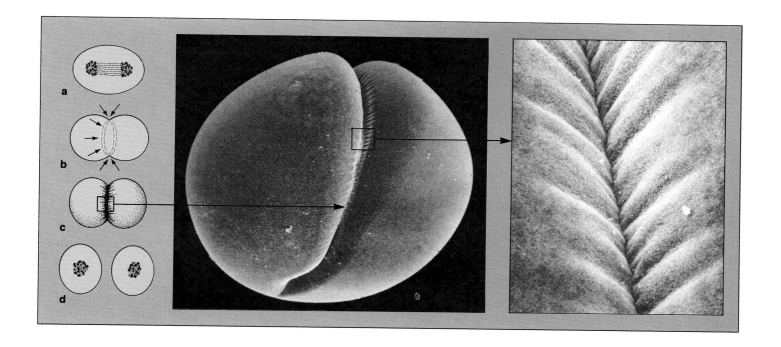

Figure 7.7 Cytokinesis in an animal cell. The scanning electron micrographs show the furrowing of the plasma membrane caused by the contraction of a microfilament ring just beneath it. (**a**) Nuclear division is complete; the spindle is disassembling. (**b**) Microfilament rings at the former spindle equator contract, like a purse string closing. (**c**) Contractions cause furrowing at the cell surface. (**d**) The cytoplasm is pinched in two.

CYTOKINESIS

Cytoplasmic division, or cytokinesis, usually coincides with the period from late anaphase through telophase. For most animal cells, deposits accumulate and form a layer around microtubules at the cell midsection. A shallow, ringlike depression appears above the layer, at the cell surface (Figure 7.7). At this depression, called a **cleavage furrow,** contractile microfilaments pull the plasma membrane inward and cut the cell in two.

A different form of cytokinesis, **cell plate formation,** occurs in most land plants. (Plant cells typically have fairly rigid walls that preclude the formation of cleavage furrows.) Vesicles filled with wall-building material fuse with remnants from the spindle, forming a disklike structure (the "cell plate"). Here, cellulose deposits form a crosswall that separates the two daughter cells (Figure 7.8).

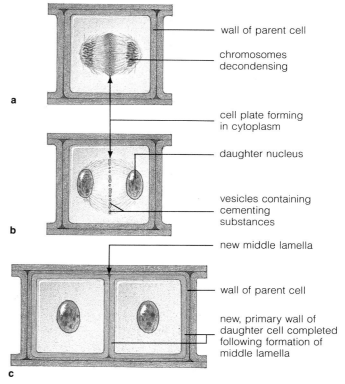

wall of parent cell

chromosomes decondensing

cell plate forming in cytoplasm

daughter nucleus

vesicles containing cementing substances

new middle lamella

wall of parent cell

new, primary wall of daughter cell completed following formation of middle lamella

Figure 7.8 Cytokinesis following mitosis in a plant cell. (**a, b**) Vesicles form at the spindle equator and gradually fuse to form a cell plate. The cell plate grows outward until it reaches the parent cell wall. The vesicles contain substances that will form the middle lamella, which will cement together the primary walls of the daughter cells. (**c**) The membrane of the vesicles is used in forming the plasma membrane on both sides of the cell plate.

SUMMARY

1. Each cell of a new generation must receive a duplicate of all the parental DNA and enough cytoplasmic machinery to start up its own operation.

2. Somatic (body) cells of multicelled eukaryotes commonly have a diploid number of chromosomes. That is, they have two of each type of chromosome characteristic of the species. The pairs of "homologous" chromosomes generally are alike in length, shape, and the heritable traits they deal with.

3. Eukaryotic chromosomes are duplicated during interphase (between cell divisions). Whereas each was one DNA molecule (and associated proteins), now it consists of two, temporarily attached as sister chromatids.

4. Eukaryotes employ different division mechanisms that serve different functions:
   a. Mitosis is a type of nuclear division that maintains the parental number of chromosomes in each of two daughter nuclei. (Thus if the parental nucleus is diploid, so will be the daughter nuclei.)
   b. Mitosis is the basis of bodily growth for multicelled eukaryotes. It also is the basis of asexual reproduction for some eukaryotes.
   c. Meiosis, a type of nuclear division that reduces the parental chromosome number by half (to the haploid number) in each of four daughter nuclei, is the basis of gamete formation. It occurs only in germ cells (a cell lineage set aside for sexual reproduction).
   d. For most organisms, actual cytoplasmic division, or cytokinesis, occurs toward the end of nuclear division or at some point afterward.

5. A cell destined to divide by mitosis spends about ninety percent of the cell cycle in interphase (a period between mitotic divisions). During interphase, the cell increases in mass, doubles its number of cytoplasmic components, and duplicates its chromosomes.

6. Mitosis proceeds through four continuous stages:
   a. Prophase. Duplicated, threadlike chromosomes condense; a microtubular spindle forms and establishes the poles for nuclear division. The nuclear envelope starts to break up late in prophase.
   b. Metaphase. Spindle microtubules harness each chromosome and orient its two sister chromatids toward opposite spindle poles. All chromosomes become aligned at the spindle equator.
   c. Anaphase. Sister chromatids of each chromosome separate to become independent chromosomes, which move to opposite poles.
   d. Telophase. Chromosomes decondense to the threadlike form, and a new nuclear envelope forms around the two parcels of chromosomes. Mitosis is completed.

Review Questions

1. Define the two types of nuclear division mechanisms that occur in eukaryotes. What is cytokinesis? 95

2. Define somatic cell and germ cell. Which type of cell can undergo meiosis? 96

3. What is a chromosome? What is a chromosome called in its unduplicated state? In its duplicated state (that is, with two sister chromatids)? 95

4. Define homologous chromosome. Do homologous chromosomes pair during mitosis, meiosis, or both? 96

5. Describe the spindle apparatus and its general function in nuclear division processes. 97

6. Name the four main stages of mitosis, and characterize each stage. 97

Self-Quiz   (Answers in Appendix IV)

1. Eukaryotic DNA is distributed to daughter cells by _mitosis_ or _meiosis_, both of which are nuclear divisions.

2. Each kind of organism contains a characteristic number of _chromosomes_ in each cell; each of those structures is composed of a _DNA_ molecule with its associated proteins.

3. A pair of chromosomes that are similar in length, shape, and the traits they govern are called _____.
   a. diploid chromosomes
   b. mitotic chromosomes
   c. homologous chromosomes
   d. germ chromosomes

4. Somatic cells of multicelled eukaryotic organisms usually have a _____ number of chromosomes, whereas gametes have a _____ number.
   a. haploid; haploid   c. diploid; diploid
   b. haploid; diploid   d. diploid; haploid

5. Interphase is the stage when _____.
   a. nothing occurs
   b. a germ cell forms its spindle apparatus
   c. a cell grows and duplicates its DNA
   d. cytokinesis occurs

6. Following mitosis, a daughter cell will end up with genetic instructions that are _____ and _____ chromosome number as the parent cell.
   a. identical to the parent cell's; the same
   b. identical to the parent cell's; one-half the
   c. rearranged; the same
   d. rearranged; one-half the

7. Cytokinesis is a term that describes _____.
   a. cell division
   b. nuclear division
   c. cytoplasmic division
   d. reducing the chromosome number

8. During interphase, a cell _____.
   a. grows
   b. doubles the number of cytoplasmic components
   c. duplicates its chromosomes
   d. all of the above

9. All of the following are stages of mitosis *except* _____.
   a. prophase
   b. interphase
   c. metaphase
   d. anaphase

10. A *pair* of duplicated homologous chromosomes is the same thing as _____.
   a. one chromatid
   b. two chromatids
   c. three chromatids
   d. four chromatids

11. Match each stage of mitosis with the following key events.
   _D_ metaphase   a. sister chromatids of each chromosome separate and move to opposite poles
   _b_ prophase
   _C_ telophase   b. threadlike chromosomes condense and a microtubular spindle forms
   _A_ anaphase
                       c. chromosomes decondense, daughter nuclei re-form
                       d. all chromosomes become aligned at spindle equator

Selected Key Terms

anaphase *100*
cell cycle *97*
cell plate formation *101*
centromere *95*
chromosome *95*
cleavage furrow *101*
cytokinesis *95*
diploid state *96*
haploid state *96*
homologous chromosome *96*

interphase *97*
meiosis *95*
metaphase *100*
mitosis *95*
prophase *100*
reproduction *94*
sister chromatid *95*
spindle apparatus *97*
telophase *100*

Readings

Alberts, B., et al. 1989. *Molecular Biology of the Cell.* Second Edition. New York: Garland Publishing.

Prescott, D. 1988. *Cells: Principles of Molecular Structure and Function.* Boston: Jones and Bartlett. Chapter 7.

Smith-Klein, C., and V. Kish. 1988. *Principles of Cell Biology.* New York: Harper & Row.

8 MEIOSIS

KEY CONCEPTS

1. Sexual reproduction of multicelled plants and animals depends on these events: meiosis, gamete formation, and fertilization. Other reproductive events, including spore formation, occur between meiosis and gamete formation in plants (page 245).

2. The cells of most animals and many plants have a diploid number of chromosomes (*two* of each type characteristic of the species), half from the mother and half from the father. In germ cells set aside for sexual reproduction, every two chromosomes that are alike (homologous) will pair with each other during meiosis.

3. Meiosis, a nuclear division process in germ cells only, reduces the diploid number by half for the forthcoming gametes. Each gamete produced is haploid, with only *one* of each pair of homologous chromosomes. The union of two gametes at fertilization restores the diploid number in the new individual.

4. During meiosis, homologous chromosomes swap segments (crossing over and recombination), then they are shuffled in ways that will give to different mixes of maternal and paternal instructions for traits in gametes.

ON ASEXUAL AND SEXUAL REPRODUCTION

A strawberry plant can do something you could not even begin to do except in your wildest imagination. It can reproduce all by itself. Through mitosis, aboveground stems called runners grow outward from the plant—and brand new plants sprout along the runners. Similarly, the entire body of a flatworm can split into two roughly equivalent parts—then, through mitosis, each part can grow into a whole flatworm.

Nature abounds with different and sometimes spectacular forms of **asexual reproduction.** By this process, *one* parent passes on a duplicate of all of its genes to offspring. ("Genes" are specific stretches of DNA, each being the inherited instructions for producing or influencing the traits of offspring.) This means, of course, that asexually produced offspring can only be genetically identical copies, or clones, of the parent.

Inheritance is much more interesting with sexual reproduction. Commonly, **sexual reproduction** involves two parents, each with two genes for every trait. Both parents pass on one of each gene to offspring by way of meiosis, gamete formation, and fertilization (page 95).

Figure 8.1 (**a**) The chin fissure, a heritable trait arising from a rather uncommon form of a gene. Actor Kirk Douglas received a gene that influences this trait from each of his parents. One gene called for a chin fissure and the other didn't, but one is all it takes in this case. (**b**) This photograph shows what Mr. Douglas' chin might have looked like if he had inherited two ordinary forms of the gene instead.

Through meiosis and fertilization, old gene combinations are broken up and new ones are put together. The immediate consequence is variation in the physical and behavioral traits of offspring. The long-term consequence can be evolutionary change.

a

b

Thus the first cell of a new individual inherits two genes for every trait—one from each parent.

If the instructions in every pair of genes were identical down to the last detail, then sexual reproduction would produce clones, also. Just imagine—you, everyone you know, every member of the entire human population might all end up looking exactly alike.

But it happens that, over evolutionary time, mutations have changed the molecular structure of many genes. In fact, any gene might come in several alternative forms, called **alleles,** that "say" slightly different things about how a trait will be expressed in offspring. (As Figure 8.1 indicates, genetic instructions about your chin can vary this way. One allele might say "put a dimple in it" and another might say "no dimple.") Different alleles govern thousands of different traits. And this brings us to a key reason why humans don't all look alike. *Sexual reproduction puts together new combinations of existing alleles in offspring.*

This chapter gives us a closer look at meiosis, the foundation for sexual reproduction. More importantly, it starts us thinking about some far-reaching consequences of the gene shufflings possible with sexual reproduction. New gene combinations among offspring lead to variations in their physical and behavioral traits. *Such variation is a testing ground for agents of selection—and a basis of evolutionary change.*

OVERVIEW OF MEIOSIS

Think "Homologues"

As we have seen, meiosis is a nuclear division mechanism that occurs in *germ cells,* a cell lineage destined to give rise to the gametes (sperm or eggs) used in sexual reproduction. Germ cells develop in a variety of reproductive structures and organs (Figure 8.2). Like somatic cells, they commonly have a diploid number of *chromosomes* (these being structures composed of DNA and proteins).

Recall that "diploid" means there are two chromosomes of each type, or 2n, in a cell. The two are **homologous chromosomes.** Generally, homologues have the same length, the same shape, and the same genes—and they line up with each other during meiosis. Only

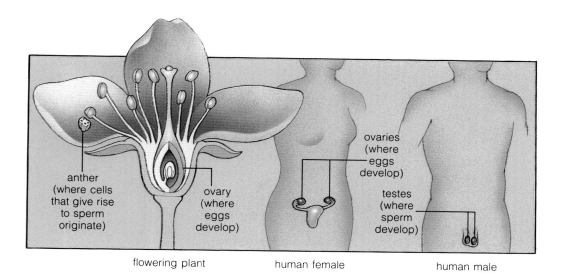

anther
(where cells
that give rise
to sperm
originate)

ovary
(where
eggs
develop)

ovaries
(where
eggs
develop)

testes
(where
sperm
develop)

flowering plant   human female   human male

Figure 8.2 Examples of the location of germ cells that give rise to sperm and eggs.

the sex chromosomes, designated X and Y, differ in form and in which genes they carry—but they still function as homologues during meiosis.

Meiosis reduces the diploid number by half, to the "haploid" number (*n*). And not just any half: *Each gamete ends up with one of each pair of homologous chromosomes.* To give an example, the diploid number for humans is 46 (that is, 23 + 23 homologues). A human gamete ends up with one of each type of 23 chromosomes. Table 8.1 gives the diploid number of chromosomes for other species. Can you say what the haploid number would be for each example?

Overview of the Two Divisions

Meiosis bears some resemblances to mitosis. While a germ cell is still in interphase, its chromosomes are duplicated (by a process called DNA replication). Each chromosome now has a duplicate of itself, and it hangs onto it at the centromere. For as long as the two parts of a duplicated chromosome remain attached to each other, they are called **sister chromatids:**

As in mitosis, microtubules that are part of a spindle apparatus harness each chromosome and move it during nuclear division.

Table 8.1	Number of Chromosomes in the Somatic Cells of Some Eukaryotes
Fruit fly, *Drosophila melanogaster*	8
Garden pea, *Pisum sativum*	14
Corn, *Zea mays*	20
Frog, *Rana pipiens*	26
Earthworm, *Lumbricus terrestris*	36
Human, *Homo sapiens*	46
Chimpanzee, *Pan troglodytes*	48
Amoeba, *Amoeba*	50
Horsetail, *Equisetum*	216

Unlike mitosis, however, there are *two divisions*, called meiosis I and II:

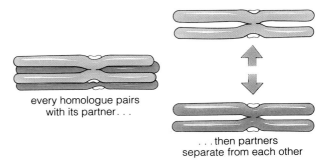

During meiosis I, each duplicated chromosome lines up with its partner, *homologue to homologue,* then the partners are separated from each other. Here we show just one pair of homologous chromosomes, but the same thing happens to all pairs in the nucleus:

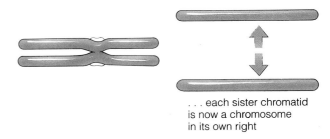

Cytokinesis typically follows. Each daughter cell has a haploid number of chromosomes—but each chromosome is still duplicated.

During meiosis II, *the sister chromatids of each chromosome are separated from each other:*

Cytokinesis typically follows this separation also. Overall, one DNA duplication during interphase, then two nuclear divisions and two cytoplasmic divisions produce four haploid cells.

STAGES OF MEIOSIS

Prophase I Activities

The first stage of meiosis, **prophase I,** is a time of major gene shufflings between homologous chromosomes. Homologues begin to pair at the onset of prophase I. It is as if they become stitched point by point along their

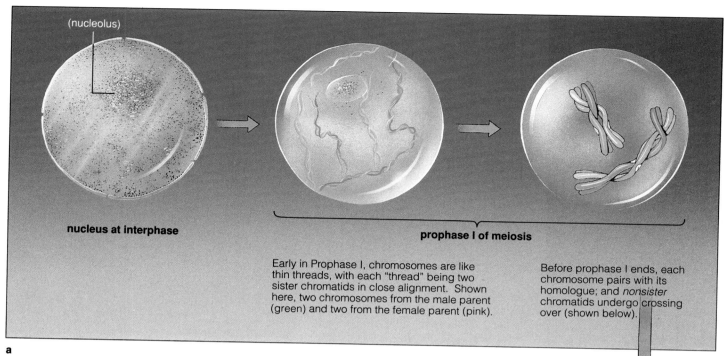

(nucleolus)

nucleus at interphase

prophase I of meiosis

Early in Prophase I, chromosomes are like thin threads, with each "thread" being two sister chromatids in close alignment. Shown here, two chromosomes from the male parent (green) and two from the female parent (pink).

Before prophase I ends, each chromosome pairs with its homologue; and *nonsister* chromatids undergo crossing over (shown below).

a

Figure 8.3 Crossing over. (**a**) Early in prophase I of meiosis, chromosomes are like thin threads, with each "thread" being two sister chromatids in close alignment. (**b, c**) Before prophase I ends, each chromosome has condensed and pairs with its homologous partner, and *nonsister* chromatids undergo crossing over.

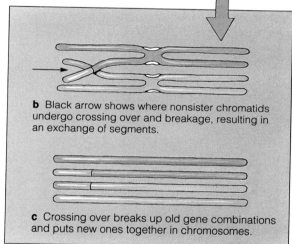

b Black arrow shows where nonsister chromatids undergo crossing over and breakage, resulting in an exchange of segments.

c Crossing over breaks up old gene combinations and puts new ones together in chromosomes.

entire length, with little space between them. (The X and Y chromosomes pair at one end only.) The arrangement favors **crossing over,** whereby *nonsister* chromatids undergo breakage at one or more sites along their length and exchange corresponding segments at the breakage points. This event is diagrammed to the right, in Figure 8.3.

Crossing over would be rather pointless if each type of gene never varied from one chromosome to the next. But, remember, a gene can come in alternative forms—alleles. You can safely bet that all the alleles on one chromosome will not be an identical match to the ones on its homologue. With each crossover, then, homologues may swap different instructions for some traits.

We will look at the mechanism of crossing over in later chapters. For now it is enough to know that crossing over leads to **genetic recombination,** which in turn leads to variation in the traits of offspring.

Crossing over is an event by which old combinations of alleles in a chromosome are broken up and new ones put together.

Figure 8.4 The 23 pairs of homologous chromosomes of humans.

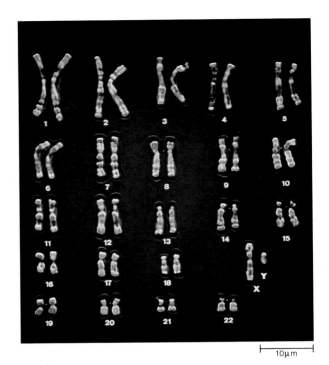

10μm

Figure 8.5 Meiosis: the nuclear division mechanism by which the parental number of chromosomes is reduced by half (to the haploid number) for forthcoming gametes. Only two pairs of homologous chromosomes are shown. The green ones are derived from one parent, and the pink ones are their homologues from the other parent.

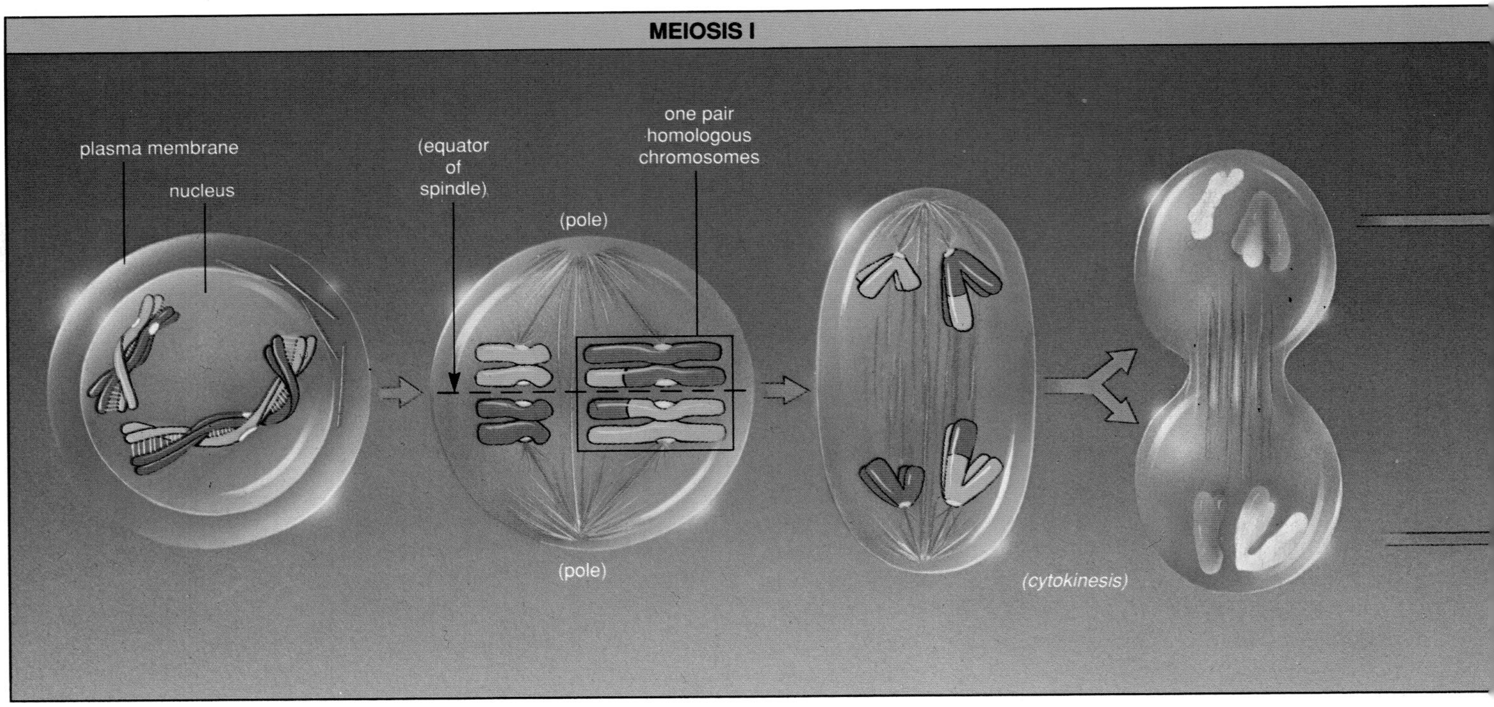

MEIOSIS I

plasma membrane

nucleus

(equator of spindle)

one pair homologous chromosomes

(pole)

(pole)

(cytokinesis)

Prophase I

Each chromosome condenses, then pairs with its homologue. Crossing over and recombination occur.

Metaphase I

Spindle apparatus forms, nuclear envelope (not shown) breaks down during transition to metaphase I. Homologues align randomly at spindle equator.

Anaphase I

Each homologue is separated from its partner, and the two are moved to opposite poles.

Telophase I

A haploid number of chromosomes (still duplicated) ends up at each pole.

Separating the Homologues

The second stage of meiosis, **metaphase I**, is a time of major shufflings of whole chromosomes, before their distribution into daughter nuclei.

Suppose the shufflings are proceeding right now in one of your germ cells. We can call that cell's homologous chromosomes "maternal" and "paternal" (one of each type was inherited from your mother and their homologues were inherited from your father). Figure 8.4 shows how many maternal and paternal chromosomes would be present in that cell, so you can get an idea of how complicated it would be to track their movements. To keep things simple, let's imagine that we are tracking only *two* pairs of those homologous chromosomes, as shown in Figure 8.5.

During metaphase I, the spindle apparatus moves all the homologous chromosomes until they are lined up at the spindle equator. Then, during anaphase I, each homologue is separated from its partner and the two are moved to opposite poles of the spindle. Are all the maternal chromosomes destined to move to one pole and the paternal chromosomes to the other? Maybe, maybe not. The metaphase I line-up and subsequent direction of movement of homologues is random. *It doesn't matter which partner moves to which pole.*

Consider Figure 8.6, which shows how just three pairs of homologues can be shuffled into any one of four possible line-ups at metaphase I. In this case, 2^3 or 8 combinations of maternal and paternal chromosomes are possible for the forthcoming gametes.

A human germ cell has 23 pairs of homologous chromosomes, not just three. So 2^{23} or *8,388,608 combinations* of maternal and paternal chromosomes are possible every time a germ cell gives rise to sperm or eggs! Are you beginning to get an idea of why such splendid mixes of traits show up even in the same family?

Because homologues align randomly at metaphase I, different gametes will end up with different mixes of maternal and paternal chromosomes.

MEIOSIS II

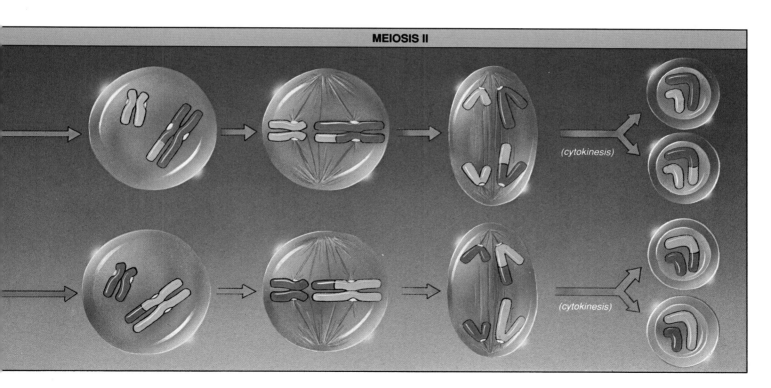

Prophase II
There is no DNA replication between divisions. Sister chromatids of each chromosome are still attached at the centromere.

Metaphase II
Each chromosome is aligned at the spindle equator.

Anaphase II
Each chromosome splits; what were once sister chromatids are now chromosomes in their own right and are moved to opposite poles.

Telophase II
Four daughter nuclei form. Following cytokinesis, each gamete has a haploid number of chromosomes, all in the unduplicated state.

four different chromosome line-ups possible:

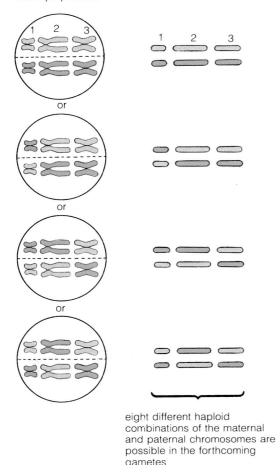

or

or

or

eight different haploid combinations of the maternal and paternal chromosomes are possible in the forthcoming gametes

Figure 8.6 Possible outcomes of the random alignment of three pairs of homologous chromosomes at metaphase I of meiosis. The three types of chromosomes are labeled 1, 2, and 3. Maternal chromosomes are shaded pink; paternal ones are shaded green.

There is no DNA duplication between the two meiotic divisions. But, remember, each chromosome was duplicated earlier (during interphase) and is still in the duplicated form when meiosis II begins.

Separating the Sister Chromatids

Meiosis II has one overriding function: separation of the two sister chromatids of each chromosome. Each duplicated chromosome is moved to the spindle equator at metaphase II. Each is split at anaphase II; its (formerly) sister chromatids are now chromosomes in their own right.

Each spindle pole is the destination of half the parental number of chromosomes. But that haploid number includes one of each type of chromosome characteristic of the species. During telophase II, new nuclear membranes form around the chromosomes after they have become clustered at the two poles. Meiosis is completed.

MEIOSIS AND THE LIFE CYCLES

Gamete Formation

The life cycle of multicelled animals proceeds from meiosis to gamete formation, fertilization, then growth by way of mitosis (Figures 8.7 through 8.9). The life cycle of pine trees, roses, and other familiar multicelled plants proceeds in the same general way, although some additional events occur between meiosis and gamete formation. (Among other things, plants form haploid spores, the consequences of which are described in Chapter 22.)

In *male* animals, meiosis and gamete formation are called **spermatogenesis** (Figure 8.8). Typically a diploid

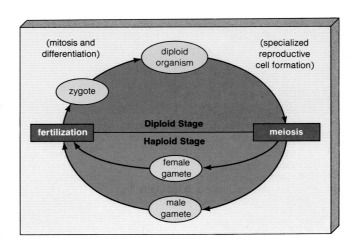

Figure 8.7 Generalized life cycle for animals.

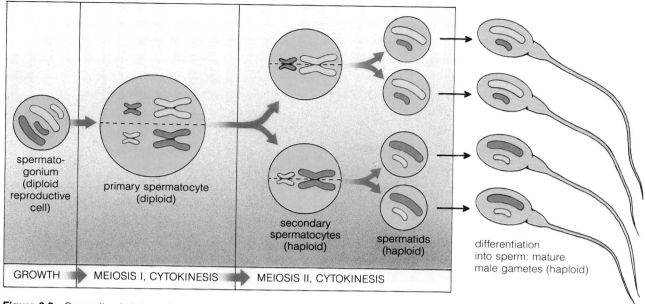

Figure 8.8 Generalized picture of spermatogenesis in male animals. (For the sake of clarity, the nuclear envelopes are not shown in Figures 8.8 through 8.11.)

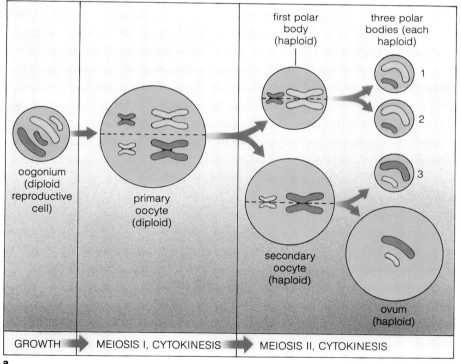

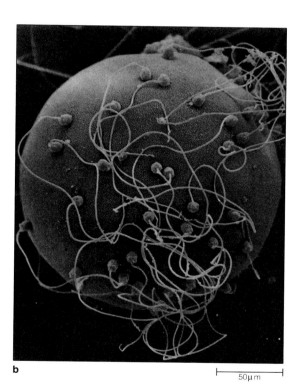

Figure 8.9 (**a**) Generalized picture of oogenesis in female animals. This sketch is not drawn to the same scale as Figure 8.8. A primary oocyte is *much* larger than a primary spermatocyte, as suggested by the scanning electron micrograph of the egg and sperm of a clam in (**b**). Also, the polar bodies are extremely small compared to an ovum, as shown in Figure 32.12.

germ cell increases in size and becomes a primary spermatocyte. This larger, immature cell undergoes meiosis and eventually four haploid spermatids are produced. The spermatids change in form, develop a tail, and become **sperm** (mature male gametes).

In *female* animals, meiosis and gamete formation are called **oogenesis.** This sequence of events differs from spermatogenesis in two important features. Compared to a primary spermatocyte, many more cytoplasmic components accumulate in a primary oocyte, the female germ cell that undergoes meiosis. Also, the cells formed after meiosis differ in size and function (Figure 8.9).

Following meiosis I, one cell (the secondary oocyte) has nearly all the cytoplasm; the other (small) cell is a "polar body." Both cells undergo meiosis II, and the outcome is one large cell and three extremely small ones (that is, three polar bodies). The large cell develops into the mature **egg,** or **ovum.** The polar bodies do not function as gametes. During meiosis they each serve as "dumping grounds" for one set of the parental chromosomes, so that the egg will end up with the necessary haploid number (page 96).

More Gene Shufflings at Fertilization

As Figure 8.10 shows, the diploid number of chromosomes is restored at **fertilization,** when the nuclei of two gametes fuse to form the zygote (the first cell of a new multicelled individual). Fertilization, too, contributes to variation in the traits of offspring.

Think about the possibilities for humans. First, genes are shuffled during prophase I, when each chromosome takes part in two or three crossovers, on the average. Second, random alignments at metaphase I lead to one of 8,388,608 possible combinations of maternal and paternal chromosomes in each gamete. Third, of all the genetically diverse male and female gametes that are produced, *which* two will get together is a matter of chance. As you can see, the sheer number of new combinations brought together at fertilization is staggering!

The mix of different combinations of alleles from two different gametes at fertilization contributes to variation in the traits of offspring.

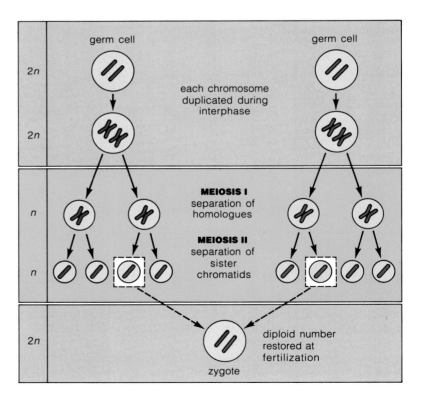

Figure 8.10 How fertilization restores a diploid (2n) number of chromosomes that has been reduced by half during meiosis.

MEIOSIS COMPARED WITH MITOSIS

In this unit, our main focus has been on mitosis and meiosis—two nuclear division mechanisms used in the reproduction of eukaryotes. Mitosis underlies asexual reproduction of the single-celled forms as well as growth of the multicelled forms. Meiosis occurs only in germ cells used in sexual reproduction. Figure 8.11 summarizes the similarities and differences between the two mechanisms.

The major difference between them is this: Mitotic cell division produces clones (genetically identical copies of the parent cell). Meiosis and fertilization give rise to novel combinations of alleles in offspring which, as a consequence, vary from the parents and one another in the details of their traits. Three events are responsible for the variation:

1. Crossing over and genetic recombination occur during prophase I of meiosis.

2. Each chromosome and its homologue align randomly at the spindle equator before their separation during metaphase I, so forthcoming gametes can end up with mixes of maternal and paternal chromosomes.

3. Fertilization is a chance mix of different combinations of alleles from two different gametes.

The variation in traits made possible by meiosis and fertilization is a testing ground for agents of selection, hence for the evolution of populations.

SUMMARY

1. The life cycle of multicelled plants and animals generally includes meiosis and gamete formation, fertilization, and growth by way of cell divisions. In animals, meiosis produces four haploid gametes (sperm in males, eggs in females). Fusion of a sperm and an egg nucleus at fertilization produces a diploid cell (zygote), which develops into the multicelled form by way of mitosis and cytokinesis.

2. Sexually reproducing organisms commonly have a diploid number of chromosomes ($2n$), or two of each type characteristic of the species. The two are "homologous" chromosomes, with the same length, shape, and gene sequence. Homologues pair with each other during meiosis. (The X and Y chromosomes differ in length, shape, and which genes they carry, but they still pair as homologues.)

3. Before meiosis, all chromosomes in a germ cell are duplicated. The duplicates remain attached (as sister chromatids) at the centromere.

4. Meiosis consists of two consecutive divisions of the chromosomes in a germ cell. In meiosis I, each chromosome pairs with and then separates from its homologue. In meiosis II, the sister chromatids of each chromosome separate from each other. In both cases, microtubules of a spindle apparatus move the chromosomes.

5. The following key events occur during meiosis I:
   a. At prophase I, homologues pair with each other. Crossing over breaks up old combinations of alleles and puts together new ones in the chromosomes. This genetic recombination leads to variation in traits among offspring.
   b. At metaphase I, each pair of homologous chromosomes aligns randomly at the spindle equator, with respect to all the other pairs, so different mixes of maternal and paternal chromosomes will end up in different gametes.
   c. At anaphase I, each chromosome is separated from its homologue and moved to the opposite spindle pole.

6. The following key events occur during meiosis II:
   a. At metaphase II, all chromosomes are moved to the spindle equator.
   b. At anaphase II, the sister chromatids of each chromosome are separated for movement to opposite poles. Once separated, they are chromosomes in their own right.

7. Following cytokinesis (cytoplasmic division), there are four haploid cells, one or all of which may function as gametes.

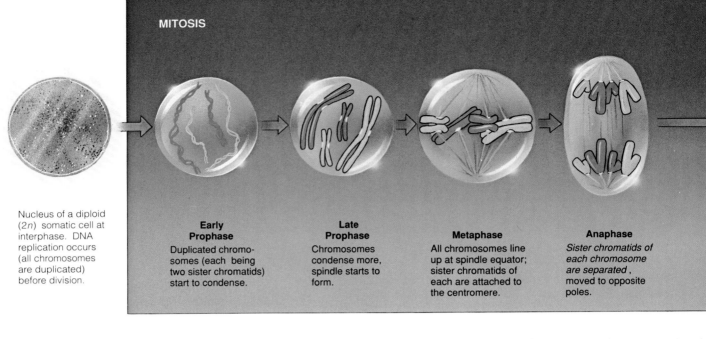

MITOSIS

Nucleus of a diploid (2n) somatic cell at interphase. DNA replication occurs (all chromosomes are duplicated) before division.

Early Prophase
Duplicated chromosomes (each being two sister chromatids) start to condense.

Late Prophase
Chromosomes condense more, spindle starts to form.

Metaphase
All chromosomes line up at spindle equator; sister chromatids of each are attached to the centromere.

Anaphase
Sister chromatids of each chromosome are separated , moved to opposite poles.

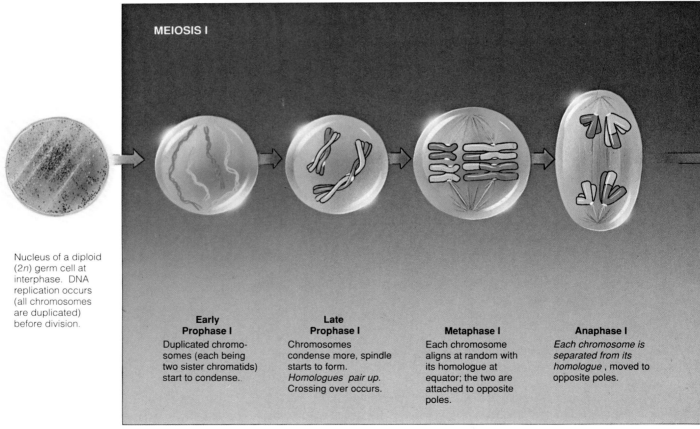

MEIOSIS I

Nucleus of a diploid (2n) germ cell at interphase. DNA replication occurs (all chromosomes are duplicated) before division.

Early Prophase I
Duplicated chromosomes (each being two sister chromatids) start to condense..

Late Prophase I
Chromosomes condense more, spindle starts to form. *Homologues pair up.* Crossing over occurs.

Metaphase I
Each chromosome aligns at random with its homologue at equator; the two are attached to opposite poles.

Anaphase I
Each chromosome is separated from its homologue , moved to opposite poles.

Figure 8.11 Summary of mitosis and meiosis, using a diploid (2n) animal cell as the example. The diagram is arranged to help you compare the similarities and differences between the two division mechanisms. (Chromosomes derived from the male parent are green; their homologues from the female parent are pink.)

Telophase
Each daughter nucleus is *diploid* (2*n*), with pairs of homologous chromosomes. Each chromosome is in the unduplicated state.

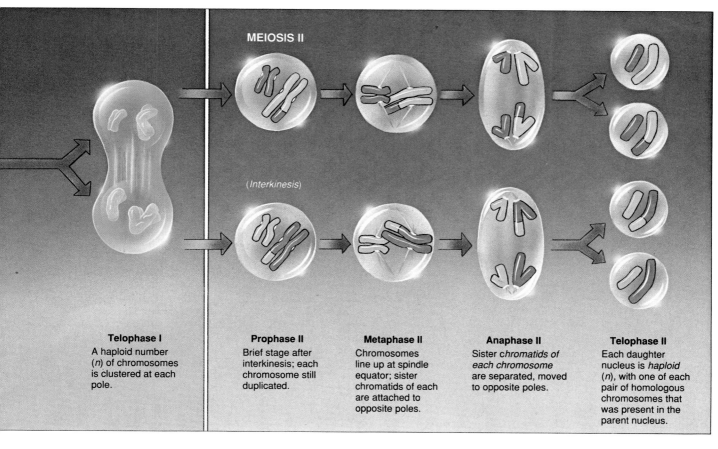

MEIOSIS II

(*Interkinesis*)

Telophase I
A haploid number (*n*) of chromosomes is clustered at each pole.

Prophase II
Brief stage after interkinesis; each chromosome still duplicated.

Metaphase II
Chromosomes line up at spindle equator; sister chromatids of each are attached to opposite poles.

Anaphase II
Sister *chromatids of each chromosome* are separated, moved to opposite poles.

Telophase II
Each daughter nucleus is *haploid* (*n*), with one of each pair of homologous chromosomes that was present in the parent nucleus.

Review Questions

1. Define sexual reproduction. How does it differ from asexual reproduction? What is one of its presumed advantages? *104*

2. The nucleus of diploid cells contains pairs of chromosomes that resemble each other in length, shape, and which genes they carry. What are the pairs called? *105*

3. Refer back to Table 8.1, which gives the diploid number of chromosomes in the body cells of a few organisms. What would be the *haploid* number for the gametes of humans? For the garden pea? *106*

4. Suppose the diploid cells of an organism have four pairs of homologous chromosomes, designated AA, BB, CC, and DD. How would its haploid number of chromosomes be designated? *105–106*

5. When, and in which type of cells, does meiosis occur? *106*

6. Define meiosis and characterize its main stages. In what respects is meiosis like mitosis? In what respects is it unique? *106, 108, 109*

7. Does crossing over occur during mitosis, meiosis, or both? At what stage of nuclear division does it occur, and what is its significance? *107*

Self-Quiz   *(Answers in Appendix IV)*

1. Life cycles of multicellular plants and animals usually include meiosis, _____ formation, fertilization, and _____ by means of cell divisions.

2. Gametes called _____ are produced in males and gametes called _____ are produced in females.

3. The somatic (body) cells of sexually reproducing organisms usually have a _____ number of chromosomes, or _____ of each type characteristic of that species.

4. Two homologous chromosomes generally contain the same _____.
   a. genes in reverse order
   b. alleles in the reverse order
   c. alleles in the same order
   d. none of the above

5. Prior to meiosis, all the chromosomes in a diploid germ cell are _____.
   a. paired   c. duplicated
   b. randomly mixed d. separated

6. Crossing over _____.
   a. alters the chromosome alignments at metaphase
   b. occurs between sperm DNA and egg DNA at fertilization
   c. leads to genetic recombination
   d. occurs only rarely

7. Because of the _____ alignment of homologous chromosomes at metaphase, gametes can end up with _____ mixes of maternal and paternal chromosomes.
   a. unvarying; different
   b. unvarying; duplicate
   c. random; duplicate
   d. random; different

8. Variation in the traits of offspring is increased by the mix of _____ allele combinations from two _____ gametes at fertilization.
   a. similar; similar
   b. different; similar
   c. different; different
   d. similar; different

9. Prior to the meiotic divisions, duplicated chromosomes remain attached as sister _____ at the area of the chromosome called the _____.
   a. chromosomes; centromere
   b. chromatids; centriole
   c. chromosomes; centriole
   d. chromatids; centromere

10. Following meiosis and cytokinesis, there are _____ haploid cells, one or all of which may function as _____.
    a. two; body cells
    b. two; gametes
    c. four; gametes
    d. four; body cells

11. The net result of meiosis is that the _____ chromosome number is _____.
    a. diploid; doubled
    b. diploid; halved
    c. haploid; doubled
    d. haploid; halved

12. Match each stage of meiosis with the correct events.
    _____ prophase I
    _____ metaphase I
    _____ anaphase I
    _____ metaphase II
    _____ anaphase II

    a. each chromosome is separated from its homologue and moves to the opposite pole
    b. two parcels of chromosomes move to the equator of two spindles
    c. sister chromatids of each chromosome are separated and move to opposite poles
    d. homologues pair with each other and crossing over occurs
    e. pairs of homologous chromosomes align randomly at the spindle equator

Selected Key Terms

allele *105*
asexual reproduction *104*
crossing over *107*
fertilization *112*
gene *104*
genetic recombination *107*
germ cell *105*

homologous chromosome *105*
oogenesis *112*
ovum (egg) *112*
sexual reproduction *104*
sister chromatid *106*
sperm *112*
spermatogenesis *110*

Readings

Cummings, M. 1988. *Human Heredity: Principles and Issues.* New York: West.

Strickberger, M. 1985. *Genetics.* Third edition. New York: Macmillan. Contains excellent introduction to chromosomes and meiosis.

OBSERVABLE PATTERNS OF INHERITANCE

KEY CONCEPTS

1. Genes, the units of instruction for producing or influencing traits in offspring, have specific locations on chromosomes. There can be alternative molecular forms of a gene (alleles) at these locations.

2. Diploid cells, which have pairs of homologous chromosomes, have pairs of genes. Mendel's monohybrid crosses of pea plants provided indirect evidence that the two genes of each pair segregate from each other during meiosis and end up in different gametes.

3. Mendel's dihybrid crosses of pea plants provided indirect evidence that a gene pair tends to assort into gametes independently of other gene pairs that are located on nonhomologous chromosomes.

4. The traits observed in offspring are influenced by dominance relations, in which one gene of a pair exerts more pronounced effects compared to its partner. Traits also are influenced by interactions among different gene pairs, by single genes that affect more than one structure or function in the body, and by environmental conditions.

MENDEL'S INSIGHTS INTO THE PATTERNS OF INHERITANCE

More than a century ago, Charles Darwin and Alfred Wallace proposed their theory of evolution by natural selection and offered a new way of looking at life's diversity. In their view, members of a population vary in heritable traits. Variations that improve chances of surviving and reproducing show up more often in each generation, those that don't become less frequent, and in time the population changes (evolves).

Not everyone accepted the theory, partly because it did not fit with a prevailing view of inheritance. It was common knowledge that instructions for heritable traits reside in sperm and eggs—but how were the instructions combined at fertilization? Many thought they blended together, like cream into coffee.

Yet if the instructions become blended, why aren't distinctive traits diluted out of a population? Why do children with freckles keep turning up among nonfreckled generations? How come the descendants of a herd of white stallions and black mares are not uniformly gray? Blending scarcely explained what people could see with their own eyes, but it was considered a rule anyway. Populations "had to be" uniform—and without variation for selective agents to act upon, evolution simply could not occur.

Even before Darwin and Wallace presented their theory, however, someone was gathering evidence that eventually would support its premise about variation in heritable traits. Gregor Mendel, a monk interested in plant breeding *and* mathematics, was beginning to identify the rules governing inheritance (Figure 9.1).

Figure 9.1 Gregor Mendel, founder of classical genetics.

Mendel's Experimental Approach

Mendel experimented with the garden pea plant, *Pisum sativum* (Figure 9.2). This plant can fertilize itself. (Its flowers produce sperm *and* eggs, and fertilization occurs in the same flower.) Some plants are **true-breeding:** successive generations are exactly like the parents in one or more traits. For example, all offspring may ''breed true'' for white flowers.

Pea plants also lend themselves to *cross-fertilization,* in which sperm from one individual fertilizes eggs from another. In some experiments, Mendel stopped plants from self-fertilizing by opening their flower buds and removing the stamens. (Stamens bear pollen grains, in which sperm develop.) Then he promoted cross-fertilization by brushing pollen from another plant on the ''castrated'' bud.

Why did Mendel tinker with plants this way? He wanted cross-fertilization to occur between two true-breeding plants having different forms of the same trait. For example, he crossed a white-flowered with a purple-flowered plant. If their offspring bore white *or* purple flowers, he could identify one plant or the other as the source of the hereditary material for that trait. If there *were* patterns in the way hereditary material is transmitted from parents to offspring, the use of variations in traits might be a way to identify them.

Some Terms Used in Genetics

Having read the chapter on meiosis, you already have insight into the mechanisms of sexual reproduction—which is more than Mendel had. He did not know about chromosomes and so could not have known that the parental chromosome number is reduced by half in gametes, then restored at fertilization. Yet Mendel had some hunches about what was going on. As we follow his thinking, let's simplify things by substituting a few modern terms used in studies of inheritance (see also Figure 9.3):

1. **Genes** are units of instructions for producing or influencing a specific trait in offspring. Each gene has its own **locus,** or particular location, on a chromosome.

2. Diploid cells have received two genes for each trait, one on each of two homologous chromosomes. Two genes at homologous loci are called a **gene pair.**

3. The two genes of a pair deal with the same trait, but they may vary in their information about it. This happens when they differ in form, as when a gene for flower color specifies ''red'' and a different molecular form of that gene specifies ''white.'' The various molecular forms of a gene are called **alleles.**

4. Gene shufflings during meiosis and fertilization can put together different mixes of alleles in offspring. Thus a pair of alleles at homologous loci might be the same molecular form or different from each other. If they are the same, this a *homozygous* condition; if different, this is a *heterozygous* condition.

5. Often one allele is ''dominant,'' meaning its effect on a trait masks the effect of any ''recessive'' allele occupying the homologous locus. Capital letters indicate dominance; lowercase letters indicate recessiveness (for example, *A* and *a*).

6. Putting this together, we say a **homozygous dominant** individual has two dominant alleles (*AA*) for the trait being studied. A **homozygous recessive** individual has two recessive alleles (*aa*). A **heterozygous** individual has two different alleles (*Aa*).

7. To keep the distinction clear between genes and the traits they specify, we use the word **genotype** for the sum total of an individual's genes (or for one gene pair at a time). We use **phenotype** for an individual's traits (the observable aspects of its form and behavior).

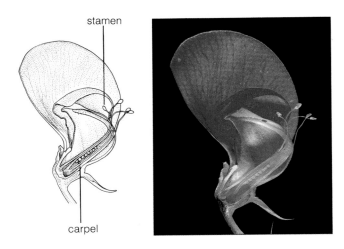

Figure 9.2 The garden pea plant (*Pisum sativum*), the focus of Mendel's experiments. A flower has been sectioned to show the location of the stamens (male reproductive structures) and the carpel (the female reproductive organ).

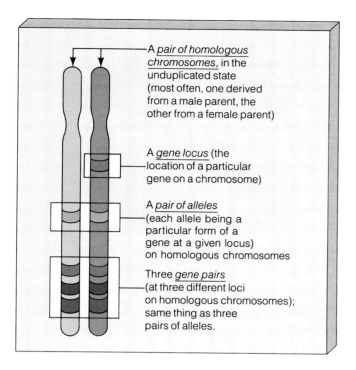

Figure 9.3 A few genetic terms illustrated. In Mendel's time, no one knew about meiosis or chromosomes, but it was clear that offspring received hereditary material from parents by way of sperm and eggs. As we now know, the hereditary material (genes) is packaged in homologous chromosomes (one from the male, one from the female parent). Thus at each gene locus along the chromosomes, one allele has come from the male parent and its partner has come from the female parent.

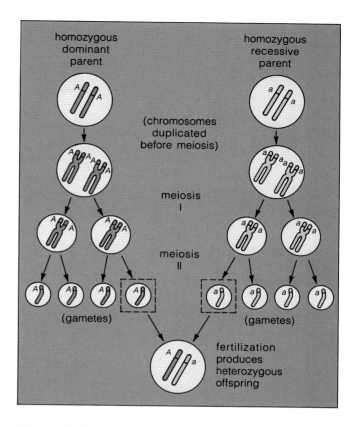

Figure 9.4 Segregation of alleles in a monohybrid cross. Two parents that are true-breeding for contrasting forms of a single trait can give rise only to heterozygous offspring.

The Concept of Segregation

Mendel's first experiments were **monohybrid crosses.** This means two parents that bred true for contrasting forms of a single trait were crossed to produce heterozygous offspring. Mendel tracked the trait through two generations of offspring, which can be designated as follows:

P parental generation

F_1 first-generation offspring

F_2 second-generation offspring

In one case, Mendel crossed a true-breeding purple-flowered plant and a white-flowered plant. *All* the F_1 offspring from that cross had purple flowers. Then he allowed F_1 plants to self-fertilize and produce seeds— and some of the F_2 offspring had white flowers!

Mendel interpreted the results this way. Each plant inherits two "units" of instructions for flower color, and those units retain their identity from one generation to the next. Each egg or sperm receives only one unit. Assuming both units were purple in one parent and white in the other, then each F_1 plant must have inherited one purple and one white unit. If that were so, then purple must be the dominant form of the trait, because it masked the white in F_1 plants.

Let's couch Mendel's interpretation in terms of what we know about meiosis and fertilization. Pea plants are diploid, with pairs of homologous chromosomes. Assume one parent is homozygous dominant (AA) for flower color and the other, homozygous recessive (aa). After meiotic cell division, each sperm or egg will carry the gene for flower color on one of its chromosomes. As Figure 9.4 shows, when sperm and eggs combine at fertilization, only one outcome is possible: $A + a = Aa$.

But what about the F_2 results? To understand this aspect of Mendel's monohybrid crosses, you have to know he crossed hundreds of plants and kept track of thousands of offspring. He also *counted* and *recorded* the number of dominant and recessive offspring for each

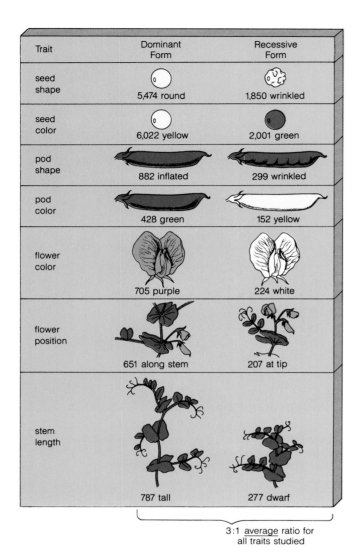

Figure 9.5 Results from Mendel's monohybrid cross experiments with the garden pea. The numbers are his counts of the F_2 plants that showed the dominant or recessive form of the trait being studied. On the average, the dominant-to-recessive ratio was 3:1.

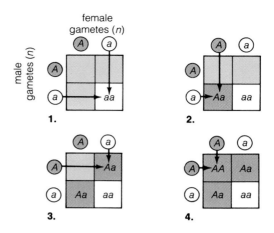

Figure 9.6 Punnett-square method of predicting the probable ratio of traits that will show up in offspring of self-fertilizing individuals known to be heterozygous (*Aa*) for a trait. The circles represent gametes. The letters inside gametes represent the dominant or recessive form of the gene being tracked. Each square depicts the genotype of one kind of offspring.

cross. An intriguing ratio emerged from his records. On the average, of every four F_2 plants, three showed the dominant form of the trait and one showed the recessive (Figure 9.5).

Mendel used his knowledge of mathematics to explain the 3:1 phenotypic ratio. He began by assuming each particular sperm is not precommitted to combining with one particular egg; fertilization has to be a chance event. This meant the monohybrid crosses could be interpreted according to rules of *probability*, which apply to chance events. ("Probability" simply means the number of times a particular outcome will occur, divided by the total number of all possible outcomes.)

The easiest way to predict the probable outcome of a cross between two F_1 plants is the **Punnett-square method** shown in Figure 9.6. Assume each F_1 plant produced two kinds of sperm (or eggs) in equal proportions: half were *A*, and half were *a*. If any sperm is equally likely to fertilize any egg, there are four possibilities for each encounter:

Possible Event:	Probable Outcome:
sperm *A* meets egg *A*	$\frac{1}{4}$ *AA* offspring
sperm *A* meets egg *a*	$\frac{1}{4}$ *Aa* $\Big\}$ or $\frac{1}{2}$ *Aa*
sperm *a* meets egg *A*	$\frac{1}{4}$ *Aa*
sperm *a* meets egg *a*	$\frac{1}{4}$ *aa*

Thus a new F_2 plant had three chances in four of carrying one or both dominant alleles. It had only one chance in

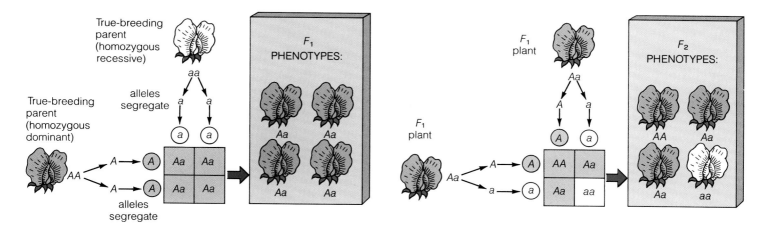

Figure 9.7 Results from one of Mendel's monohybrid crosses. Notice that the dominant-to-recessive ratio is 3:1 for the second-generation (F_2) plants.

As you can tell from the numerical results in Figure 9.5, however, the F_2 ratios weren't *exactly* 3:1. Almost certainly, Mendel's reliance on a large number of crosses and his understanding of probability kept him from being confused by minor deviations from the predicted results. To see why, flip a coin a few times. We all know that a flipped coin is just as likely to end up heads as tails. But often it ends up heads, or tails, several times in a row. When you flip the coin only a few times, the actual ratio may differ greatly from the predicted ratio (1:1). Only when you flip the coin *many* times will you come close to the predicted ratio.

four of carrying two recessive alleles. That is a probable phenotypic ratio of 3:1 (see Figure 9.7).

Results from his monohybrid crosses led Mendel to formulate a principle:

Mendelian principle of segregation. **Diploid organisms inherit a pair of genes for each trait (on a pair of homologous chromosomes). The two genes segregate from each other during meiosis, so each gamete formed will end up with one or the other gene, but not both.**

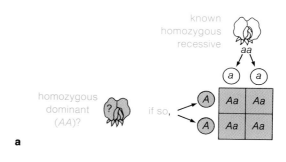

Testcrosses

Mendel gained support for his concept of segregation through the **testcross**. In this type of experimental cross, F_1 hybrids are crossed to an individual known to be true-breeding for the same recessive trait as the recessive parent.

For example, Mendel crossed purple-flowered F_1 plants with true-breeding, white-flowered plants. If his concept were correct, there would be about as many recessive as dominant plants in the offspring from the testcross. That is exactly what happened (Figure 9.8). As predicted, about half the testcross offspring were purple-flowered (*Aa*) and half were white-flowered (*aa*).

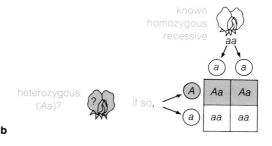

Figure 9.8 Punnett-square method of predicting the outcomes of a testcross between an individual known to be homozygous recessive for a trait (here, white flower color) and an individual that shows the dominant form of the trait. (**a**) If the individual of unknown genotype is homozygous dominant, all offspring will show the dominant form of the trait. (**b**) If the individual is heterozygous, about half the offspring will show the recessive form.

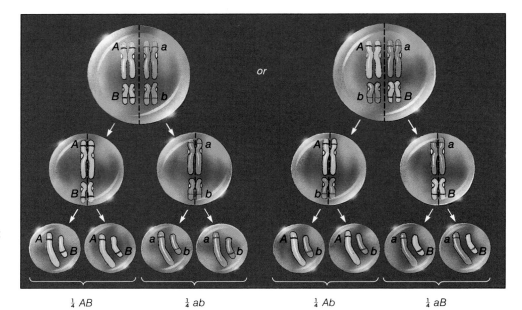

Figure 9.9 Example of independent assortment, showing just two pairs of homologous chromosomes. The different combinations of alleles possible in gametes arise through the random alignment of homologues during metaphase I of meiosis (page 109).

$\frac{1}{4}$ AB $\frac{1}{4}$ ab $\frac{1}{4}$ Ab $\frac{1}{4}$ aB

The Concept of Independent Assortment

In another series of experiments, Mendel crossed true-breeding pea plants having contrasting forms of two traits. In such **dihybrid crosses,** the F_1 offspring inherit two gene pairs, neither of which consists of identical alleles. In the following example of a dihybrid cross, *A* and *B* stand for dominance in flower color and height; *a* and *b* stand for recessiveness:

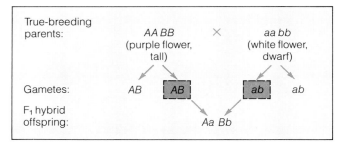

Mendel anticipated (correctly) that all the F_1 offspring of the cross would be purple-flowering and tall. But what would happen when those *Aa Bb* offspring formed sperm and eggs of their own? Would a gene for flower color and a gene for height *travel together or independently of each other* into the gametes?

As we now know, a gene pair tends to segregate into gametes independently of other gene pairs—when the others are located on *non*homologous chromosomes. In this case, four combinations of alleles are possible:

$$\frac{1}{4}AB \qquad \frac{1}{4}Ab \qquad \frac{1}{4}aB \qquad \frac{1}{4}ab$$

These possibilities arise through the random alignment of maternal and paternal chromosomes during metaphase I of meiosis (page 109 and Figure 9.9).

Now think of the way alleles can be mixed at fertilization. Simple multiplication (four kinds of sperm times four kinds of eggs) tells us that *sixteen* combinations of alleles are possible in dihybrid F_2 plants. Figure 9.10 lays out the possibilities, using the Punnett-square method. When we add up the possible combinations, we get $\frac{9}{16}$ tall purple-flowered, $\frac{3}{16}$ dwarf purple-flowered, $\frac{3}{16}$ tall white-flowered, and $\frac{1}{16}$ dwarf white-flowered plants. That is a probable phenotypic ratio of 9:3:3:1.

Results from all of Mendel's dihybrid F_2 crosses were close to a 9:3:3:1 ratio. This evidence that gene pairs tend to travel independently into gametes led to the formulation of another principle:

Mendelian principle of independent assortment. **Each gene pair tends to assort into gametes independently of other gene pairs located on nonhomologous chromosomes.**

VARIATIONS ON MENDEL'S THEMES

It was Mendel's genius to limit his studies to clearly dominant or recessive phenotypes. But the phenotypes of many other traits are not as straightforward, as the following examples will demonstrate.

Dominance Relations

Different degrees of dominance may exist between a pair of alleles. One or both may be fully dominant, or one may be incompletely dominant over the other.

In **incomplete dominance,** the phenotype is intermediate between the homozygous dominant or recessive

Figure 9.10 Results from Mendel's dihybrid cross between true-breeding parent plants differing in two traits (flower color and height). Here, *A* and *a* represent the dominant and recessive alleles for flower color. *B* and *b* represent the dominant and recessive alleles for height. On the average, the phenotypic combinations in the F_2 generation occur in a 9:3:3:1 ratio. Keep in mind that working a Punnett square is really a way to show *probabilities* of certain combinations occurring as a result of allele shufflings during meiosis and fertilization.

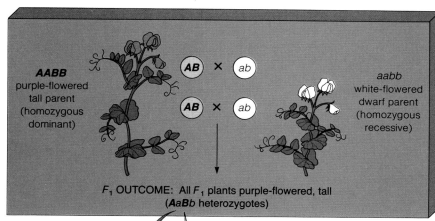

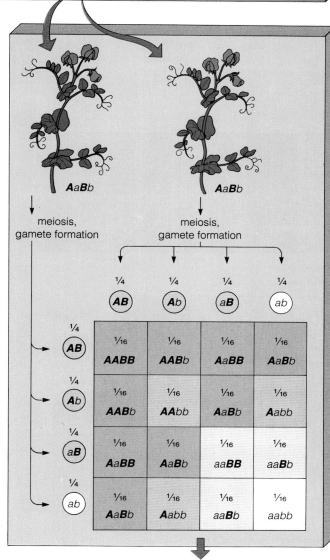

ADDING UP THE F_2 COMBINATIONS POSSIBLE:

- 9/16 or 9 purple-flowered, tall
- 3/16 or 3 purple-flowered, dwarf
- 3/16 or 3 white-flowered, tall
- 1/16 or 1 white-flowered, dwarf

types. Suppose you cross true-breeding red-flowered and white-flowered snapdragons. All of the F_1 offspring will have *pink* flowers. Suppose you then cross two of the F_1 plants. You will discover that the "red" allele was not blended away, because the F_2 offspring will have red, pink, or white flowers in a predictable ratio. (The dominant allele calls for a red pigment, and it takes two of them to color the flowers red. With their single dominant allele, heterozygotes can produce only enough pigment to color the flowers pink.)

In **codominance,** a pair of nonidentical alleles gives rise to two different phenotypes. Neither allele dominates expression of the other in heterozygotes. You probably have heard of **ABO blood typing.** It refers to a type of recognition protein at the surface of red blood cells. Two alleles for that protein are codominant. They give rise to different forms of the protein, called A and B. A third allele for the protein is recessive.

Four blood types are possible, depending on which two alleles are present in a person's cells. Using the symbols I^A and I^B for the codominant alleles and i for the recessive allele,

Allelic Combination:	Blood Type Produced:
$I^A I^A$ or $I^A i$	A
$I^B I^B$ or $I^B i$	B
$I^A I^B$	AB
ii	O

Red blood cells of homozygous recessive people are neither A nor B; that is what the "O" means.

Interactions Between Different Gene Pairs

One gene pair can influence other gene pairs, with their combined activities producing some effect on phenotype.

In **epistasis,** one gene pair masks the expression of another and some expected phenotypes do not appear at all. Such interactions can affect the color of fur or skin in mammals. The different colors result from vari-

Figure 9.11 A rare albino rattlesnake, showing the pink eyes and white coloration of animals that are unable to produce the pigment melanin. (Eyes are pink because the absence of melanin allows red light to be reflected from blood vessels in the snake's eyes.)

Surface coloration in birds and mammals is due almost entirely to the color of feathers or fur. In fishes, amphibians, and reptiles, it is due to color-bearing cells in the skin. Some of the cells contain melanin (a brownish-black pigment) or red to yellow pigments. Others contain crystals that reflect light and alter the effect of other pigments present.

The mutation affecting melanin production in the snake shown here had no effect on the production of yellow-to-red pigments and light-reflecting crystals. So the snake's skin appears iridescent yellow as well as white.

ations in the amount and distribution of a brownish-black pigment called melanin. One gene locus affects how much melanin will be produced; a *B* allele (for black) at that locus is dominant to *b* (for brown). However, another gene locus controls whether there will be any melanin at all. *CC* or *Cc* individuals can produce melanin. But a *cc* individual cannot; it will be an *albino*, a phenotype arising from the absence of melanin (Figure 9.11).

Multiple Effects of Single Genes

A single gene can exert effects on seemingly unrelated aspects of an individual's phenotype. This form of gene expression is called **pleiotropy**. The *Commentary* gives a good example of such effects.

Environmental Effects on Phenotype

The external environment has profound effects on gene expression. For example, exposure to sunlight can affect genes that control the amount of melanin deposited in a given region of skin. Tanning is one outcome; freckles are another.

The internal environment (that is, conditions within the body) also influences gene expression. At puberty, for example, the male body steps up its production of testosterone (a sex hormone). Among other things, increased levels of testosterone affect genes that govern the development of cartilage in the larynx. The voice deepens as a result of those changes. Without the hormonal increase, the voice will remain higher pitched later in life.

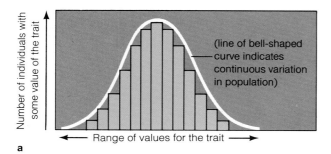

a

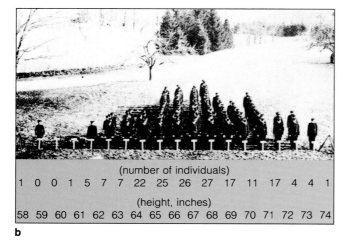

(number of individuals)																
1	0	0	1	5	7	7	22	25	26	27	17	11	17	4	4	1
(height, inches)																
58	59	60	61	62	63	64	65	66	67	68	69	70	71	72	73	74

b

Figure 9.12 (a) Generalized bell-shaped curve typical of populations showing continuous variation in some trait. (b) Example of continuous variation in a population sample: height distribution in a group of 175 U.S. Army recruits about the turn of the century.

Sickle-Cell Anemia

A certain allele gives rise to a defective form of hemoglobin, the oxygen-transporting pigment in red blood cells. When the defective pigment molecules give up the oxygen, they stack together like rigid poles. The stacking often deforms red blood cells so that they clump together in tiny blood vessels. The impaired oxygen flow causes severe damage to many tissues and organs. The symptoms of this genetic disorder are called *sickle-cell anemia*. Heterozygotes still have one functional allele and show few symptoms. But homozygous recessives show severe phenotypic consequences.

Figure a shows the pleiotropic effects possible in sickle-cell homozygotes. The color-enhanced scanning electron micrographs show (b) normal red blood cells and (c) cells characteristic of sickle-cell anemia. Because of their abnormal, asymmetrical shape, sickle cells do not flow smoothly through blood capillaries. They pile up in clumps that impede blood flow. Tissues served by the capillaries become starved for oxygen and nutrients even as they become saturated with waste products.

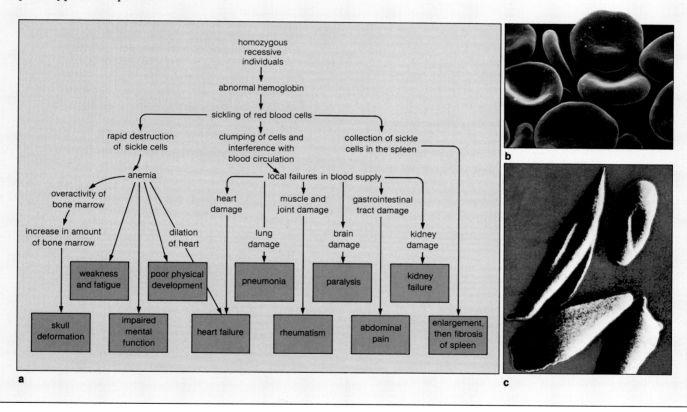

Continuous Variation in Traits

Mendel's pea plants were tall *or* dwarfed, green- *or* yellow-seeded, and so on. For most traits, however, humans and other organisms show **continuous variation.** This means that small degrees of phenotypic variation occur over a more or less continuous range, as from tall to short (Figure 9.12).

For example, eye color as well as skin color depends on the additive effects of genes controlling the melanin production and distribution. Black eyes have abundant melanin deposits in the iris; dark brown eyes have less, and light brown or hazel eyes have still less. Green, gray, and blue eyes don't have green, gray, or blue pigments. They have so little melanin that we readily see the blue wavelengths being reflected from the iris.

SUMMARY

1. Mendel's hybridization studies with garden pea plants demonstrated that diploid organisms have two units of hereditary material (genes) for each trait, and that genes retain their identity when passed on to offspring.

2. Mendel conducted monohybrid crosses (between two true-breeding plants having contrasting forms of a single trait). The crosses indicated that there can be alternative forms (alleles) of a gene, some of which are dominant over other, recessive forms.

3. Homozygous dominant individuals have two dominant alleles (AA) for the trait being studied. Homozygous recessives have two recessive alleles (aa). Heterozygotes have two different alleles (Aa).

4. In Mendel's monohybrid crosses ($AA \times aa$), all F_1 offspring were Aa. Crosses between F_1 plants produced these combinations of alleles in F_2 offspring:

$$A \times A = AA \text{ (dominant)}$$
$$A \times a = Aa \text{ (dominant)}$$
$$a \times A = Aa \text{ (dominant)}$$
$$a \times a = aa \text{ (recessive)}$$

This produced an expected phenotypic ratio of 3:1.

5. Results from such monohybrid crosses supported the Mendelian principle of segregation: Diploid organisms have a pair of genes for each trait, on a pair of homologous chromosomes. The two genes segregate from each other during meiosis, such that each gamete formed will end up with one or the other, but not both.

6. Mendel also performed dihybrid crosses (between two true-breeding plants having contrasting forms of two different traits). Results from many experiments were close to a 9:3:3:1 phenotypic ratio:

> 9 dominant for both traits
> 3 dominant for A, recessive for b
> 3 dominant for B, recessive for a
> 1 recessive for both traits

7. On the basis of such dihybrid crosses, the Mendelian principle of independent assortment was formulated: Each gene pair tends to assort into gametes independently of other gene pairs located on *non*homologous chromosomes.

8. Since Mendel's time, we have learned that (1) degrees of dominance may exist between some gene pairs, (2) gene pairs can interact to produce some positive or negative effect on phenotype, (3) a single gene can have effects on many seemingly unrelated traits, and (4) the internal and external environments influence gene expression.

Review Questions

1. State the Mendelian principle of segregation. Does segregation occur during mitosis or meiosis? *121*

2. Distinguish between the following terms: (a) gene and allele, (b) dominant trait and recessive trait, (c) homozygote and heterozygote, (d) genotype and phenotype. *118, 119*

3. Give an example of a self-fertilizing organism. What is cross-fertilization? *118*

4. Distinguish between monohybrid and dihybrid crosses. What is a testcross, and why is it valuable in genetic analysis? *119, 121, 122*

5. State the Mendelian principle of independent assortment. Does independent assortment occur during mitosis or meiosis? *122*

Self-Quiz *(Answers in Appendix IV)*

1. Alleles are _____.
   a. alternative molecular forms of a gene
   b. alternative molecular forms of a chromosome
   c. self-fertilizing, true-breeding homozygotes
   d. self-fertilizing, true-breeding heterozygotes

2. A heterozygote is _____.
   a. one of at least two forms of a gene
   b. a condition in which both alleles are the same
   c. a condition in which both alleles are different
   d. a haploid condition in genetic terms

3. The observable traits of an organism are called its _____.
   a. phenotype   c. genotype
   b. sociobiology   d. pedigree

4. In the monohybrid cross $AA \times aa$, the F_1 offspring are _____
   a. all AA d. $\frac{1}{2} AA$ and $\frac{1}{2} aa$
   b. all aa e. none of the above
   c. all Aa

5. The second generation of offspring from a genetic cross is called the _____.
   a. F_1 generation
   b. F_2 generation
   c. hybrid generation
   d. none of the above

6. In the genetic cross $Aa \times Aa$ involving complete dominance, the next generation will show a phenotypic ratio of _____.
   a. 1:2:1
   b. 1:1:1:1
   c. 3:1
   d. 9:3:3:1
   e. none of the above

7. Which of the following statements most accurately explains Mendel's principle of segregation?
   a. particular units of heredity are transmitted to offspring
   b. two genes on a pair of homologous chromosomes segregate from each other in meiosis
   c. members of a population become segregated
   d. pairs of factors are inherited completely independent of each other

8. Dihybrid crosses between two true-breeding organisms with contrasting forms of two traits (as in *Aa Bb × Aa Bb*) produce offspring phenotypic ratios close to _____.
   a. 1:2:1
   b. 1:1:1:1
   c. 3:1
   d. 9:3:3:1

9. "Each pair of genes tends to assort into gametes independently of other gene pairs located on nonhomologous chromosomes" is a statement of Mendel's _____.
   a. principle of dominance
   b. principle of segregation
   c. principle of independent assortment
   d. none of the above

10. Match each genetic term appropriately.
   __C__ dihybrid cross
   __d__ monohybrid cross, two heterozygotes
   __e__ homozygous condition
   __b__ heterozygous condition
   __a__ true-breeding parents

   a. *AA × aa*
   b. *Aa*
   c. *Aa Bb × AaBb*
   d. *Aa × Aa*
   e. *aa × AaBb*

Genetics Problems   *(Answers in Appendix III)*

1. One gene has alleles *A* and *a*; another gene has alleles *B* and *b*. For each of the following genotypes, what type(s) of gametes will be produced? (Independent assortment is expected.)
   a. *AA BB*   c. *Aa bb*
   b. *Aa BB*   d. *Aa Bb*

2. Still referring to the preceding problem, what genotypes will be present in the offspring from the following matings? (Indicate the frequencies of each genotype among the offspring.)
   a. *AA BB × aa BB*   c. *Aa Bb × aa bb*
   b. *Aa BB × AA Bb*   d. *Aa Bb*

3. In one experiment, Mendel crossed a true-breeding pea plant having green pods with a true-breeding pea plant having yellow pods. All of the F_1 plants had green pods. Which trait (green or yellow pods) is recessive? Can you explain how you arrived at your conclusion?

4. Being able to curl up the sides of your tongue into a U-shape is under the control of a dominant allele at one gene locus. (When there is a recessive allele at this locus, the tongue cannot be rolled.) Having free earlobes is a trait controlled by a dominant allele at a different gene locus. (When there is a recessive allele at this locus, earlobes are attached at the jawline.) The two genes controlling tongue-rolling and free earlobes assort independently. Suppose a woman who has free earlobes and who can roll her tongue marries someone who has attached earlobes and who cannot roll his tongue. Their first child has attached earlobes and cannot roll the tongue.
   a. What are the genotypes of the mother, the father, and the child?
   b. If this same couple has a second child, what is the probability that it will have free earlobes and be unable to roll the tongue?

5. In addition to the two genes mentioned in Problem 1, assume you now study a third gene having alleles *C* and *c*. For each of the following genotypes, indicate what type (or types) of gametes will be produced:
   a. *AA BB CC*   c. *Aa BB Cc*
   b. *Aa BB cc*   d. *Aa Bb Cc*

6. Recall that Mendel crossed a true-breeding tall, purple-flowered pea plant with a true-breeding dwarf, white-flowered plant. All the F_1 plants were tall and purple-flowered. If an F_1 plant is now self-pollinated, what is the probability of obtaining an F_2 plant heterozygous for the genes controlling height and flower color?

7. Assume that a new gene was recently identified in mice. One allele at this gene locus produces a yellow fur color. A second allele produces a brown fur color. Suppose you are asked to determine the dominance relationship between these two alleles. (Is it one of simple dominance, incomplete dominance, or codominance?) What types of crosses would you make to find the answer? On what types of observations would you base your conclusions?

8. The ABO blood system has often been employed to settle cases of disputed paternity. Suppose, as an expert in genetics, you are called to testify in a case where the mother has type A blood, the child has type O blood, and the alleged father has type B blood. How would you respond to the following statements:
   a. *The attorney of the alleged father:* "Since the mother has type A blood, the type O blood of the child must have come from the father, and since my client has type B blood, he obviously could not have fathered this child."
   b. *The mother's attorney:* "Further tests revealed that this man is heterozygous and therefore he must be the father."

9. In mice, at one gene locus, the dominant allele (*B*) produces a dark brown pigment; and the recessive allele (*b*) produces a tan pigment. An independently assorting gene locus has a dominant allele (*C*) that permits the production of all pigments. Its recessive allele (*c*) makes it impossible to produce any pigment at all. The pigmentless condition is called "albino."
   a. A homozygous *bb cc* albino mouse mates with a homozygous *BB CC* brown mouse. In what ratios would the phenotypes and genotypes be expected in the F_1 and F_2 generations?
   b. If an F_1 mouse from part (a) above were backcrossed to its albino parent, what phenotypic and genotypic ratios would be expected?

Selected Key Terms

allele *118*	homozygous recessiveness *118*
codominance *123*	incomplete dominance *122*
continuous variation *125*	independent assortment *122*
dihybrid cross *122*	locus *118*
gene *118*	monohybrid cross *119*
gene pair *118*	phenotype *118*
genotype *118*	Punnett-square method *120*
heterozygous condition *118*	segregation *121*
homozygous condition *118*	testcross *121*
homozygous dominance *118*	true-breeding organism *118*

Readings

Dunn, L. 1965. *A Short History of Genetics*. New York: McGraw-Hill.

Mendel, G. 1959. "Experiments in Plant Hybridization." Translation in J. Peters (editor), *Classic Papers in Genetics*. Englewood Cliffs, New Jersey: Prentice-Hall.

Suzuki, D., et al. 1989. *An Introduction to Genetic Analysis*. Fourth edition. New York: Freeman.

10 CHROMOSOME VARIATIONS AND HUMAN GENETICS

KEY CONCEPTS

1. Genes are arranged linearly along chromosomes. Although genes on the same chromosome tend to stay together during meiosis, crossing over disrupts such linkages. The farther apart two genes are along a chromosome, the greater will be the frequency of crossing over and recombination between them.

2. Crossing over during meiosis leads to genetic variation—hence to variation in traits among offspring. So does independent assortment of homologous chromosomes during meiosis, the outcome of which is the random mix of maternal and paternal chromosomes in gametes.

3. Abnormal changes in the structure or number of chromosomes cause many genetic disorders. Together with crossing over and independent assortment, these events play roles in evolution. They lead to different combinations of traits in offspring, and the differences can be acted upon by agents of selection.

RETURN OF THE PEA PLANT

The year was 1884. A paper Mendel had written on his hybrid crosses of pea plants had been gathering dust in a hundred libraries for nearly two decades, and Mendel himself had just passed away. Ironically, the experiments described in that forgotten paper were about to be devised all over again.

Improvements in microscopy had rekindled efforts to locate the cell's hereditary material, and researchers were zeroing in on the nucleus. By 1882, Walther Flemming had observed threadlike bodies—chromosomes—in the nuclei of dividing cells. By 1884, a question was taking shape: Could those threadlike chromosomes be the hereditary material?

Then researchers realized each gamete has half the number of chromosomes of a fertilized egg. In 1887, August Weismann proposed that a special division process must reduce the chromosome number by half before gametes form. Sure enough, in that same year meiosis was discovered. Weismann now began to pro-

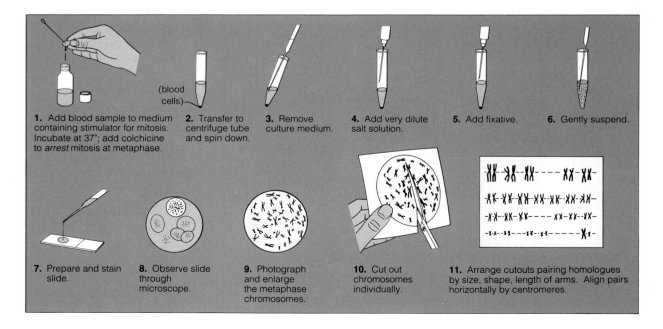

1. Add blood sample to medium containing stimulator for mitosis. Incubate at 37°; add colchicine to *arrest* mitosis at metaphase.

2. Transfer to centrifuge tube and spin down.

3. Remove culture medium.

4. Add very dilute salt solution.

5. Add fixative.

6. Gently suspend.

7. Prepare and stain slide.

8. Observe slide through microscope.

9. Photograph and enlarge the metaphase chromosomes.

10. Cut out chromosomes individually.

11. Arrange cutouts pairing homologues by size, shape, length of arms. Align pairs horizontally by centromeres.

Figure 10.1 Simplified picture of karyotype preparation. Karyotype of a human male. Human cells have a diploid chromosome number of 46. The nucleus contains 22 pairs of autosomes and 1 pair of sex chromosomes (X and Y). Each chromosome of a given type has already undergone replication.

mote his theory of heredity: The chromosome number is halved during meiosis, then restored when sperm and egg combine at fertilization; thus half the hereditary material in offspring is paternal in origin, and half is maternal. His views were hotly debated, and the debates drove researchers into testing the theory. Throughout Europe there was a flurry of experimental crosses—just like the ones Mendel had carried out.

Finally, in 1900, researchers came across Mendel's paper while checking for literature related to their own hybridization studies. To their chagrin, their results merely confirmed what Mendel already had said: Diploid cells have two units of instruction (genes) for each heritable trait, and the units segregate prior to gamete formation.

Autosomes and Sex Chromosomes

In the early 1900s, microscopists discovered a chromosomal difference between the sexes. Most of the chromosomes *are* the same number and the same type in both sexes; these were named **autosomes.** However, one or two chromosomes are *not* the same in males and females; these were named **sex chromosomes.** In humans and most other species, females have two sex chromosomes designated "X." Males have one X chromosome and another, physically different chromosome designated "Y."

Today, autosomes and sex chromosomes can be precisely characterized at metaphase, when they are in their most condensed form. At metaphase, each type has a certain length, banding pattern, and so on. (The bands appear because some regions take up more stain than others.) These features are used to create a **karyotype:** a visual representation in which the chromosomes of a cell are arranged in order, from largest to smallest (Figure 10.1). Figure 8.4 shows a professionally prepared karyotype.

Through microscopic studies, we know that each gamete produced by a female (XX) normally carries an X chromosome. Half the gametes produced by a male (XY) carry an X and half carry a Y chromosome. When a sperm and an egg both carry an X chromosome and combine at fertilization, the new individual will be female. When the sperm carries a Y chromosome, the new individual will be male (Figure 10.2).

The Y chromosome of humans carries very few genes, but among them is a "sex-determining gene." The presence or absence of its product (testis determining factor, or TDF) dictates whether a new individual will develop testes *or* ovaries. The hormones secreted from those organs trigger the development of male *or* female characteristics.

The human X chromosome carries at least 139 and possibly more than 300 genes. Like all the other chromosomes, it carries some genes associated with sexual traits (such as distribution of hair and body fat). But the X chromosome also carries many genes concerned with *nonsexual* traits, such as eye color.

Any gene on the X or Y chromosome may be called a "sex-linked gene." However, researchers now use the more precise designations, **X-linked** or **Y-linked genes.**

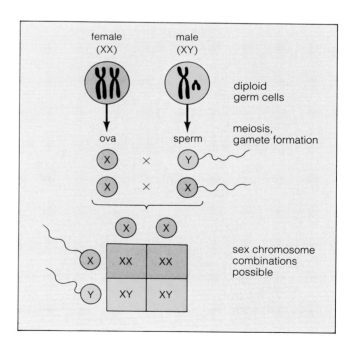

Figure 10.2 Sex determination in humans. This same pattern occurs in many animal species. Only the sex chromosomes, not the autosomes, are shown. Males transmit their Y chromosome to their sons, but not to their daughters. Males receive their X chromosome only from their mother.

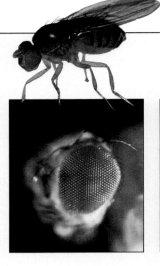

(a) Wild-type eye color for *Drosophila* and (b) white-eye phenotype.

(c) Right: Correlation between sex and eye color in *Drosophila*. Given the genetic makeup of the F_2 generation, the recessive allele (depicted here by the white dot) must be carried on the X chromosome only.

Figure 10.3 X-linked genes: clues to patterns of inheritance.

In the early 1900s, the embryologist Thomas Hunt Morgan began work to explain the apparent connection between gender and certain nonsexual traits. For example, blood clotting in humans occurs in both males and females. Yet for centuries, it was known that hemophilia (a blood-clotting disorder) shows up most often in the males, not females, of a family lineage. This phenotypic outcome was not like anything Mendel saw in his hybrid crosses between pea plants. In those crosses, either one parent plant or the other could carry a recessive allele. It made no difference *which* parent carried it; the phenotypic outcome was the same.

Morgan studied eye color and other nonsexual traits in the fruit fly, *Drosophila melanogaster*. These small flies can be grown in bottles on bits of cornmeal, molasses, and agar. A female lays hundreds of eggs in a few days, and her offspring can reproduce in less than two weeks. Morgan could track hereditary traits through nearly thirty generations of thousands of flies in a year's time.

At first, all the flies were wild-type for eye color; they had brick-red eyes, as in (a). ("Wild-type" simply means the normal or most common form of a trait in a population.) Then, through an apparent mutation in a gene controlling eye color, a *white-eyed* male appeared (b).

Morgan established true-breeding strains of white-eyed males and females. Then he did a series of *reciprocal crosses*. (These are pairs of crosses. In the first, one parent displays the trait in question; in the second, the other parent displays the trait.)

White-eyed males were mated with true-breeding (homozygous) red-eyed females. All the F_1 offspring of the cross had red eyes—but of the F_2 offspring, only some of the *males* had white eyes. Then white-eyed females were mated with true-breeding red-eyed males. Of the F_1 offspring of that second cross, half were red-eyed females and half were white-eyed males. Of the F_2 offspring, $\frac{1}{4}$ were red-eyed females, $\frac{1}{4}$ were white-eyed females, $\frac{1}{4}$ were red-eyed males, and $\frac{1}{4}$ were white-eyed males!

The seemingly odd results indicated the gene for eye color was related to gender. Probably it was located on one of the sex chromosomes. But which one? Since females (XX) could be white-eyed, the recessive allele would have to be on one of their X chromosomes. Suppose white-eyed males (XY) also carry the recessive allele on their X chromosome—*and suppose there is no corresponding eye-color allele on their Y chromosome*. Those males would have white eyes because the recessive allele would be the only eye color gene they had!

In (c) are the results we can expect when the idea of an X-linked gene is combined with Mendel's concept of segregation. By proposing that a specific gene occurs on the X but not the Y chromosome, Morgan was able to explain the outcome of his reciprocal crosses. The results of the experiments matched the predicted outcomes.

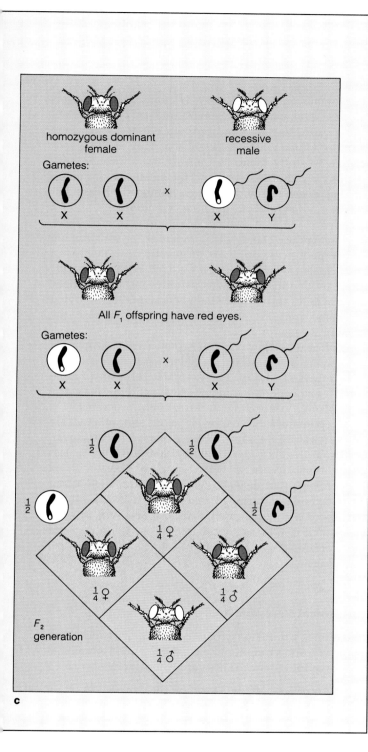

homozygous dominant female

recessive male

Gametes:

X X   X Y

All F₁ offspring have red eyes.

Gametes:

X X   X Y

½   ½

½   ½

¼ ♀

¼ ♀   ¼ ♂

F₂ generation

¼ ♂

c

In the 1900s, Thomas Hunt Morgan and his coworkers performed a series of hybridization experiments with the fruit fly, *Drosophila melanogaster*. As described in Figure 10.3, their work led to the discovery of X-linked genes, and it reinforced a major concept: *Each gene is located on a specific chromosome.*

Linkage and Crossing Over

Through their studies of the fruit fly, researchers came to realize that many traits were being inherited *as a group* from one parent or the other. They identified four groups that apparently corresponded to the haploid number of chromosomes (four) in the fruit fly's gametes. The genes on any one of those chromosomes were probably being inherited as a block and were not segregating independently of one another (as genes on nonhomologous chromosomes do).

The term **linkage** is now used to describe the tendency of genes located on the same chromosome to end up together in the same gamete. Linkage is not inevitable, however. It can be disrupted by **crossing over**: the breakage and exchange of segments between homologous chromosomes (page 107).

Think about the location of any two genes on the same chromosome. The probability of a crossover occurring between those two genes is proportional to the distance separating them along the chromosome. Suppose two genes *A* and *B* are twice as far apart as two other genes, *C* and *D*:

We would expect crossing over to disrupt the linkages between the first two much more frequently.

Two genes located physically close together on a chromosome nearly always end up in the same gamete; they are very closely linked (Figure 10.4a). Two genes relatively far apart are more vulnerable to crossing over and recombination, compared to closely linked genes (Figure 10.4b). Two genes very far apart on the same chromosome are affected by crossing over so often that they may appear to assort independently.

Crossing over plays a role in evolution. It introduces variation in genotypes (the genetic makeup of individuals in a population) and so gives rise to differences in phenotype (physical and behavioral traits). Selective agents in the environment sift through the differences, so to speak, and tend to favor genotypes that make individuals well adapted to a given environment.

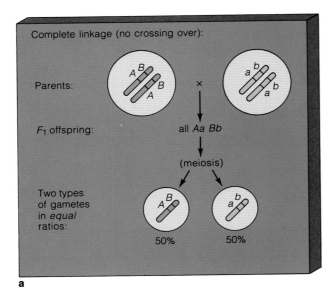

Complete linkage (no crossing over):

Parents:

$A\ B$ $\times$ $a\ b$
$B\ A$ $b\ a$

F_1 offspring: all *Aa Bb*

(meiosis)

Two types
of gametes
in *equal*
ratios:

$A\ B$ $a\ b$

50%   50%

a

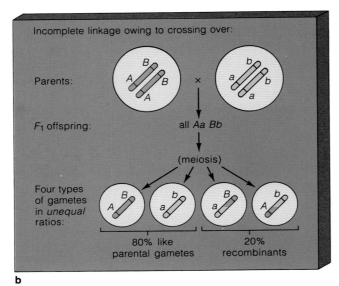

Incomplete linkage owing to crossing over:

Parents:

B B $\times$ b b
A A a a

F_1 offspring: all *Aa Bb*

(meiosis)

Four types
of gametes
in *unequal*
ratios:

$A\ B$ $a\ b$ $a\ B$ $A\ b$

80% like   20%
parental gametes   recombinants

b

Figure 10.4 How crossing over can affect gene linkage, using two gene loci as the example.

CHROMOSOME VARIATIONS IN HUMANS

Pea plants and fruit flies lend themselves to genetic analysis. They grow and reproduce rapidly in small spaces, under controlled conditions. Because a geneticist lives much longer than they do, their traits can be tracked through many generations in relatively little time.

Humans are another story. Humans live under variable conditions in diverse environments. Typically they find a mate by chance and reproduce if and when they want to. Human subjects live just as long as the geneticists who study them, so tracking traits through generations is rather tedious. And the small size of human families doesn't provide enough numbers for easy statistical inferences about inheritance patterns.

Even so, human genetics is a rapidly growing field. Researchers use standardized methods for constructing **family pedigrees**, or charts of genetic relationships of individuals in families (Figure 10.5). With pedigrees, they can identify inheritance patterns and track genetic abnormalities through several generations. By studying the same trait in many families, they increase the numerical base for analysis.

The next section of this chapter gives examples of the genetic disorders we deal with as individuals and as members of society. The examples serve as a framework for considering some practical and ethical aspects of screening, counseling, and treatment programs.

Keep in mind that "disorder" and "abnormality" are not necessarily the same thing. *Abnormal* means deviation from the average. An abnormality is a rare or less common occurrence, as when a person is born with six toes on each foot instead of five. Whether such a trait is viewed as disfiguring or merely interesting is subjective; there is nothing inherently life-threatening or even ugly about it. Other abnormalities cause mild to severe medical problems, and *disorder* is the more appropriate word here.

Autosomal Recessive Inheritance

Sometimes a mutation produces a recessive allele on an autosome and gives rise to a condition called **autosomal recessive inheritance.** These are its characteristics:

1. Males or females can carry the recessive allele on an autosome (not on a sex chromosome).

2. Heterozygotes are symptom-free. Recessive homozygotes are affected.

3. When both parents are heterozygous, there is a fifty percent chance each child born to them will be heterozygous also, and a twenty-five percent chance it will be homozygous recessive (Figure 10.6). When both parents

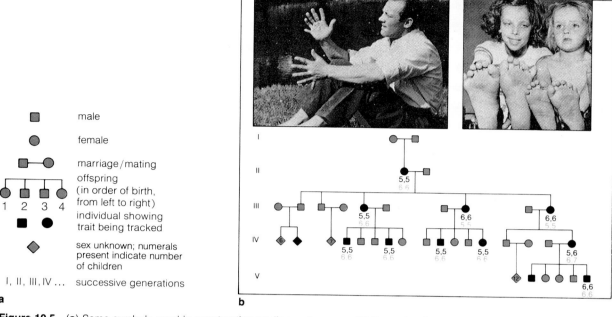

Figure 10.5 (**a**) Some symbols used in constructing pedigree diagrams. (**b**) Example of a pedigree for *polydactyly*, a condition in which an individual has extra fingers, extra toes, or both. The number of fingers on each hand is shown in black numerals, and the number of toes on each foot is shown in blue. The phenotype of female 1 is uncertain.

Polydactyly is an example of how gene expression can vary. As a human embryo develops, a dominant allele *D* controls how many sets of bones will form within the body regions destined to become hands and feet. The *Dd* genotype varies in how it is expressed. The pedigree shown here indicates the kind of variation possible.

Symbol legend:
- ■ male
- ● female
- ■—● marriage/mating
- offspring (in order of birth, from left to right)
- 1 2 3 4
- ■ ● individual showing trait being tracked
- ◆ sex unknown; numerals present indicate number of children
- I, II, III, IV ... successive generations

a

---

are homozygous recessive, all of their children will be affected.

Galactosemia, an autosomal recessive condition, arises when a breakdown product of lactose (milk sugar) cannot be metabolized. It occurs in about 1 in 100,000 newborns. Early symptoms include malnutrition, diarrhea, and vomiting. Without treatment, galactosemics usually die in childhood.

Normally, lactose is converted first to galactose and ultimately to glucose-1-phosphate, which can be degraded by glycolysis. But galactosemics cannot produce molecules of an enzyme required for one of the conversion steps (they have two defective, recessive alleles coding for that enzyme):

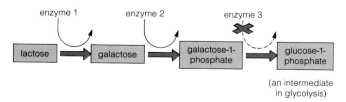

Galactose builds up in the blood, and in large concentrations it can damage the brain, liver, and eyes. Un-

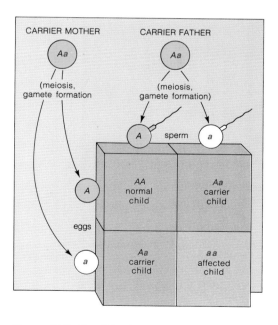

Figure 10.6 Possible phenotypic outcomes for autosomal recessive inheritance when both parents are heterozygous carriers of the recessive allele (in red here).

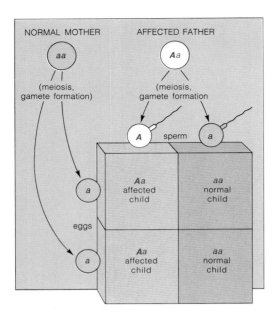

Figure 10.7 Possible phenotypic outcomes for autosomal dominant inheritance, assuming the dominant allele is fully expressed in the carriers. (The dominant allele is shaded red.)

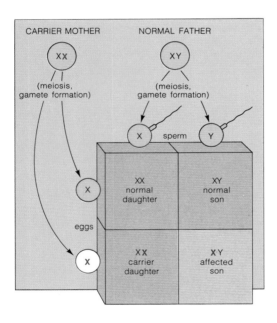

Figure 10.8 Possible phenotypic outcomes for X-linked inheritance when the mother carries a recessive allele on one of her X chromosomes (shaded red here).

usually high concentrations can be detected in urine samples of homozygous recessive infants. If the disorder is detected early enough, infants can be put on a diet that includes milk substitutes and can grow up symptom-free.

Autosomal Dominant Inheritance

Recessive alleles that cause genetic disorders can persist at fairly high frequencies, because heterozygotes usually survive and reproduce. (Their one normal allele may yield enough of the required gene product). But what if a *dominant* allele causes the disorder? In **autosomal dominant inheritance,** a dominant allele on an autosome is always expressed to some extent. If its expression reduces the chance of surviving and reproducing, its frequency among individuals in the population will decrease.

Even so, a few dominant alleles that cause pronounced disorders do remain in populations. Mutations can replenish the supply of defective alleles, so to speak. Also, *some dominant alleles do not affect reproduction or they are only expressed after reproductive age.* Figure 10.7 shows an inheritance pattern for an autosomal dominant condition.

One autosomal dominant allele causes *achondroplasia,* a type of dwarfism, in about 1 in 10,000 individuals. When long bones develop in affected children, cartilage forms in ways that lead to disproportionately short arms and legs. Affected persons are less than four feet, four inches tall. The dominant allele has no other phenotypic effects in heterozygotes, who normally are fertile. Homozygous fetuses usually are stillborn.

Huntington's disorder, a rare form of autosomal dominant inheritance, causes progressive degeneration of the nervous system. In about half the cases, symptoms emerge from age forty onward—after most people have already had children. In time, movements become convulsive, brain function deteriorates rapidly, and death follows.

X-Linked Recessive Inheritance

Some genetic disorders fall in the category of **X-linked recessive inheritance,** which has these characteristics:

1. The mutated gene occurs on the X (not the Y) chromosome.

2. Heterozygous females are phenotypically normal; the nonmutated allele on their other X chromosome covers the required function. Males typically are affected; they have only one allele for the trait, and it is recessive.

3. When the male is normal but the female is heterozygous, there is a fifty percent chance each daughter

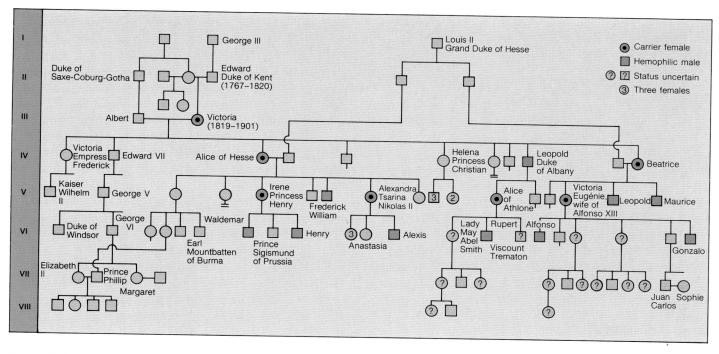

Figure 10.9 Descendants of Queen Victoria, showing carriers and affected males that possessed the X-linked gene conferring the disorder hemophilia A. Many individuals of later generations are not shown in this family pedigree.

born to them will be a carrier, and a fifty percent chance each son will be affected (Figure 10.8). When the female is homozygous recessive and the male is normal, all daughters will be carriers and all sons will be affected.

Hemophilia A is an example of X-linked recessive inheritance. Normally, a blood clotting mechanism quickly stops bleeding from an injury. Some people bleed for an unusually long time because the mechanism is defective. The reactions leading to clot formation depend on the products of several genes. If any of the genes is mutated, its defective product can cause one of several bleeding disorders.

A mutated form of the gene for "clotting factor VIII" gives rise to hemophilia A. Males with a recessive allele on their X chromosome are always affected. They run the risk of dying from untreated bruises, cuts, or internal bleeding. Blood-clotting time is more or less normal in heterozygous females. The nonmutated gene on their other X chromosome produces enough factor VIII to cover the required function.

Hemophilia A affects only about 1 in 7,000 human males. The frequency of the recessive allele was unusually high among the royal families of nineteenth-century Europe, whose members often intermarried. Queen Victoria of England was a carrier, as were two of her daughters (Figure 10.9). At one time, eighteen of her sixty-nine descendants were affected males or female carriers.

Changes in Chromosome Structure

On rare occasions, chromosome structure becomes abnormally rearranged. Deletions, duplications, inversions, and translocation are examples of such rearrangements.

A **deletion** is a loss of a chromosome segment. It happens when a segment of a chromosome is lost or when viral attack, irradiation, or chemical action cause breaks in a chromosome. The loss almost always means problems, for genes influencing one or more traits may be missing. For example, one deletion from human chromosome 5 leads to mental retardation and a malformed larynx. When affected infants cry, the sounds produced are more like meowing—hence the name of the disorder, *cri-du-chat* (meaning cat-cry).

A **duplication** is a gene sequence in excess of its normal amount in a chromosome. This happens, for example, when a deletion from one chromosome is inserted into its homologue. Duplications probably have been important in evolution. Cells require specific gene products, so mutations of most genes would be selected against. But *duplicates* of a gene could evolve through mutation (the normal gene would still provide the required product), and eventually they could yield products with related or even new functions. This apparently happened in chromosome regions that code for polypeptide chains of the hemoglobin molecule (Figure 2.30).

In humans and other primates, those regions have strikingly similar gene sequences and they produce whole families of proteins with slight differences in structure and sometimes function.

An **inversion** is a chromosome segment that separated from the chromosome and then was inserted at the same place—but in reverse. The reversal alters the position and sequence of its genes.

A **translocation** is the transfer of part of one chromosome to a *non*homologous chromosome. In some types of human cancer, a segment of chromosome 8 has been transferred to chromosome 14. Genes on that segment had been precisely regulated at their normal chromosomal location, but controls apparently are lost at the new location (page 160).

Chromosomes underwent structural changes during the evolution of humans and their closest primate relatives. Eighteen of the twenty-three pairs of human chromosomes are identical to their counterparts in chimpanzees and gorillas. However, karyotype analysis of banding patterns shows that inversions and translocations occurred in the others.

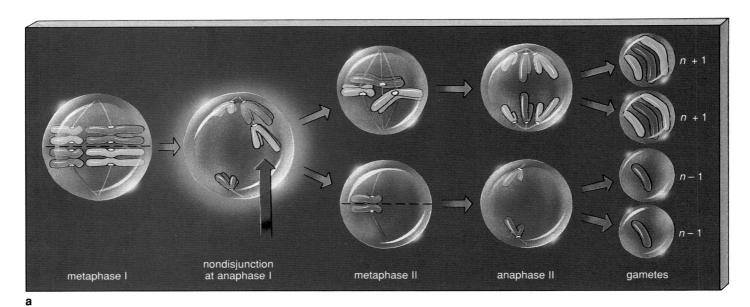

a

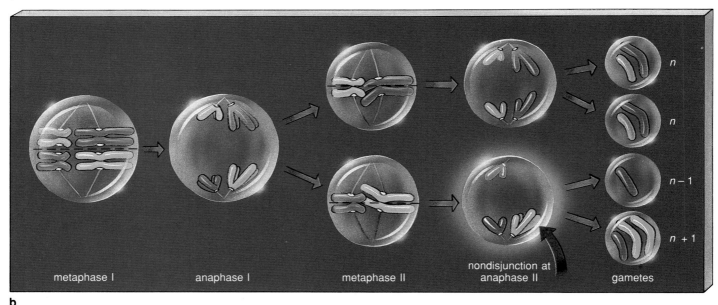

b

Figure 10.10 Two examples of nondisjunction, an event that can change the chromosome number in gametes (hence in offspring).

Changes in Chromosome Number

Sometimes new individuals end up with the wrong chromosome number. This happens as a result of non-disjunction or some other abnormal cellular event. In **nondisjunction,** one or more chromosomes fail to separate during meiosis. Perhaps a chromosome does not separate from its homologue at anaphase I. Or perhaps sister chromatids of a chromosome do not separate at anaphase II. As Figure 10.10 shows, some or all of the gametes can end up with one extra chromosome or less than the parental number.

Suppose a human gamete has an *extra* chromosome ($n + 1$). If it combines with a normal gamete at fertilization, the diploid cells of the new individual will have three of one type of chromosome ($2n + 1$). Such a condition is called **trisomy.** If the gamete is *missing* a chromosome, the new individual will have a chromosome number of $2n - 1$. One chromosome in its diploid cells will not have a homologue—a condition called **monosomy.**

Nondisjunction commonly triggers miscarriage (the spontaneous expulsion of the fetus from the uterus during pregnancy). Nondisjunction also can lead to Down syndrome and other human genetic disorders.

Down Syndrome. Sometimes the cells of an individual have three copies of chromosome 21. That condition, called trisomy 21, leads to Down syndrome. ("Syndrome" means a set of symptoms characterizing a particular disorder; typically the symptoms occur together.) Figure 10.11 shows a few children who are affected by the syndrome. Figure 10.12a shows a karyotype of a girl with Down syndrome. Figure 10.12b charts the relationship between the frequency of the disorder and the mother's age.

Figure 10.11   Children with Down syndrome.

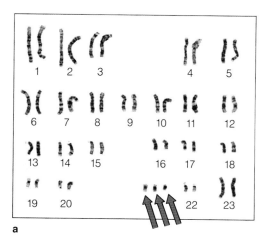

a

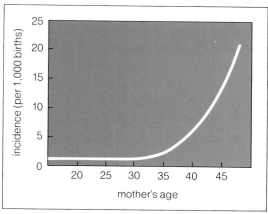

b

Figure 10.12 (**a**) Karyotype of a girl with Down syndrome; red arrows identify the trisomy of chromosome 21. (**b**) Relationship between the frequency of Down syndrome and the mother's age. Results are from a study of 1,119 children with the disorder who were born in Victoria, Australia, between 1942 and 1957.

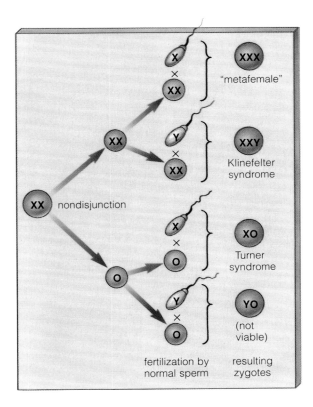

Figure 10.13 Genetic disorders that result from nondisjunction of X chromosomes followed by fertilization involving normal sperm.

Trisomic 21 embryos are often miscarried, but about 1 of every 1,000 liveborns in North America alone will develop the disorder. Most affected children show moderate to severe mental retardation, and about forty percent have heart defects. Skeletal development is slower than normal and muscles are rather slack. Older children are shorter than normal and have distinguishing facial features, including a small skin fold over the inner corner of the eyelid. With special training, affected children often can participate in normal activities, and they can enjoy life to the fullest extent allowed by their condition (Figure 10.11).

Turner Syndrome. About 1 in every 5,000 newborns is destined to have Turner syndrome. Through a nondisjunction, affected individuals have a chromosome number of 45 instead of 46 (Figure 10.13). They are missing a sex chromosome (XO), this being a type of **sex chromosome abnormality.**

Turner syndrome occurs less often than other sex chromosome abnormalities, probably because most XO embryos are miscarried early in pregnancy. Affected persons have a distorted female phenotype. Their ovaries are nonfunctional, they are sterile, and secondary sexual

Prospects and Problems in Human Genetics

Chances are, you know of someone who has a genetic disorder. Of all newborns, possibly 1 percent will have pronounced problems arising from a chromosomal aberration. Between 1 and 3 percent more will have problems because of mutant genes that produce defective proteins or none at all. Of all patients in children's hospitals, 10 to 25 percent are treated for problems arising from genetic disorders.

Human geneticists work to diagnose and treat heritable disorders. However, we apparently cannot approach the disorders in the same way we approach infectious diseases (such as influenza, measles, and polio). Infectious agents are enemies from the environment, so to speak. We have had no qualms about mounting counterattacks with immunizations and antibiotics that can eliminate or control the agents. With genetic disorders, the problem is inherent in the chromosomes of individual human beings.

How do we attack an "enemy" within? Do we institute regional, national, or global programs to identify affected persons? Do we tell them they are "defective" and run a risk of bestowing their disorder on their children? Who decides which alleles are "harmful"? Should society bear the cost of treating disorders such as Down syndrome? If so, should society also have some say in whether affected fetuses will be born at all, or aborted? These questions are only the tip of an ethical iceberg, and answers have not been worked out in universally acceptable ways.

traits fail to develop at puberty. Often they age prematurely and have shortened life expectancies.

Klinefelter Syndrome. Nondisjunction can give rise to XXY males who show Klinefelter syndrome (Figure 10.13). Males affected by this sex chromosome abnormality are sterile and often mentally retarded. Their testes are much smaller than normal, body hair is sparse, and there may be some breast enlargement. Injections of the hormone testosterone can reverse the feminized phenotype but not the sterility or mental retardation.

Phenotypic Treatments

Genetic disorders cannot be permanently cured, but sometimes we can get around their phenotypic consequences. Treatments include diet modifications, environmental adjustments, surgery, and chemical modification of gene products.

Controlling the diet can suppress or minimize the outward symptoms of several disorders. Galactosemia is controlled this way (page 133). So is *phenylketonuria*, or PKU. Normally, a gene product (an enzyme) converts one amino acid to another (phenylalanine to tyrosine). The first amino acid builds up in people who are homozygous recessive for a mutated form of the gene. If the excess is diverted into other metabolic pathways, phenylpyruvic acid and other compounds may be produced. At high levels, phenylpyruvic acid can lead to mental retardation. A diet that provides the minimum required amount of phenylalanine will alleviate the symptoms of PKU. When the body is not called upon to dispose of excess amounts, affected persons can lead normal lives.

Environmental adjustments can alleviate the outward symptoms of other genetic disorders. For example, true albinos can avoid direct sunlight. Individuals affected by sickle-cell anemia can avoid strenuous activity when oxygen levels are low, as at high altitudes.

Surgical reconstructions can correct or minimize many phenotypic defects. One type of *cleft lip* is an abnormality of the upper lip. A vertical fissure occurs at the lip midsection and often extends into the roof of the mouth. Surgery can usually correct the lip's appearance and function.

Phenotypic treatments also include chemical modification of gene products. *Wilson's disorder* arises from an abnormal accumulation of copper in the body. The body requires trace amounts of copper, which serves as a cofactor for several enzymes. Excess copper can damage the brain and liver, leading to convulsions and death. One drug binds with the copper, and the excess is eliminated by the urinary system.

Genetic Screening

In some cases, genetic disorders can be detected early enough to start preventive measures *before* symptoms can develop. In other cases, carriers who show no outward symptoms can be identified before giving birth to affected children. "Genetic screening" refers to large-scale programs to detect affected persons or carriers in a population. Most hospitals in the United States routinely screen all newborns for PKU, for example, so it is becoming less common to see people who show symptoms of the disorder.

Genetic Counseling

Sometimes prospective parents suspect they are very likely to produce a severely afflicted child. Either their

XYY Condition. About 1 in every 1,000 males has one X and two Y chromosomes, this being called an XYY condition. It is probably inappropriate to apply the term "syndrome" to this sex chromosome abnormality. XYY males tend to be taller than average and some may show mild mental retardation, but most are phenotypically normal. They are fertile and usually have genetically normal (XX or XY) children.

At one time, XYY males were thought to be genetically predisposed to become criminals. But a comprehensive study in Denmark showed that the number who do end up in prison is no more notable than the percentage of other tall men. Compared to normal (XY) males, their rate of conviction was indeed greater (41.7 percent compared to 9.3 percent). But this is not necessarily proof of a predisposition to crime. With their moderately impaired mental ability, XYY males simply may have been easier to catch. More importantly, over 99 percent of XYY individuals have no criminal record at all and lead normal lives.

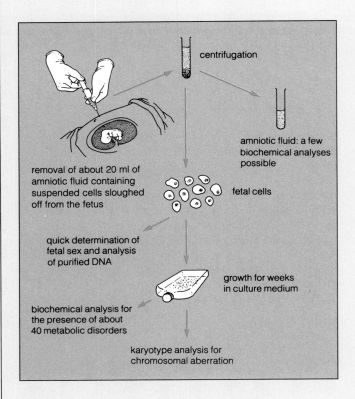

a Steps in amniocentesis, a procedure used in prenatal diagnosis of many genetic disorders.

Image labels:
- centrifugation
- amniotic fluid: a few biochemical analyses possible
- removal of about 20 ml of amniotic fluid containing suspended cells sloughed off from the fetus
- fetal cells
- quick determination of fetal sex and analysis of purified DNA
- growth for weeks in culture medium
- biochemical analysis for the presence of about 40 metabolic disorders
- karyotype analysis for chromosomal aberration

first child or a close relative shows an abnormality and they now wonder if future children will be affected the same way. In such cases, clinical psychologists, geneticists, social workers, and other consultants may be brought in to give emotional support to parents at risk.

What happens if an embryo is diagnosed as having a severe disorder? Unfortunately, there are no known cures for changes in chromosome number or structure. Prospective parents might decide on an *induced abortion* (an induced expulsion of the embryo from the uterus). Such decisions are bound by ethical considerations. We can expect the medical community to provide prospective parents with some of the information they need to make their own choice. That choice must be consistent with their own values, within the broader constraints imposed by society.

Genetic counseling begins with accurate diagnosis of parental genotypes; this may reveal the potential for a specific disorder. Biochemical tests can be used to detect many metabolic disorders. Detailed family pedigrees can be constructed to aid the diagnosis.

For disorders showing simple Mendelian inheritance patterns, it is possible to predict the chances of having an affected child—but not all disorders follow Mendelian patterns. Even ones that do can be influenced by other factors, some identifiable, others not. Even when the extent of risk has been determined with some confidence, prospective parents must know the risk is the same for *each* pregnancy. For example, if the child has one chance in four of being affected by a genetic disorder, the next child will have one chance in four also.

Prenatal Diagnosis

What happens when a woman is already pregnant? Suppose a woman forty-five years old wants to know if the child she is bearing will suffer from Down syndrome. Through prenatal diagnosis, this and about 450 other genetic disorders can be detected.

One detection procedure is based on *amniocentesis:* sampling the contents of the fluid-filled sac (amnion) that contains the fetus in the mother's uterus (Figure a). During the fourteenth to sixteenth week of pregnancy, the thin needle of a syringe is inserted through the mother's abdominal wall and into the amnion. Epidermal cells shed from the fetus float about in the amniotic fluid. The syringe withdraws some fluid—along with its sample of fetal cells. The cells are cultured and allowed to undergo mitosis. Abnormalities can be diagnosed by karyotype analysis and other tests that can be completed within weeks. Cells obtained by amniocentesis also can be tested for biochemical defects, such as the one causing sickle-cell anemia.

Amniocentesis carries a small risk: care must be taken not to puncture the fetus or cause infection.

Chorionic villi sampling (CVS), a newer procedure, uses cells drawn from the chorion (a membranous sac surrounding the amnion). This procedure can be used earlier in pregnancy (by the eighth week), and results often are available in one or two days. However, a greater risk is associated with CVS than with amniocentesis.

SUMMARY

1. Genes (units of instruction for heritable traits) are arranged one after the other along a chromosome.

2. Sexual reproduction begins with meiosis and gamete formation and ends at fertilization. By this process, offspring receive the same number and type of chromosomes as those found in the body cells of their parents.

3. Human diploid cells have two chromosomes of each type (one from the mother, one from the father). The two are homologues; they pair at meiosis. All but two are autosomes; they are the same in males and females. The two exceptions are sex chromosomes (XX in females and XY in males).

4. Genes on one chromosome segregate from ther partners on the homologous chromosome and end up in separate gametes.

5. Homologues segregate during meiosis. That is, the maternal *or* paternal chromosome of each pair align randomly at the spindle equator before they are separated from each other and shuffled at random into a gamete.

6. Genes on the *same* chromosome tend to stay together during meiosis and end up in the same gamete. But crossing over can disrupt such linkages. The farther apart two genes are on a chromosome, the greater will be the frequency of crossing over and recombination between them.

7. Chromosome *structure* can be altered by deletions, duplications, inversions, or translocations. Chromosome *number* can be altered by nondisjunction (the failure of chromosomes to separate during meiosis, so that gametes end up with one extra or one missing chromosome). Gene mutations, changes in chromosome structure, and changes in chromosome number cause many genetic disorders.

8. Variation in a population arises not only through mutation, but also through independent assortment, crossing over, and changes in the structure or number of chromosomes. These events play roles in evolution. They change the genotypes (genetic makeup of individuals) and so lead to differences in phenotype (physical and behavioral traits) upon which selective agents can act.

Self-Quiz   (Answers in Appendix IV)

1. _____ segregate during _____.
   a. Homologues; mitosis
   b. Genes on one chromosome; meiosis
   c. Homologues; meiosis
   d. Genes on one chromosome; mitosis

2. Two genes of a pair on homologous chromosomes end up in separate _____.
   a. body cells
   b. gametes
   c. nonhomologous chromosomes
   d. offspring
   e. both b and d are possible

3. Genes on the same chromosome tend to remain together during _____ and end up in the same _____.
   a. mitosis; body cell
   b. mitosis; gamete
   c. meiosis; body cell
   d. meiosis; gamete
   e. both a and d

4. The probability of a crossover occurring between two genes on the same chromosome is _____.
   a. unrelated to the distance between them
   b. increased if they are closer together on the chromosome
   c. increased if they are farther apart on the chromosome
   d. impossible

5. Chromosome structure can be altered by _____.
   a. deletions
   b. duplications
   c. inversions
   d. translocations
   e. all of the above

6. Nondisjunction is caused by _____.
   a. crossing over in meiosis
   b. segregation in meiosis
   c. failure of chromosomes to separate during meiosis
   d. multiple independent assortments

7. A gamete affected by nondisjunction would have _____.
   a. a change from the normal chromosome number
   b. one extra or one missing chromosome
   c. the potential for a genetic disorder
   d. all of the above

8. Genetic disorders can be caused by _____.
   a. gene mutations
   b. changes in chromosome structure
   c. changes in chromosome number
   d. all of the above

9. Which of the following does not contribute to variation in a population?
   a. independent assortment
   b. crossing over
   c. mitosis
   d. changes in chromosome structure and number

10. Match the chromosome terms appropriately.
    _____ crossing over
    _____ deletion
    _____ nondisjunction
    _____ translocation
    _____ gene mutation

    a. a change in DNA which may affect genotype and phenotype
    b. movement of a chromosome segment to a nonhomologous chromosome
    c. disrupts gene linkages during meiosis
    d. causes gametes to have abnormal chromosome numbers
    e. loss of a chromosome segment

Genetics Problems   *(Answers in Appendix III)*

1. Recall that human sex chromosomes are XX for females and XY for males.
   a. Does a male child inherit his X chromosome from his mother or father?
   b. With respect to an X-linked gene, how many different types of gametes can a male produce?
   c. If a female is homozygous for an X-linked gene, how many different types of gametes can she produce with respect to this gene?
   d. If a female is heterozygous for an X-linked gene, how many different types of gametes can she produce with respect to this gene?

2. One human gene, which may be Y-linked, controls the length of hair on men's ears. One allele at this gene locus produces nonhairy ears; another allele produces rather long hairs (hairy pinnae).
   a. Why would you *not* expect females to have hairy pinnae?
   b. If a man with hairy pinnae has sons, all of them will have hairy pinnae; if he has daughters, none of them will. Explain this statement.

3. Suppose that you have two linked genes with alleles *A,a* and *B,b* respectively. An individual is heterozygous for both genes, as in the following:

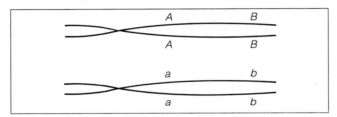

If the crossover frequency between these two genes is zero percent, what genotypes would be expected among gametes from this individual, and with what frequencies?

4. In hemophilia A, the body's blood-clotting mechanism is defective. This condition has been traced to a recessive allele of an X-linked gene. Refer now to Figure 10.9. Why are only the females shown as carriers of the recessive allele?

5. Huntington's disorder is due to a dominant autosomal allele. Usually this disorder does not manifest itself until after age thirty-five. Individuals having Huntington's disorder are almost always heterozygous. As a genetic counselor, you are visited by a twenty-year-old woman. Her mother has Huntington's disorder but her father is normal. What is the probability that this woman will develop Huntington's disorder as she grows older? Suppose, at her present age, she marries someone with no family history of the disorder. If they have a child, what is the probability that it will have Huntington's disorder?

6. Color blindness is an X-linked trait. A woman heterozygous for color blindness (*Gg*) marries someone who has normal color vision. What is the probability that their first child will be color blind? Their second child? If they have two children only, what is the probability that both will be color blind?

7. A person affected by Turner syndrome has only a single sex chromosome (X only), yet may survive. In contrast, a person having a single Y chromosome and no X chromosome cannot survive. What does this tell you about the genetic content of the X and Y chromosomes?

8. Fertilization of a normal egg by a sperm that has no sex chromosomes (male nondisjunction) can lead to Turner syndrome. Also, fertilization of an egg that has no sex chromosomes (female nondisjunction) by a sperm carrying one X chromosome can lead to the same disorder. Suppose a hemophilic male and a carrier (heterozygous) female have a child. The child is nonhemophilic and is affected by Turner syndrome. In which parent did nondisjunction occur?

9. The trisomic XXY condition is also called Klinefelter syndrome. How could this syndrome arise if nondisjunction occurred in the female parent of an affected individual? How could it arise if nondisjunction occurred in the male parent?

10. If nondisjunction occurs for the X chromosomes during oogenesis, then some eggs having two X chromosomes and others having no X chromosomes are produced at about equal frequencies. If normal sperm fertilize these two types of eggs, what genotypes are possible?

11. Phenylketonuria (PKU) is an autosomal recessive condition. About 1 of every 50 persons who inherit the gene responsible for the disorder is heterozygous and displays no symptoms of the disorder.
    a. If you select a symptom-free female at random from the population, what is the probability that she will be heterozygous?
    b. If you select a symptom-free male and a symptom-free female at random, what is the probability that both will be heterozygous? What is the probability that they could have a child with PKU?

Selected Key Terms

autosomal dominant inheritance *134*	karyotype *129*
	linkage *131*
autosomal recessive inheritance *132*	monosomy *137*
	nondisjunction *137*
autosome *129*	sex chromosome *129*
chromosomal deletion *135*	sex chromosome abnormality *138*
chromosomal duplication *135*	trisomy *137*
chromosomal inversion *136*	X-linked gene *129*
chromosomal translocation *136*	X-linked recessive inheritance *134*
crossing over *131*	
family pedigree *132*	Y-linked gene *129*

Readings

Edlin, G. 1988. *Genetic Principles: Human and Social Consequences.* Second edition. Portola Valley, California: Jones & Bartlett.

Fuhrmann, W., and F. Vogel. 1986. *Genetic Counseling.* Third edition. New York: Springer-Verlag.

Holden, C. 1987. "The Genetics of Personality." *Science* 237:598–601. For students interested in human behavioral genetics.

Patterson, D. August 1987. "The Causes of Down Syndrome." *Scientific American* 257(2):52–60.

Weiss, R. November 1989. "Genetic Testing Possible Before Conception." *Science News* 136(21):326.

DNA STRUCTURE AND FUNCTION · 11

1. Hereditary instructions are encoded in the linear sequence of nucleotides that make up DNA molecules. The four kinds of nucleotides in DNA differ in which nitrogen-containing base they contain (adenine, guanine, thymine, or cytosine).

2. In each DNA molecule, two strands of nucleotides are twisted together like a spiral staircase; they form a double helix. Hydrogen bonds occur between the bases of the two strands. As a rule, adenine pairs only with thymine, and guanine only with cytosine.

3. Before a cell divides, its DNA is replicated with the help of enzymes and other proteins. Each double-stranded DNA molecule starts unwinding, and a new, complementary strand is assembled on the exposed bases of each parent strand according to base-pairing rules.

4. There is only one DNA molecule in each chromosome. Except in bacterial chromosomes, it is tightly bound with many histones and other proteins that have roles in the structural organization of DNA.

For much of human history, we have been dealing with a world that is often harsh. In a bad year, an influenza virus might strike hundreds of thousands of us. In any year, wheat rusts and other pathogenic (disease-causing) microbes might destroy millions of acres of crops and contribute to starvation on a global scale. Every year, some of our kind are born with crippling genetic disorders.

And yet, through research that began only a few decades ago, we soon may have more control over our individual and combined destinies. A newly developing technology is giving us the means to alter, to our advantage, the DNA molecules of pathogens, crop plants—and human beings. How, if at all, will the technology be used? As a member of society, you will be dealing with this question for years to come. You owe it to yourself and others to gain some understanding of the possible benefits and risks of the new technology. This topic is addressed in Chapter 13, but it begins here, with the structure and function of DNA itself.

DISCOVERY OF DNA FUNCTION

In the spring of 1868, Johann Friedrich Miescher discovered a previously unknown substance in the cell nucleus. In time the substance came to be known as deoxyribonucleic acid, or DNA. No one knew what its function might be. In fact, seventy-five years passed before DNA was even recognized as the master blueprint of inheritance in all organisms. Clues about its function turned up through many seemingly unrelated studies.

In 1928, for example, Fred Griffith attempted to create a vaccine against a bacterium that causes a type of pneumonia in humans. He isolated two strains of the bacterium (designated R and S) and used them in four experiments.

First, laboratory mice were injected with live R cells. The R strain proved to be harmless, because the mice did not develop pneumonia. Second, mice were injected with live S cells. The mice died, and blood samples from them teemed with live S cells. The S strain was pathogenic. Third, S cells were killed by exposure to high temperature. Mice that were injected with the heat-killed

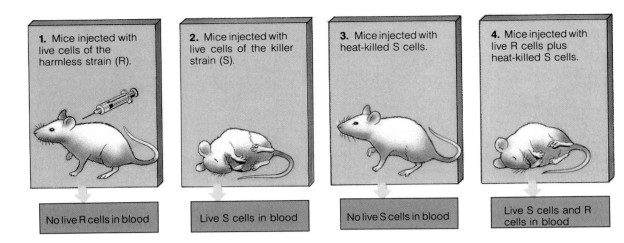

Figure 11.1 Summary of the results from Griffith's experiments with harmless (R) strains and disease-causing (S) strains of *Streptococcus pneumoniae*, as described in the text. You may be wondering why the S form is deadly and the R form harmless. The disease-causing strain produces a thick external capsule that protects the bacterial cells from attack by the host's immune system. Cells of the R strain form no such capsule; the host's defense system has the chance to destroy those cells before they can cause disease.

Figure 11.2 Hershey-Chase bacteriophage studies pointing to DNA as the substance of heredity.

Bacteriophages, a class of viruses that infect bacteria, consist only of protein and DNA (Figure 11.3). When bacteriophages attach to a host cell, they inject their DNA into it. Soon the cell starts making viral nucleic acids and proteins (including enzymes) necessary to build new bacteriophages. Then viral enzymes degrade the bacterial cell wall. The cell bursts, releasing the new infectious generation. In 1952, Alfred Hershey and Martha Chase devised a way to track the viral DNA and proteins during the infections.

(**a**) Bacteriophage proteins contain sulfur but no phosphorus—and the DNA contains phosphorus but no sulfur. In one experiment, bacteriophages were labeled with a radioactive isotope of sulfur (^{35}S) to tag their proteins. In a second experiment, they were labeled with a radioactive isotope of phosphorus (^{32}P) to tag their DNA.

(**b**) Labeled bacteriophages were allowed to infect unlabeled cells suspended in fluid. Hershey and Chase whirred the fluid in a kitchen blender to cleave the bacteriophage bodies from the cells. (**c**) Labeled protein remained in the fluid; it was associated with the bacteriophage bodies. Labeled DNA remained with the bacterial cells—it had to contain the hereditary instructions for producing new bacteriophages.

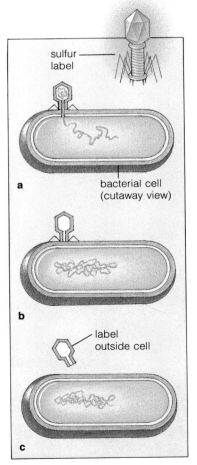
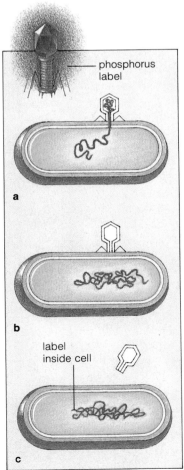

FIRST EXPERIMENT

SECOND EXPERIMENT

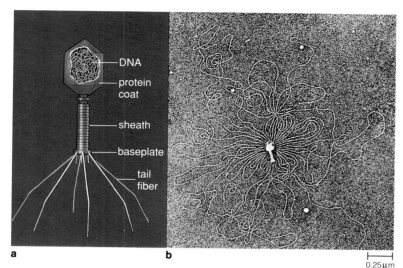

a

b

0.25 μm

Figure 11.3 (**a**) Component parts of one type of bacteriophage. (**b**) Electron micrograph of a broken bacteriophage, with a single DNA molecule released from the protein coat.

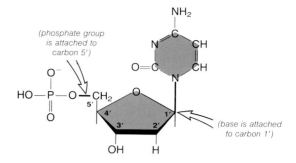

(phosphate group is attached to carbon 5')

(base is attached to carbon 1')

Figure 11.4 Example of one of the four nucleotides of DNA. Each nucleotide has a sugar (deoxyribose), a phosphate group, and one of four bases (shown here, cytosine). The small numerals identify the carbon atoms to which other atoms or molecules are attached.

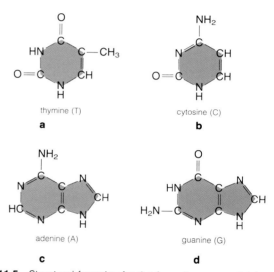

thymine (T)

a

cytosine (C)

b

adenine (A)

c

guanine (G)

d

Figure 11.5 Structural formulas for the four nitrogen-containing bases of DNA. (**a, b**) The single-ring pyrimidines and (**c, d**) the double-ring purines.

cells did not die. Fourth, live R cells were mixed with heat-killed S cells and injected into mice. Oddly, the mice died, and blood samples from them teemed with live S cells as well as R cells. The results of this series of experiments are summarized in Figure 11.1.

What was going on in the fourth experiment? The harmless cells apparently had picked up the instructions for causing infection and had been transformed into pathogens. The transformation was permanent—that is, *heritable*—for hundreds of generations of bacteria descended from the transformed cells also caused disease!

A few years later, researchers found that *extracts* of heat-killed S cells also could transform harmless R cells. However, by adding a certain enzyme to such extracts, Oswald Avery and his coworkers were able to block the transformation of harmless bacterial cells. That enzyme degrades DNA molecules but has no effect on proteins, which at the time were the leading candidates in the search for the hereditary molecule. In contrast, protein-degrading enzymes had no effect at all on the transformation. In 1944, Avery reported that DNA was probably the substance of heredity.

By the early 1950s, experiments with bacteriophage, a class of viruses that infect bacteria, were providing more evidence that DNA functions as the molecule of heredity (see, for example, Figures 11.2 and 11.3). That was the decade in which James Watson and Francis Crick deciphered the structure of DNA and thereby ushered in the golden age of molecular genetics.

DNA STRUCTURE

Components of DNA

A DNA molecule is composed of four kinds of **nucleotides,** the subunits of nucleic acids (page 33). A nucleotide consists of a five-carbon sugar (deoxyribose), a phosphate group, and one of the following nitrogen-containing bases:

| adenine | guanine | thymine | cytosine |
| (A) | (G) | (T) | (C) |

The nucleotides in DNA are structurally similar to one another (Figure 11.4). But notice, in Figure 11.5, that thymine and cytosine are smaller, single-ring structures

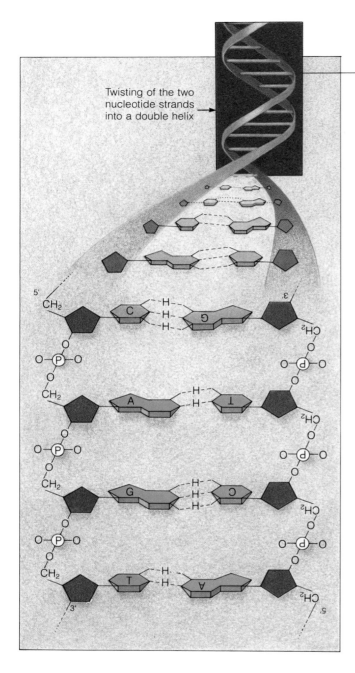

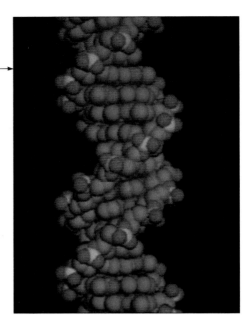

Twisting of the two
nucleotide strands
into a double helix

Figure 11.6 Arrangement of bases (blue) in the DNA double helix.

(pyrimidines). Adenine and guanine are larger, double-ring structures (purines).

The four kinds of nucleotide bases making up a DNA molecule differ in relative amounts from species to species. But the amount of adenine always equals the amount of thymine (A = T), and the amount of guanine equals the amount of cytosine (G = C).

Patterns of Base Pairing

How are nucleotide bases arranged in DNA? The first convincing evidence came from Rosalind Franklin, who developed the best x-ray diffraction images of DNA. (Crystallized preparations of DNA disperse x-rays in a regular pattern. Franklin used the pattern to calculate the positions of atoms in the crystal.) By her calculations, nucleotides had to be arranged into a long, thin molecule of uniform diameter. And bonding patterns among the nucleotides might make the molecule twist helically, like a circular stairway.

Using these and other clues, Watson and Crick put together a model of DNA structure. As they alone perceived, DNA consists of *two* strands of nucleotides, twisted together into a *double* helix (Figure 11.6). Hydrogen bonds join the bases of one strand with bases of the other. For the entire length of a DNA molecule, adenine always pairs with thymine, and cytosine always pairs with guanine. In other words, only two kinds of **base pairs** occur in DNA:

$$A--T \quad \text{and} \quad G--C$$

However, the *order* of bases in a nucleotide strand can vary greatly. In even a tiny stretch of DNA from a rose, gorilla, human, or any other organism, the base sequence might be:

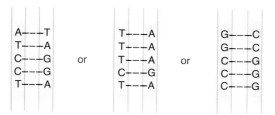

As you can see, DNA molecules show constancy and variation in their molecular structure. This is the molecular foundation for the unity and diversity of life.

Base pairing between the two nucleotide strands in DNA is *constant* **for all species (adenine to thymine, guanine to cytosine).**

The base sequence (that is, which base follows the next in a nucleotide strand) is *different* **from species to species.**

DNA REPLICATION

Assembly of Nucleotide Strands

The process by which DNA is duplicated prior to cell division is called **replication.** The two nucleotide strands making up the double helix unwind from each other and their bases are left exposed. Cells have stockpiles of free nucleotides, and these pair with exposed bases on the unwound strands. Each parent strand remains intact and a companion strand is assembled on each one according to the base-pairing rule.

As replication proceeds, each parent strand is twisted into a double helix with its new, partner strand. Because the parent strand is conserved, each "new" DNA molecule is really half-old, half-new. Figure 11.7 is a diagram of this mode of replication.

Prior to cell division, the double-stranded DNA molecule unwinds and is replicated. Each parent strand remains intact —it is conserved—and a new, complementary strand is assembled on each one.

Replication Enzymes

A DNA double helix does not unwind all by itself during replication. Enzymes and other proteins unwind the molecule, keep the two strands separated, and assemble a new strand on each one.

DNA polymerases are major replication enzymes. They govern nucleotide assembly on a parent strand. They also "proofread" the growing strands for mis-

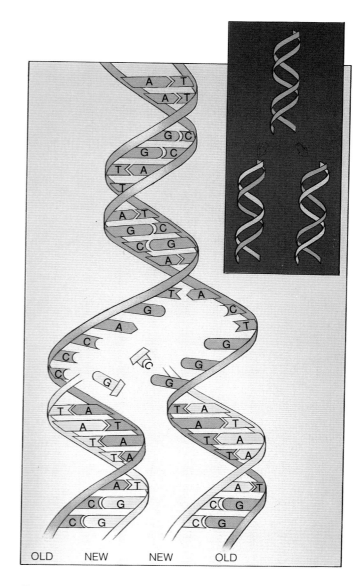

Figure 11.7 DNA replication. The original two-stranded DNA molecule is shown in blue. A new strand (yellow) is assembled on each of the two original strands. The two molecules resulting from the replication process are "half-old, half-new."

matched base pairs, which are replaced with correct bases. The proofreading function is one reason why DNA is replicated with such accuracy. On the average, for every 100 million nucleotides added to a growing strand, only *one* mistake slips through the proofreading net.

ORGANIZATION OF DNA IN CHROMOSOMES

There is one DNA molecule in each chromosome. If you could line up all forty-six human chromosomes end to end, they would extend about a meter. Obviously, all that DNA would become a tangled mess if it were not organized in some way.

The DNA of humans and all other eukaryotes is tightly bound with many proteins, including **histones.** Some histones are like spools for winding up small stretches of DNA. Each histone-DNA spool is a **nucleosome:**

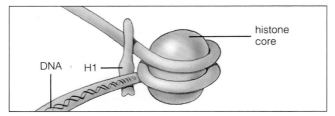

Another histone (H1) stabilizes the arrangement and plays a role in higher levels of organization. The chro-

mosome becomes coiled repeatedly through interactions between the histones and DNA; this process greatly increases its diameter (Figure 11.8). Further folding results in a series of loops. The looped regions vary in size, and each may contain one or (at most) a few gene sequences. What are the functions of such packing variations along the length of a chromosome? As you will see in the next chapter, some are known to influence the activity of different genes at different times in the life of an organism.

SUMMARY

1. Deoxyribonucleic acid, or DNA, is the master blueprint of hereditary instructions in cells. It is assembled from small organic molecules called nucleotides.

2. All nucleotides have a five-carbon sugar (deoxyribose) and a phosphate group. They also have one of four nitrogen-containing bases: adenine, thymine, guanine, or cytosine.

3. In a DNA molecule, two nucleotide strands are twisted together into a double helix. The bases of one strand pair with bases of the other strand (by hydrogen bonding).

4. There is constancy in base pairing in a DNA molecule. Adenine always pairs with thymine (A = T), and guanine always pairs with cytosine (G = C).

5. There is variation in the *sequence* of base pairs in the DNA of different species.

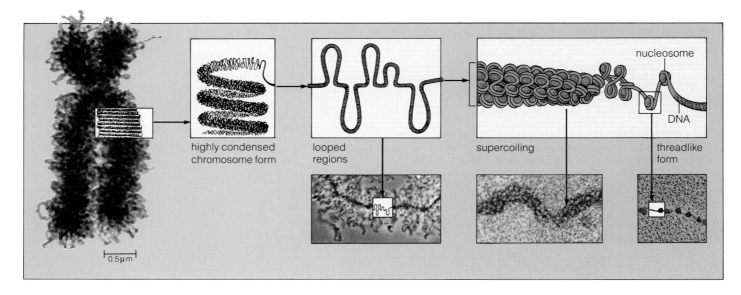

Figure 11.8 Organization of DNA in chromosomes, based on studies of electron micrographs (such as those in the bottom row). The nucleosome is the basic packing unit. It consists of a double loop of DNA around a core of histone molecules.

6. In DNA replication, the two strands of the double helix unwind from each other and a new strand of complementary sequence is assembled on each one. Two double-stranded molecules result, in which one strand is "old" (it is conserved) and the other is "new." Replication requires many enzymes and other proteins.

7. In each eukaryotic chromosome, a DNA molecule is bound tightly with many proteins. Interactions between the DNA and proteins give rise to the structural organization that is characteristic of the metaphase chromosome.

Self-Quiz *(Answers in Appendix IV)*

1. _____ bonds hold the bases of one nucleotide strand to bases of the other nucleotide strand of a DNA double helix.

2. Which of the following is *not* a nitrogenous base of DNA?
   a. adenine
   b. thymine
   c. guanine
   d. cytosine
   e. uracil

3. Base pairing in the DNA molecule follows which configuration?
   a. A--G, T--C
   b. A--C, T--G
   c. A--U, C--G
   d. A--T, C--G

4. A single strand of DNA with the base sequence C–G–A–T–T–G would be complementary to the sequence _____.
   a. C–G–A–T–T–G
   b. G–C–T–A–A–G
   c. T–A–G–C–C–T
   d. G–C–T–A–A–C

5. The DNA of one species differs from others in its _____.
   a. sugars
   b. phosphate groups
   c. base pair sequence
   d. all of the above

6. When DNA replication begins, _____.
   a. the two strands of the double helix unwind from each other
   b. the two strands condense tightly for base-pair transfers
   c. two DNA molecules bond
   d. old strands move to find new strands before bonding

7. DNA replication produces _____.
   a. two half-old, half-new double-stranded molecules
   b. two double-stranded molecules, one with the old strands and one with newly assembled strands
   c. three new double-stranded molecules, one with both strands completely new and two that are discarded
   d. none of the above

8. The process of DNA replication requires _____.
   a. a supply of new nucleotides
   b. forming of new hydrogen bonds
   c. many enzymes and other proteins
   d. all of the above

9. DNA polymerase has _____ functions.
   a. strand assembly
   b. phosphate attachment
   c. proofreading
   d. a and c are both correct

10. Match these DNA concepts appropriately.
   _____ base pair sequences
   _____ metaphase chromosome
   _____ constancy in base pairing
   _____ replication
   _____ double helix

   a. two nucleotide strands twisted together
   b. A = T, G = C
   c. one strand old (conserved), the other new
   d. accounts for differences among species
   e. structure results from interactions between DNA and proteins

Review Questions

1. DNA is composed of four different kinds of nucleotides. Name the three molecular parts of a nucleotide. Name the four different kinds of nitrogen-containing bases that may occur in the nucleotides of DNA. *145*

2. What kind of bond holds two DNA chains together in a double helix? Which nucleotide base-pairs with adenine? Which pairs with guanine? *146*

3. The four bases in DNA may differ greatly in relative amounts from one species to the next—yet the relative amounts are always the *same* among all members of a single species. How does the concept of base pairing explain these twin properties—the unity and diversity—of DNA molecules? *147*

Selected Key Terms

adenine (A) *145*
base pair *146*
cytosine (C) *145*
DNA polymerase *147*
DNA replication *147*
guanine (G) *145*
histone *148*
nucleosome *148*
nucleotide *145*
thymine (T) *145*

Readings

Alberts, B., et al. 1989. *Molecular Biology of the Cell.* Second edition. New York: Garland Publishing.

Cairns, J., G. Stent, and J. Watson (editors). 1966. *Phage and the Origins of Molecular Biology.* Cold Spring Harbor, New York: Cold Spring Harbor Laboratories. Collection of essays by the founders of and converts to molecular genetics. Gives a sense of history in the making—the emergence of insights, the wit, the humility, the personalities of the individuals involved.

Radman, M., and Wagner. August 1988. "The High Fidelity of DNA Duplication." *Scientific American* 259(2).

Watson, J. 1978. *The Double Helix.* New York: Atheneum. Highly personal view of scientists and their methods, interwoven into an account of how DNA structure was discovered.

12

FROM DNA TO PROTEINS

KEY CONCEPTS

1. Life cannot exist without enzymes and other proteins. Instructions for building proteins are encoded in genes, which are stretches of DNA molecules. The "code words" of a gene are sequences of nucleotide bases, read three at a time.

2. The path from genes to proteins has two steps. In transcription, an RNA strand is assembled on exposed bases of an unwound gene region according to base-pairing rules. In translation, the code-word sequence of DNA (and then RNA) is converted into the amino acid sequence of a polypeptide chain. One or more such chains form a protein molecule.

3. Although replication and repair enzymes work to preserve DNA instructions over time, on rare occasions one to several bases in a gene sequence are permanently deleted, added, or replaced. Such gene mutations are the original source of genetic variation in populations.

4. Which gene products appear in a cell, at what times, and in what amounts is governed by control elements built into DNA molecules as well as by regulatory proteins, enzymes, and hormones.

DNA is like a book of instructions in each cell. The alphabet used to create the book is simple enough: A, T, G, and C. But how is the alphabet arranged into the sentences (genes) that become expressed as proteins? How does a cell skip through the book, reading only those genes that will provide specific proteins at specific times? Answers to these questions begin with the nucleotide sequence of DNA.

PROTEIN SYNTHESIS

As we saw in the preceding chapter, a DNA molecule consists of two long strands of nucleotides. When DNA is replicated, the two strands unwind from each other, and their exposed bases serve as a structural pattern, or *template*, upon which a complementary strand is built. At other times in a cell's life, however, DNA regions are unwound so that enzymes can gain access to specific genes.

Each **gene** is a stretch of nucleotides in DNA that calls for the assembly of different amino acids into a polypeptide chain. Such chains are the basic structural units of proteins.

The path from genes to proteins has two steps, called transcription and translation. Think about how the path is followed in a eukaryotic cell. In **transcription,** molecules of ribonucleic acid, or **RNA,** are produced on DNA

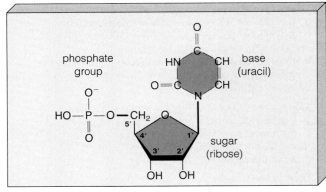

a

templates in the nucleus. In **translation,** RNA molecules shipped from the nucleus into the cytoplasm are used as the templates for assembling polypeptide chains. Following translation, one or more chains become folded into protein molecules. Some of the proteins serve as structural materials in cells; many serve as enzymes in the synthesis of carbohydrates, lipids, and other substances that are essential for the structure and function of cells.

Transcription

A strand of RNA is almost, but not quite, like a single strand of DNA. Its nucleotides consist of a sugar (ribose), a phosphate group, and a nitrogen-containing base. Its bases are adenine, cytosine, guanine, and **uracil** (Figure 12.1). Like the thymine in DNA, uracil can base-pair with adenine.

During transcription, an RNA strand is assembled on a DNA template according to the rule of base pairing:

$$
\begin{array}{ccc}
\text{DNA} & \text{C---G} & \text{RNA} \\
& \text{G---C} & \\
& \text{T---A} & \\
& \text{A---U} &
\end{array}
$$

Transcription differs from DNA replication in two key respects. First, only one region of one DNA strand (not the whole strand) is used as the template. Second, different enzymes (RNA polymerases) are involved.

Transcription starts at a **promoter,** a base sequence that signals the start of a gene. After an RNA polymerase binds with a promoter, it moves along the DNA template and joins nucleotides into an RNA *transcript* (Figure

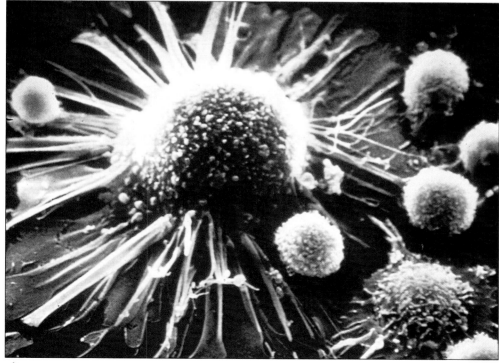

b

Figure 12.1 (**a**) Example of one of the four nucleotides in RNA, a type of molecule that is absolutely central to the synthesis of proteins according to genetic messages. Such gene expression must be executed with controlled precision, otherwise cell structure, function, or both may suffer. Cancer cells, such as the one shown in (**b**), are one outcome of abnormal gene expression. We will return to this topic at the chapter's end.

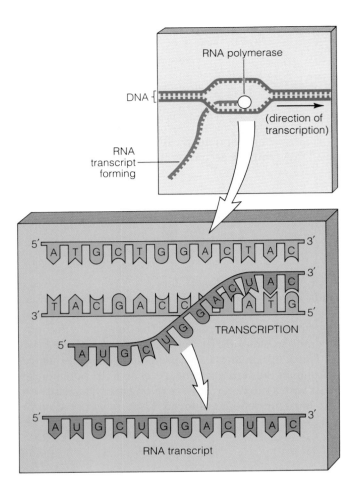

Figure 12.2 Transcription: the synthesis of an RNA molecule on a DNA template. Following transcription, the unwound region of DNA winds up again into a double helix.

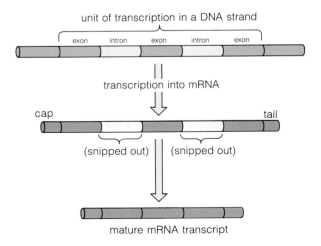

12.2). When the enzyme reaches the end of the gene region, another signal causes the transcript to be released from the template.

Classes of RNA

Three types of RNA molecules are transcribed from different DNA regions. All three take part in translation, the second step in protein synthesis:

ribosomal RNA (rRNA)	a type of molecule that combines with certain proteins to form the *ribosome* (the structural "workbench" on which a polypeptide chain is assembled)
messenger RNA (mRNA)	the "blueprint" (a linear sequence of nucleotides) delivered to the ribosome for translation into a polypeptide chain
transfer RNA (tRNA)	an adaptor molecule; it can pick up a specific amino acid *and* pair with an mRNA code word for that amino acid

Only mRNA carries protein-building instructions out of the nucleus. And it does not get shipped without alterations. Just as a dressmaker might snip off some threads or bows on a dress before it leaves the shop, so does a cell tailor its mRNA. For example, newly formed transcripts contain **introns** (noncoding portions) and **exons** (portions that do get translated into proteins). As Figure 12.3 shows, the introns are snipped out and the exons are spliced together to form the mature mRNA transcript.

Translation

The Genetic Code. Like a DNA strand, an mRNA transcript is a linear sequence of nucleotides. So we are still left with a central question: What are the protein-building "words" encoded in that sequence?

Francis Crick, Sidney Brenner, and others came up with the answer. They deduced the nature of the **genetic**

Figure 12.3 (*Left*) Transcript processing. Before an mRNA transcript leaves the nucleus, noncoding portions of the genetic message (introns) are snipped out, and a nucleotide cap is added at one end and a tail at the other.

code—that is, how the nucleotide sequence of DNA and then mRNA corresponds to the amino acid sequence of a polypeptide chain. The bases of the nucleotides are read three at a time, and each base triplet calls for an amino acid (Figure 12.4). A start signal built into the DNA, and therefore into the mRNA copy, establishes the correct "reading frame" for blocking out every three nucleotides in the sequence.

Hereditary instructions for building proteins are encoded in the nucleotide sequence of DNA and mRNA.

Every three nucleotides (base triplet) specifies an amino acid, which becomes linked with others into a polypeptide chain.

The genetic code consists of sixty-four different base triplets. (Sixty-one actually specify amino acids. The other three act like stop signs in protein synthesis; they signify that no more amino acids are to be added to the polypeptide chain.) Each base triplet in mRNA is called a **codon.** As you can see from Figure 12.4, many amino acids can be specified by two or more different codons.

With few exceptions, the genetic code is universal for all forms of life. A codon that calls for a certain amino acid in bacteria calls for the same amino acid in protistans, fungi, plants, and animals.

Codon-Anticodon Interactions. So now we have an mRNA molecule, with its string of codons, arriving in the cytoplasm of a eukaryotic cell. Here it will interact in precise ways with its molecular relatives, the tRNAs and rRNAs.

Each kind of tRNA has an **anticodon,** a sequence of three nucleotide bases that can pair with a specific mRNA codon. Each tRNA also has a molecular "hook," an attachment site for an amino acid. To keep things simple in illustrations to follow, we can portray codon-anticodon interactions in this fashion:

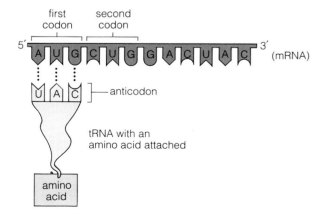

First Letter	Second Letter				Third Letter
	U	C	A	G	
U	phenylalanine	serine	tyrosine	cysteine	U
	phenylalanine	serine	tyrosine	cysteine	C
	leucine	serine	stop	stop	A
	leucine	serine	stop	tryptophan	G
C	leucine	proline	histidine	arginine	U
	leucine	proline	histidine	arginine	C
	leucine	proline	glutamine	arginine	A
	leucine	proline	glutamine	arginine	G
A	isoleucine	threonine	asparagine	serine	U
	isoleucine	threonine	asparagine	serine	C
	isoleucine	threonine	lysine	arginine	A
	(start) methionine	threonine	lysine	arginine	G
G	valine	alanine	aspartate	glycine	U
	valine	alanine	aspartate	glycine	C
	valine	alanine	glutamate	glycine	A
	valine	alanine	glutamate	glycine	G

Figure 12.4 The genetic code. The codons in an mRNA molecule are nucleotide bases, read in blocks of three. Each of those base triplets will call for a specific amino acid during mRNA translation. In this diagram, the first nucleotide of any triplet is given in the left column. The second is given in the middle columns; the third, in the right column. Thus we find (for instance) that trytophan is coded for by the triplet ⬚U⬚G⬚G. Phenylalanine is coded for by both ⬚U⬚U⬚U and ⬚U⬚U⬚C.

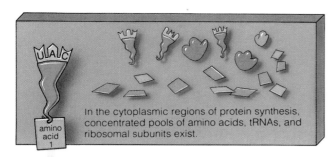

In the cytoplasmic regions of protein synthesis, concentrated pools of amino acids, tRNAs, and ribosomal subunits exist.

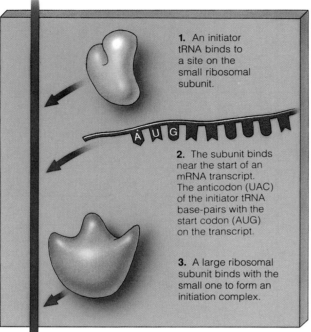

1. An initiator tRNA binds to a site on the small ribosomal subunit.

2. The subunit binds near the start of an mRNA transcript. The anticodon (UAC) of the initiator tRNA base-pairs with the start codon (AUG) on the transcript.

3. A large ribosomal subunit binds with the small one to form an initiation complex.

a Initiation

Stages of Translation.

Translation proceeds through three stages called initiation, elongation, and termination. All three stages take place on the surface of ribosomes. As Figure 12.5 shows, each ribosome is composed of two separate subunits that come together only during initiation of translation.

In *initiation*, an "initiator" tRNA able to start translation binds to a small ribosomal subunit, which in turn binds to the transcript. Next, a large ribosomal subunit joins with the small one. We now have an intact ribosome, an mRNA transcript, and an initiator tRNA. Chain elongation can begin.

In *chain elongation*, a start codon on the mRNA defines the reading frame for assembling amino acids in sequence. A series of tRNAs deliver amino acids to the ribosome. A peptide bond forms between the growing polypeptide chain and each amino acid added.

With *chain termination*, a stop codon is reached, and the ribosome and polypeptide chain are detached from the mRNA transcript. The chain joins the pool of free proteins in the cytoplasm or enters the cytomembrane system for further processing. Here you may wish to refer to page 50, which outlines the final destinations for new proteins.

Figure 12.5 Simplified picture of protein synthesis.

initiation complex (an mRNA transcript loaded on an intact ribosome)

binding site for mRNA

first binding site for tRNA

second binding site for tRNA

mRNA

tRNA

1. This diagram shows the relative positions of the binding sites for tRNAs and the mRNA transcript on an intact ribosome.

2. Once the initiator tRNA is aligned in the first binding site, another tRNA can occupy the adjacent binding site. Which particular tRNA binds is dictated by the next mRNA codon.

3. The anticodon of the second tRNA base-pairs with the second codon. As it does, its attached amino acid aligns with the amino acid of the initiator tRNA.

b Chain elongation

Mutation and Protein Synthesis

In general, the base sequence in DNA must be preserved from one generation to the next, otherwise offspring might not be able to synthesize properly all the proteins necessary for their own growth, development, and reproduction. Yet changes do occur in the DNA.

As we saw in the preceding chapters, crossing over and recombination put new mixes of alleles in chromosomes. We also saw that the structure and number of chromosomes can change through nondisjunction or some other abnormal event.

Another kind of change is called the **gene mutation.** It is a deletion, addition, or substitution of one to several bases in the nucleotide sequence of a gene. Gene mutations are rare, chance events. On the average, the mutation rate for a gene is only one in a million replications.

Some gene mutations are induced by **mutagens,** environmental agents that can attack a DNA molecule and modify its structure. Some viruses, ultraviolet radiation, and certain chemicals are examples of mutagens. Other gene mutations are spontaneous rather than induced. For example, if A accidentally becomes paired with A during replication, enzymes with proofreading

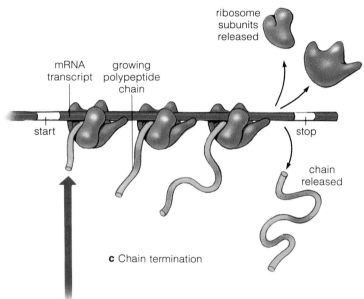

ribosome subunits released

mRNA transcript   growing polypeptide chain

start   stop

chain released

c Chain termination

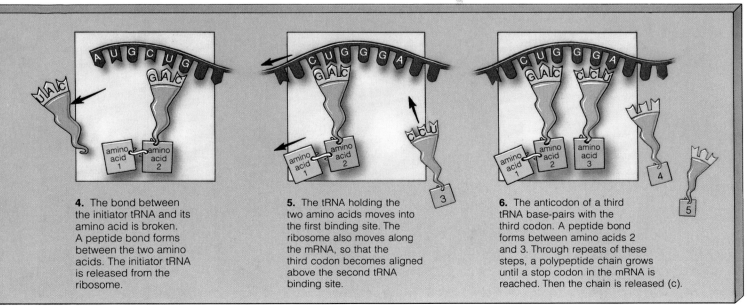

4. The bond between the initiator tRNA and its amino acid is broken. A peptide bond forms between the two amino acids. The initiator tRNA is released from the ribosome.

5. The tRNA holding the two amino acids moves into the first binding site. The ribosome also moves along the mRNA, so that the third codon becomes aligned above the second tRNA binding site.

6. The anticodon of a third tRNA base-pairs with the third codon. A peptide bond forms between amino acids 2 and 3. Through repeats of these steps, a polypeptide chain grows until a stop codon in the mRNA is reached. Then the chain is released (c).

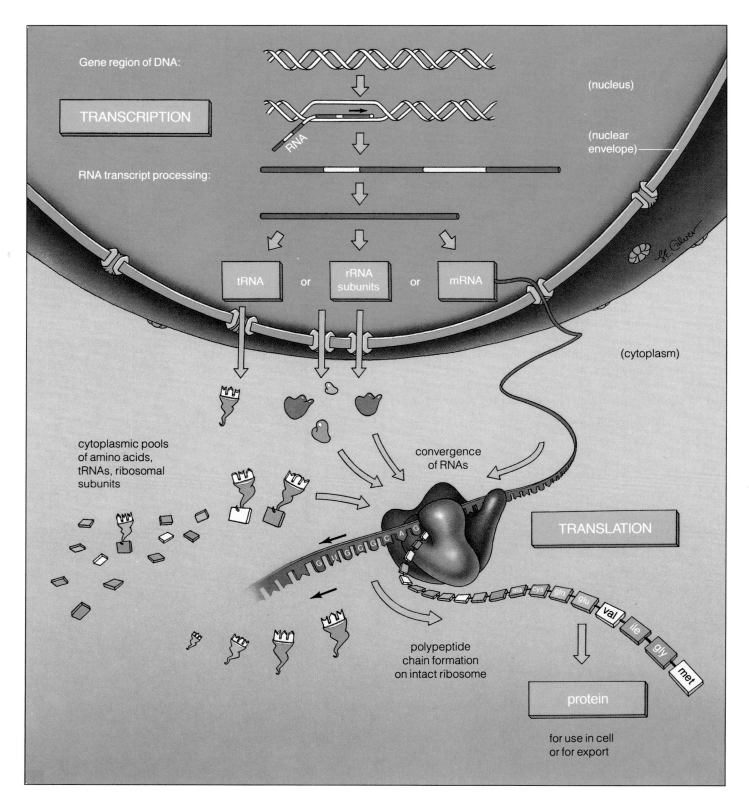

Figure 12.6 Summary of the flow of genetic information in protein synthesis in eukaryotic cells.

functions might detect the mistake—but which A will they "fix"?

You might be thinking that a change in a single base pair is insignificant. But sickle-cell anemia has been traced to a single mutation in the DNA strand coding for the beta chain of hemoglobin. Only one amino acid is substituted for another in the resulting chain—yet the substitution can have severe consequences (see the *Commentary* on page 125).

Barbara McClintock discovered another type of spontaneous gene mutation. Through studies of Indian corn (maize), she realized that certain DNA regions frequently "jump" to new locations in the same DNA molecule or in a different one. These *transposable elements* often inactivate the genes into which they become inserted and give rise to changes in phenotype (Figure 12.8).

Let's turn now to controls over gene activity. Before doing so, you may wish to review Figure 12.6, which summarizes the flow of information from genes to proteins.

CONTROLS OVER GENE ACTIVITY

All the different cells of your body carry the same genes, and they use most of them to synthesize proteins that are basic to any cell's structure and functioning. That is why proteins of the cytoskeleton are the same from one cell to the next, as are many of the enzymes used in metabolism.

Yet each type of cell also uses a small fraction of genes in highly specialized ways. Even though they all carry the genes for hemoglobin, only red blood cells activate those genes. Even though they carry the genes for antibodies (protein "weapons" against specific agents of disease), only certain white blood cells activate them. *These and all other living cells control which genes are active and which gene products appear, at what times, and in what amounts.*

The best-understood gene controls deal with the *rate* of transcription (how many RNA transcripts peel off a gene in a given period). Transcription, recall, starts at a promoter (a base sequence at the start of genes). The enzyme RNA polymerase must bind with a promoter before it can transcribe genes. Some genes are transcribed more often simply because their promoters bind the enzyme strongly, compared to genes with weak promoters.

The transcription rate also can be increased or slowed down through the action of regulatory proteins, such as repressors. Some regulatory proteins interact with a promoter. Others interact with an **operator,** a short base sequence between a promoter and the start of a gene.

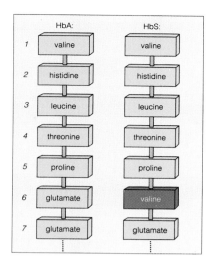

Figure 12.7 A substitution of one amino acid for another in a polypeptide chain. In sickle-cell hemoglobin, valine is substituted for glutamate at position 6 of the beta chain (see above). This mutation puts a "sticky" (hydrophobic) patch on the surface of a hemoglobin molecule. When oxygen concentrations in the blood are low, the sticky patches interact and hemoglobin molecules aggregate into rods, distort red blood cells, and cause them to clump in blood vessels (page 125).

Figure 12.8 Color variations in kernels of Indian corn. All the cells in a kernel have pigment-coding genes, so you might expect the whole kernel to be the same color. Some are indeed fully colored, but others are spotted or entirely colorless. Early in plant growth, transposable elements (movable DNA regions) jumped about in the chromosomes of some cells. Their insertions into new chromosomal locations produced mutations, some of which inactivated genes that affected pigment synthesis. The mutated genes of an affected cell were expressed in all of its daughter cells in the kernel's tissues.

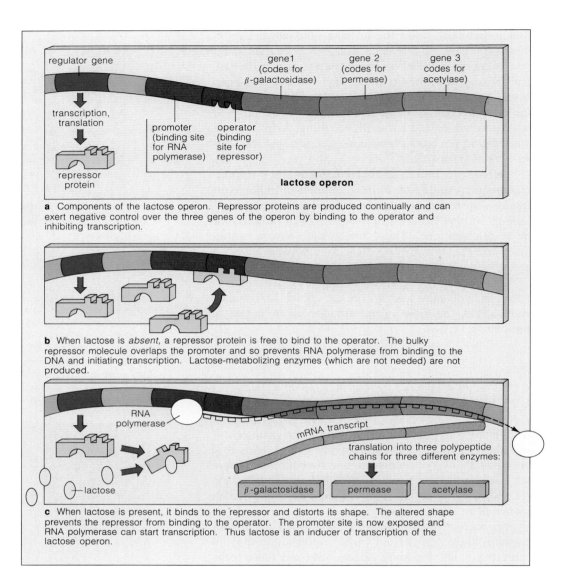

a Components of the lactose operon. Repressor proteins are produced continually and can exert negative control over the three genes of the operon by binding to the operator and inhibiting transcription.

b When lactose is *absent*, a repressor protein is free to bind to the operator. The bulky repressor molecule overlaps the promoter and so prevents RNA polymerase from binding to the DNA and initiating transcription. Lactose-metabolizing enzymes (which are not needed) are not produced.

c When lactose is present, it binds to the repressor and distorts its shape. The altered shape prevents the repressor from binding to the operator. The promoter site is now exposed and RNA polymerase can start transcription. Thus lactose is an inducer of transcription of the lactose operon.

Figure 12.9 Negative control of the lactose operon. The first gene of the operon codes for an enzyme that splits lactose into two subunits (glucose and galactose). The second one codes for an enzyme that transports lactose molecules across the plasma membrane and into the cytoplasm. The third plays a complex role in lactose metabolism.

Gene Control in Prokaryotes

Studies of *Escherichia coli*, a type of bacterial cell living in the gut of mammalian hosts, yielded the first insights into gene controls. Here we will look briefly at how lactose metabolism is controlled in those cells.

After you drink a glass of milk, *E. coli* rapidly transcribes three adjacent genes that allow it to use milk sugar (lactose) as an energy source. Two binding sites (a promoter and an operator) precede those genes, and it is here that transcription is controlled. Any gene (or group of genes), together with its promoter and operator sequence, is called an **operon**.

A repressor protein affords negative control over the lactose operon. It binds with the operator whenever lactose concentrations are low. Being a rather large molecule, the repressor overlaps the promoter and so blocks RNA polymerase's access to the genes. Thus, lactose-degrading enzymes are not produced when they are not needed (Figure 12.9).

The repressor does not block transcription when lactose is present. A lactose molecule binds with and alters the shape of the repressor. In its altered shape, the repressor cannot bind to the operator, so RNA polymerase has access to the genes. Thus lactose-degrading enzymes are produced when required.

Gene Control in Eukaryotes

Selective Gene Expression. Compared to *E. coli*, much less is known about gene controls in multicelled eukaryotes. The main reason is that patterns of gene expression vary within and between different body tissues. Remember that all the cells in your body have the same genes (they are descended from the same fertilized egg). But cells of your brain, liver, and other tissues are **differentiated**—they are specialized in composition, structure, and function.

Differentiation arises through *selective* gene expression in different cells. Depending on the cell type and the control agents acting on it, some genes might be turned on only at one particular stage of the life cycle. Others might be left on all the time or never activated at all. Still other genes might be switched on and off throughout the individual's life. We will return to this topic in Chapter 32.

Cell differentiation occurs in multicelled eukaryotes as a result of *selective gene expression*. Although all the cells in the body inherit the same genes, they activate or suppress some fraction of those genes in different ways to produce pronounced differences in cell structure or function.

X Chromosome Inactivation. One kind of control over gene expression is exerted through the packaging of DNA in chromosomes. Consider that each cell in a mammalian female has two X chromosomes. The gene products of only one are required for normal cell functioning; it may even be that a double dose of gene products from two X chromosomes would prove lethal.

As a female embryo develops, one X chromosome in each cell becomes condensed and transcription of most of its genes is permanently suppressed (Figure 12.10). Which of the two becomes condensed is a matter of chance.

The embryo grows through cell divisions, and the condensed chromosome is replicated and passed on to all the daughter cells. Because the paternal X chromosome was randomly inactivated in some cells and the maternal X chromosome was inactivated in others, every adult female is a "mosaic" of X-linked traits, with patches of tissue in which a paternal *or* maternal allele is being expressed.

The mosaic effect is evident in human females affected by *anhidrotic ectodermal dysplasia*. A mutant allele on one X chromosome gives rise to this skin disorder, which is characterized by an absence of sweat glands. In patches of defective skin, the X chromosome bearing the normal allele has been inactivated and the X chromosome bearing the mutant allele is functional (Figure 12.11).

Gene Control of Cell Division. Every second of the day, millions of cells in different parts of your body divide and replace their worn-out, dead, and dying

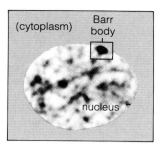

Figure 12.10 Inactivated X chromosome in the nucleus of a cell from a mammalian female. Such chromosomes are called Barr bodies.

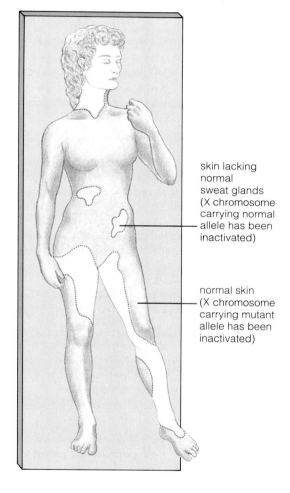

skin lacking normal sweat glands (X chromosome carrying normal allele has been inactivated)

normal skin (X chromosome carrying mutant allele has been inactivated)

Figure 12.11 Pattern of gene expression in a woman affected by anhidrotic ectodermal dysplasia, a disorder in which patches of skin do not have normal sweat glands. Such "mosaic" tissue effects were discovered by Mary Lyon. They arise through random inactivation of the X chromosome during embryonic development.

Cancer—When Gene Controls Break Down

On rare occasions, gene controls over cell division become permanently altered. A cell divides again and again, until its offspring begin to crowd surrounding cells and interfere with tissue functioning. The alteration has spawned a **tumor:** a tissue mass composed of cells that are dividing at an abnormally high rate.

The problem is not that tumor cells divide at a horrendous rate; normal cells divide much faster when they replace a surgically removed portion of the liver. Rather, tumor cells have lost the controls telling them when to stop. They will not stop as long as conditions for growth remain favorable.

When a tumor is *benign,* it may continue to grow more rapidly than normal but it remains in the same place in the body. Surgical removal of the tissue mass removes its threat to health. When a tumor is *malignant,* its cells can migrate and then grow and divide in other organs.

Normally, recognition proteins allow cells to bind together in tissues and organs (page 55). When genes for those proteins are altered or suppressed, the cell can leave its proper place and travel (in blood or lymph) to other tissues, where it can form a new growth. This process of invasion is called **metastasis.**

All of the many types of malignant tumors are grouped into the general category of **cancer.** At the minimum, all cancer cells have these characteristics:

1. *Profound changes in the plasma membrane and cytoplasm.* Membrane permeability is amplified. Some membrane proteins are lost or altered, and new ones appear. The cytoskeleton shrinks, becomes disorganized, or both. Enzyme activity shifts (as in an amplified reliance on glycolysis).

2. *Abnormal growth and division.* Inhibitors of overcrowding in tissues are lost. Cell populations increase to unusually high densities. New proteins trigger an abnormal increase in small blood vessels that service the growing cell mass.

3. *Weakened capacity for adhesion.* Cells cannot become properly anchored in the parent tissue.

4. *Lethality.* Unless cancer cells are eradicated, they will kill the individual.

Any gene having the potential to induce cancerous transformations is called an **oncogene.** When introduced into a normal cell, an oncogene transforms it into a tumor cell. Such genes were first identified in infectious agents called retroviruses (page 224). But similar gene sequences *also* occur in the normal DNA of many species and rarely trigger cancer! These sequences are called **proto-oncogenes.**

Proto-oncogenes code for proteins necessary in normal cell functioning. They may become cancer-causing genes only on rare occasions, when specific mutations alter their structure or their expression. In other words, the *normal* expression of proto-oncogenes is vital—even though their abnormal expression may be lethal.

Insertion of viral DNA into the DNA of a host cell can skew transcription of a proto-oncogene. **Carcinogens** may do the same thing. Carcinogens include many natural and synthetic compounds (such as asbestos and certain components of cigarette smoke), x-rays, gamma rays, and ultraviolet radiation.

Yet cancer seems to be a multistep process, with mutations in more than one proto-oncogene required to bring it about. Look again at the characteristics of cancer cells. Now think about some of the products of proto-oncogenes: *growth factors (signals sent by one cell to trigger growth in other cells), regulatory proteins involved in cell adhesion, and protein signals for cell division.*

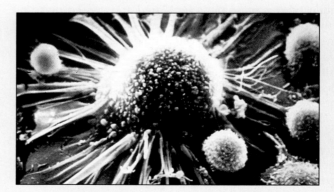

A cancer cell (center), displaying some of the gross abnormalities arising from alterations in genes.

predecessors. The cells do not all divide at the same rate. Some, including nerve cells, have long-term roles and are arrested at interphase. Others have short-term roles and divide rapidly, as do cells in the protective epithelium near the stomach wall, which is constantly exposed to the corrosive effects of gastric fluids.

No one knows exactly how, but genes govern cell growth and division. When most of a rat's liver is surgically removed, protein synthesis and cell divisions accelerate phenomenally: *4 billion* replacement cells are produced in four days. Then brakes are applied and the division rate slows; by the seventh day, most of the missing liver tissue has been replaced.

On rare occasions, a cell loses control over division. We refer to this loss as cancerous transformation, an event described in the *Commentary.* Possibly more than any other example, this type of transformation brings home the critical extent to which you and all other organisms depend on controls over gene expression.

SUMMARY

1. The path leading from DNA to proteins can be summarized this way:

The path has two steps. In *transcription,* an exposed region of one strand of the DNA double helix serves as the template for assembling an RNA strand. In *translation,* three classes of RNA molecules (mRNA, tRNA, and rRNA) interact to convert the message originally encoded in DNA into a polypeptide chain.

2. The relationship between a sequence of DNA or mRNA and a sequence of amino acids in a polypeptide chain is called the genetic code. The code words are a sequence of nucleotides that are read in blocks of three (base triplets). Each base triplet in mRNA is a codon; the complementary base triplet in tRNA is an anticodon.

3. Transcription follows the same base-pairing rule that applies to DNA replication, but uracil (not thymine) pairs with the adenine present in the DNA template strand:

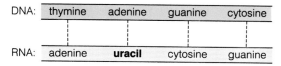

DNA:	thymine	adenine	guanine	cytosine
RNA:	adenine	**uracil**	cytosine	guanine

4. Translation begins with the convergence of two ribosomal subunits, an initiator tRNA, and an mRNA transcript. A polypeptide chain forms as tRNAs deliver amino acids to the ribosome. The chain detaches from the ribosome when a stop codon is reached in the mRNA.

5. Overall, the protein-building instructions in DNA are preserved through the generations. But crossing over and recombination, changes in chromosome structure or number, and gene mutations can change parts of those instructions.

6. Gene expression is controlled by many interacting elements, including control sites built into DNA molecules, regulatory proteins, enzymes, and hormones. Their interactions govern which gene products appear, at what times, and in what amounts.

Review Questions

1. Are the products specified by eukaryotic DNA assembled *on* the DNA molecule? If so, state how. If not, tell where they are assembled, and on which molecules. *151*

2. Figure 12.6 shows the steps by which hereditary instructions are transcribed from DNA into RNA, which is then translated into proteins. Study this figure and then, on your own, write a description of this sequence, taking care to define the terms *transcription* and *translation. 156*

3. Define *genetic code.* Is the same basic genetic code used for protein synthesis in all living organisms? *153*

4. Define the three types of RNA. What is a codon? An anticodon? *152, 153*

5. Cells depend on controls over which gene products are synthesized, at what times, at what rates, and in what amounts. Describe one type of control over transcription in *E. coli,* a type of prokaryote. *158*

6. A plant, fungus, or animal is composed of diverse cell types. How might this diversity arise, given that all of the body cells in each organism inherit the *same* set of genetic instructions? *158, 159*

7. Somatic cells of human females have two X chromosomes. During what developmental stage are genes on *both* chromosomes active? Explain what happens to one of those chromosomes after that stage. *159*

8. What are the characteristics of cancer cells? Explain the difference between a benign tumor and one exhibiting metastasis. *160*

Self-Quiz *(Answers in Appendix IV)*

1. Nucleotide bases, read _____ at a time, serve as the "code words" of genes.

2. Genetic information in DNA is transferred to RNA strands during _____, the first step in protein synthesis.
   a. replication   c. multiplication
   b. duplication   d. transcription

3. The RNA molecule is _____.
   a. a double helix
   b. usually a double nucleotide strand
   c. always a double nucleotide strand
   d. usually a single nucleotide strand

4. During transcription, base pairing is similar to that of DNA replication except _____.
   a. cytosine in DNA pairs with guanine in RNA
   b. adenine in DNA pairs with uracil in RNA
   c. thymine in DNA pairs with adenine in RNA
   d. guanine in DNA pairs with cytosine in RNA

5. _____ starts when two ribosomal subunits, an initiator tRNA, and an mRNA transcript come together.
   a. Transcription
   b. Replication
   c. Subduction
   d. Translation

6. The coded genetic instructions for forming polypeptide chains are carried to the ribosome by _____.
   a. DNA
   b. rRNA
   c. mRNA
   d. tRNA

7. The function of tRNA is to _____.
   a. deliver amino acids to the ribosome
   b. pick up genetic messages from rRNA
   c. synthesize mRNA
   d. all of the above

8. How many amino acids are coded for in this mRNA sequence: CGUUUACACCGUCAC?
   a. three
   b. five
   c. six
   d. seven
   e. more than seven

9. An anticodon pairs with the nitrogen-containing bases of _____.
   a. mRNA codon
   b. DNA codons
   c. tRNA anticodon
   d. amino acids

10. Match the terms related to protein-building appropriately.

____ disrupts genetic instructions	a. interacting DNA control sites, regulatory proteins, enzymes, and hormones
____ genetic code word	b. RNAs convert genetic messages into polypeptides
____ transcription	c. series of nucleotide bases
____ translation	d. one DNA strand serves as the template
____ gene expression	e. gene recombination, changes in chromosome structure and number, gene mutation

11. Using the genetic code shown in Figure 12.4, translate the mRNA sequence UAUCGCACCUCAGGAUGAGAU. Which of the following polypeptide chains does this sequence specify?
   a. tyr-arg-thr-ser-gly-stop-asp...
   b. tyr-arg-thr-ser-gly...
   c. tyr-arg-tyr-ser-gly-stop-asp...
   d. none is correct

Selected Key Terms

anticodon 153
cancer 160
codon 153
differentiation 158
exon 152
gene 150
gene mutation 155
genetic code 152
intron 152
messenger RNA 152
metastasis 160
mutagen 155

oncogene 160
operator 157
operon 158
promoter 151
proto-oncogene 160
ribosomal RNA 152
ribosome 152
selective gene expression 158
transcription 150
transfer RNA 152
translation 151
uracil 151

Readings

Darnell, J. October 1985. "RNA." *Scientific American* 253(4):68–78.

Feldman, M., and L. Eisenbach. November 1988. "What Makes a Tumor Cell Metastatic?" *Scientific American* 259(5):60–85.

Kupchella, C. 1987. *Dimensions of Cancer.* Belmont, California: Wadsworth.

Prescott, D. 1988. *Cells.* Boston: Jones & Bartlett. Chapter 8 contains an excellent introduction to protein synthesis.

Ptashne, M. January 1989. "How Gene Activators Work." *Scientific American* 260(1):41–47.

Weintraub, H. January 1990. "Antisense RNA and DNA." *Scientific American* 262(1):40–46.

RECOMBINANT DNA AND GENETIC ENGINEERING

KEY CONCEPTS

1. Genetic experiments have been occurring in nature for billions of years as a result of gene mutations, crossing over and recombination, and other events. Humans are now engineering genetic changes by way of recombinant DNA technology.

2. With recombinant DNA technology, DNA molecules are cut into fragments, inserted into cloning tools such as plasmids, then amplified rapidly to form amounts suitable for research and practical applications.

3. The new technology raises social, legal, ecological, and ethical questions regarding its benefits and risks.

For more than 3 billion years, nature has been conducting genetic experiments through mutation, chromosomal crossing over, and other events. Genetic messages have changed countless times, and this is the source of life's diversity.

We humans have been changing the genetic character of species for thousands of years. Through artificial selection, we coaxed modern crop plants and new breeds of cattle, birds, dogs, and cats from wild ancestral stocks. We developed meatier turkeys, sweeter oranges, seedless watermelons, and flamboyant ornamental plants. We produced the tangelo (tangerine × grapefruit) and the mule (donkey × horse). Through surgery, vaccinations, drug therapy, and other medical practices, we have preserved vulnerable genetic messages in the population.

Today we are "engineering" genetic changes through **recombinant DNA technology.** With this technology, DNA from different species can be cut, spliced together, then inserted into bacteria or other types of rapidly dividing cells—which multiply in quantity the recombinant DNA molecules. Genes can be isolated, modified, and reinserted into the organism (or transplanted into a different one). In many cases, the engineered genes produce functional proteins.

Recombinant DNA technology has staggering potential for medicine, agriculture, and industry. It does not come without risks. With this chapter, we consider some basic aspects of the new technology. We also address ecological, social, and ethical questions related to its application.

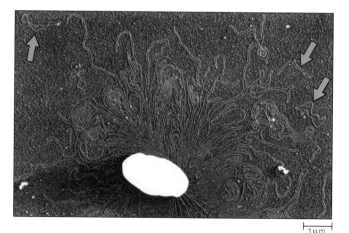

Figure 13.1 A ruptured bacterial cell (*Escherichia coli*). Notice the larger bacterial chromosome and the smaller plasmids (blue arrows).

RECOMBINANT DNA TECHNOLOGY

Plasmids

Recombinant DNA technology grew out of experiments with bacteria. All bacteria have a single chromosome, a circular DNA molecule that contains all the genes necessary for normal growth and development. Many species also have smaller circles of "extra" DNA called plasmids (Figure 13.1). **Plasmids** are circular DNA molecules that carry only a few genes and that replicate *independently* of the chromosome.

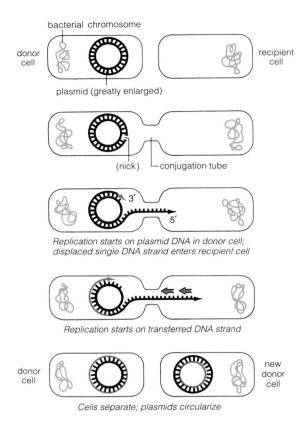

Figure 13.2 Transfer of a plasmid between two bacterial cells during conjugation. For clarity, the bacterial chromosomes are not drawn full-size. (Each bacterial chromosome actually contains about forty times more genetic information than the largest plasmids do.)

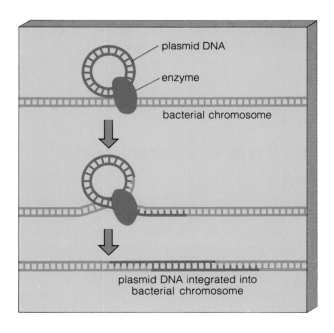

Genes carried on some plasmids permit **bacterial conjugation,** a process by which a donor cell transfers DNA to a recipient cell (Figure 13.2). On rare occasions, the plasmid becomes integrated into the recipient's chromosome. Like all cells, bacteria have enzymes capable of cutting DNA strands during normal repair operations. One such enzyme recognizes a short nucleotide sequence that happens to occur on the plasmid as well as on the chromosome. It cuts both molecules at the sequence, then splices their cut ends together (Figure 13.3).

In the late 1960s and early 1970s, researchers learned how to use different bacterial enzymes to cut DNA into fragments. They also learned how to package the fragments into other types of plasmids for insertion into host cells. They developed ways to pinpoint specific DNA fragments in cells. They also started to identify nucleotide sequences of individual genes and to map the *genome* (all the DNA in a haploid set of chromosomes) for different species.

Producing Restriction Fragments

Different species of bacteria produce **restriction enzymes,** the sole function of which is to cut apart foreign DNA molecules that may get into the cell. Several hundred different restriction enzymes have been identified. Each makes its cut only at sites having a short, specific nucleotide sequence. Researchers can select the ones that will produce a DNA fragment containing a particular gene.

Restriction enzymes are useful in another respect. Many produce staggered cuts, so both ends of a DNA fragment end up with short, single-stranded tails:

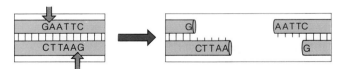

These "sticky ends" of the fragment can base-pair with any other DNA molecule cut by the same restriction enzyme.

Figure 13.3 (*Left*) Integration of a plasmid into a bacterial chromosome through recombination. Only a small stretch of the circular bacterial chromosome is shown.

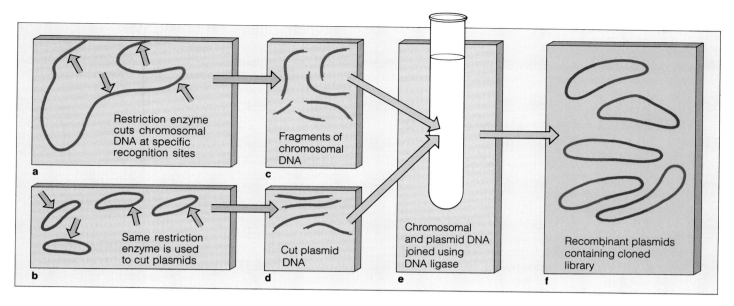

Figure 13.4 Formation of a DNA library.

DNA Amplification

Suppose we use the same restriction enzyme to cut plasmids and a chromosome. When the chromosomal fragments and cut plasmids are mixed together, they base-pair at random at their sticky ends. By using another enzyme, **DNA ligase,** the base pairing can be made permanent (Figure 13.4).

We now have a **DNA library**—a collection of DNA fragments produced by restriction enzymes and incorporated into plasmids. We can insert the DNA library into bacteria or other host cells for amplification. In other words, after repeated replications and divisions of the host cells, we end up with **cloned DNA**—multiple, identical copies of DNA fragments from the original chromosome.

DNA amplification can occur through other methods besides cloning. For example, single-stranded mRNA can serve as a template for assembling a DNA strand that is identical in sequence to some desired gene. The assembly requires a special viral enzyme, **reverse transcriptase.** After the "hybrid" DNA/RNA molecule is assembled, the RNA strand is degraded. Other enzymes convert the remaining strand of DNA to double-stranded form (Figure 13.5). Any DNA molecule "copied" from mRNA is known as **cDNA.**

Today, the **polymerase chain reaction** is the most commonly used method of DNA amplification. With this method, the gene of interest is split into two single strands, which enzymes then copy over and over to produce millions of copies of DNA containing that gene.

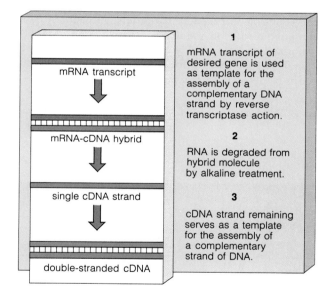

Figure 13.5 Formation of double-stranded cDNA from an mRNA transcript.

COMMENTARY

RFLPs and Genetic Fingerprinting

Some aspects of recombinant DNA technology are already affecting our lives. Among them is the work with *restriction fragment length polymorphisms*, or RFLPs, in human chromosomes.

"Restriction fragment length" refers to the use of restriction enzymes to cut human DNA into fragments of specific lengths. The fragments are separated by special procedures to identify the ones of interest. "Polymorphism" refers to the fact that each person's DNA has a slightly unique pattern of sites where the enzymes make their cuts. Thus each person has a slightly different pattern of DNA fragments. (The only exceptions are identical twins, who have identical DNA.)

The pattern variations can be detected by using radioactive probes (page 19). It turns out that mutant alleles responsible for genetic disorders sometimes have a unique restriction site that is *not* present in the DNA of normal persons. Sickle-cell anemia is an example; a mutant allele coding for a defective hemoglobin chain can be detected in this manner (see Figure 12.7).

Another kind of RFLP may revolutionize criminal investigations. Each person has a *genetic fingerprint*, a unique array of RFLPs inherited from each parent in a Mendelian pattern. Already, we can resolve paternity cases by comparing the genetic fingerprint of the child with that of the disputed parent.

Today, murderers have become much easier to identify if they leave even a few drops of their blood at the scene of the crime or if a few drops of the victim's blood is found on their clothing. The bloodstain will contain enough DNA for its genetic fingerprint to be compared with a suspect's DNA. Rapists also may be deterred by the new technology. A genetic fingerprint from just a drop or two of semen recovered from a victim can identify a rapist. Courts have already ruled for conviction when a suspect's genetic fingerprint has matched that found in semen samples. Although genetic fingerprinting has been disputed in some cases, the accuracy of the laboratory work, not the approach itself, is being questioned.

Table 13.1	Examples of Cloned Human Gene Products Approved for Use or Under Development	
Protein	**Used in Treating**
Insulin	Diabetes
Somatotropin (growth hormone)	Pituitary dwarfism
Erythropoetin	Anemia
Factor VIII	Hemophilia
Factor IX	Hemophilia
Interleukin-2	Cancer
Tumor necrosis factor	Cancer
Interferons	Some cancers, viral infections
Monoclonal antibodies	Infectious diseases
Atrial natriuretic factor	High blood pressure
Tissue plasminogen factor	Heart attack, stroke

Regardless of the method used, the resulting DNA fragments can be separated according to their sizes. The separation procedure spreads out the fragments so that particular ones can be identified or isolated. The *Commentary* describes only one of the ways DNA fragments can be used.

Expressing a Cloned Gene

Even when DNA fragments containing a desired gene have been isolated and amplified, it may not be possible for a host cell to transcribe and translate them into functional protein. For example, human genes contain noncoding regions (introns) as well as coding regions (exons). The genes cannot be translated unless the introns are spliced out and the exons spliced together into a mature mRNA transcript (Figure 12.3). Bacterial host cells don't have the proper splicing enzymes. That is why researchers working with human genes typically use cDNA (synthesized from mature mRNA transcripts from which the introns have already been removed).

Despite these and other obstacles, several human gene products already are being mass-produced or developed. Table 13.1 lists some of these products.

Mapping the Human Genome

We now can map the exact nucleotide sequence of cloned genes. Machines have been developed that can sequence 10,000 nucleotides of DNA on a daily basis. To sequence the estimated 3 billion nucleotides in the twenty-three types of human chromosomes, many researchers must collectively devote at least ten to twenty years of effort. Success depends on advances in speeding up the sequencing process. Many laboratories are collaborating in the mapping attempt.

GENETIC ENGINEERING: RISKS AND PROSPECTS

Paul Berg and his colleagues were the first to insert foreign DNA into a bacterial plasmid. Their pioneer work in the 1970s brought a question into sharp focus: Is it dangerous to transfer genes between different species? Let's look at the work going on today with bacteria, plants, and animals.

Genetically Engineered Bacteria

The bacterial strains used in many genetic engineering experiments are harmless to begin with, and they are also modified (by mutation) so they cannot survive outside the laboratory. Even so, there is some concern about possible risks of introducing genetically engineered bacteria into humans or the environment.

Consider what happened when Steven Lindow genetically engineered a strain of *Pseudomonas syringae.* This common bacterium lives on leaves and stems, and it happens to make many crop plants susceptible to frost damage. Proteins on the bacterial cell surface enhance the formation of ice crystals and so cause plant damage. Lindow excised the "ice-forming" gene from some *P. syringae* cells. The modified cells could not synthesize the ice-forming protein. For one experiment, these so-called *ice-minus bacteria* were to be sprayed on strawberry plants in an isolated field just before a frost. The experiment would indicate whether the plant cells would then resist freezing.

The proposed experiment only involved an organism from which a harmful gene had been *deleted,* yet it triggered a bitter legal debate on the risks of deliberately releasing genetically engineered microbes in the environment. The courts finally ruled in favor of allowing the experiment to proceed, and a small patch of strawberries was sprayed (Figure 13.6). As predicted, there was no ecological disaster—but a few environmental activists entered the patch at night and pulled up the plants.

The lessons of the ice-minus controversy are important. Rules governing the release of genetically engineered organisms have since been clarified. Environmental impact reports are filed first, and biotechnologists have learned they must communicate effectively with the public about their work. We must weigh carefully the potential for ecological problems against the public good that any particular experiment may produce.

Figure 13.6 Spraying an experimental strawberry patch in California with "ice-minus" bacteria. (Government regulations required that the sprayer use elaborate protective gear.)

Figure 13.7 Crown gall tumors on a willow tree. The tree was infected by *Agrobacterium tumefaciens*, which carries tumor-inducing genes on a plasmid.

Figure 13.8 Ten-week-old mouse littermates, the one on the left weighing 29 grams, and the one on the right, 44 grams. The larger mouse grew from a fertilized egg into which the gene for human somatotropin (growth hormone) had been inserted.

Genetically Engineered Plants

The first successful attempt to genetically engineer plants occurred in the early 1980s. Researchers inserted DNA fragments into a type of plasmid from a bacterium that infects many plants. Some genes carried on the plasmid cause the formation of crown gall tumors (Figure 13.7). Normally, the tumor-inducing genes become integrated into the DNA of infected plants. Researchers first removed the harmful genes and substituted other, desired genes. Then they inserted the plasmid into plant cells. In some cases, the foreign genes were expressed normally in the plant tissues.

Genetically engineered crop plants may allow us to increase global food production. Millions of humans die each year from starvation. Simply increasing food production to keep pace with the burgeoning human population is not a solution to the problem, but it is one of the few short-term options available.

For example, very few crop plants can grow in salty soil. This is a serious problem, because most croplands do not get enough rainfall and must be irrigated. Yet irrigation water brings in enormous amounts of salts, which build up in the soil. All over the world, croplands are rapidly being "salted out" and can no longer support existing crop plants. Some researchers are attempting to recombine the genes of sugar beets and a few other moderately salt-tolerant species with conventional crop plants. Genetic engineering may produce salt-resistant, high-yield strains.

Genetically Engineered Animals

In 1982, Ralph Brinster and Richard Palmiter introduced the gene for somatotropin (growth hormone) from rats into fertilized mouse eggs. When the mice grew, it became clear that the rat gene had become integrated into the mouse DNA and was being expressed. The mice grew much larger than their normal littermates. Their cells had up to thirty-five copies of the gene, and blood concentrations of the hormone were several hundred times higher than normal values. More recently, the gene for human somatotropin was successfully introduced and expressed in mice (Figure 13.8).

Similar experiments with large domesticated animals have not been successful. For example, when the somatotropin gene was inserted into pigs, it was expressed—but the pigs developed arthritis-like symptoms and other disorders.

Gene modification in animals is extremely difficult. New or modified genes must be inserted in the body cells of an animal or in gametes. They also must end up in specific locations in chromosomes so that their expres-

COMMENTARY

Human Gene Therapy

Inserting one or more normal genes into the body cells of an organism to correct a genetic defect is called **gene therapy.** The idea of doing this to offer relief from severe genetic disorders seems to be socially acceptable at present, even though we are not close to having the technology by which gene therapy might be accomplished.

In contrast, inserting genes into a normal human (or sperm or egg) in order to modify or enhance a particular trait is called many things, including **eugenic engineering;** but mostly it is called a horrifying idea. Who decides which traits are "desirable"? What if all prospective parents start picking the sex of their children through genetic engineering? (Three-fourths of one recently surveyed group said they would choose a boy. What would be the long-term social implications of a drastic shortage of girls?) If it is okay to engineer taller or blue-eyed individuals, would it be okay to engineer "superhuman" offspring with exceptional strength or intelligence? Fortunately, perhaps, intelligence and most other traits arise through complex interactions among many genes; these qualities will be outside the reach of genetic manipulation for some time.

This chapter only touched on some of the social and ethical issues raised by recombinant DNA technology and genetic engineering. Some individuals say that the DNA of each species never should be altered. But as the earlier discussion should make clear, nature itself alters DNA much of the time. The real argument, of course, is whether we have the wisdom to bring about beneficial changes without causing harm to ourselves or the environment.

The challenge of manipulating the human genome reflects our very human tendency to leap before we look. And the pressure to restrict all genetic engineering recalls an old attitude: "If God had wanted us to fly, he would have given us wings." And yet, something about the human experience gave us the *capacity* to imagine wings of our own making—and that capacity carried us to the frontiers of space.

Where are we going from here with this new product of our imagination, recombinant DNA technology? To gain perspective on the question, spend some time reading the history of our species. It is a history of survival of all manner of threats, expansions, bumblings, and sometimes large-scale disasters. It is also a story of increasingly intertwined interactions with the environment and with one another. The questions confronting you today are these: Should we be more cautious, believing that one day the risk takers may go too far? And what do we as a species stand to lose if the risks are *not* taken?

---

sion will be properly regulated. And they must not disrupt the function of other genes.

For example, genes can be injected into a sperm nucleus just after it penetrates an egg. But the eggs are so vulnerable to being poked that the procedure has a high rate of failure. Even when gene delivery is successful, researchers still cannot control *where* in the DNA the inserted gene will end up. In its new location, will the inserted gene alter a proto-oncogene, with its potential to cause cancer (page 160)? Or cause a mutation? Or alter expression of related genes?

SUMMARY

1. Genetic "experiments" have been occurring in nature for billions of years. Mutation, crossing over and recombination at meiosis, and other natural events have all contributed to the current diversity that we see among organisms.

2. Humans have been manipulating the genetic character of different species for thousands of years. The emergence of recombinant DNA technology in the past few decades has enormously expanded our capacity to cause genetic change. Recombinant DNA technology is founded on procedures by which DNA molecules can be cut into fragments, inserted into plasmids or some other cloning tool, then propagated in a population of rapidly dividing cells.

3. A DNA clone is any DNA sequence that has been amplified in dividing cells. DNA sequences also can be amplified in test tubes by the polymerase chain reaction.

4. Recombinant DNA technology and genetic engineering have enormous potential for research and applications in medicine, agriculture, and home and industry. As with any new technology, potential benefits must be weighed against potential risks, including ecological and social disruptions.

5. Although the new technology has not developed to the extent that human genes can be modified, the social, legal, ecological, and ethical questions should be explored in detail before such an application is possible.

Review Questions

1. What is a plasmid? What is a restriction enzyme? Do such enzymes occur naturally in organisms? 163, 164

2. Recombinant DNA technology involves the following:
   a. Producing DNA restriction fragments. 164
   b. Cloning those restriction fragments into a suitable vector (generating a DNA library). 165
   c. Selecting the clone containing the restriction fragments you may wish to study. 165

Briefly describe one of the methods used in each of these categories.

3. Having read about examples of genetic engineering in this chapter, can you think of some additional potential benefits of this technology? Can you envision other potential problems?

Self-Quiz   (Answers in Appendix IV)

1. Gene mutations, crossing over and recombination during meiosis, and other natural events are the basis of the _____ observed in present-day organisms.

2. Causing genetic change by deliberately manipulating DNA is known as _____.

3. _____ are small circles of bacterial DNA that are separate from the bacterial chromosome.

4. Genetic researchers use plasmids as _____.

5. Rejoining cut DNA fragments from any organism is best known as _____.
   a. cloning genes
   b. mapping genes
   c. recombinant DNA technology
   d. conjugating DNA

6. Using the metabolic machinery of a bacterial cell to produce multiple copies of genes carried on hybrid plasmids is _____.
   a. a way to create a DNA library
   b. bacterial conjugation
   c. mapping a genome
   d. DNA amplification

7. Any DNA sequence that has been amplified in dividing cells is a _____.
   a. DNA clone
   b. DNA library
   c. chunk of foreign DNA
   d. gene map

8. The polymerase chain reaction _____.
   a. is a natural reaction in bacterial DNA
   b. cuts DNA into fragments
   c. amplifies DNA sequences in test tubes
   d. inserts foreign DNA into bacterial DNA

9. Which may benefit from recombinant DNA technology?
   a. households
   b. industry
   c. medicine
   d. agriculture
   e. all of the above

10. Match the recombinant DNA information appropriately.
    _____ DNA clone
    _____ bacterial plasmid
    _____ natural genetic "experiments"
    _____ polymerase chain reaction
    _____ modification of human genes

    a. mutation and meiotic recombination
    b. raises social, legal, and ethical questions
    c. a method of test tube gene amplification
    d. any cellular amplification of DNA sequences
    e. a cloning tool

Selected Key Terms

bacterial conjugation 164
cDNA 165
cloned DNA 165
DNA library 165
DNA ligase 165
eugenic engineering 169
gene therapy 169
genetic fingerprint 166

plasmid 163
polymerase chain reaction 165
recombinant DNA technology 163
restriction enzyme 164
reverse transcriptase 165
RFLP 166

Readings

Alberts, B., et al. 1989. *Molecular Biology of the Cell*. Second edition. New York: Garland Publishing. Chapter 5 has excellent coverage of DNA recombination and genetic engineering.

Anderson, W.F. 1985. "Human Gene Therapy: Scientific and Ethical Considerations." *Journal of Medicine and Philosophy* 10:274–291.

Brill, W. 1985. "Safety Concerns and Genetic Engineering in Agriculture." *Science* 227:381–384.

Guyer, R., and D. Koshland, Jr. 22 December 1989. "The Molecule of the Year." *Science* 246(4937):1543–1546.

Palmiter, R., et al. 1983. "Metallothionein–Human GH Fusion Genes Stimulate Growth of Mice." *Science* 222:809–814. Report on landmark experiments in mammalian gene transfers.

White, R., and J. Lalouel. February 1988. "Chromosome Mapping with DNA Markers." *Scientific American* 258(2):40–48.

FACING PAGE: *Millions of years ago, a bony fish died, and sediments gradually buried it. Today its fossilized remains are studied as one more piece of the evolutionary puzzle.*

MICROEVOLUTION

1. Individuals of a population show variations in traits, corresponding to differences in the relative abundances of alleles for most gene locations on chromosomes. *Microevolution* refers to changes in the frequency of those alleles over time.

2. Allele frequencies change as a result of mutation, genetic drift, gene flow, and natural selection. *New* alleles arise only through mutation; *existing* alleles are shuffled into, through, or out of populations by the other processes.

3. Natural selection is not an "agent," combing actively and purposefully through populations for the "best" individuals. It is simply *a measure of the difference in survival and reproduction* that has occurred among individuals who differ from one another in one or more traits.

4. For sexually reproducing organisms, speciation may occur when populations of a species become reproductively isolated from one another and differences build up in their pools of alleles. The divergence between populations can become pronounced, so that their members no longer can interbreed and produce fertile offspring under natural conditions.

EMERGENCE OF EVOLUTIONARY THOUGHT

When we hear the word "evolution," many of us think of Charles Darwin and the five-year voyage around the world that started him thinking about the meaning of life's diversity (Figure 14.1). But the history of evolutionary thought began long before Darwin's time. Centuries ago, scholars in the West believed that each kind of living thing, or "species," as they called it, had remained the same since the time of creation. They viewed all species as unchanging links in a great chain, extending from the lowest forms of life to humans and on to spiritual beings. Once all the links had been discovered and described, they thought, the meaning of life would be revealed.

With the global explorations of the sixteenth century, however, scholars were soon overwhelmed with descriptions of thousands upon thousands of exotic species from Asia, Africa, the Pacific islands, and the Americas. Where did all those species "fit" in the great chain? And why were so many living in one part of the world but not others?

More questions emerged during the eighteenth century. Anatomists had been comparing the body plans of different mammals (such as humans, whales, and bats) and other major animal groups. They found some striking similarities. For example, the arms of humans, flippers of whales, and wings of bats differ in size, shape, and function. Yet why are the number of bones and the connections among them so much alike? And

a

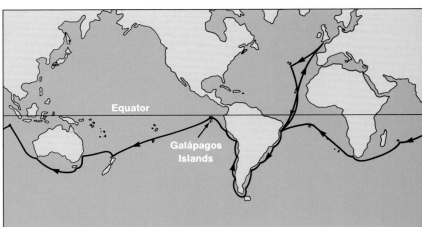

b

Figure 14.1 (**a**) Charles Darwin a few years after he returned from his five-year voyage aboard H.M.S. *Beagle* (**b**). A replica of the *Beagle* is shown in (**c**) sailing off the coast of South America. Inland from the coast, Darwin explored parts of the Andes, where he observed fossils of marine organisms in rock layers (**d**) that were 12,000 feet above sea level. Scenes from the Galápagos Islands, about 600 miles off the coast of Ecuador: a blue-footed booby (**e**), marine iguanas (**f**), and tropical plants of the highlands of Santa Cruz Island (**g**). The diversity of life on such islands influenced Darwin's thinking about the evolution of species.

why do they form in similar ways in the animal embryo? According to one explanation, some body plans were so perfect there was no need to come up with new ones for each organism at the time of creation.

Yet if that were true, how could there be body parts with no apparent function? For example, some snakes have bones corresponding to a pelvic girdle, a set of bones to which *hind limbs* attach (Figure 14.2). What were the bones doing there if snakes had been created in a state of limbless perfection? Similarly, humans have bony parts exactly like the bones in a tail. What were parts of a tail doing in a perfectly designed human body?

By the mid-eighteenth century, geologists had already started to map the stratification (horizontal layering) of sedimentary rocks beneath the earth's surface.

Such layers were deposited very slowly, one above the other, over time. And different layers held different kinds of **fossils,** the remains or body impressions of once-living organisms. (For example, fossils of simple marine organisms are restricted to deep rock layers in many parts of the world. Fossils in rock layers above them are similar, but some are more complex in structure. And fossils in the uppermost layers closely resemble living marine organisms.)

Many naturalists tried to reconcile the unmistakable layering of fossils with the traditional view of creation. Georges-Louis Leclerc de Buffon, for example, argued that if all species were created at the same time and place, they would not now be dispersed throughout the world; mountain barriers or oceans would have stopped them. *Perhaps species originated in more than one place.* Also, the "imperfections" in body plans and the fossil sequences could be interpreted to mean that species were not unalterably perfect. *Perhaps species became modified over time.* Awareness of **evolution**—changes within lines of descent over time—was in the wind.

"Evolution" was already being discussed in 1831, when Darwin was twenty-two years old and wondering what to do with his life. Although he had just earned a degree in theology from Cambridge University, he was happiest when studying natural history. It was John Henslow, a Cambridge botanist, who perceived Darwin's real interests. Henslow arranged for Darwin to become ship's naturalist aboard H.M.S. *Beagle* (Figure 14.1).

The *Beagle* sailed for South America to complete work on mapping the coastline. During its five-year voyage, the ship stopped at islands, near mountain ranges, and along rivers, so Darwin had ample opportunity to study the native plants and animals. He also had time to mull over an idea advanced by the geologist Charles Lyell.

According to Lyell, volcanic activity, mountain formation, and erosion occurred in the past just as they do

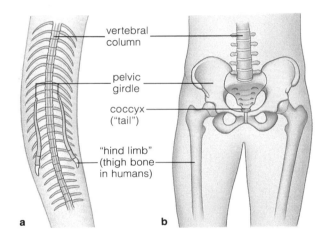

Figure 14.2 Bony parts of a python (**a**) that correspond to the pelvic girdle of other vertebrates, including humans (**b**). Small "hind limbs" protrude through the skin on the underside of the snake.

Figure 14.3 The armadillo (right) and a reconstruction of an extinct animal, the glyptodont (above). Resemblances between these unique, scaly animals and their similar, restricted geographic distribution provided Darwin with a clue that helped him develop his theory of how evolution occurs.

in the present—and because those processes occur so slowly, it must have taken millions of years to form the present landscape. According to nearly everyone else, the earth was less than 6,000 years old, and if that were true, it was hard to imagine that the many thousands of known species had evolved in so little time. If Lyell were correct, however, *there had been time enough for evolution.*

Neither Darwin nor anybody else had ever seen one species evolve into another. But among the fossils Darwin collected in Argentina were the remains of extinct animals that looked suspiciously like living armadillos (Figure 14.3). Nothing else in the world resembled either animal, and Darwin later wondered whether one form gave rise to the other.

It is one thing to say evolution may be possible, but quite another to say *how* it occurs. In time, Darwin came to suspect that geographic isolation is related to evolution and species diversity. For example, the Galápagos Is-

lands are about 1,000 kilometers off the coast of Ecuador (Figure 14.1). Every island or cluster of islands has diverse species, including distinct species of finch (Figure 14.4). Yet all the finch species closely resemble a finch living on the western coast of South America and nowhere else. Each species has a distinct type of beak suitable for crushing seeds, spearing insects, or securing some other foodstuff available on a given island. Perhaps all the species descended from the mainland finch, then changed slightly after becoming isolated on different islands.

But *how* could such changes occur? A clue came from an essay by Thomas Malthus, a clergyman and economist. In Malthus' view, any human population tends to outgrow its resources, and its members must compete for what is available. Darwin thought about all the populations he had observed during his voyage. The individual members of those populations had varied in body size, form, coloring, and other traits. It dawned

a

e

b

c

d

Figure 14.4 Examples of variation in beak shape among different finch species of the Galápagos Islands. These and eight other species apparently are descended from a common ancestor, a seed-eating ground finch.

(**a**) *Certhidea olivacea*, a tiny tree-dwelling finch that resembles a warbler in song and behavior; it uses its slender beak to probe for insects.

(**b**) *Geospiza scandens* and (**c**) *G. conirostris*, two species with a beak adapted for eating cactus flowers and fruits. (The former is shown in the midst of dinner in **e**.) Other species in the same genus have thick, strong beaks adapted for crushing cactus seeds.

(**d**) *Camarhynchus pallidus*, a finch that feeds on wood-boring insects such as termites. It swings its small body like a woodpecker does, to hammer at bark. It does not have the woodpecker's long, probing tongue, but it has learned to break cactus spines to appropriate lengths, then hold the "tools" in its beak and use them as probes.

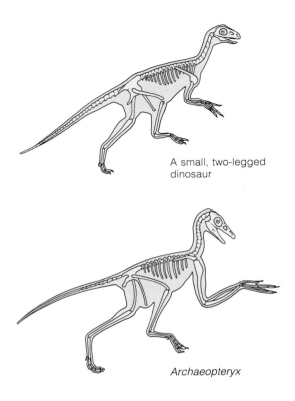

A small, two-legged dinosaur

Archaeopteryx

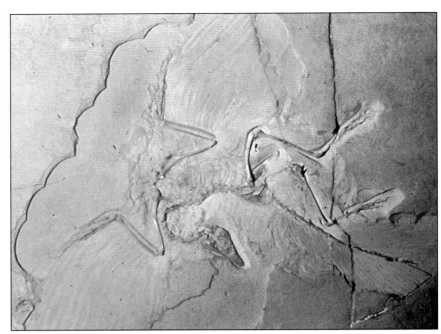

Figure 14.5 One of six fossils of *Archaeopteryx* from limestone deposits that are more than 140 million years old. *Archaeopteryx* was a transitional form on (or very near) the evolutionary road leading from reptiles to birds. In fact, the fossils might have been classified as reptilian, had it not been for the clear imprints of feathers in the finely grained limestone.

Archaeopteryx had teeth and hind limbs like its reptilian ancestors, but in other features (such as its feathers and wishbone) it was already like modern birds. It did not have the strong muscle attachments and bony framework of modern birds, but it certainly could glide and perhaps it could fly feebly.

on him that some traits—beak size and shape, for example—could lead to differences in the ability to secure resources.

If there were "struggles for existence" (competition) within a population, then individuals born with a stronger seed-crushing beak or some other favorable trait might have an edge in surviving and reproducing. *Nature would select individuals with competitive traits and eliminate others— and so a population could change.* Favored individuals would pass on the useful traits to offspring, their offspring would do the same, and so on. In time, descendants of the favored individuals would make up most of the population, and less favored individuals might have no descendants at all.

And so Darwin saw "natural selection" among variant individuals as a mechanism of evolution. He did not formally announce his theory right away; he wanted first to sift through the evidence for flaws in his reasoning. Then, in 1858, he received a paper from the naturalist Alfred Wallace, who had arrived at the same exact conclusion!

Darwin's colleagues prevailed upon him to present a paper along with Wallace's. The next year Darwin's detailed evidence in support of the theory was published in book form.

Darwin's theory still faced a crucial test. If evolution occurs, there should be evidence of one kind of organism changing into another. The fossil record seemed to contain no transitional forms, the so-called "missing links" between major groups of organisms. Oddly, two years after Darwin's book was published such a fossil did turn up, but few paid much attention to it. *Archaeopteryx* resembled reptiles *and* birds (Figure 14.5). Like fossils of small two-legged reptiles, it had teeth and a long, bony tail. Like modern birds, its body was covered with feathers!

No other evidence turned up in Darwin's time. Almost seventy years passed before advances in genetics led to widespread acceptance of his theory of natural selection. In the meantime, his name was associated mostly with the idea that life evolves—something others had proposed before him.

Figure 14.6 Variation in shell color and banding patterns among populations of one species of snails found on islands of the Caribbean. All individuals of a species have the same number of genes on their chromosomes, but alternative forms (alleles) are possible at most gene locations along the chromosomes. Variation in traits arises from the enormously different combinations of alleles carried by different members of the population.

MICROEVOLUTIONARY PROCESSES

Variation in Populations

As Darwin perceived, individuals don't evolve; *populations* do. By definition, a **population** is a group of individuals occupying a given area and belonging to the same species. All individuals of the same population have a certain number of traits in common. They have the same overall form and appearance, as when all members of a bird population have blue feathers, a slender beak, a gizzard, a four-chambered heart, and so on. These are *morphological traits* (*morpho-* means form). The physical and chemical operation of various cells and body parts proceeds in much the same way for all individuals during their growth, development, day-to-day housekeeping activities, and reproduction. These are *physiological traits* (they relate to body functioning). All individuals of a population make similar basic responses to stimuli, as when humans reflexively yank their hands away from a spider on a light switch or rummage through a refrigerator when they feel hungry. These are *behavioral traits*.

In most natural populations, the manifestations of different traits are not quite the same from one individual to the next. One member of a flock of blue, straight-feathered birds might have white, frazzled feathers. A few members might be more sensitive to winter cold or more adept at attracting a mate with a dazzling courtship display. Humans normally have hair, but the hair differs in color, texture, amount, and distribution from one person to the next. Figure 14.6 shows a stunning example of variation in the color and banding patterns of snails, yet even this example only hints at the immense variation in all individuals of all populations, past and present, on earth.

Sources of Variation. Variations in traits are partly *heritable.* Children resemble their parents more than they resemble anyone else. Information about heritable traits resides in hundreds or thousands of genes (specific regions of DNA molecules). All individuals inherit the same number and kinds of genes characteristic of their species. But some number of the genes in one individual have a slightly different molecular structure than their counterparts in another individual. This genetic variation results in variation in traits, or different **phenotypes.** Whether your hair is black, brown, red, or white depends on *which* molecular forms of certain genes you happened to inherit from your mother and father.

Variation in traits is an outcome of certain events that were introduced in earlier chapters. These events fall into five broad categories:

1. Gene mutation

2. Abnormal changes in chromosome structure or number

3. Crossing over and genetic recombination at meiosis

4. Independent assortment of chromosomes at meiosis

5. Fertilization between genetically different gametes

Hardy-Weinberg Principle:

The genotypic frequencies for a population in equilibrium will fit the formula

$$AA \quad Aa \quad aa$$
$$p^2 + 2pq + q^2$$

where p = the frequency of allele A and q = the frequency of allele a.

The allele frequencies and the genotypic frequencies will be stable from generation to generation if the following assumptions are true:

There is no mutation, the population is infinitely large and is isolated from other populations, mating is random, and all genotypes are equally viable and fertile.

Figure 14.7 Hardy-Weinberg equilibrium. To prove the validity of the Hardy-Weinberg rule stated above, let's follow the course of two alleles, A and a, through succeeding generations.

For all members of the population, the gene locus must be occupied by either A or a. In mathematical terms, the frequencies of A and a must add up to 1. For example, if A occupies half of all the gene loci and a occupies the other half, then $0.5 + 0.5 = 1$. If A occupies ninety percent of all the gene loci, then a must occupy the remaining ten percent ($0.9 + 0.1 = 1$). No matter what the proportions of alleles A and a,

$$p + q = 1$$

You know that during sexual reproduction of diploid organisms, the two alleles at a gene locus segregate and end up in separate gametes. Thus p is also the proportion of gametes carrying the A allele, and q the proportion carrying the a allele. To find the expected frequencies of the three possible genotypes (AA, Aa, and aa) in the next generation, we can construct a Punnett square:

	$p\,\textcircled{A}$	$q\,\textcircled{a}$
$p\,\textcircled{A}$	AA (p^2)	Aa (pq)
$q\,\textcircled{a}$	Aa (pq)	aa (q^2)

Because the frequency of genotypes must add up to 1,

$$p^2 + 2pq + q^2 = 1$$

To see how these calculations can be applied, let's follow the allele frequencies for a population of 1,000 diploid individuals made up of the following genotypes:

$$450 \ AA$$
$$500 \ Aa$$
$$\underline{50 \ aa}$$
$$1{,}000 \text{ individuals (or 2,000 alleles)}$$

Theoretically, of every 1,000 gametes produced, the frequency of A will be $450 + \frac{1}{2}(500) = 700$, or $p = 0.7$. The frequency of a will be $\frac{1}{2}(500) + 50 = 300$, or $q = 0.3$. Notice that

$$p + q = 0.7 + 0.3 = 1$$

After one round of random mating, the frequencies of the three genotypes possible in the next generation will be as follows:

$$AA = p^2 \quad = 0.7 \times 0.7 = 0.49$$
$$Aa = 2pq = 2 \times 0.7 \times 0.3 = 0.42$$
$$aa = q^2 \quad = 0.3 \times 0.3 = 0.09$$

and

$$p^2 + 2pq + q^2 = 0.49 + 0.42 + 0.09 = 1$$

Notice that the allele frequencies have not changed:

$$A = \frac{2 \times 490 + 420}{2{,}000 \text{ alleles}} = \frac{1{,}400}{2{,}000} = 0.7 = p$$

$$a = \frac{2 \times 90 + 420}{2{,}000 \text{ alleles}} = \frac{600}{2{,}000} = 0.3 = q$$

The genotypic frequencies have changed initially. However, given that the distribution of genotypes fits the equation $p^2 + 2pq + q^2$, the genotypic frequencies will be stable over succeeding generations. You can verify this by calculating the most probable allele frequencies for gametes produced by the second-generation individuals:

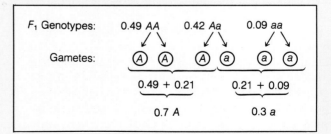

which is back where we started from. Because the allele frequencies are exactly the same as those of the original gametes, they will yield the same frequencies of genotypes as in the second generation.

You could go on with the calculations until you ran out of paper, or patience. As long as the population adheres to the conditions stated in the boxed inset for the Hardy-Weinberg principle, you would end up with the same results. When the frequencies of different alleles and different genotypes remain constant through successive generations, the population is in Hardy-Weinberg equilibrium: it is not evolving.

Of these events, only mutation *creates* new forms of a gene (page 155). The rest shuffle *existing* genes into new combinations in new individuals. But what a shuffle! By one estimate, more than 10^{600} combinations of genes are possible in human gametes, yet there are not even 10^{10} humans alive today. So unless you have an identical twin, it is extremely unlikely that another person with your exact genetic makeup has ever lived—or ever will.

A population is a group of individuals occupying a given area and belonging to the same species.

Far more genetic variation is possible than can ever be expressed in the individuals of any population.

Genetic Equilibrium. Think about a population of snails or some other kind of sexually reproducing, diploid animal. Each snail has two chromosomes of each type, so it has two genes (not one) for every locus. Recall that the different molecular forms of a gene are also called *alleles* (page 105). Now, suppose one snail allele codes for yellow shells and another, for white. A count of the yellow shells and white shells suggests that more snails probably carry the yellow allele. Through genetic analysis, you might be able to determine the actual **allele frequencies:** the abundance of each kind of allele in the whole population.

When a population is evolving, its allele frequencies are changing through successive generations. A formula called the Hardy-Weinberg principle is used to establish a reference point for measuring the rates of evolutionary change (Figure 14.7). At this point, **genetic equilibrium,** allele frequencies for a trait remain stable through the generations; there is zero evolution. Genetic equilibrium would be possible under these conditions:

1. No mutations are occurring.

2. The population is very, very large.

3. The population is isolated from other populations of the same species.

4. All members survive, mate, and reproduce (no selection).

5. Mating is completely random.

Remember, though, genetic equilibrium is a reference point only; a natural population never meets all these conditions. Mutation changes its allele frequencies over long spans of time. Three other forces—genetic drift, gene flow, and natural selection—also can drive the population away from genetic equilibrium, even in the space of a few generations. Changes in allele frequencies brought about by mutation, genetic drift, gene flow, and natural selection are called **microevolution,** and they are directly observable aspects of evolutionary change. These microevolutionary processes, listed in Table 14.1, will now be described.

Mutation

A **mutation** is a heritable change in the kind, structure, sequence, or number of component parts of DNA. Mutations are the original source of phenotypic variation. They are random in terms of which gene will be affected and, most importantly, *whether they will be harmful or beneficial to the individual.*

Most mutations are harmful. Why? The effect of a mutated gene depends on how its protein product is received in the environment and on how it meshes with the coordinated workings of the whole body. No matter what the species, each individual inherits a combination of many genes that are already fine-tuned by selection processes for a given range of operating conditions. The protein product of a mutant gene is likely to be less functional, not more so, under those conditions.

Yet, every so often, a mutation must have provided its bearer with advantages. Suppose the enzyme product of a mutant gene operates only at a higher temperature. Suppose an individual carrying the mutant gene lives in an environment that is becoming warmer, with temperatures higher than members of its species normally encounter. Here the mutation might be a boon, not a bust.

Also, some mutations have neutral effects. Sometimes the gene product is not essential for survival (as when it specifies white or yellow shells). Sometimes the functional allele on the other chromosome masks its effects (page 118). Finally, some mutated genes might be closely linked to highly adaptive genes in the chro-

Table 14.1	Major Microevolutionary Processes
Mutation	*A heritable change in the kind, structure, sequence, or number of component parts of DNA*
Genetic drift	*A random fluctuation in allele frequencies over time, due to chance occurrences alone*
Gene flow	*A change in allele frequencies as individuals leave a population or as individuals enter it*
Natural selection	*A change or stabilization of allele frequencies due to the differential reproduction of variant members of a population*

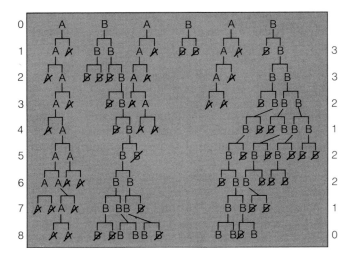

Figure 14.8 Illustration of genetic drift. Each individual who carries allele *A* is represented in this diagram as A; each individual who carries allele *B* is represented as B. They all produce two identical offspring in each generation. Half the offspring die before reproductive age (population size remains constant) but which ones die is random. The relative abundance of the two types of individuals fluctuates until *A* no longer is represented in the population and *B* becomes fixed in the population.

Numbers to the left of the diagram signify the sequential generations; numbers to the right signify the number of individuals bearing the allele *A* who survived in each generation. (Which individuals were to die in this example was determined by tossing a coin.)

mosome. It could hitchhike with adaptive genes through meiosis—and perhaps turn out to be beneficial under new environmental tests.

Genetic Drift

Sometimes allele frequencies change randomly through the generations because of chance events alone. This process is called **genetic drift**. Genetic drift is most rapid when population size is small.

Suppose only some members of a small population carry an allele *A*, and suppose none of them reproduces over several generations. Whether you call it chance or bad luck, some do not mate and others fall ill or accidentally die early. The *A* allele would disappear from the population (Figure 14.8). It would do so even if it were the "best" (most advantageous) allele for a given trait. Such a run of bad luck would be less likely in a large population, where hundreds might carry the *A* allele.

During extreme cases of genetic drift (founder effects and bottlenecks), a population originates or is rebuilt

from very few individuals. With the *founder effect*, a few individuals leave a population and manage to establish a new one. Simply by chance, allele frequencies for many traits may not be the same as they were in the original population—and a different range of phenotypes will become available for agents of selection. The founder effect probably played a role in establishing the variation among finch species on the Galápagos Islands.

With *bottlenecks*, disease, starvation, or some other stressful situation nearly wipes out a large population, and even though the population recovers, its relative abundances of alleles have been altered at random. Cheetahs may have gone through a severe bottleneck, for there is unusual allelic uniformity among them. Their extreme uniformity concerns many conservationists. No matter how large a genetically uniform population may be, it is extremely vulnerable to extinction.

Gene Flow

Allele frequencies change when individuals leave a population (immigration) or new individuals enter it (emigration). This microevolutionary process is called **gene flow**. The physical flow of alleles tends to minimize genetic variation between populations. For example, genes flow rather freely between baboon troops in Africa. Male baboons periodically wander off or are driven out of one troop, and sometimes they join up with another troop some distance away. When they mate, they have a homogenizing effect on the genetic character of the population.

Natural Selection

Natural selection is the most important microevolutionary process, along with genetic drift. Darwin gained insight into this process when he correlated his observations of inheritance with certain features of populations and the environment. Here we outline the key correlations:

First Correlation
Observation: Natural populations have enormous reproductive potential—yet, over time, population size tends to remain much the same. (To give a simple example, a single sea star can release 2,500,000 eggs every year, but the oceans obviously do not fill with sea stars.)

Observation: In natural environments, food supplies and other resources do not increase explosively; in fact, they remain much the same over time.

Observation: The limited availability of resources puts limits on population growth. (There is only so much water, nutrients, and growing space for a plant population; only so many plants to feed a bighorn sheep population; only so many sheep to feed a mountain lion population; and so on.)

Inference: When a population outstrips the supplies of necessary resources, there must be competition among its members for the resources that *are* available. Because of this competition, only some of the individuals who were born will themselves reproduce.

Second Correlation

Observation: The members of a natural population show great variation in their traits, and much of the variation is passed on through generations (it has a heritable, or genetic, basis).

Inference: Some heritable traits must be more adaptive than others; they give the individual a competitive edge in surviving and reproducing in the environment.

Inference: Over the generations, then, bearers of adaptive traits leave more offspring. There is differential reproduction (in each new generation, individuals with the adaptive traits make up more of the population's reproductive base).

Inference: Over the generations, the character of the population changes—it evolves—as some traits increase in frequency and others decrease or disappear.

With these correlations, we are now ready for a more precise definition of this major microevolutionary process. **Natural selection** is the *differential survival and reproduction* of individuals of a population—individuals that differ in one or more traits.

Evolution by natural selection is now a well-documented phenomenon in nature. Hundreds of studies of plants, animals, and microorganisms have demonstrated the validity of Darwin's correlations and inferences. Here we will consider a few examples.

EVIDENCE OF NATURAL SELECTION

Natural selection may have stabilizing, directional, or disruptive effects on the range of phenotypes in a population. Figure 14.9 is a generalized overview of these effects.

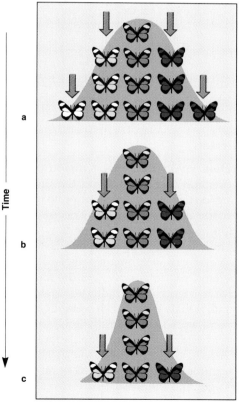

Stabilizing Selection

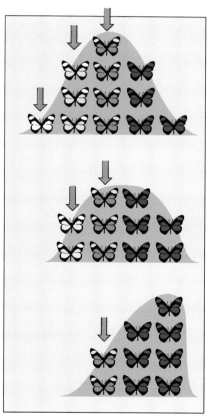

Directional Selection

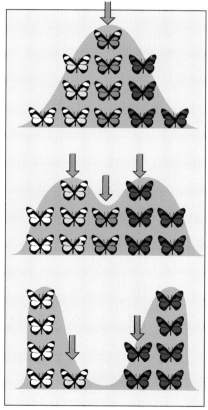

Disruptive Selection

Figure 14.9 Three modes of natural selection, using the phenotypic variation of a small population of butterflies as the example. The bell-shaped curve represents the range of continuous variation in wing color. The most common forms (powder blue) occur between extreme forms of the trait (white at one end of the curve, deep purple at the other). Arrows signify which forms are being selected against over time (charts **a** down through charts **c**).

Sickle-Cell Anemia—Lesser of Two Evils?

Often two or more forms of a trait persist over the generations at a frequency too high to be maintained by mutation alone. This type of stabilizing selection is called "balanced polymorphism." It results when heterozygotes for a trait have a competitive edge over homozygotes. In some environmental contexts, one allele is favored—but in other contexts, its nonidentical partner is favored. *The survival value of any allele must be weighed in the context of the environment in which it is being expressed.*

The sickle-cell trait in humans is an example. Sickle-cell anemia, a genetic disorder, is caused by a mutant allele that codes for a defective form of hemoglobin (page 125). In tropical and subtropical regions of Africa, the mutant allele (HbS) is maintained at a high frequency relative to the normal one (HbA). Homozygotes often die in their early teens or early twenties, but heterozygotes make up nearly a third of the population.

Where the sickle-cell trait is most prevalent, so also is malaria. The parasite that causes malaria is transmitted to humans by a type of mosquito that evolved in the tropics and subtropics.

Individuals who do not carry the mutant allele have a far greater chance of surviving and reproducing than individuals who do—provided they don't get malaria. HbS/HbA heterozygotes happen to have greater resistance to malaria and are more likely to survive severe infections! So the persistence of the harmful sickle-cell trait becomes a matter of relative evils.

In Central Africa, malaria has been an agent of selection for less than 2,000 years. In tropical and subtropical regions of the Middle East and Asia, it has been around for much longer. Even though the sickle-cell trait occurs at high frequencies in these regions also, the symptoms are not as pronounced as they are in Africa. Apparently, mutations in other alleles at other loci have reduced the serious effects of the HbS allele.

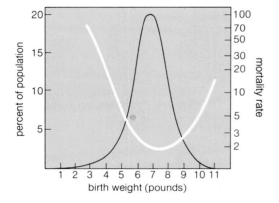

Figure 14.10 Weight distribution for 13,730 newborns (black curve) and their survival rate (white curve). Here, stabilizing selection favors a birth weight between 7½ and 8 pounds and operates against newborns whose weight is significantly higher or lower than this.

Stabilizing Selection

Stabilizing selection favors the most common phenotypes in the population (Figure 14.9 and the *Commentary*). By allowing certain allele frequencies to persist over time, this mode of selection tends to counter the effects of mutation, genetic drift, and gene flow.

For example, humans weigh about 7 pounds at birth, on the average. Newborns weighing significantly more or less than this tend not to survive. As the shape of the survival curve in Figure 14.10 indicates, stabilizing selection favors individuals with a birth weight between 7½ and 8 pounds and works against individuals at either extreme. Studies in widely different populations yield much the same results.

As another example, nautiloids evolved hundreds of millions of years ago, and the fossil record reveals very little change from the ancestral forms to the single remaining species. A similar example of stabilizing selection can be drawn from the horsetails of the plant kingdom (Figure 14.11). In such extreme cases of stabilizing selection, a single body plan has been conserved in successive species through time.

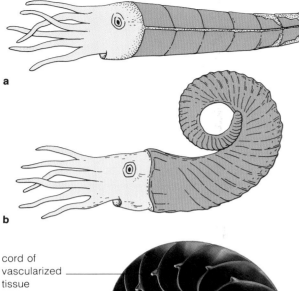

a

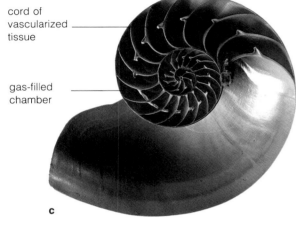

b

cord of
vascularized
tissue

gas-filled
chamber

c

d

Figure 14.11 (**a, b**) Diagrams of a type of tentacled mollusk called nautiloids. The one in (**a**) flourished about 450 million years ago; the one in (**b**), about 400 million years ago. The chambered shell of the only existing representative of the nautiloid lineage, shown in (**c**) and (**d**), is sliced in half lengthwise to show the intricate partitions. These partitions have persisted over time and may be viewed as an extreme case of stabilizing selection.

(**e, f**) The sphenophyte lineage extends back more than 380 million years. Its only existing members, the horsetails (of the genus *Equisetum*), retain traits of their ancient relatives.

e Fossil

f Living representatives of lineage

Directional Selection

Directional selection shifts allele frequencies in a consistent direction in response to a new environment or a directional change in the old one. By this process, forms at one end of the range of phenotypic variation become more common than intermediate forms (Figure 14.9b).

Peppered Moths. Directional selection has been documented in populations of the peppered moth (*Biston betularia*) in England. At one time, a speckled light-gray form was common and a dark-gray form was extremely rare. Between 1848 and 1898, the dark form increased in frequency. For example, all but two percent of the moth population near one city had become dark gray. (Much later, researchers identified two genes that code for wing and body color. So the trait definitely has a heritable basis and is subject to natural selection.)

Peppered moths are active at night but rest during the day on tree trunks—where they are vulnerable to bird predators. Before the industrial revolution, tree trunks were cloaked with light-gray speckled lichens. Light-gray speckled moths resting on the lichens were

a

b

Figure 14.12 An example of variation that is subject to directional selection in changing environments. (**a**) The light- and dark-colored forms of the peppered moth are resting on a lichen-covered tree trunk. (**b**) This is how they appear on a soot-covered tree trunk, which was darkened by industrial air pollution.

camouflaged from the birds, but dark moths stood out like sore thumbs (Figure 14.12). Then soot and other pollutants from factories started killing the lichens and darkening the trunks. Now the rare dark moths blended with the changing background and the light moths did not. Dark moths survived and reproduced more, so allele frequencies in the populations changed.

About a hundred different moth species underwent directional selection in response to pollution in industrial regions throughout Great Britain. Strict pollution controls went into effect in 1952, however, so lichens have made a comeback and tree trunks are largely free of soot. As you might predict, the frequency of dark moths is now declining.

Insecticide Resistance. Increasing resistance to insecticides is another example of directional selection. An initial insecticide application kills most of the targeted insects. However, a few insects may survive (some aspect of their structure, physiology, or behavior allows them to resist the pesticide's chemical effects). If the resistance has a genetic basis, there will be more resistant individuals in the next generation.

Often, farmers have resorted to heavier and more frequent insecticide applications to counter the increased numbers of resistant forms. The insecticides have become selective agents; they actually favor the resistant forms. Crop damage from insects is now greater than it was before the widespread use of insecticides.

Disruptive Selection

Disruptive selection favors forms at both ends of the phenotypic range and operates against intermediate forms (Figure 14.9). An example comes from a study of a small population of Galápagos finches during a severe dry season, when seeds and a few wood-boring insects were the only types of food available. The finches varied in beak size and shape. At one end of the phenotypic range were birds with longer beaks, which could be used to open cactus fruits and expose the seeds. At the other end were birds with the deepest, widest beaks. They were able to crack hard cactus seeds on the ground and strip away tree bark to get at insects. In this case, more birds with extreme beak variations survived than birds in between (Figure 14.4).

Sexual Selection

Natural selection has been a major force behind differences in appearance between males and females of a species. Among birds and mammals, for example, the males are often larger, more varied in color and patterning, and more aggressive than females (Figure 14.13). Usually the females are the agents of selection; they

a

b

c

Figure 14.13 Some results of sexual selection. (**a**) Male bird of paradise (*Paradisaea raggiana*) engaged in a spectacular courtship display that has caught the eye (and, perhaps, sexual interest) of the female. Males of this species compete fiercely for females, which serve as selective agents. (**b**) Northern fur seals mate only on small islets and rocky beaches of North America. Males large enough to command the rocks enjoy mating privileges with about ten to twenty females. Males not able to secure a territory do not mate at all, so they contribute nothing to the allele pool of the next generation. As an outcome of this sexual selection, males weigh about twice as much as females. (**c**) Brilliantly hued male sugarbird with a subdued-hued female.

choose their mates and so are directly associated with reproductive success. **Sexual selection** is based on any trait that gives the individual a competitive edge in mating and producing offspring. We will return to this topic in Chapter 38.

SPECIATION

Defining the Species

As we have seen, changes in allele frequencies are not irreversible. Selection may be reversed if the environment changes in an appropriate direction. Even if an allele is lost from a population (through genetic drift or selection), it may be reinstated by mutation or gene flow. So what prevents a population from cycling its alleles, over and over again?

An irreversible step, acting like a ratchet, causes genetically isolated populations to branch in different evolutionary directions. This step, "evolution's ratchet," is part of the process by which species originate—a process called **speciation.** Let's now consider this process, using the following definition as our point of reference for all sexually reproducing organisms:

A species is one or more populations of individuals that can interbreed under natural conditions and produce fertile offspring, and that are reproductively isolated from other populations.

Divergence and Isolation

Divergence. No matter how diverse the individuals of a population become, they remain members of the same species as long as they continue to interbreed successfully and share a common pool of alleles. But sometimes barriers arise between parts of the population and create local breeding units. Then, two or more pools of alleles may exist where there had been only one.

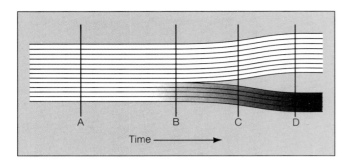

Figure 14.14 Divergence leading to speciation. Because evolution is gradual here, we cannot identify the moment in time that one species becomes two. Each vertical line represents a different population. In A, there is only one species. In D, there are two. In B and C, the divergence has begun but is far from complete.

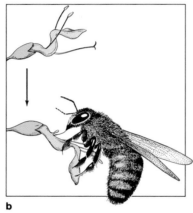

a   b

Figure 14.15 Mechanical isolation between two species of sage (*Salvia mellifera* and *S. apiana*). The first species (**a**) has a small floral landing platform for its small or medium-size pollinators. The second species (**b**) has a large landing platform and long stamens, which extend some distance away from the nectary. Even though small bees can land on this larger platform, they can do so without brushing against the pollen-bearing stamens. It takes larger pollinators to do this. Hence the small pollinators of *S. mellifera* are mostly incapable of spreading pollen to flowers of *S. apiana;* and the large pollinator of *S. apiana* cannot land on and cross-pollinate *S. mellifera*. The plants and their pollinators are all drawn to the same scale.

Suppose, over time, there is no gene flow between the local units, and selection, genetic drift, and mutation operate independently in each one. The units are now reproductively isolated populations; over time, differences in allele frequencies will accumulate between them. This process is called **divergence**. When divergence becomes great enough, members of the two populations will not be able to interbreed successfully under natural conditions; the populations have become separate species (Figure 14.14).

Geographic Barriers. The populations of most species are not strung out continuously, with one merging into the others. Most often they are isolated geographically to some extent, with gene flow being more of an intermittent trickle than a steady stream. But sometimes barriers form and shut off even the trickles. This can happen rapidly, as when a major earthquake changed the course of the Mississippi River in the 1800s and isolated populations of insects that couldn't swim or fly.

Geographic isolation also happens very slowly, as when gradual shifts in climate cause the breakup of formerly continuous environments. Millions of years ago, for example, extensive forests in Africa gave way to grasslands with isolated stands of trees as a result of long-term shifts in rainfall.

Once geographic isolation is absolute, genetic drift or selection may lead to divergence, then to speciation.

Reproductive Isolating Mechanisms. Any aspect of structure, function, or behavior that prevents interbreeding is a **reproductive isolating mechanism.** Such mechanisms prevent movement of alleles (gene flow) between populations, as the following examples will illustrate.

Two populations may be *mechanically isolated,* as when differences in reproductive organs prevent interbreeding. For example, two species of sage plants differ in the size and arrangement of their floral parts. Both reproduce with the aid of pollinators, which carry pollen from plant to plant. But each sage plant species has a "landing platform" that only one type of pollinator fits on, so cross-fertilization between the plant species is inhibited (Figure 14.15).

Two populations also may remain isolated through *hybrid inviability and infertility.* Suppose fertilization occurs between the gametes of different species. The resulting embryo usually dies because of physical or chemical incompatibilities with the mother. Hybrids that do live are commonly weak and their survival chances are not good. In a few cases, hybrid offspring are vigorous but sterile. A cross between a female horse and a male donkey produces a mule—a hybrid that is fully functional *except* in its capacity to reproduce.

In addition, *behavioral isolation* works to prevent interbreeding among related species in the same territory. Complex courtship rituals often precede mating. The song, head bobbing, wing spreading, and dancing by a male bird of one species may stimulate a female of his species—but the female of a related species would not recognize his behavior as a sexual overture.

Polyploidy. Speciation also may occur by way of **polyploidy.** In this condition, offspring end up with three or more of each type of chromosome characteristic

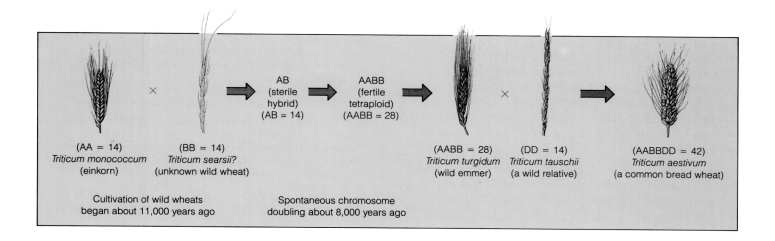

Figure 14.16 Proposed speciation in wheat by way of polyploidy and hybridizations.

of the parental stock. Polyploidy can arise through complete nondisjunction at meiosis, followed by the formation of diploid instead of haploid gametes (page 96). It also can arise when germ cells duplicate their DNA but fail to divide.

About half of all flowering plants are polyploid species. The ability of many flowering plants to reproduce asexually or to undergo self-fertilization probably accounts for the widespread occurrence of this condition. (Because the polyploid individual need not wait for a sexual partner with the same chromosome number, speciation is instantaneous. That individual can give rise to a whole population just like itself.) Polyploidy is less common in animals, probably because it upsets the balance between autosomes and sex chromosomes—a balance that is crucial for animal development and reproduction.

Speciation also occurs when polyploidy is followed by successful hybridization. Most hybrids between two species are sterile because they have different numbers or types of chromosomes. (This usually prevents homologous pairing at meiosis.) But if polyploidy happens to occur in the hybrid's germ cells, then the "extra" set of chromosomes can pair with the original ones at meiosis, and viable gametes can form. Wheat is an example of a polyploid hybrid (Figure 14.16).

Wheat grains dating from 11,000 B.C. have been found in the Near East. Several species of wild diploid wheat still grow there. They have 14 chromosomes (two sets of 7, designated AA). Also growing in the region is a wild grass with 14 chromosomes, designated BB. They differ from the A chromosomes, judging from their failure to pair with them at meiosis. One tetraploid wheat species has 28 chromosomes; analysis during meiosis shows that they are AABB. The A chromosomes pair with As, and the B chromosomes pair with Bs. A hexaploid wheat has 42 chromosomes (six sets of seven). Its chromosomes are AABBDD, the last set (DD) coming from *Triticum taushchii*, another wild grass.

SUMMARY

1. Individuals of a population show great variation in their traits, corresponding to differences in the relative abundances of alleles for most gene locations on chromosomes.

2. Microevolution means changes in allele frequencies as a result of mutation, genetic drift, gene flow, and natural selection. Mutation is the only source of new alleles; the other processes simply shuffle existing alleles into, through, or out of the population.

3. Mutations arise randomly and infrequently, and most are harmful. Individuals inherit a combination of many genes already fine-tuned for a given range of operating conditions, and a mutation usually disrupts interdependencies among gene products. But if environmental conditions change, the product of a mutated gene may prove beneficial.

4. Genetic drift is the chance increase or decrease in the relative abundances of alleles in a population. Gene flow is a change in allele frequencies following the movement of individuals from one population to another and their subsequent reproduction.

5. Natural selection is the differential survival and reproduction of variant members of a population. Agents of selection can have stabilizing, directional, or disruptive effects on the range of phenotypes.

6. A species is one or more populations whose members are able to interbreed under natural conditions and produce fertile offspring, and are reproductively isolated from other populations. (This definition applies only to sexually reproducing species.)

7. Sometimes two populations of a species (or local breeding units within a population) become reproductively isolated from one another, and divergence occurs. Divergence is a buildup of differences in allele frequencies between reproductively isolated populations. When divergence is great enough, successful interbreeding is no longer possible under natural conditions; speciation has occurred.

8. Geographic barriers as well as mechanical, physiological, or behavioral barriers can keep different populations reproductively isolated from each other.

Review Questions

1. What is the Hardy-Weinberg baseline against which changes in allele frequencies may be measured? *178, 179*

2. Changes in allele frequencies may be brought about by mutation, genetic drift, gene flow, and selection pressure. Define these occurrences, then describe the way each one can send allele frequencies out of equilibrium. *179*

3. What implications might the effect of genetic drift hold for an earlier concept of "survival of the fittest"? *180*

4. Define stabilizing, directional, and disruptive forms of selection and give a brief example of each. *181–184*

5. Give two examples of reproductive isolating mechanisms, and outline what they accomplish. *186*

Self-Quiz   *(Answers in Appendix IV)*

1. A _____ is a group of individuals occupying a given area and belonging to the same species.

2. Variation in the individuals of a population corresponds to _____ in the relative abundances of alleles for most gene locations along the chromosomes.

3. Allele frequencies change as a result of _____.
   a. mutation   d. natural selection
   b. genetic drift   e. all of the above
   c. gene flow

4. The only source of new alleles is _____.
   a. mutation   d. natural selection
   b. genetic drift   e. all of the above
   c. gene flow

5. Existing alleles are shuffled into, through, or out of populations by _____.
   a. mutation   d. natural selection
   b. genetic drift   e. b, c, and d only
   c. gene flow

6. Which of the following statements about mutation is *not* true?
   a. they arise randomly   c. they arise rather infrequently
   b. most are harmful   d. most are beneficial

7. Speciation is _____.
   a. the extinction of a distinct population
   b. the accumulation of environmental factors that cause geographic isolation
   c. the process whereby different species originate
   d. a means of altering gene frequencies in a population

8. The sickle-cell trait evolved in tropical and subtropical regions of Asia, the Middle East, and Africa, then it appeared in the United States population with the influx of individuals who were forcibly brought over from Africa prior to the Civil War. In terms of microevolution, this is an example of _____.
   a. mutation   c. gene flow
   b. genetic drift   d. natural selection

9. Divergence is defined by _____.
   a. accumulated differences in allele frequencies between reproductively isolated populations
   b. a loss in the ability of two populations to interbreed
   c. eventual occurrence of speciation
   d. all of the above

10. Match the evolution concepts appropriately.
   _____ gene flow
   _____ sexually reproducing species
   _____ natural selection
   _____ mutation
   _____ genetic drift

   a. the sole source of new alleles
   b. chance increase or decrease in relative allele frequencies of a population
   c. one or more populations whose members interbreed and produce fertile offspring
   d. change in allele frequencies in a population due to immigration, emigration
   e. differential survival and reproduction of variant members of population

Selected Key Terms

allele frequency *179*
divergence *185–186*
evolution *174*
fossil *174*
gene flow *180*
genetic drift *180*
genetic equilibrium *179*
microevolution *172*
mutation *179*
natural selection *172, 181*
phenotype *177*
polyploidy *186*
population *177*
reproductive isolating mechanism *186*
speciation *185*

Readings

Cook, L., G. Mani, and M. Varley. 1986. "Postindustrial Melanism in the Peppered Moth." *Science* 231:611–613.

Dobzhansky, T. 1973. "Nothing in Biology Makes Sense Except in the Light of Evolution." *The American Biology Teacher* 35(3):125–129. Personal views of a leading geneticist, who argues that the principle of evolution does not clash with religious faith.

Futuyma, D. 1987. *Evolutionary Biology.* Second edition. Sunderland, Massachusetts: Sinauer.

Moorhead, A. 1969. *Darwin and the Beagle.* New York: Harper & Row.

MACROEVOLUTION **15**

KEY CONCEPTS

1. Speciation occurs through microevolution—changes in allele frequencies within reproductively isolated populations of a species. In contrast, macroevolution refers to the large-scale patterns, trends, and rates of change in *groups of species* over time. The patterns include the retention of certain traits and the modification of others within major groups.

2. Two trends—mass extinction and adaptive radiation—have changed the course of biological evolution many times. The pace of evolution has been gradual in some lineages, but in others it has proceeded with bursts of speciation events followed by long periods of little change.

3. Evidence of macroevolution comes from the fossil record in combination with the geologic record, radioactive dating methods, comparative morphology, and comparative biochemistry.

4. The evolution of life has been linked, from the time of origin to the present, to the physical and chemical evolution of the earth.

The history of life spans nearly 4 billion years. It is a story of how species originated, persisted or became extinct, and stayed put or radiated into new environments. **Macroevolution** refers to the large-scale patterns, trends, and rates of change among groups of species. Those groupings (the so-called higher taxa), include all the different genera, families, phyla or divisions, and so on up to the most inclusive groups of species, the kingdoms.

Given our knowledge of natural selection and other microevolutionary processes, we can interpret the evidence of large-scale patterns and trends by starting with a simple fact: *Evolution proceeds by modifications of organisms that already exist.* In other words, "new" species don't appear out of thin air. They emerge as mutation, natural selection, and genetic drift change the allele frequencies in reproductively isolated populations of an existing species. Given this fact, there must be underlying threads of relatedness connecting all species since the origin of life. Attempts to reconstruct the past and to classify life's diversity take this relatedness into account.

Because evolution proceeds by modification of already existing species, there is a continuity of relationship among all species that have ever appeared on earth.

EVIDENCE OF MACROEVOLUTION

The Fossil Record

"Fossil" comes from a Latin word for something that has been "dug up." As generally used, however, a **fossil** is recognizable evidence of an organism that lived long ago. Most of what we know of life's history comes from fossilized skeletons, shells, leaves, seeds, and tracks. To be preserved as fossils, body parts or impressions must be buried before they decompose, and the rock layers in which they are entombed must not be disturbed much over time.

The fossil record is uneven in terms of which organisms and environments are represented. For example, fossils of mollusks and other animals with hard shells

a

b

Figure 15.1 Two pieces of the macroevolutionary puzzle. (**a**) Fossilized leaves from *Archaeopteris*, which was probably on the evolutionary road leading to gymnosperms and flowering plants. Some *Archaeopteris* trees were more than twenty meters tall. (**b**) Complete fossil of a female ichthyosaur, about 200 million years old, that died while giving birth. The extinction of these dolphin-sized marine reptiles coincided with the rise of modern sharks, which were more efficient at feeding and swimming.

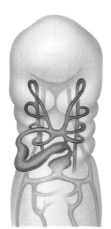

ADULT SHARK   HUMAN EMBRYO
(3 millimeters long)

Figure 15.2 Evidence of evolution from comparative embryology. Aortic arches, a two-chambered heart, and other structures develop in a fish embryo. These structures, which persist in adult fishes, function in respiration in aquatic environments. Fishlike respiratory structures also develop in amphibian embryos. Used during aquatic larval stages, some disappear and others become altered for adult life on land.

Similarly, the early embryos of reptiles, birds, and mammals also develop fishlike structures. Shown here, the aortic arches *(red)*, a two-chambered heart *(orange)*, and certain veins *(blue)* in an early human embryo. Notice how these structures resemble those of an adult shark. They provide evidence of the retention of basic developmental processes during the evolution of different groups of vertebrates.

or skeletons are abundant, but fossils of jellyfishes and other soft-bodied animals are not. Seafloors, floodplains, swamps, and natural traps such as caves and tar pits favor fossilization; rapidly eroding hills do not.

The fossil record also is uneven in terms of the quality of specimens. Most fossils are broken, incomplete, and often crushed or deformed. A few are spectacularly complete and well preserved (Figure 15.1). They tell us a great deal about the structure, function, and behavior of extinct organisms.

Some parts of the record are gone forever. Other parts are yet to be discovered. We have fossils of about 250,000 species, dating mostly from the past 600 million years. Although that is not much, compared to the estimated 4 to 10 million existing species, it's better than nothing at all. Besides, the fossil record for some lineages *is* nearly complete. This is especially true of shallow seas in which sediments were steadily deposited.

The completeness of the fossil record varies as a function of the kinds of organisms represented, where they lived, and the stability of the region since the time of fossilization.

Dating Fossils

When geologists of the mid-1800s started mapping the layers of sedimentary rock beneath the earth's surface, they discovered that certain fossils occur in the same kinds of layers over vast areas, even on different continents. Sedimentary rocks form by a gradual "rain" of erosion products and skeletons of tiny marine organisms, so it was logical to assume that the deepest layers formed first and the ones closest to the surface formed last. By Darwin's time, the layered pattern of fossils was used as the basis of a time scale for biological events—but it was only a relative time scale with no dates.

Earth history simply was divided into four great eras, based on abrupt transitions in the fossil record. The four divisions, the *Proterozoic, Paleozoic, Mesozoic,* and the "modern" era, the *Cenozoic,* are still in use. Research based on radioactive dating methods finally revealed the actual time spans between their boundaries. These methods are based on the unvarying rates of decay of certain radioisotopes, as described in the *Commentary* on page 19. It turned out that the Proterozoic began 2.5 billion years ago. Further research revealed that the origin and early evolution of the earth itself occurred during an even more ancient era, now called the *Archean,* which began more than *4.5 billion years ago.*

We are accustomed to thinking of a year as a long time. Few of us will last a century. Yet 1 million years is 10,000 centuries end to end. And that is still only $\frac{1}{3,500}$ of the history of life. Think about that for a minute.

Comparative Morphology

Evidence of macroevolution comes from detailed comparisons of body form and structural patterns in major groups. This work is called **comparative morphology.**

Stages of Development. At certain stages of development, the embryos of different organisms within a major group bear striking resemblance to one another, and often the similarities indicate evolutionary relationships. For example, certain structures such as those shown in Figure 15.2 appear in all vertebrate embryos at corresponding stages of their development. Such structural similarities are one of the reasons why fishes, amphibians, reptiles, birds, and mammals are said to belong to the same subphylum (Vertebrata) despite the large variation among adult forms.

Why do similar embryonic stages persist in different vertebrates? Cells have specific positions and roles in the tiny embryo, and interactions among their hormones and other gene products are vital for normal growth and development. Most gene mutations affecting embryonic cells usually have devastating effects on later stages (Chapter 32). Thus, nearly all gene mutations affecting the embryonic stages were probably selected against during vertebrate history.

Then why are there such variations in form among adult vertebrates? At least some variations probably arose through mutations in **regulatory genes** that control the *rates of growth* of different body parts.

Consider how the skull bones of newborn chimpanzees and humans are proportionally alike (Figure 15.3). From infancy onward, changes in skull proportions are dramatic for chimps but almost nonexistent for humans. We cannot account for the difference by assuming chimps and humans have different sets of genes; their genes are very nearly identical. But suppose certain regulatory genes underwent mutation in an isolated population of a species that was ancestral to both chimps and humans. Suppose the mutations blocked the rapid growth required for proportional changes in the skull bones. That microevolutionary event could have put the early ancestors of humans on a separate evolutionary road.

Homologous Structures. Different species may have homologous structures, meaning they resemble one another in body form or patterning due to descent from a common ancestor. In **morphological divergence,** one

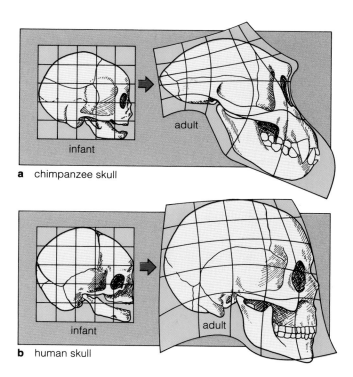

a chimpanzee skull

b human skull

Figure 15.3 Comparison of the proportional changes in a chimpanzee skull and a human skull, both of which are remarkably similar in infants. (Imagine that these representations of the infant skulls are paintings on a blue rubber sheet divided into a grid. Stretching the sheet deforms the squares of the grid, and the resulting changes between the two adult skulls reflect differences in the growth patterns in each square of the grid.)

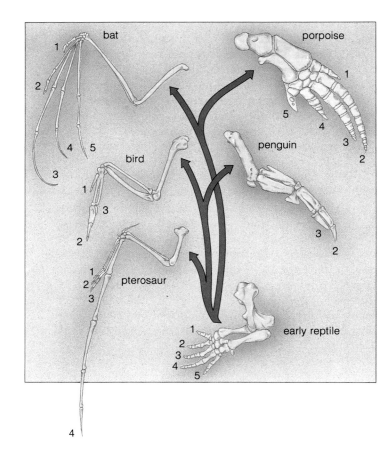

Figure 15.4 Morphological divergence in the vertebrate forelimb, starting with the generalized form of ancestral early reptiles. Diverse forms have emerged even while many similarities in the number and position of bones have been preserved.

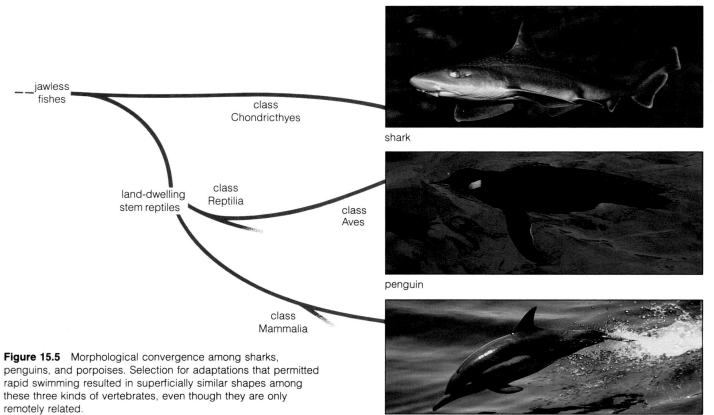

Figure 15.5 Morphological convergence among sharks, penguins, and porpoises. Selection for adaptations that permitted rapid swimming resulted in superficially similar shapes among these three kinds of vertebrates, even though they are only remotely related.

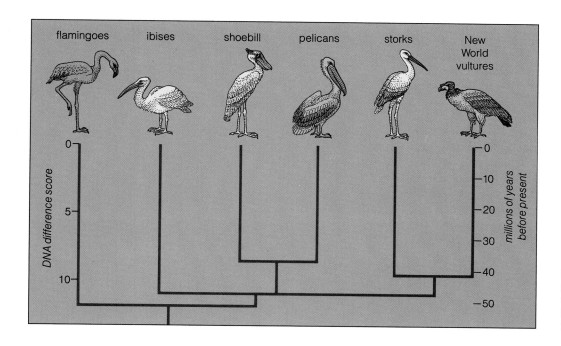

Figure 15.6 Relationships among some New World vultures, storks, and other birds, as indicated by DNA hybridization techniques.

or more homologous structures have departed in appearance, function, or both from the ancestral form.

The vertebrate forelimb is an example of divergence (Figure 15.4). Most land-dwelling vertebrates have a five-toed limb. Such a limb was the point of departure for the evolution of wings in three groups of vertebrates (pterosaurs, birds, and bats). All three types of wings have the same component parts, as do the flippers of porpoises. Similarly, the five-toed limb was the forerunner of the long, one-toed limbs of modern horses, the stubby limbs of moles and other burrowing mammals, and the pillarlike limbs of elephants.

Analogous Structures. In **morphological convergence,** dissimilar and only distantly related species adopt a similar way of life, and body parts that take on similar functions end up resembling one another. Similar body parts used for similar functions in evolutionarily remote lineages are said to be "analogous" to one another.

The forelimbs of sharks, penguins, and porpoises provide examples of convergence (Figure 15.5). Fishes ancestral to sharks used fins as body stabilizers, and sharks do the same. Birds ancestral to penguins could fly, but penguins use their "wings" as fins. Four-legged reptiles ancestral to porpoises returned to the seas, and their descendants use their "front legs" as fins. All three types of fins have become similar in proportion, position, and function. The overall shape of penguins and porpoises also converged on that of the shark. All three vertebrates have a streamlined shape that reduces drag during rapid swimming.

Comparative Biochemistry

The genetic makeup of different species contains information about evolutionary relationships. For example, a significant number of mutations occurred in different lineages since life began some 3.8 billion years ago.

Some apparently were **neutral mutations** (or nearly so), meaning that the mutated allele had no more measurable effect on survival and reproduction than did other alleles for the trait. This means neutral mutations cannot be selected against, and so they have tended to accumulate in the DNA. It seems neutral mutations have accumulated at regular rates in different classes of proteins. They can be used as the units of time in a "molecular clock" for dating the divergence of two species from a common ancestor.

Neutral mutations are being identified by methods of the sort described next. In general, they have helped answer some thorny questions about evolutionary relationships.

DNA Hybridization Studies. Nucleotide sequences of DNA also reveal information about evolutionary relationships. In DNA hybridization studies, segments of DNA from two species are converted to single-stranded form, then allowed to recombine into "hybrid" double helices. The extent to which they recombine is a direct measure of the similarity between them. Figure 15.6 shows a chart of evolutionary relationships as suggested by DNA hybridization studies.

Immunological Comparisons. When foreign proteins enter the body's internal environment, they trigger the production of antibodies. An antibody is a chemical weapon; it binds with specific foreign agents and marks them for destruction by the body's immune system (page 394).

When a protein from one animal species (say, a type of frog) is injected into a rabbit, the rabbit produces antibodies to it. The antibodies can be recovered and mixed with the same protein from a *different* frog species. When mixed, the protein molecules and antibody molecules bind as large aggregates. The same test can be repeated with the same protein from many other frog species. The closer the proteins are in structure, the stronger the reactions to them will be. A strong reaction means the frog is closely related to the original species tested; a weaker reaction means it is more distantly related.

MACROEVOLUTION AND EARTH HISTORY

The fossil record, comparative morphology, and comparative biochemistry provide evidence of the evolution of life on a grand scale. When this diverse evidence is carefully pieced together, an important fact emerges: *The evolution of life has been linked, from its origin to the present, to the physical and chemical evolution of the earth.*

Origin of Life

Early Earth and Its Atmosphere. Billions of years ago, explosions of dying stars ripped through our galaxy and left behind a dense cloud of dust and gas that extended trillions of kilometers in space. As the cloud cooled, countless bits of matter gravitated toward one another. By 4.6 billion years ago, the cloud had flattened into a slowly rotating disk. Our sun was born at the extremely dense, hot center of the disk. There, thermonuclear reactions began that would feed themselves for the next 10 billion years.

Farther out from the center, the earth was forming along with other planets. By 3.8 billion years ago, it was hurtling through space as a thin-crusted inferno (Figure 15.7). An early atmosphere developed when gases that had formed in the earth's molten interior (or were trapped below the crust) were forced to the outside. This first atmosphere had little free oxygen—and its near-absence favored the origin of life. (Free oxygen disrupts the structure of amino acids, nucleotides, and other biological molecules exposed to it.)

Early on, water vapor must have been released from the breakdown of rocks during volcanic eruptions, but it would have evaporated in the intense heat blanketing the crust. In time the crust cooled, and rains started stripping mineral salts from the parched rocks. Salt-laden waters collected in depressions in the crust and formed the early seas.

If the earth had formed as a smaller planet, it would not have had enough gravitational mass to hold onto an atmosphere. If it had settled into an orbit closer to the sun, its surface would have remained too hot for rain to form. If the orbit had been more distant, its surface would have become too cold, locking up any water as ice. *Because of its size and distance from the sun, the early earth retained liquid water on its surface.* Without liquid water, life as we know it never would have originated.

The first living cells emerged between 4 billion and 3.8 billion years ago. There is no record of the event. Most rocks from that period melted, solidified, and remelted many times because of large-scale movements in the earth's mantle and crust. Some rocks were buried so deeply that heat and compression altered any clues they might have held.

Even so, we can gain insight into the manner in which life originated by considering four questions:

1. What were physical and chemical conditions like at the time of origin?

Figure 15.7 Representation of the primordial earth, about 4 billion years ago. Within another 500 million years, living cells would be present on the surface. (During its formation, the moon presumably was closer to the earth. Here it looms on the horizon.)

2. Based on known physical, chemical, and evolutionary principles, could life have originated spontaneously under those conditions?

3. Can we postulate a sequence of events by which the first living systems developed?

4. Can we devise experiments to test whether that sequence could indeed have taken place?

Synthesis of Biological Molecules. All the components found in biological molecules were present on the early earth. We know this from rock samples of meteorites, the earth's moon, and Mars that were formed 4.5–4.6 billion years ago. We also know that lightning, hot volcanic ash, even shock waves have enough energy to drive the synthesis of biological molecules under abiotic conditions. (*Abiotic* means not involving or produced by organisms.) For example, Stanley Miller mixed hydrogen, methane, ammonia, and water in a reaction chamber (Figure 15.8). He recirculated the mixture and kept bombarding it with a spark discharge to simulate lightning. Within one week, many amino acids and other organic compounds had formed.

Self-Replicating Systems. During the 300 million years after the first rains began, organic compounds accumulated in the shallow waters of the earth. The first self-replicating systems emerged in this organic "soup."

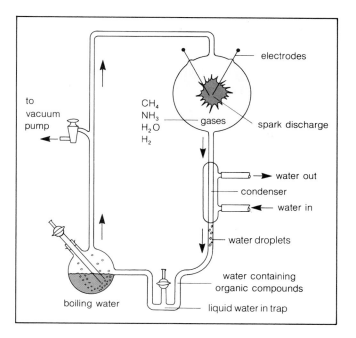

Figure 15.8 Stanley Miller's apparatus used in studying the synthesis of organic compounds under conditions believed to have been present on the early earth. (The condenser cools the circulating steam and causes water to condense into droplets.)

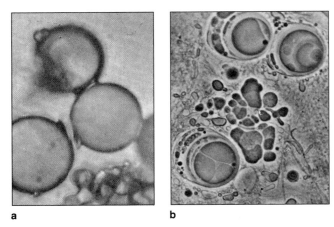

Figure 15.9 Microscopic spheres of (**a**) proteins and (**b**) lipids that self-assembled under abiotic conditions.

By *systems* we mean the following interacting molecules:

DNA — *protein-building instructions*

RNA — *transcribers and translators of DNA*

proteins — *including enzymes required for DNA replication and protein synthesis*

Clay crystals at the bottom of tidal flats and estuaries may have been the first templates (structural patterns) for protein synthesis. Iron, zinc, and other metal ions often are embedded in such crystals. (If you have ever baked meat in a clay or metal pot, you know that bits of protein stick to both kinds of heated surfaces.) Amino acids could have joined up randomly in the water, but clay templates would have allowed longer chains to form in less time with less chance of breaking apart.

Suppose amino acids became linked together on a clay template. Suppose the resulting protein's shape and chemical behavior allowed it to function as a weak enzyme in hastening the linkage between other amino acids. Clay templates promoting such linkages would have had selective advantage over other templates. There would have been *chemical competition* for available amino acids—and selection for the first large molecules characteristic of living cells.

Now suppose a clay template (or amino acids stuck to it) also attracted nucleotides. Intriguingly, most existing enzymes are assisted by small molecules—coenzymes—some of which are structurally identical to RNA nucleotides. Simple systems of enzymes, coenzymes, and RNA have been created in the laboratory. Experiments with these systems suggest RNA could have replaced clay as a template for protein synthesis.

At this point we have no idea how DNA entered the picture. We do know this: The reactions leading to the first self-replicating systems could not have been purely random. *Physical and chemical conditions of the early earth made some reactions more probable than others.* For example,

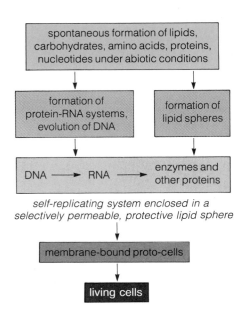

Figure 15.10 One proposed sequence of events that led to the first self-replicating systems and, later, to the first living cells.

amino acids can exist in two forms, like mirror-images of two hands. Almost all living things have "lefthanded" amino acids. In contrast, meteorites contain disorganized arrays of lefthanded and righthanded forms. When the disorganized arrays are exposed to clay crystals, the clay attracts only lefthanded amino acids!

The First Plasma Membranes. The plasma membrane of all living cells is a lipid bilayer with its associated proteins. Metabolism cannot proceed—life cannot exist—without this vital barrier against random chemical fluctuations in the outside world. How did plasma membranes originate?

Most likely, the first living cells were little more than membrane-bound sacs around some nucleic acids that served as templates for protein synthesis. Experiments show that such sacs can form spontaneously. For example, Sidney Fox heated amino acids under dry conditions to form protein chains, which he placed in hot water. After the chains were allowed to cool, they self-assembled into small, stable spheres (Figure 15.9). The spheres were selectively permeable to some substances. They also tended to pick up lipids from the water and so formed a lipid-protein film at their surface. In other experiments, lipids alone self-assembled into small, water-filled sacs that displayed many properties characteristic of cell membranes.

And so self-replicating systems and the membranes required to protect them from the environment could have arisen through spontaneous chemical events. Figure 15.10 summarizes a sequence of events that could have led to the first cells.

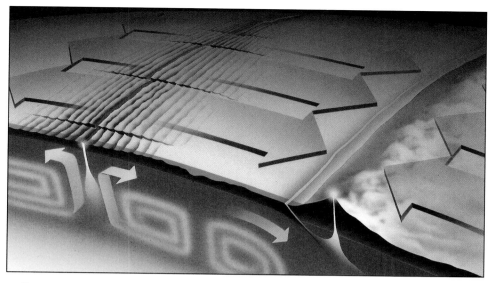

a

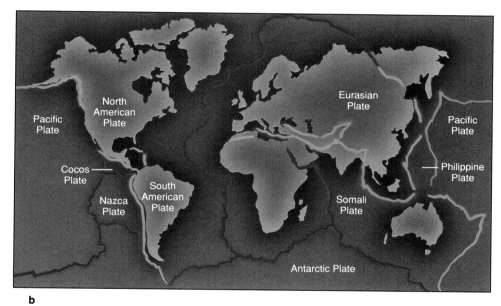

b

Figure 15.11 Plate tectonics, the arrangement of the earth's crust in rigid plates that split apart, move about, and collide with one another.

(**a**) The seafloor is slowly spreading from mid-oceanic ridges, measurably displacing the continents away from those ridges. Hot, molten material deep inside the earth slowly wells up and spreads out laterally beneath the crust, much as hot air rises from a stove and spreads out at the ceiling. Oceanic ridges are places where the material has ruptured the crust. As cooled material moves away from the ridges, it acts like a conveyor belt, carrying older continental crust with it.

Similarly, heat builds up beneath the continents, causing deep rifts that eventually cause land masses to split apart. We see this happening today at the Great Rift Valley in eastern Africa.

The major plates of the earth's crust are shown in (**b**). As the plates push against a continental margin, they are often thrust beneath it. The thrusting causes the crumpling and upheavals that have created most major mountain ranges.

Drifting Continents and Changing Seas

The first living cells did not emerge in a wonderfully stable environment, and things haven't settled down much for their single-celled and multicelled descendants. Even today, living things contend with more than volcanic eruptions, earthquakes, and other local upheavals. They contend with long-term consequences of drifting and colliding continents, with newly forming and disappearing ocean basins, and with bombardments from outer space!

Plate tectonics refers to the arrangement of the earth's outer layer (lithosphere) in slablike plates, all in motion and floating on a hot, plastic layer of the underlying mantle. Driven by a process called seafloor spreading

(Figure 15.11), the plates move a few centimeters a year, on the average. The positions of the continents and oceans change as a result of these movements.

For much of earth history, continents have collided to form supercontinents, which later split open at deep rifts that eventually became new ocean basins. During the Paleozoic, for example, an early continent (Gondwana) drifted southward from the tropics, across the south polar region, then northward. Later, Gondwana and other land masses crunched together to form Pangea. This single world continent extended from pole to pole, and an immense ocean spanned the rest of the globe (Figure 15.12). Pangea started breaking up during the Mesozoic, and the drifting and collisions among its fragments continues today.

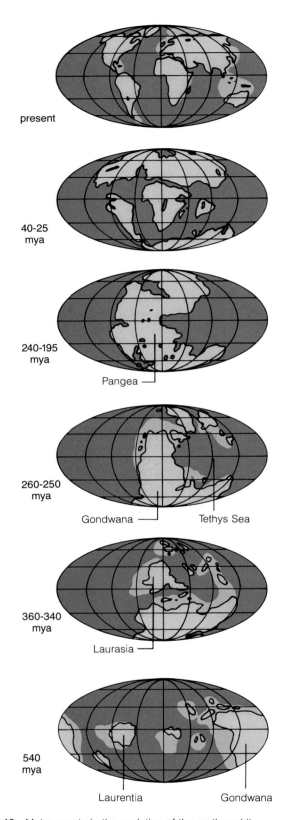

present

40-25 mya

240-195 mya

Pangea

260-250 mya

Gondwana — Tethys Sea

360-340 mya

Laurasia

540 mya

Laurentia   Gondwana

Figure 15.12 Major events in the evolution of the earth and its organisms. The time spans of the different eras are not to scale; if they were, the chart would run off the page. Think of the spans as minutes on a clock, with life originating at midnight (near the dawn of the Proterozoic). The Paleozoic would not even begin until 10:04 a.m. The Mesozoic would begin at 11:09 a.m., the Cenozoic at 11:47 a.m., and the Recent epoch of the Cenozoic would begin during the last 0.1 *second* before noon.

Era	Period	Epoch	Millions of Years Ago (mya)
CENOZOIC	Quaternary	Recent	0.01-
		Pleistocene	1.65
	Tertiary	Pliocene	5
		Miocene	25
		Oligocene	38
		Eocene	54
		Paleocene	65
MESOZOIC	Cretaceous	Late	
			100
		Early	
			138
	Jurassic		
			205
	Triassic		
			240
PALEOZOIC	Permian		
			290
	Carboniferous		
			360
	Devonian		
			410
	Silurian		
			435
	Ordovician		
			505
	Cambrian		
			550
PROTEROZOIC			
			2,500
ARCHEAN			

Range of Global Diversity (marine and terrestrial)	Times of Major Geologic and Biological Events

1.65 mya to present. Major glaciations. Modern humans emerge and begin what may be greatest **mass extinction** of all time on land, starting with Ice Age hunters.

65-1.65 mya. Unprecedented mountain building as continents rupture, drift, collide. Major climatic shifts; vast grasslands emerge. Major **radiations** of flowering plants, insects, birds, mammals. Origins of earliest human forms.

65 mya. Asteroid impact? **Mass extinction** of all dinosaurs and many marine organisms.

135-65 mya. Pangea breakup continues, broad inland seas form. Major **radiations** of marine invertebrates, fishes, insects, dinosaurs. Origin of angiosperms (flowering plants).

181-135 mya. Pangea breakup begins. Rich marine communities. Major **radiations** of dinosaurs.

205 mya. Asteroid impact? Mass extinction of many organisms in seas, some on land; dinosaurs, mammals survive.

240-205 mya. Recovery, **radiations** of marine invertebrates, fishes, dinosaurs. Gymnosperms the dominant land plants. Origin of mammals.

240 mya. Mass extinction. Nearly all species in seas and on land perish.

280-240 mya. Pangea, worldwide ocean form; shallow seas squeezed out. Major **radiations** of reptiles, gymnosperms.

360-280 mya. Tethys Sea forms. Recurring glaciations. Major **radiations** of insects, amphibians. Spore-bearing plants dominate; gymnosperms present. Origin of reptiles.

370 mya. Mass extinction of many marine invertebrates, most fishes.

435-360 mya. Laurasia forms, Gondwana moves north. Vast swamplands, early vascular plants. **Radiations** of fishes continue. Origin of amphibians.

435 mya. Glaciations as Gondwana crosses South Pole. **Mass extinction** of many marine organisms.

500-435 mya. Gondwana moves south. Major **radiations** of marine invertebrates, early fishes.

550-500 mya. Land masses dispersed near equator. Simple marine communities. Origin of animals with hard parts.

700-550 mya. Supercontinent Laurentia breaks up; widespread glaciations.

2,500-570 mya. Oxygen present in atmosphere. Origin of aerobic metabolism. Origin of protistans, algae, fungi, animals.

3,800-2,500 mya. Origin of photosynthetic bacteria.

4,600-3,800 mya. Formation of earth's crust, early atmosphere, oceans. Chemical evolution leading to origin of life (anaerobic bacteria).

4,600 mya. Origin of earth.

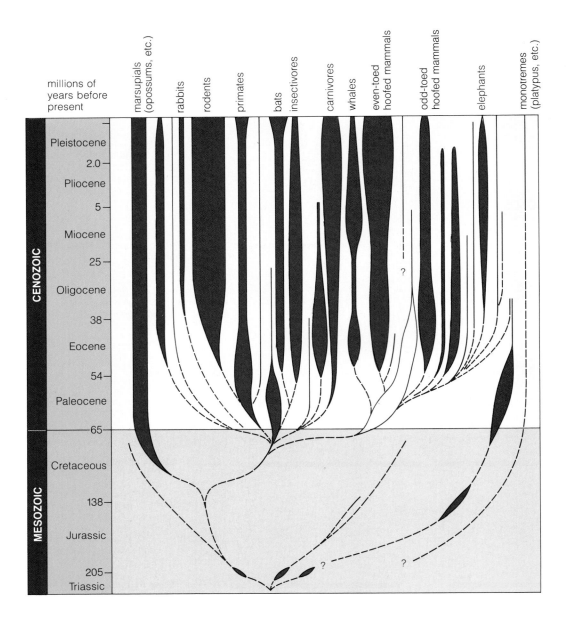

Figure 15.13 The great adaptive radiation of mammals in the first 10–12 million years of the Cenozoic Era. This radiation is thought to have resulted from the invasion of adaptive zones vacated by dinosaurs at the close of the Mesozoic.

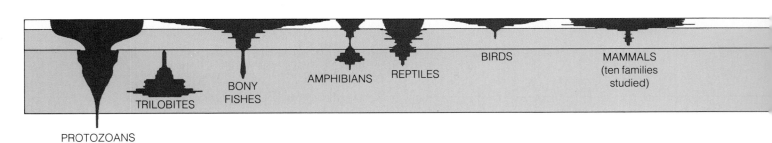

Figure 15.14 A limited representation of the evolutionary histories of organisms. Notice the diverse patterns of radiation and extinction for the sample shown here. Notice also the spectacular current success of the insects. (The yellow band represents the Cenozoic to the present; green, the Mesozoic; and blue, the Paleozoic. The widths of the lineages, all shown in rust, represent the approximate numbers of *families* in the case of plants and animals.)

Such changes in land masses, shorelines, and oceans had profound effects on the evolution of life. When land masses were widely dispersed, populations were isolated and speciation was favored, especially in warm coastal waters. When land masses collided, shorelines disappeared and volcanic activity was intense. Habitats were lost and species diversity tended to decline. Also, with each major shift in land masses, warm or cold ocean currents changed directions and climates changed drastically. Immense glaciers formed many times and tied up great volumes of ocean water, so shallow seas disappeared. When the glaciers melted, continental margins were flooded. Once again there were repercussions for the evolution of life on land and in the seas.

As if these planetary insults were not enough, asteroids or some other huge extraterrestrial objects have repeatedly struck the earth's surface. Some of the impacts almost certainly changed the course of biological evolution.

Extinctions and Adaptive Radiations

Figure 15.12 is an overview of two trends that have dominated the history of life. There have been recurring *mass extinctions,* followed by periods of recovery and *adaptive radiations.*

Some number of species within a lineage inevitably disappear as local conditions change. The rather steady rate of their disappearance over time is called "background extinction." In contrast, a **mass extinction** is an abrupt rise in extinction rates above the background level. It is a catastrophic, global event in which not just one species but major groups of species are wiped out simultaneously.

Major groups tend to survive episodes of mass extinction when their members are widely dispersed in different regions. The ones hit hardest generally are adapted for specialized ways of life in tropical regions. Most importantly, luck is the rule of the game during mass extinctions. For example, what *had* been the most adaptive traits may make no difference whatsoever if an asteroid the size of Vermont hits the earth.

In an **adaptive radiation,** a lineage fills the environment with new species through bursts of microevolutionary activity. Adaptive radiations are common features during the first few millions of years following a mass extinction. For example, most orders of living mammals appeared during the 12 million years after the mass extinction that marks the boundary between the Mesozoic and Cenozoic eras (Figure 15.13).

Radiations take place when there are unfilled *adaptive zones.* These are most easily defined as ways of life, such as "burrowing in the seafloor" or "catching insects in the air at night." Before a lineage can successfully occupy an adaptive zone, it must have physical, ecological, and evolutionary access to it.

Physical access may mean colonizing a new habitat or geographic region for the species. (For example, the ancestors of the Galápagos finches probably reached their new habitat because they were able to ride the winds that carried them away from the mainland.) Ecological access can occur if an adaptive zone is unoccupied or if the invading species can outcompete the resident species.

Evolutionary access may result when a key innovation develops in a species. A *key innovation* is a modification in structure or function that permits an individual to exploit the environment in an improved or novel way. For example, the modification of forelimbs into wings opened new adaptive zones to the ancestors of birds and bats.

Keep in mind that Figure 15.12 is only a generalized picture of life's history. It shows the five greatest mass extinctions—but there were others in between. It shows the shrinking and expanding range of species diversity for all the major groups combined. But within that overall pattern, each major group has its own distinctive history of extinctions and radiations (Figure 15.14).

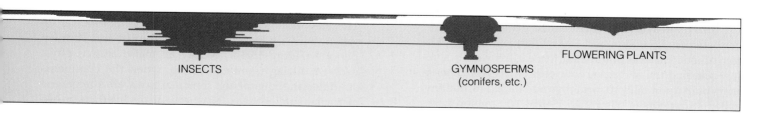

INSECTS · GYMNOSPERMS (conifers, etc.) · FLOWERING PLANTS

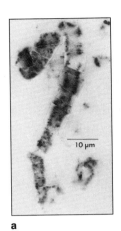

a **b**

Figure 15.15 (a) One of the oldest known fossils, dated at 3.5 billion years old. It is a filament formed of walled cells. (b) Stromatolites in Western Australia, which formed between 2,000 and 1,000 years ago in shallow seawater. Calcium deposits preserved their structure. They are identical to stromatolites more than 3 billion years old.

With these qualifications in mind, let's discuss briefly the macroevolutionary trends that dominated the five eras of geologic time—the Archean, Proterozoic, Paleozoic, Mesozoic, and Cenozoic eras. This discussion will set the stage for the next unit of the book, which describes the history and current range of diversity for all five kingdoms.

The Archean and Proterozoic Eras

Until about 3.7 billion years ago, the earth's crust was highly unstable. Even after somewhat rigid land masses had formed, they were fringed with active volcanoes. Yet rocks 3.5 billion years old contain fossils of well-developed prokaryotic cells that probably lived in tidal mud flats. Those cells resembled the simple bacteria found today in mud flats, bogs, and pond mud where oxygen is absent. Fermentation, the most common anaerobic pathway, was probably the first ATP-generating pathway to evolve. Before the close of the Archean era, the first photosynthetic bacteria had evolved.

Between 2.5 billion and 700 million years ago, photosynthetic bacteria dominated the earth's warm, shallow seas and the first eukaryotic cells emerged. Bacterial mats built upon mats to form mound-shaped formations called stromatolites (Figure 15.15). By 600 million years ago, the mat-formers had almost disappeared. Animals had appeared; instead of making their own food, they began grazing on the photosynthesizers.

The accumulation of oxygen in the Proterozoic atmosphere had two irreversible effects on the course of evolution. First, the abundance of free oxygen prevented further synthesis of organic compounds outside of living cells; the chemical origin of life would be a one-time event. Second, the oxygen-rich atmosphere was an adaptive zone of global dimensions. In some bacterial lineages, metabolic machinery became modified in ways that permitted aerobic respiration—a key innovation that foreshadowed the evolution of multicelled plants, fungi, and animals.

The earliest plants were green algae, known from fossils over 1.2 billion years old. Rock formations in Australia, dated at 900 million years, have yielded fossils of at least fifty-six species of green and red algae, plant spores, and fungi. About 750 million years ago, at the Proterozoic–Paleozoic boundary, animals with shells, spines, and armor plates appeared abruptly. A rock formation 530 million years old provides a remarkable window on the early evolution of animals (Figure 15.16).

The Paleozoic Era

The Paleozoic is subdivided into the Cambrian, Ordovician, Silurian, Devonian, Carboniferous, and Permian periods (Figure 15.12). Nearly all major animal phyla evolved during Cambrian times, when the continents straddled the equator and their margins were submerged in warm, shallow seas. Most organisms lived on or just beneath the seafloor, feeding on organic debris or suspended particles. Trilobites were a dominant group (Figure 15.16). They were nearly wiped out during the late Cambrian, apparently as a result of an abrupt and widespread mixing of warm, oxygen-rich waters of shallow seas with cold, oxygen-poor waters from the

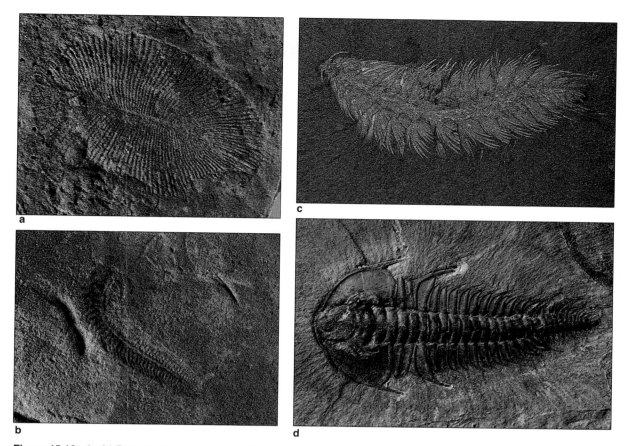

Figure 15.16 (a, b) From Australia's Ediacara Hills, fossils of two Cambrian animals, about 600 million years old. (c) From the Burgess Shale of British Columbia, a remarkably well-preserved fossil of a marine worm of Cambrian times. Notice the bundles of fine bristles that may have functioned in locomotion. (d) One of the earliest Cambrian trilobites.

deep oceans. Trilobite populations adapted to cold water at the fringes of continents survived—but those in the shallow seas were devastated.

During the Ordovician, Gondwana drifted southward and seas flooded more of the land. Major adaptive radiations were favored as vast, shallow marine environments opened up. Reef organisms flourished, and fast-swimming predators (nautiloids) dominated the evolutionary stage. By the late Ordovician, Gondwana straddled the South Pole. Immense glaciers formed on its surface and the shallow seas were drained (Figure 15.17). This first ice age may have triggered the first global mass extinction; reef life everywhere collapsed.

Gondwana drifted northward during the Silurian and on into the Devonian. Reef organisms recovered, and there was a major radiation of predatory fishes equipped with armor plates and massive jaws. The invasion of land was beginning rather inconspicuously. Small-stalked plants became established along the muddy margins of the land (Figure 15.18). In time, lobe-finned

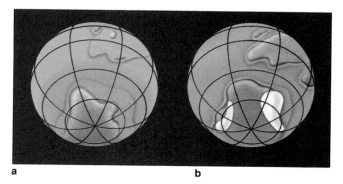

Figure 15.17 Plate tectonics and global changes in climate. Gondwana drifted southward during the Ordovician and vast, warm, shallow seas inundated most of the land. This was a time of major adaptive radiations in marine environments. Late in the Ordovician, Gondwana straddled the pole (a), and within 5 million years, huge ice sheets had formed over it, locking up enough ocean water to cause a drop in sea level. (b) Colder waters moved into the tropics and large land areas were drained, with dire consequences for many families of marine organisms.

a   b

Figure 15.18 Fossils of plants that pioneered the invasion of land during Silurian times. (**a**) Stems of the oldest known plant, *Cooksonia*, were less than 7 centimeters tall. (**b**) Fossils of a type of Devonian plant *(Psilophyton)* that may have been the earliest ancestors of seed-bearing plants.

fishes ancestral to amphibians started to move onto land. Like many existing fishes of stagnant waters, they were able to breathe air.

Another global mass extinction occurred at the Devonian–Carboniferous boundary, when sea levels swung wildly. Nearshore species were washed out to sea and became entombed in deep ocean sediments, and huge boulders were ripped from the coasts. The cause of the Devonian extinction is not known.

Major adaptive radiations of plants and insects occurred during the Carboniferous, when land masses were submerged and drained no less than fifty times. Immense swamp forests with large, scaly-barked trees became established along the coasts. Over time the seas moved in and buried the forests in sediments and debris; they became reestablished as the seas moved out—and they were submerged again. (The organic mess left behind has since compacted into the extensive coal deposits of Britain and of the Appalachians of North America.)

Insects, amphibians, and early reptiles flourished in the vast swamp forests of the Permian era. As the Permian drew to a close, nearly all species on land and in the seas perished in the greatest of all mass extinctions. This happened at a time when all land masses collided to form a vast continent, Pangea, surrounded by a single world ocean. Changes in the distribution of water, land area, and land elevation had catastropic effects on world temperature and climate. To the north, arid lowlands and humid uplands emerged. To the south, near the pole, glaciers advanced over the land. The small reptilian ancestors of dinosaurs, birds, snakes, and lizards met the challenge of drastic swings in climate and survived.

The Mesozoic Era

The Mesozoic lasted about 175 million years; it is subdivided into Triassic, Jurassic, and Cretaceous periods. The most ecologically specialized lineages had been hit hardest during the Permian extinction, particularly in tropical reefs. Recovery occurred through adaptive radiations of new lineages. This time the increase in diversity did not level off, as it did in Paleozoic times; it continued to increase to the present day.

The major radiation of reptiles had ended abruptly before the Mesozoic, but divergences in a few lineages now gave rise to mammals and some wonderful reptilian monsters, including the dinosaurs.

Many marine organisms perished during a mass extinction of unknown causes at the Triassic–Jurassic boundary. On land, some dinosaurs and small mammals were among the survivors. During the late Jurassic and early Cretaceous, dinosaurs emerged as rulers of the evolutionary stage. Did they die out with a whimper or a bang at the close of the Mesozoic? This question is hotly debated (see *Commentary*).

Dinosaurs so dominate our thinking about the Mesozoic that we often overlook one of the most spectacular events of all. Early in the Cretaceous, flowering plants emerged and underwent a major radiation in the evolutionarily small space of about 10 million years. In most habitats they overwhelmed the already declining gymnosperms. The radiation continues today, fueled by a speciation rate exceeding that of any other plant lineage.

The Cenozoic Era

The world's land masses were undergoing major reorganization at the dawn of the Cenozoic. Lava poured through faults and fissures in the earth's crust. Coastlines fractured; intense volcanic activity and uplifting produced mountains along the margins of massive rifts and plate boundaries. The Alps, Andes, Himalayas, and Cascade Range were born through these upheavals.

As the continents assumed their current configurations, major shifts in climate affected the further evolution of life. For example, vast, semiarid, cooler grasslands emerged—new adaptive zones into which plant-eating mammals and their predators radiated.

Today the continents are dispersed nearly to their maximum separation, and they intersect the tropics and

The Dinosaurs—A Tale of Global Impacts, Radiations, and Extinctions

By 200 million years ago, near the close of the Triassic, the first dinosaurs had evolved. They were not much larger than a wild turkey. Most had high metabolic rates—some researchers say they were warmblooded—and many ran swiftly on two legs. The Triassic dinosaurs were not the dominant lineage on land; center stage belonged to *Lystrosaurus* and other plant-eating, mammal-like reptiles that were too large to be bothered by most predators. Then *Lystrosaurus* ran out of luck and adaptive zones opened up for the dinosaurs, perhaps when an extraterrestrial object struck the earth. There is a crater about the size of Rhode Island in central Quebec (Figure a), and even if it is not *the* impact crater, it certainly tells us that huge impacts have occurred.

The blast wave and resulting firestorm, earthquakes, and possibly lava flows from an impact that size would have been extremely destructive. The atmospheric distribution of rocks and water vaporized during the impact could have caused global darkening, and would have been followed by months or years of acid rain. The animals that survived this time of extinction and later ones tended to be small and metabolically active—and less vulnerable to long-term swings in climate, whatever their cause.

The surviving dinosaurs underwent a major adaptive radiation. Over the next 140 million years, their descendants were the ruling reptiles. Some, including the "ultrasaurs," reached monstrous proportions; they weighed 70 tons and towered 50 feet above the ground (Figure b). Many dinosaurs perished during a mass extinction at the end of the Jurassic, then during a pulse of extinctions during the Cretaceous. Each time, some lineages recovered and new forms replaced them on the evolutionary stage. By the late Cretaceous, perhaps a hundred different species had emerged. Duckbilled dinosaurs appeared in forests and swamps. Tanklike

a Manicougan crater, Quebec—an example of earth's past encounters with extraterrestrial objects. To the right, an artist/astronomer's interpretation of what might have happened during the last few minutes of the Cretaceous.

Triceratops and other plant eaters flourished in more open regions; they were prey for the agile, fearsomely toothed *Tyrannosaurus rex*.

The final blow came at the Cretaceous–Tertiary (K–T) boundary. Perhaps long-term changes in climate or other planetary events had been winnowing out some of the dinosaurs for a few million years before then. Or perhaps a shower of comets bombarded the earth during that same time span. Whatever the cause, there was a pulse of extinctions culminating with the disappearance of all (or nearly all) of the remaining dinosaurs.

The impact associated with the K–T boundary probably occurred on land. A thin, worldwide layer of iridium-rich rock has been dated precisely at that boundary. Iridium is rare on the earth's surface but common in asteroids. A few asteroids still remain in earth-crossing orbits, and perhaps there was one more 63 million years ago than there is now. The thin, iridium-rich layer may have formed from dust settling after a collision.

So where is the crater? India is one possibility; immense flows of basaltic lava occurred precisely at the K–T boundary, and the crater may be masked by the basalt. A small farm town in Iowa is another possibilty. It is built over a huge crater that was later filled in with glacial sediments. The crater also has been dated at the K–T boundary.

Even more exotic forces may be at work in and around the solar system. There is some evidence that mass extinctions have occurred at regular intervals of 26–30 million years. Were dinosaurs and other major lineages the victims of a cosmic timetable? Consider that comets and other debris form a dense cloud at the fringes of our solar system. Does the orbit of an unknown star swing near that cloud to the extent that comets are hurled toward the earth?

Intriguingly, 2.3 millon years ago, a huge object from space hit the Pacific Ocean. About 2.3 million years ago, vast ice sheets started forming abruptly in the Northern Hemisphere. Long-term shifts in climate may have been ushering in this most recent ice age, but a global impact might have accelerated the process. (Water vaporized during the impact would have formed a cloud cover that prevented sunlight from reaching the earth's surface.) The formation of ice sheets following the global impact is one more bit of information that compels us to look skyward, also, in our attempts to piece together the evolutionary story of life.

b And how might the small, unobtrusive mammals of this time ever have ventured out from under the shrubbery if dinosaurs had not disappeared? Would *you* even be here today?

confine the polar regions. These conditions have favored a period of unparalleled species richness. The vast island chains of the tropical Pacific—remnants and omens of tectonic movements—may well be the richest ecosystems ever to appear on earth, and the vast tropical forests of South America, Africa, and Southeast Asia are probably not far behind.

This is the geologic stage for what may turn out to be the greatest mass extinction of all time. Beginning about 50,000 years ago, early humans started following the migrating herds of wild animals around the Northern Hemisphere. Within a few thousand years, major groups of large mammals, including the woolly mammoth, had disappeared. The pace of extinction has been accelerating ever since, as humans hunt animals for food, fur, feathers, or fun; as they destroy habitats to clear land for cattle or crops. The repercussions are global in scope; we will return to this topic in Chapter 37.

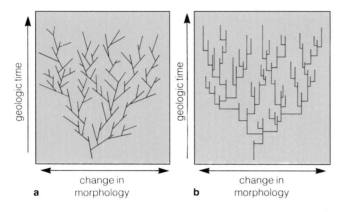

Figure 15.19 Two interpretations of how morphological diversity develops in a lineage. Each short vertical line represents a single species. In a *gradual* model (**a**), changes occur more or less steadily but at different rates among species. In a *punctuational* model (**b**), rapid morphological change is associated with speciation, as indicated by the horizontal lines that signify branch points, and the morphologies of established species remain constant through time.

ORGANIZING THE EVIDENCE—PHYLOGENETIC CLASSIFICATION

We have considered some evidence of the appearances and disappearances of diverse species on a changing geologic stage. Let's now consider their **phylogeny**—their evolutionary relationships, starting with the most ancestral species and including all the branches leading to all of its descendants.

Relationship by way of descent resembles a tree, with living species forming a huge canopy of leaves that are related by collections of twigs, branches, and major limbs leading to a common ancestral trunk. Environmental agents have shaped the tree, pruning and nourishing the canopy into its present form. But *species* are the only real parts of the tree of descent. The twigs, branches, and limbs are simply categories of relationship that *we* perceive among species. Each twig or branch in a tree simply represents a single line of descent, or **lineage.** The branch points are speciation events.

When the branchings are drawn with shallow angles, they mean rapid change; steeper angles mean gradual change from the ancestral species form. According to one model (gradualism), most morphological change within a species occurs very slowly, through genetic drift and other processes. According to another model (punctuation), most changes take place rapidly during speciation, with stabilizing selection working against major shifts in form once the species is established. Figure 15.19 shows how both models might be used to interpret the morphological diversity within a family of organisms.

Figure 15.19 shows branchings in the tree of descent for a family only. Just imagine how many branchings there must be in the whole forest of life, with its many millions of species! Cataloging all of it is a task in itself; assigning the entries to categories in a classification system is another.

Centuries ago, Carolus Linnaeus devised a two-name system for cataloging life's diversity. The first was the genus name (plural, genera); the second was the species name. Species that share certain physical characteristics were grouped into the same genus. Later, classification schemes were developed with increasingly inclusive categories, based on perceived similarities and differences among them. Similar genera were grouped into families, then similar families grouped into orders, and so on up to the category of kingdoms (Figure 15.20). The categories are still in use, although researchers now consider phylogeny as well as shared characteristics when they construct classification schemes.

This book uses a modified version of Robert Whittaker's phylogenetic system of classification. In this scheme, the major groups of species are assigned to one of five kingdoms (Figure 15.21 and Appendix II).

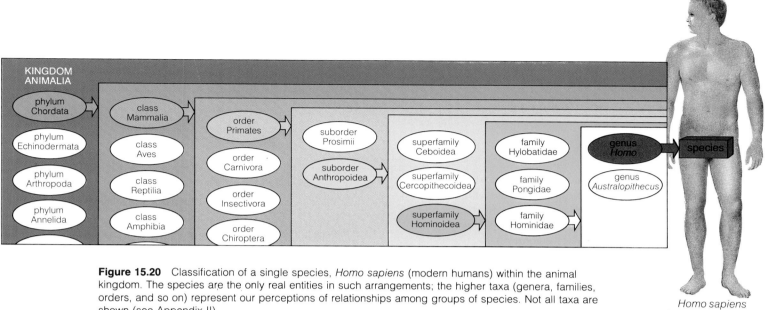

Figure 15.20 Classification of a single species, *Homo sapiens* (modern humans) within the animal kingdom. The species are the only real entities in such arrangements; the higher taxa (genera, families, orders, and so on) represent our perceptions of relationships among groups of species. Not all taxa are shown (see Appendix II).

Homo sapiens
(only living species
of this genus)

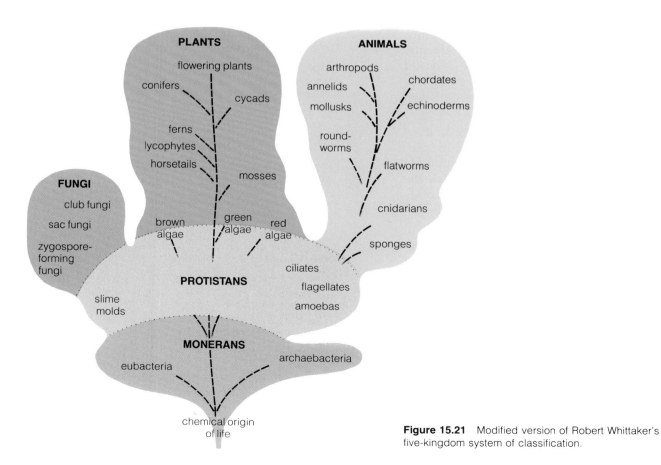

Figure 15.21 Modified version of Robert Whittaker's five-kingdom system of classification.

Regardless of its strengths, no classification system should be viewed as *the* system. As long as new observations are made, different people will interpret relationships among species in different ways. The system used here helps summarize knowledge about the world of life but it, too, is subject to modification as new evidence turns up.

SUMMARY

1. Because evolution proceeds by modifications of already existing species, there is a continuity of relationship among all species. The large-scale patterns, trends, and rates of change among groups of species (higher taxa) over time are called macroevolution.

2. Evidence of evolutionary relationships comes from the fossil record and earth history, radioactive dating methods, comparative morphology, and comparative biochemistry. The evidence reveals similarities and differences in form, function, behavior, and biochemistry. It can be used to construct evolutionary trees in which the branch points are speciation events and each branch, a lineage (a single line of descent).

3. Biologists arrange species into groupings that are meant to reflect how they are related by descent. Very closely related species are grouped into the same genus. Closely related genera are grouped into the same family, and so on up through orders, classes, and phyla (or divisions). Five kingdoms—the monerans, protistans, fungi, plants, and animals—are generally accepted as the most inclusive groupings.

4. The fossil record varies as a function of the kinds of organisms represented, where they lived, and the geologic stability of the region since the time of fossilization. Together with the geologic record, it shows that the evolution of life has been linked, since the time of its origin to the present, to the evolution of the earth.

5. Comparative morphology often reveals similarities in embryonic development stages that indicate evolutionary relationships. It shows evidence of divergences and convergences in body parts among certain major groups.

6. Comparative biochemistry is based on mutations that have accumulated in the DNA of different species. The mutations serve as a molecular clock for dating the time of divergence between two species from a common ancestor. DNA hybridization studies and immunological reactions reveal similarities and differences in genes and gene products.

7. Over geologic time, tectonic movements of the earth's crust have caused profound shifts in land masses, shorelines, and oceans. The course of biological evolution has been redirected many times by these events.

8. The two main macroevolutionary trends are mass extinction (a catastrophic, global event in which major groups of species perish abruptly) and adaptive radiation (bursts of speciation events by which a lineage fills the environment with new species).

Review Questions

1. What is the difference between microevolution and macroevolution? *189–190*

2. What factors influence the completeness of the fossil record? *190*

3. Explain the difference between: *191–193*
   a. homologous and analogous structures
   b. morphological divergence and convergence

4. How is an adaptive radiation defined, and what criteria must be met before a radiation can occur? *201*

5. Describe the chemical and physical characteristics of the earth 4 billion years ago. How do we know what it was like? *194–196*

6. Describe the experimental evidence for the spontaneous origin of large organic molecules, the self-assembly of proteins, and the formation of organic membranes and spheres, under conditions similar to those of the early earth. *195, 196*

7. How does continental drift occur, and in what ways does this process influence changes in biological communities? *197, 201*

8. When did plants, insects, and vertebrates invade the land? *203*

9. The Atlantic Ocean is widening, and the Pacific and Indian oceans are closing. Write a short essay on the possible biological consequences of the forthcoming formation of a second Pangea.

Self-Quiz   *(Answers in Appendix IV)*

1. Large-scale patterns, trends, and rates of changes in *groups of species* over time are called _____.

2. Two trends have changed the course of biological evolution repeatedly; they are _____ and _____.

3. The rate of evolution has been _____ in some lineages but in others it has been much more rapid with _____ of speciation events followed by lengthy periods with little evidence of change.

4. The fossil record of evolution correlates with evidence from _____.
   a. the geologic record
   b. radioactive dating
   c. comparative morphology
   d. comparative biochemistry
   e. all of the above

5. The branch points on an evolutionary tree represent _____.
   a. a lineage or single line of descent
   b. areas in time that were rich in fossils
   c. speciation events
   d. the points in time at which groups of genera arose

6. Biologists generally agree that there (is) are _____ kingdom(s) of life.
   a. one
   b. two
   c. three
   d. four
   e. five

7. Evidence of similarities and differences in genes between members of a major group of organisms comes from which type of study?
   a. fossil evidence
   b. embryonic development
   c. DNA hybridization
   d. comparative morphology

8. Forces believed to have profoundly affected the evolution of life are _____.
   a. movements of the earth's land masses
   b. shifts in shorelines
   c. shifts in oceans
   d. all of the above

9. The evolution of life forms on earth has been dependent on _____.
   a. organisms now known only as fossils
   b. organisms that have changed little through long time periods
   c. the physical and chemical evolution of earth
   d. the rate of change occurring in different earth environments

10. Match the evolution terms appropriately.

    _____ main evolutionary trends
    _____ fossils
    _____ comparative biochemistry
    _____ comparative morphology
    _____ macroevolution

    a. evidence of evolution based on DNA mutations
    b. evidence of evolution based on studies of body forms and structural patterns in major groups
    c. study of changes in groups of species
    d. mass extinctions and adaptive radiations
    e. recognizable signs of life in the past

Readings

Bambach, R., C. Scotese, and A. Ziegler. 1980. "Before Pangea: The Geographies of the Paleozoic World." *American Scientist* 68(1): 26–38.

Cech, T. November 1986. "RNA as an Enzyme." *Scientific American* 255(5):64–75.

Gore, R. June 1989. "The March Toward Extinction." *National Geographic* 175(6):662–699.

Grieve, R. April 1990. "Impact Cratering on the Earth." *Scientific American* 262(4):66–73.

Margulis, L. 1982. *Early Life.* Boston: Science Books International. Easy-to-read introduction to the origin and evolution of prokaryotes and eukaryotes. Paperback.

Simpson, G. 1983. *Fossils and the History of Life.* New York: Scientific American Books.

Stanley, S. 1987. *Extinction.* New York: Scientific American Books.

Selected Key Terms

adaptive radiation *201*
analogous structure *193*
Archean Era *202*
Cenozoic era *204*
comparative morphology *191*
fossil *189*
gradualism *208*
homologous structure *191*
lineage *208*
macroevolution *189*

mass extinction *201*
Mesozoic Era *204*
morphological convergence *193*
morphological divergence *191*
neutral mutation *193*
Paleozoic Era *202*
phylogeny *208*
plate tectonics *197*
Proterozoic Era *202*
punctuation *208*

16 HUMAN EVOLUTION: A CASE STUDY

KEY CONCEPTS

1. The fossil record gives evidence of evolutionary trends among the primates.

2. Among those trends were skeletal modifications that began when four-legged primates were adapting to life in the trees. Those primates and their presumed descendants foreshadowed the evolution of humanlike forms that were capable of walking upright, on two legs.

3. Modifications in the hands led to increased dexterity and manipulative skills. There was less reliance on the sense of smell and more reliance on daytime vision. The brain became larger and more complex; among humans and their predecessors, this trend was accompanied by refined technologies and cultural evolution.

4. All of the trends in primate evolution were the foundation for a remarkable characteristic of humans, their plasticity (their capacity to remain flexible and adapt to a wide range of challenges imposed by unpredictable, complex environments).

In the preceding chapter, we likened the history of life to a great tree. Each branch of the tree represents a line of descent, and the branch points represent divergences leading to new species. There are more than 40,000 existing species of fishes, amphibians, reptiles, birds, and mammals on the vertebrate branch of that tree. As we turn to the evolution of any one of those species, it helps to keep a key point in mind. At each crossroad leading to a new species, complex traits were already in place and functioning—*and new traits emerged only through modification of what went before.* The evolution of the human species speaks eloquently of this principle.

THE MAMMALIAN HERITAGE

Humans are members of the class Mammalia. Like other vertebrates, mammals have an internal skeleton with two key features: a nerve cord within a column of bones (backbone), and a skull that houses sense organs and a three-part brain (hindbrain, midbrain, and forebrain). Two mammalian features are central to the story of

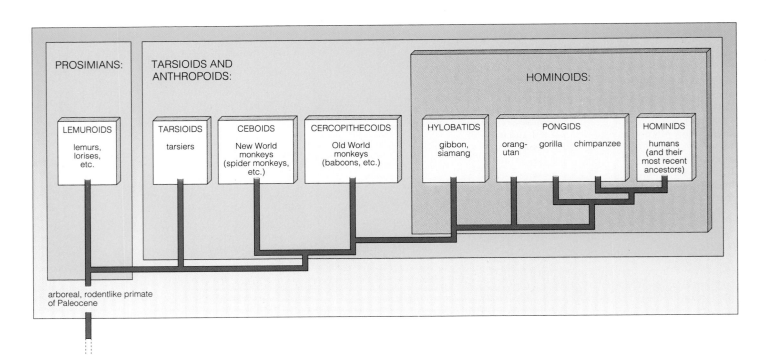

PROSIMIANS:

LEMUROIDS

lemurs, lorises, etc.

TARSIOIDS AND ANTHROPOIDS:

TARSIOIDS

tarsiers

CEBOIDS

New World monkeys (spider monkeys, etc.)

CERCOPITHECOIDS

Old World monkeys (baboons, etc.)

HOMINOIDS:

HYLOBATIDS

gibbon, siamang

PONGIDS

orang-utan   gorilla   chimpanzee

HOMINIDS

humans (and their most recent ancestors)

arboreal, rodentlike primate of Paleocene

human evolution. The first is their **dentition**—the type, number, and size of teeth. Mammals have four types of upper and lower teeth that match up and work together to crush or cut food:

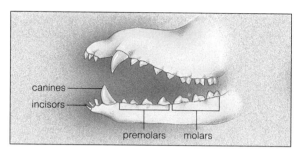

Teeth tell us quite a bit about what an animal can eat, hence something of its life-style. *Incisors*, shaped like flat chisels or cones, are used to nip or cut food. They are pronounced in horses and other grazing animals. Pointed *canines* are used in biting and piercing. Carnivores use long canines to hook into prey; some monkeys and apes use them to split open tough plant parts. *Premolars* and *molars* (cheek teeth) are a platform for food, and their surface bumps (cusps) are used in grinding and shearing. Plant-eating animals have large cheek teeth.

Teeth also happen to fossilize very well. Many inferences about the life-styles of human ancestors are drawn from fragments of jaws and teeth they left behind. Teeth provide clues about diet—and diet in turn provides clues about early habitats.

The second feature characterizing mammals is their extended period of **infant dependency** and **learning**. Young mammals depend on adults for nourishment and

Figure 16.1 (*Left*) Simplified family tree for primates. The two highest groupings (the prosimians and the tarsioids and anthropoids) are shown in green. Major groups of living primates are indicated by the gold boxes; representative members are indicated. All monkeys, apes, and humans are anthropoids. Apes, humans, and recent human ancestors are further classified as hominoids; their evolutionary relationships are the focus of this chapter.

protection and as models for behavior. They rely on a limited set of social behaviors that can be learned and repeated. Some also show a capacity to add quickly to their behavioral repertoire. This behavioral flexibility has reached its fullest expression among the primates.

THE PRIMATES

Trends in Primate Evolution

Figure 16.1 shows a family tree for **primates**. Its members are the prosimians (including lemurs) and the anthropoids (including monkeys, apes, and humans). Apes, humans, and recent human ancestors are classified as **hominoids**. Only the members of the human lineage are classified as **hominids**.

Most primates live in tropical or subtropical forests, woodlands, or savannas, which are open grasslands with a few stands of trees. Like their ancient forerunners, the vast majority of primates are tree-dwellers. As a group, they are difficult to define, for no one feature sets them apart from other mammals. Perhaps the primates are best defined in terms of the following *trends in their evolution*, many of which began when ancestral forms were adapting to life in the trees:

1. Change in overall skeletal structure and mode of locomotion

2. Modification of the hands, this leading to increased dexterity and manipulation

3. Less reliance on the sense of smell and more reliance on daytime vision, including color and depth perception

4. Change in dentition, toward fewer, less specialized teeth

5. Brain expansion and elaboration

6. Behavioral evolution

These trends did not proceed at the same time or same pace in different lineages. And some trends proceeded in a few groups only. Taken as a whole, however, they give us a sense of the adaptive potential that is uniquely "primate."

Figure 16.2 Comparison of the skeletal organization and stance of monkeys, apes (the gorilla is shown here), and humans. Modifications of the basic mammalian plan have allowed three distinct modes of locomotion. The quadrupedal monkeys climb and leap, and apes climb and swing by their forelimbs. Both modes of locomotion are well suited for life in the trees. Humans are habitual two-legged walkers.

monkey

gorilla

human

Skeletal Changes. A monkey skeleton is suitable for life in the trees. It allows rapid climbing, leaping, and running along branches (Figures 16.2 and 16.3). For example, the armbones are shorter than the legbones, so monkeys can run palms-down. (Try doing this yourself.) An ape skeleton is suitable for climbing through trees and using the longer arms to carry some of the body weight. The shoulder blade positioning allows apes to swivel their arms freely above the head. (Monkeys cannot do this.)

Of all the primates, only humans can stride freely on two legs for long periods of time. Their habitual two-legged gait, called **bipedalism,** emerged through skeletal reorganizations in primates ancestral to humans.

Modification of Hands. The first (four-legged) mammals spread their toes apart to help support body weight as they walked or ran. (Primates still can spread their toes or fingers apart.) Many mammals also make cupping motions, as when squirrels bring food to the mouth. Two other hand movements developed among ancestral tree-dwelling primates. Modifications in the handbones allowed fingers to be wrapped around objects (*prehensile* movements), and the thumb and tip of each finger to touch each other (*opposable* movements).

Among the early tree-dwelling primates, hands began to be freed from their earlier role in supporting body weight. Much later, on the evolutionary road leading to humans, refinements in hand movements led to the precision grip and power grip:

These hand positions gave early humans the capacity to make and use tools—which were the foundation for unique technologies and cultural development.

Enhanced Daytime Vision. Early primates had an eye socket on each side of the skull. Later ones had forward-directed eyes, so they had an overlapping visual field and depth perception. Over time, they became quite good at discerning shape, movement, and variations in color and in light intensity (dim to bright). These visual stimuli are typical of life in the trees.

Changes in Dentition. Monkeys have long canines and rather rectangular jaws. Human teeth are smaller and more uniform in length, and the jaw is bow-shaped. During the evolution of forms leading to humans, the jaws and teeth became modified in ways that reflect a

a

b

c

Figure 16.3 Representative primates. Gibbons (**a**) have limbs and a body adapted for brachiation (swinging arm over arm through the trees). Monkeys are quadrupedal (four-legged) climbers, leapers, and runners, as the spider monkey in (**b**) demonstrates. Tarsiers (**c**) are vertical clingers and leapers.

shift from eating insects, then fruits and leaves, and on to a mixed diet.

Brain Expansion and Elaboration. During primate evolution, the brain increased in mass and complexity. The largest brain size in living apes is 650 cubic centimeters (for the gorilla). The smallest known brain size in modern humans is 855 (the average is 1,350).

Behavioral Evolution. By using existing primates as our model, we can identify a trend toward longer life spans, longer spans between pregnancies, single births (rather than litters), and longer periods of infant dependency. The increased parental investment in fewer offspring requires strong bonds between parents and offspring, intense maternal care, and longer periods of learning.

The capacity for learning expanded further during human evolution. Brain modifications and behavioral complexity were interlocked, with new developments in one stimulating development of the other. Nowhere is this more evident than in the parallel evolution of human culture and the human brain. Here we define **culture** as the sum total of behavior patterns of a social group, passed between generations by learning and by symbolic behavior, especially language. It is the culmination of a long evolutionary history.

Primate Origins

Primates evolved from ancestral mammals more than 60 million years ago. The first known primates resembled small rodents or tree shrews (Figure 16.4). Like tree shrews, they probably had huge appetites and foraged at night for insects, seeds, buds, and eggs. They had a long snout and a well-developed sense of smell, useful for detecting food or predators. They could claw their way upward through the shrubbery, though not with much speed or grace.

Between 54 million and 38 million years ago (the Eocene), some of their descendants were living in the trees. Fossils give evidence of increased brain size, a shorter snout, enhanced daytime vision, and refined grasping movements. How did these traits emerge?

Figure 16.4 One of the night-foraging tree shrews of Indonesia.

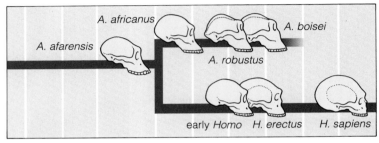

millions of years before present

One interpretation...

...and another

Figure 16.5 Two of several proposed phylogenetic trees for the hominids. There is general agreement that the hominids emerged during the late Miocene, between 6 and 4 million years ago. The *Homo* lineage is well documented; the australopith connections are not yet understood.

Consider that the trees were a promising adaptive zone that provided new food sources and safety from ground-dwelling predators. Yet the trees also were visually complex. Imagine the dappled sunlight, boughs swaying in the wind, colorful fruit, darting insects, perhaps predatory birds. A well-developed snout would not have been of much use (air currents disperse odors). But individuals with a refined sense of color, shape, and movement would have a competitive edge. So would individuals skilled at running, swinging, and leaping (especially!) from branch to branch. Selective agents surely favored the ones with enough brains to assess distance, body weight, winds, and suitability of the destination—and to compensate quickly for miscalculations.

By 35 million years ago, during Oligocene times, the tree-dwelling ancestors of monkeys and apes had emerged. Some forms lived above humid swamps lining the rivers of tropical rain forests. Given the altogether nightmarish predatory reptiles that inhabited the swamps, perhaps we can sense why the Oligocene primates rarely ventured to the ground—and why it became imperative to think fast and grip strongly. Slip-ups were always possible; a surprising number of primates still fall out of the trees.

An adaptive radiation of apelike forms—the first hominoids—occurred between 23 million and 20 million years ago. The continents were beginning to assume their current positions, and climates were becoming cooler and drier (page 204). Forests gave way to grasslands, and perhaps speciation was favored as subpopulations of apes became reproductively isolated within the shrinking stands of trees. This was the time of origin for a series of Miocene apes called the "dryopiths." Between 10 million and 6 millon years ago, some of their descendants gave rise to the forerunners of modern gorillas, chimpanzees—and humans.

THE HOMINIDS

The first hominids probably emerged between 6 million and 4 million years ago, during the late Miocene or early Pliocene. Fossils of these humanlike forms have been discovered in Africa and are known to be 4 million years old. There were many different hominids, but they all had three features in common: *bipedalism, omnivorous feeding behavior,* and *further brain expansion and reorganization.* These features probably correlated with the emerg-

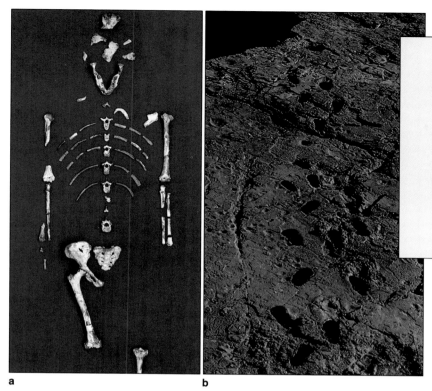

Following the path produces, at least for me, a kind of poignant time wrench. At one point, and you need not be an expert tracker to discern this, the traveler stops, pauses, turns to the left to glance at some possible threat or irregularity, then continues to the north. This motion, so intensely human, transcends time. Three million seven hundred thousand years ago, a remote ancestor—just as you or I—experienced a moment of doubt.

—Mary Leakey

Figure 16.6 (**a**) Fossil remains of Lucy, one of the earliest known australopiths. Lucy was only 1.1 meters (3 feet 8 inches) tall. The density of her limb bones is indicative of very strong muscles.
(**b**) Footprints made in soft, damp volcanic ash 3.7 million years ago at Laetoli, Tanzania, as discovered by Mary Leakey. The arch, big toe, and heel marks are those of upright early hominids.

a   b

ence of cooler, drier climates. When the African rain forests started shrinking, some hominids made the transition to life in mixed woodlands and in the vast grasslands that were opening up.

Rather than speculate on how bipedal, omnivorous, and large-brained hominids evolved during the transition, think about the plasticity inherent in those three features. Here, **plasticity** means the ability to be flexible and to adapt to a wide range of demands. The first hominids were faced with a new, complex, and unpredictable world. Yet they had the use of hands freed from their load-bearing functions. They could survive on more than one type of food. And they had the brains to *learn* how to adapt to a changing environment. Thus, the emergence of hominids occurred through modifications of traits observed among the other primates; *it was based on the primate heritage.*

Australopiths

The earliest known hominids, the **australopiths,** can be grouped into two broad categories:

1. Gracile forms (slightly built), currently going by the names *Australopithecus afarensis* and *A. africanus*

2. Robust forms (muscular, heavily built), including *A. boisei* and *A. robustus*

We don't really know how the early hominids were related. Apparently this was a "bushy" period of evolution, with many divergent forms; and inferences about their relatedness must be drawn from a limited number of fossil fragments. Figure 16.5 focuses on two of several interpretations of the fossil record.

Hipbones 4 million years old show muscle insertions typical of two-legged walkers. The hipbone of one skeleton (from a female dubbed "Lucy") allowed body weight to be centered directly over the hindlimbs; this is a hallmark of bipedalism. Besides this evidence, the australopiths left footprints (Figure 16.6).

In other traits, the australopiths were transitional between the Miocene apes and later hominids. Bipedalism had freed their hands—and fossil handbones suggest they were quite dexterous. Like small apes, their cranial capacity was about 400 cubic centimeters. Unlike apes, their jaw was slightly bow shaped. And their dentition suggests that some were omnivores and others, vegetarians.

For example, *A. africanus* was probably omnivorous; the cheek teeth formed a platform that could grind plants, but the incisors were relatively large. *A. robustus* had strong jaw muscles and a large grinding platform typical of plant-eaters. *A. boisei* had very large, heavily cusped molars; it may have specialized in chewing seeds, nuts, and other tough plant material.

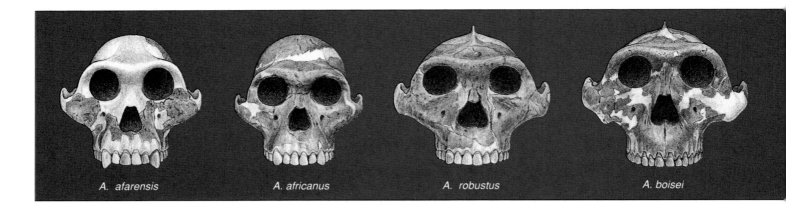

| A. afarensis | A. africanus | A. robustus | A. boisei |

Stone Tools and Early *Homo*

By about 2.5 million years ago, hominids had started to make stone tools. (Before then, they could have used sticks and other perishable materials as tools, much like chimpanzees do today.) The first toolmaker may have been the hominid called "early *Homo*" in one classification scheme (and "late australopith" or *H. habilis* in others). Compared to australopiths, early *Homo* had a smaller face, more generalized teeth, and a larger brain (Figure 16.7). This hominid apparently was a scavenger, hunter, or both; its remains have been found with bones of frogs, pigs, lizards, birds, fish, and small antelopes. And it may have been ancestral to modern humans.

At some point, early *Homo* began using rocks to crack open animal bones and expose the soft, edible marrow. Individuals started using sharp-edged flakes, formed naturally (and not too often) as rocks tumbled through a river or down a hill. They used such flakes to scrape flesh from animal bones. At some point, they started making flakes themselves.

The earliest known "manufactured" tools were crudely chipped pebbles discovered by Mary Leakey at Olduvai Gorge (Figure 16.8). This African gorge cuts through a sequence of sedimentary deposits, with the more recently deposited layers containing ever more sophisticated tools. The sequence gives insight into the increasingly refined exploitation of a major food source—game herds of the open grasslands.

Olduvai may also give evidence of our early social and cultural development. Nearly 1.9 million years ago, *our ancestors had a concept of "home."* We have evidence of a brushwood shelter, supported by a circle of small rocks, where early humans gathered to eat captured or scavenged animals. It was an extraordinary development among primates. Other primates forage together, stay together overnight, then move on—but they don't bring back food to a home base. Also, early *Homo* traveled several miles from home and back to get special rocks (lava, quartzite) that were best for making certain kinds of tools. *Early on, the human brain was assessing distant horizons.*

Homo erectus

Between 1.5 million and 300,000 years ago, during the Pleistocene, extreme shifts in climate put the adaptive potential of the early humans to the test. Vast glacial sheets formed and sea levels fell; then the glaciers melted and sea levels rose. Land bridges and coastal regions were alternately submerged and exposed. During repeated interglacials, a larger brained human species, *Homo erectus*, migrated out of Africa and into Southeast Asia, China, and Europe.

Compared with modern humans, *H. erectus* had a long, thick-walled skull (Figure 16.7). But there was practically no difference in brain size between a large *H. erectus* and a small modern human. The increased brain size correlates with the spectacular travels of *H. erectus;* no other hominids had ranged so far from their original environment. It correlates also with refinements in toolmaking. And *H. erectus* seemed to have made controlled use of fire—and how advantageous campfires would have been, during the moves to colder regions.

Homo sapiens

Somewhere between 300,000 and 200,000 years ago, modern humans *(H. sapiens)* apparently arose from *H. erectus* stock. There was a transitional period, often with different geographic forms coexisting in time, then *H. erectus* gradually disappeared. Early *H. sapiens* had a rounder, higher skull, with a face more delicately structured. The teeth and jaw were rather small; many forms had a chin (their predecessors did not). Not until about 40,000 years ago did anatomically modern humans, *H. sapiens sapiens*, emerge. Before then, a number of archaic groups lived in Europe, the Near East, and China. Among them were the Neanderthals. Neanderthals had projecting facial bones, large incisors, and often large browridges. They had a slightly larger brain than we

Figure 16.7 Comparison of skull shapes of the early hominids relative to modern humans (*Homo sapiens sapiens*). In general, australopiths had a small brain and large face, compared with the larger brained, smaller faced forms of the genus *Homo*. The drawings are not all to the same scale. White areas of skulls are reconstructions.

do—ranging between 1,300 and 1,750 cubic centimeters. And they were proficient hunters and gatherers in diverse environments. They lived in caves, rock shelters, and open-air camps, in varied climates. Yet they disappeared suddenly about 35,000 or 40,000 years ago. Did they become extinct or evolve rapidly into modern forms? We do not know.

Since then, human evolution has been almost entirely cultural rather than biological, and so we leave the story. From the biological perspective, however, we can make these concluding remarks: Humans have spread throughout the world by rapidly devising the cultural means to deal with a broad range of environmental conditions. Compared with their predecessors, modern humans have developed spectacularly rich and varied cultures, moving from "stone-age" technology to the age of "high tech." Yet hunters and gatherers persist in parts of the world, attesting to the great plasticity and depth of human adaptations.

SUMMARY

1. Primates include prosimians (lemurs and related forms) and anthropoids (monkeys, apes, and humans). Apes and humans are hominoids; and only humans and their recent ancestors are further classified as hominids.

2. The following evolutionary trends for primates as a whole are related largely to their tree-dwelling ancestry:
   a. From a four-legged gait to bipedalism (habitual free-striding, two-legged gait). This trend involved changes in the shoulders, backbone, pelvic girdle, legs, and feet. The changes evolved in the trees, where primates were largely freed from the selective pressures imposed by ground-dwelling predators but were challenged by a visually complex habitat.
   b. Increased manipulative skills owing to modification of the hands, which began to be freed from their load-bearing function among tree-dwelling primates.

chopper   chopper   bola (?)

hand ax

cleaver

Figure 16.8 Representatives of the more than 37,000 stone tools recovered from Olduvai Gorge. *Upper row:* A crude chopper and a more advanced form having a sharp edge. The stone ball may represent a transition from passive to aggressive tool use. It resembles the Argentine bolas, which are strung together on lengths of hide and thrown at animals to entangle the legs and bring them down. *Lower row:* A hand ax and a cleaver.

   c. Less reliance on the sense of smell and more reliance on enhanced daytime vision, including color vision and depth perception.
   d. From specialized to omnivorous eating habits.
   e. Brain expansion and reorganization. Larger, more complex brains are correlated with increasingly sophisticated technology (from simple to refined tools) and with social development.
   f. All of these trends were the foundation for the remarkable *plasticity* of the hominids (their capacity

to remain flexible and to adapt to a wide range of demands imposed by a complex environment).

3. The oldest known primates date from the Paleocene. The first hominids (australopiths) emerged between 6 million and 4 million years ago. All were bipedal, with a larger brain than their predecessors. Some were slightly built omnivores; others were heavily built and well-muscled vegetarians.

4. Fossils of early *Homo*, the first known representative of the human lineage, date from 2 million years ago. Early *Homo* was omnivorous, larger brained, and taller than its predecessors, used simple tools, and showed some social development.

5. *Homo erectus* fossils associated with abundant cultural artifacts date from about 1.5 million to 300,000 years ago. This form was adapted to a wide range of habitats. Fully human forms *(H. sapiens)* emerged between 300,000 and 200,000 years ago, possibly from *H. erectus* stock. By 40,000 years ago, modern forms *(H. sapiens sapiens)* had evolved. From that point on, cultural evolution has outstripped biological evolution of the human species.

Review Questions

1. What are the general evolutionary trends that occurred among the primates as a group? What way of life apparently was the foundation for these trends? *213*

2. What conditions seem to have been responsible for the great adaptive radiation of apelike forms during the Miocene? *216*

3. What is the difference between "hominoid" and "hominid"? Are we hominoids, hominids, or both? *213*

4. Describe the key characteristics of hominid evolution. How do they relate to the concept of plasticity? *216–217*

Self-Quiz   *(Answers in Appendix IV)*

1. Primates include _____.
   a. lemurs   c. apes   e. all of the above
   b. monkeys   d. humans

2. _____ are hominoids; only _____ are hominids.
   a. lemurs; monkeys and their immediate ancestors
   b. apes and humans; humans and their recent ancestors
   c. monkeys; apes and their recent ancestors
   d. monkeys, apes, and humans; apes and their recent ancestors

3. The key trends in primate evolution began in the _____.
   a. savanna   c. trees
   b. water   d. forest floor

4. Which of the following was not a trend in primate evolution?
   a. bipedalism
   b. hand modification that increased manipulative skills
   c. shift from omnivorous to specialized eating habits
   d. less reliance on smell, more on vision
   e. brain expansion and reorganization
   f. all were evolutionary trends

5. Early hominids displayed great plasticity. This means that _____.
   a. they were adapted to a wide range of demands in complex environments
   b. they had flexible bones that cracked easily
   c. they were limber enough to swing through the trees
   d. they were adapted for a narrow range of demands in complex environments

6. The oldest known primates date from the _____.
   a. Miocene (25 million to 13 million years ago)
   b. Paleocene (65 million to 54 million years ago)
   c. Oligocene (38 million to 25 million years ago)
   d. Pliocene (13 million to 2 million years ago)

7. The first known hominids are generally classified as _____.
   a. *Homo*   c. cercopiths
   b. dryopiths   d. australopiths

8. The first known hominids were _____.
   a. bipedal and with larger brains than their predecessors
   b. slightly built omnivores
   c. heavily built, muscular vegetarians
   d. all of the above

9. Fossils of the early *Homo*, the first known representative of the human line, date from _____ million years ago.
   a. 8   c. 4
   b. 6   d. 2

10. Match the early primates with their descriptions.
   _____ australopiths   a. first known representatives of the
   _____ *Homo*   human line
   _____ anthropoids   b. only humans and their recent
   _____ *Homo sapiens*   ancestors
         sapiens   c. fully modern humans
   _____ hominids   d. the first hominids
                            e. monkeys, apes, humans

Selected Key Terms

australopith *217*   hominoid *213*
bipedalism *216*   *Homo erectus 218*
culture *215*   *Homo sapiens 218*
dentition *213*   plasticity *217*
early *Homo 218*   primate *213*
hominid *213*

Readings

Reader, J. 1981. *Missing Links.* Boston: Little, Brown. Readable, exquisitely illustrated historical account of discoveries concerning human evolution.

Rensberger, R. 1981. "Facing the Past." *Science 81* 2(8):41–50. Intriguing look at how artists' reconstructions can bias our perceptions of what the early hominids looked like.

Weiss, M., and A. Mann. 1989. *Human Biology and Behavior: Anthropological Perspective.* Fifth edition. Boston: Little, Brown.

FACING PAGE: *Patterns of diversity in nature, here represented by different species of plants and fungi.*

IV

17

VIRUSES, MONERANS, AND PROTISTANS

KEY CONCEPTS

1. Living organisms have the metabolic means to maintain, grow, and reproduce themselves. Viruses are infectious agents that are not alive; it takes the metabolic machinery of a living host cell to synthesize the viral nucleic acids and proteins necessary to construct a new virus particle.

2. Prokaryotic cells (bacteria) are not structurally complex; they do not have the specialized organelles of eukaryotic cells. However, they are metabolically diverse and many make complex behavioral responses to their surroundings.

3. More than 2.5 billion years ago, three prokaryotic lineages diverged from a common ancestor. Their modern descendants are the archaebacteria, eubacteria, and eukaryotes.

4. The protistan kingdom is dominated by eukaryotic organisms that are alike in being single-celled but quite diverse in their modes of nutrition. Protistans may be reminiscent of single cells that gave rise to the kingdoms of multicelled eukaryotes—the fungi, plants, and animals.

In the preceding unit, we considered some of the observations and methods being used to discern evolutionary relationships among organisms. We turn now to the existing spectrum of life, with two kingdoms of single-celled forms—the monerans and protistans—the main focus of this chapter. Most monerans and protistans are *microbes*, meaning they are too small to be seen without a microscope (Figure 17.1). Before we begin, however, let's consider the viruses, which also are microscopically small. Viruses are *not* alive, but they do have impact on organisms in all five kingdoms.

VIRUSES

In ancient Rome *virus* meant "poison." In the late 1800s, this rather nasty word was bestowed on a newly discovered class of infectious agents, smaller than the microbes being studied by Louis Pasteur and others of that era. Many viruses deserve their name, for they can have devastating effects on humans, cats, cattle, insects, crop plants, fungi, bacteria, and even other viruses. You name it, and there probably are one or more kinds of viruses that can infect it.

 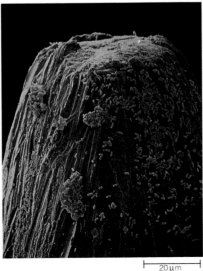 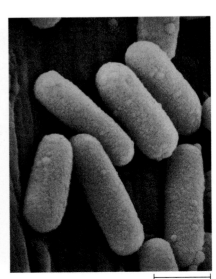

100 µm   20 µm   1 µm

Today we define a **virus** as a noncellular infectious agent having two characteristics. *First*, a virus consists of DNA or RNA surrounded by a protective protein coat, which is sometimes enclosed in a lipid envelope. *Second*, a virus can replicate only after its DNA or RNA enters a host cell and subverts the cell's biosynthetic machinery. When outside of a host, a virus is no more alive than a chromosome is alive.

A virus can lock onto a specific molecular group at the surface of a cell that can serve as its host. Once bound, either the whole virus or its nucleic acid alone enters the cell cytoplasm. At some point, the viral DNA or RNA is replicated—and its genes are transcribed and translated into viral proteins. Then the replicated nucleic acids and new proteins are assembled into new virus particles. Usually, many new particles are assembled and released together, either by budding from the host cell or through a rupturing of its plasma membrane (lysis).

Figure 17.2 shows the cycle of replication for one of the *bacteriophages,* a class of viruses that infect bacterial cells. Although bacteriophages have bad effects on their hosts, they have been put to good use in genetics research.

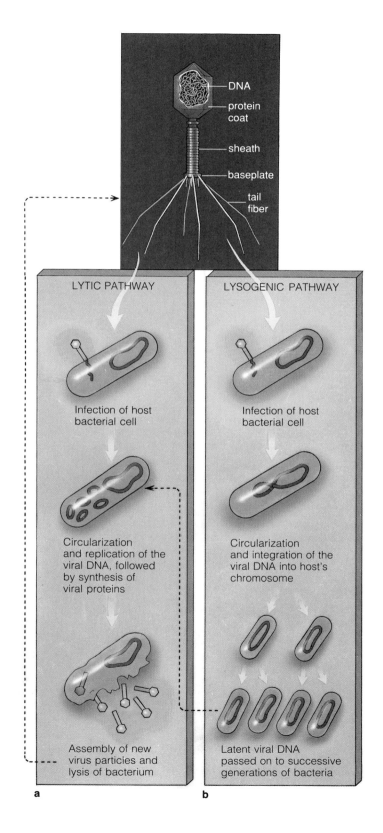

Figure 17.1 (*Left*) How small are bacteria? Shown here, *Bacillus* cells on the tip of a pin.

Figure 17.2 (*Right*) Replication of a bacteriophage. Different bacteriophages infect different types of bacteria. (**a**) In a *lytic pathway* of infection, the virus replicates its genetic material inside a host cell and commandeers the cell's metabolic machinery, causing it to make viral proteins. The replicated genetic material and proteins are used to assemble new virus particles, which are released when the infected cell ruptures. (**b**) In a *lysogenic pathway*, a host cell is infected but not killed, and the viral genetic material becomes integrated into the bacterial chromosome by natural recombination mechanisms (page 163). Integration is straightforward for DNA viruses. RNA viruses rely on *reverse transcription*. A DNA transcript of the viral RNA is synthesized, then inserted into the bacterial chromosome. When the host cell is about to reproduce, it replicates the foreign DNA along with its own DNA, so both are passed on to subsequent bacterial generations. Later, the viral DNA may leave the chromosome, whereupon the lytic pathway is followed.

Figure labels: DNA, protein coat, sheath, baseplate, tail fiber

LYTIC PATHWAY
- Infection of host bacterial cell
- Circularization and replication of the viral DNA, followed by synthesis of viral proteins
- Assembly of new virus particles and lysis of bacterium

a

LYSOGENIC PATHWAY
- Infection of host bacterial cell
- Circularization and integration of the viral DNA into host's chromosome
- Latent viral DNA passed on to successive generations of bacteria

b

Table 17.1 Classification of Animal Viruses

DNA Viruses	Some Diseases Produced	RNA Viruses	Some Diseases Produced
Adenoviruses	Acute respiratory tract infections; under some circumstances can cause malignant tumors in hamsters	Enteroviruses	Diarrhea; polio; aseptic meningitis
		Rhinoviruses	Common colds
Parvoviruses	Some types of gastroenteritis (diarrhea, vomiting)	Togaviruses	Yellow fever; German measles (rubella); equine encephalitis
Papovaviruses	Warts in humans, rabbits, dogs; some cancers in mice, hamsters	Influenza viruses	Influenzas
Herpes viruses	Fever blisters; chickenpox; shingles; certain genital infections with neurological effects; some induce cancers; one (Epstein-Barr virus) implicated in infectious mononucleosis	Paramyxoviruses	Mild respiratory disorders; Newcastle disease; measles; mumps
		Rhabdoviruses	Rabies
		Arenaviruses	Meningitis; hemorrhagic fevers
		Coronaviruses	Upper respiratory disease
Poxviruses	Smallpox; cowpox; formation of fibromas (nodules or benign tumors)	Retroviruses	Certain tumors (sarcomas); AIDS; leukemia
		Reoviruses	Mild respiratory disorders; diarrhea

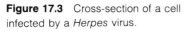

an adenovirus

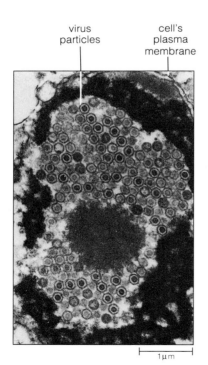

virus particles   cell's plasma membrane

1 μm

Figure 17.3 Cross-section of a cell infected by a *Herpes* virus.

Figure 17.4 Effect of a viral infection on Rembrandt tulips. These and other types of flowers are variegated as a result of a relatively benign viral infection that affects pigment formation in different tissue regions. Virus particles are passed on to new generations of plants, which also produce variegated flowers.

Different *animal viruses* infect invertebrates and vertebrates. Among them are the *Herpes* viruses (page 487 and Figure 17.3) and the viruses that cause recurring, worldwide epidemics of influenza. Between 1918 and 1920 alone, a Spanish influenza virus killed more than 20 million people. Influenza viruses infect tissues of the upper respiratory tract. Often the weakened tissues are susceptible to agents of pneumonia, a complication that accounts for a large number of the deaths associated with flu epidemics. Table 17.1 lists the DNA and RNA viruses responsible for some troubling diseases, including forms of cancer.

Plant viruses are known to cause more than a thousand diseases among plants. Typically, sucking insects (such as aphids) or other invertebrates transmit the virus from an infected plant to another plant. Viral diseases can greatly reduce the yields of a variety of crops, including potatoes, tomatoes, cauliflower, cucumbers, turnips, and barley. Figure 17.4 shows the visible effects of one viral disease.

Viruses may not be the most stripped-down disease agents. For example, ''viroids'' are naked strands or circles of RNA (with no protein coat) that depend on a host cell for replication. Although viroids are mere snippets of genes, some may cause certain forms of cancer; others wipe out huge crops of citrus, avocados, potatoes, and other plants.

MONERANS

Bacteria, the sole members of the kingdom Monera, are the most abundant and far-flung microbes. Different species live in boiling mud, hot springs, snow, deserts, ''pristine'' lakes, and deep oceans. There may be tens of billions of them in a handful of rich soil—several times the number of people on earth. Many bacteria live in or on other organisms. The ones in your gut and on your skin outnumber the cells of your body. Fortunately, animal cells are much larger than bacteria, so humans are only a few percent ''bacterial'' by weight.

Characteristics of Bacteria

Bacterial cells are prokaryotic; they have no nucleus or other membrane-bound organelles (Figure 17.5 and Table 17.2). Many metabolic reactions take place at the plasma membrane—the bacterial equivalent of organelle membranes. Proteins are synthesized rapidly at the numerous ribosomes distributed through the cytoplasm or attached to the plasma membrane.

Nearly all bacteria have a semirigid cell wall, which is usually a tough mesh of polysaccharide-protein mol-

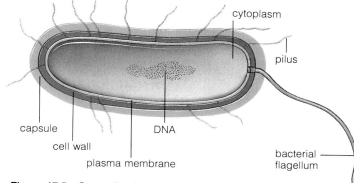

Figure 17.5 Generalized body plan of a bacterium.

Table 17.2   Characteristics of Bacterial Cells
1. Bacterial cells are prokaryotic (they have no nucleus or other membrane-bound organelles in the cytoplasm).
2. Bacterial cells have a single chromosome (a circularized DNA molecule); many species also have plasmids.
3. Most bacteria have a cell wall composed of peptidoglycan.
4. Most bacteria reproduce by binary fission.
5. Collectively, bacteria show great diversity in their modes of metabolism.

ecules (peptidoglycans). In some species, a sticky capsule or slime layer covers the wall and allows the bacterium to adhere to rocks, plant parts, teeth, and other surfaces. There may also be one or more **bacterial flagella** anchored to both the wall and plasma membrane (Figure 17.5). A bacterial flagellum rotates like a propeller and so moves the cell through the fluid environment. The cell wall commonly imparts one of three shapes to the bacterium:

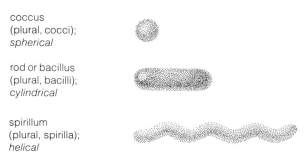

coccus
(plural, cocci);
spherical

rod or bacillus
(plural, bacilli);
cylindrical

spirillum
(plural, spirilla);
helical

Some bacteria are **photosynthetic autotrophs**; they produce their own organic compounds from simple inorganic substances, using sunlight as an energy source. A few are **chemosynthetic autotrophs**; they also produce

Table 17.3 Some Major Groups of Bacteria

Group	Main Habitats	Characteristics	Representatives
Archaebacteria			
Methanogens	Anaerobic sediments of lakes, swamps; also animal gut	Chemosynthetic; methane producers; used in sewage treatment facilities	*Methanobacterium*
Extreme halophiles	Brines (extremely salty water)	Heterotrophic; also have photosynthetic machinery of a unique sort	*Halobacterium*
Thermoacidophiles	Acidic soil, hot springs, hydrothermal vents on seafloor	Heterotrophic or chemosynthetic; use inorganic substances such as sulfur as a source of electrons for ATP formation	*Sulfolobus, Thermoplasma*
Photosynthetic eubacteria			
Cyanobacteria	Mostly lakes, ponds; some marine, terrestrial	In photosynthesis, water is electron donor, oxygen a by-product; some fix nitrogen	*Anabaena, Nostoc*
Purple sulfur bacteria	Anaerobic, organically rich muddy soils, and sediments of aquatic habitats	Photosynthetic; use reduced sulfur compounds as electron donors	*Chromatium*
Chemosynthetic eubacteria			
Nitrifying bacteria	Soil, freshwater, marine habitats	Major ecological role (nitrogen cycle)	*Nitrosomonas, Nitrobacter*
Heterotrophic eubacteria			
Spirochetes	Aquatic habitats; parasites of animals	Helically coiled, motile; free-living and parasitic species; some major pathogens	*Spirochaeta, Treponema*
Gram-negative, aerobic rods and cocci	Soil, aquatic habitats; parasites of animals, plants	Some major pathogens; some (e.g., *Rhizobium*) fix nitrogen	*Pseudomonas, Neisseria, Rhizobium, Agrobacterium*
Gram-negative, facultative anaerobic rods	Soil, plants, animal gut	Many are major pathogens; one (*Photobacterium*) is bioluminescent	*Salmonella, Shigella, Proteus, Escherichia, Photobacterium*
Rickettsias and chlamydias	Host cells of insects, other animals	Intracellular parasites; many pathogens	*Rickettsia, Chlamydia*
Gram-positive cocci	Soil; skin and mucous membranes of animals	some major pathogens	*Staphylococcus, Streptococcus*
Endospore-forming rods and cocci	Soil; animal gut	Some major pathogens	*Bacillus, Clostridium*
Gram-positive nonsporulating rods	Fermenting plant, animal material; gut, vaginal tract	Some important in dairy industry, others serious contaminators of milk, cheese	*Lactobacillus, Listeria*
Actinomycetes	Soil; some aquatic habitats	Include anaerobes and strict aerobes; major producers of antibiotics	*Actinomyces, Streptomyces*

their own organic compounds, using simple inorganic substances as the energy source. But the vast majority of bacteria are **heterotrophs** of one sort or another. They must use organic compounds produced by other organisms for energy.

Most bacterial cells reproduce by way of fission; they divide into two daughter cells following DNA replication (Figure 17.6). Daughter cells may stick together in pairs, clusters, or chains when they fail to separate completely after division. In addition to the bacterial chromosome, which is a circular DNA molecule, many bacterial species have "extra" DNA in the form of plasmids.

Major Groups of Bacteria

How are the many thousands of known bacterial species related to one another? Traditionally, bacteria are characterized according to overall form, wall properties, growth patterns, and tolerance of environmental conditions such as pH and temperature. (For example, cell walls of "Gram-positive" bacteria retain a deep purple stain used in microscopy, but the walls of "Gram-negative" bacteria lose the color.)

But characterizing a given bacterium is not the same thing as classifying it. Bacteria are not well represented

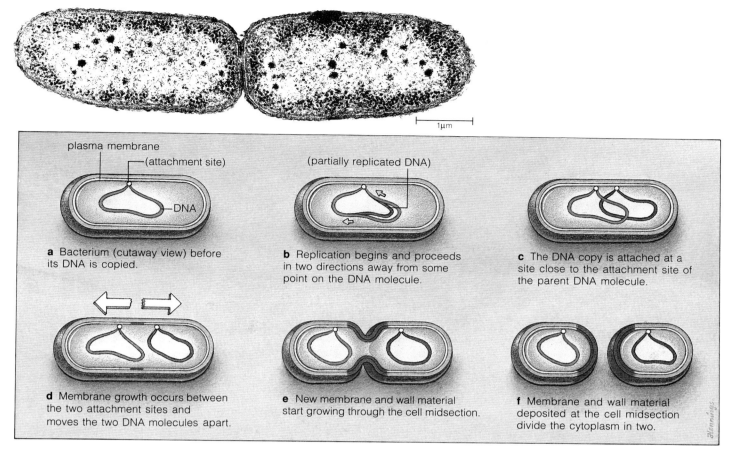

Figure 17.6 Bacterial reproduction by binary fission, a cell division mechanism. The micrograph shows the bacterium *Pseudomonas* after new membrane and wall material divided the cytoplasm of the parent cell in two.

Within the figure:

plasma membrane

(attachment site)

(partially replicated DNA)

DNA

1μm

a Bacterium (cutaway view) before its DNA is copied.

b Replication begins and proceeds in two directions away from some point on the DNA molecule.

c The DNA copy is attached at a site close to the attachment site of the parent DNA molecule.

d Membrane growth occurs between the two attachment sites and moves the two DNA molecules apart.

e New membrane and wall material start growing through the cell midsection.

f Membrane and wall material deposited at the cell midsection divide the cytoplasm in two.

in the fossil record, and until recently, they kept their evolutionary secrets to themselves. Comparative biochemistry studies have now shown there are two major groups, the archaebacteria and eubacteria.

Archaebacteria. Three types of archaebacteria (the methanogens, extreme halophiles, and thermoacidophiles) are closely related to one another but not to any other bacteria (Table 17.3). They all live in harsh settings reminiscent of the ancient environments in which life began (hence the name, *archae*bacteria). They are genetically and structurally distinct from all other organisms.

Methanogens live in swamps, mud, and the gut of cows, humans, and other animals. Through their metabolic activities, these anaerobic bacteria produce large amounts of natural gas as well as the "marsh gas" of swamps and sewage treatment facilities. Because they produce about 2 billion tons of methane gas (CH_4) annually, they play a major role in the global cycling of carbon (page 534).

Extreme halophiles live in salt lakes and brackish seas. They can spoil salted fish, salted animal hides, and commercially produced sea salt. Most are heterotrophs and use aerobic pathways for ATP formation. But when oxygen supplies are low, some switch to photosynthesis by using a light-trapping pigment (bacteriorhodopsin) embedded in their plasma membrane. The thermoacidophiles live in hot springs, acidic soils, even near volcanic vents at the ocean floor.

Photosynthetic Eubacteria. There are thousands of diverse eubacteria. Here we will consider a few examples, using the mode of nutrition (photosynthetic, chemosynthetic, and heterotrophic) to distinguish among them.

Like plants, some eubacteria use sunlight energy for photosynthetic reactions leading to ATP formation. Cyanobacteria (which also are called blue-green algae) are an example. You can see them at the surface of freshwater ponds. Many species grow as chains of cells

a

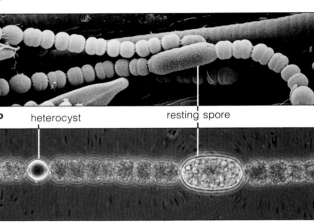

b   heterocyst   resting spore

c

Figure 17.7 (**a**) Slimy mat formed by chains of cyanobacterial cells floating near the surface of a nutrient-enriched pond. (**b**) Closer view of cells in the chains. Resting spores form along the chain when conditions do not favor growth; they germinate and give rise to a new chain when favorable conditions return. (**c**) The heterocysts are cells in the chain that have become specialized for nitrogen fixation.

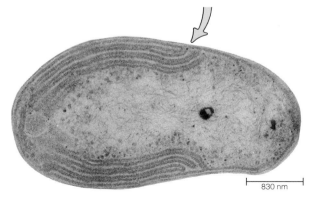

830 nm

Figure 17.8 *Nitrobacter*, one of the nitrifying bacteria. The arrow points to a site where the plasma membrane folds into the cytoplasm. The membrane infoldings greatly increase the membrane surface area available for metabolic reactions.

that surround themselves with a mucous sheath. Often the chains are intertwined into dense, slimy mats.

Some cyanobacteria, including *Anabaena*, produce a nitrogen-fixing enzyme when environmental supplies of nitrogen-containing compounds run low. Some cells in the chains develop into **heterocysts,** and these cells produce the enzyme molecules. Their newly formed nitrogen compounds move through cell junctions and are shared with photosynthetic cells; carbohydrates formed in the photosynthetic cells are shared with the heterocysts (Figure 17.7).

Chemosynthetic Eubacteria. Among the most important chemosynthetic forms are the nitrifying bacteria. They have roles in the global cycling of nitrogen, a component of all amino acids and proteins. Nitrifying bacteria attack ammonia or nitrite in soil or water, stripping electrons from them for use in reactions leading to ATP formation. In some species, the reactions are carried out across extensive infoldings of the plasma membrane (Figure 17.8).

Heterotrophic Eubacteria. Many notorious eubacteria fall in this category, including the agents of syphilis and gonorrhea (page 486). Some are transmitted from one host to another by ticks or insects. For example, ticks transmit a spirochete *(Borrelia burgdorferi)* from deer and other wild animals to humans; it is the agent of Lyme disease. Infected persons develop a rash around the tick bite, then severe headaches, backaches, chills, and fatigue. Without prompt treatment, they can develop arthritis-like symptoms and other disorders.

Some heterotrophic eubacteria can go dormant by forming a resistant body, an **endospore,** around their genetic material and a bit of cytoplasm (Figure 17.9). Endospores resist heat, drying, radiation, and other environmental insults; they germinate and give rise to new bacterial cells when favorable conditions return. One endospore-former, *Clostridium botulinum*, produces a deadly toxin. Cattle and birds die after eating fermented grains tainted with the toxin. *C. botulinum* also can

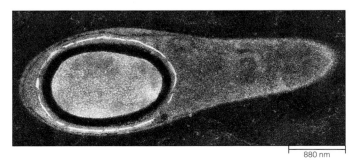

880 nm

Figure 17.9 Transmission electron micrograph of a developing spore of the bacterium *Clostridium tetani*.

produce toxins in improperly sterilized and sealed cans or jars. Eating the tainted food can lead to botulism, a form of poisoning that affects muscle activity; death can follow as a result of respiratory failure.

Escherichia coli is another heterotrophic eubacterium. It normally benefits its human hosts. *E. coli* produces vitamin K and compounds useful in fat digestion, and its activities help prevent many disease-causing microbes from colonizing the gut. Even so, some *E. coli* strains produce a potent toxin and cause serious diarrhea. *E. coli* diarrhea is the leading cause of infant mortality for most of the human population.

Ironically, heterotrophic eubacteria called actino-mycetes are major producers of chemical weapons (antibiotics) against many of their own dangerous relatives. Other beneficial eubacteria are *Azotobacter* and *Rhizobium;* these nitrogen-fixing forms enhance soil fertility and plant growth. *Lactobacillus* breaks down milk sugar and helps newborn mammals digest milk. Some species help make cheese, sour cream, yogurt, and other fermented milk products.

About the "Simple" Bacteria

Bacteria are small. They are not as structurally complex as eukaryotic cells. *But bacteria are not simple.* For example, bacteria deal with the environment in sophisticated ways. Photosynthetic species sense the intensity of light and move toward its source (or away if it is too bright). Heterotrophic species sense and move toward higher concentrations of nutrients. Aerobes move toward oxygen, anaerobes move away from it. Many bacteria can detect and avoid toxins.

As another example, even though bacteria are not multicellular, some species give a good imitation of it. Millions of cells of *Myxococcus xanthus* form "predatory" colonies that trap cyanobacteria and other microbes. Their enzyme secretions degrade the "prey" that becomes stuck to the colony, and the *M. xanthus* cells absorb the breakdown products. More than this, migrating *M. xanthus* cells change direction and *move as a single unit* toward what may be food, then move away if their target is not chemically "tasty."

THE RISE OF EUKARYOTIC CELLS

The "records" built into living bacteria tell us something about the origin of eukaryotes. Comparisons of nucleotide sequences of the DNA and RNA from different bacteria suggest that archaebacteria, eubacteria, and the forerunners of eukaryotes diverged from a common ancestor long before fossil evidence of eukaryotes started accumulating (page 202).

Single-celled protistans and multicelled plants, fungi, and animals are eukaryotes. They are alike (and *unlike* bacteria) in having a nucleus and other membrane-bound organelles that separate different metabolic activities in the space of the cell cytoplasm. The simplest eukaryotes are soft-bodied protistans called amoebas. Because amoebas have no hard parts, they vanish quickly after they die. If the ancestors of eukaryotes also were soft-bodied, fossils of them may never be found. At present, we have no direct evidence of how eukaryotes arose, although the kind of speculation described in the *Commentary* gives us interesting things to think about.

PROTISTANS

By definition, **protistans** are singled-celled eukaryotic organisms, but the boundaries of their kingdom are poorly defined. Remember, the boundaries of major groups of organisms are artificial. *We simply impose them on the continuous threads of descent that tie together all species, past and present.* Without doubt, the protistan lineages spill over into the kingdoms of plants, fungi, and animals. (Thus, what may be "a photosynthetic protistan" to one biologist may be "a plant" to another.) With this qualification in mind, let's consider the characteristics of the protistan groups listed in Table 17.4.

Table 17.4 Classification of Protistans	
Phylum*	Common Name
Gymnomycota	Slime molds
Euglenophyta	Euglenids
Chrysophyta	Yellow-green algae Golden algae Diatoms
Pyrrophyta	Dinoflagellates
Sarcomastigophora	
Subphylum Mastigophora	Flagellated protozoans

Subphylum Sarcodina	Amoeboid protozoans:
	  Amoebas
	  Foraminiferans
	  Heliozoans
	  Radiolarians

Apicomplexa	Sporozoans and kin
Ciliophora	Ciliated protozoans

*Also called Division in some classification schemes; the terms are equivalent.

Speculations on the Origin and Evolution of Eukaryotes

About 2.5 billion years ago, divergences from an ancestral prokaryote marked the beginning of three major lineages: archaebacteria, eubacteria, and the forerunner of eukaryotes. A family tree showing the presumed relationships among these lineages is shown in Figure a; it is based on recent work in comparative biochemistry.

Origin of Organelles

Archaebacteria and eubacteria are fundamentally different in wall characteristics and many other properties. And eukaryotes differ from both of those prokaryotic lineages in having membrane-bound organelles. How did organelles emerge in the cells ancestral to eukaryotes? The most conspicuous organelle, the nucleus, may have evolved through modification of infoldings of the plasma membrane. Such infoldings occur in many bacteria (Figure 17.8 shows an example).

According to Lynn Margulis and many other biologists, some organelles had symbiotic origins. *Symbiosis* ("living together") refers to interactions in which one species serves as host to another species (the guest).

Such partnerships can be observed today (Figures b and c). Mitochondria, recall, are organelles specialized for aerobic respiration. They are uncannily like bacteria in size and structure, their DNA is replicated independently of the cell's DNA, and their genetic code is slightly different from that of the cell.

Perhaps the forerunners of mitochondria were free-living, predatory cells, similar to existing amoebalike forms that are weakly tolerant of free oxygen. Suppose they preyed on aerobic bacteria—some of which resisted digestion. We can imagine that the guest found itself in a protected, nutrient-rich environment; and the host was supplied with "extra" ATP that could be channeled into growth, greater activity, and the assembly of more structures, such as hard body parts. In time, the bacteria became modified as their increasingly complex host cells performed some functions for them. They became mitochondria, incapable of independent existence.

Similarly, chloroplasts are like photosynthetic bacteria in their metabolism and overall DNA sequences, and they can replicate somewhat independently of the cell. Chloroplasts may have evolved a number of times, in a

(**a**)   Proposed relationships among prokaryotes and eukaryotes.

(**b**)   A predatory bacterium (small, dark oval) which has ended up living in the space between the cell wall and plasma membrane of a larger bacterium.

(**c**)   One of nature's experiments—*Plakobranchus*, a marine mollusk that feeds on algae, the chloroplasts of which become incorporated in its tissues and continue functioning, providing the animal with oxygen.

(**d**)   *Trichoplax adhaerens*, one of the simplest multicelled animals, being little more than a flattened ball of ciliated cells. This tiny animal was discovered crawling about in a seawater aquarium.

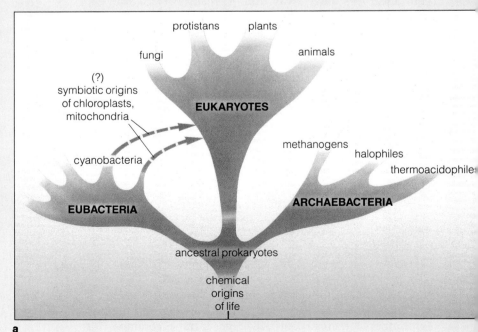

number of different lineages. (Chloroplasts of existing eukaryotic species vary in shape and their array of light-absorbing pigments, just as different species of photosynthetic bacteria do.)

Beginnings of Multicellularity

Multicellularity signifies interdependence and division of labor among specialized cell types. Think about the advantages it bestows—say, on one type of organism that is tasty to another. Compared to single cells, a multicelled organism might be too large to eat. It may have muscles and other specialized tissues that provide the means for rapid response to danger. Each tissue or organ performs a single function very well, and it depends on other tissues or organs to help maintain favorable operating conditions for the whole organism.

Among the simple eukaryotes are forms that may be reminiscent of nature's early experiments in multicellularity. Think about the division of labor in *Volvox*, a photosynthetic, colonial organism (Figure 4.1). Some *Volvox* colonies are single-layered, hollow spheres of as many as *600,000* flagellated cells. The sphere can be moved forward when the flagella beat in unison. Reproduction is assigned to a few cells; they divide and give rise to daughter colonies that float inside the parent sphere. The daughter cells are released after they pro-duce and secrete certain enzymes that dissolve the jellylike secretions holding the parent colony together. Some species even show sexual differentiation (certain cells produce eggs, sperm, or both).

Think about the green algae that are merely chains or sheets of cells. Tiny multicelled plants that first invaded the land were not much more complex than this. Or think about how a tiny marine animal (*Trichoplax*) is no more than a flattened ball of ciliated cells, half a millimeter across (Figure d). It has no right side, left side, front, or back. It simply moves in any direction, amoebalike. *Trichoplax* may resemble some of the first multicelled animals.

Finally, think about just one of your own multicelled systems—the calcium-containing bones of your skeleton. Given the central roles of calcium in cell division and cell movements, the earliest eukaryotes must have had proteins that served as calcium storage centers. (Such proteins could latch onto calcium entering the cell, much like calcium-binding proteins do in existing eukaryotes.) In time, proteins of this sort must have become incorporated into shells and internal structures. Certainly by 750 million years ago, simple multicelled animals with hard parts exploded onto the evolutionary stage. In the increasingly elaborate shells and internal skeletons of the early animals, we have hints of the origins of bones—of our own skeletal system and calcium reservoirs.

b   1 µm   c

body folds pushed back to show chloroplasts

d

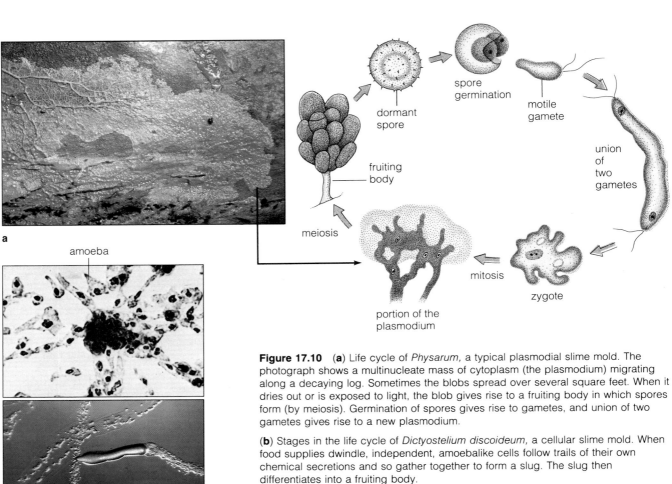

a

amoeba

b

Figure 17.10 (**a**) Life cycle of *Physarum*, a typical plasmodial slime mold. The photograph shows a multinucleate mass of cytoplasm (the plasmodium) migrating along a decaying log. Sometimes the blobs spread over several square feet. When it dries out or is exposed to light, the blob gives rise to a fruiting body in which spores form (by meiosis). Germination of spores gives rise to gametes, and union of two gametes gives rise to a new plasmodium.

(**b**) Stages in the life cycle of *Dictyostelium discoideum*, a cellular slime mold. When food supplies dwindle, independent, amoebalike cells follow trails of their own chemical secretions and so gather together to form a slug. The slug then differentiates into a fruiting body.

Slime Molds

Slime molds are phagocytic cells. They live mainly on decaying plant material, such as rotting logs, where they engulf bacteria and other microbes for food.

In many respects, slime molds resemble organisms of other kingdoms. Like certain bacteria, the cells of some slime mold species gather together, then differentiate to form spore-bearing structures called fruiting bodies. Each structure is a stalk with reproductive cells (spores) at its tip. Certain spores have thick walls and are dispersed by air currents, just like fungal spores.

Slime molds spend part of their life creeping about like animals. Cells of "plasmodial" slime molds migrate as a single blob of cytoplasm, leaving a slimy track behind. Amoebalike cells of *Dictyostelium discoideum* and other "cellular" slime molds retain their separate identity when clustered together (Figure 17.10).

Euglenids

Ponds and lakes, especially stagnant ones, are home to euglenids. For their size, these flagellated, photosynthetic protistans show great complexity (Figure 17.11).

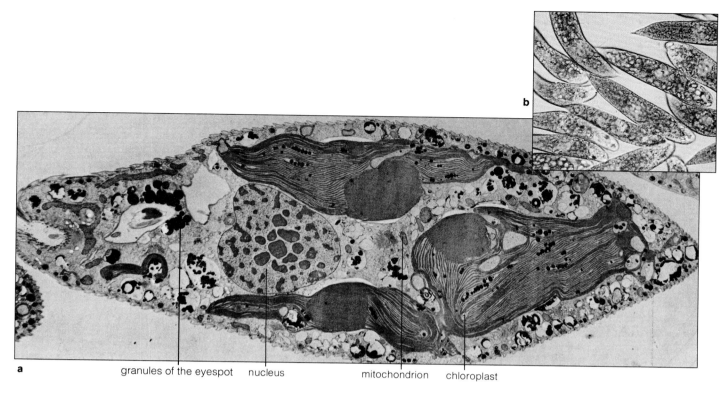

a granules of the eyespot nucleus mitochondrion chloroplast

Figure 17.11 Example of the complexity possible in a eukaryotic cell. (**a**) This long-section of *Euglena* shows the profusion of internal organelles. (**b**) Light micrograph of living *Euglena* cells.

One feature is an **eyespot,** which contains granules of light-absorbing pigments. It senses sunlight intensity and is used to help position the body for favorable light exposure. Some *Euglena* species can survive without photosynthesis if nutrients are abundant (they switch to other ATP-generating pathways).

Chrysophytes

The photosynthetic protistans called chrysophytes include yellow-green algae, golden algae, and diatoms. They abound in freshwater and marine habitats. Other pigments mask the chlorophyll and make these organisms look brownish-yellow.

Most chrysophytes have silica hard parts. The thin, two-part silica wall of diatoms fits together like a pillbox. Exchanges between the plasma membrane and the environment occur through numerous wall perforations (Figures 3.1 and 17.12).

The silica shells of diatoms have been accumulating for 100 million years, and extensive deposits of the resulting fine, crumbly material occur in different regions. Many abrasives, filtering materials, and insulating materials are made of "diatomaceous earth."

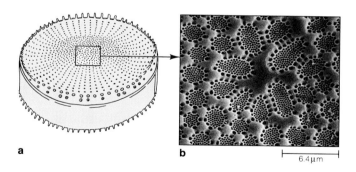

a b 6.4 μm

Figure 17.12 (**a**) Diatom shells. (**b**) Closer view of a shell, showing the intricate perforations.

a

Figure 17.13 (**a**) Source of red tides along the Florida coast—the dinoflagellate *Ptychodiscus brevis*. (**b**) Part of a fish kill resulting from a dinoflagellate "bloom."

b

a

b

c

Figure 17.14 Flagellated and amoeboid protozoans. (**a**) The trypanosome responsible for African sleeping sickness. It is transmitted to humans by the tsetse fly, a biting insect. (**b**) Foraminiferan shells, former homes of amoeboid protozoans. (**c**) *Amoeba proteus*, crawling about in a water droplet.

Dinoflagellates

Nearly all dinoflagellates are photosynthetic members of marine plankton. ("Plankton" refers to communities of aquatic organisms, mostly microscopic, that drift or swim weakly through the water.) Some have flagella that fit like ribbons in grooves between stiff cellulose plates at the body surface (Figure 17.13).

Dinoflagellates appear yellow-green, brown, or red, depending on the predominant photosynthetic pigments. Every so often, red dinoflagellates undergo population explosions and color the seas red or brown. Because some forms produce a neurotoxin, the resulting *red tides* can have devastating effects. Hundreds of thousands of fish that feed on plankton may be poisoned and wash up along the coasts. The neurotoxin does not affect clams, oysters, and other mollusks, but it builds up in their tissues. Humans who eat the tainted mollusks may die.

Flagellated and Amoeboid Protozoans

Some protistans are called protozoans ("first animals"), a name that refers in part to their predatory or parasitic habits. There are more than 65,000 protozoan species. Fewer than two dozen cause diseases in humans, yet their influence is staggering. In any year, hundreds of millions suffer protozoan infections!

Flagellated Protozoans. The flagellated protozoans include free-living and parasitic forms. Some trypanosomes, for example, are harbored in the salivary glands of biting insects that can infect new hosts (Figure 17.14).

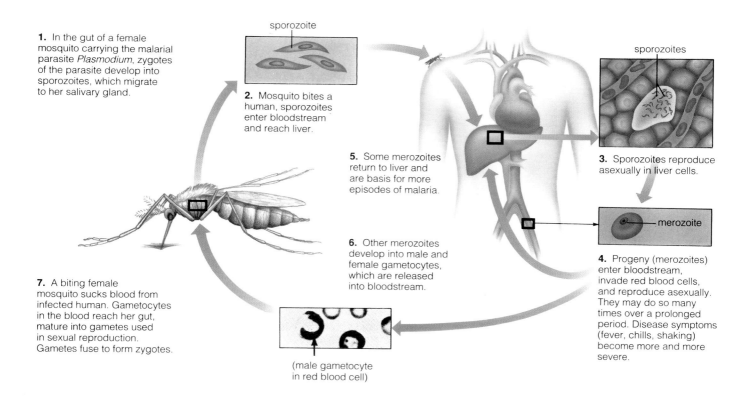

1. In the gut of a female mosquito carrying the malarial parasite *Plasmodium*, zygotes of the parasite develop into sporozoites, which migrate to her salivary gland.

sporozoite

2. Mosquito bites a human, sporozoites enter bloodstream and reach liver.

sporozoites

3. Sporozoites reproduce asexually in liver cells.

5. Some merozoites return to liver and are basis for more episodes of malaria.

merozoite

4. Progeny (merozoites) enter bloodstream, invade red blood cells, and reproduce asexually. They may do so many times over a prolonged period. Disease symptoms (fever, chills, shaking) become more and more severe.

6. Other merozoites develop into male and female gametocytes, which are released into bloodstream.

7. A biting female mosquito sucks blood from infected human. Gametocytes in the blood reach her gut, mature into gametes used in sexual reproduction. Gametes fuse to form zygotes.

(male gametocyte in red blood cell)

Figure 17.15 Life cycle of the sporozoan *Plasmodium*, which causes the disease malaria. The life cycle unfolds in the human body and in an insect (the female *Anopheles* mosquito), which transfers the sporozoan to new hosts during bites.

Also in this group are the trichomonads. *Trichomonas vaginalis*, a worldwide nuisance, is transferred to new human hosts during sexual intercourse; without treatment, trichomonad infection damages membranes in the urinary and reproductive tracts.

Infection by *Giardia intestinalis* usually causes mild intestinal disturbances, but sometimes it can be fatal. This flagellated protozoan forms **cysts** (walled, resting structures) that leave the body in feces. It infects a new host who ingests food or water contaminated with the feces. Even the water of remote mountain streams can contain the cysts and should be boiled before drinking.

Amoeboid Protozoans. Among the members of this group are the foraminiferans and amoebas. Adult forms move or capture prey by sending out pseudopods ("false feet," which really are temporary extensions of the cell body). Most feed on algae, bacteria, and other protozoans.

The amoebas live in freshwater, seawater, and soil. Foraminiferans live mostly in the seas. The hard shells of many species are peppered with hundreds of thousands of holes through which pseudopods extend (Figure 17.14b). They include *Amoeba proteus* of biology laboratory fame (Figure 17.14c). A parasitic form, *Entamoeba histolytica*, causes a severe intestinal disorder (amoebic dysentery). It travels in cysts within feces, which may contaminate water and soil in regions with inadequate sewage treatment.

Sporozoans

The sporozoans are parasites. All have an infectious, sporelike stage (sporozoites) that often is transmitted to new hosts by insects. Some *Plasmodium* species cause malaria (Figure 17.15). Mosquitoes transmit them to bird or human hosts. About 150 million people are stricken with malaria each year, mostly in tropical and subtropical regions. Infection persists for long periods and generally is not fatal, unless the species *P. falciparum* is the causative agent. People traveling through countries with high rates of malaria are advised to use antimalarial drugs.

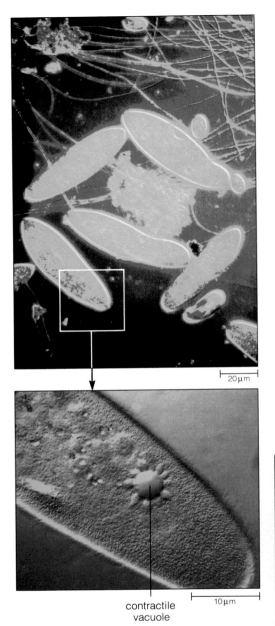

contractile
vacuole

|— 10 µm —|

|— 20 µm —|

Ciliated Protozoans

The ciliated protozoans demonstrate the structural complexity that is possible in a single cell. There are 8,000 freshwater and marine species, including fast-swimming, predatory forms with poison-charged harpoons and numerous cilia at the body surface (Figures 17.16 and 17.17). The cilia beat in synchrony, propelling the body through the water. Most forms, including *Paramecium*, have rows of cilia that beat bacteria and other food into a gullet (a cavity leading into the body). Food is digested in cytoplasmic vesicles, and wastes are eliminated at an anal pore. Like amoeboid protozoans, *Paramecium* uses contractile vacuoles to dispose of excess water.

This concludes our survey of the kingdoms of single-celled organisms. When we think about redwoods, whales, and other complex organisms, it seems almost incomprehensible that organisms as simple as the protistans gave rise to the multicelled kingdoms. Yet as the *Commentary* in this chapter suggests, it may be that we are forgetting to consider what must have been an immense evolutionary parade of intermediate forms.

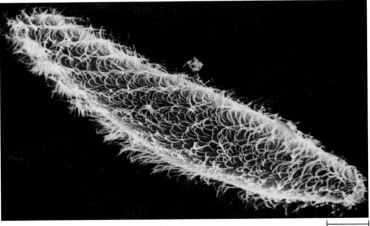

|— 20 µm —|

Figure 17.16 *Paramecium*, a ciliated protozoan of freshwater habitats. Its single-celled body contains more solutes than the surroundings, and water tends to move into it by osmosis (page 42). What keeps the cell body from bloating and rupturing the plasma membrane? Like some other protistans, *Paramecium* has a contractile vacuole. Tubelike extensions of this organelle extend through the cytoplasm, collecting water and draining it into a central vacuolar space. When filled, the vacuole contracts and the water is forced into a small pore that drains to the outside.

SUMMARY

1. In the microbial world are two kingdoms of single-celled organisms: the monerans (bacteria) and protistans. Viruses also are present in this world, but they are not alive.

2. A virus is a noncellular infectious agent with two characteristics. First, it consists only of a nucleic acid core and a protein coat that sometimes is enclosed in a lipid envelope. Second, it can replicate only after its genetic material enters a host cell and directs the cellular machinery into synthesizing the materials necessary to produce new virus particles.

3. All bacteria are prokaryotic, with no organelles. All have one chromosome (a circle of DNA); many also have plasmids (smaller, extra circles of DNA). Most reproduce by binary fission. Bacteria are metabolically diverse, with many different types of photosynthesizers, chemosynthesizers, and heterotrophs. They also show varied behavior.

4. Three prokaryotic lineages diverged from a common ancestor during the Archean Era: archaebacteria, eubacteria, and the forerunners of eukaryotes. Archaebacteria (the methanogens, extreme halophiles, and thermoacidophiles) thrive in harsh settings, much like the environments in which life probably originated. Eubacteria (all other existing bacteria) are fundamentally different from the other lineage in wall characteristics and other properties.

5. Protistans are single-celled eukaryotes (that have membrane-bound organelles and diverse modes of nutrition. The boundaries of their kingdom are poorly defined, with some lineages extending into the kingdoms of multicelled eukaryotes: plants, fungi, and animals.

6. The slime molds (phagocytic) and euglenids, chrysophytes, and dinoflagellates (mostly photosynthetic) are protistans. So are the flagellated protozoans (mostly motile predators or parasites); sporozoans and kin (parasites); and ciliated protozoans (mostly motile predators).

Review Questions

1. What is a virus? Why is a virus considered to be no more alive than a chromosome? *223*

2. Outline the replication cycle of a virus that enters the lytic pathway. *223*

3. Describe the key characteristics of a bacterium. What are some differences between archaebacteria and eubacteria? *225–229*

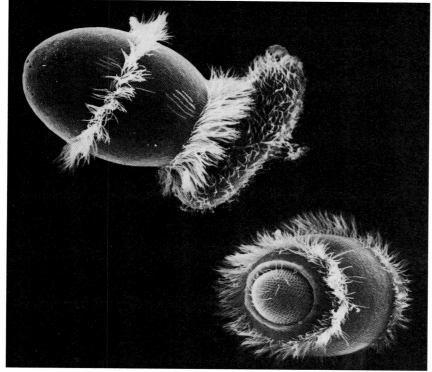

Figure 17.17 Mealtime for *Didinium*, a ciliated protozoan with a big mouth. Dinner in this case is *Paramecium*, poised at the mouth (upper left) and swallowed (lower right).

20 µm

4. Name a few photosynthetic, chemosynthetic, and heterotrophic eubacteria. Describe some that are likely to give you the most trouble recreationally (if you enjoy water sports), medically, and ecologically. *226* (if you are worried about the greenhouse effect, for example; see page *536*).

5. What is an endospore? Are all endospore-forming bacteria dangerous? *228–229*

6. Name the main categories of protistans. Think about where most of them live. Can you draw a few correlations between biotic and abiotic conditions in their environment and their structural characteristics? *229–236*

Self-Quiz   *(Answers in Appendix IV)*

1. Viruses cannot reproduce without the metabolic machinery of _____.

2. Viruses are _____.
   a. the simplest living organisms
   b. agents of infection
   c. nonliving
   d. a and b above
   e. b and c above

3. Viruses infect _____.
   a. bacteria
   b. plants
   c. animals
   d. all of the above

4. The two main structural features of viruses are a _____ and a _____.
   a. DNA core; protein coat
   b. nucleic acid core; plasma membrane
   c. DNA-containing nucleus; lipid envelope
   d. nucleic acid core; protein coat

5. The two kingdoms of single-celled organisms are the _____.
   a. animals and plants
   b. fungi and plants
   c. monerans and protistans
   d. viruses and monerans
   e. fungi and protistans

6. All bacteria have _____ circular chromosome(s) and may have smaller, extra circles of _____ known as plasmids.
   a. one; RNA
   b. two; RNA
   c. one; DNA
   d. two; DNA

7. Among the diverse bacteria are _____.
   a. photosynthetic autotrophs
   b. chemosynthetic autotrophs
   c. heterotrophs
   d. two of the above
   e. a, b, and c above

8. Archaebacteria are thought to be like ancient prokaryotes because they _____.
   a. have RNA rather than DNA as their hereditary material
   b. live in places reminiscent of conditions on early earth
   c. photosynthesize by mechanisms similar to that of their ancestors
   d. all of the above

9. Bacteria reproduce by _____.
   a. mitosis
   b. meiosis
   c. binary fission
   d. use of elaborate sexual systems

10. Match the groups to their descriptions.
   _____ monerans and   a. cannot reproduce without
           protistans   pirating metabolic machinery of
   _____ archaebacteria   living cells
   _____ protistans   b. all bacteria except archaebacteria
   _____ viruses   c. all single-celled organisms
   _____ eubacteria   d. single-celled eukaryotes
                              e. methanogens, extreme halo-
                                 philes, and thermoacidophiles

Selected Key Terms

archaebacterium *227*   heterocyst *228*
bacterial flagellum *225*   heterotroph *226*
bacteriophage *223*   microbe *222*
binary fission *227*   multicellularity *231*
chemosynthetic autotroph *225* photosynthetic autotroph *225*
endospore *228*   protistan *229*
eubacterium *227*   symbiosis *230*
eyespot *233*   virus *223*

Readings

Brock, T., and M. Madigan. 1988. *Biology of Microorganisms*. Fifth edition. Englewood Cliffs, New Jersey: Prentice-Hall.

Frankel-Conrat, H., P. Kimball, and J. Levy. 1988. *Virology*. Second edition. Englewood Cliffs, New Jersey: Prentice-Hall.

Frazier, W., and D. Westoff. 1988. *Food Microbiology*. Fourth edition. New York: McGraw-Hill. Good reference on the microbes that have major effects on our food supplies.

Margulis, L., and K. Schwartz. 1988. *Five Kingdoms*. New York: Freeman. An illustrated guide to the diversity of life. Paperback.

Stanier, R., et al. 1986. *The Microbial World*. Fifth edition. Englewood Cliffs, New Jersey: Prentice-Hall.

Woese, C. 1981. "Archaebacteria." *Scientific American* 244(6):98–125.

KEY CONCEPTS

1. Fungi are heterotrophs. Their mode of nutrition (extracellular digestion and absorption) is essential in the decomposition of organic material and the cycling of nutrients in nature.

2. Like plants, fungi apparently evolved first in aquatic environments. Many existing fungi have symbiotic relationships with plants. This type of association may have been important when ancient plants and fungi invaded the land.

3. Plants are mostly multicelled, photosynthetic autotrophs that directly or indirectly nourish almost all other forms of life.

4. During the transition to life on land, plant parts became specialized for life in new environments. The specializations included root systems, vascular tissues, a waxy cuticle, and specialized means of nourishing, protecting, and dispersing gametes.

How often, if we think of them at all, do we think of plants and fungi as nothing more than part of the scenery—a stationary backdrop for the riveting activities of animals? To be sure, plants and fungi don't crawl, run, leap, or fly about. Yet many millions of years ago, simple aquatic plants managed to make the transition to life on land—and their descendants came to cloak the earth, inch by inch, continent by continent. Their evolution, while not smacking of the drama of, say, the rise and fall of dinosaurs, was nevertheless remarkable.

What adaptations allowed those pioneers to spread through a world vastly different from the lagoons and pools they left behind? Here was a world of dry winds and often pronounced seasonal shifts in temperature and rainfall. Water was not always available to provide them with dissolved nutrients and to serve as a medium for dispersing the new generations.

Spores, seeds, roots and shoots containing internal pipelines for water and nutrients—these were some of the passports to life on higher and drier land. Be glad ancient plants made the journey. Without them, and without the decomposing activities of fungi that accompanied them onto land, we humans and all other terrestrial animals never would have evolved.

Figure 18.1 Plants at the boundary between two different worlds, one aquatic, one terrestrial. More than 400 million years ago, the ancestors of existing land plants crossed the boundary and made their successful bids for life on land. Shown here, one of the brown algae (sea palm) of the intertidal zone. Of all the aquatic algae, it appears that only the descendants of certain green algae made it onto dry land.

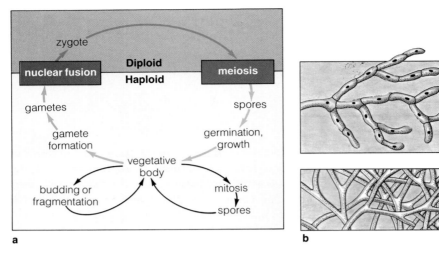

a

Figure 18.2 (**a**) Generalized life cycle for many fungi. Black arrows show the main asexual modes of reproduction, which typically dominate the life cycle. Blue arrows show the sexual mode of reproduction. Notice how the haploid stage dominates fungal life cycles. (**b**) Example of the filaments (hyphae) of a mycelium, the meshlike vegetative body that functions in food absorption.

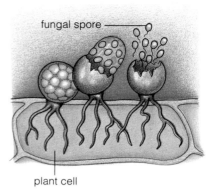

Figure 18.3 A chytrid, one of the microscopically small aquatic fungi. The sketch shows how motile spores are released from one type of chytrid that has parasitized a plant cell.

PART I. KINGDOM OF FUNGI

General Characteristics of Fungi

Mode of Nutrition. Fungi are heterotrophs that survive by decomposing living or nonliving organic matter. They can break down just about anything organic— nature's garbage (including dead plants and animal wastes), your groceries, clothing, paper, photographic film, leather, and paint.

Most fungi are **saprobes,** meaning they get nutrients from nonliving organic matter. Some are **parasites;** they get nutrients directly from the tissues of a living host. All fungi rely on extracellular digestion and absorption. Enzymes secreted by their cells break down large organic molecules into smaller bits, which the cells then absorb. The fungal mode of nutrition is vital for the world of life. In autumn, for example, one deciduous elm tree alone can shed 400 pounds of withered leaves. Without fungi and other decomposers, natural communities would become buried in their own garbage, nutrients would not be recycled, and life could not go on.

Body Plan. The vast majority of fungi are multicelled, with a body adapted for absorbing nutrients and reproducing. The food-absorbing part of the body is the **mycelium** (plural, mycelia), a mesh of tiny, branching filaments that grow over or into organic matter. Each filament is a **hypha** (plural, hyphae). Most often a hypha is composed of elongated cells having walls reinforced with chitin, a nitrogen-containing polysaccharide. The

mycelium extends through or over a large area of its surroundings and so provides the fungus with a favorable means of absorbing nutrients.

In many multicelled fungi, some hyphae become interwoven and modified, forming reproductive structures in which gametes or spores develop. (A mushroom is such a structure.) The **fungal spore** of land-dwelling species is a walled cell, resistant to dry or cold spells, that can be dispersed from the parent body. When it germinates, it gives rise to a new mycelium. As Figure 18.2 shows, fungal spores can be produced as part of the sexual cycle (by way of meiosis) or as part of a nonsexual cycle (by mitosis).

Major Groups of Fungi

About 60,000 species of fungi have been identified. The main groups are chytrids, water molds, zygospore-forming fungi, sac fungi, club fungi, and "imperfect" fungi (Table 18.1). Only chytrids and water molds produce motile spores. Spores of other fungi are dispersed by air currents or other means.

Chytrids and Water Molds. Muddy or aquatic habitats are home to microscopic chytrids. Most chytrids feed on decaying plants; some parasitize living plants, animals, and fungi. In many single-celled species, motile spores swim through the water and settle onto a host cell, germinate, then grow into a globe-shaped cell having rootlike absorptive structures (Figure 18.3). Most

Table 18.1 Major Groups of Fungi

Common Name	Typical Habitats
Chytrids	Aquatic (mud, decaying plants or animals); some parasitic
Water molds and related forms	Aquatic; some parasitic
Zygospore-forming fungi	Soil, decaying plant parts; some parasitic
Sac fungi	Soil, decaying plant parts; many pathogens of plants
Club fungi	Soil, decaying plant parts; many pathogens
Imperfect fungi	Diverse (e.g., soil, grains, human body)

a

b

500 µm

c 10 µm   d

water molds are major decomposers of aquatic habitats. Some parasitize algae or aquatic animals. The cottony growths you may have seen on goldfish or tropical fish are mycelia of a parasitic water mold. A few water molds can devastate certain crops (see *Commentary*).

Zygospore-Forming Fungi. Like fungi generally, the members of this group rely mostly on asexual reproduction (by spore formation). When they reproduce sexually, their zygotes develop a thick wall and so become zygospores, which give rise to stalked, spore-producing structures upon germination. Among these fungi are the black bread molds, notorious spoilers of baked goods (Figure 18.4a). Winds have dispersed the spores of one species *(Rhizopus stolonifer)* just about everywhere, including the North Pole. Another zygospore former, *Pilobolus*, disperses its spores in a wonderful blast (Figure 18.4b).

Sac Fungi. Possibly the most famous sac fungi are the single-celled yeasts, which normally live in the nectar of flowers and on fruits and leaves. Bakers use the carbon dioxide products of a yeast to make bread dough rise (Figure 18.4c); vintners and distillers use the ethanol product of various yeasts for alcoholic beverages. In nature, the multicelled sac fungi are far more numerous than yeasts, and they are important decomposers. Their spores occur in or on structures shaped like globes or cups (Figure 18.4d). This fungal group contains edible species (such as morels and truffles) and disease-causing species (see *Commentary*).

Figure 18.4 Representative zygospore-forming fungi (**a,b**) and sac fungi (**c,d**).

Spores of the black bread mold, *Rhizopus stolonifer* (**a**), give moldy bread a thoroughly unappetizing appearance. *Pilobolus* (**b**) grows on animal feces. The dark sacs above the swollen portion of each stalk contain spores. Water pressure builds up inside the swollen portion so that the spore sac can be blasted 2 meters away—a remarkable feat, considering the stalk is less than 10 millimeters tall!

(**c**) Cells of *Saccharomyces cerevisiae*, the yeast that makes bread dough rise. These single-celled fungi reproduce mainly by budding. (**d**) Scarlet cup fungus *(Sarcoscypha)*. Spore-bearing sacs occur within the cup, which is composed of tightly interwoven hyphae.

A Few Fungi We Would Rather Do Without

You know you are a serious student of biology when you can view organisms objectively in terms of their roles in nature, not in terms of the impact they have on humans generally and yourself in particular. As a student you can indeed respect that saprobic fungi are vital decomposers in the web of life, and that many parasitic fungi help keep populations of destructive insects and weeds in check.

The true test is when you raid the refrigerator for a bowl of exorbitantly priced raspberries and discover a fungal species beat you to them, or when another fungus has begun feeding on the warm, damp tissues between your toes.

Who among us can praise the fungal species that cause athlete's foot and similar diseases that make the skin redden, crack, or turn scaly? Which home gardeners can wax poetic about black spot or powdery mildew on their roses and other ornamental plants? Which farmers happily lose millions of dollars each year to invasions of their crops by rusts or smuts?

Some fungi have even influenced the course of human history. More than a century ago, Irish peasants cultivated potatoes as their main food source. Between 1845 and 1860, growing seasons were cool and damp, year after year. The cool conditions encouraged the rapid spread of *Phytophthora infestans,* a type of water mold responsible for *late blight* (the rotting of potato and tomato plants). The fungus produced abundant spores, spore dispersal through the watery film on plants went unimpeded, and destruction was widespread. During this fifteen-year period, a third of Ireland's population starved to death, died in the outbreak of typhoid fever that followed as a secondary effect, or fled to the United States and other countries.

And what about *Claviceps purpurea,* which is parasitic on rye and other grains? Alkaloid by-products of its activities have some medical uses. They are used to treat migraine headaches and to return the uterus to pre-pregnancy size and prevent hemorrhaging after childbirth. But when the alkaloids are eaten in large amounts, as might happen when rye flour is contaminated, *ergotism* develops. Symptoms of this disease include hysteria, hallucination, convulsions, vomiting, diarrhea, dehydration, and often gangrenous limbs. Severe cases are fatal.

Some Fungal Pathogens	
Water molds:	
Phytophthora infestans	Late blight of potato, tomato
Plasmapara viticola	Downy mildew of grapes (below)
Zygospore-forming fungi:	
Rhizopus	Food spoilage
Sac fungi:	
Ophiostoma ulmi	Dutch elm disease
Cryphonectria parasitica	Chestnut blight
Venturia inequalis	Apple scab
Claviceps purpurea	Ergot of rye, ergotism
Monilinia fructicola	Brown rot of stone fruits
Club fungi:	
Puccinia graminis	Black stem wheat rust
Ustilago maydis	Smut of corn
Amanita (some)	Severe or fatal food poisoning
Imperfect fungi:	
Verticillium	Plant wilt
Microsporum, *Trichophyton,* *Epidermophyton*	Various species cause ringworms, including athlete's foot
Candida albicans	Infection of mucous membranes

Ergotism epidemics were common in Europe in the Middle Ages, when rye was a major crop. Ergotism thwarted Peter the Great, the Russian czar who was obsessed with conquering ports along the Black Sea for his vast and nearly landlocked empire. The soldiers laying siege to the ports ate mostly rye bread and fed rye to their horses. The former went into convulsions and the latter, into "blind staggers." Quite possibly, outbreaks of ergotism were an excuse to launch the Salem witch-hunts in colonial Massachusetts.

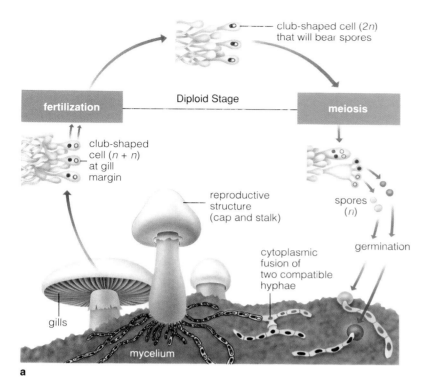

club-shaped cell (2n)
that will bear spores

Diploid Stage

fertilization

meiosis

club-shaped
cell (n + n)
at gill
margin

reproductive
structure
(cap and stalk)

spores
(n)

cytoplasmic
fusion of
two compatible
hyphae

germination

gills

mycelium

a

b

c

d

Figure 18.5 Generalized life cycle for club fungi (**a**) and some representatives. (**b**) The fly agaric mushroom *(Amanita muscaria)*, which causes hallucinations when eaten. Other species of *Amanita* are usually fatal when eaten. (**c**) A coral fungus. (**d**) A shelf fungus growing outward from a tree trunk.

Club Fungi. Figure 18.5 illustrates some of the diversity that exists among the club fungi. Although most species are important decomposers, the rusts and smuts cause serious plant diseases.

When most people hear the word fungus, they think of a mushroom—which is produced only by certain club fungi. Each mushroom is a short-lived reproductive structure consisting of a stalk and a cap (Figure 18.5a and b). Its spores are borne at the end of club-shaped cells, which are arranged along sheets of tissue (gills) in the cap. The rest of the fungus is an extensive mycelium growing through organic material in soil or decaying wood.

Imperfect Fungi. Fungi are classified mainly on the basis of their reproductive modes. A sexual phase is absent (or undetected) in many species, so they are lumped together in this informal category. Many imperfect fungi cause diseases. Some show predatory behavior. Others, including *Penicillium* species, are commercially and medically important (Figure 18.6 and *Commentary*).

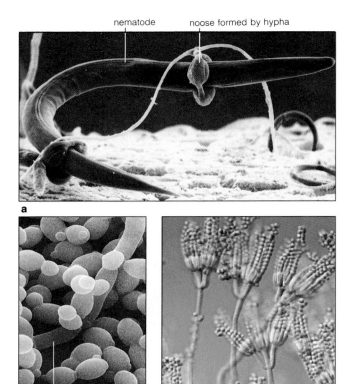

nematode | noose formed by hypha

a

b hypha   budding cell   **c**

Figure 18.6   Representative imperfect fungi.

(**a**) The trapping mechanism of *Arthrobotrys dactyloides*, a predatory fungus. The fungal hyphae form nooselike rings that can swell rapidly with incoming water after the hyphal cell walls are stimulated (as when a nematode brushes past). Within a tenth of a second, the increased turgor pressure shrinks the "hole" in the noose and captures the nematode. Hyphae grow into the animal's body and release digestive enzymes.

(**b**) *Candida albicans*, which causes "yeast infections" of the mucous membranes of the mouth and vaginal tract. A long hypha and budding cells are shown.

(**c**) *Penicillium*. This phase-contrast micrograph shows rows of spores, called conidia (singular, conidium), and the structures that produce them. Spores of this type are common among fungi.

Lichens and Other Symbionts

Many land-dwelling photosynthesizers and fungi enter into symbiotic relationships, in which one or both species come to depend permanently on the other. An example is the **lichen,** which is a fungus and a captive photosynthetic partner such as a green alga (Figure 18.7). Lichens often live in inhospitable places, including bare rock and wind-whipped tree trunks. Their secretions slowly break down rock and convert it to soil that can support larger plants. In the arctic tundra, where large plants are scarce, reindeer and musk oxen can survive on lichens. Near cities, air pollution is often monitored by observing lichens, which cannot grow in heavily polluted air.

Often the fungal member of a lichen absorbs all but twenty percent of the carbohydrates produced by its partner. It is difficult to see what the partner gets out of being enslaved, given how the drain on nutrients stunts its growth. Only a few green algae truly benefit from the relationship. They grow very slowly and cannot compete successfully with other organisms on their own, but they thrive in the shelter provided by a lichen.

Benefits flow both ways in a **mycorrhiza,** or "fungus-root." In this symbiotic relationship, fungal hyphae thread densely around the roots of forest trees and shrubs (page 309). The fungus gets carbohydrates from the plant. It also conserves dissolved mineral ions when they are plentiful and releases them to the plant when they are scarce in the soil. Many plants do not readily absorb mineral ions in the absence of mycorrhizae—which happen to be highly susceptible to acid rain. This susceptibility is having repercussions in the world's forests (page 572).

The fossil record suggests that many aquatic fungi and plants had entered into symbiotic partnerships before the invasion of the land, many millions of years ago. The partnership certainly would have increased chances for survival for both organisms in the new environment.

Figure 18.7 (**a**) Cross-section through one type of lichen, showing how fungal hyphae form layers above and below algal cells. (**b**) A lichen (*Cladonia rangiferina*), sometimes called reindeer "moss."

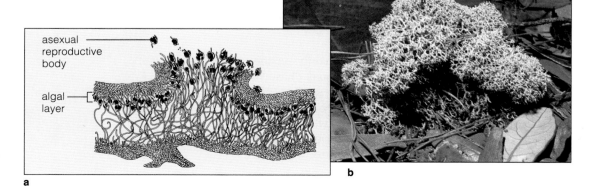

asexual reproductive body

algal layer

a

b

PART II. KINGDOM OF PLANTS

General Characteristics of Plants

We turn now to the kingdom dominated by multicelled, photosynthetic autotrophs—to the green plants that use sunlight energy, water, and dissolved minerals to produce their own food. Together with photosynthetic bacteria and protistans, plants directly or indirectly nourish almost all other forms of life on earth.

There are more than 280,000 species of plants living in freshwater and seawater, on land, even high above forest floors (attached to other plants). In size alone, they range from microscopic algae to giant redwoods. Most are **vascular plants**. They have a well-developed *root system* that absorbs water and nutrients from the soil. They also have a well-developed *shoot system* (stems, leaves, and other structures) that functions in photosynthesis and reproduction. And they have specialized tissues that transport water and solutes through the roots, stems, and leaves.

There are about 13,000 species of **red, brown,** and **green algae.** Nearly all are multicelled, aquatic species. On land, there are fewer than 16,000 species of bryophytes, which are **nonvascular plants:** they have very simple internal transport systems or none at all. Strictly speaking, the bryophytes do not have true roots, stems, and leaves, for these structures are partly defined by the vascular tissues inside them.

Evolutionary Trends Among Plants

Plants have a long evolutionary history. Multicelled green algae were flourishing in aquatic habitats 700 million years ago, long before animals with hard parts evolved. About 300 million years later, simple stalked plants that presumably descended from them had ventured onto land, in partnership with mycorrhizal fungi. After another 55 million years had passed, forests of tall, woody-stemmed plants with large, fernlike fronds were established. Fossil evidence of these plants gives us insight into some overall trends in their evolution.

From Nonvascular to Vascular Plants. During the transition to life on land, underground plant parts became specialized for absorbing water and nutrients from a large volume of soil. Aboveground parts became adapted for exploiting an abundant resource—sunlight energy. In some lineages, complex vascular tissues evolved within the increasingly extensive roots and shoots. One vascular tissue, *xylem*, transported water and dissolved mineral ions throughout the plant body. Another vascular tissue, *phloem*, transported the products of photosynthesis.

Structural support tissues also evolved; probably they proved adaptive in allowing some plants to grow taller and avoid being shaded by their neighbors. Stems and leaves became protected by a waxy coat (cuticle) that restricted water loss in dry environments; evaporation was controlled at tiny openings (stomata) across the cuticle. These tissue specializations are seen in nearly all existing land plants (page 293).

Sexual Reproduction on Land. Sexual reproduction typically involves fusion of a sperm and egg. In aquatic habitats, motile sperm swim to the eggs or are carried to them by currents. When plants began invading the land, they initially were confined to muddy or moist regions; their sperm and eggs could not get together elsewhere. The move to higher, drier land meant that gametes would have to be nourished, protected, and dispersed under hostile conditions. The mechanisms for doing so emerged through a shift from haploid to diploid dominance in the life cycle.

On page 96, we saw how the life cycle of sexually reproducing organisms alternates between diploid ($2n$) and haploid (n) phases. For most land plants, a **sporophyte** (spore-producing body) develops during the diploid phase; then a **gametophyte** (gamete-producing body) develops during the haploid phase. The first phase begins at fertilization, with the formation of a zygote. The zygote grows into a sporophyte. The second phase begins in the mature sporophyte, when some cells (touchingly called "mother cells") undergo meiosis and give rise to haploid spores. A **plant spore** is usually a single haploid cell that develops into a gametophyte, which produces gametes in its sex organs. The second phase of the life cycle ends at fertilization (Figure 18.8).

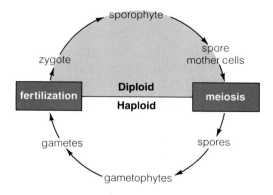

Figure 18.8 Generalized life cycle of plants, showing how a diploid sporophyte phase alternates with a haploid gametophyte phase. (This is often called "alternation of generations.") Only specialized parts of the plant produce spores and gametes.

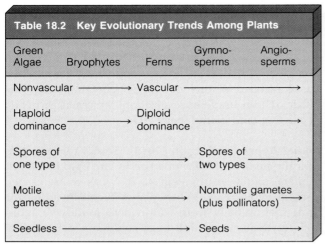

Table 18.2 Key Evolutionary Trends Among Plants

Green Algae	Bryophytes	Ferns	Gymno- sperms	Angio- sperms
Nonvascular ⟶	Vascular	⟶		
Haploid dominance ⟶	Diploid dominance	⟶		
Spores of one type	⟶	Spores of two types	⟶	
Motile gametes	⟶	Nonmotile gametes (plus pollinators) ⟶		
Seedless	⟶	Seeds	⟶	

a

b

Given enough sunlight, nutrients, and the ever-present liquid water, simple aquatic plants spend most of the time reproducing, so the haploid phase often dominates the life cycle. The diploid phase dominates the life cycle of complex land plants. Most activities are channeled into the growth and development of a large, multicelled sporophyte, such as a pine tree. With its roots and shoots the sporophyte is adapted for obtaining nutrients and water, even through dry spells.

Over evolutionary time, the sporophytes of some lineages began holding onto their spores and gameto-phytes, nourishing and protecting them until conditions favored dispersal. Also, two kinds of spores evolved, and these gave rise to more specialized male and female gametophytes. The sperm-producing male gameto-phytes became packages (pollen grains) for getting sperm to the eggs without the need for liquid water. Finally, among the most complex vascular plants, the embryo sporophyte came to be nourished and protected inside a seed coat. Seeds are ideal packages for surviving hostile conditions. It is probably no coincidence that the dominant seed plants arose during the climatic extremes of Permian times (page 204).

Table 18.2 summarizes the key evolutionary trends among plants. With these trends in mind, let's turn to the spectrum of plant diversity, beginning with the algae.

Red, Brown, and Green Algae

"Algae" originally meant simple aquatic plants, but the organisms once grouped under the term are now classified in different kingdoms. For example, we now classify "blue-green algae" (cyanobacteria) as monerans and "golden algae" as protistans. Only the red, brown, and green algae are still viewed as members of the plant kingdom.

Red Algae. "Red" algae actually appear green, red, purple, or greenish-black, depending on the types and abundances of different photosynthetic pigments. These pigments include phycobilins, which can trap blue-green light in deep water. Figure 18.9 shows two representatives of this group. Red algae are major producer organisms of marine communities. Some species have stone-like cell walls in which calcium carbonate has been deposited; they are among the builders of coral reefs (page 565). Several species of red algae are used to make agar (a laboratory culture medium), a moisture-retaining agent in baked goods, and a setting agent in gelatin desserts. In Japan especially, a red alga is commercially grown for food.

Brown Algae. This group includes many olive-green, golden, and dark-brown species (Figures 18.1 and 18.9d). Xanthophylls and other pigments give them their distinct color. Most brown algae live offshore or in intertidal zones, attached to submerged rocks by rootlike structures. Some species have a stemlike structure, leaflike blades, and gas-filled "floats" that help hold the plant upright. Tall, underwater forests of giant kelps are examples of the sporophytes; the gametophytes are microscopically small.

When you eat commercial ice cream, pudding, salad dressing, canned and frozen foods, jelly beans, or beer, and when you use cough syrup, toothpaste, a variety of cosmetics, paper, textiles, ceramics, or floor polish, thank the brown algae. Some species produce algin, a substance we use as a thickening, emulsifying, or suspension agent.

c

d

Figure 18.9 Red algae, one (**a**) showing a common, finely branched growth pattern and another (**b**) showing sheetlike growth. (**c**) A green alga showing branching growth. (**d**) A brown alga. This variety of kelp is usually submerged, even at low tide; its long stalks are anchored to rocks below.

Figure 18.10 Life cycle of *Chlamydomonas,* a single-celled green alga. Asexual reproduction is most common, but sexual reproduction also can occur between different mating strains.

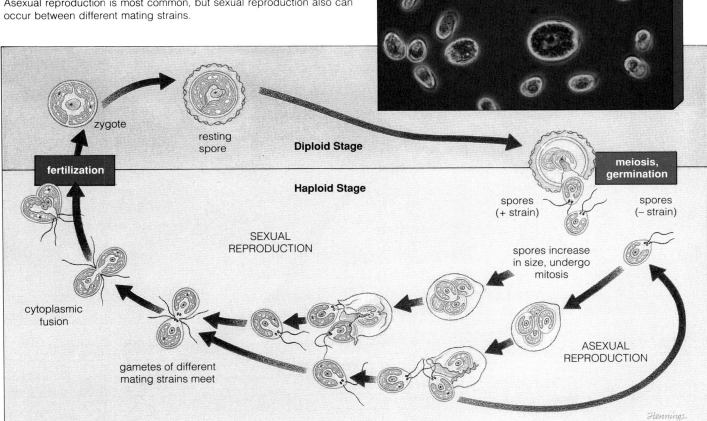

Green Algae. Many green algae live in the seas, others live in fresh water, and a few even grow on snow, soil, and tree trunks. Figure 18.9c shows a representative green alga. Another green alga (*Ulva,* or sea lettuce) is being harvested for food in some Asian countries; still another *(Chlorella)* is a potential source of oxygen for space stations and submarines.

Like complex land plants, green algae have a distinct array of photosynthetic pigments (chlorophylls *a* and *b,* carotenoids, and xanthophylls), their cells have cellulose walls, and they store carbohydrates as starch. Figures 18.10 and 18.11 show just two of the many patterns of sexual and asexual reproduction among these mostly aquatic plants.

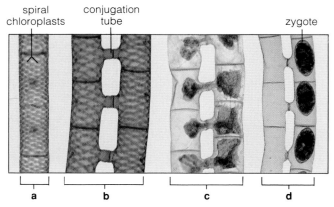

spiral chloroplasts conjugation tube zygote

a   b   c   d

Figure 18.11 One mode of sexual reproduction in watersilk *(Spirogyra)*, a filamentous green alga with ribbonlike, spiral chloroplasts (**a**). A conjugation tube forms between cells of adjacent, haploid filaments of different mating strains (**b**). The cellular contents of one strain pass through the tubes into cells of the other strain, where zygotes form (**c,d**). The zygotes will develop thick walls, and meiosis will occur in them when they germinate to form new haploid filaments.

Figure 18.12 Life cycle of a moss, a representative bryophyte. Notice how the sporophyte is attached to (and dependent upon) the gametophyte.

Bryophytes

Bryophytes are nonvascular land plants. Like their vascular relatives, they display three features that surely were adaptive during the transition to land. First, above-ground parts usually have a cuticle with numerous stomata. Second, protective tissue layers surround reproductive cells and help keep them from drying out. The *sporangium* (plural, sporangia) houses spores; the *archegonium* (archegonia) and *antheridium* (antheridia) house the eggs and sperm, respectively. Third, the embryo sporophyte begins its early development *inside* the female gametophyte.

Bryophytes include mosses, liverworts, and hornworts. All are small plants, generally less than 20 centimeters (8 inches) long. Although they have leaflike, stemlike, and rootlike parts (Figures 18.12 and 18.13), they do not have xylem or phloem. Most bryophytes grow in regions that are moist for much of the year.

Mosses are the most common bryophytes. Their spores give rise to threadlike gametophytes that grow into the well-known moss plants. Generally, eggs and sperm develop at the shoot tips, inside protective layers

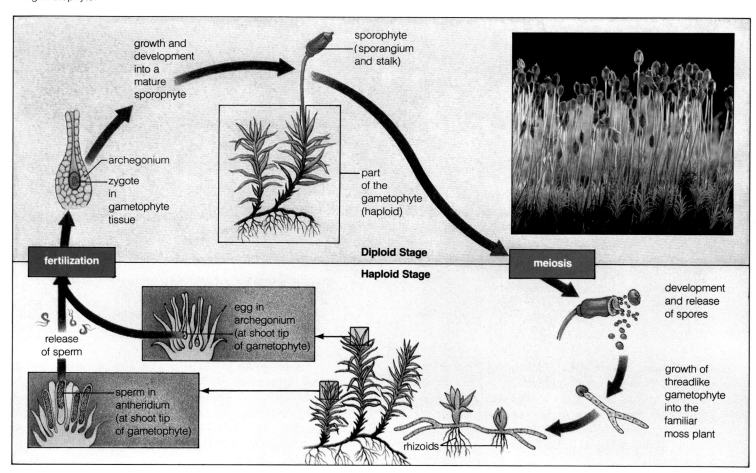

of the gametophytes. Sperm reach the eggs by swimming through a film of water on plant parts. After fertilization, the zygote develops into a sporophyte, which depends on the gametophyte for nutrients and water. The bryophytes are distinct from all other land plants in having an independent gametophyte and a dependent sporophyte.

Lycophytes, Horsetails, and Ferns

Seedless vascular plants once dominated the land (Figure 18.14a). Existing members of this group include the lycophytes, horsetails, and ferns. Their sporophytes, which develop independently of the gametophyte, have well-developed vascular tissues. Although these plants are adapted for life on land, they are confined largely to wet, humid regions (their gametophytes have no vascular tissues for water transport). Also, the sperm must travel through some water to reach the eggs.

Lycophytes. About 350 million years ago, some lycophytes were tree-sized. Among their descendants are the small club mosses *(Lycopodium)* found on forest floors. Spores often are borne on leaves that are sometimes grouped into conelike clusters. After germinating, the spores give rise to small, free-living gametophytes. Figure 18.14b shows the sporophytes of one club moss.

Horsetails. The ancient relatives of horsetails included tall, treelike forms. A single genus, *Equisetum*, has survived to the present (Figure 18.14c). Horsetails grow in moist soil along streams and in disturbed habitats, such as vacant lots, roadsides, and beds of railroad tracks. Their sporophytes typically have underground stems (rhizomes) and scalelike leaves on an aboveground, photosynthetic stem. Pioneers of the American West used horsetails to scrub their cooking pots; the

a

b c

d

Figure 18.14 Seedless vascular plants. (**a**) Reconstruction of some of the vascular plants of the Carboniferous era. Most of the tall, treelike forms are extinct. The descendants of such forms include the modern-day lycophytes, horsetails, and ferns. (**b**) A lycophyte *(Lycopodium)* and (**c**) a horsetail *(Equisetum)*. Notice the conelike spore-bearing structures of their sporophytes. (**d**) Ferns in a Tasmanian rain forest.

Figure 18.13 A liverwort *(Marchantia)*, showing the leaflike structure of its gametophytes. The cups hold asexually produced vegetative bodies that can grow into new plants.

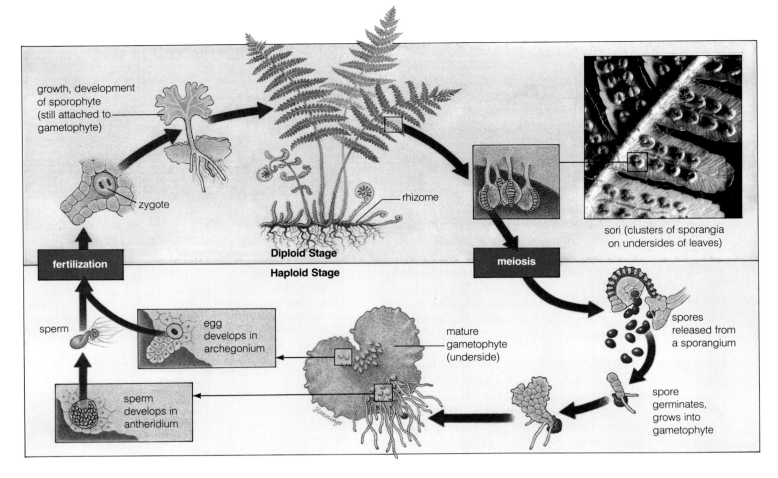

Figure 18.15 Fern life cycle.

walls of stem cells contain silica, giving the stems a sandpaper quality (see also Figure 14.11e).

All horsetail spores form in protective layers at the shoot tips, then they are dispersed by air currents. They must germinate within a few days to produce gametophytes, which are free-living plants about the size of a small pea.

Ferns. Among the 12,000 or so species of ferns are some of the most popular nursery plants. Most ferns are native to tropical and temperate regions (Figure 18.14d). Some floating species are less than 1 centimeter (0.4 inch) across; some tropical tree ferns are 25 meters (82 feet) tall. Except for tropical tree ferns, the stems are mostly underground. Fern leaves (fronds) are usually finely divided, like feathers, but extraordinary diversity exists on the basic plan.

You may have noticed rust-colored patches on the lower surface of many fern fronds. Each patch is a **sorus** (plural, sori), a cluster of sporangia (Figure 18.15). Each sporangium is microscopic and looks rather like a baby rattle. At dispersal time, the sporangium snaps open and causes the spores to catapult through the air. A germinating spore develops into a small gametophyte,

the most common being the green, heart-shaped type shown in Figure 18.15.

We now leave the simple vascular plants. Take a moment to review Table 18.3, which compares their characteristics with those of the remaining major plant groups.

Existing Seed Plants

In terms of sheer numbers and distribution, the most successful vascular plants are the seed-bearing species. Their ancestors first appeared during Devonian times. Fossil evidence of one species, a "progymnosperm," was shown earlier in Figure 15.1. Plants of this type were on the evolutionary road leading to both **gymnosperms** and **angiosperms.**

The word gymnosperm is derived from the Greek *gymnos* (meaning naked), and *sperma* (which is taken to mean seed). As the name implies, gymnosperm seeds are rather unprotected; they are perched at the surface of reproductive parts. The word angiosperm is derived from the Greek *angeion* (meaning vessel) and *sperma*. The "vessel" refers to floral parts that surround and protect the seeds during development.

Table 18.3 The Major Groups of Plants and Their Characteristics

Division	Number of Species	Some General Characteristics
Simple, Mostly Aquatic Plants		
Red algae	4,000	Single-celled to multicelled aquatic plants with branching filaments or fan shapes. Most in warm marine habitats; some deep-water species; some reef builders. Abundant phycobilin pigments.
Brown algae	1,500	Multicelled aquatic plants, including kelps. Almost all marine, many in colder seas. Some with structures that resemble roots, stems, leaves. Abundant xanthophyll pigments. Diploid dominance in many species.
Green algae	7,000	Single-celled to multicelled aquatic plants, most freshwater, wet soil; many in warm seas. Filaments, colonies, sheetlike forms. Pigments like land plants.
Nonvascular Plants		
Bryophytes	16,000	Mosses, liverworts, hornworts. Underground "stems," rootlike, leaflike structures. Many moist, humid habitats. Water required for fertilization. Diploid sporophyte dependent on dominant (haploid) gametophyte.
Seedless Vascular Plants		
Lycophytes	1,000	Club mosses the most familiar. Mostly wet or shady habitats; water required for fertilization. Cuticle, stomata. Diploid dominance. Some with two types of spores.
Horsetails	15	One existing genus. Swamps, moist woodlands, lake edges, railroad beds; water required for fertilization. Hollow photosynthetic stem with spore-bearing and scalelike leaves. Diploid dominance.
Ferns	12,000	Mostly tropical, temperate plants. Wet, humid habitats. Finely divided leaves typical; rhizomes. Tree ferns have woody stems. Diploid dominance.
Vascular Plants with "Naked" Seeds (Gymnosperms)		
Conifers	550	Mostly evergreen, woody trees and shrubs with needlelike or scalelike leaves. Widespread in Northern, Southern hemispheres. Seed- and pollen-bearing cones. Diploid dominance.
Cycads	100	Palmlike, woody stem, very slow growth. Tropics, subtropics. Pollen- and seed-bearing cones on different plants. Diploid dominance.
Ginkgos	1	Woody stem, deciduous fan-shaped leaves. Microspores, ovules on different plants. Fleshy seeds. Diploid dominance.
Gnetophytes	70	Some woody branching shrubs; one has strappy leaves. Pollen- and seed-bearing cones on different plants. Diploid dominance.
Vascular Plants with Flowers, Protected Seeds (Angiosperms)		
Flowering Plants:	235,000	Woody and nonwoody (herbaceous) plants. Flowering, seed-bearing. Depend on wind or animal pollinators. Almost every land region, some aquatic. Diploid dominance.
Monocots	65,000	Grasses, palms, lilies, orchids, onions, etc. Floral parts often arranged in threes or multiples of three; one cotyledon (seed leaf); parallel-veined leaves common.
Dicots	170,000+	Most fruit trees, roses, cabbages, melons, beans, potatoes, etc. Floral parts often arranged in fours, fives, or multiples of these; two seed leaves; net-veined leaves common.

Figure 18.16 Lesser known gymnosperms. (**a**) A cycad and its cones. (**b**) The fleshy seeds of a ginkgo tree. (**c,d**) A gnetophyte (*Welwitschia*) and its female cones.

Gymnosperms

There are over 700 species of gymnosperms. The sporophytes of nearly all species are conspicuous trees or shrubs. The small gametophytes are not free living; they are enclosed by sporophyte tissues. The most familiar gymnosperms are the conifers. Seed-bearing plants called cycads, ginkgos, and gnetophytes also belong to this group (Figure 18.16). All three types are restricted in their native distribution and numbers of species.

Conifers generally are woody trees and shrubs with needlelike or scalelike leaves. Familiar examples are the pines, spruces, firs, hemlocks, junipers, cypresses, and redwoods. Most are evergreen species; although they shed old leaves throughout the year, they retain enough leaves to distinguish them from deciduous species. All conifers have cone-shaped clusters of leaves bearing the sporangia (hence the name conifer, which means cone bearing). Seeds develop on scales of female cones.

We can use the pine as a general example of conifer life cycles (Figure 18.17). The pine tree produces two kinds of spores in two kinds of cones. The scales of male cones bear sporangia (pollen sacs), in which mother cells undergo meiosis and give rise to haploid **microspores.** The spores develop into pollen grains, each containing a male gametophyte. The scales of female cones bear **ovules** (sporangia with tissue layers that will become a seed coat). Inside each ovule, a mother cell undergoes meiosis. Only one of the resulting haploid spores, the *megaspore*, survives and develops into a female gametophyte.

Each spring, air currents lift millions of pollen grains off their cone perches. The extravagant numbers ensure that at least some pollen grains will land on female cones. The arrival of a pollen grain on female reproduc-

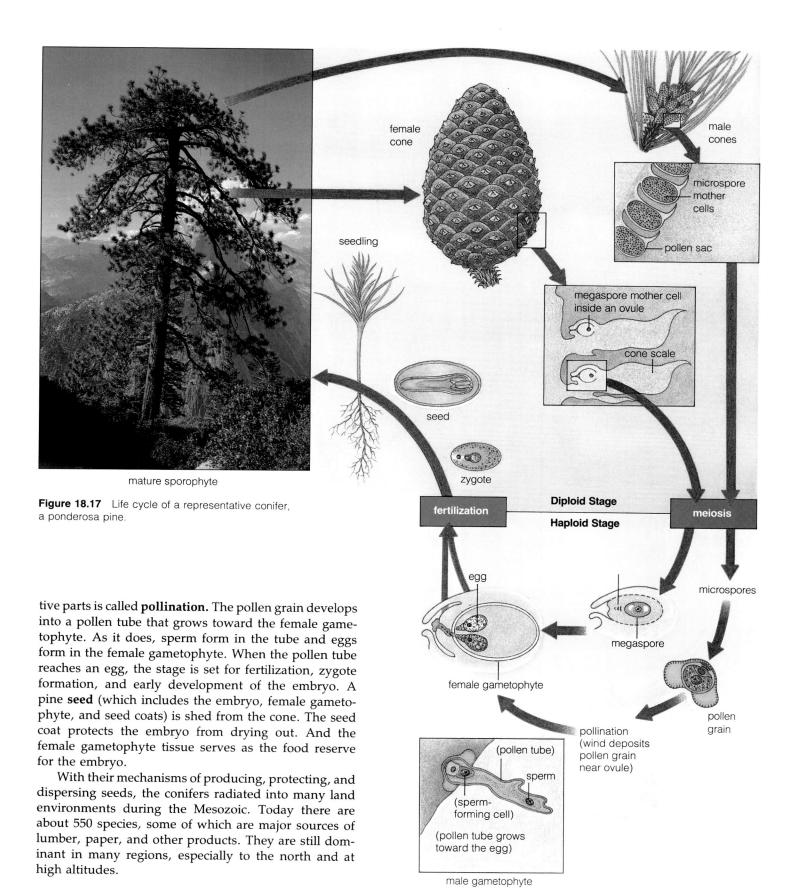

female cone

seedling

seed

zygote

mature sporophyte

Figure 18.17 Life cycle of a representative conifer, a ponderosa pine.

male cones

microspore mother cells

pollen sac

megaspore mother cell inside an ovule

cone scale

Diploid Stage

Haploid Stage

fertilization

meiosis

egg

microspores

megaspore

female gametophyte

pollen grain

pollination (wind deposits pollen grain near ovule)

(pollen tube)

sperm

(sperm-forming cell)

(pollen tube grows toward the egg)

male gametophyte

tive parts is called **pollination.** The pollen grain develops into a pollen tube that grows toward the female gametophyte. As it does, sperm form in the tube and eggs form in the female gametophyte. When the pollen tube reaches an egg, the stage is set for fertilization, zygote formation, and early development of the embryo. A pine **seed** (which includes the embryo, female gametophyte, and seed coats) is shed from the cone. The seed coat protects the embryo from drying out. And the female gametophyte tissue serves as the food reserve for the embryo.

With their mechanisms of producing, protecting, and dispersing seeds, the conifers radiated into many land environments during the Mesozoic. Today there are about 550 species, some of which are major sources of lumber, paper, and other products. They are still dominant in many regions, especially to the north and at high altitudes.

Angiosperms—The Flowering Plants

Of all the divisions of plants, the angiosperms (flowering plants) are the most successful. They have dominated the land for more than 100 million years. Today there are about 235,000 known species, and new ones are being discovered almost daily in previously unexplored regions of the tropics.

Angiosperms are also the most diverse types of plants. They range in size from tiny duckweeds (about a millimeter long) to *Eucalyptus* trees more than 100 meters tall. Most are free living and photosynthetic, but some are saprobic or parasitic. Diverse species are found on dry land and in wetlands, freshwater, and seawater (Figure 18.18).

There are two classes of angiosperms, the **monocots** and **dicots.** The monocots include grasses, palms, lilies, and orchids. The main crop plants (wheat, corn, rice, rye, and barley) are domesticated grasses—all monocots. Dicots are more diverse; there are nearly 200,000 species.

They include most of the familiar shrubs and trees other than conifers, most nonwoody (herbaceous) plants, the cacti, and the water lilies. For comparative purposes, we show the life cycle of a monocot in Figure 18.19. The life cycle of a typical dicot is described in detail in the next unit of the book, which focuses on the structure and function of flowering plants.

As you will see in the next unit, many factors contributed to the adaptive success of angiosperms. As with other seed plants, the diploid sporophyte dominates the life cycle. It retains and nourishes the gametophyte, and it is well adapted for life on land. The embryos are nourished by a unique tissue (endosperm) within the seed. Also, the seeds are packaged in fruits, which function in protection and dispersal. Above all, angiosperms have unique reproductive structures called **flowers.** Many diverse floral structures have coevolved with animal pollinators—insects, bats, birds, and rodents. This innovation probably figured in the rise of angiosperms and the gradual decline of gymnosperms in so many regions over the past 100 million years.

a

b

c

d

Figure 18.18 Representative flowering plants. Diverse species are beautifully adapted to nearly all environments, ranging from desert floors (**a**) to the snowline of high mountains (**b**). All use pollinators. Many, including the sugarcane plants in (**c**), are wildly successful in their numbers and distribution, being planted, tended, and harvested by humans. Some flowering plants, including water lilies (**d**), live in water and so overlap the domain of their simpler aquatic relatives, the algae.

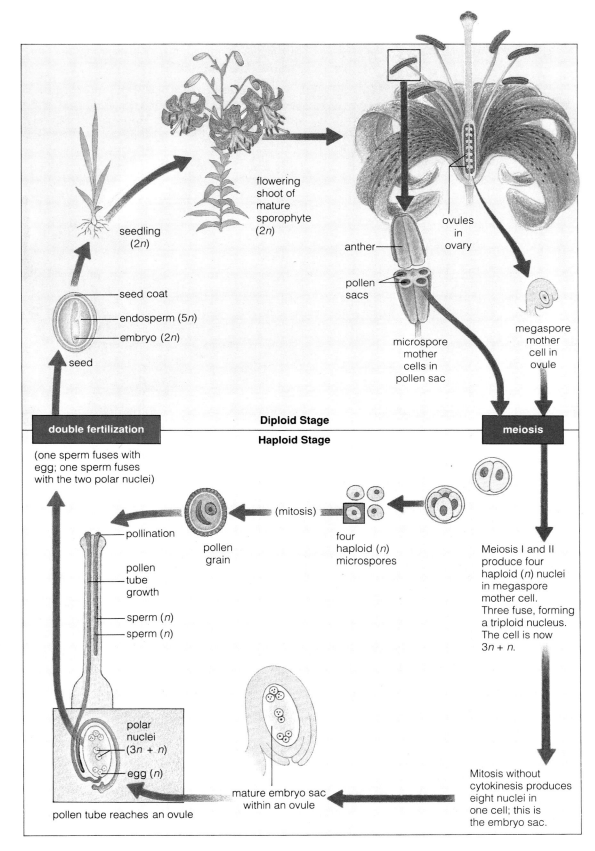

Figure 18.19 Life cycle of a monocot *(Lilium)*. Although about seventy percent of the known flowering plant species follow the pattern shown in Figure 22.4, microscopic slides abound of lilies, and you are likely to encounter these slides in the classroom.

seedling
(2n)

flowering
shoot of
mature
sporophyte
(2n)

ovules
in
ovary

anther

pollen
sacs

seed coat

endosperm (5n)

embryo (2n)

seed

microspore
mother
cells in
pollen sac

megaspore
mother
cell in
ovule

Diploid Stage

double fertilization

Haploid Stage

meiosis

(one sperm fuses with
egg; one sperm fuses
with the two polar nuclei)

(mitosis)

pollination

pollen
grain

four
haploid (n)
microspores

Meiosis I and II
produce four
haploid (n) nuclei
in megaspore
mother cell.
Three fuse, forming
a triploid nucleus.
The cell is now
$3n + n$.

pollen
tube
growth

sperm (n)

sperm (n)

polar
nuclei
($3n + n$)

egg (n)

mature embryo sac
within an ovule

pollen tube reaches an ovule

Mitosis without
cytokinesis produces
eight nuclei in
one cell; this is
the embryo sac.

SUMMARY

1. Fungi are heterotrophs that are important decomposers for the world of life. Many also are serious pathogens of plants and animals. Fungal species are either saprobic (feeding on nonliving organic matter) or parasitic (feeding on living organisms). Fungal cells secrete digestive enzymes that break down food into nutrient molecules, which are absorbed across the fungal plasma membrane.

2. Most fungi are multicelled, with haploid and diploid stages alternating in the life cycle. The multicelled body (mycelium) is composed of microscopic filaments (hyphae). Fungi can reproduce asexually (as by spore formation) or sexually.

3. The main groups of fungi are the chytrids, water molds, zygospore-forming fungi, sac fungi, club fungi, and "imperfect fungi."

4. The plant kingdom includes aquatic and terrestrial species, nearly all of which are photosynthetic autotrophs. Land plants are believed to have evolved from multicelled green algae over 400 million years ago.

5. These were the major evolutionary trends among plants:
   a. Structural adaptations to dry periods, especially vascular tissues (xylem and phloem).
   b. A shift from haploid (gametophyte) dominance to diploid (sporophyte) dominance.
   c. A shift from spores of the same type to spores of two different types. This paved the way for the evolution of male gametophytes that became specialized for dispersal without liquid water, and female gametophytes that became specialized for holding onto, protecting, and nourishing the embryo sporophytes (in seeds).

6. The red, brown, and green algae are mostly multicelled aquatic plants. Existing nonvascular land plants (those without well-developed xylem and phloem) include the bryophytes (mosses, liverworts, and hornworts). Existing vascular land plants include the lycophytes, horsetails, ferns, gymnosperms, and angiosperms (flowering plants).

7. Land plants typically have a cuticle and stomata that help control water loss. They have sporangia (protective tissue layers around their spores). They have ovules or similar protective and nutritive tissue layers around their gametophytes. The embryo sporophyte begins its development *within* gametophyte tissues.

8. Bryophytes are restricted to moist habitats; water is required for fertilization. The simplest vascular plants (lycophytes, horsetails, and ferns) also require the presence of water for fertilization.

9. The most complex vascular plants (gymnosperms and angiosperms) have escaped from dependency on liquid water for reproduction. Their gametophytes are attached to and protected by the sporophyte, which (with its root and shoot systems) is well adapted to conditions on dry land. Their male gametophytes and seeds protect and disperse the new generation and help it through hostile conditions.

Review Questions

1. Name and describe a member of each group of fungi. *240–243*

2. What is a lichen? A mycorrhiza? *244*

3. Define sporophyte, gametophyte, spore, and gamete. Which are haploid? Which are diploid? *245*

4. Describe the evolutionary trends among plants that figured in the invasion of land. *245, 246*

5. What are some differences between bryophytes and the vascular plants? What are gymnosperms? Angiosperms? *248–251*

6. Choose a garden plant, crop plant, or weed that grows in your neighborhood. Make a diagram of it, labeling its parts. Attempt to correlate some of its structures with seasonal variations in temperature, moisture, and other local environmental conditions.

Self-Quiz   *(Answers in Appendix IV)*

1. Fungi are _____.
   a. photosynthetic autotrophs
   b. chemosynthetic autotrophs
   c. heterotrophs
   d. a and b above

2. The fungal mode of nutrition involves _____.
   a. light-dependent and light-independent reactions
   b. phagocytosis and intracellular digestion
   c. extracellular digestion and absorption
   d. cycles of starvation and nutrient abundance

3. Fungi are major _____ of nature.
   a. producers
   b. saprobes and parasites
   c. decomposers
   d. b and c above

4. A mycorrhiza is a _____
   a. fungal disease of the foot
   b. fungus-plant relationship
   c. type of parasitic water mold
   d. type of fungus endemic to barnyards

5. The _____ stage dominates the life cycle of vascular plants.
   a. diploid
   b. haploid

6. Which of the following is *not* characteristic of gymnosperms and angiosperms (flowering plants)?
   a. vascular tissues
   b. diploid dominance
   c. single spore type
   d. nonmotile male gametes

7. Which of the following statements is *not* true?
   a. Red, brown, and green algae are mostly multicelled aquatic plants.
   b. The bryophytes are nonvascular plants.
   c. Lycophytes, horsetails, ferns, gymnosperms, and angiosperms are vascular plants.
   d. Horsetails and gymnosperms are the simplest vascular plants.

8. Evolution of spores of different types led to _____.
   a. convergence of haploid and diploid functions
   b. male gametophytes that could be dispersed without water
   c. seeds
   d. b and c above

9. A seed has the following parts:
   a. female gametophyte   c. seed coats
   b. embryo   d. all of the above

10. The gametophytes of gymnosperms and angiosperms are attached to and nourished by the _____.
    a. cuticle
    b. sorus
    c. sporophyte
    d. flower

11. Match the plant groups appropriately.
    _____ lycophytes, horse-   a. gametophytes attached to
           tails, ferns   sporophytes
    _____ fungi   b. major decomposers
    _____ bryophytes   c. mosses, liverworts, hornworts
    _____ angiosperms   d. simplest vascular plants
    _____ gymnosperms,   e. flowering plants
           angiosperms

12. Match each structure with its function.
    _____ cuticle, stomata   a. reproduction
    _____ seed   b. produce gametes
    _____ sporophyte   c. produce spores
    _____ ovule   d. control water loss
    _____ gametophyte   e. disperse embryo sporophyte
    _____ flower   f. protect, nourish gametophytes

Selected Key Terms

alga *245*
angiosperm *250*
dicot *254*
flower *254*
gametophyte *245*
gymnosperm *250*
hypha *240*
lichen *244*
megaspore *252*
microspore *252*
monocot *254*
mycelium *240*

mycorrhiza *244*
nonvascular plant *245*
ovule *252*
parasite *240*
plant spore *245*
pollination *253*
saprobe *240*
seed *253*
sorus *250*
sporophyte *245*
vascular plant *245*

Readings

Bold, H., and J. LaClaire. 1987. *The Plant Kingdom*. Fifth edition. Englewood Cliffs, New Jersey: Prentice-Hall. Paperback.

Gensel, P., and H. Andrews. 1987. ''The Evolution of Early Land Plants.'' *American Scientist* 75:478–489.

Moore-Landecker, E. 1982. *Fundamentals of the Fungi*. Second edition. Englewood Cliffs, New Jersey: Prentice-Hall. Well-written introduction to the kingdom of fungi.

Raven, P., R. Evert, and S. Eichhorn. 1986. *Biology of Plants*. Fourth edition. New York: Worth. Lavishly illustrated.

Stern, K. 1988. *Introductory Plant Biology*. Fourth edition. Dubuque, Iowa: Boston. Paperback.

19 ANIMALS

KEY CONCEPTS

1. There are probably more than 2 million species of animals—multicelled, motile heterotrophs that pass through a period of embryonic development during their life cycle. They range from simple sponges to vertebrates.

2. By comparing the body plans of existing animal groups and integrating this information with what can be seen in the fossil record, we can identify major trends that occurred in animal evolution.

3. The most revealing aspects of the animal body plan are its type of symmetry, gut, and cavity (if any) between the gut and body wall; whether it has a distinct head end; and whether it is divided into a series of segments.

Quick! What is an animal? Most likely the "animal" that just leaped into your mind was a mammal or a bird, reptile, amphibian, or fish (but probably a mammal). These are all **vertebrates,** the only animals with a "backbone" and the ones most familiar to us. Yet, of 2 million or more species of animals, fewer than 50,000 are vertebrates! Among the **invertebrates** (animals without a backbone) are lesser known but spectacularly diverse species, including the feathery-headed tube worm shown in Figure 19.1.

Suppose we take representatives of each animal phylum and arrange them in sequence, from the structurally simplest to the most complex. Sponges would be near one end of the spectrum and vertebrates near the other. If we accept that new species arise only from preexisting species (page 174), then all the phyla in our arrangement are related by way of descent—some closely, others distantly. Thus, comparing the similarities and differences among them may allow us to identify some major trends in animal evolution.

Figure 19.1 Guess which one of these fits our stereotype of what "an animal" is. To the left, a tube-dwelling marine worm, with mucus-coated featherlike structures that trap food particles suspended in water. To the right, a furry little arctic fox in its winter coat.

OVERVIEW OF THE ANIMAL KINGDOM

General Characteristics of Animals

There are more than thirty phyla of animals, but we can follow several evolutionary trends without considering every single one. Our sampling will include sponges, cnidarians, flatworms, roundworms, rotifers, mollusks, annelids, arthropods, echinoderms, and chordates (Table 19.1). Like other animals, they are defined by the following characteristics:

1. Animals are multicellular, and except for sponges, their cells form tissues. The tissues usually are arranged into organs and organ systems.

2. Animals are heterotrophs (they cannot produce their own food). They eat or absorb nutrients produced by other organisms.

3. Animals are diploid organisms that reproduce sexually and, in many cases, asexually.

4. Animal life cycles include a period of embryonic development. In brief, cell divisions transform the zygote into a multicelled embryo. Then the cells become arranged into germ layers (*ectoderm*, *endoderm*, and, in most species, *mesoderm*), and these give rise to all tissues and organs of the adult (page 475).

5. Most animals are motile, at least during part of the life cycle.

Body Plans

We can use five body features to track the increasing complexity among different animal groups. These are *body symmetry, cephalization, type of gut, type of body cavity*, and *segmentation*.

Body Symmetry and Cephalization. Nearly all animals have a radial or bilateral body plan. A hydra, for example, has body parts arranged radially, like spokes of a bike wheel (Figure 19.2a). A slice down its center divides it into equal halves; another slice at right angles to the first divides the hydra into equal quarters. You cannot get the same result with a crayfish, which shows

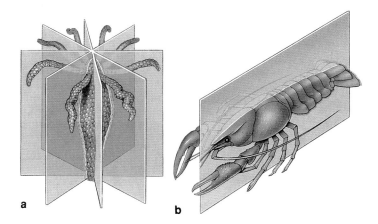

Figure 19.2 Planes of radial symmetry in a hydra (**a**) and of bilateral symmetry in a crayfish (**b**).

Table 19.1 Animal Phyla Described in This Chapter		
Phylum	Some Representatives	Number of Known Species
Porifera (poriferans)	Sponges	8,000
Cnidaria (cnidarians)	Hydrozoans, jellyfishes, corals, sea anemones	11,000
Platyhelminthes (flatworms)	Turbellarians, flukes, tapeworms	15,000
Nematoda (roundworms)	Pinworms, hookworms	20,000
Rotifera (rotifers)	Species with crown of cilia	1,800
Mollusca (mollusks)	Snails, slugs, clams, squids, octopuses	110,000
Annelida (segmented worms)	Leeches, earthworms, polychaetes	15,000
Arthropoda (arthropods)	Crustaceans, spiders, insects	1,000,000
Echinodermata (echinoderms)	Sea stars, sea urchins	6,000
Chordata (chordates)	Invertebrate chordates: Tunicates, lancelets	2,100
	Vertebrates:	
	Fishes	21,000
	Amphibians	3,900
	Reptiles	7,000
	Birds	8,600
	Mammals	4,500

Figure 19.3 Body plans of bilateral animals. (**a**) Acoelomate, or without a body cavity, as in flatworms. (**b**) Pseudocoelomate, without a continuous peritoneal lining; various organs, especially of the reproductive system, occupy the pseudocoel. (**c,d**) Coelomate, a plan typical of vertebrates and several invertebrate groups (including annelids and echinoderms).

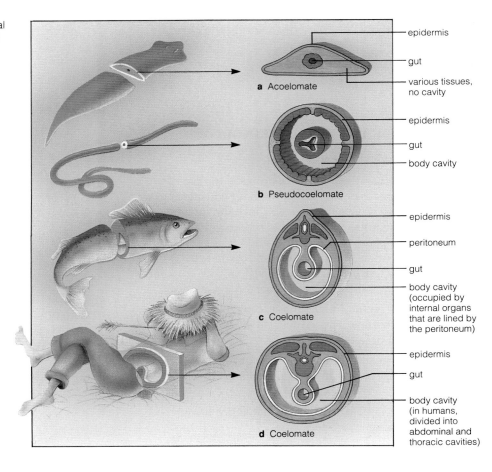

a Acoelomate
- epidermis
- gut
- various tissues, no cavity

b Pseudocoelomate
- epidermis
- gut
- body cavity

c Coelomate
- epidermis
- peritoneum
- gut
- body cavity (occupied by internal organs that are lined by the peritoneum)

d Coelomate
- epidermis
- gut
- body cavity (in humans, divided into abdominal and thoracic cavities)

Figure 19.4 (*Below*) Key trends in animal evolution, identified by comparing the body plans of major groups of existing animals.

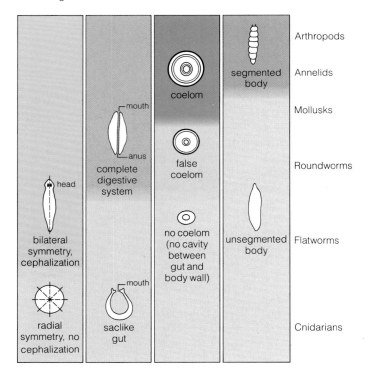

bilateral symmetry, cephalization

radial symmetry, no cephalization

mouth / anus — complete digestive system

mouth — saclike gut

coelom

false coelom

no coelom (no cavity between gut and body wall)

segmented body

unsegmented body

Arthropods

Annelids

Mollusks

Roundworms

Flatworms

Cnidarians

bilateral symmetry. Basically, its right and left halves are mirror images of each other (Figure 19.2b). It has a head end and a tail end (*anterior* and *posterior*). And it has a back side and a front or belly side (*dorsal* and *ventral*).

Radial animals are aquatic, and their body plan allows them to respond to food or danger floating past in any direction. Their ancestors originated before the first bilateral animals, which crawled on the seafloor. The forward end of the crawlers was the first to encounter food and other stimuli, so there surely were selective advantages in having sensory structures and nerve cells concentrated up front, in a head (cephalization). We can indeed identify trends toward bilateral symmetry and cephalization; they involved the development of paired muscles, paired sensory structures, paired nerves, and paired brain regions.

Type of Gut. A **gut** is a body region where food is digested and absorbed. Some guts are saclike, with only one opening (a mouth) for taking in food and expelling residues. Many are part of a tubelike system with openings at both ends (mouth and anus). These "complete" digestive systems are more efficient at processing nutrients. (Different tube regions are specialized for different tasks, such as preparing, digesting, and storing food.) The evolution of a complete gut helped pave the way for increases in body size and activity.

Body Cavities. A body cavity occurs between the gut and the body wall of most bilateral animals. The most common type, the **coelom,** has a lining called the *peritoneum* (Figure 19.3). The lining also covers organs inside the coelom and helps hold them in place. Like modern-day flatworms, the early bilateral animals did not have a coelom; their tissues were packed solidly between the gut and body wall. In some lineages, a "false" coelom evolved; this cavity forms in a distinct way and is not lined with peritoneum. However, a true coelom apparently was a prerequisite for the evolution of large, complex animals. Being cushioned and protected inside a body cavity, the organs of early "coelomate" animals had the potential for increased size and activity.

Segmentation. A "segmented" animal has a series of body units that may or may not be alike. For example, an earthworm has many similar-looking segments. Insect segments are grouped in such a way that they form three regions (head, thorax, and abdomen). In many lineages, segmentation foreshadowed the development of increasingly specialized head parts, legs, wings, and other appendages.

Not all of the body features just described appeared in every lineage, as Figure 19.4 shows. Their absence does not mean an animal is "primitive" or evolutionarily stunted; even the simplest animals are exquisitely adapted to their environment. Taken together, however, the similarities and differences among animals help us perceive broad evolutionary relationships (Figure 19.5). Let's turn now to specific examples from each major phylum.

SPONGES

Simple as they are, sponges are one of nature's success stories. They have been among the most abundant aquatic animals ever since Cambrian times. Some are as small as a fingernail; a few are big enough to sit in. Sponges have no body symmetry, no organs, and even their "tissues" are not like those of other animals. Flattened cells line the outer surface and parts of the internal cavities. Between the linings is a semifluid matrix with needles or fibers; both elements provide structural support (Figures 19.6 and 19.7).

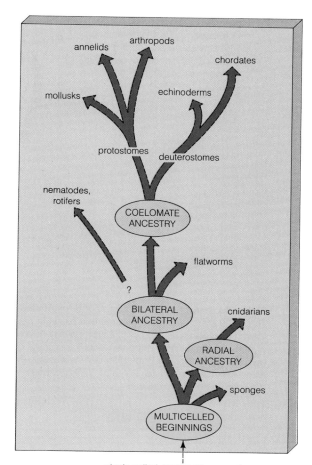

single-celled, protistanlike ancestors

Figure 19.5 A family tree for animals, showing the broad evolutionary relationships among major groups. All groups shown were established by the time the Cambrian drew to a close, some 570 million years ago.

Figure 19.6 Sponges in their marine habitat.

Figure 19.7 (**a**) Body plan of a simple sponge. Dark purple shows the location of "food-trapping" collar cells. (**b**) Small section through the body wall. (**c**) Structure of a collar cell.

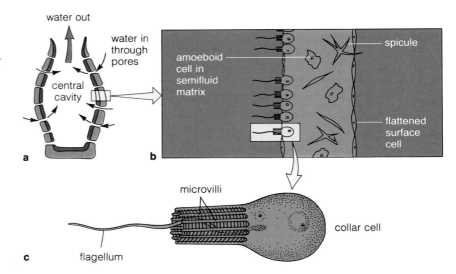

water out

water in through pores

central cavity

a

amoeboid cell in semifluid matrix

spicule

flattened surface cell

b

microvilli

flagellum

collar cell

c

Figure 19.8 (**a**) Sea nettle *(Chrysaora),* one of the large jellyfishes. The frilled structures are oral arms that assist in capturing and ingesting prey. Jellyfishes of this type usually have an abundance of nematocysts which, in the sea nettle and some other species, give painful stings. (**b**) Body plan of a hydrozoan *(Hydra),* showing its tissue organization. (**c**) Polyp of *Hydra,* capturing and then digesting a small crustacean.

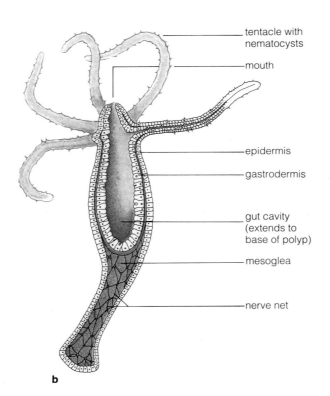

tentacle with nematocysts

mouth

epidermis

gastrodermis

gut cavity (extends to base of polyp)

mesoglea

nerve net

b

a

c

Sponges survive by moving large volumes of water through pores and chambers in their body wall and trapping bacteria or food particles suspended in the water. Currents are created by the beating flagella of thousands or millions of *collar cells* in the interior linings. Food becomes trapped on the "collars" of microvilli on these cells, which may transfer some of the food to amoebalike cells inside the matrix for further digestion, storage, and distribution.

Sponges reproduce sexually, and the young sponges pass through a **larval stage**—a sexually immature, free-living form—before growing into adults. Some sponges also reproduce asexually by fragmentation (small fragments break away from the parent body and give rise to new sponges).

CNIDARIANS

The aquatic animals called cnidarians include hydrozoans, jellyfishes, corals, and sea anemones (Figures 19.8 through 19.10). Two body forms are common; both are radial and wonderfully suitable for capturing prey from any direction. One, the *medusa* (plural, *medusae*), floats like a tentacle-fringed bell or upside-down saucer in the water. The other form, the tubelike *polyp*, is usually attached to rocks:

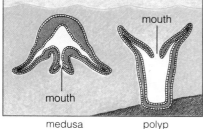

medusa   polyp

The main body cavity of medusae or polyps is a saclike gut with one opening (a mouth). This is a more efficient

a

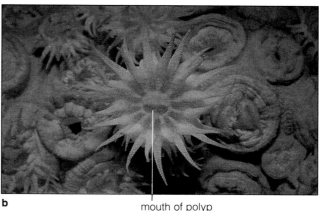

b   mouth of polyp

Figure 19.9 Corals from the Great Barrier Reef, Australia. (**a**) Individual polyps of this soft coral (*Telesto*) lack an external skeleton. (**b**) Hard corals, such as *Tubastraea*, have polyps with a calcium-reinforced external skeleton. Most of the massive, reef-forming hard corals are colonial.

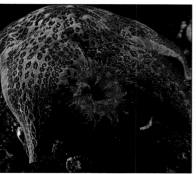
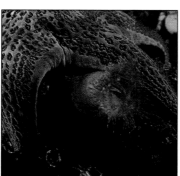

Figure 19.10 A bright-orange sea anemone escaping from a predatory sea star. It closes its mouth, so water inside its gut can serve as a "skeleton" (a hydrostatic skeleton) that its muscles can act against. Muscle contractions allow the sea anemone to detach from rocks and make thrashing movements. (Only a few sea anemones can do this.)

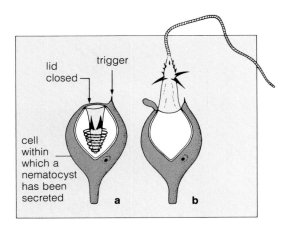

Figure 19.11 One type of nematocyst, a capsule with an inverted thread inside. This one has a bristlelike trigger (**a**). When prey (or predators) touch the trigger, the capsule becomes more "leaky" to water. As water diffuses inward, pressure inside the capsule increases and the thread is forced to turn inside out (**b**). The thread's tip may penetrate the prey, releasing a toxin as it does this.

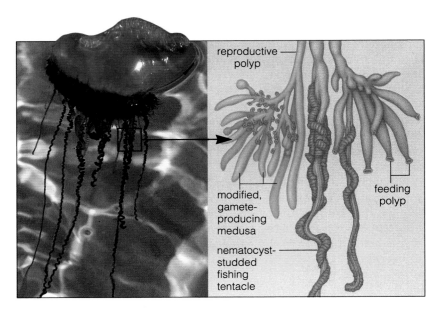

Figure 19.12 A floating colonial hydrozoan, the Portuguese man-of-war (*Physalia*). The colony has many modified polyps and medusae.

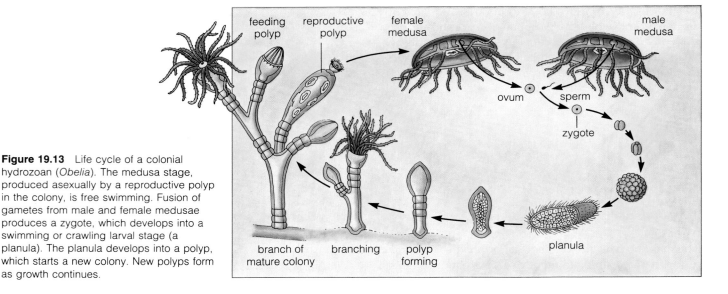

Figure 19.13 Life cycle of a colonial hydrozoan (*Obelia*). The medusa stage, produced asexually by a reproductive polyp in the colony, is free swimming. Fusion of gametes from male and female medusae produces a zygote, which develops into a swimming or crawling larval stage (a planula). The planula develops into a polyp, which starts a new colony. New polyps form as growth continues.

feeding arrangement, compared to sponges, for cnidarians can eat big rather than bacteria-sized prey (Figure 19.8). Also, they have true tissues, including sheetlike coverings called **epithelia** (singular, epithelium) and *mesoglea* (a secreted, jellylike layer). They have a *nerve net*, a system of nerve cells that work with sensory cells and contractile cells to bring about movement, shape changes, and feeding. And of all the animals, only cnidarians produce *nematocysts,* capsules that discharge prey-capturing threads (Figure 19.11). The toxin-filled ones can give humans a painful sting.

Most corals have calcium-reinforced external skeletons that are the main building material for reefs. Their most extravagant accomplishment is the Great Barrier Reef, which parallels the Australian coast for about 2,000 kilometers. Most of the reef-forming hard corals live as colonies; they are composed of many interconnected polyps. Different members of other colonial forms are specialized for feeding, defense, and reproduction (Figure 19.12).

All cnidarian life cycles include both sexual and asexual modes of reproduction. The zygote resulting from sexual reproduction nearly always develops into a planula, a swimming or creeping larva (Figure 19.13). Animals resembling planulas may have given rise to the lineages of bilateral animals.

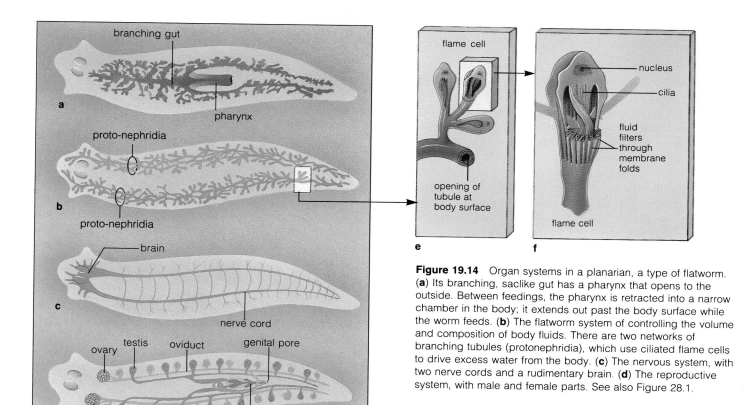

Figure 19.14 Organ systems in a planarian, a type of flatworm. (**a**) Its branching, saclike gut has a pharynx that opens to the outside. Between feedings, the pharynx is retracted into a narrow chamber in the body; it extends out past the body surface while the worm feeds. (**b**) The flatworm system of controlling the volume and composition of body fluids. There are two networks of branching tubules (protonephridia), which use ciliated flame cells to drive excess water from the body. (**c**) The nervous system, with two nerve cords and a rudimentary brain. (**d**) The reproductive system, with male and female parts. See also Figure 28.1.

FLATWORMS

This phylum includes turbellarians, flukes, and tapeworms. All but the tapeworms have a saclike gut, with food entering through a pharynx (a muscular tube). Unlike sponges and cnidarians, the flatworms are bilateral, cephalized, and equipped with organ systems (but no coelom). Also, three germ layers form in their embryos.

The embryonic midlayer (mesoderm), which gives rise to muscles and reproductive structures, was pivotal in the evolution of complex animals. It allowed contractile cells to evolve independently of the other layers, and it became the embryonic source of blood, bones, and other tissues and organs of complex animals.

Turbellarians

Turbellarians typically feed on whole small animals or suck tissues from dead or wounded ones. Most live in the seas; a few, including planarians, live in freshwater. Planarians have an organ system for regulating the volume and composition of body fluids. The system has many small, branched tubes (proto-nephridia) that extend from the body surface to *flame cells* in the body

(Figure 19.14e and f). Flame cells get their name from a tuft of cilia that "flickers" inside them and drives excess fluid out of the flatworm body.

Planarians commonly reproduce by dividing in half, with each half then regenerating the missing portion. Flatworms generally are hermaphrodites, meaning one individual has both female and male reproductive organs, including a penis. Sexual reproduction usually involves the simultaneous exchange of sperm between two individuals.

Flukes

The flukes (trematodes) are **parasites;** they obtain nutrients from a living host, usually without killing it. Some flukes suck in cells, tissue juices, or blood at the body surface of vertebrate hosts; others take up residence inside the host's body. The internal parasites generally have complex life cycles that require a "definitive" host (in which the parasite matures and reproduces) and an "intermediate" host (in which some development, but no sexual reproduction occurs). For example, humans are the definitive host for several notorious, disease-causing flukes, some of which use insects as intermediate hosts (see *Commentary*).

Tapeworms

Tapeworms (cestodes) are intestinal parasites of vertebrates. Ancestral tapeworms apparently became so good
at letting their hosts provide for them that their own
gut became superfluous and disappeared. Present-day
species simply absorb soluble nutrients from their hosts.

A tapeworm attaches to the intestinal wall by a *scolex*,
a structure with suckers, hooks, or both (Figure 19.15).
Just behind the scolex is a region where new tapeworm
body units, the *proglottids*, form by budding. The *Commentary* shows one of these units (Figure b). Each
proglottid is almost like an individual, for it has a
hermaphroditic reproductive system. Proglottids mate
and transfer sperm, and fertilized eggs may accumulate
in older proglottids. Sooner or later, the oldest ones
leave the body by way of feces and so carry the eggs to
the outside, where they become available to infect a
new host.

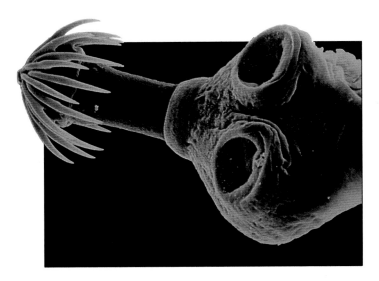

Figure 19.15 Scanning electron micrograph of a scolex of a
tapeworm that parasitizes shorebirds.

ROUNDWORMS (NEMATODES)

Roundworms (nematodes) are truly abundant, yet most
people have never seen one. They live just about everywhere, from snowfields to deserts and hot springs. A
cupful of rich soil has thousands of them, and a dead
earthworm or a fruit rotting on the ground will almost
certainly have an interesting variety of scavenging types.
Roundworms also parasitize plants and animals. Humans alone are infected by about thirty species, including
hookworms (see *Commentary*). One very long type of
hookworm lives just under the skin and lifts it in thin,
serpentlike ridges. Thousands of years ago, healers
removed the "serpents" by winding them out very
slowly, around a stick. The symbol of our medical
profession continues to be a serpent wound around a
staff.

Roundworms are bilateral, and their cylindrical body
is usually tapered at both ends. They have a complete
digestive system (Figure 19.16). Their false coelom contains reproductive organs, and fluid in the cavity circulates nutrients through the body. These worms have
a *cuticle*, a tough, flexible body covering. Even though
different roundworms live in diverse habitats, their body
plan remains much the same. It is as if the plan worked
well under many different conditions, so they just stayed
with it.

ROTIFERS

Of the rotifers, we might say this: Seldom has so much
been packed in so little space. Most rotifers are less than
a millimeter long, but Figure 19.17 will give you an idea
of the structural complexity possible in animals with a
false coelom. Most rotifers prowl about in lakes and
ponds or on wet mosses, feeding on bacteria, and other
animals feed on them. They are often abundant and
important in food webs.

Figure 19.16 Body plan of a
roundworm, which is specialized for
feeding and reproducing, and little
else. (Gametes as well as digestive
residues leave the body by way of
the cloaca, the last organ of
the gut.)

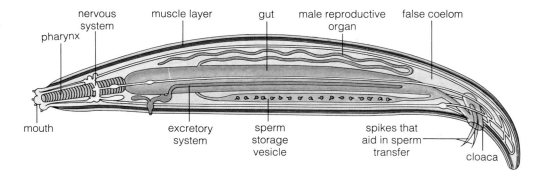

A Rogues' Gallery of Parasitic Worms

Many parasitic flatworms and roundworms call the human body home, much to our enormous discomfort. In any given year, about 200 million people house blood flukes responsible for *schistosomiasis*. Figure a shows how the life cycle of a Southeast Asian blood fluke (*Schistosoma japonicum*) requires a human host, a watery medium for swimming larval stages, and an intermediate host (an aquatic snail). Fluke eggs hatch into ciliated, swimming larvae (1). These burrow into an intermediate host, where they multiply asexually and develop into fork-tailed larval forms (2, 3). These larvae leave the snails and actively swim about until they encounter human skin (4). They bore inward and migrate to the circulatory system, where they mature. Sexual reproduction (5) occurs in small, thin-walled veins in the intestines, where the female fluke lays her eggs. The eggs leave the body by way of feces, and the cycle begins anew.

Infected humans typically mount an immune response to the masses of fluke eggs being produced in their body. White blood cells and other immune fighters infiltrate the infected areas, and grainy masses form in tissues. In time, the liver, spleen, bladder, and kidneys deteriorate and malfunction.

Tapeworms also do damage to humans. One kind uses pigs as intermediate hosts; another uses cattle. Humans become infected when they eat insufficiently cooked pork or beef (Figure b). Or consider how the larvae of one tapeworm are eaten by copepods (tiny relatives of crabs). The larvae avoid digestion, then they develop further in fishes that eat the copepods. Humans who eat infected fishes that are raw, improperly pickled, or insufficiently cooked can become hosts for the adult tapeworms.

Then there are the roundworms called pinworms and hookworms. The pinworm *Enterobius vermicularis* infects humans (especially children) in temperate regions. It lives in the large intestine. At night, the centimeter-long female pinworms migrate to the anal region to lay eggs. Their presence at the body surface causes itching, and scratchings made in response will transfer the eggs to other objects. Newly laid eggs contain embryos, but within a few hours they are juveniles and ready to hatch if the eggs are inadvertently ingested by another human.

Hookworms can be an awful problem for humans in the tropics or subtropics. Adult hookworms live in the small intestine, where they feed on blood and other tissues. Teeth or sharp ridges bordering their mouth can cut into the intestinal wall. Adult females, about a centimeter long, can release a thousand eggs daily. These leave the body by way of feces, then hatch into juveniles. A juvenile penetrates the skin of a barefoot person. Inside a host, the parasite travels the bloodstream to the lungs, where it works its way into the air spaces. After moving up the windpipe, it is swallowed. Soon it is in the small intestine, where it may mature and live for several years.

Another roundworm, *Trichinella spiralis*, causes painful and sometimes fatal symptoms. The adults live in the lining of the small intestine. Female worms release

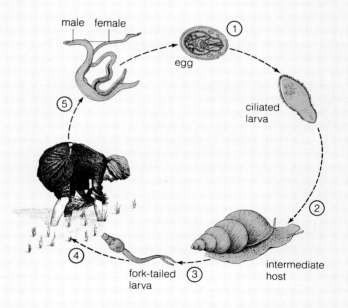

a Life cycle of a blood fluke, *Schistosoma japonicum.*

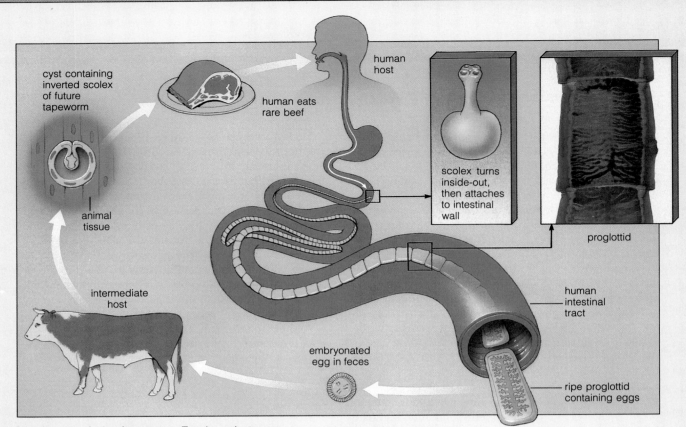

cyst containing inverted scolex of future tapeworm

animal tissue

human host

human eats rare beef

scolex turns inside-out, then attaches to intestinal wall

proglottid

intermediate host

human intestinal tract

embryonated egg in feces

ripe proglottid containing eggs

b Life cycle of a beef tapeworm, *Taenia saginata*.

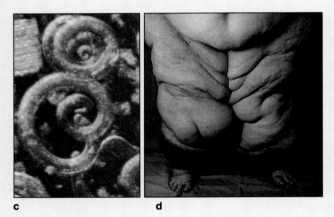

c

d

(c) Juveniles of a roundworm, *Trichinella spiralis*, inside muscle tissue. (d) Elephantiasis in a woman, caused by the roundworm *Wuchereria bancrofti*.

juveniles (Figure c), and these work their way into blood vessels and travel to muscles. There they become encysted (they produce a covering around themselves and enter a resting stage). Humans usually become infected by eating insufficiently cooked meat from pigs or certain game animals. The presence of encysted juveniles cannot easily be detected when fresh meat is examined, even in a slaughterhouse.

Finally, Figure d shows the results of prolonged and repeated infections by the roundworm *Wuchereria bancrofti*. Adult worms live in the lymph nodes, where they can obstruct the flow of lymph that normally is returned to the bloodstream (page 387). The obstruction causes fluid to accumulate in the legs and other body regions, which undergo grotesque enlargement, a condition called *elephantiasis*. A mosquito is the intermediate host. The female roundworms produce active young that travel the bloodstream at night. If a mosquito sucks blood from a human, the juveniles can enter the insect's tissues. After some growth, they move to the insect's sucking device, where they are ready to enter a new definitive host when the mosquito gets hungry again.

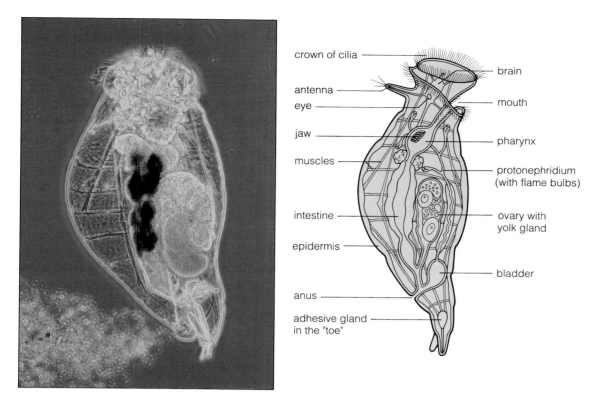

Figure 19.17 Lateral view of a rotifer, which is busily laying eggs. In many rotifer species, males are unknown; females produce diploid eggs that develop into more diploid females. In other species, females can reproduce diploid eggs that develop into females *or* they can produce haploid eggs that develop into haploid males. If the haploid eggs happen to get fertilized by a male, they develop into females. The male rotifers appear only occasionally, and they are dwarfed and short-lived.

TWO MAIN EVOLUTIONARY ROADS

Bilateral animals not much more complex than flatworms emerged during Cambrian times. A major divergence occurred not long afterward, and it led to two distinctly different groups of animals, the **protostomes** and **deuterostomes**. As shown in Figure 19.5, mollusks, annelids, and arthropods are protostomes; echinoderms and chordates are deuterostomes.

Both types of animals usually have a coelom and a complete digestive system. But their embryos develop in different ways. For example, early cell divisions differ. In deuterostomes, the divisions cut the fertilized egg as you might cut an apple into four wedges, from top to bottom. In protostomes, the "wedges" would be cut at a slight angle, not straight up and down. An indentation forms in the early ball-shaped embryo for both types of animals. In deuterostomes, this first indentation develops into the anus, and a second one becomes the mouth. In protostomes, it develops into the mouth; the anus

forms elsewhere. Finally, the coelom begins as outpouchings of the gut wall in deuterostomes, but it arises from a split in tissue at the sides of the gut in protostomes.

MOLLUSKS

"Mollusk" means a soft-bodied animal. Snails, clams, and octopuses are familiar examples. Mollusks have a head, foot, and usually a calcified shell. The ones with a well-developed head have distinctive eyes and tentacles. Mollusks also have a *mantle*, a tissue fold that hangs down like a skirt around the body. Food entering the gut is often preshredded, having been processed by a tonguelike organ (radula) with hard teeth. Besides these features, mollusks also have a heart, complex reproductive and excretory organs, and respiratory structures called *gills*.

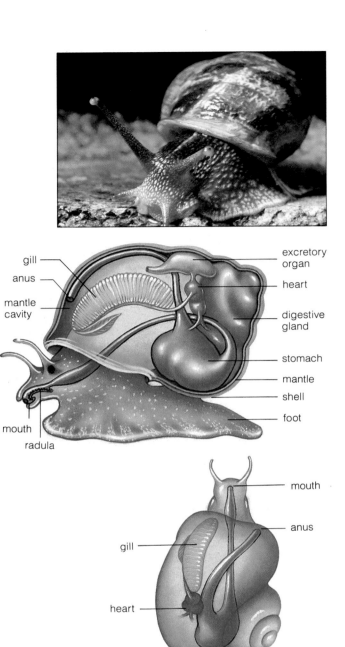

Figure 19.18 Body plan of a familiar mollusk, the land snail.

Figure 19.19 Diversity on the basic molluscan plan. Walk along a rocky shore and you will probably see chitons. The ones shown in (**a**) and (**b**) live in the intertidal zone of Monterey Bay, California. Members of this class of mollusks show beautiful variations in the color and patterns of their elongated shells. (**c**) One of the soft-bodied gastropods called sea slugs.

Gastropods

Snails and slugs make up the largest group of mollusks, the gastropods ("belly foots"). A gastropod is so named because it is largely a stomach and digestive gland sitting atop a muscular foot (Figure 19.18). Many snails have a coiled or cone-shaped shell. Coiling is a way of compacting the organs into a mass that can be balanced above the rest of the body, much as you would balance a backpack full of books. In sea slugs (nudibranchs) and

sea hares, the shell is reduced or lost, as evident in Figure 19.19c.

As most gastropod larvae develop, some body parts undergo a strange internal realignment (torsion). The gills, anus, and kidney openings that were at the back of the animal end up over the head. This could create something of a sanitation problem, what with wastes being dumped near the respiratory structures and the mouth. But gastropods use cilia to create currents that sweep away the wastes.

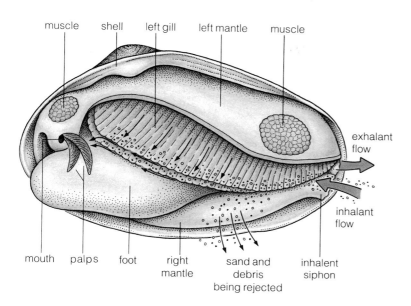

Figure 19.20 Anatomy of a clam. The left shell has been lifted off for this diagram. Food trapped in mucus on the gills is sorted by the palps, and suitable particles are swept by cilia to the mouth.

Figure 19.21 An octopus, one of the cephalopods. Notice the suction cups on its tentacles and its well-developed eyes.

Bivalves

Clams, scallops, oysters, and mussels are well-known bivalves (animals with a "two-valved shell"). Humans have been eating one type or another since prehistoric times. Some bivalves are only 1 or 2 millimeters across. A few giant clams of the South Pacific are more than a meter across and weigh 225 kilograms (close to 500 pounds). The bivalve head is not much to speak of, but the foot is usually large and specialized for burrowing. Nearly all bivalves use their gills for "suspension feeding" as well as respiration. Cilia on the gills draw water into and out of the mantle cavity. Mucus on the gills traps tiny bits of food that are suspended in the water (Figure 19.20). Bivalves hunkered in beneath mud or sand have tubes (siphons) that deliver water and suspended food to the mantle cavity and carry out wastes.

Cephalopods

The cephalopods include squids, octopuses, nautiluses, and cuttlefish, all fast-swimming predators of the seas. Some are only a few centimeters long. The giant squid can grow to 18 meters (about 60 feet); it is the largest invertebrate known.

Instead of a foot, cephalopods have tentacles, usually equipped with suction pads (Figure 19.21). They use tentacles to capture prey and a beaklike pair of jaws to bite or crush it. Venomous secretions often speed the captive's death. Cephalopods move rapidly by a type of *jet propulsion*. When muscles in the mantle relax, water is drawn into the mantle cavity. When other muscles contract, a jet of water is squeezed out through a funnel-shaped siphon. By adjusting the siphon, the animal partly controls which way it moves.

Being highly active, cephalopods have great demands for oxygen, and they alone among the mollusks have a closed circulation system. (Blood flows rapidly from a main heart through the gills, each of which has an accessory heart that boosts oxygen uptake and carbon dioxide removal.) Their nervous system is well developed, and the brain is large and complex. Giant nerve fibers connect the brain with muscles used in jet propulsion, making it possible for a cephalopod to respond quickly to food or danger. Cephalopod eyes resemble yours. Finally, cephalopods can learn. Keep giving an octopus a mild electric shock after showing it an object with a distinctive shape, and it will eventually learn to avoid the object. In terms of memory and learning, cephalopods are the world's most complex invertebrates.

a

b

Figure 19.22 Terrestrial leech attached to a human host, shown before feeding (**a**) and afterward, when it is engorged with blood (**b**). In the past, especially in the early nineteenth century, doctors sometimes were unrestrained in using leeches to "cure" disorders ranging from nosebleeds to obesity; they often applied fifty leeches at a time to patients. Nowadays, leeches are used more selectively, as in relieving congested tissue grafts.

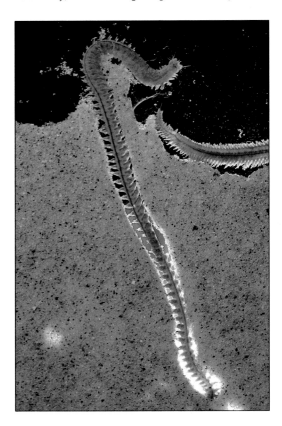

Figure 19.23 A marine polychaete inside its burrow. The reinforced bristles (setae) along the length of its body aid in crawling movements and in gripping the burrow wall.

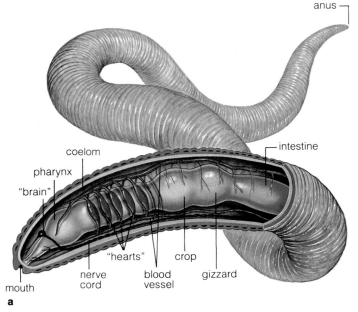

anus

coelom

intestine

pharynx

"brain"

mouth

"hearts"

nerve
cord

blood
vessel

crop

gizzard

a

Figure 19.24 (**a**) Arrangement of nerves, blood vessels, and the digestive system of an earthworm. All extend through one coelomic chamber after another. (**b**) The earthworm system of controlling the volume and composition of body fluids. The functional unit is the nephridium.

ANNELIDS

The annelids, or segmented worms, include the leeches, polychaetes, and earthworms. Leeches live in fresh water, the sea, and moist tropical regions. Most swallow small animals or kill them and suck out their juices. The ones most people have heard about feed on vertebrate blood (Figure 19.22). The leech gut has many side branches for storing food from a big meal, this being handy for an animal that may have long waits between meals. The leech uses suckers at opposite ends of its muscular body to inch forward over objects.

The polychaetes are a diverse group of mostly marine worms. Different kinds burrow, crawl, swim, or become attached to substrates. They dine on small animals, algae, or organic debris in sediments. Some polychaetes live in tubes of their own making; most of these have tentacles or featherlike structures originating near the mouth (Figure 19.1). Polychaetes use chitin-reinforced bristles, called *setae*, for crawling about or holding onto the walls of burrows or tubes (Figure 19.23).

Earthworms (a type of oligochaete) also have setae. Earthworms are scavengers. They burrow in moist soil or mud, where they feed on decomposing plant material and other organic matter. In compacted soil, they literally eat dirt as they burrow. Earthworms aerate soil, to the benefit of many plants; they also make nutrients available to other organisms by carrying subsoil to the surface. Figure 19.24 shows the complex organ systems of this annelid.

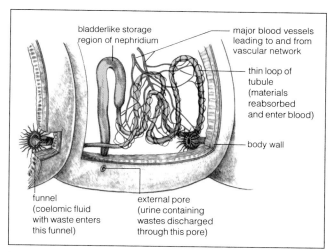

bladderlike storage region of nephridium

major blood vessels leading to and from vascular network

thin loop of tubule (materials reabsorbed and enter blood)

body wall

funnel (coelomic fluid with waste enters this funnel)

external pore (urine containing wastes discharged through this pore)

b

What other features besides setae characterize the annelids? Their segmentation is pronounced. Partitions usually separate the segments into a series of coelomic chambers, and a complete digestive system extends through all of them. Annelids have a closed circulatory system and *nephridia* (singular, *nephridium*), kidneylike organs that help control the volume and composition of body fluids (Figure 19.24). A thin, flexible cuticle covers the body and permits bending as well as gas exchange. Annelids have clusters of nerve cells in each segment and a rudimentary brain—meaning they have both peripheral and centralized regions of neural control. Annelids probably were an early offshoot of the protostome branch, and they remind us of the developments that led to increased size and more complex organs.

ARTHROPODS

Of all animals, the arthropods show the most diversity; more than a million species are known. Ancestral annelids or animals like them apparently gave rise to four different arthropod groups:

Trilobites	*Now extinct (page 202)*
Chelicerates	*Horseshoe crabs, spiders, scorpions, ticks, mites*
Crustaceans	*Copepods, crabs, lobsters, shrimps, barnacles*
Uniramians	*Centipedes, millipedes, insects*

Arthropod Adaptations

Arthropods are spectacularly abundant, they are distributed through more habitats than any other animals,

they eat more kinds of foods (and more of it), and they are very good at defending themselves against predators. Five adaptations in particular have contributed to their success:

1. Hardened exoskeleton
2. Division of labor in the life cycle
3. Jointed appendages
4. Specialized respiratory systems
5. Efficient nervous system and sensory organs

The Arthropod Exoskeleton. Think back on the annelid worm's cuticle. Its thin cuticle has a drawback, for predators can pierce right through it. When arthropods were evolving, the pressures of predation must have favored thicker, hardened cuticles. The cuticle of existing arthropods is actually an external skeleton, or **exoskeleton**. It contains protein and chitin, a combination that is flexible, lightweight, and protective. Some cuticles also have calcium deposits; they are less flexible but they are harder, like armor plates.

Although exoskeletons probably evolved as a defense against predators, they took on other functions when aquatic arthropods began invading the land. *An arthropod exoskeleton is a superb barrier to evaporative water loss, and it can support a body deprived of water's buoyancy.*

Exoskeletons do restrict increases in size. But arthropods grow in spurts and molt between these spurts. For arthropods, **molting** means that they shed their exoskeleton and grow a new one at different stages of the life cycle. Aquatic arthropods swell up a bit with water and so enlarge before their new exoskeleton hardens. Land-dwelling arthropods swell up with air.

Division of Labor in the Life Cycle. Some arthropods grow and molt several times, without much change in body form, before becoming sexually mature adults. Other arthropods go through sexually immature stages (larvae) that molt and *change* as they grow. Often, the change requires massive tissue reorganization and drastic remodeling. We saw an example of this in Figure 1.4, which showed the emergence of an adult moth. Transformation of a larva into the adult form is called **metamorphosis**.

Metamorphosing insects show a division of labor between the sexually immature stages and the adult. For example, the moth caterpillar specializes in *feeding and growth*. The adult specializes in *dispersal and reproduction*; it does not grow, and much of its energy is channeled into producing gametes and encountering a mate. With this division of labor, the insect is highly adapted to seasonal changes in food sources and environmental conditions.

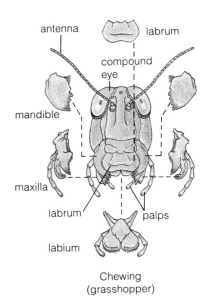

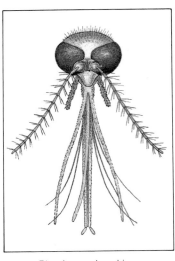

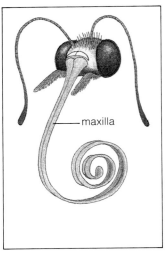

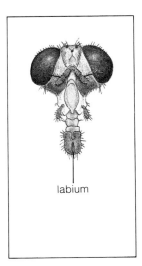

antenna	labrum
compound eye	
mandible	
maxilla	
labrum	palps
labium	

Chewing
(grasshopper)

Piercing and sucking
(mosquito)

maxilla

Siphoning tube
(butterfly)

labium

Sponging
(housefly)

Figure 19.25 Examples of the specialized appendages of arthropods. Shown here, insect headparts, adapted for feeding in specialized ways.

Jointed Appendages. As arthropods evolved, body segments became more specialized, reduced in number, and grouped or fused together in a variety of ways. In some lineages, segments became organized into a head, thorax, and abdomen; in others, they formed a forebody and hindbody. Also, appendages became highly specialized in different body regions. For example, the arthropod head has pairs of appendages used in feeding and sensing information (Figure 19.25). Other appendages perform such tasks as walking, swimming, flying, sperm transfer, and spinning silk.

Respiratory Systems. Of all the respiratory structures, the tubes called *tracheas* contributed most to arthropod diversity, especially among the insects. Tracheas begin at pores on the body surface and branch to supply oxygen directly to body tissues (Figure 28.3). Tracheas support high metabolic rates in small-bodied insects. In evolutionary terms, they were the foundation for such energy-consuming activities as insect flight.

Specialized Sensory Structures. The arthropod eye and other sensory organs contribute immensely to the success of these animals. Many arthropods have a wide angle of vision, and they can process visual information from many directions. That is partly why dragonflies are so good at capturing other insects in midair; their eyes contain more than *30,000* photoreceptor units. We will consider these and other sensory structures in Chapter 30.

Chelicerates

Of the familiar chelicerates—the spiders, scorpions, ticks, and mites—we might say this: Never have so many been loved by so few. Spiders especially have a bad reputation, even though many do good work for us by killing uncounted numbers of insect pests. But a few venomous species deserve the reputation. So do ticks, blood-sucking parasites that often transmit disease-causing microbes to their hosts (including the ones that cause Rocky Mountain spotted fever). Most mites are free-living scavengers, but some parasitic forms live in such interesting places as the vertebrate ear. A number are major agricultural pests.

Spiders generally are eight-legged, many-eyed predators (Figure 19.26). Their forebody has two pairs of appendages for subduing and handling prey. Small hindbody appendages spin out threads of silk when the spider makes prey-ensnaring webs and egg cases.

Crustaceans

Shrimps, crayfishes, lobsters, crabs, copepods, and pill bugs belong to this group. All are important components of food webs, and many edible varieties are commercially important. Most crustaceans live in the seas, but there are freshwater and land-dwelling species.

Crustaceans commonly have many segments, which take diverse forms. For example, crabs and lobsters have a shieldlike cover (carapace) over some or all segments.

a

b

c

Figure 19.26 Representative spiders. (**a**) The brown recluse, the bite of which can be severe to fatal for humans. This North American spider lives under bark and rocks, and in and around buildings. It can be identified by the violin-shaped mark on its forebody. (**b**) A female black widow, the bite of which can be painful and sometimes dangerous. (**c**) Wolf spider. Like most spiders, it is harmless to humans and plays a major role in keeping insect pests in check.

a

b

c

Figure 19.27 Some crustaceans. (**a**) A land crab. (**b**) A marine copepod. These crustaceans typically are a few millimeters long at most. The free-swimming, herbivorous species are ecologically important; they represent a good part of the diet of fishes and other carnivores in aquatic communities. (**c**) An American blue lobster, which walks about on the seafloor preying on snails, clams, and small fishes. Humans prey on it; lobster is considered to be one of the tastiest shellfishes.

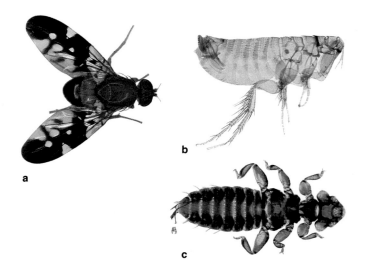

d

Figure 19.29 Representatives of some orders of insects.

(**a**) Mediterranean fruit fly (order Diptera). The larval forms destroy citrus fruit and other valuable crops. (**b**) Flea (order Siphonaptera), with big strong legs, excellent for jumping onto and off animal hosts. (**c**) Duck louse (order Mallophaga), which dines on particles of feathers and bits of skin.

(**d**) Scarab beetle, a member of the largest order (Coleoptera) in the animal kingdom. There are more than 300,000 known species. (**e**) Also in this order are ladybird beetles, which are raised commercially. They are released in great numbers as biological controls of aphids and other insect pests.

(**f**) Stinkbugs (order Hemiptera), newly hatched. (**g**) European earwig (order Dermaptera), a common household pest.

(**h**) Luna moth (order Leptidoptera), a beautiful flying insect of North America. Like most other moths and butterflies, its wings and body are covered with microscopic scales. (**i**) Dragonfly (order Odonata). This remarkable aerialist captures and eats other insects in midflight.

b

Figure 19.28 (**a**) One of the mild-mannered millipedes, which scavenge on decaying plant parts. (**b**) One of the not-so-mild-mannered centipedes of Southeast Asia, an aggressive predator that can bring down small frogs and lizards.

Their strong claws are used in collecting and tearing food, and in defense. The head has paired eyes, sensory appendages (antennae), jawlike mandibles, and structures used for handling food. The weakly swimming copepod has a single eye in the middle of the head (Figure 19.27).

Insects and Their Relatives

Millipedes and Centipedes. These close relatives of the insects are notable for having numerous paired legs (Figure 19.28). Millipedes are slow-moving, nonaggressive scavengers of decaying plant material. Centipedes are fast-moving, aggressive carnivores, equipped with fangs and venom glands. They live under rocks, logs, or forest litter and prey on insects, earthworms, and snails, although some can subdue small lizards and toads. One harmless centipede hides in houses and eats insect pests; its unexpected appearance is often startling.

e

f

g

h

i

Insects. In terms of sheer diversity, insects are the most successful invertebrates; there are more than 800,000 known species. Small size, widespread reliance on metamorphosis, and wings contribute to their success. Being small, insects can grow and reproduce in great numbers on a single plant that might be only an appetizer for another animal. Metamorphosing insects can use different resources at different times. And the insect's wings allow movement to widely scattered food sources, even to migrate with the changing seasons. (Unlike a bird wing, the insect wing develops as a lateral fold of the exoskeleton.) Figure 19.29 shows members of major orders of insects.

The factors that contribute to their success also make insects our most aggressive competitors. They eat veg-etable crops and stored food; they destroy wool, paper, and timber. They bite and suck blood from us and our pets, and they transmit diseases. On the bright side, some insects pollinate certain crop plants, and many "good" insects attack or parasitize the ones we call "bad" insects.

An insect typically has a head, thorax, and abdomen. The head has paired sensory antennae and mouthparts that are modified for biting, chewing, puncturing, or sucking (Figure 19.25). The thorax has three pairs of legs and often two pairs of wings. Digestion proceeds mostly in the midgut region, and water is reabsorbed through small tubes, the *Malpighian tubules* that join the midgut. Nitrogen-containing wastes (from protein break-down) also diffuse into the tubes, where they are

tube feet   spine

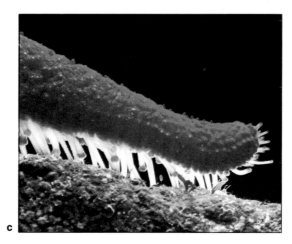

d

Figure 19.30 Some echinoderms: sea urchin (**a**), sea star (**b**), and brittle stars (**d**), which move by rapid, snakelike action of their arms. Sea stars use tube feet for locomotion (**c**), sea urchins use their spines and tube feet. Spines also protect sea urchins, and so do the venom glands of some species. With or without the venom, a spine that punctures and breaks off beneath human skin can inflame tissues.

converted into harmless crystals for elimination with feces. This system allows land-dwelling insects to get rid of potentially toxic wastes without losing precious water.

Some insects have a staggering reproductive capacity. By one estimate, if all the progeny of a single female fly were to survive and reproduce through only six more generations, that fly would have more than 5 trillion descendants!

ECHINODERMS

We turn now to the deuterostome lineage, beginning with the echinoderms (''spiny-skinned'' animals). Figure 19.30 shows some members of this phylum, which includes sea stars, sea urchins, brittle stars, and sea cucumbers.

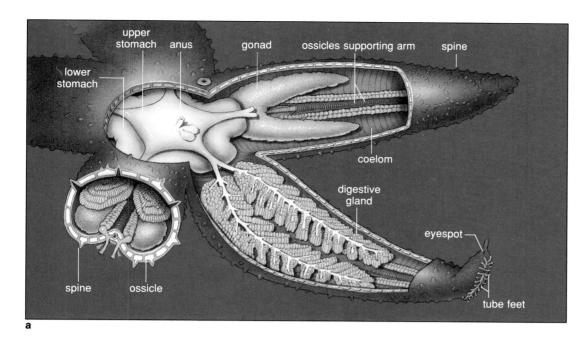

a

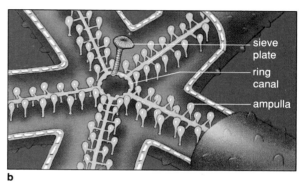

b

Figure 19.31 (**a**) Body plan of a sea star. (**b**) The water-vascular system, including tube feet.

Adult echinoderms have a radial body plan with bilateral features. (For example, some ancient echinoderms were bilateral, and so are echinoderm larvae.) The nervous system is decentralized; there is no brain. This allows, say, any arm of a sea star to become the leader, with the rest of the body moving in the direction that it is moving.

If you turn a sea star over, you will see *tube feet*, fluid-filled, muscular structures with suckerlike adhesive disks (Figure 19.30c). Tube feet are used for walking, burrowing, clinging to a rock, or gripping a clam or snail about to become dinner. They are part of a *water-vascular system* unique to echinoderms (Figure 19.31).

In sea stars, the water-vascular system includes a main canal in each arm. Short side canals extend from them and deliver water to the tube feet. Each tube foot has an ampulla, a fluid-filled, muscular structure something like the rubber bulb on a medicine dropper. When the ampulla contracts, it forces fluid into the foot, which thereby lengthens. Tube feet change shape constantly as muscle action redistributes fluid through the water-vascular system.

Some sea stars swallow their prey whole, but most push part of their stomach outside the mouth and start digesting prey even before swallowing it. They get rid of residues through the mouth. Most do have a small anus, but this is of no help in getting rid of empty clam or snail shells.

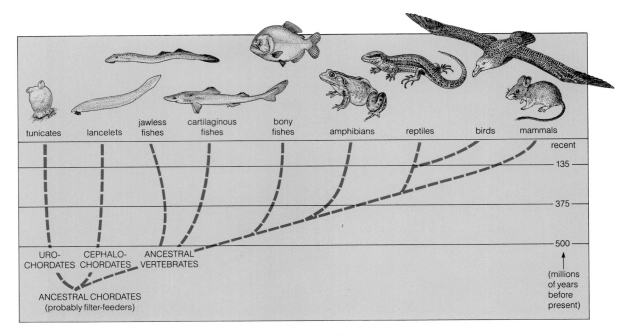

Figure 19.32 A proposed family tree for chordates, these being the vertebrates and their backboneless relatives, the tunicates and lancelets.

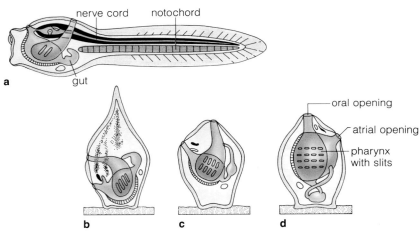

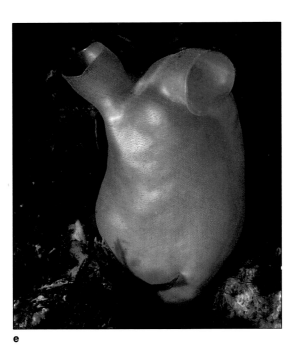

Figure 19.33 Tunicate body plan, from the larval stage (**a**) to the adult (**d,e**). The tadpole-like larva swims for only a few minutes or days until it locates a suitable living site. It attaches its head to a substrate (**b**), and metamorphosis begins; the tail, notochord, and most of the nervous system are resorbed (recycled to form new tissues). The slits in the pharynx multiply (**c**). Organs become rotated until the openings through which water enters and leaves are directed away from the substrate (**d**).

CHORDATES

The bilateral animals called chordates are shown in Figure 19.32. More than 47,000 are vertebrates—animals with a cartilaginous or bony backbone. Only about 2,100 are invertebrate chordates (those without a backbone). Four features distinguish the chordates from other ani-mals: a notochord; a dorsal, tubular nerve cord; a pharynx with slits in the wall; and a tail that extends past the anus.

A **notochord** is a long rod of stiffened tissue (not cartilage or bone) that serves as a supporting structure for the body. A tubular **nerve cord** lies above the notochord and gut. A muscular **pharynx,** the entrance

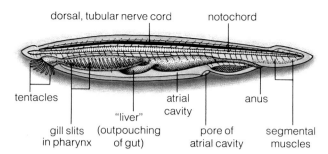

Figure 19.34 Cutaway view and photograph of a lancelet, showing the position of its nerve cord and flexible notochord.

Figure 19.35 *(Right)* Proposed evolution of gill-supporting structures, as found in jawless fishes (**a**), into the hinged vertebrate jaw (**c**). The first gill opening of the mud-dwelling jawless fishes was converted into a spiracle through which water could be drawn. The first in the series of gill-supporting structures became enlarged and equipped with teeth.

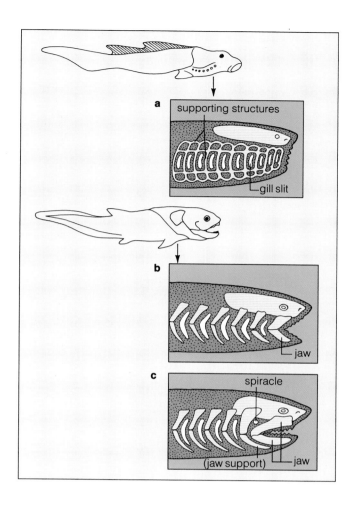

to the digestive tract, has *gill slits* for filter feeding or gas exchange. These features are evident in all chordate embryos and often persist into adulthood.

Invertebrate Chordates

Tunicates and lancelets are invertebrate chordates. All tunicates are confined to the seas. They secrete a gelatinous or leathery "tunic" around themselves (hence the name). Unlike the other chordates, they have no coelom. A bilateral larval stage resembling a frog tadpole bears a notochord (Figure 19.33). The tadpole attaches to a rock, shell, or wood piling, then it undergoes drastic metamorphosis. Its tail and notochord disappear, and the tunic thickens. The pharynx enlarges and develops many slits. In adults, the pharynx is a sieve for collecting diatoms and other bits of food suspended in water. It also is an efficient respiratory organ; it is thin walled and has a good blood supply.

There are only a few species of lancelets, fishlike animals that are tapered at both ends (Figure 19.34). Lancelets are rarely more than 5 centimeters long, but they have well-developed muscles and can bury them-

selves quickly in sand or mud. A buried lancelet keeps its mouth at the surface of mud or sand, and cilia drive water through the gill slits along the pharynx. Food sieved by the pharynx is delivered to the rest of the gut.

Evolutionary Trends Among the Vertebrates

There are seven classes of existing vertebrates—the jawless fishes, cartilaginous fishes, bony fishes, amphibians, reptiles, birds, and mammals. Unit VI describes their structure and function. Here, we will survey their major features, beginning with the evolutionary trends that brought them about.

A key evolutionary trend was the replacement of the notochord by a column of hard, bony units. The vertebral column proved to be a strong internal skeleton for the body's muscles to work against. Some bony units also proved to have enormous potential when they became modified into powerful jaws (Figure 19.35). *The vertebral column was the foundation for fast-moving, predatory fishes— some of which were ancestral to all other vertebrate animals.*

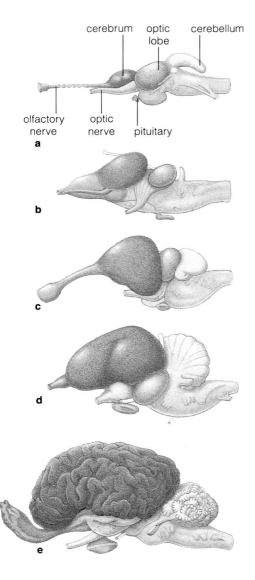

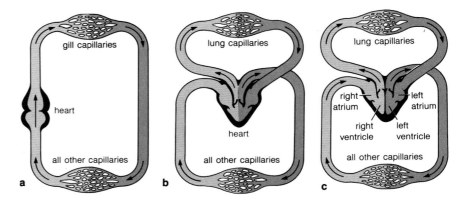

Figure 19.37 Circulatory system of fishes, amphibians, and mammals. All vertebrates have a closed circulatory system with a heart divided into chambers (atria and ventricles).

In fishes (**a**), a two-chambered heart pumps blood to the gills, where it picks up oxygen; the blood delivers oxygen to the rest of the body, then finally returns to the heart. In amphibians (**b**), the atrium is divided. In their three-chambered heart, oxygen-poor blood is sent to the lungs, and blood that is moderately well oxygenated is sent along a separate route to the rest of the body. In some reptiles and all birds and mammals (**c**), the separation is complete. Their four-chambered heart is the basis for two separate and efficient circulation routes, which are described on page 378.

Figure 19.36 Evolutionary trend toward an expanded, increasingly complex brain, as suggested by comparing the brain of existing vertebrates: (**a**) codfish; (**b**) frog, an amphibian; (**c**) alligator, a reptile; (**d**) bird; and (**e**) horse, a mammal. The drawings are not to the same scale.

A related trend was the regional expansion of the nerve cord into a spinal cord and complex brain. Consider that the evolution of jaws opened up many possibilities for feeding. Yet it certainly intensified the competition for prey, and it increased the prospect of one fish being eaten by another. Fishes equipped to recognize food or predators from a distance were more likely to survive and reproduce. Over time, their equipment came to include better senses of smell and vision; and the brain expanded in its programs for survival (Figure 19.36). *The trend toward more complex sense organs and nervous systems began in the fishes and continued among vertebrates that invaded the land.*

The pharynx was the starting point for another evolutionary trend. As fishes increased in size and became more active, they required more efficient ways to absorb and distribute oxygen. Gill slits became increasingly important in respiration. But gill slits won't work on land (unless they are kept moist, they become badly stuck together). During the Devonian, some fishes developed outpockets of the pharynx wall—and these evolved into lungs. In a related development, the heart became a more complex pump for moving oxygen and carbon dioxide to and from the lungs (Figure 19.37). Thus, *the vertebrate lungs and heart evolved by way of modification of the pharynx.*

Finally, a key trend began with the paired fins of certain fishes. Over time, their fins became lobed and equipped with internal supports of bone or cartilage. The "lobe-finned" fishes were the forerunners of land-

Figure 19.38 (**a**) Coelacanth (*Latimeria*), a "living fossil" that resembles the early lobe-finned fishes. Evolution of bony or cartilaginous structures in the lobed fins of certain fishes (**b**) into the limbs of early amphibians (**c**).

a

b

c

dwelling vertebrates. Among their few living descendants are the coelacanths (Figure 19.38). *The fins of certain fishes were the starting point for the legs, arms, and wings seen among modern amphibians, reptiles, birds, and mammals.*

Fishes

The first fishes emerged more than 450 million years ago, probably from filter-feeding chordates. They included the agnathans, which had a food-straining pharynx, no jaws, and no vertebral column (a notochord provided support). They also included the placoderms, with jaws and external body plates of bony material. Placoderms were bottom-dwelling scavengers or predators that ambushed mostly the bottom-dwelling agnathans. They became extinct about 325 million years ago, but the agnathans managed to leave a few descendants (the lampreys and hagfishes).

Today, fishes are the most numerous and diverse vertebrates. Their overall body form tells us something about their watery world. Being about 800 times denser than air, water resists movement through it. And the fish body typically is streamlined, with muscles organized for forward motion. It typically has *fins*, appendages that help propel, stabilize, and guide the body through water. Often there are protective *scales*. The scales covering the body of most free-swimming fishes, such as sardines, are small; they provide protection without weighing the fish down.

Figure 19.39 A sea lamprey, here pressing its toothed oral disk to the glass wall of an aquarium. Like other living agnathans, it has a slender, rounded, eel-like body with skin that is soft and lacks scales. The skeleton is cartilaginous.

Jawless Fishes. Today, living jawless fishes are limited to seventy or so species of lampreys and hagfishes. These have cylindrical, eel-like bodies, a notochord, and a cartilaginous skeleton. They do not have paired fins. Lampreys have a suckerlike oral disk with horny, toothlike plates that can rasp the flesh from prey (Figure 19.39). Some species give the whole group a poor reputation; they have a habit of latching onto salmon, trout, and other commercially valuable fishes, then sucking out body juices and tissues. About the turn of

caudal fin

dorsal fin

pelvic fin (paired)

pectoral fin (paired)

Figure 19.40 Two cartilaginous fishes: the blue-spotted reef ray (**a**) and a shark (**b**). Two bony fishes: the soldier fish (**c**) and a sea horse (**d**).

the century, lampreys began invading the Great Lakes of North America. Populations of lake trout and other large fishes collapsed. The populations are recovering since the development of a chemical that poisons lamprey larvae, but the battle goes on.

Hagfishes look like large worms with "feelers" around the mouth. They are not the favorite of fishermen. Not only do they burrow into fishes trapped by setlines or nets, but the ones that end up on deck secrete copious amounts of sticky, slimy mucus.

Cartilaginous Fishes. Sharks, skates, and rays belong to this group. Most are specialized predators with a streamlined body and paired fins that enhance maneu-

verability and stability. All types have cartilage skeletons, scales, and usually five to seven gill slits on both sides of the pharynx (Figure 19.40). Their teeth (modified scales) are shed and replaced continually. Some sharks are 15 meters long; they are among the largest living vertebrates. Sharks that eat large fishes and marine mammals use their sharp, triangular teeth to capture and rip off chunks of flesh. Skates and rays are mostly bottom dwellers with flattened teeth for crushing hard-shelled invertebrates. They have distinctive, enlarged fins extending onto the side of the head (Figure 19.40).

Bony Fishes. Long ago, this group diverged into at least four lineages: the lungfishes, crossopterygians (the

Figure 19.41 Two amphibians: (**a**) the American toad. (**b**) The terrestrial stage in the life cycle of a red-spotted salamander.

coelacanth is one of these), bichirs, and ray-finned fishes. Most of the 21,000 existing species are ray-finned fishes. Of these, the ones called teleosts are the most successful, and among them are the fish we prize most for food. They include salmon, tuna, rockfish, catfish, perch, minnows, and eels, as well as deep-sea luminescent varieties. They generally have highly maneuverable fins and light, flexible scales; they can make rapid, complex movements. Their respiratory system delivers plenty of oxygen to metabolically active tissues. There are exceptions to this general list of features, including the sea horse (Figure 19.40d).

Amphibians

Amphibians arose from lobe-finned fishes of the sort shown in Figure 19.38. By the time amphibians were established on land, they were encountering new sensory information. The brain underwent dramatic development, particularly in regions concerned with processing signals related to vision and balance.

Today's amphibians include salamanders, frogs, toads, and wormlike apodans; Figure 19.41 shows examples. All have a mostly bony endoskeleton and, usually, four legs. When salamanders walk, they bend from side to side, much like a fish moving through water. The first four-legged animals probably walked this way. Adult amphibians eat just about anything they can catch. Their head size dictates the upper limit of the size of prey. (One frog has such a large head it is called the walking mouth.) Nearly all amphibians must lay their eggs in water or damp places.

Frogs and toads have a stiffened skeleton, long hindlimbs, and powerful muscles. They can catapult themselves into the air or propel themselves forcefully through water. Their sticky-tipped tongue is usually attached at the *front* of the mouth and flips out to capture prey.

In water dwellers, gills and the skin (which is thin and vulnerable to drying out) serve in respiration. Land dwellers rely on lungs, skin, and the pharynx. Some skin glands produce toxins, and the most noxious species are brightly colored, the better to inform predators of their inedibility. The skin of South African clawed frogs (*Xenopus laevis*) contains chemicals that defend the body against microbes, which flourish in the frog's habitat. Researchers are hoping these chemicals can be used as new antibiotics.

Reptiles

In the Late Carboniferous, insects began a major adaptive radiation into the lush habitats on land, and this event probably triggered the evolution of reptiles from certain amphibians. Like their modern descendants, those amphibious forms almost certainly were carnivores. And almost certainly, the abrupt expansion in the quantity and choices of insects represented a major food source, ready for exploitation. The jaw became adapted for feeding on insects; limb bones became more efficient for moving on land.

A key adaptation to life on land was the increased reliance on internal fertilization and the development of

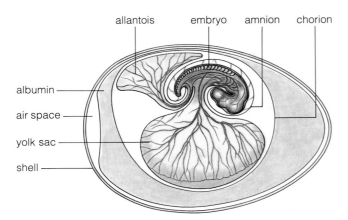

Figure 19.42 Generalized structure of the amniotic egg. Its membranes—the allantois, amnion, and chorion—protect the embryo and provide it with metabolic support (page 476).

Figure 19.43 (*Right*) A gallery of reptiles. (**a**) Galápagos tortoise, highly specialized for life in a mobile home. It carts its rigid, domed shell around on sturdy legs. Its marine relatives, the sea turtles, have flipperlike legs protruding from a more streamlined shell. (**b**) One of the crocodiles, which differ mainly in the shape of the snout. Differences relate to specialized diets. (**c**) Green python, which first suffocates and then eats its prey. (**d**) Tuatara, a "living fossil." The body plan of this lizard has not changed much since the age of dinosaurs. (**e**) Defensive behavioral display of a chameleon caught crossing open ground in Kenya.

a

b

the **amniotic egg** (Figure 19.42). This type of egg is typical of reptiles, birds, and mammals—but not amphibians. Often a leathery or calcified shell protects the embryo inside while permitting the exchange of oxygen, carbon dioxide, and water. It departs from amphibian eggs in an important respect. The amniotic egg contains membranes that protect the embryo and provide it with metabolic support (page 476).

Reptiles underwent an adaptive radiation in the Mesozoic. Dinosaurs and related forms emerged; they ruled the land for the next 125 million years. Their domination ended abruptly when the Cretaceous drew to a close (page 205). Today's reptiles include turtles, crocodilians, and snakes and lizards (Figure 19.43). Their diversity is greatest in the tropics. More than ninety percent of the known reptiles are snakes and lizards, and most of those are snakes. Lizards typically are small, but the largest (the Komodo monitor lizard) is large enough to hunt deer and even water buffalo.

Living reptiles have a bony skeleton and skin that resists drying out; most also have hardened scales. Reptiles depend entirely on lungs, with air being sucked in (not forced in by mouth muscles, as in amphibians). Most rely on internal fertilization, and most lay leathery

eggs. They often adjust internal body temperature by behavioral and physiological means. And they may show complex social behavior, as when male and female parents assist newly hatched crocodiles in their move out of the egg and into the water. Young alligators of the Gulf Coast swamps remain with the mother for two years after hatching.

Birds

Birds apparently descended from crocodilianlike reptiles that ran around on two legs during Jurassic times, some 160 million years ago. The oldest known bird (*Archaeopteryx*, shown in Figure 14.5) resembled those reptiles, especially in its limb bones. Birds still resemble reptiles in many of their internal structures, their scaly legs, and their habit of laying eggs. Figure 19.44 shows two representatives of this class of animals.

Birds vary tremendously in size, body proportions, coloration, and capacity for flight. Their songs and other social behaviors are remarkably complex, as you will see in later chapters. The largest bird, the ostrich, weighs about 150 kilograms (330 pounds); a small hummingbird

c

d

e

a

b

humerus

breastbone

c

Figure 19.44 Birds showing two characteristics of the class: flight (**a**) and dedicated parental behavior toward the nestlings (**b**). The diagram (**c**) shows a generalized skeletal system for birds.

Figure 19.45 Representative mammals. (**a**) Duck-billed platypus, an egg-laying mammal. (**b**) Opossum and (**c**) kangaroo, both pouched mammals. Their young emerge in somewhat unfinished form and undergo further development in a pouch at the surface of the mother's body, as illustrated by the young kangaroo in (**c**). Like all other mammals, placental mammals (including humans) nourish their young with milk from mammary glands (**d**).

a

b

c

barely tips the scales at 2.25 grams (0.08 ounce). Yet in basic body plan, nearly all birds meet the two key requirements for flight: low weight and great power.

The **bird wing** is a forelimb constructed of feathers, powerful muscles, and lightweight bones. Contraction of muscles attached to the humerus (Figure 19.44) is the basis of the powerful downstroke required for flight. Bird bones are strong and yet weigh very little because of air cavities in the bony tissue. The endoskeleton of a frigate bird, which has a wingspan of 7 feet (over 2 meters), weighs only 4 ounces (less than 12 grams). That is less than the feathers weigh! All birds have a large, strong, four-chambered heart and an efficient system for oxygen delivery, as shown in Figure 28.4.

Mammals

The characteristics and evolutionary trends among mammals were described in Chapter 16. A key point to

remember is this: It was not until the rise of modern mammals at the dawn of the Cenozoic that the brain began to reveal its true potential. Especially among the primates, it began expanding to include larger, interconnected masses of information-encoding and information-processing cells. This development was the foundation for our own remarkable capacity for memory, learning, and conscious thought.

There are three major groups of living mammals: the egg-laying, pouched, and placental mammals. Representatives are shown in Figure 19.45. Embryos develop internally in nearly all types, and the liveborn are nourished by milk-secreting glands of the mother. Most mammals are covered with hair, although whales are not, and reptilian scales persist on the tails of beavers and rats. Adults typically have a permanent set of teeth (reptiles have successive sets). Mammals have lungs, a four-chambered heart, and a well-developed cerebral cortex. Their structure and function will occupy our attention in Unit VI.

d

SUMMARY

1. Animals are multicelled, motile heterotrophs that reproduce sexually and, often, asexually. They pass through a period of embryonic development during the life cycle. Except for sponges, animals have true tissues, and most have organs and organ systems. But they vary enormously in their degree of structural and functional complexity.

2. Several trends are apparent in animal evolution, even though all trends did not proceed in every group:

a. From the development of a bilaterally symmetrical, cephalized body to paired nerves and brain regions, paired muscles, paired sensory structures (such as eyes), and paired appendages (such as fins and legs).

b. From a saclike gut to a complete digestive system, with mouth and anus. This permitted regional specialization of the gut, which meant energy and nutrients could be obtained more efficiently from food.

c. From a body packed rather solidly with tissues to one with a coelom—a cavity between the gut and body wall, lined with peritoneum. In this cavity, organs could increase in size and complexity.

d. From an unsegmented to a segmented body. This permitted complex movements and led to appendages becoming specialized for different tasks in different segments.

3. The arthropods are the most abundant, widely distributed animals. They eat more kinds and amounts of food, and they successfully resist predation. Their success derives from five adaptations: a hardened exoskeleton, jointed appendages, specialized respiratory systems, and a well-developed nervous system and sensory organs. Many insects benefit further from small size, reliance on metamorphosis, and wings.

4. Many flatworms and roundworms parasitize humans; several species cause serious diseases.

5. Chordates are characterized by a notochord; a dorsal, tubular nerve cord; a pharynx with gill slits; and a tail that extends past the anus. The first three features were starting points for key trends in vertebrate evolution. These were the emergence of a vertebral column; powerful muscles and jaws; increasingly complex brain, sensory structures, lungs, and heart; and limbs (from lobed fins of certain fishes).

Review Questions

1. What are the key differences between invertebrates and vertebrates? *258*

2. Choose one group of invertebrates and describe its distinguishing features. Do the same for one group of vertebrates.

3. Name some paired body parts that evolved only in bilaterally symmetrical, cephalized animals. *259, 260*

4. Which animals have a saclike gut? A complete gut? In evolutionary terms, what was the advantage of having a complete gut? *260, 261*

5. What is a coelom? Why was its emergence in some animal lineages so important to their further evolution? *261*

6. What kinds of specialized body parts evolved following the emergence of segmentation in some animal lineages? *261*

Self-Quiz *(Answers in Appendix IV)*

1. The five body features that help us identify major trends in animal evolution are _____, _____, _____, _____, and _____.

2. Which of the following would *not* describe the nutritional mode of any animal?
   a. parasitic or predatory autotroph
   b. parasitic or predatory heterotroph
   c. nutrient ingestion or absorption

3. Which of the following is *not* a general characteristic of animals?
   a. multicellularity, with cells forming tissues and usually organs
   b. exclusive reliance on sexual, not asexual, reproduction
   c. motility at some stage of the life cycle
   d. all of the above are characteristic of animals

4. Animals generally show radial or bilateral symmetry. Which ones have both radial *and* bilateral features?
   a. cnidarians
   b. mollusks
   c. arthropods
   d. echinoderms
   e. this is a trick question; no animal shows both

5. In terms of sheer numbers and distribution, _____ are the most successful animals.
   a. arthropods
   b. sponges
   c. snails and clams
   d. sea stars
   e. vertebrates

6. Jointed appendages, compound eyes, specialized respiratory system, and hardened exoskeleton are characteristic of _____.
   a. arthropods
   b. sponges
   c. snails and clams
   d. sea stars
   e. vertebrates

7. Which animals are notorious for causing serious diseases in humans?
   a. cnidarians
   b. flatworms and roundworms
   c. segmented worms
   d. chordates

8. The _____ is the body cavity of more complex animals; it is located between the gut and body wall.
   a. pharynx
   b. peritoneum
   c. coelom
   d. archenteron

9. Chordate features that figured prominently in vertebrate evolution were _____.
   a. vertebral column, nerve cord, tail
   b. notochord, nerve cord, pharynx with gill slits, tail
   c. notochord, muscles, jaws, tail
   d. vertebral column, muscles, jaws, pharynx with gill slits

10. Match the animals with their characteristics.
   _____ arthropods   a. some cause serious diseases
   _____ roundworms,   b. well-developed tissues are
        flatworms   absent
   _____ insects   c. metamorphosis and wings
   _____ chordates   d. most abundant and widely
   _____ sponges   distributed animals
                             e. notochord, nerve cord, pharynx
                                with gill slits, tail

Selected Key Terms

amniotic egg 286
coelom 261
cuticle 266
deuterostome 269
ectoderm 259
endoderm 259
epithelium 264
exoskeleton 273
gill 269
gill slit 281
gut 260
invertebrate 258
mantle 269
medusa 263
mesoderm 259
metamorphosis 273

molting 273
nematocyst 264
nephridium 273
nerve cord 280
nerve net 264
notochord 280
peritoneum 261
pharynx 280
polyp 263
protostome 269
trachea 274
vertebral column 281
vertebrate 258
water-vascular system 279

Readings

Hickman, C. P., and L. S. Roberts. 1988. *Integrated Principles of Zoology.* Eighth edition. St. Louis: Mosby.

Kozloff, E. 1990. *Invertebrates.* Philadelphia: Saunders. Excellent, authoritative book.

Mitchell, L., J. Mutchmor, and W. Dolphin. 1988. *Zoology.* Menlo Park, California; Benjamin/Cummings. Clearly written, beautifully illustrated survey of the animal kingdom.

Moyle, P., and J. Cech, Jr. 1988. *Fishes: An Introduction to Ichthyology.* Second edition. Englewood Cliffs, New Jersey: Prentice-Hall.

Romer, A. S., and T. S. Parsons. 1986. *The Vertebrate Body.* Sixth edition. Philadelphia: Saunders.

Welty, J., and L. Baptista. 1988. *The Life of Birds.* Fourth edition. New York: Saunders.

FACING PAGE: *A flowering plant (Prunus) busily doing what it does best: producing flowers for the fine art of reproduction.*

20 PLANT TISSUES

KEY CONCEPTS

1. Most of the 275,000 plant species are flowering plants. Despite their diversity, flowering plants are composed of only three kinds of tissues. Dermal tissues form a protective cover for all parts of the root and shoot systems. Ground tissues make up the bulk of the plant body. Strands of vascular tissues thread through the ground tissues; they move water, dissolved minerals, and products of photosynthesis through the roots, stems, and leaves.

2. Plants grow at the tips of their roots and shoots. At each tip, some of the cells produced in a region of undifferentiated cells (apical meristem) divide repeatedly, elongate, and develop into the specialized cell types characteristic of the three categories of plant tissues. The growth originating at root and shoot tips produces the primary tissues of the plant body.

3. Many plants also show secondary growth, which increases the diameter of roots and stems to produce the tissues we call wood. Secondary growth originates with undifferentiated cells called lateral meristems. A fully developed lateral meristem is like a long cylinder that produces new cells along the length of an older root or stem.

In the spring of 1980, Mount St. Helens in southwestern Washington exploded violently. Within minutes, hundreds of thousands of mature trees near the volcano's northern flank were blown down like matchsticks. Thick ashes and pumice turned the previously forested region into a scarred, barren sweep of land. Events of this magnitude dramatize what the world would be like without plants, reminding us that we could no more survive without them here than on the rock-strewn surface of the moon.

What characterizes the world's plants, which directly or indirectly nourish other organisms and make the land

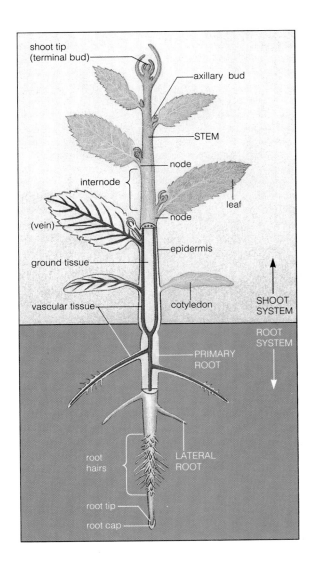

Figure 20.1 Body plan for one type of flowering plant. Some vascular tissues by which water and nutrients move through the plant body are shown in red. Notice how they thread through the ground tissue (shaded yellow), which makes up the bulk of the plant body. Dermal tissue (epidermis) forms a covering for the root and shoot systems shown here.

habitable? Can we identify patterns of structural organization among them? Do plants, like animals, have internal systems for transporting and distributing materials among their component cells? How do plants reproduce? What governs their growth and development? These questions are addressed in this unit.

THE PLANT BODY: AN OVERVIEW

As we saw in the preceding unit, there are more than 275,000 species of plants, and no one species can be used as a "typical example" of their body plans. Here our main focus will be on the most familiar types, the flowering plants (angiosperms).

Shoot and Root Systems

Flowering plants typically have well-developed shoot and root systems (Figure 20.1). The **shoot system** (stems and leaves) has internal pipelines that conduct water, minerals, and organic substances among roots, leaves, and other plant parts. The stems serve as frameworks for upright growth. With upright growth, photosynthetic tissues in leaves are favorably exposed to light and the flowers are favorably exposed to pollinators. Some parts of the shoot system store food.

The **root system,** which usually grows below ground, absorbs water and dissolved minerals from soil and conducts water and solutes to aboveground plant parts. The root system also stores food, anchors the plant, and sometimes structurally supports it.

Plant Tissues

Three kinds of tissues predominate in root and shoot systems. The bulk of the plant body consists of different types of *ground* tissues, which have *vascular* tissues dispersed through them (Figure 20.1). *Dermal* tissues serve as a protective covering for the plant body.

Ground Tissues. As Figure 20.2 suggests, cells of the ground tissues differ in their wall structure. The most abundant cell type, *parenchyma*, generally has thin walls. In stems, roots, leaves, flowers, and the flesh of fruits,

parenchyma cells are massed together with ample air spaces between them. Different types of parenchyma cells take part in photosynthesis, storage, secretion, and other tasks.

Two other ground tissues, *collenchyma* and *sclerenchyma*, have cells with thickened walls that are either rigid or pliable. Both tissues offer mechanical support for plant parts. For example, collenchyma is a component of the strong yet pliable "strings" of celery stalks. Hemp and flax (used in producing rope, paper, and thread) have long, tapered sclerenchyma cells called fibers. Coconut shells have extremely thick-walled sclerenchyma cells called sclereids.

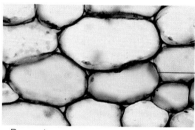

a Parenchyma

thin
primary
wall

irregularly
thickened
primary
wall

b Collenchyma

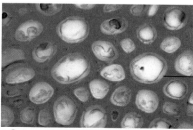

thick
secondary
wall

c Sclerenchyma

Figure 20.2 Examples of ground tissues, which make up the bulk of the plant body. All three are cross-sections from the stem of a sunflower plant (*Helianthus*).

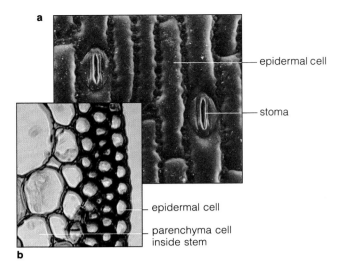

Figure 20.3 (**a**) Surface view of leaf epidermis from a corn plant (*Zea mays*). (**b**) Cross-section through the stem of a corn plant, showing part of the epidermis.

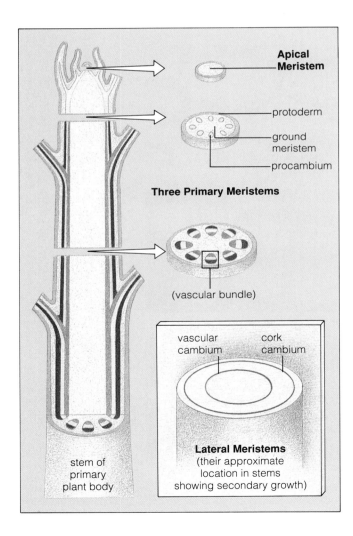

Vascular Tissues. Flowering plants have two kinds of vascular tissues—xylem and phloem—that serve different functions. **Xylem** conducts water and dissolved minerals absorbed from the soil; it also mechanically supports the plant. Its main water-conducting cells are dead at maturity; all that remains are walls with recesses, pits, or open ends. The cells overlap or are joined at their ends to form the continuous pipelines through roots, stems, and leaves.

Phloem is the vascular tissue by which sugars and other solutes are rapidly transported through the plant body. Its main conducting cells (sieve tube members) are alive at maturity. Their walls have clusters of pores through which the cytoplasmic contents of adjacent cells are connected. "Companion cells" in the phloem help the sieve tube members move sugars from regions of photosynthesis to other plant parts.

Dermal Tissues. A continuous layer of tightly packed cells, the **epidermis,** covers the primary plant body. Waxes and a fatty substance (cutin) that have been deposited on the outer surface of epidermal cell walls form a surface coating called a **cuticle.** The cuticle restricts water loss and offers some resistance to microbial attack. For the most part, water vapor as well as carbon dioxide and oxygen move across the epidermis at openings (stomata) between pairs of specialized epidermal cells (Figure 20.3). A protective cover, the periderm, replaces epidermis when roots and stems increase in diameter and become woody.

How Plant Tissues Arise: The Meristems

When a new plant starts to grow, cells divide and lengthen at the tip of the first root and shoot. Each tip has a dome-shaped mass of cells called the **apical meristem.** Descendants of some of those cells develop into the specialized tissues of the lengthening root and stem. (Other cells in the mass do nothing more than divide and so perpetuate the apical meristem even as

Figure 20.4 Locations of primary meristems (yellow) and lateral meristems (red) in plants that show both primary and secondary growth.

Descendants of some of the cells at apical meristems give rise to three kinds of primary meristems. In turn, those three meristems give rise to epidermis, the ground tissues, and vascular tissues of the primary plant body.

Woody plants put on new primary growth each season. They also undergo secondary growth, which leads to increases in the diameter of older stems and roots. Secondary growth originates at two types of lateral meristems, called vascular cambium and cork cambium. The first type gives rise to secondary vascular tissues. The second type gives rise to periderm, the protective covering that replaces epidermis.

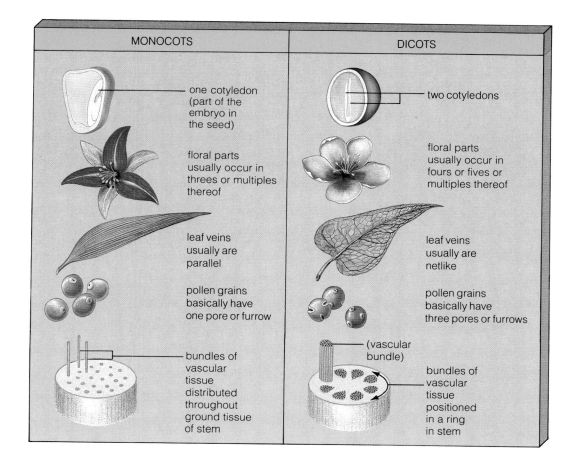

MONOCOTS	DICOTS
one cotyledon (part of the embryo in the seed)	two cotyledons
floral parts usually occur in threes or multiples thereof	floral parts usually occur in fours or fives or multiples thereof
leaf veins usually are parallel	leaf veins usually are netlike
pollen grains basically have one pore or furrow	pollen grains basically have three pores or furrows
bundles of vascular tissue distributed throughout ground tissue of stem	(vascular bundle) bundles of vascular tissue positioned in a ring in stem

Figure 20.5 Main differences between monocots and dicots. See also Figures 20.6 and 20.7.

tissues are developing behind it.) Growth originating at root and shoot tips is *primary* growth. Many plants, including buttercups and corn, die after only one season of primary growth.

Plants with a woody body show *secondary* growth at regions other than root and shoot tips. Secondary growth originates at self-perpetuating tissue masses called **lateral meristems,** and it increases the diameter of older roots and stems. Each spring, for example, a maple tree undergoes primary growth at its root and shoot tips, and secondary growth adds to its woody parts. Figure 20.4 shows the location of meristems in one type of flowering plant stem.

Monocots and Dicots Compared

As we have seen, there are two classes of flowering plants, and they are referred to informally as **monocots** and **dicots** (page 254). Grasses, lilies, orchids, irises, cattails, and palms are examples of monocots. Nearly all familiar trees and shrubs (other than conifers) are dicots.

Monocots and dicots are similar in structure and function, but they differ in some distinctive ways. For example, monocot seeds have one cotyledon and dicot seeds have two. A "cotyledon" is a leaflike structure originating in the seed, as part of the plant embryo. After the seed germinates, the cotyledons may unfurl somewhere along the length of the tiny seedling. Figure 20.5 shows other differences between monocots and dicots.

SHOOT SYSTEM

Arrangement of Vascular Bundles

The primary tissues of monocot and dicot stems usually are organized in one of two patterns, based on the distribution of vascular bundles. A **vascular bundle** is a strandlike arrangement of primary xylem and phloem. Commonly, some fibers and parenchyma cells form a sheath around the strand. The stems of most monocots and some dicots have the vascular bundles distributed throughout the ground tissue. Figure 20.6 shows an example of this. The stems of most dicots and conifers have the vascular bundles arranged as a ring that divides the ground tissue into **cortex** and **pith** (Figure 20.7).

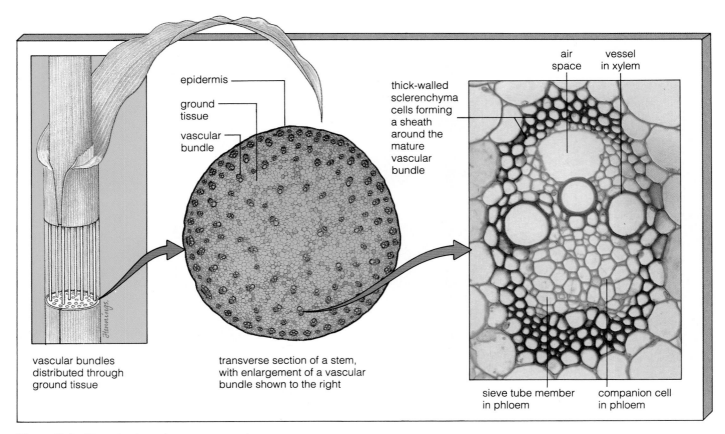

Figure 20.6 Stem structure of corn, a monocot.

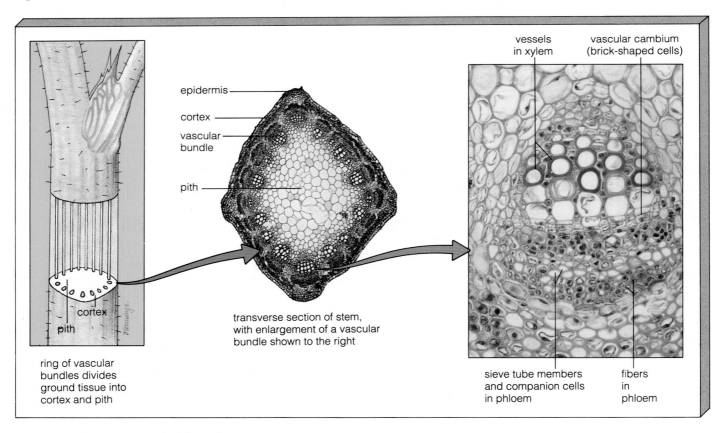

Figure 20.7 Stem structure of alfalfa, a dicot.

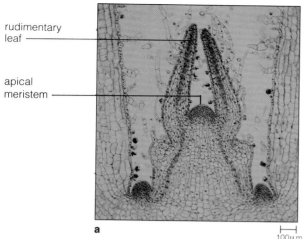

rudimentary leaf

apical meristem

a

100μm

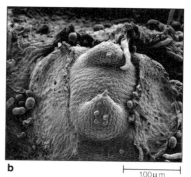

b

100μm

c

100μm

Figure 20.8 (**a**) Leaf formation at the shoot tip of *Coleus*, as seen in thin section. (**b,c**) Scanning electron micrographs of the same type of shoot tip.

Arrangement of Leaves and Buds

For most vascular plants, **leaves** are the main sites of photosynthesis. Leaves develop on the flanks of the tip of a main stem or its branchings. Each starts out as a slight bulge from apical meristem and enlarges into a thin, rudimentary leaf (Figure 20.8). The bulges are close together at first, but as plant growth continues, the leaves that form from the bulges become spaced at intervals along the length of the stem. The point on the stem where one or more leaves are attached is a **node,** and each stem region between two successive nodes is an **internode** (Figure 20.1).

Look at a twig of a walnut tree in winter, when it is devoid of leaves (Figure 20.9a). At the shoot tip is a **bud,** an undeveloped shoot of mostly meristematic tissue that is often protected by a covering of modified leaves. In addition to this ''terminal bud,'' other buds occur at regular intervals along the length of the stem. These ''lateral buds'' form in the upper angle where a leaf is attached to the stem.

Buds give rise to leaves, flowers, or both (Figure 20.9b). Depending on the species, each node has one, two, or three or more leaves and buds.

Leaf Structure

Leaf Shapes. Many dicot leaves have a broad blade attached by a stalk (petiole) to the stem. Many monocot leaves are not stalked; the blade simply encircles the stem, forming a sheath. In some species, the blade is deeply lobed or divided into smaller leaflets (Figures

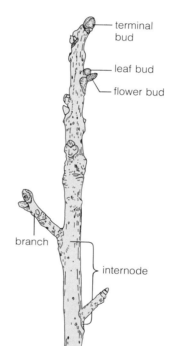

terminal bud

leaf bud

flower bud

branch

internode

a

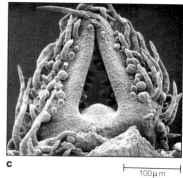

b

c

d

Figure 20.9 (**a**) Arrangement of buds on a three-year-old twig of a walnut tree. (**b–d**) Formation of leaves at a terminal bud of a dogwood tree.

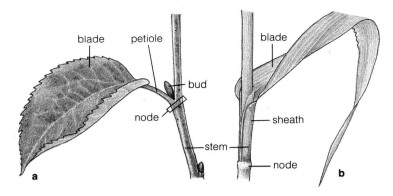

Figure 20.10 General form of leaves commonly seen in dicots (a) and monocots (b).

20.10 and 20.11). There are many variations on the basic leaf plans. For example, some leaves have hairs and scales, others have hooks that impale predators.

Most leaves are short-lived. In "deciduous" species such as birches, the leaves drop away from the stem as winter approaches. Species such as camellias also drop leaves, but they appear "evergreen" because the leaves do not all drop at the same time.

Leaf Internal Structure. The leaves just described have a large external surface area exposed to sunlight and carbon dioxide in the air. Photosynthetic parenchyma cells are organized inside those leaves, between the upper and lower epidermis. Extensive air spaces between the cells enhance the uptake of carbon dioxide and the release of oxygen during photosynthesis (Figure

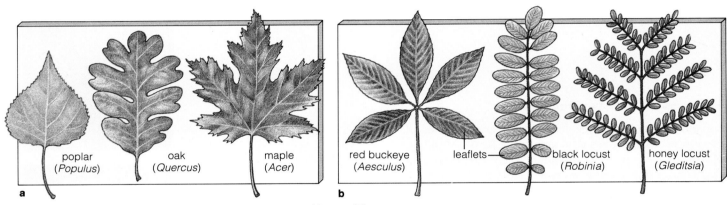

Figure 20.11 Examples of simple leaves (a) and compound leaves (b).

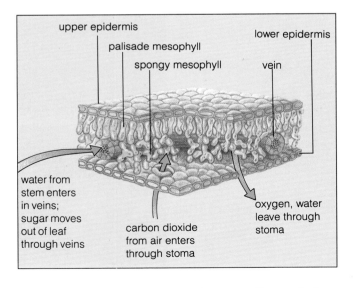

Figure 20.12 Example of leaf internal structure. Photosynthetic parenchyma cells make up the stacked palisade mesophyll and the more loosely organized, spongy mesophyll.

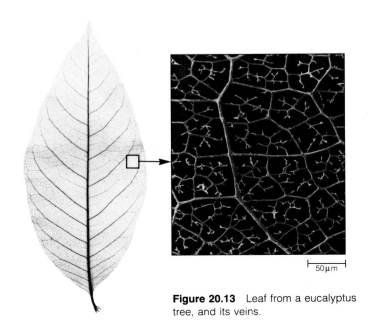

Figure 20.13 Leaf from a eucalyptus tree, and its veins.

Uses and Abuses of Leaves

Imagine a meal without leaves. Our plates would not be graced with lettuces, parsley, spinach, or chard; our senses would not be tweaked by splendid spices and seasonings from the leaves of parsley, sage, rosemary and thyme. There would be no more of the teas brewed from a variety of leaves.

But leaves have uses beyond the table. Oils from the leaves of orange trees and lavender find their way into perfumes and scented soaps. Oils of eucalyptus, camphor, and other leaves find their way into medicine chests. Doctors prescribe drugs isolated from belladona, an extract of leaves of nightshade plants. They use digitalis, isolated from foxglove leaves, to help regulate heartbeat and blood circulation. They treat skin that has been damaged from the sun or other forms of radiation with juices from the leaves of *Aloe vera*.

Landscape architects and imaginative homeowners use the leaves of different trees and shrubs much as an artist uses different colors and textures. Ecologically aware people plant shrubs next to buildings and so help keep them from heating up in summer and losing heat in winter. We get twine and rope from leaves of century plants (*Agave*), cords and textiles from leaf fibers of Manila hemp, hats from leaves of Panamanian palms, and thatched roofs from the leaves of palms and grasses.

We kill cockroaches, fleas, lice, and flies with insecticides derived from Mexican cockroach plants. We kill over a hundred kinds of insects, mites, and nematodes with extracts of Neem tree leaves—without killing off the natural predators of all those far-too-common pests.

Some people also smoke, chew, or tuck onto their mouth the leaves of tobacco plants—and so become candidates for the hundreds of thousands of deaths each year from lung, mouth, and throat cancers. Cocaine, derived from leaves of the coca plant, is used medicinally and abused by increasing numbers of individuals—with devastating social and economic effects, as described in the *Commentary* on page 439.

20.12). Vascular bundles called *veins* form a network through the leaf (Figure 20.13). The veins move water and solutes to the photosynthetic cells and carry products away from them.

Usually the lower epidermis of a leaf contains most of the *stomata* (singular, stoma). Carbon dioxide enters the plant and oxygen and water vapor leave it by way of these tiny openings (Figure 20.12). The next chapter will describe how stomata open and close in response to environmental conditions.

ROOT SYSTEM

Plants must absorb enough water and dissolved minerals for growth and maintenance, and it takes a tremendous root surface area to do this. For example, the root system of a rye plant only four months old had a surface area of 639 square meters (764 square yards), about 130 times more than its shoot system!

Taproot and Fibrous Root Systems

A root system penetrates downward and spreads out laterally, anchoring the aboveground parts. The roots of carrots, sugar beets, and most other plants also store photosynthetically produced food, some to be used by root cells and some to be transported later to aboveground parts.

In most dicot seedlings, the first (primary) root increases in diameter and grows downward. Branchings called *lateral roots* emerge sideways along its length (Figure 20.14a). The youngest branchings are found near the root tip. A primary root and its lateral branchings represent a **taproot system.** A carrot has a taproot system. So does a pine tree, the roots of which can penetrate the soil to depths of 6 meters and more.

Generally, the primary root is short-lived in monocots such as grasses. In its place, numerous *adventitious roots* arise from the stem of the young plant. (The term "adventitious" refers to any structure arising at an unusual location, such as roots that grow from stems or leaves.) Adventitious roots and their branchings are all somewhat alike in length and diameter, and they form a **fibrous root system** (Figure 20.14b). Generally, fibrous root systems do not penetrate the soil deeply.

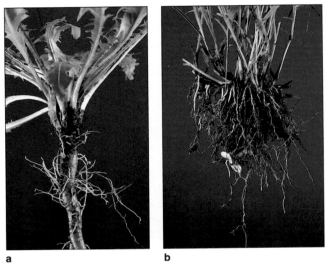

a   b

Figure 20.14 (**a**) Taproot system of a dandelion. (**b**) Fibrous root system of a grass plant.

Root Structure

Cells of root tips divide at the apical meristem and in a limited zone behind it, where primary tissues start to differentiate. Cells elongate in the next few millimeters and they mature beyond that region.

Root Cap. A root tip has a dome-shaped cell mass, the *root cap* (Figure 20.15). Root apical meristem produces the cell mass and in turn is protected by it. The root cap is pushed forward as the root grows, and some of its cells are torn loose. The slippery remnants lubricate the cap and enhance movement through the soil.

Root Epidermis. Behind the root cap, the epidermis, ground tissue, and vascular tissues form. Root epidermis is the absorptive interface with the environment. Some of the epidermal cells send out long extensions called *root hairs* (Figure 21.4). Root hairs greatly increase the surface available for taking up water and solutes. That is why you should never yank a plant out of the ground when transplanting it; too much of the fragile absorptive surface would be torn off.

Vascular Column. Most often, the vascular tissues of a root are arranged as a central cylinder, or **vascular column**. The column is surrounded by ground tissue

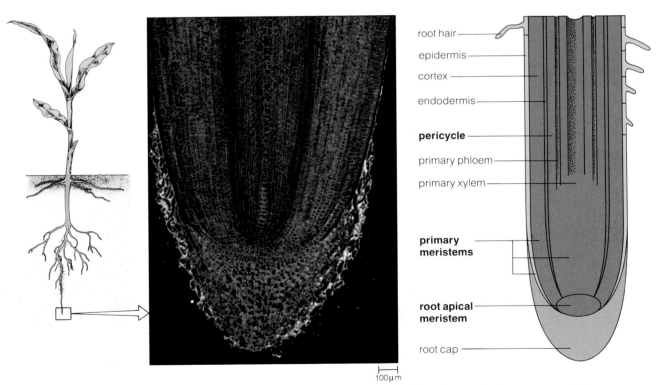

Figure 20.15 Generalized root tip, sliced lengthwise. The micrograph shows a corn root tip.

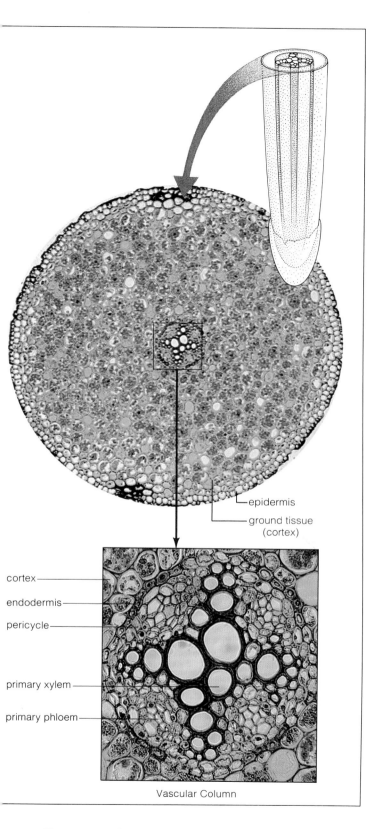

Figure 20.16 Section through a young root from a buttercup (*Ranunculus*). The closeup shows details of its vascular column.

cortex

endodermis

pericycle

primary xylem

primary phloem

Vascular Column

epidermis

ground tissue (cortex)

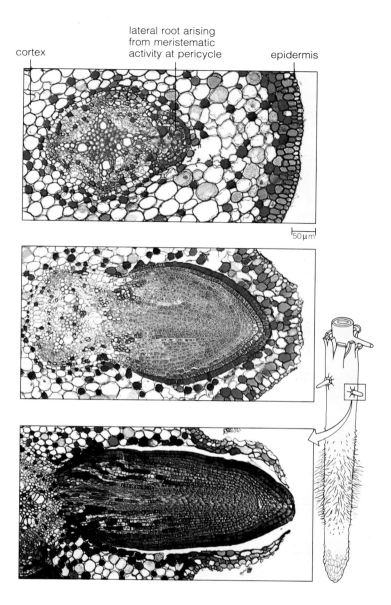

cortex

lateral root arising from meristematic activity at pericycle

epidermis

50 µm

Figure 20.17 Lateral root formation in a willow (*Salix*).

called the root cortex (Figure 20.16). Abundant air spaces in the ground tissue allow oxygen to reach the living root cells, which depend on oxygen for aerobic respiration. Also, many cell junctions (called plasmodesmata) connect the cytoplasm of adjacent cells of the cortex. Water entering the root moves from cell to cell until it reaches the *endodermis*, the innermost cell layer of the root cortex. As you will see in the next chapter, the endodermis helps control the movement of water and dissolved minerals into the vascular column.

Just inside the endodermis is the *pericycle*. This part of the vascular column has one or more layers of cells that give rise to lateral roots, which grow out through the cortex and epidermis (Figure 20.17).

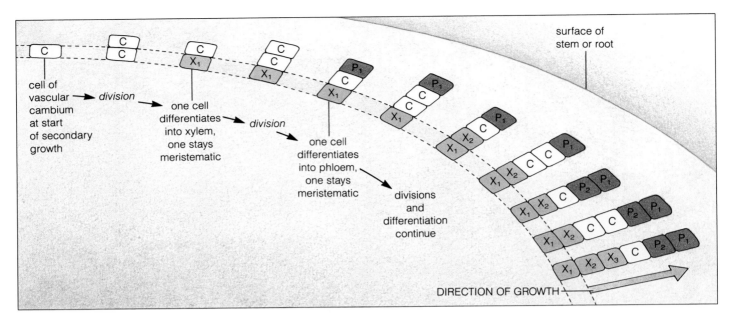

Figure 20.18 Relationship between the vascular cambium and its derivative cells (secondary xylem and phloem). This is a composite drawing of growth in a stem through successive seasons. Notice how the ongoing divisions displace the cambial cells, moving them steadily outward even as the core of xylem increases the stem or root thickness.

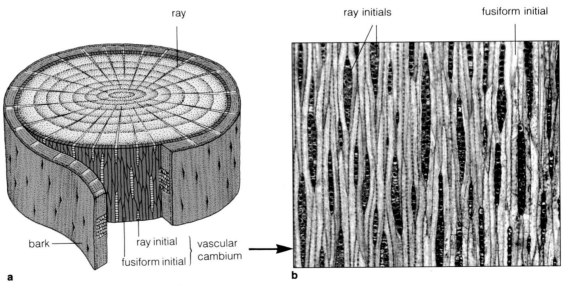

Figure 20.19 (a) Location of vascular cambium in an older stem showing secondary growth. Some of the meristematic cells of the vascular cambium (the fusiform initials) produce the secondary xylem and phloem that conduct water and food vertically through the stem. Other meristematic cells (ray initials) produce parenchyma and other cells that serve as lateral channels for water and as food storage centers. (b) Vascular cambium from the trunk of an apple tree (*Malus*).

WOODY PLANTS

Herbaceous and Woody Plants Compared

The life cycle of flowering plants extends from seed germination to seed formation, then eventual death. Most monocots and some dicots are nonwoody, or "herbaceous," plants; they show little or no secondary growth during the life cycle. In contrast, many dicots and all gymnosperms are "woody" plants; they show secondary growth during two or more growing seasons. Herbaceous and woody plants are characterized as follows:

annuals	*Life cycle completed in one growing season; little (if any) secondary growth. Examples: snap beans, corn, marigolds.*
biennials	*Life cycle completed in two growing seasons (root, stem, leaf formation the first season; flowering, seed formation, death the second). Example: carrots.*
perennials	*Vegetative growth and seed formation continue year after year. Some have secondary tissues, others do not. Examples: the herbaceous cacti, woody shrubs (roses), vines (ivy, grape), and trees (apples, elms, magnolias).*

Tissue Formation During Secondary Growth

The older stems and roots of many plants become more massive and woody through the activity of two types of lateral meristems. As indicated in Figure 20.4, these meristems are called **vascular cambium** and **cork cambium.**

When fully developed, the zone of vascular cambium is like a cylinder, one or a few cells thick. Its meristematic cells give rise to secondary xylem and phloem tissues that conduct substances up, down, and horizontally through the enlarging stem or root. Xylem forms on the inner face of the vascular cambium, and phloem forms on the outer face (Figures 20.18 and 20.19).

The mass of xylem increases season after season, and it usually crushes the thin-walled phloem cells from the preceding growth period. New phloem cells are added each year, outside the growing inner core of xylem.

In time, the mass of new tissue inside a stem or root causes the cortex and outer phloem to rupture. Parts of the cortex split away and carry epidermis with them. But cork cambium is produced by meristematic cells. In turn, the cork cambium produces the *periderm*, a corky replacement for the epidermis. "Cork" refers to the cells produced by the cork cambium. "Bark" refers to all living and nonliving tissues between the vascular cambium and the stem or root surface.

As you can see from Figure 20.20, living phloem in older trees is confined to a thin zone beneath the periderm. Stripping off a band of phloem all the way around a tree's circumference will kill the tree. (When the vertical phloem pipelines are broken, there is no way to transport photosynthetically derived food to the roots, which will die.) This is a common practice in countries where tropical rain forests are being destroyed; it kills the trees before the land is cleared for agriculture (page 577).

Early and Late Wood

In regions having prolonged dry spells or cool winters, the vascular cambium of stems and roots becomes inactive during parts of the year. The first xylem cells produced at the start of the growing season tend to have large diameters; they represent *early wood*. As the season progresses, the cell diameters become smaller; these cells represent the *late wood*.

The last-formed, small-diameter cells of late wood end up next to the first-formed, large-diameter cells of the next season's growth. When you look at a full-diameter slice from an old tree trunk, you won't see the individual cells. But the early and late wood will reflect the light differently, and you can identify them as light

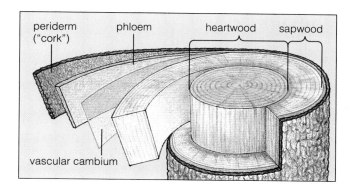

Figure 20.20 Structure of a woody stem showing extensive secondary growth. Heartwood, the core of the mature tree, is devoid of living cells. Sapwood is the cylindrical zone of xylem between the heartwood and vascular cambium; it also contains living parenchyma cells. All the living and nonliving tissues between the vascular cambium and the surface of a woody stem are often called bark.

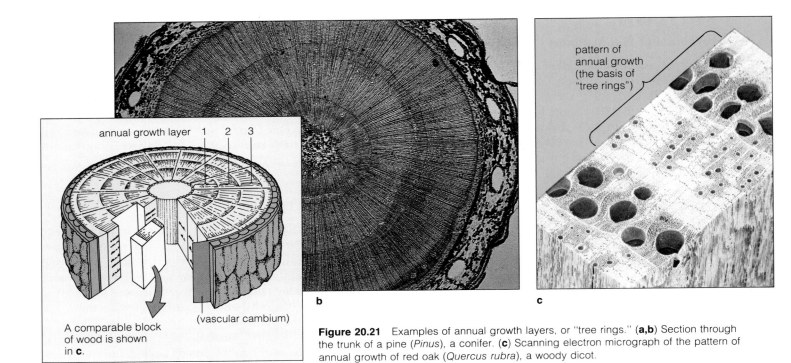

annual growth layer 1 2 3

A comparable block
of wood is shown
in **c**.

(vascular cambium)

a

b

c

pattern of
annual growth
(the basis of
"tree rings")

Figure 20.21 Examples of annual growth layers, or "tree rings." (**a,b**) Section through the trunk of a pine (*Pinus*), a conifer. (**c**) Scanning electron micrograph of the pattern of annual growth of red oak (*Quercus rubra*), a woody dicot.

bands and dark bands (Figure 20.21). The alternating bands, each with early and late wood, represent annual growth layers, or "tree rings."

SUMMARY

1. There are two classes of flowering plants (angiosperms), informally called monocots and dicots. The root systems and shoot systems (stems, leaves) of monocots and dicots are composed of three kinds of tissues (dermal, ground, and vascular tissues).

2. Dermal tissues include epidermis (which covers and protects the surfaces of primary plant parts) and periderm (which replaces epidermis on plants showing secondary growth).

3. There are three types of ground tissues. One (parenchyma) makes up the bulk of fleshy plant parts; its generally thin-walled, living cells function in photosynthesis, storage, and other tasks. The other two types (collenchyma and sclerenchyma) have cells with thickened walls, and both provide mechanical support for growing plant parts.

4. Vascular tissues include xylem and phloem. Xylem contains cells (actually the water-permeable walls of cells dead at maturity) that interconnect to form pipelines for conducting water and dissolved minerals. Phloem, the food-conducting tissue, contains living cells joined end to end to form tubes.

5. The primary growth of roots and shoots originates at apical meristems, the undifferentiated tissue of self-perpetuating cells at root and shoot tips. Descendants of some cells of the apical meristem produce the primary tissues. Nonwoody annual plants, which live one season, show little or no secondary growth.

6. Lateral meristems (vascular cambium and cork cambium) increase the diameter of stems and roots of plants showing secondary growth. Perennial plants that live for many seasons show extensive secondary growth.

7. Stems give photosynthetic tissues favorable exposure to light and display the flowers. Their vascular tissues distribute substances to and from roots, leaves, and other plant parts. Monocot stems have vascular bundles distributed throughout the ground tissue; dicot stems have vascular bundles arrayed as a cylinder that separates the ground tissue into cortex and pith.

8. Photosynthetic parenchyma cells are organized between the upper and lower epidermis of leaves, with abundant air spaces around them. Numerous openings (stomata) in the epidermis allow water vapor and gases to be exchanged efficiently between the leaf interior and the environment.

9. Roots absorb water and dissolved minerals from the surroundings and conduct them to aerial plant parts; they anchor and sometimes support the plant and often have food storage regions.

Review Questions

1. List some of the functions of the root system and the shoot system of a flowering plant. *293*

2. What are the three main types of tissues in flowering plants, and what are the functions of each? *293, 294*

3. What are some of the differences between monocots and dicots? *295* Which of the following stem sections is typical of most monocots? *295* Of most dicots? Label the main tissue regions of each: *294, 295*

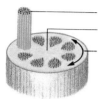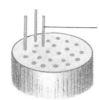

4. Where are apical meristems located, and what kinds of tissues originate from them? *294, 295*

5. Where are the two types of lateral meristems located, and what kinds of tissues originate from each? *295*

6. How are annual growth layers formed in woody stems? If you were to strip away a band of phloem all the way around a tree's circumference, what would happen to the tree? Why? *303, 294*

Self-Quiz   *(Answers in Appendix IV)*

1. In vascular plants, _Dermial_ tissues cover and protect the root and shoot systems; _ground_ tissues make up the greater portion of any plant body; and strands of _Vascular_ tissues conduct food, water, and minerals.

2. The three basic kinds of plant tissues arise from cell division and elongation in regions called _Apical_ meristems at the tips of roots and shoots. Increases in root or stem diameter occur at _Lateral_ meristems.

3. The two classes of flowering plants are _____.
   a. angiosperms and gymnosperms
   b. monocots and dicots
   c. shrubs and trees
   d. herbs and shrubs

4. One type of ground tissue, _____, makes up the bulk of fleshy plant parts.
   a. parenchyma   c. collenchyma
   b. sclerenchyma   d. epidermis

5. _____ cells are thin-walled, alive at maturity, and function in photosynthesis, storage, and other tasks.
   a. Parenchyma   c. Collenchyma
   b. Sclerenchyma   d. Epidermis

6. Cells of the ground tissues _____ and _____ have thick walls and help mechanically to support the plant parts.
   a. parenchyma; collenchyma
   b. collenchyma; sclerenchyma
   c. parenchyma; sclerenchyma
   d. parenchyma; epidermis

7. _____ is a vascular tissue that conducts water and minerals; _____ is a vascular tissue that conducts food.
   a. Phloem; xylem
   b. Vascular cambium; phloem
   c. Xylem; phloem
   d. Xylem; vascular cambium

8. Herbaceous annual plants show little or no _____ growth; perennial plants show considerable _____ growth.
   a. secondary; secondary   c. secondary; primary
   b. primary; secondary   d. primary; primary

9. _____ stems have vascular bundles scattered throughout the ground tissue; _____ stems have vascular bundles arranged as a cylinder that separates ground tissue into cortex and pith.
   a. Dicot; dicot   c. Monocot; monocot
   b. Dicot; monocot   d. Monocot; dicot

10. Match these plant parts with their structure or function.
   _c_ roots
   _D_ leaves
   _f_ dermal tissue
   _a_ stems
   _e_ ground tissues
   _b_ vascular tissues

   a. expose photosynthetic tissues to light, display flowers; distribute materials to and from roots, leaves
   b. contain xylem and phloem
   c. absorb, conduct water and minerals, anchor the plant, store food
   d. has photosynthetic cells between upper, lower epidermis
   e. contain parenchyma, collenchyma, sclerenchyma
   f. contain epidermis, periderm

Selected Key Terms

apical meristem *294*
bud *297*
cork cambium *303*
cortex *295*
cuticle *294*
dicot *295*
epidermis *294*
internode *297*
lateral meristem *295*
leaf *297*

monocot *295*
node *297*
phloem *294*
pith *295*
root system *293*
shoot system *293*
vascular bundle *295*
vascular cambium *303*
vascular column *300*
xylem *294*

Readings

Bold, H., C. Alexopoulos, and T. Delevoryas. 1987. *Morphology of Plants and Fungi*. Fifth edition. New York: Harper & Row.

Raven, P., R. Evert, and S. Eichhorn. 1986. *Biology of Plants*. Fourth edition. New York: Worth. Exquisite color micrographs and illustrations of plant cells and tissues.

Rost, T., et al. 1984. *Botany: An Introduction to Plant Biology*. Second edition. New York: Wiley.

Stern, K. 1988. *Introductory Plant Biology*. Fourth edition. Dubuque, Iowa: Brown. Beautifully illustrated, accessible introduction to plant structure and function. Paperback.

PLANT NUTRITION AND TRANSPORT

1. Land plants are adapted for absorbing water and dissolved mineral ions that may not always be plentiful in soil; conserving water; and conducting water, minerals, and photosynthetic products through the plant.

2. Evaporation at the leaf surface pulls up water in xylem, a vascular tissue that extends from roots, through stems, to the leaves. The cumulative strength of hydrogen bonds between water molecules allows water to be pulled upward as continuous fluid columns.

3. Water escapes mainly from small passageways (stomata) across the cuticle-covered epidermis of leaves, but the carbon dioxide required for photosynthesis must diffuse into the plant through the same passageways. Stomata open and close at different times and so control water loss and carbon dioxide uptake.

4. Differences in water pressure between sites of photosynthesis and metabolically active, growing tissues drive the distribution of organic materials through the plant. The organic materials move through the phloem.

It took you eighteen years or so to grow to your present height. A corn plant grows more than that in three months! Yet how often do we stop to think that plants do anything impressive? Being endowed with great mobility, intelligence, and rich emotions, we tend to be fascinated more with ourselves than with immobile, expressionless plants.

With this chapter we turn to some adaptations by which plants function in the environment. You know from previous chapters that plants are photosynthetic autotrophs; they require only sunlight, water, carbon dioxide, and some minerals to nourish themselves. But plants (like people) do not have unlimited supplies of resources. Most soils are frequently dry. The air is only 340 parts carbon dioxide to a million parts of everything else. And soil water does not hold lavish amounts of dissolved minerals. As you will see, *many aspects of plant structure and function are responses to low concentrations of environmental resources.*

Table 21.1 Role of Mineral Elements in Plant Function

Element	Some Known Functions	Some Deficiency Symptoms
Macronutrients:		
Nitrogen	Component of proteins, nucleic acids, coenzymes	Stunted growth; light green older leaves; older leaves yellow and die (chlorosis)
Potassium	Activation of enzymes, role in maintaining water-soluble balance*	Reduced growth; mottled, spotted, or curled older leaves; burned leaf margins; weakened roots and stems
Calcium	Roles in cementing cell walls, formation of mitotic spindle	Leaves deformed; terminal buds die; reduced root growth
Magnesium	Component of chlorophyll; activation of enzymes	Chlorosis; drooped leaves
Phosphorus	Component of nucleic acids, ATP, phospholipids	Purplish veins in older leaves; fewer seeds and fruits; stunted growth
Sulfur	Component of most proteins, two vitamins	Light green or yellow leaves; reduced growth

*All mineral elements contribute to the water-solute balance, but potassium is notable because there is so much of it.

NUTRITIONAL REQUIREMENTS

Plants generally require sixteen essential elements. They use three of those elements—oxygen, carbon, and hydrogen—as the main building blocks for their carbohydrates, lipids, proteins, and nucleic acids. The plants get oxygen from water, gaseous oxygen (O_2), and carbon dioxide (CO_2) in the air. They get hydrogen from water molecules.

The other thirteen elements become available to plants as dissolved salts; they are "mineral ions." As Table 21.1 shows, six of the mineral ions are *macronutrients,* meaning a significant fraction of each becomes incorporated in plant tissues. The rest are *micronutrients;* only small traces occur in plant tissues. Micronutrients as well as macronutrients play vital roles in photosynthesis and other metabolic events. Plants cannot grow well when they are deprived of any essential element.

In many agricultural regions, for example, crop harvests suffer from nitrogen scarcity. There actually is plenty of gaseous nitrogen ($N\equiv N$) in the air, but plants do not have the metabolic means to break apart the three covalent bonds in each molecule. Crop yields often depend on applications of nitrogen-rich fertilizers or on the activity of "nitrogen-fixing" bacteria present in the soil. These bacteria convert nitrogen to forms they—and the plants—can use.

Clover, beans, and other legumes have an advantage in this respect. Nitrogen-fixing bacteria reside in localized swellings (nodules) on the plant roots (Figure 21.1). The bacteria feed on some of the plant's organic molecules. But they also provide the plant with usable nitrogen. Thus, the plants get more of a scarce resource through a type of *symbiotic* relationship with bacteria; the two species interact on a permanent basis in a mutually beneficial way.

Element	Some Known Functions	Some Deficiency Symptoms
cronutrients:		
lorine	Role in root, shoot growth; role in photolysis	Wilting; chlorosis; some leaves die
n	Role in chlorophyll synthesis; role in electron transport	Chlorosis; yellow and green striping in grasses
ron	Roles in flowering, germination, fruiting, cell division, nitrogen metabolism	Terminal buds, lateral branches die; leaves thicken, curl, become brittle
nganese	Role in chlorophyll synthesis; coenzyme activity	Light green leaves with green major veins; leaves whiten and fall off
c	Role in formation of auxin, chloroplasts, and starch; enzymes component	Chlorosis; mottled or bronzed leaves; abnormal roots
pper	Component of several enzymes	Chlorosis; dead spots in leaves; stunted growth; terminal buds die
lybdenum	Component of enzyme used in nitrogen metabolism	Possible nitrogen deficiency; pale green, rolled or cupped leaves

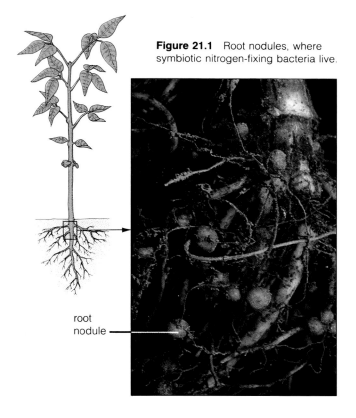

Figure 21.1 Root nodules, where symbiotic nitrogen-fixing bacteria live.

root nodule

Figure 21.2 (**a**) Specialized leaves of the Venus flytrap, a carnivorous plant. This plant grows in nitrogen-poor soil. Its two-lobed leaves open and close like a clamshell. When an insect brushes against the leaf "hairs" (long extensions of specialized epidermal cells), the movement triggers cellular changes at the leaf midrib. The lobes of the leaf close, and the spines fringing the leaf margins intermesh to become a cage around the insect (**b,c**). Glandlike epidermal cells secrete enzymes that digest proteins of the trapped insect body. The plant uses nitrogen released from the proteins in biosynthesis.

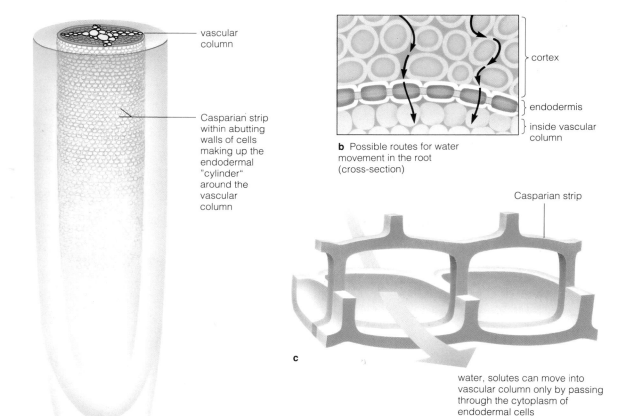

vascular
column

Casparian strip
within abutting
walls of cells
making up the
endodermal
"cylinder"
around the
vascular
column

cortex

endodermis

inside vascular
column

b Possible routes for water
movement in the root
(cross-section)

Casparian strip

water, solutes can move into
vascular column only by passing
through the cytoplasm of
endodermal cells

Figure 21.3 (**a**) Location of the vascular column in a root. (**b**) Possible routes for water movement in the root, cross-section. Water moving into the root travels along the walls of cells making up the cortex or through the spaces between the cells. (**c**) Water moves into the vascular column only through the cytoplasm of endodermal cells; it cannot penetrate their walls at the Casparian strip.

The Venus flytrap is another example of how plants are adapted to low concentrations of an environmental resource. This plant grows well in nitrogen-poor soil, but only because it gets nitrogen from a rather unusual source (Figure 21.2).

WATER ABSORPTION

Water availability profoundly affects root development, hence growth of the entire plant. Roots branch out in some locations and are replaced by roots that branch into different locations as conditions change. It is not that roots "explore" the soil for resources. Outward root growth simply is stimulated in regions where water and dissolved mineral ions are abundant.

Refer to Figure 20.16, which shows the structure of a typical root. Once water has been absorbed from the surrounding soil, it moves through the root cortex until it reaches the endodermis of the vascular column. There, a water-repellent "Casparian strip" prevents water from moving past the abutting walls of the endodermal cells (Figure 21.3). Water cannot enter the vascular column unless it moves across the plasma membrane and through the cytoplasm of those cells. Plasma membranes, recall, selectively move some substances but not others across the lipid bilayer. *Membrane transport mechanisms help control the types of absorbed solutes that will become distributed through the plant.*

Water absorption is enhanced by **root hairs,** each of which is an extension of a specialized epidermal cell (Figure 21.4). Root hairs greatly increase the surface area available for absorbing water and solutes. A single root system might develop millions or billions of root hairs.

Mycorrhizae, or "fungus-roots," also enhance water absorption. A **mycorrhiza** is a symbiotic association between a fungus and a young root (Figure 21.5). One type of fungus grows as a mat of thin filaments around roots and several centimeters into the soil. Collectively, the filaments have a tremendous surface area for absorbing mineral ions from a large volume of soil. The fungus uses some of the root's sugars and nitrogen-containing compounds. As it grows, the root uses some of the minerals secured by the fungus (Figure 21.6).

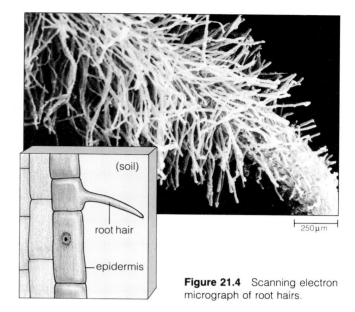

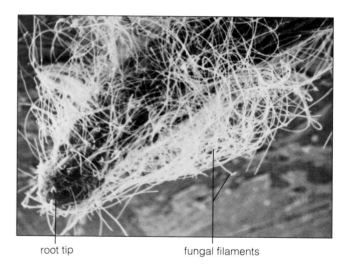

Figure 21.4 Scanning electron micrograph of root hairs.

root tip   fungal filaments

Figure 21.5 Mycorrhiza (fungus-root) of a lodgepole pine tree.

Figure 21.6 Effect of mycorrhizal fungi on plant growth. The six-month-old juniper seedlings on the left were grown in sterilized, low-phosphorus soil with a mycorrhizal fungus. The seedlings on the right were grown in soil without the fungus.

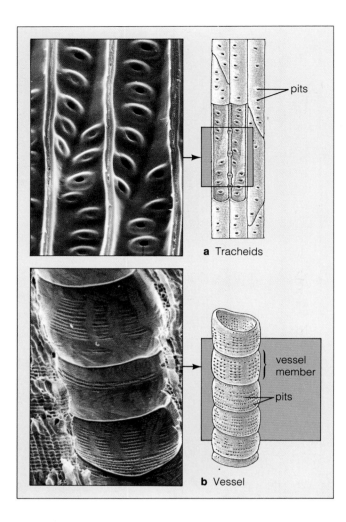

Figure 21.7 Examples of the main cells in xylem that conduct water and dissolved mineral ions through the plant body. The tracheids are from a pine tree stem; the vessel from a red oak stem. Water flows as continuous columns through the tubes formed by adjacent cell walls. Compare Figure 21.8.

In figure, labels: pits; **a** Tracheids; vessel member; pits; **b** Vessel

WATER TRANSPORT AND CONSERVATION

Transpiration

Water moves from roots to stems, then into leaves. Only about one percent is used in growth and metabolism; the rest evaporates into the air. Water evaporation from stems, leaves, and other plant parts is called **transpiration.**

How does water get all the way to the top of plants, including trees that are 100 meters tall? Water moves through the water-conducting cells of xylem (Figures 21.7 and 21.8). The cells are dead at maturity and only their walls remain, so the cells themselves are not pulling water "uphill." Rather, *water is pulled up by the drying power of air, which creates continuous negative pressures (tensions) that extend downward from leaves to roots.* This is the key point of the **cohesion theory of water transport:**

1. The drying power of air causes transpiration (evaporation from plant parts exposed to air).

2. The pulling action of transpiration puts the water in xylem in a state of tension that is continuous from leaves, down through the stems, to roots.

3. As long as water molecules continue to escape from the plant, molecules of soil water move into roots (following the osmotic gradient) and are pulled up as replacements.

4. Columns of water are pulled up by the collective strength of hydrogen bonds between water molecules, which are confined in the narrow, tubular xylem cells.

5. Hydrogen bonds are enough to hold water molecules together in the xylem and to hold them to cell walls, but they are not strong enough to prevent them from breaking away from each other during transpiration.

Control of Water Loss

Only a small fraction of the water moving into a leaf is used in photosynthesis, growth, and other activities. When water loss by transpiration exceeds water uptake by roots, however, plants can wilt and die.

Transpiration occurs mostly through **stomata** (singular, stoma), the small passageways across the cuticle-covered epidermis of leaves and stems (pages 77 and 294). So does the inward diffusion of carbon dioxide. When stomata are open, carbon dioxide can be absorbed for photosynthesis. But when they are open, water nearly always moves out! *Stomata must open and close at different times to control water loss and carbon dioxide uptake.*

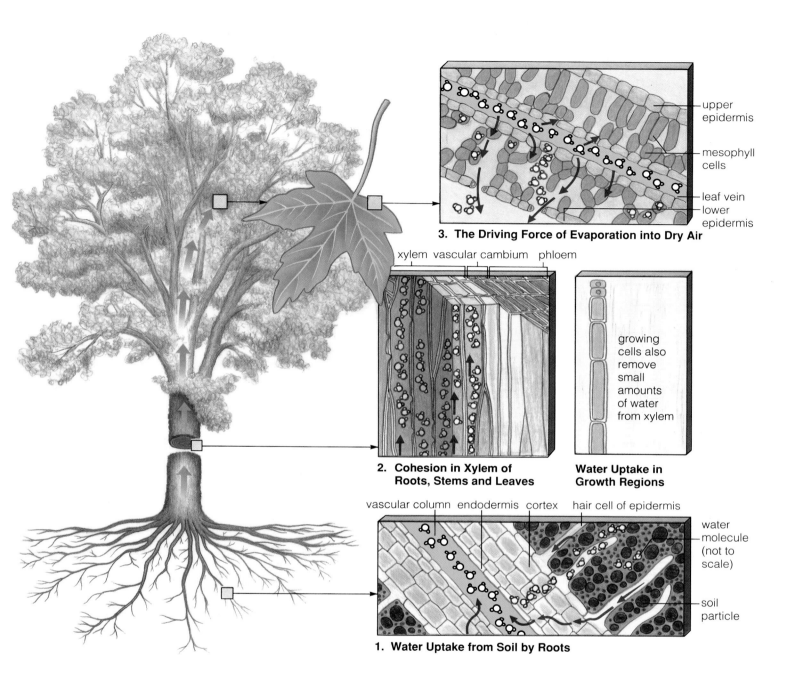

3. The Driving Force of Evaporation into Dry Air

- upper epidermis
- mesophyll cells
- leaf vein
- lower epidermis

xylem vascular cambium phloem

2. Cohesion in Xylem of Roots, Stems and Leaves

growing cells also remove small amounts of water from xylem

Water Uptake in Growth Regions

vascular column endodermis cortex hair cell of epidermis

- water molecule (not to scale)
- soil particle

1. Water Uptake from Soil by Roots

Figure 21.8 Cohesion theory of water transport. Tensions in water in the xylem extend from leaf to root. These tensions are caused mostly by transpiration (the evaporation of water from plant parts). As a result of the tensions, columns of water molecules that are hydrogen-bonded to one another are pulled upward.

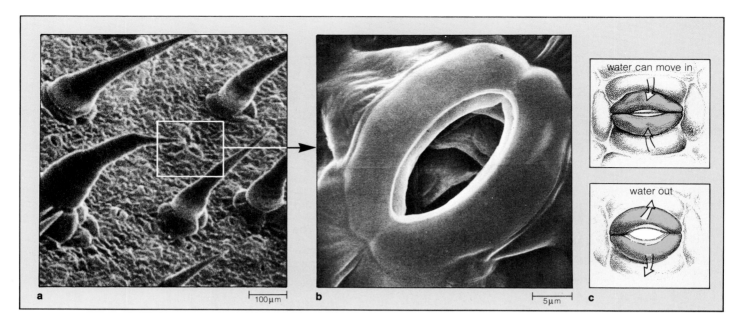

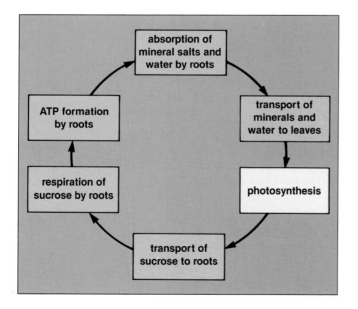

Figure 21.9 (**a,b**) Stomata on the lower epidermis of a cucumber leaf. Look closely and you can see several stomata among the hairlike structures at the leaf's surface. The guard cells flanking the stoma in (**b**) are clearly visible; inside, you can see parts of the photosynthetic mesophyll cells. (**c**) A stoma is closed when water pressure in the guard cells is low; it is open when the guard cells are swollen with water. When water moves out, the cells will collapse against each other, eliminating the gap between them.

Figure 21.10 Interrelated processes that influence the coordinated growth of roots, stems, and leaves. When one process is rapid, the others also may speed up. Any environmental factor limiting one process eventually slows growth of all plant parts.

A stoma opens and closes according to how much water and carbon dioxide are present in the two **guard cells** flanking it (Figure 21.9). When the sun comes up, light acts directly on cells, including the guard cells. It triggers the active transport of potassium ions into guard cells, followed by an inward movement of water by osmosis. As water pressure builds up inside, the guard cells become swollen and move apart, producing the stomatal opening. Thus, water is lost and carbon dioxide moves into the leaf during the day.

Photosynthesis stops when the sun goes down, but carbon dioxide accumulates in cells as a by-product of aerobic respiration. Potassium in the guard cells moves out now, followed by water. The guard cells collapse against each other, closing the gap between them. Thus, transpiration is reduced and water is conserved during the night.

Stomata remain open during daylight, when photosynthesis occurs. Water is lost, but carbon dioxide can enter the leaves.

Stomata remain closed during the night, when carbon dioxide accumulates through aerobic respiration. Then, water is conserved.

MINERAL UPTAKE AND ACCUMULATION

As Figure 21.10 shows, solute absorption and accumulation are coordinated throughout the plant body in ways that affect growth. Cells expend energy and actively accumulate solutes, especially dissolved mineral ions. Energy from ATP drives the membrane pumps by which solutes are actively transported into cells (page 80).

In photosynthetic cells, ATP necessary for the membrane pump operation is formed during both photosynthesis and aerobic respiration. What about nonphotosynthetic cells, such as those in roots? How do they get all the ATP necessary for active transport? Here, ATP is formed almost entirely through aerobic respiration in the mitochondria of individual cells.

TRANSPORT THROUGH THE PHLOEM

Storage and Transport Forms of Organic Compounds

Sucrose and other organic compounds resulting from photosynthesis are used throughout the plant. Leaf cells use some of the compounds and the rest are transported to roots, stems, buds, flowers, and fruits. Carbohydrates become stored as starch in most plant cells. Quantities of fats become stored in many seeds and fruits, including the avocado. Proteins become stored in many seeds.

Starch molecules are too large to cross cell membranes, so they cannot leave the cells in which they are formed. They also are too insoluble to be transported to other regions of the plant body. Fats are largely insoluble in water, and they cannot be transported out of their storage sites. Storage proteins do not lend themselves to transport, either.

Storage forms of organic compounds are converted to transportable forms through hydrolysis and other reactions. For example, hydrolysis of starch liberates the glucose units, which combine with fructose. The resulting molecule, sucrose, is the main form in which sugars are transported through plants.

Storage starch, fats, and proteins are converted to smaller subunits that are soluble and transportable through the plant body.

Translocation

Sucrose and other organic compounds are distributed through the plant by **translocation,** a process that occurs in phloem. Phloem, a vascular tissue, contains interconnecting tubes formed by the walls of living cells. The tubes lie side by side in overlapping array within vascular bundles, and they extend from leaf to root. Water and organic compounds are transported rapidly through their numerous wall perforations (Figure 21.11).

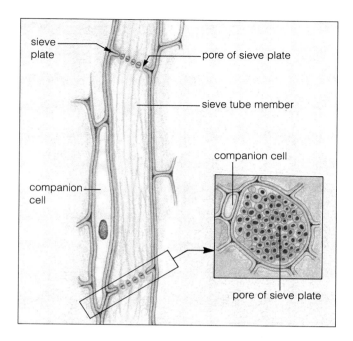

Figure 21.11 The main cell type in phloem, sliced lengthwise. The boxed inset shows how the wall of this cell, called a sieve tube member, is perforated at plates between adjacent cells. Companion cells metabolically support the sieve tube members and may direct their activities.

Figure 21.12 Honeydew droplet exuding from the tail end of an aphid feeding on the sugars present in the phloem of a plant.

Small insects called aphids tell us something about translocation. An aphid forces its mouthpart into the phloem cells and feeds on their dissolved sugars. The contents of the cells are under high pressure, often five times as much as in an automobile tire. This pressure forces the fluid through the aphid gut and out the other end as "honeydew" (Figure 21.12). Park your car under trees being attacked by aphids and it might get a spattering of sticky honeydew droplets, thanks to fluid pressure in the phloem.

Pressure Flow Theory

The movement of organic molecules in phloem follows a "source-to-sink" pattern. The main **source regions** are the sites of photosynthesis in leaves. A **sink region** is any plant part that requires organic compounds to meet its nutritional needs or that stockpiles them for later use. Growing leaves, fruits, seeds, and roots are examples of sink regions.

According to the **pressure flow theory,** translocation depends on pressure gradients between source and sink regions. Consider what happens when sucrose and other organic compounds move into a phloem cell in a leaf (a source region). Solute concentrations increase in the cell, followed by an inward movement of water by osmosis. Fluid pressure builds inside as the water encounters resistance from the cell wall, so some water is forced outward. Then water and its dissolved components move to regions in the tube system where water pressure is lower (Figure 21.13).

What maintains the low pressure at sink regions? There, organic compounds are unloaded from the tubes and actively transported into sink cells (which use them in wall building, starch formation, aerobic respiration, and other activities). When the solutes move in, water follows by osmosis.

Movement of organic compounds through the phloem is driven by differences in water pressure between the source regions and sink regions.

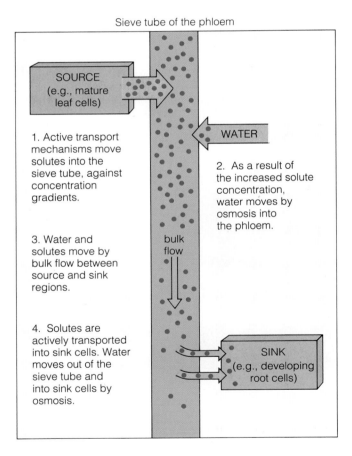

Sieve tube of the phloem

SOURCE
(e.g., mature leaf cells)

1. Active transport mechanisms move solutes into the sieve tube, against concentration gradients.

WATER

2. As a result of the increased solute concentration, water moves by osmosis into the phloem.

3. Water and solutes move by bulk flow between source and sink regions.

bulk flow

4. Solutes are actively transported into sink cells. Water moves out of the sieve tube and into sink cells by osmosis.

SINK
(e.g., developing root cells)

Figure 21.13 Proposed mechanism of pressure flow in the phloem of vascular plants.

SUMMARY

1. Plants have a large surface area that favors the uptake of water and nutrients, which often are present in low concentrations in the surrounding soil.

2. Plants require oxygen, carbon, hydrogen, and sixteen essential mineral ions. The mineral ions function in metabolic activities and in establishing solute concentration gradients across cell membranes. Such gradients are the basis for water movement into cells, and they help maintain cell shape and growth.

3. In the water-conducting cells of xylem, continuous negative pressures (tensions) extend down from leaves to roots. Transpiration (the evaporation of water from leaves and other plant parts exposed to air) causes the tension.

   a. When water molecules vacate transpiration sites, replacements are pulled under tension to the site.

   b. The cumulative strength of hydrogen bonds between water molecules allows water to be pulled upward as continuous fluid columns. This is the key point of the cohesion theory of water transport.

4. The plant loses water but takes up carbon dioxide during the day, when its numerous stomata remain open. It conserves water and restricts the inward movement of carbon dioxide during the night, when its stomata close. The controlled opening and closing of stomata balance carbon dioxide requirements (for photosynthesis) with the need for water conservation.

5. Organic compounds are distributed through the plant by translocation, which is driven by differences in water pressure between source regions (photosynthesis or storage sites) and sink regions (any metabolically active or growing tissue). The process takes place in the vascular tissue called phloem.

Review Questions

1. Give examples of the features that enable land plants to absorb water and nutrients from their surroundings, which have low concentrations of these required substances. *307–309*

2. Which three elements are the main building blocks of a plant? *307* Which six elements are considered macronutrients for land plants? *306*

3. Describe some of the specific roles that mineral ions play in plant functioning. How do solute absorption and accumulation affect plant growth? *306, 307, 309*

4. Define mycorrhiza. Why do you think it is important to include some of the native soil around roots when transplanting a plant from one place to another? *309*

5. Describe transpiration. State how the cohesion theory of water transport helps explain what is going on in this form of water movement. *310*

6. Transpiration competes with other water-requiring cell processes. Can you name some of these processes?

7. Sucrose transport from one plant organ to another is called translocation. Can you explain how it works in terms of the four key points of the pressure flow theory? *313, 314*

Self-Quiz   *(Answers in Appendix IV)*

1. Water can be pulled upward in a plant due to the cumulative strength of ___Hydroge Bonds___ between water molecules.

2. Water leaves and carbon dioxide enters the plant through ___Stoma___, which are tiny openings in the leaf epidermis.

3. The *uptake* of water and nutrients is enhanced by ___C___.
   a. a large number of leaves
   b. a tall stem structure
   c. a large surface area
   d. none of the above

4. Besides mineral ions, plants require ___D___.
   a. calcium, oxygen, and hydrogen
   b. carbon, potassium, and oxygen
   c. oxygen, carbon, and nitrogen
   d. carbon, oxygen, and hydrogen

5. In plant metabolism, mineral ions have roles in ___C___.
   a. metabolic activities
   b. establishing solute concentration gradients across cell membranes
   c. water movement into cells
   d. maintaining cell shape and growth
   e. all of the above

6. The loss of water molecules from leaves is known as ___E___.
   a. respiration   c. transpiration
   b. expiration   d. tension

7. The key point of the cohesion theory of water transport is ___B___.
   a. the volume of water lost from leaf surfaces
   b. the cumulative strength of hydrogen bonds between water molecules
   c. the width of the stem
   d. all of the above

8. During the day, plants lose ___d___ and take up _____.
   a. carbon dioxide; water
   b. water; oxygen
   c. oxygen; water
   d. water; carbon dioxide

9. At night, plants conserve ___d___ and restrict _____ intake.
   a. carbon dioxide; water
   b. water; oxygen
   c. oxygen; water
   d. water; carbon dioxide

10. Translocation of organic compounds in phloem is between photosynthesis or storage regions and regions where metabolism is high. These two regions are known respectively as ___d___.
    a. leaves and stems
    b. sink and source
    c. stems and leaves
    d. source and sink

11. Match the concepts of plant nutrition and transport.

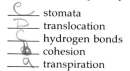

___e___ stomata
___d___ translocation
___c___ hydrogen bonds
___b___ cohesion
___a___ transpiration

a. theory of water transport in plants
b. evaporation from stems and leaves
c. creation of a continuous water column in a plant
d. transport of organic molecules from source to sink region
e. balancing of carbon dioxide uptake with water conservation

Selected Key Terms

Casparian strip *309*
cohesion theory of
  water transport *310*
guard cell *312*
mycorrhiza *309*
pressure flow theory *314*
root hair *309*
sink region *314*
source region *314*
stomata *310*
symbiotic relationship *307*
translocation *313*
transpiration *310*

Readings

Epstein, E. 1973. "Roots." *Scientific American* 228(5):48–58.

Galston, A., P. Davies, and R. Satter. 1980. *The Life of a Green Plant.* Englewood Cliffs, New Jersey: Prentice-Hall. A simplified treatment of much of plant physiology.

Hewitt, E., and T. Smith. 1975. *Plant Mineral Nutrition.* New York: Wiley.

Hitz, W., and R. Giaquinta. 1987. "Sucrose Transport in Plants." *Bioessays* 6:217–221.

Salisbury, F., and C. Ross. 1985. *Plant Physiology.* Third edition. Belmont, California: Wadsworth.

22   PLANT REPRODUCTION AND DEVELOPMENT

KEY CONCEPTS

1. A sporophyte with flowers as well as roots, stems, and leaves dominates the life cycle of angiosperms (flowering plants). The flowers of many species coevolved with insects, birds, and other agents that function in pollination and seed dispersal.

2. Haploid microspores form in male floral structures and give rise to male gametophytes (pollen grains), in which sperm develop. Haploid megaspores form in female floral structures and give rise to female gametophytes, in which eggs develop.

3. Pollen grains are released from the parent sporophyte and are adapted for traveling to the eggs. Female gametophytes remain attached to the parent plant and are nourished by it.

4. Following fertilization, seeds develop. Each seed consists of an embryo sporophyte as well as tissues that function in its nutrition, protection, and dispersal. From the time a seed germinates, its growth and development are influenced by plant hormones, the secretions of which are linked to seasonal changes and other environmental cues.

REPRODUCTIVE MODES

Although it probably is not something you think about very often, flowering plants engage in sex. As in humans, they have elaborate reproductive systems that produce, protect, and nourish sperm and egg cells. As in humans, female structures of flowering plants house the embryo during its early development. Flowers serve as exclusive or open invitations to third parties—pollinators—that function in getting sperm and egg together. Long before humans ever thought of it, flowering plants were using tantalizing colors and fragrances in improving the odds for sexual success.

Plants also do something humans cannot do (at least not yet). They can reproduce asexually. Recall that *sexual* reproduction requires the formation of gametes, followed by fertilization (Chapter 8). This means two sets of genetic instructions (from two gametes) are present in the fertilized egg. *Asexual* reproduction occurs by way of mitosis, so individuals of the new generation form a clone (they all are genetically identical to the parent plant).

What we usually think of as "the plant" is the **sporophyte,** a vegetative body commonly composed of diploid cells (Figure 18.8). Radish plants, cactus plants, and elm trees are examples. The sporophyte produces the reproductive shoots called **flowers.** Some cells in the flowers divide by meiosis and give rise to **gametophytes,** which are composed of haploid cells. Male gametophytes produce sperm; female ones produce eggs.

The female gametophytes are not free-living plants; they are usually tiny multicelled bodies embedded within floral tissues. The male gametophytes are released from flowers (as small pollen grains); they are like shipping crates for sperm-producing cells until they actually land on a female flower part.

Sporophytes also can be reproduced asexually by several means. For example, strawberry plants send out horizontal aboveground stems (runners), and new roots and shoots develop at every other node along the stems. Buds on short underground stems of onions or lilies grow into new plants. And new plants grow at nodes of underground horizontal stems (rhizomes) of Bermuda grass. Asexual reproduction also occurs with a little help from humans. Whole orchards of pear trees, for example, have been grown from cuttings or buds of a parent tree. The entire navel-orange industry of southern California forms a huge clone that came from a single parent tree, still growing in Riverside.

Sexual reproduction dominates the life cycle of most flowering plants, and it will be our focus here.

GAMETE FORMATION IN FLOWERS

Floral Structure

Figure 22.1 shows the parts of a flower (the carpels, stamens, petals, and sepals). Most or all are attached to a receptacle, a modified end of the floral shoot.

A flower has one or more female parts, which are closed vessels called **carpels.** The chamber within a carpel is the **ovary;** here, eggs develop, fertilization takes place, and seeds mature. In many plants, the carpel narrows into a slender column (style) between the ovary and the landing platform for pollen (the stigma).

Flowers have male parts called **stamens.** Often, a stamen consists of a slender stalk capped by an anther. Anthers contain **pollen sacs,** the chambers in which pollen grains develop.

"Perfect" flowers have both male and female parts. "Imperfect" flowers have male *or* female parts, but not both. Some species, including oaks, have male and female flowers on the same plant. Other species, including willows, have them on separate plants.

Most flowers have petals, which are attached to the receptacle below the male and female flower parts. Collectively, the petals form a "corolla." The color, pattern, and shape of a corolla often attract bees and other pollinators. Sepals are the outermost, leaflike parts. The green sepals enclosing an unopened rosebud are an example.

Microspores to Pollen Grains

Let's now turn to pollen grain formation, which begins while an anther is growing inside a flower bud. Mitotic divisions produce four masses of "mother" cells, each destined to divide by meiosis and so give rise to four haploid **microspores.** Several layers of cells form a walled chamber around each cell mass; this is the "sac" in which pollen develops.

Each microspore in a pollen sac undergoes mitotic cell division. The resulting two-celled, haploid body is a **pollen grain** (Figures 22.2 and 22.3). One cell will produce the sperm. The other will develop into a **pollen tube,** which will grow through tissues of a carpel and so transport sperm to the ovary.

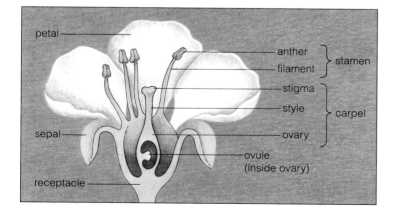

petal — anther ⎫
filament ⎬ stamen
stigma ⎫
style ⎬ carpel
sepal — ovary ⎭
ovule (inside ovary)
receptacle —

Figure 22.1 Common arrangement of floral appendages. Shown here, a cherry (*Prunus*) flower, with a single carpel that will form the cherry.

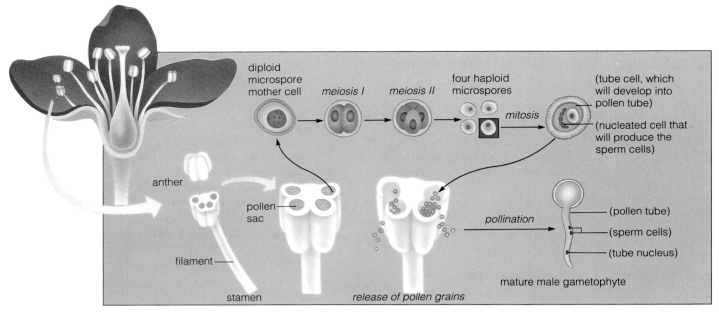

Figure 22.2 Stages in the development of a male gametophyte.

Figure 22.3 Scanning electron micrographs of pollen grains from (**a**) day lily and (**b**) ragweed.

Megaspores to Eggs

Meanwhile, in the carpel of a flower, one or more dome-shaped cell masses have been developing on the inner wall of the ovary. In some carpels, only one mass develops; in others, hundreds or thousands form. Each mass is the start of an **ovule** which, if all goes well, will become a seed.

As the cell mass grows, one or two protective layers (integuments) form around it. Within the mass, a diploid mother cell divides by meiosis and four haploid **megaspores** form. Typically, three of the four disintegrate. The one remaining undergoes mitosis three times *without* cytoplasmic division, so at first it is a single cell with eight nuclei (Figure 22.4). The cytoplasm divides only after each nucleus migrates to a specific location in the cell. The result is a seven-celled **embryo sac,** the female

gametophyte. One cell, the "endosperm mother cell," has two nuclei. It will help form **endosperm,** a nutritive tissue around the forthcoming embryo. Another cell is the egg.

POLLINATION AND FERTILIZATION

Pollination

The transfer of pollen grains to a stigma is called **pollination.** Insects, birds, or other agents are often required for the transfer. The relationship between flowering plants and their pollinators is one of the most intriguing of all evolutionary stories. It is the topic of the *Commentary* on pages 320–322.

Fertilization and Endosperm Formation

Once a pollen grain is deposited on the proper stigma, the pollen tube starts its journey to an ovule. Two sperm cells form inside the tube before or during its growth through the stigma and style. Upon reaching the ovary, the pollen tube grows toward an ovule. When it penetrates the embryo sac, its tip ruptures and the two sperm are released (Figure 22.4).

"Fertilization" generally means the fusion of a sperm nucleus and an egg nucleus. **Double fertilization** occurs in flowering plants. Commonly, one sperm nucleus fuses with that of the egg, forming a *diploid* (2n) zygote. Meanwhile, the other sperm nucleus and both nuclei of the endosperm mother cell all fuse together, forming a

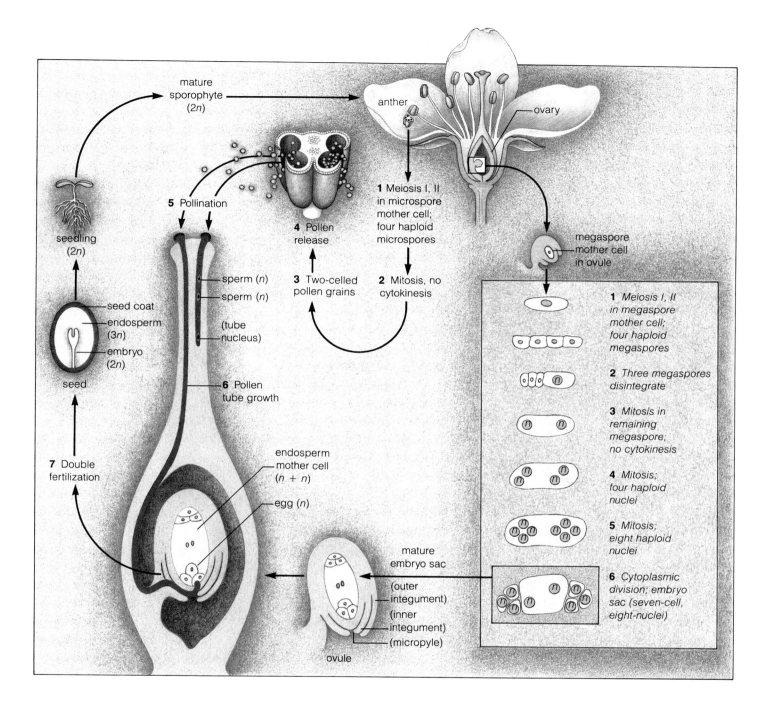

Figure 22.4 Generalized life cycle of a flowering plant, showing details of embryo sac formation within an ovary that has one ovule. Compare this illustration with Figure 22.2.

triploid (3*n*) nucleus. This produces a "primary endosperm cell." Tissues derived from this cell will nourish the seedling (the young sporophyte) until its leaves form and photosynthesis is under way.

In double fertilization, one sperm nucleus fuses with the egg nucleus, and a diploid zygote results. The other sperm nucleus fuses with the two nuclei of the endosperm mother cell, which gives rise to triploid (3*n*) nutritive tissue.

Coevolution of Flowering Plants and Their Pollinators

An astonishing variety of flowering plants can be found almost everywhere, from snow-covered flanks of mountains to low deserts, from freshwater ponds to the surface of open oceans. How did this distribution and diversity come about, given that plants (unlike animals) cannot just pick up and move to new places?

For the answer, we must go back 400 million years, to the time when plants began invading the land. Insects that could scavenge on moist, decaying plant parts and tiny spores were probably right behind them. Fossils provide us with evidence of the evolution of stronger stems and taller plants. We also have fossils of scavengers that had become adapted to withstanding exposure to air. The absence of competition for edible but aerial plant parts seems to have favored a variety of feeding adaptations (such as sucking, piercing, and chewing mouth-

parts). It also favored the development of insect wings.

The first *seed*-bearing plants made their entrance in humid coastal forests, some 395 million years ago. They were ancestral to the gymnosperms and flowering plants. Often their ovules and pollen sacs were located in conelike formations of modified leaves, and it seems the pollen grains simply drifted on air currents to the ovules.

Pollen grains happen to be rich sources of protein. Suppose some insects came to associate "cone" with "food source." *They would have begun serving as pollinating agents.* Some of the dustlike pollen would be eaten, but some would cling to the insect body and be transported to ovules. Insects clambering about reproductive cones would not be precision pollinators—but they would be more effective than air currents alone. Pollen would be delivered right to the door, so to speak. The tastier the

(**a-c**) Flowers with red and yellow components, colors that attract bird pollinators.

a

b

c

pollen, the more home deliveries, and the more seeds formed. *And the greater the number of seeds formed, the greater the reproductive success.*

What we are describing here is a case of **coevolution.** The word means the joint evolution of two (or more) species interacting in close ecological fashion. When one species evolves, the change affects selection pressures operating between the two species and so the other also evolves.

In our evolutionary story, there was natural selection of variant plants able to attract beneficial insects. At the same time, there was selection of pollinator insects. Because of their ability to recognize a particular food and locate it quickly, the pollinators were able to outcompete other foraging insects.

Another, perhaps related change should be mentioned. Existing pollen-eating beetles have strong mouthparts, and many chew on the ovules they pollinate. *Chewing behavior may have been a selective force in the evolution of floral structure.* At one time, ovules were naked and vulnerable on cone scales. The ovules of today's flowering plants are protected inside closed carpels—away from hungry insects.

Many aspects of floral structure can be correlated with specific pollinators. An example is the glorybower, which has some bright red petals. Its petals form a tube for nectar. The flower does not visually attract beetles or honeybees; they cannot detect red wavelengths. Beetles and honeybees are attracted to strong odors, but the glorybower is odorless. (In flowers with large, deep tubes, insects also face the distinct possibility of drowning in the nectar.)

Birds pollinate the glorybower and other flowers with red and yellow components (Figures a–c). Birds have a keen sense of vision and can detect the color red. They also have a poor sense of smell, so floral odors would be wasted on them. Some birds have beaks as long as the floral tube, and the flower's anthers and stigmas are located where the bird's head will brush against them. Birds require considerable energy to power their flights. They visit the nectar cups of many plants of the same species—and so they promote cross-pollination.

Some flowers pollinated by beetles (and flies) have strong and seemingly awful odors that smell like decaying meat or moist dung. Perhaps such odors originally resembled the smells of decaying matter in the forest litter, where beetles first evolved. In contrast, bees are attracted to strong, sweet odors and to flowers with bright yellow, blue, purple, or ultraviolet components (Figures d–f). The plumelike hairs on most bees retain

d

e

Nectar guides for insect pollinators. (**d**) The nectar guide of the marsh marigold (*Caltha palustris*) is visible to bees. Unlike the human eye, the bee eye can detect ultraviolet light. Although the petals of this flower appear solid yellow to us, its distinctive markings become apparent when the flower is exposed to ultraviolet light. (**e**) Bees also use the close-range guide of the passion flower (*Passiflora caerula*).

pollen from the flowers being visited. The landing platforms of some bee-pollinated flowers favorably position the bee body for brushing against pollen-laden anthers (Figure f).

Butterflies forage by day and often are attracted to sweet-smelling, red, and upright flowers having a more or less horizontal landing platform. Most moths forage by night. They pollinate flowers with strong, sweet odors and white or pale-colored petals, which are more visible in the dark (Figure g). Butterflies and moths have long, narrow mouthparts, corresponding to narrow floral tubes or spurs. When uncoiled, the mouthpart of the Madagascar hawkmoth is a record 22 centimeters long—the same length as the floral tube of an orchid (*Angraecum sesquipedale*)! Hawkmoths do not require a landing platform; they hover in front of a floral tube.

(**f**) Coevolutionary match between a flower and its pollinator. The color of Scotch broom attracts bees. Some of the petals serve as a landing platform that corresponds to the size and shape of the bee body. The pollinator's weight on the platform forces petals apart. The pollen-laden stamens, which are positioned to strike against the bee, are thereby released. The pollen brushes against dense hairs covering the bee body. Periodically, the bee grooms itself, packing pollen for its trip back to the hive. The orange-colored mass of pollen visible here has been packed in a pollen basket formed by stiff hairs on the outer hind leg.

(**g**) Stephanotis, like other night-flowering plants, has no distinctive color pattern. Its white petals and strong scent attract moth pollinators (white or pale colors reflect more light and so are more visible at night).

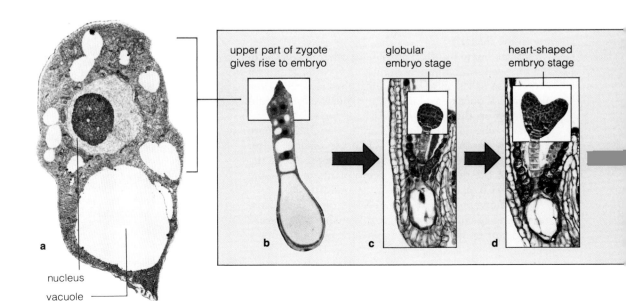

nucleus

vacuole

EMBRYONIC DEVELOPMENT

Seed Formation

When the zygote first forms, it is attached to the parent plant. It undergoes some development even before the mitotic divisions begin that will lead to the mature embryo (Figure 22.5). Some of the first divisions produce a simple row of cells by which nutrients will be transferred from the parent plant to the embryo. Other cell divisions and differentiation produce the embryonic tissues.

Cotyledons, or "seed leaves," develop as part of the embryo (Figures 22.5 and 22.6). Many plants have large cotyledons that absorb the endosperm and function in food storage. Other plants have thin cotyledons that may produce enzymes for transferring stored food from the endosperm to the germinating seedling.

As the embryo continues to develop, the endosperm expands and integuments of the ovule harden and thicken. A fully mature ovule is a **seed;** its integuments become the seed coat.

Fruit Formation and Seed Dispersal

The ovary containing the ovule (or ovules) develops into a **fruit.** A fruit is any matured or ripened ovary. Many fruits, including apples and tomatoes, are juicy and fleshy. Others, including grains and nuts, are dry. The fruit wall of a walnut is dry and intact at maturity.

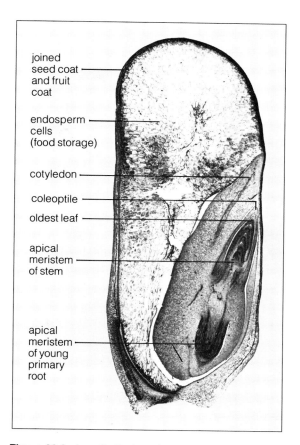

Figure 22.6 Longitudinal section through a grain of corn (*Zea mays*).

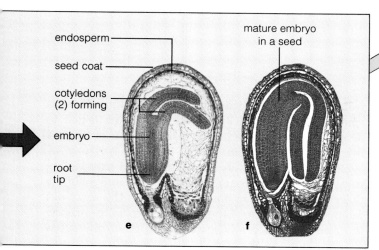

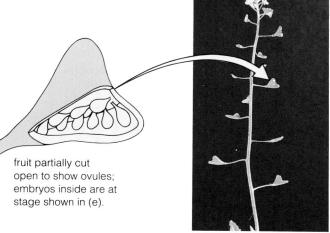

fruit partially cut open to show ovules; embryos inside are at stage shown in (e).

Figure 22.5 Some stages in the development of shepherd's purse (*Capsella bursa-pastoris*), a dicot. The micrographs are not to the same scale. (**a**) Internal organization of the zygote. The organization is heritable and contributes to the shaping and cell differentiations of the early embryo, identified by the yellow boxes in (**b-d**). The embryo in (**e**) is well differentiated; the one in (**f**) is mature.

a Apple blossoms

b Petals fallen away

c Enlarging ovaries

Figure 22.7 Fruit formation on an apple tree. Petals dropping away from the flower usually signify that fertilization has been successful. After this, the ovary and receptacle expand (**b**). Sepals and stamens can still be observed on the immature fruit (**c**).

Table 22.1	Kinds of Fruits of Some Flowering Plants	
Type	Characteristics	Some Examples
Simple (formed from single carpel, or two or more united carpels of one flower)	1. Fruit wall *dry; split* at maturity	Pea, magnolia, tulip, mustard
	2. Fruit wall *dry; intact* at maturity	Sunflower, wheat, rice, maple
	3. Fruit wall *fleshy*, sometimes with leathery skin	Grape, banana, lemon, cherry
Aggregate (formed from numerous but separate carpels of single flower)	*Aggregate* (cluster) of matured ovaries (fruits), all attached to receptacle (modified stem end)	Blackberry, raspberry
Multiple (formed from carpels of several associated flowers)	*Multiple* matured ovaries, grown together into a mass; may include accessory structures (such as receptacle, sepal, and petal bases)	Pineapple, fig, mulberry
Accessory (formed from one or more ovaries *plus* receptacle tissue that becomes fleshy)	1. *Simple:* a single ovary surrounded by receptacle tissue	Apple, pear
	2. *Aggregate:* swollen, fleshy receptacle with dry fruits (seeds) on its surface	Strawberry

Multiple flowers remain clustered together in a pineapple. Figures 22.7 through 22.9 show examples of different fruits. Table 22.1 gives a listing of the main types of fruits.

Fruits function in seed protection and dispersal in specific environments. For example, maple fruits have winglike extensions (Figure 22.8b). When the fruit drops, the wings cause it to spin sideways. With such spinnings, seeds can be dispersed to new locations, where they will not have to compete with the parent plant for soil water, minerals, and sunlight. Many fruits have hooks, spines, hairs, and sticky surfaces. They are taxied to new locations by adhering to feathers or fur of animals that brush against them.

Fleshy fruits such as strawberries and cherries are tasty to many animals and are adapted for surviving the digestive enzymes in the animal gut. The enzymes remove just enough of the hard seed coats to increase the chance of successful germination when the seeds are expelled from the body.

floral remnants

ovary

a

seed (in carpel)

wing

b

receptacle

ovary

c

Figure 22.8 (**a**) Pineapple, a multiple fruit; (**b**) maple, a dry fruit; and (**c**) strawberry, an aggregate fruit.

a

b

Figure 22.9 Two economically important crop plants. (**a**) The seeds of nutmeg (*Myristica fragrans*) are ground to produce one of the most popular spices in the world. The spice called mace comes from the fleshy seed coat, which appears as bands of red tissue in this photograph. (**b**) The pods (fruits) of cacao (*Theobroma cacao*) contain seeds that are processed into chocolate and cocoa.

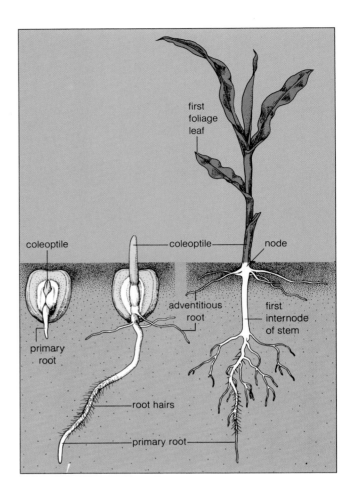

Figure 22.10 Some stages in the development of a corn plant, a monocot. The shoot's young leaves are enclosed in a coleoptile, which protects them during upward growth through the soil. The first node develops at the coleoptile's base, and adventitious roots develop here. When a corn seed is planted deeply, the first internode of the stem elongates as shown. When a seed is planted close to the soil surface, light inhibits elongation of the first internode. In this case, the adventitious roots and the first root look as if they originate in the same region (just below the soil surface).

PATTERNS OF GROWTH AND DEVELOPMENT

At some point following its dispersal from the parent plant, a mature embryo grows into a seedling, which in turn develops into a mature sporophyte. Flowers, fruits, and new seeds form. Often, old leaves drop away from the plant in autumn. What kinds of internal mechanisms govern the plant's growth and development? Do environmental signals set them in motion? Let's consider some answers (and best guesses) to these questions.

Seed Germination and Early Growth

Before or after seed dispersal, embryonic growth idles. Then, at **germination,** the embryo absorbs water, resumes growth, and breaks through the seed coat. Environmental factors, including water, oxygen, temperature, light, and daylength, influence germination. Most mature seeds do not contain enough water for cell expansion or metabolism. Usually, water availability is seasonal and germination coincides with the return of spring rains. As more and more water molecules move inside, the seed swells and the coat finally ruptures.

Once the seed coat splits, oxygen moves in more easily from the surrounding air. Cells of the embryo switch to aerobic respiration and metabolism moves into high gear. Cells divide and elongate continuously to produce the first root of the seedling. Germination is completed when the first root visibly protrudes from the seed coat.

In seedling tissues, cells divide in different planes and expand in different directions. These differences are genetically controlled, and they lead to tissues and structures having specific shapes. Figures 22.10 and 22.11 show the patterns of growth and development characteristic of a corn plant (a monocot) and a soybean plant (a dicot).

Once the first root begins its downward growth, the plant cannot pick up and move on if conditions are not ideal. Yet the plant has a genetically based capacity to respond to a certain range of environmental conditions. Through interactions with the environment, plants can adjust their patterns of growth. Suppose a seed germinates alongside a highway, and a paper bag tossed out

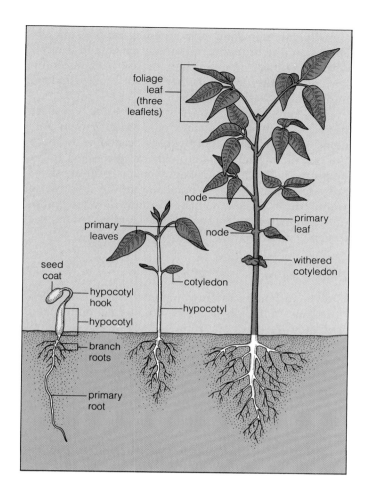

Figure 22.11 Some stages in the development of a soybean plant, a dicot. Food-storing cotyledons are lifted above the soil surface when cells of the hypocotyl (the stem *below* the cotyledon) elongate. The hypocotyl becomes hook shaped and forces a channel through the soil as it grows. The cotyledons can be pulled up through the channel without being torn apart. At the soil surface, light causes the hook to straighten. For several days, cells of the cotyledons function in photosynthesis; then the cotyledons wither and fall off. Photosynthesis is taken over by the first leaves that develop along the stem and later by foliage leaves, each divided into three leaflets. Flowers develop in buds at the nodes shown here.

of a passing car comes to rest on top of it. The first shoot of the plant will bend and grow out from under the bag, toward the light. Later in the year, the plant will respond to environmental rhythms, including seasonal change. As you will now see, plant hormones are involved in these and many other growth responses.

Plant growth and development depend on hormonal responses to signals from the environment.

Effects of Plant Hormones

Hormones influence plant growth and development in powerful but poorly understood ways. **Hormones** are signaling molecules; they are secreted from certain cell types, and they trigger changes in the activities of target cells. Any cell having receptors for a given hormone is a "target." The known plant hormones are auxins, gibberellins, cytokinins, abscisic acid, and ethylene (Table 22.2).

Table 22.2	Main Plant Hormones and Some Known (or Suspected) Effects
Auxins	*Promote cell elongation in coleoptiles and stems; long thought to be involved in phototropism and gravitropism*
Gibberellins	*Promote stem elongation; might help break dormancy of seeds and buds; stimulate breakdown of starch*
Cytokinins	*Promote cell division; promote leaf expansion and retard leaf aging*
Abscisic acid	*Promotes stomatal closure; might trigger bud and seed dormancy*
Ethylene	*Promotes fruit ripening; promotes abscission of leaves, flowers, and fruits*
Florigen (?)	*Arbitrary designation for as-yet-unidentified hormone (or hormones) thought to cause flowering*

Auxins promote stem elongation (Figure 22.12). Some synthetic auxins are used as herbicides, which selectively kill certain plants. One type (2,4-D) is used extensively to kill weeds in cereal crops. The **gibberellins** are also known to promote stem elongation.

Cytokinins stimulate cell division, promote leaf expansion, and retard leaf aging. **Abscisic acid** promotes seed and bud dormancy. It also helps restrict water loss from plants during short periods of drought. (It acts on guard cells in ways that cause stomata to close.)

Ethylene stimulates fruit ripening, and it triggers abscission (the dropping of flowers, fruits, and leaves). Ancient Chinese ripened their picked fruit faster by burning incense, although they didn't know ethylene in the smoke was responsible.

There are other hormones besides the five types just described. Root and leaf cells produce their own hormones. An unknown hormone associated with shoot tips can inhibit lateral bud growth, an effect called **apical dominance.** (As gardeners know, pinching off the shoot tips encourages increased branching of the stem below and produces bushier plants.) Also, the flowering process may be triggered by an as-yet-unidentified hormone, designated "florigen."

Plant Tropisms

Growth responses to environmental factors are called "tropisms." Plant physiologists do not fully understand how they work.

Phototropism. When light strikes one side of a plant more than the other, the stem will curve toward the light (Figure 22.13). Also, the leaves will turn until their flat surface is facing the light. When a plant adjusts its direction and rate of growth in response to light, we call this **phototropism.**

Coleoptiles, for example, show phototrophic responses. (A coleoptile is the hollow, cylindrical structure that protects the tender young leaves of grass seedlings.) As Frits Went recognized in the 1920s, coleoptile tips contain a growth-promoting substance (he named it auxin). The auxin moves down from the tip into cells less exposed to light. Those cells are stimulated to grow longer, and this bends the tip toward the light.

Gravitropism. After seeds germinate, the first root always curves down and the coleoptile or stem always curves up. This is an example of **gravitropism,** a growth response to the earth's gravitational force. Figure 22.14

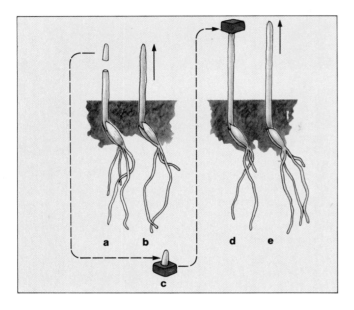

Figure 22.12 Experiment demonstrating that an auxin present in the tip of a grass coleoptile promotes elongation of cells below. (**a**) An oat coleoptile tip is cut off. Compared with a normal coleoptile (**b**), the stump does not elongate much. When the tip is placed on a tiny block of agar for several hours (**c**), auxin moves into the agar. When the agar is placed on another de-tipped coleoptile (**d**), elongation proceeds about as rapidly as in an intact coleoptile (**e**).

Figure 22.13 Phototropism in bean, pea, and oat seedlings. These plants were grown in darkness, then exposed to light from the right side for a few hours before they were photographed.

bean   pea   oat

shows what happens when a potted seedling is turned on its side in a dark room. Cell elongation on the upper side of the horizontal stem decreases markedly but increases on the lower side. With the adjusted growth, the stem curves upward, even in the absence of light.

What roles, if any, do plant hormones play in gravitropism? A horizontal stem of a seedling might respond as a result of changing cell sensitivity to auxin. Gravitropism of roots may be triggered by some type of growth-inhibiting hormone in the root cap. A horizontally positioned root will not curve downward if its root cap has been surgically removed, but it will do so if the cap is put back on.

Thigmotropism. Peas, beans, and many other plants with long, slender stems are climbing vines; they generally do not grow upright without physical support. These plants show **thigmotropism,** or unequal growth resulting from physical contact with solid objects in their surroundings. Suppose one side of a stem grows against a fencepost. Cells stop elongating on the side of the stem making contact, and within minutes the stem starts to curl around the post. It might do so several times before cells on both sides start elongating at about the same rate once again. Auxin and ethylene may be involved in this response.

Biological Clocks and Their Effects

Like all other organisms, plants have internal time-measuring mechanisms called **biological clocks.** The clocks have roles in adjusting daily activities. They also have roles in seasonal adjustments in the plant's patterns of growth, development, and reproduction.

Circadian Rhythms. Some plant activities occur regularly in cycles of about 24 hours, even when environmental conditions remain constant. These are called "circadian" rhythms. For example, many plants hold their leaves in horizontal positions during the day but fold them closer to the stem at night (Figure 22.15). Even when you keep one of those plants in constant light or darkness for a few days, it folds its leaves into the "sleep" position in the evening anyway! In some way, the plant measures time without light-on (sunrise) and light-off (sunset) signals.

Photoperiodism. Many plants show **photoperiodism,** a seasonal response made when days or nights become longer or shorter than some set length. Apparently, a blue-green molecule called **phytochrome** serves as the plant's "eye."

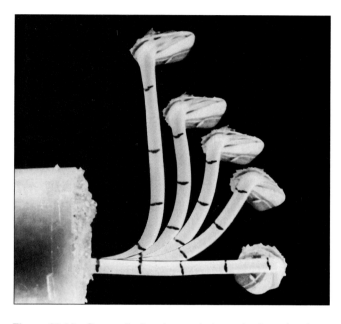

Figure 22.14 Composite time-lapse photograph of gravitropism in a dark-grown sunflower seedling. In this plant, the two cotyledons emerge aboveground because the stem portion just below the cotyledons elongates. Just before this five-day-old plant was positioned horizontally, it was marked at 0.5-centimeter intervals. After 30 minutes, upward curvature was detectable. The most upright position shown was reached within 2 hours.

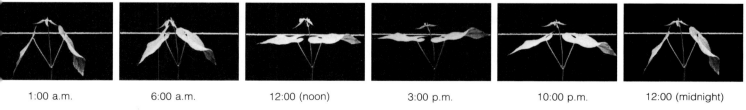

| 1:00 a.m. | 6:00 a.m. | 12:00 (noon) | 3:00 p.m. | 10:00 p.m. | 12:00 (midnight) |

Figure 22.15 Leaf movements in a bean plant. This plant was kept in constant darkness, and its leaf movements continued independently of sunrise (6 a.m.) and sunset (6 p.m.). Folding the leaves closer to the stem may prevent phytochrome from being activated by bright moonlight, which could interrupt the dark period necessary to trigger flowering. Or perhaps it helps slow heat loss from leaves otherwise exposed to the cold night sky above.

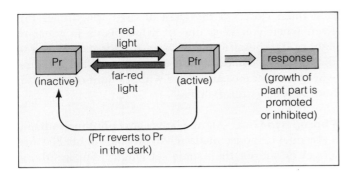

Figure 22.16 Interconversion of phytochrome from the active form (Pfr) to the inactive form (Pr). This pigment is part of a switching mechanism that can promote or inhibit growth of different plant parts.

Figure 22.17 Effect of daylength on spinach, a long-day plant. The plant on the left was grown under short-day conditions; the one on the right, under long-day conditions.

Phytochrome can absorb light of red or far-red wavelengths, with different results. It is converted to an active form (Pfr) at sunrise, when red wavelengths dominate the sky. It reverts to inactive form (Pr) at night and during the day (Figure 22.16). It does this also at sunset, when far-red wavelengths predominate, and in the shade of other leaves (which absorb the red). Botanists suspect that Pfr controls which types of enzymes are being produced in particular cells—*and different enzymes are required for different growth responses.*

Pfr influences seed germination, stem elongation and branching, leaf expansion, and the formation of flowers, fruits, and seeds. We see the effects of Pfr deficiency when plants are grown from seed in darkness. The plants put more resources into stem elongation and less into leaf expansion or stem branching.

Plants show responses to changes in light and darkness. Phytochrome serves as a switching mechanism in the biological clock governing these responses.

The Flowering Process. By measuring the lengths of day and night, plants control the flowering process, which is keyed to changes in daylength through the year. (In a given geographic region, the length of any day remains much the same from one year to the next.)

For example, in late summer and early fall, many plants form flowers, produce seeds, and shed leaves in response to shorter daylengths. These responses are so predictable that flowering plants can be categorized in terms of daylength. "Long-day plants" flower in spring when daylength becomes *longer* than some critical value.

"Short-day plants" flower in late summer or early autumn when daylength becomes *shorter* than some critical value. "Day-neutral plants" flower whenever they become mature enough to do so.

Figure 22.17 shows what happens when spinach (a long-day plant) grows under short-day conditions and long-day conditions. A spinach plant will not flower and produce seeds unless it is exposed to 14 hours of light each day for two weeks. These conditions do not exist in the tropics—which is why spinach is a poor choice for a seed farm in the tropics.

One short-day plant, the cocklebur, measures time with great sensitivity. A single night longer than $9\frac{1}{2}$ hours triggers flowering—but if artificial light interrupts that dark period for even a minute or two, the plant won't flower! Red wavelengths are the most disruptive for cockleburs and other short-day plants. They affect phytochrome in such a way that flowering is inhibited.

At least one hormone probably interacts with phytochrome to influence flowering. The elusive hormone thought to control flowering is designated "florigen," and it must be produced in leaves and transported to newly forming buds. Trim all but one leaf from a cocklebur plant, cover the remaining leaf with black paper for more than $9\frac{1}{2}$ hours, and the plant will flower. Cut off the leaf immediately after its dark period and the plant will *not* flower.

Senescence. Plants make a major investment in reproduction. They actually withdraw nutrients from leaves, roots, and stems and distribute them to newly forming flowers, fruits, and seeds. Most plants end up with tan-colored, dead leaves. The deciduous ones (which shed leaves at the end of each growing season) transport

From Embryo to Mature Oak

Where the ocean breaks along the central California coast, the land rolls inward as steep and rounded hills. These sandstone hills started forming 65 million years ago, when violent movements in the earth's crust caused parts of the submerged ocean floor to crumple upward into a jagged new coastal range. Since then, rains and winds played across the land, softening the stark contours and sending mineral-laden sediments into the canyons. In time, grasses cloaked the inland hills, and their organic remains accumulated and enriched the soil. On these hillsides, in these canyons, the coast live oak (*Quercus agrifolia*) began to evolve more than 10 million years ago.

Quercus agrifolia is a long-lived giant; some trees are known to be 300 years old. They can grow more than 100 feet tall, and their evergreen branches can spread even wider. In early spring, clusters of male flowers develop and form golden catkins among the leaves. Wind carries pollen grains from the catkins to female flowers near the branch tips of the same or neighboring trees. After pollination, a sperm-bearing pollen tube grows through the style and into the ovary, where fertilization takes place. The newly formed zygote undergoes repeated cell divisions, which give rise to root and shoot tips and to cotyledons. Integuments of the ovule form the seed coat, and ovary walls develop into a shell. By early fall, the seed reaches maturity and is shed from the tree as a hard-shelled acorn.

Three centuries ago, long before Gaspar de Portola sent landing parties ashore to found colonies throughout upper California, oaks were shedding the seeds of a new generation. Suppose it was then that a scrubjay, foraging at the foot of a hillside, came across a worm-free acorn. In storing away food for leaner days, the bird used its beak to scrape a small crater in the soil, then dropped in the acorn and covered it with decaying leaves. Although a scrubjay remembers its hiding places most of the time, this particular acorn lay forgotten. The next spring, it germinated.

The oak seed embarked on a journey of continued growth. Cells divided repeatedly, grew longer, and increased in diameter. Water pressure forced the enlargement—water taken up osmotically as ions accumulated in the seedling's first root. A root cap formed and protected the root as it grew downward through the soil. Cell differentiations gave rise to the cortex,

epidermis, and a vascular column through which water and ions would flow. Lateral roots emerged, probably under the influence of hormones. As new roots grew longer, their absorptive surfaces increased. When the first shoot began its upward surge, separate vascular bundles began forming; eventually they would form a continuous cylinder of secondary xylem and phloem.

In parenchyma cells of the new roots, stems, and leaves, central vacuoles grew and pressed the cytoplasm outward until it became only a thin zone against the cell wall. The cell's surface area increased relative to the volume of cytoplasm, and this enhanced absorption. Water, mineral ions, and carbon dioxide could be har-

acorn

vested rapidly in spite of their dilute concentrations in soil and air. Mycorrhizae, the symbiotic association of roots and fungi, further enhanced absorption. Stomata developed in leaves and began to control water and carbon dioxide movements.

As the oak seedling grew, vascular pipelines served as functional links among all plant parts. Through xylem, water and minerals moved from roots to stems and leaves. Through phloem, organic compounds were shuttled from one region of the oak to another.

At the whim of a scrubjay, the seed had sprouted in a well-drained, sunlit basin at the foot of a canyon. Rainwater accumulated each winter, keeping the soil moist enough to encourage luxuriant growth through spring and the dry summers. Out in the open, red wavelengths of sunlight activated phytochrome in the seedling, triggering hormonal events that encouraged stem branching and leaf expansion. All the while, delicate hormone-mediated responses were being made to the winds, the sun, the tug of gravity, the changing seasons.

And so the oak increased in size. Every season, year after year, century after century, roots continued to develop and snake through a tremendous volume of the moist soil. Branches continued to spread beneath the sun. Leaves proliferated—leaves where the oak put together its own food with sunlight energy, water, carbon dioxide, and the few simple minerals it mined from the soil.

On their way to the gold fields, prospectors of the California Gold Rush rested in the shade of the oak's immense canopy. The great earthquake of 1906 scarcely disturbed the giant, anchored as it was by a root system extending 80 feet through the soil. By chance, the brush fires that periodically sweep through California's coastal canyons did not seriously damage the tree. Fungi that could have rotted its roots never took hold; the soil was too well drained and the water table too deep. Leaf-chewing insects were kept in check by protective chemicals in the leaves and by predatory birds living in the canyon.

During the 1960s, human population growth surged in California. The land outside cities began to show the effects of population overflow as native plants gave way to suburban housing. A developer turned his tractors into the canyon but was so impressed with the giant oak that the tree was not felled. Death came later.

The new homeowners were not aware of the ancient, delicate relationships between the giant tree and the land that sustained it. Soil was graded between the trunk and the drip line of the overhanging canopy. Flower beds were mounded against the trunk. Lawns were planted beneath the branches and sprinklers installed. Overwatering in summer created standing water next to the great trunk—and the oak root fungus (*Armillaria*) that had been so successfully resisted until then became established. With its roots rotting away, the oak began to suffer the effects of massive disruption to the feedback relationships among its roots, stems, and leaves. Eventually it had to be cut down. In their fifth winter, in their red brick fireplace, the homeowners began burning three centuries of firewood.

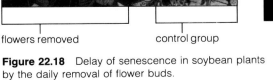

flowers removed   control group

Figure 22.18 Delay of senescence in soybean plants by the daily removal of flower buds.

Figure 22.19 Effect of the relative length of day and night on Douglas fir plant growth. The plant at the left was exposed to 12-hour light and 12-hour darkness for a year; its buds became dormant because daylength was too short. The plant at the right was exposed to 20-hour light and 4-hour darkness; buds remained active and growth continued. The middle plant was exposed to 12-hour light, 11-hour darkness, and 1-hour light in the middle of the dark period. This light interruption of an otherwise long dark period also prevented bud dormancy.

nutrients to storage sites in twigs, stems, and roots before the leaves wither and fall off.

The dropping of leaves, flowers, fruits, or other plant parts is called **abscission.** Ethylene formed in cells near the break points may trigger the process. Abscisic acid may also play a role, perhaps because it stimulates ethylene production in cells that are near the point where the plant part will break off.

The sum total of processes leading to the death of plant parts or the whole plant is called **senescence.** The funneling of nutrients into reproductive parts may be one cue for senescence of leaves, stems, and roots. When the drain of nutrients is stopped by removing each newly emerging flower, the leaves and stems stay green and healthy much longer than they otherwise would (Figure 22.18). Gardeners routinely remove flower buds from many plants to maintain vegetative growth.

Senescence requires other cues, however. When a cocklebur is induced to flower, its leaves yellow regardless of whether the nutrient-demanding young flowers are left on or pinched off. It is as if a "death signal" forms during short days and leads to both flowering and senescence. The signal counteracts cytokinins, which delay senescence.

Dormancy. As autumn approaches and days grow shorter, growth slows or stops in many plants. It stops even if temperatures are still moderate, the sky is bright, and water is plentiful. When a perennial or biennial plant stops growing under conditions that seem (to us) quite suitable for growth, it has entered a state of **dormancy.** Ordinarily, its buds will not resume growth until early spring.

Short days and long nights are strong cues for dormancy. When a short period of red light interrupts the long dark period for Douglas firs, the plants respond as if nights are shorter and days are longer. They continue to grow taller (Figure 22.19). In this case, phytochrome conversions during the dark period prevent dormancy.

A dormancy-breaking process is at work between fall and spring. It probably involves gibberellins and abscisic acid, and it requires exposure to low winter temperatures for specific periods. For example, Delicious apples grown in Utah require 1,230 hours near 43°F (6°C); apricots grown there require 720 hours. Generally, the varieties of trees that grow in the southern United States require less cold exposure than those growing in northern states and Canada.

Through interactions between hormones and biological clocks, plants make seasonal adjustments in their patterns of growth, development, and reproduction.

SUMMARY

1. The sphorophyte of flowering plants is a vegetative body with roots, stems, leaves, and flowers. In sexual reproduction, reproductive cells in its flowers divide by meiosis to form haploid spores that will divide by mitosis to produce gametophytes. Male gametophytes give rise to sperm, female gametophytes to eggs.

2. A flower has one or more carpels (female parts), stamens (male parts), petals, and sepals, most or all of which are attached to the receptacle. The anther of each stamen contains pollen sacs in which pollen grains develop. A pollen grain is a male gametophyte. The ovary inside each carpel is a closed vessel in which eggs develop, fertilization takes place, and seeds mature. The carpel's stigma is a landing platform for pollen grains.

3. Pollination is the transfer of pollen grains to an appropriate stigma. (Insects, birds, and other agents coevolved with many flowering plants and serve as pollen carriers.) As a pollen tube grows through tissues of the carpel, two sperm cells form within it. At double fertilization, one sperm nucleus fuses with an egg nucleus to form a diploid ($2n$) zygote. Commonly, the other sperm nucleus and two nuclei of an endosperm mother cell all fuse together, forming a cell that will give rise to triploid ($3n$) endosperm, a nutritive tissue.

4. Seeds are mature ovules. One or more ovules form inside a carpel. They start out as tissue masses on the inner wall of the ovary; they become surrounded by cell layers (integuments) and they house the egg. After double fertilization and while the embryo undergoes initial development, the ovule expands. Endosperm accumulates inside or is absorbed by the embryo, and its integuments harden and thicken to become the seed coat. Simultaneously, the ovary develops into the fruit, which serves to protect and disperse seeds.

5. Following dispersal from the parent plant, seeds germinate (the embryo absorbs water, resumes growth, and breaks through the seed coat). Germination is completed when the first (primary) root protrudes from the seed coat. The plant increases in volume and mass. Dermal, ground, and vascular tissues form; later, flowers, fruits, and new seeds form, then older leaves drop off. Hormones and environmental signals interact to bring about these developmental events.

6. There are five known categories of plant hormones. Auxins and gibberellins promote stem elongation. Cytokinins stimulate cell division, promote leaf expansion, and retard leaf aging. Abscisic acid limits water loss (by triggering stomatal closure) and may have roles in seed and bud dormancy. Ethylene promotes fruit ripening and abscission.

7. Plants make seasonal adjustments in their patterns of growth, development, and reproduction in response to changing environmental conditions. They do this with the aid of biological clocks (internal time-measuring mechanisms). Phytochrome pigments serve as a switching mechanism in one type of biological clock.

Review Questions

1. Distinguish between these terms:
   a. Megaspore and microspore *317*
   b. Pollination and fertilization *318, 319*
   c. Pollen grain and pollen tube *317*
   d. Ovule and female gametophyte *318, 316*

2. Observe the kinds of flowers growing in the area where you live. On the basis of what you have read about the likely coevolutionary links between flowering plants and their pollinators, can you perceive what kinds of pollinating agents your floral neighbors might depend upon? (Refer to page 320)

3. Give some specific examples of adaptations that enhance seed protection and dispersal. *324*

4. List the five known plant hormones (or groups of hormones) and the main functions of each. *327*

5. What is phytochrome, and what is its role in plant growth? *329, 330*

Self-Quiz   *(Answers in Appendix IV)*

1. The *sporophyte/gametophyte* (pick one) bears flowers, roots, stems, and leaves; it dominates the life cycle of flowering plants.

2. Many species of flowering plants coevolved with insects, birds, and other agents that function as _____.

3. A flower may have _____.
   a. sepals
   b. petals
   c. carpels
   d. all of the above

4. Male gametophytes eventually produce _____ and female gametophytes produce _____.
   a. spores; eggs
   b. sperm; spores
   c. eggs; sperm
   d. sperm; eggs

5. *Double fertilization* refers to formation of the _____.
   a. zygote
   b. endosperm tissue
   c. endosperm tissue and zygote
   d. two zygotes

6. Seeds are mature _____ and fruits are mature _____.
   a. ovaries; ovules
   b. ovules; stamens
   c. ovules; ovaries
   d. stamens; ovaries

7. Seed germination is over when the _____.
   a. embryo absorbs water
   b. embryo resumes growth
   c. primary root protrudes from seed coat
   d. none of the above

8. Which of the following statements is false?
   a. auxins and gibberellins promote stem elongation
   b. cytokinins stimulate cell division, promote leaf expansion, and retard leaf aging
   c. abscisic acid promotes water loss
   d. ethylene promotes fruit ripening and abscission

9. The various plant hormones _____.
   a. are independent in their activities
   b. interact with environmental signals to produce their effects
   c. are active only in plant embryos within seeds
   d. are active only in adult plants

10. Match the plant reproduction and development terms.

    _____ immature fruit
    _____ pollination
    _____ phytochrome
    _____ double fertilization
    _____ immature seed
    _____ major parts of a seed

    a. one sperm nucleus fuses with egg nucleus, the other with two nuclei of endosperm mother cell
    b. ovule
    c. pigment in biological clock
    d. transfer of pollen grains to stigma
    e. ovary
    f. embryo plant, endosperm, seed coats

Selected Key Terms

abscission *333*	flower *316*	phototropism *328*
biological clock *329*	fruit *323*	plant hormone *327*
carpel *317*	gametophyte *316*	pollen grain *317*
coevolution *321*	germination *326*	pollination *318*
cotyledon *323*	megaspore *318*	seed *323*
dormancy *333*	microspore *317*	senescence *333*
double fertilization *318*	ovary *317*	sporophyte *316*
embryo sac *318*	ovule *318*	stamen *317*
endosperm *318*	photoperiodism *329*	

Readings

Bowley, J.D., and M. Black. 1985. *Seeds: Physiology of Development and Germination*. New York: Plenum Press.

Nickell, L. 1982. *Plant Growth Regulators: Agricultural Uses*. New York: Springer-Verlag. Concise explanations of agricultural practices that include use of growth regulators.

Proctor, M., and P. Yeo. 1973. *The Pollination of Flowers*. New York: Taplinger. Beautifully illustrated introduction to pollination.

FACING PAGE: *How many and what kinds of body parts does it take to function as a lizard in a tropical forest? Make a list of what comes to mind as you start reading Unit VI, then see how resplendent the list can become at the unit's end.*

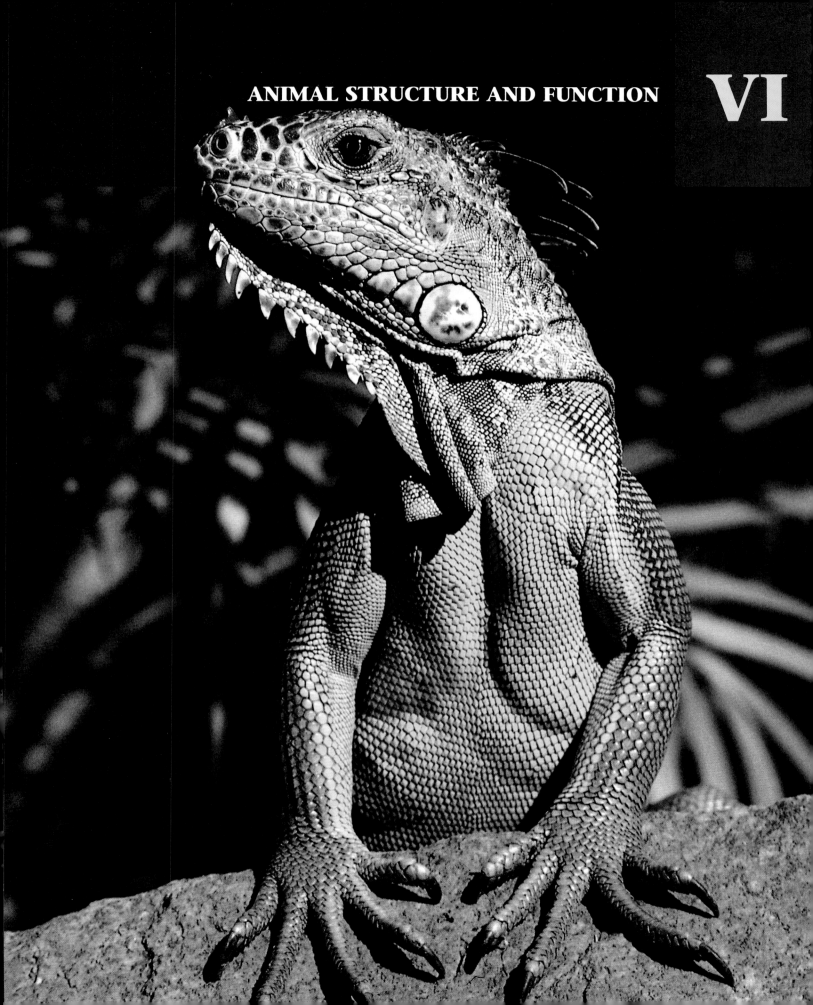

23

ANIMAL TISSUES, ORGAN SYSTEMS, AND HOMEOSTASIS

KEY CONCEPTS

1. The cells of most animals interact at three levels of organization—in *tissues,* many of which are combined in *organs,* which are components of *organ systems.*

2. Most animals are constructed of only four types of tissues: epithelial, connective, nervous, and muscle tissues.

3. Each cell engages in basic metabolic activities that ensure its own survival. At the same time, cells of a tissue or organ perform activities that contribute to the survival of the animal as a whole.

4. The combined contributions of cells, tissues, organs, and organ systems help maintain a stable "internal environment" that is required for individual cell survival. This concept is central to understanding the functions of any organ system, regardless of its complexity.

After a cold night in the Kalahari Desert of Africa, animals small enough to fit in the pocket of an overcoat emerge from their underground burrows. These animals, called meerkats, stand on their hind legs and face eastward, exposing a large surface area of their chilled bodies to the warm rays of the morning sun (Figure 23.1).

The meerkats don't know it, but they are working to ensure good operating conditions for their enzymes. The animal body depends on enzymes, and if its temperature exceeds or falls below a tolerable range, enzyme activity and metabolism itself suffer. Like many other animals, meerkats rely on behavioral adjustments to help maintain "inside" temperatures even though outside temperatures change.

With this unit, we turn to the structure of the animal body (its anatomy) and the mechanisms by which the body functions in the environment (its physiology). Each animal is structurally and physiologically adapted for

Figure 23.1 Meerkats of the Kalahari Desert, working as they do every morning to maintain their internal body temperature. How animals function in their environment is the subject of this unit.

several essential tasks, which may be summarized in the following way:

1. Maintain internal "operating conditions" within some tolerable range even though external conditions change.

2. Locate and take in nutrients and other raw materials, distribute them through the body, and dispose of wastes.

3. Protect itself against injury or attack from viruses, bacteria, and other foreign agents.

4. Reproduce, and often help nourish and protect the new individuals during their early development.

Complex animals consist of millions, even trillions of cells organized into tissues. A **tissue** is a group of cells and intercellular substances that function together in one or more of the specialized tasks listed above. Tissues split up the work, so to speak, in ways that contribute to the survival of the animal as a whole.

Different tissues become organized in specific proportions and patterns to form **organs,** such as the stomach. In **organ systems,** two or more organs interact chemically, physically, or both in performing a common task. For example, different organs of your digestive system ingest and prepare food for absorption by cells, then eliminate food residues.

OVERVIEW OF ANIMAL TISSUES

Even the most complex animal is constructed of only four kinds of tissues, called epithelial, connective, muscle, and nervous tissues. Figures 23.2 through 23.6 show examples of all four. Nearly all of the cells in these body tissues are collectively called **somatic cells.** (Somatic comes from the Greek *soma,* meaning body.) The exceptions are the **germ cells,** the only cells in the animal body that give rise to sperm and eggs.

Epithelial Tissue

Your skin is an example of **epithelium** (plural, epithelia), a type of tissue in which cells are linked tightly together, with little space or extracellular material between them. The cells are organized as one or more layers. One surface of the epithelium is free (it does not have overlying tissue), and the other adheres to an underlying layer (a "basement membrane").

Skin is one of several types of epithelia that cover and protect external body surfaces. Other epithelia line the body's cavities, ducts, or tubes, and substances diffuse or are actively transported across them. Still others contain glands, which function in secretion. (Endocrine glands secrete hormones or other substances that are distributed to other cells by the bloodstream. Exocrine glands secrete saliva, digestive enzymes, milk, or other substances through ducts to a free epithelial surface.) Figures 23.2 and 23.3 show examples of epithelia.

Connective Tissue

There are four categories of connective tissue, called connective tissue proper, cartilage, bone, and blood. The first two types have cells and fibers scattered through a watery or jellylike material, the "ground substance." The fibers contain collagen, elastin, or both. Collagen is a many-stranded protein that imparts strength to the fibers; elastin, another protein, gives them elasticity.

Connective Tissue Proper. As Figure 23.4 shows, dense, loose, and fatty tissues fall in this category. *Dense* connective tissue provides strong connections between different body parts. For example, it is a component of tendons, which attach muscle to bone. When a muscle is stretched, the collagen fibers of this tissue resist being

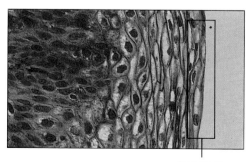

free surface

Figure 23.2 Example of stratified epithelium, which functions in protecting underlying tissues.

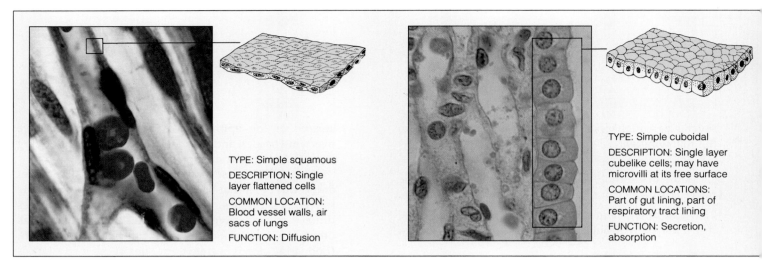

Figure 23.3 Examples of simple epithelium, showing the three basic cell shapes in this type of tissue.

TYPE: Simple squamous

DESCRIPTION: Single layer flattened cells

COMMON LOCATION: Blood vessel walls, air sacs of lungs

FUNCTION: Diffusion

TYPE: Simple cuboidal

DESCRIPTION: Single layer cubelike cells; may have microvilli at its free surface

COMMON LOCATIONS: Part of gut lining, part of respiratory tract lining

FUNCTION: Secretion, absorption

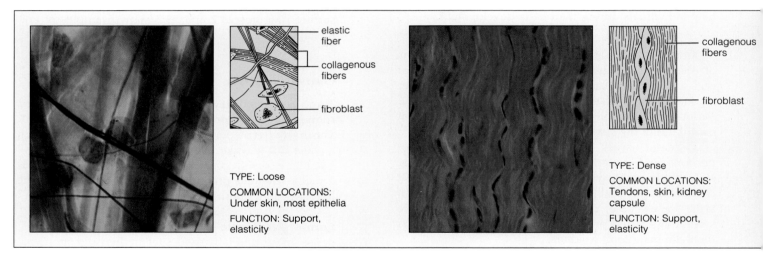

Figure 23.4 Examples of connective tissue.

elastic fiber

collagenous fibers

fibroblast

TYPE: Loose

COMMON LOCATIONS: Under skin, most epithelia

FUNCTION: Support, elasticity

collagenous fibers

fibroblast

TYPE: Dense

COMMON LOCATIONS: Tendons, skin, kidney capsule

FUNCTION: Support, elasticity

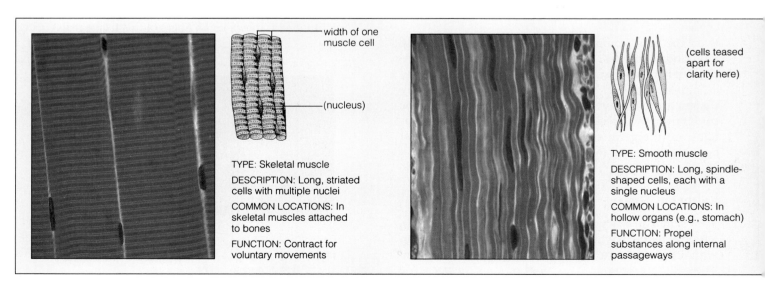

Figure 23.5 Examples of skeletal, smooth, and cardiac muscle tissues.

width of one muscle cell

(nucleus)

TYPE: Skeletal muscle

DESCRIPTION: Long, striated cells with multiple nuclei

COMMON LOCATIONS: In skeletal muscles attached to bones

FUNCTION: Contract for voluntary movements

(cells teased apart for clarity here)

TYPE: Smooth muscle

DESCRIPTION: Long, spindle-shaped cells, each with a single nucleus

COMMON LOCATIONS: In hollow organs (e.g., stomach)

FUNCTION: Propel substances along internal passageways

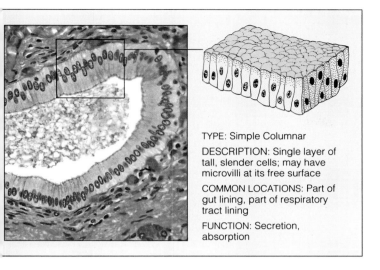

TYPE: Simple Columnar

DESCRIPTION: Single layer of tall, slender cells; may have microvilli at its free surface

COMMON LOCATIONS: Part of gut lining, part of respiratory tract lining

FUNCTION: Secretion, absorption

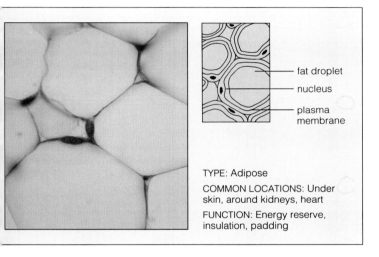

fat droplet

nucleus

plasma membrane

TYPE: Adipose

COMMON LOCATIONS: Under skin, around kidneys, heart

FUNCTION: Energy reserve, insulation, padding

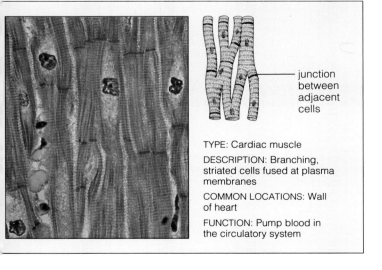

junction between adjacent cells

TYPE: Cardiac muscle

DESCRIPTION: Branching, striated cells fused at plasma membranes

COMMON LOCATIONS: Wall of heart

FUNCTION: Pump blood in the circulatory system

pulled apart under the tension. *Loose* connective tissue, which has both collagen and elastin fibers, supports epithelia and many organs, and it surrounds blood vessels and nerves. *Adipose* tissue is an energy reserve and also pads some organs. Its large, densely clustered cells are specialized for fat storage.

Cartilage. This connective tissue cushions and helps maintain the shape of many body parts. With its jellylike ground substance and dense array of fibers, cartilage can resist compression and stay resilient. Cartilage is located at the ends of many bones, in parts of the nose, and in the external ear. Cartilage built into disks in the backbone serves as shock pads.

Bone. Bone tissue is organized as flat plates, cylinders, and other bony structures that support and protect softer tissues and organs. The body's arm and leg bones act with muscles; they form a leverlike system for movement. Bone tissue also stores mineral salts, and marrow cavities of some types contain bone marrow that produces blood cells.

Although bone tissue is hardened by mineral salts, it is not completely solid. Spaces of different types extend through the hardened portions. Some spaces are chambers for living bone cells (page 349).

Blood. The connective tissue called blood transports oxygen to cells and wastes away from them; it also transports hormones and enzymes. Some of its components protect against blood loss (through clotting mechanisms), and others defend against disease-causing agents. Chapter 26 describes this tissue and its complex functions.

Muscle Tissue

In **muscle tissue,** specialized cells contract (shorten) in response to stimulation, then passively lengthen and so return to their resting state. Muscle tissue helps move the whole body as well as its individual parts. The three categories of muscle tissues (smooth, skeletal, and cardiac) are illustrated in Figure 23.5.

Skeletal muscle tissue contains many long, cylindrical cells. Typically, a number of skeletal muscle cells are bundled together, then several bundles are enclosed in a tough connective tissue sheath to form "a muscle," such as the biceps.

Smooth muscle tissue consists of spindle-shaped cells held together by connective tissue. It occurs in walls of blood vessels, the stomach, and other internal organs. In vertebrates, smooth muscle is said to be "involuntary," because the animal usually cannot directly control its contraction.

Figure 23.6 A sampling of the millions of neurons that form communication lines within and between different regions of the human body. Shown here, motor neurons, which relay signals from the brain or spinal cord to muscles and glands. Collectively, these and other neurons sense environmental change, integrate a great number and variety of signals about those changes, and initiate appropriate responses.

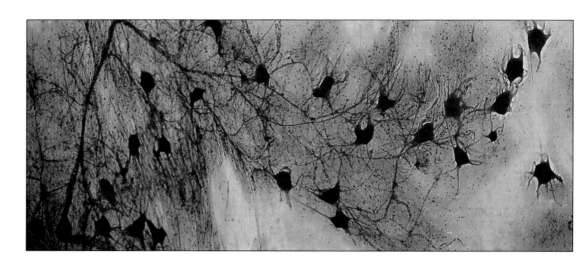

Figure 23.7 Organ systems of the human body. All vertebrates have the same types of systems, serving similar functions.

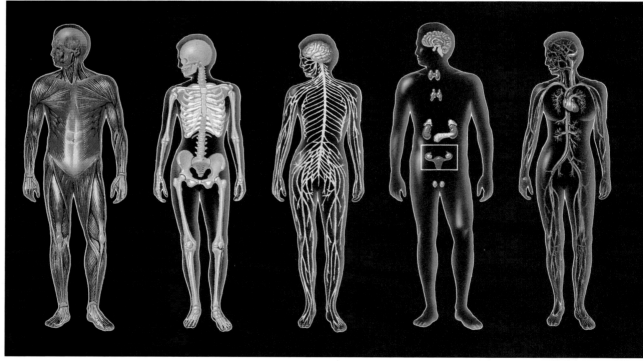

INTEGUMENTARY SYSTEM	MUSCULAR SYSTEM	SKELETAL SYSTEM	NERVOUS SYSTEM	ENDOCRINE SYSTEM	CIRCULATORY SYSTEM
Protection from injury and dehydration; body temperature control; excretion of some wastes; reception of external stimuli; defense against microbes.	Movement of internal body parts; movement of whole body; maintenance of posture; heat production.	Support, protection of body parts; sites for muscle attachment, blood cell production, and calcium and phosphate storage.	Detection of external and internal stimuli; control and coordination of responses to stimuli; integration of activities of all organ systems.	Hormonal control of body functioning; works with nervous system in integrative tasks.	Rapid internal transport of many materials to and from cells; helps stabilize internal temperature and pH.

Cardiac muscle tissue is the contractile tissue of the heart. The plasma membranes of adjacent cardiac muscle cells are fused together. Cell junctions at these fusion points allow the cells to contract as a unit. When one muscle cell receives a signal to contract, its neighbors are also stimulated into contracting.

Nervous Tissue

In **nervous tissue,** many cells called neurons are organized as lines of communication that extend throughout the body. Some types of neurons detect specific changes in environmental conditions; some coordinate the body's immediate and long-term responses to change. Still others relay signals to muscles and glands that can carry out responses. Figure 23.6 shows a sampling of this last type; they are called motor neurons, and they will occupy our attention in Chapter 30.

MAJOR ORGAN SYSTEMS

The tissues just described become organized into the same kinds of organ systems in all vertebrates (Figure 23.7). You might think we are stretching things a bit when we say that each of those systems contributes to the survival of all living cells in the body. After all, what could the body's skeleton and musculature have to do with the life of a tiny cell?

Yet interactions between the skeletal and muscular systems give the individual a means of moving about—toward sources of nutrients and water, for example. Some of their components help circulate blood through the body, as when contractions of certain leg muscles help move blood in veins back to the heart. Through blood circulation, nutrients and other substances are transported to individual cells, and wastes are carried away from them.

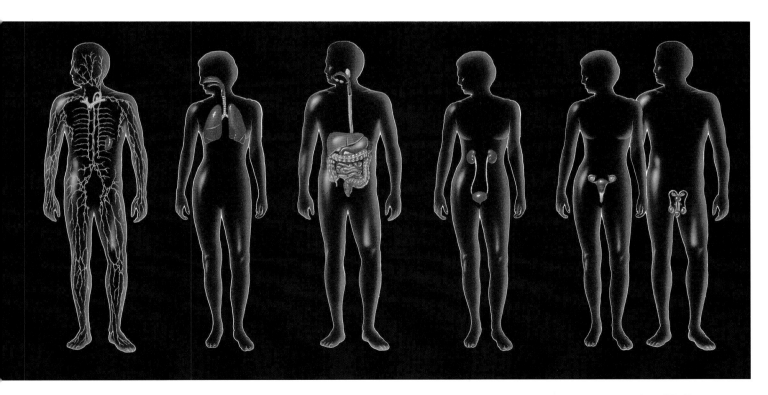

LYMPHATIC SYSTEM	RESPIRATORY SYSTEM	DIGESTIVE SYSTEM	URINARY SYSTEM	REPRODUCTIVE SYSTEM
Return of some tissue fluid to blood; roles in immunity (defense against specific invaders of the body).	Provisioning of cells with oxygen; removal of carbon dioxide wastes produced by cells; pH regulation.	Ingestion of food, water; preparation of food molecules for absorption; elimination of food residues from the body.	Maintenance of the volume and composition of extracellular fluid. Excretion of blood-borne wastes.	Male: production and transfer of sperm to the female. Female: production of eggs; provision of a protected, nutritive environment for developing embryo and fetus. Both systems have hormonal influences on other organ systems.

HOMEOSTASIS AND SYSTEMS CONTROL

The Internal Environment

To stay alive, your body's cells must be continually bathed in fluid that supplies them with nutrients and carries away metabolic wastes. In this they are no different from an amoeba or any other free-living, single-celled organism. However, many *trillions* of cells are crowded together in your body—and they all must draw nutrients from and dump wastes into the same fluid.

The fluid *not* inside cells is called **extracellular fluid,** and it amounts to about 17 liters in the average human. Much of the fluid is *interstitial,* meaning it occupies the spaces between cells and tissues. The rest is *plasma,* the fluid portion of blood. Interstitial fluid exchanges substances with blood and with the cells it bathes.

In functional terms, the extracellular fluid is continuous with the fluid inside cells. That is why drastic changes in its composition and volume have drastic effects on cell activities. Its concentrations of hydrogen, potassium, calcium, and other ions are especially important in this regard. Those concentrations must be maintained at levels that are compatible with the survival of the body's individual cells. Otherwise, the animal itself cannot survive.

It makes no difference whether the animal is simple or complex. *The component parts of any animal work together to maintain the stable fluid environment required by its living cells.* This concept, which is absolutely central to our understanding of the structure and function of animals, may be summarized this way:

1. Each cell of the animal body engages in basic metabolic activities that ensure its own survival.

2. Concurrently, the cells of a given tissue typically perform one or more activities that contribute to the survival of the whole organism.

3. The combined contributions of individual cells, organs, and organ systems help maintain the stable internal environment—that is, the extracellular fluid—required for individual cell survival.

Mechanisms of Homeostasis

The word **homeostasis** refers to stable operating conditions in the internal environment. In this state, the operation of the body is proceeding within the limits set by controlling mechanisms. Interactions among those mechanisms keep physical and chemical aspects of the body within tolerable ranges.

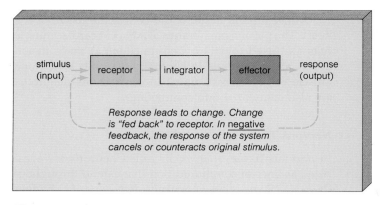

Figure 23.8 Components necessary for negative feedback at the organ level.

In a **negative feedback mechanism,** detection of a change in the internal environment brings about a response that tends to return conditions to the original state. Think about a furnace with a thermostat. The thermostat senses the air temperature and "compares" it to a preset point on a thermometer built into the furnace control system. When the temperature falls below the preset point, the thermostat signals a switching mechanism that turns on the heating unit. When the air becomes heated enough to match the prescribed level, the thermostat signals the switching mechanism, which shuts off the heating unit.

Similarly, meerkats and many other animals rely on feedback mechanisms to raise or lower body temperature. In humans, that temperature is maintained near 37°C (98.6°F), even during extremely hot or cold weather.

Under some circumstances, **positive feedback mechanisms** set in motion a chain of events that *intensify* the original condition. Positive feedback is associated with instability in a system. For example, sexual arousal leads to increased stimulation, which leads to more stimulation, and so on until an explosive, climax level is reached (page 473). As another example, during childbirth, pressure of the fetus on the uterine walls stimulates production and secretion of the hormone oxytocin. Oxytocin causes muscles in the walls to contract, this increases pressure on the fetus, and so on until the fetus is expelled from the mother's body.

Homeostatic control mechanisms require three components: receptors, integrators, and effectors (Figures 23.8 and 23.9). **Receptors** can detect a specific change in the environment. For example, when someone kisses you, there is a change in pressure on your lips. Receptors in the skin of your lips translate the stimulus energy into a signal, which can be sent to the brain. Your brain is an **integrator,** a control point where different bits of information are pulled together in the selection of a

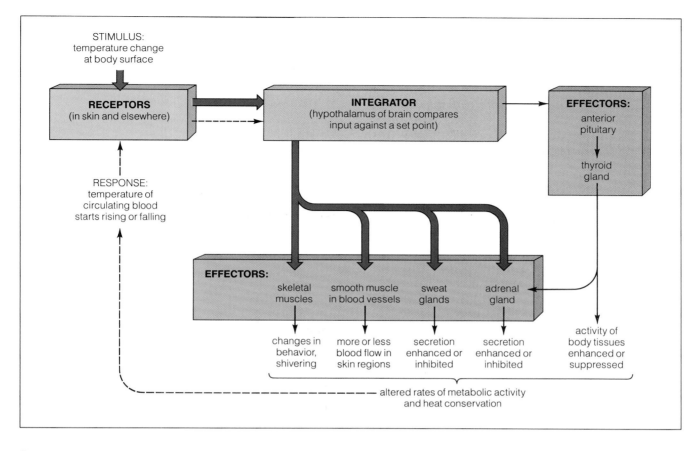

Figure 23.9 Homeostatic controls over the internal temperature of the human body. The dashed line shows how the feedback loop is completed. The blue arrows indicate the main control pathways.

response. It can send signals to your muscles or glands (or both), which are **effectors** that carry out the response. In this case, the response might include flushing with pleasure and kissing the person back. Or it might include flushing with rage and shoving the person away from your face.

Your brain continually receives information about how things *are* operating (the information from receptors) and also how they *should be* operating (information from "set points," which are often built into the brain itself). When conditions deviate significantly from a set point, the brain functions to bring them back to the most effective operating range. It does this by way of signals that cause specific muscles and glands to increase or decrease their activity.

Homeostatic control mechanisms maintain physical and chemical aspects of the internal environment within ranges that are most favorable for cell activities.

What we have been describing here is a general pattern of monitoring and responding to a constant flow of information about the external and internal environments. During this activity, organ systems operate together in coordinated fashion. Throughout this unit, we will be asking the following questions about their operation:

1. What physical or chemical aspect of the internal environment are organ systems working to maintain as conditions change?

2. By what means are organ systems kept informed of change?

3. By what means do they process incoming information?

4. What mechanisms are deployed in response?

As you will see, the operation of all organ systems is under neural and endocrine control.

SUMMARY

1. A tissue is an aggregation of cells and intercellular substances united in the performance of a specialized activity. An organ is a structural unit in which tissues are combined in definite proportions and patterns that allow them to perform a common task. An organ system has two or more organs interacting chemically, physically, or both in ways that contribute to the survival of the animal as a whole.

2. Tissues, organs, and organ systems work in ways that help maintain a stable internal environment (the extracellular fluid) required for individual cell survival.

3. Epithelial tissues cover external body surfaces, line internal cavities and tubes, and form the secretory portion of glands. Connective tissues support or bind together other tissues; they include connective tissue proper, cartilage, bone, and blood. Nervous tissue detects and coordinates information about change in the internal and external environments, and it controls responses to those changes. Muscle tissue is the basis of contraction, which underlies internal movements of body parts and movement through the environment.

4. Organ systems work largely through homeostatic control mechanisms that help maintain a stable internal environment. In negative feedback (the most common mechanism), the response to a disturbance in a system decreases the original disturbance. In positive feedback, a response intensifies the original disturbance.

5. Control of the internal environment depends on the body's receptors, integrators, and effectors. Receptors detect stimuli, which are specific changes in the internal or external environment. Integrating centers (such as the brain) receive information from receptors about how some aspect of the body *is* operating and compare it to a "set point" about how it *should be* operating. On the basis of this information, signals are relayed to effectors (muscles, glands, or both), which carry out the appropriate response.

Review Questions

1. What is an animal tissue? An organ? An organ system? List the major organ systems of the human body, along with their functions. *337, 340, 341*

2. Define extracellular fluid, interstitial fluid, and blood plasma. *342*

3. Perhaps the most important concept in animal physiology relates the functioning of cells, organs, and organ systems to the internal environment. Can you state the three main points of this concept? *342*

4. Epithelial tissue and connective tissue differ from each other in overall structure and function. Describe how. *337*

5. State the overall functions of (a) muscle tissue and (b) nervous tissue. *339, 341*

6. Define homeostasis. What are the three components necessary for homeostatic control over the internal environment? *342*

7. What are the differences between negative feedback and positive feedback mechanisms? *342*

Self-Quiz   (Answers in Appendix IV)

1. The four types of tissues in most animals are _Epithia,_ _Connective_, _nervous_, and _muscle_

2. Animals are structurally and functionally adapted for these tasks:
   a. maintenance of the internal environment
   b. nutrient acquisition, processing, distribution, and waste disposal
   c. self-protection against injury or attack
   d. reproduction
   e. all of the above

3. _____ tissues cover external body surfaces, line internal cavities and tubes, and some form the secretory portions of glands.
   a. Muscle
   b. Nervous
   c. Connective
   d. Epithelial

4. Most _____ tissues bind or mechanically support other tissues; but one type functions in physiological support of other tissues.
   a. muscle
   b. nervous
   c. connective
   d. epithelial

5. _____ tissues detect and coordinate information about environmental changes and control responses to those changes.
   a. Muscle
   b. Nervous
   c. Connective
   d. Epithelial

6. _____ tissues contract and make possible internal body movements as well as movements through the external environment.
   a. Muscle
   b. Nervous
   c. Connective
   d. Epithelial

7. Cells in the animal body _____.
   a. engage in metabolic activities that ensure their survival
   b. perform activities that contribute to the survival of the animal
   c. contribute to maintaining the extracellular fluid
   d. all of the above

8. In a state of _____, physical and chemical aspects of the body are being kept within tolerable ranges by controlling mechanisms.
   a. positive feedback
   b. negative feedback
   c. homeostasis
   d. metastasis

9. In negative feedback mechanisms, _____.
   (a.) a detected change brings about a response that tends to return internal operating conditions to the original state
   b. a detected change suppresses internal operating conditions to levels below the set point
   c. a detected change raises internal operating conditions to levels above the set point
   d. less solutes are fed back to the affected cells

10. _Receptors_ detect specific environmental changes, an _Integrator_ pulls different bits of information together in the selection of a response, and _effectors_ carry out the response.

11. Match the concepts.
   _d_ muscles and glands
   ___ positive feedback
   _c_ body receptors
   _b_ negative feedback
   _a_ brain

   a. integrating center
   b. the most common homeostatic mechanism
   c. eyes and ears
   d. effectors
   e. chain of events intensifies the original condition

Selected Key Terms

adipose tissue *339*
blood *339*
bone *339*
cardiac muscle *341*
cartilage *339*
dense connective tissue *337*
effector *343*
epithelium *337*
extracellular fluid *342*
germ cell *337*
homeostasis *342*
integrator *342*

loose connective tissue *339*
muscle tissue *339*
negative feedback mechanism *342*
nervous tissue *341*
organ *337*
organ system *337*
positive feedback mechanism *342*
receptor *342*
skeletal muscle *339*
smooth muscle *339*
somatic cell *337*
tissue *337*

Readings

Bloom, W., and D.W. Fawcett. 1986. *A Textbook of Histology.* Eleventh edition. Philadelphia: Saunders. Outstanding reference text.

Leeson, C.R., T. Leeson, and A. Paparo. 1985. *Textbook of Histology.* Philadelphia: Saunders.

Ross, M., and E. Keith. 1985. *Histology: A Text and Atlas.* New York: Harper & Row.

Vander, A., J. Sherman, and D. Luciano. 1990. "Homeostatic Mechanisms and Cellular Communication" in *Human Physiology.* Fifth edition. New York: McGraw-Hill.

Vaughan, T. 1986. *Mammalogy.* Third edition. Philadelphia: Saunders. Chapter 22 is an excellent introduction to temperature regulation in mammals.

24 PROTECTION, SUPPORT, AND MOVEMENT

KEY CONCEPTS

1. The integumentary system (the skin and its component structures) is not merely a bag for bones and other internal body parts. It protects the body from abrasion, ultraviolet radiation, bacterial attack, and other environmental insults. It contributes to the functions of the rest of the body, as when it synthesizes the vitamin D necessary for calcium metabolism.

2. Skeletal and muscular systems work together to move the body through the environment and to change the positions of body parts. Bones function not only in movement but also in protection and support for soft organs, in mineral storage, and in blood cell formation.

3. When adequately stimulated, all three types of muscle tissue (smooth, cardiac, and skeletal) can contract (develop force and shorten) and then relax. Skeletal muscle is the only type that interacts with the skeleton to bring about locomotion and positional changes.

In the image you hold of yourself, you are tall or short, pale or dark, sparsely or profusely haired, taut-skinned or flabby, slow moving or always on the move (or somewhere in between). You have three organ systems to thank for your body's shape, its superficial features, and its positional changes and movements. These are the integumentary, skeletal, and muscular systems (Figure 24.1).

INTEGUMENTARY SYSTEM

Functions of Skin

No garment ever made approaches the qualities of the one covering your body—*your skin.* What besides skin maintains its shape in spite of repeated stretchings and washings, kills many bacteria on contact, screens out harmful rays from the sun, is waterproof, repairs small cuts and burns on its own, and with a little care, will last as long as you do?

Together, the skin and structures derived from it (such as hair, nails, oil glands, and sweat glands) are called the **integumentary system.** Keep in mind that skin is much more than "a covering" (which is what integument means), just as a Ferrari is more than a hunk of metal. Skin protects the rest of the body from abrasion and bacterial attack. It helps control internal temperature. Its profusion of blood vessels serves as a reservoir for blood that can be tapped and shunted to metabolically active regions, such as muscles being given a workout. Skin produces the vitamin D required for calcium metabolism. And signals from its receptors help the brain assess what is happening in the surrounding world.

Structure of Skin

Skin has two distinct regions, a covering called the **epidermis** and an underlying **dermis** (Figure 24.2). The dermis is mostly dense connective tissue that cushions

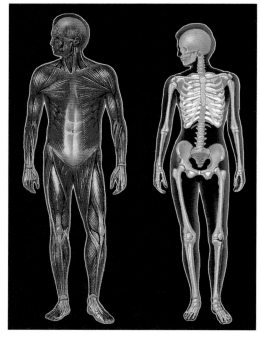

Figure 24.1 Starting point for a tour of the structure and function of vertebrate organ systems. Traveling from the outside in, this chapter describes skin and its derivatives (integumentary system) and interactions between the skeletal and muscular systems.

the body from everyday stresses and strains. Beneath this is the hypodermis, a tissue that anchors the skin and yet allows it some freedom of movement. Fat stored in the hypodermis insulates the body against cold and protects some organs, such as the uterus, against injury.

Skin structure varies considerably. Most of your own skin is as thin as a paper towel; it thickens only on the soles of your feet and in other regions subjected to pounding or abrasion. Some fishes have hard dermal scales, others have bare skin coated with slimy mucus. The scales, feathers, hair, beaks, hooves, horns, claws, nails, and quills of different vertebrates are all produced by cell differentiation in epidermal tissues (Figure 24.3).

Each day, millions of dead cells at your skin's surface are worn off by abrasion. Fortunately, you get new top layers of skin every thirty-five to forty-five days. In the process of **keratinization,** cells in mid-epidermal regions die and become dead bags with a lot of keratin (a water-insoluble protein) inside. Rapid cell divisions below them push the bags toward the skin surface. The rapid divisions contribute to skin's capacity to mend itself after abrasion, cuts, or burns. Keratinized cells at the surface form a barrier against dehydration, bacteria, and many toxic substances.

a

b

Figure 24.3 Feathers (**a**) and hair (**b**)—two kinds of structures arising from cell differentiations in the epidermis. The scanning electron micrograph in (**b**) shows the overlapping cells of the outer layers of a hair shaft, here emerging from the epidermal surface of skin.

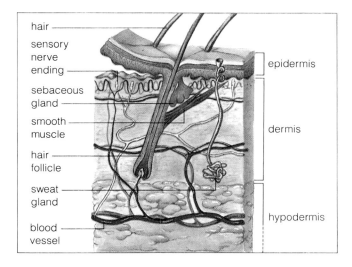

hair

sensory nerve ending

sebaceous gland

smooth muscle

hair follicle

sweat gland

blood vessel

epidermis

dermis

hypodermis

Figure 24.2 Section through vertebrate skin (epidermis and dermis). Also shown is the hypodermis, a tissue layer below the skin.

Figure 24.4 Sunlight and the skin. Melanin-producing cells of the epidermis are stimulated by exposure to ultraviolet radiation. With prolonged sun exposure, melanin levels increase and light-skinned people become tanned (visibly darkened). Tanning provides some protection against ultraviolet radiation, but prolonged exposure can damage the skin. Over the years, tanning causes elastin fibers of the dermis to clump together. The skin loses its resiliency and begins to look like old leather.

Prolonged exposure to ultraviolet radiation also suppresses the immune system. Certain phagocytes and other specialized cells in the epidermis defend the body against specific viruses and bacteria. Sunburns interfere with the functioning of these cells. This may be why sunburns can trigger the small, painful blisters called "cold sores." The blisters are a symptom of a viral infection. Nearly everyone harbors this virus (*Herpes simplex*); usually it becomes localized in a nerve ending near the skin surface, where it remains dormant. Stress factors—including sunburn—can activate the virus and trigger the skin eruptions.

Ultraviolet radiation also can activate proto-oncogenes in skin cells (page 160). Epidermal skin cancers start out as scaly, reddened bumps. They grow rapidly and can spread to adjacent lymph nodes unless they are surgically removed. Basal cell

carcinomas start out as small, shiny bumps and slowly grow into ulcers with beaded margins. Their threat to the individual ceases when they are surgically removed, provided they are removed in time.

Three pigments—melanin, hemoglobin, and carotene—contribute to skin color. Cells in deep epidermal layers produce melanin, a brownish-black pigment that protects cells against ultraviolet radiation. Suntanned skin has increased melanin concentrations in the epidermis (Figure 24.4). Pale skin has very little melanin. Its pinkish cast results from hemoglobin, a pigment giving red blood cells their color. Red shows through thin-walled blood vessels and through the epidermis, both of which are transparent. Carotene, a yellow-orange pigment, occurs in the uppermost layers of epidermis; it is abundant in the skin of most Asians.

Dense connective tissue makes up most of the dermis, and it cushions the body against everyday stretching and other mechanical insults. The dermis tears when skin is stretched too much during pregnancy, and this causes white scars ("stretch marks"). Although hairs are derived from epidermal cells, each grows out of a follicle (a husklike cavity) that is largely embedded in the dermis (Figure 24.2). A hair consists of dead keratinized cells, overlapping one another like roof shingles (Figure 24.3). The most abused cells tend to frizz out near the end of the hair shaft; we call these "split ends."

The average scalp has about 100,000 hairs, but nutrition and hormones influence hair growth and density. Protein deficiency causes hair to thin (hair cannot grow without the amino acids required for keratin synthesis). Severe fever, emotional stress, and excessive vitamin A intake also cause hair thinning. Excessive hairiness in women may result from abnormal production of testosterone, a hormone that influences patterns of hair growth and other secondary sexual traits.

As we age, the rate of epidermal cell division slows, and our skin becomes thinner and more susceptible to injury. Glandular secretions that kept the skin soft and moistened start dwindling. Collagen and elastin fibers in the dermis break down and become more sparse, so the skin loses its elasticity and wrinkles deepen. Excessive tanning, prolonged exposure to cold winds, and tobacco smoke accelerate the aging processes.

SKELETAL SYSTEM

Functions of Bones

Just as skin is more than a baglike covering, so is the skeletal system more than a framework of dried-out bones. Its component parts function in these tasks:

1. *Movement:* The action of skeletal muscle on bones moves the body.

2. *Protection:* Bones serve as hard compartments that enclose and protect the brain, spinal cord, heart, lungs, and other vital organs.

3. *Support:* Bones of the skeletal system support and anchor the skin and all soft organs.

4. *Mineral Storage:* Bone tissue serves as a "bank" for calcium, phosphorus, and other mineral ions. The body

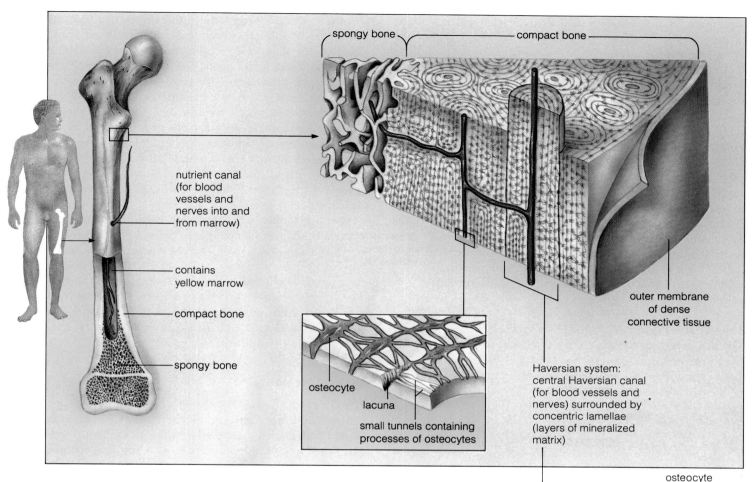

spongy bone

compact bone

nutrient canal
(for blood
vessels and
nerves into and
from marrow)

contains
yellow marrow

compact bone

spongy bone

osteocyte

lacuna

small tunnels containing
processes of osteocytes

outer membrane
of dense
connective tissue

Haversian system:
central Haversian canal
(for blood vessels and
nerves) surrounded by
concentric lamellae
(layers of mineralized
matrix)

makes deposits and withdrawals of these reserves, depending on metabolic needs.

5. *Blood Cell Formation:* Parts of some bones (such as the breastbone) are sites of red blood cell production.

Characteristics of Bone

In size and shape, human bones range from a pea-like wrist bone to thighbones much longer than they are wide. Bones are classified as long, short (or cubelike), flat, and irregular. Here we will focus mainly on the long bones of the body's limbs.

Bone Structure. Bone, a connective tissue, has living cells and collagen fibers distributed through a ground substance. The tissue is hardened by calcium salts deposited around the fibers, but it is not entirely solid (Figure 24.5). Tiny, needlelike hard parts make up "spongy bone," which actually is quite firm and strong. In many bones, **red marrow** fills the spongy tissue, which serves as a major site of blood cell formation.

osteocyte

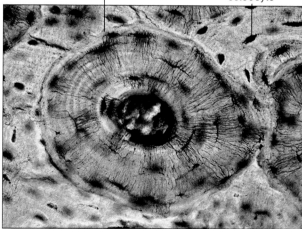

75 μm

Figure 24.5 Structural organization of the long bones of mammals. The micrograph shows several Haversian systems. Nutrients and hormones reach living bone cells (osteocytes) by way of blood vessels present in these systems. The osteocytes reside in lacunae (spaces in the bone tissue). Small tunnels connect neighboring spaces.

Most mature bones have yellow marrow in their interior cavities. Yellow marrow also produces red blood cells when blood loss from the body is severe.

Dense, "compact" bone surrounds the spongy tissue. It forms the shaft of long bones and helps them withstand mechanical shocks. Figure 24.5 shows how the tissue has thin, concentric layers around small canals. These "Haversian canals" form interconnected channels for blood vessels and nerves that service the living cells.

How Bones Develop. Long bones form in cartilage models in the embryo. Bone-forming cells (osteoblasts) secrete material inside the shaft of the cartilage model. Calcium becomes deposited in the cartilage, which breaks down into cavities that merge to become the marrow cavity (Figure 24.6). The bone-forming cells continue to secrete bone tissue along the cavity walls and eventually become trapped by their own secretions. Then, they are called osteocytes (living bone cells).

Bone Tissue Turnover. Bone tissue is like a bank from which minerals are constantly deposited and withdrawn. The turnover is important for bone remodeling programs that occur when young individuals are growing. For

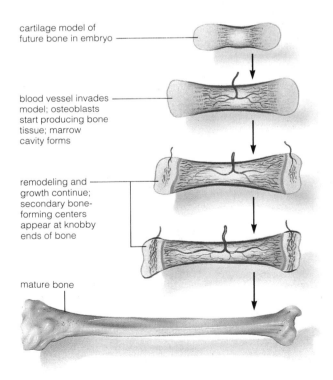

cartilage model of future bone in embryo

blood vessel invades model; osteoblasts start producing bone tissue; marrow cavity forms

remodeling and growth continue; secondary bone-forming centers appear at knobby ends of bone

mature bone

Figure 24.6 Long bone formation, beginning with osteoblast activity in a cartilage model that has already formed in the animal embryo. The bone-forming cells are active in the shaft region first; their activities are repeated in the knobby bone ends until only cartilage is left in the joints at both ends of the shaft.

example, the diameter of the thighbone increases as certain bone cells deposit minerals at the surface of the shaft. At the same time, other bone cells destroy a small amount of bone tissue inside the shaft. Thus the thighbone becomes thicker and stronger—but not too heavy.

Bone turnover also helps maintain calcium levels for the body as a whole. Consider how the body resorbs calcium. First, bone cells secrete enzymes that break down bone tissue. As the component minerals dissolve, the released calcium enters interstitial fluid; from there, it is taken up by the blood. This resorption activity is central to the hormonal control of calcium balance, as described on page 456.

Bone turnover can deteriorate with increasing age, especially among older women. The bone mass decreases in the backbone, legs, feet, and elsewhere. The backbone can collapse and curve so much that the ribcage position is lowered, leading to complications in internal organs. The syndrome is called *osteoporosis*. Decreasing osteoblast activity, calcium and sex hormone deficiencies, excessive protein intake, and decreased physical activity may contribute to osteoporosis.

Skeletal Structure

The skeleton of humans and most other living vertebrates contains both bone and cartilage (Figure 24.7). Humans started walking on their hind legs about 3 million years ago and haven't stopped since. The upright posture puts the skeleton's backbone into an S-shaped curve, which is not an ideal arrangement. (The older we get, the longer we have resisted the pull of gravity in an imperfect way, and the more lower back pain we have.) But nature works through modifications of preexisting structures, and the skeleton of ancestral four-legged vertebrates was one of them.

You have 206 bones, distributed in two different portions of your skeleton (Figure 24.8). The **axial skeleton** includes the skull, vertebral column (backbone), ribs, and sternum (the breastbone). The **appendicular skeleton** includes the bones of arms, hands, legs, feet, pelvic girdle (at the hips), and pectoral girdle (at the shoulders).

The flexible, curved backbone is held upright by large muscles and straplike ligaments. It extends from the skull to the pelvic girdle, where it transmits the weight of your torso to the lower limbs. The delicate spinal cord threads through a cavity formed by bony parts of the vertebrae, which are arranged one above the other. **Intervertebral disks,** which contain cartilage, are shock absorbers and flex points between the vertebrae (Figure 24.9). Severe or rapid shocks may cause a disk to herniate (slip out of place). The protruding disk may press against neighboring nerves or the spinal cord and cause excruciating pain.

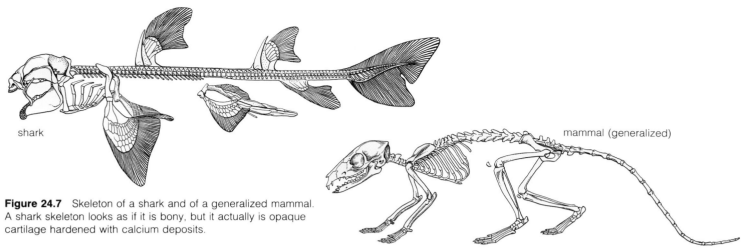

Figure 24.7 Skeleton of a shark and of a generalized mammal. A shark skeleton looks as if it is bony, but it actually is opaque cartilage hardened with calcium deposits.

shark

mammal (generalized)

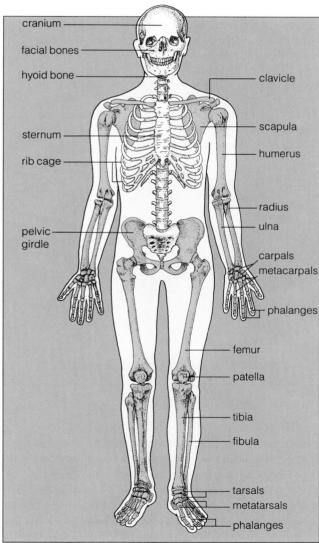

cranium
facial bones
hyoid bone
clavicle
sternum
scapula
rib cage
humerus
radius
ulna
pelvic girdle
carpals
metacarpals
phalanges
femur
patella
tibia
fibula
tarsals
metatarsals
phalanges

Figure 24.8 The human skeleton, with the axial portion in yellow and the appendicular portion in tan. Can you identify similar structures in the endoskeleton of the generalized mammal shown in Figure 24.7?

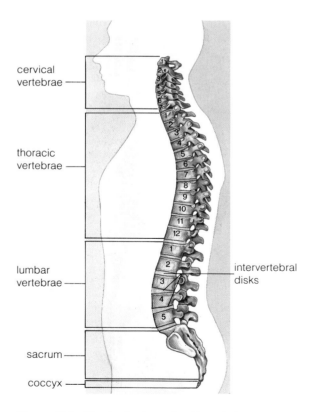

cervical vertebrae
thoracic vertebrae
lumbar vertebrae
intervertebral disks
sacrum
coccyx

Figure 24.9 Vertebral column (backbone) of the human skeleton, side view. The cranium balances on the uppermost vertebra of the column.

On Runner's Knee

When you run, one foot and then the other is pounding hard against the ground. Each time a foot hits the ground, the knee joint above it must absorb the full force of your body weight. The knee joint allows us to do many things. It allows the leg bones beneath it to swing and, to some degree, to bend and twist. And the joint can absorb a force nearly seven times the body's weight—but there is no guarantee that it can do so repeatedly. Nearly 5 million of the 15 million joggers and runners in the United States alone suffer from "runner's knee," which refers generally to various disruptions of the bone, cartilage, muscle, tendons, and ligaments at the knee joint.

Like most joints, the knee joint permits considerable movement. The two long bones joined here (the femur and tibia) are actually separated by a cavity. They are held together by ligaments, tendons, and a few fibers that form a capsule around the joint. A membrane that lines the capsule produces a fluid that lubricates the joint, and where the bone ends meet, they are capped with a cushioning layer of cartilage.

Between the femur and tibia are wedges of cartilage that add stability and act like shock absorbers for the weight placed on the joint. Here also are thirteen fluid-filled sacs (bursae) that help cut down friction.

When the knee joint is hit hard or twisted too much, its cartilage can be torn. Once cartilage is torn, the body often cannot repair the damage. Orthopedic surgeons usually recommend removing most or all of the torn tissue; otherwise it can cause arthritis. Each year, more than 50,000 pieces of torn cartilage are surgically removed from the knees of football players alone. Football players,

tennis players, basketball players, weekend joggers—all are helping to support the burgeoning field of "sports medicine."

The seven ligaments that strap the femur and tibia together are also vulnerable to injury. A ligament is not meant to be stretched too far, and blows to the knee during collision sports (such as football) can tear it apart. A ligament is composed of many connective tissue fibers. If only some of the fibers are torn, it may heal itself. If the ligament is severed, however, it must be surgically repaired. (Edward Percy likens the surgery to sewing two hairbrushes together.) Severed ligaments must be repaired within ten days. The fluid that lubricates the knee joint happens to contain phagocytic cells that remove the debris resulting from day-to-day wear and tear in the joint. The cells will also go to work indiscriminately on torn ligaments and turn the tissue to mush.

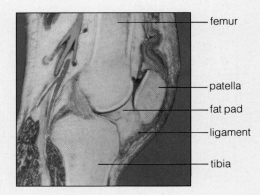

Longitudinal section through the knee joint.

Joints

Areas of contact (or near-contact) between bones are called joints. The most common type, **synovial joints,** are freely movable. A flexible capsule of dense connective tissue holds the bones near each other. Cells of a membrane that lines the capsule interior secrete a lubricating fluid into the joint. Unfortunately, freely movable joints sometimes move too freely and their structural organization is disrupted (see the *Commentary*).

Over time, the cartilage at the bone ends of freely movable joints may simply wear away, a condition called *osteoarthritis.* In contrast, *rheumatoid arthritis* is a degen-erative disorder with a genetic basis. The synovial membrane thickens and becomes inflamed, cartilage degenerates, and bone becomes deposited in the joint.

At **cartilaginous joints,** cartilage fills the space between bones and permits only slight movements. Such joints occur between vertebrae and between the breastbone and ribs. At **fibrous joints,** fibrous tissue unites the bones and no cavity is present. Fibrous joints loosely connect the flat skull bones of a fetus. At childbirth, the bones slide over each other and help prevent skull fractures. A newborn still has "soft spots" (fontanels), but the fibrous tissue hardens completely during adulthood, so the skull bones become fused together.

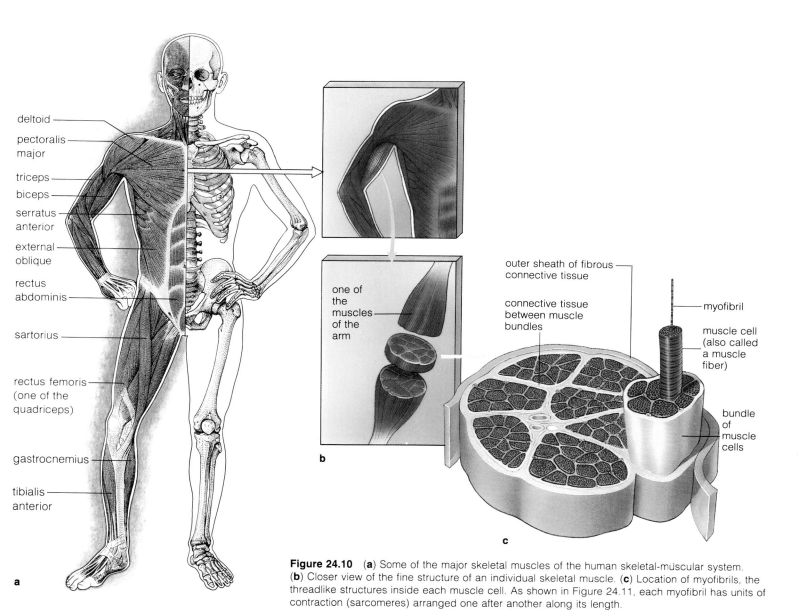

deltoid

pectoralis
major

triceps

biceps

serratus
anterior

external
oblique

rectus
abdominis

sartorius

rectus femoris
(one of the
quadriceps)

gastrocnemius

tibialis
anterior

a

one of
the
muscles
of the
arm

b

outer sheath of fibrous
connective tissue

connective tissue
between muscle
bundles

myofibril

muscle cell
(also called
a muscle
fiber)

bundle
of
muscle
cells

c

Figure 24.10 (**a**) Some of the major skeletal muscles of the human skeletal-muscular system. (**b**) Closer view of the fine structure of an individual skeletal muscle. (**c**) Location of myofibrils, the threadlike structures inside each muscle cell. As shown in Figure 24.11, each myofibril has units of contraction (sarcomeres) arranged one after another along its length.

MUSCULAR SYSTEM

Comparison of Muscle Tissues

There are three types of muscle tissues, called skeletal, cardiac, and smooth muscle (page 339). In general, the cells of all three types have three properties in common. First, they show *excitability*. All cells have an electric gradient across the plasma membrane (the outside is more positive than the inside). In excitable cells, the gradient reverses suddenly and briefly in response to adequate stimulation. Such sudden reversals in charge are "action potentials." Second, muscle cells can *contract* (shorten) in response to action potentials. Third, muscle cells are *elastic*; after contracting, they can lengthen.

The smooth muscle of vertebrates occurs mostly in the wall of internal organs. For example, smooth muscle in the stomach and intestinal walls propels substances through the digestive tract. Cardiac muscle occurs only in the heart walls, and its action will be described in a later chapter. Here we focus on skeletal muscle—the only type that interacts with the skeleton to bring about locomotion and positional changes of body parts.

Skeletal Muscle Contraction

Figure 24.10 shows a few of the main skeletal muscles of the human body. They are composed of a few hundred to hundreds of thousands of muscle cells. Each muscle

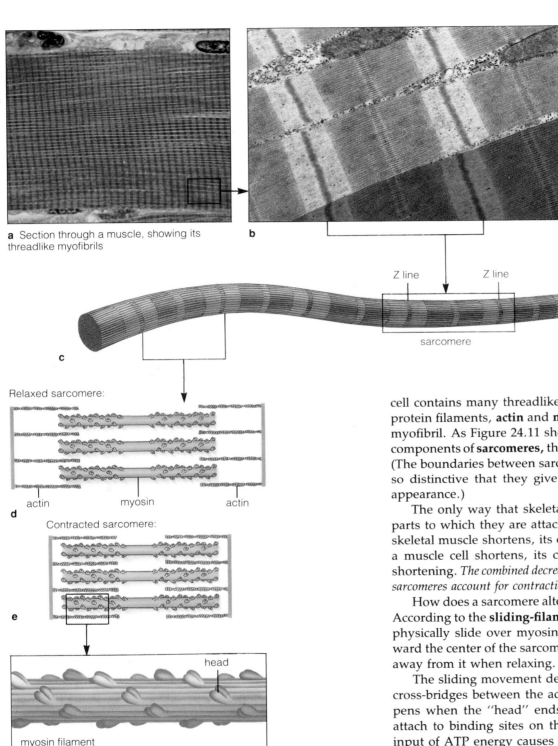

a Section through a muscle, showing its threadlike myofibrils

b

Z line   Z line

sarcomere   myofibril

c

Relaxed sarcomere:

actin   myosin   actin

d

Contracted sarcomere:

e

head

myosin filament

actin filament

f

Figure 24.11 Fine structure of a muscle cell (**a,b**). Interactions between actin and myosin filaments in each sarcomere (**c-f**) are the basis of skeletal muscle contraction.

cell contains many threadlike **myofibrils.** Two types of protein filaments, **actin** and **myosin,** are packed in each myofibril. As Figure 24.11 shows, actin and myosin are components of **sarcomeres,** the basic units of contraction. (The boundaries between sarcomeres, called Z lines, are so distinctive that they give skeletal muscle a striped appearance.)

The only way that skeletal muscles can move body parts to which they are attached is to shorten. When a skeletal muscle shortens, its cells are shortening. When a muscle cell shortens, its component sarcomeres are shortening. *The combined decreases in length of the individual sarcomeres account for contraction of the whole muscle.*

How does a sarcomere alternately contract and relax? According to the **sliding-filament model,** actin filaments physically slide over myosin filaments. They move toward the center of the sarcomere during contraction and away from it when relaxing.

The sliding movement depends on the formation of cross-bridges between the actin and myosin. This happens when the "head" ends of the myosin molecules attach to binding sites on the actin (Figure 24.12). An input of ATP energy causes the myosin heads to tilt in a short power stroke, toward the center of the sarcomere. The actin filaments attached to them also move toward the center. Then the myosin heads detach, reattach at the next actin binding site in line, and move the actin filaments a bit more. A single contraction takes a whole series of power strokes by myosin heads in each sarcomere.

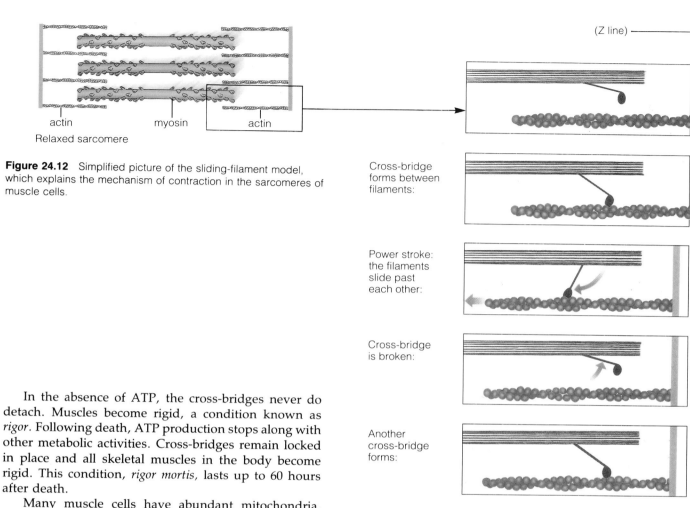

actin

myosin

actin

Relaxed sarcomere

(Z line)

Cross-bridge forms between filaments:

Power stroke: the filaments slide past each other:

Cross-bridge is broken:

Another cross-bridge forms:

Another power stroke:

(toward center of sarcomere)

Figure 24.12 Simplified picture of the sliding-filament model, which explains the mechanism of contraction in the sarcomeres of muscle cells.

In the absence of ATP, the cross-bridges never do detach. Muscles become rigid, a condition known as *rigor.* Following death, ATP production stops along with other metabolic activities. Cross-bridges remain locked in place and all skeletal muscles in the body become rigid. This condition, *rigor mortis,* lasts up to 60 hours after death.

Many muscle cells have abundant mitochondria, which produce ATP through aerobic respiration. Such cells can contract for extended periods, provided enough oxygen is transported to them by the circulation. ATP *can* be replaced by rapidly breaking down glycogen through glycolysis, the energy-releasing pathway described on page 81. (Glycogen is the main fuel supply for cells that consume ATP rapidly.) During strenuous activity, when oxygen levels decline in muscle tissues, glycolysis is critical for ATP synthesis. Glycolysis cannot continue for long, and muscle cells fatigue quickly when glycogen is depleted.

Control of Contraction

Skeletal muscle receives signals to contract from the nervous system. Appropriate signals can trigger action potentials that travel along the plasma membrane of the muscle cell, then invade the cell interior by way of membranous tubes (Figure 24.13). The tubes connect with a membrane system surrounding the cell's myofibrils. This system, the **sarcoplasmic reticulum,** stores

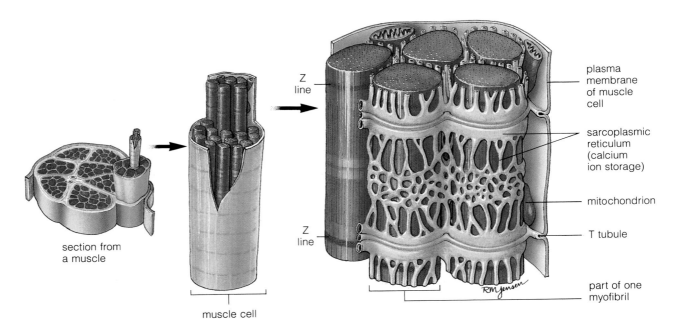

Z line

plasma membrane of muscle cell

sarcoplasmic reticulum (calcium ion storage)

mitochondrion

Z line

T tubule

section from a muscle

muscle cell

RMJensen

part of one myofibril

Figure 24.13 Membrane systems of a muscle cell. The plasma membrane encloses the cell's myofibrils, each with many sarcomeres. The plasma membrane is continuous with membranous tubes (T tubules) that thread inward, near each Z line. The tubes connect with a calcium-storing system (sarcoplasmic reticulum). Action potentials travel along the plasma membrane, stimulate the T tubules, and then trigger calcium release from the sarcoplasmic reticulum. Without calcium ions, actin and myosin filaments in the myofibrils cannot interact to bring about contraction.

calcium ions and releases them in response to action potentials. *Calcium ions bind to sites on the actin filaments, and this allows cross-bridges to form.*

A muscle relaxes when calcium ions are actively taken up after contraction and stored in the sarcoplasmic reticulum. It contracts when calcium ions are released from the sarcoplasmic reticulum. By controlling the frequency with which action potentials reach the sarcoplasmic reticulum in the first place, the nervous system controls calcium ion levels in muscle tissue—and so exerts control over muscle contraction.

SKELETAL-MUSCULAR INTERACTIONS

The human body has more than 600 muscles, arranged as pairs or groups. Some work together to promote the same movement. Others work in opposition (antagonistically), so that the action of one opposes or reverses the action of another. When they contract, these muscles transmit force to the bones to which they are attached. Together, the skeleton and the muscles attached to it are like a system of levers in which rigid rods (bones)

Athletes and Anabolic Steroids

(From Elaine Marieb, Human Anatomy and Physiology, 1989, Benjamin/Cummings Publishing Company)

American society loves a winner and its top athletes reap large social and monetary rewards. Thus, it is not surprising that some will grasp at anything that will increase their performance—including anabolic steroids. Anabolic steroids, variants of testosterone engineered by pharmaceutical companies, were introduced in the 1950s to treat victims of certain muscle-wasting diseases and anemia, and to prevent muscle atrophy in patients immobilized after surgery. Testosterone is responsible for the increase in muscle and bone mass and other physical changes that occur during puberty and convert boys into men. Convinced that megadoses of the steroids could produce enhanced masculinizing effects in grown men, many athletes were using the steroids by the early 1960s; the practice is still going on today.

It has been difficult to determine the incidence of anabolic steroid use among athletes because the use of drugs has been banned by most international competitions, and users (and prescribing physicians or drug dealers) are naturally reluctant to talk about it. Nonetheless, there is little question that many professional bodybuilders and athletes competing in events that require great muscle strength (e.g., shot put, discus throwing, and weight lifting) are heavy users. Sports figures such as football players have also admitted to using steroids as an adjunct to training, diet, and psychologic preparation for games. Advantages of anabolic steroids cited by athletes include increased muscle mass and strength, increased oxygen-carrying capability owing to enhanced red blood cell volume, and an

increase in aggressive behavior (i.e., the urge to "steam-roller" the other guy).

Typically bodybuilders who use steroids combine high doses (up to 200 milligrams a day) with heavy resistance training. Intermittent use begins several months before an event and both oral and injected steroid doses (a method called stacking) are increased gradually as the competition nears.

But do the drugs do all that is claimed for them? Research studies have reported increases in isometric strength and body weight increases in steroid users. While these are results weight lifters dream about, there is a hot dispute over whether this also translates into athletic performance requiring the fine muscle coordination and endurance needed by runners, for instance. Present evidence indicates that it does not, and nonusers handily outperform users in such athletic endeavors.

Do the seemingly minimal advantages conferred by steroid use outweigh the risks? This too is doubtful. Physicians say they cause bloated faces, shriveled testes, and infertility; damage the liver and promote liver cancer; and cause changes in blood cholesterol levels (which may predispose long-term users to coronary heart disease). Additionally, it now appears that the psychiatric hazards of anabolic steroids may be equally threatening. Recent studies have indicated that one-third of anabolic steroid users have serious mental problems. Manic behavior in which the users undergo Jekyll-Hyde personality swings and become extremely violent is common, as are depression and delusions.

The question of why athletes use these drugs is easy to answer. Some admit to a willingness to do almost anything to win, short of killing themselves. Are they unwittingly doing this as well?

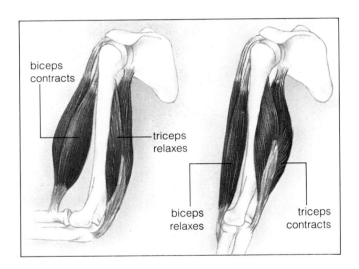

Figure 24.14 Antagonistic muscle pair, showing how two muscles can produce movement in opposite directions.

move about at fixed points (the joints). Most attachments are close to joints. This means a muscle has to contract only a small distance to produce a large movement of some body part.

A limb can be extended and rotated around a joint because of arrangements between pairs or groups of muscles. Muscles in body limbs are often arranged in antagonistic pairs, such as the biceps and triceps shown

in Figure 24.14. Notice how these muscles bridge both the elbow and shoulder joints. When one member of this antagonistic pair (the biceps) contracts, the elbow joint flexes (bends). As it relaxes and its partner (the triceps) contracts, the limb extends and straightens.

SUMMARY

1. The integumentary system (skin and its derivatives) protects the rest of the body from abrasion, bacterial attack, ultraviolet radiation, and dehydration. It helps control internal temperature and pH, and it serves as a blood reservoir for the rest of the body. Its receptors are essential in detecting environmental stimuli.

2. Bones are the structural elements of vertebrate skeletons. They function in movement (by interacting with skeletal muscles to which they are attached), in protecting and supporting other body parts, in mineral storage, and in blood cell formation.

3. The human skeleton has an axial portion (skull, backbone, ribs, and breastbone) and an appendicular portion (limb bones, pelvic girdle, and pectoral girdle). Intervertebral disks are shock pads and flex points in the backbone.

4. Smooth, cardiac, and skeletal muscle tissue all show excitability, contraction, and elasticity. Only skeletal

muscle interacts with the skeleton to bring about movement of the body through the environment or positional changes of its parts.

5. Each muscle cell contains many threadlike myofibrils, which contain actin and myosin filaments. The filaments are organized into sarcomeres, the basic units of contraction.

6. Sarcomeres contract when action potentials trigger the release of calcium ions from a membrane system (sarcoplasmic reticulum) in the muscle cell. Calcium binding alters the actin filaments so that the heads of adjacent myosin molecules can bind to them and form cross-bridges. ATP drives the power strokes that cause actin filaments to slide past the myosin filaments and so shorten the sarcomere.

7. In combination with skeletal muscles, the skeleton works like a system of levers in which rigid rods (bones) move about at fixed points (joints). A limb can be extended and rotated around a joint because of the way pairs or groups of muscles are arranged relative to joints.

Review Questions

1. What are some of the functions of skin and its derivative structures? *346*

2. What are some of the functions of bone tissue? *348, 349*

3. Name the three properties that all three muscle tissues (smooth, cardiac, and skeletal) have in common. *353*

4. Look at Figure 24.10. Then, on your own, sketch and label the fine structure of a muscle, down to one of its individual myofibrils. Can you identify the basic unit of contraction in a myofibril? *353*

5. How do actin and myosin interact in a sarcomere to bring about muscle contraction? What role does ATP play? *354*

Self-Quiz   *(Answers in Appendix IV)*

1. The __integumentary__ system protects the body from abrasion, ultraviolet radiation, bacterial attack, and other environmental stresses.

2. __Skeletal__ and __muscular__ systems work together to move the body and specific body parts.

3. Besides functioning in movement, various bones __protect__ and __support__ soft organs, store __minerals__, and are production sites for __blood cells__.

4. The three types of muscle tissue are __smooth__, __cardiac__, and __skeletal__.

5. Which of the following is *not* a function of the integumentary system?
   a. protect the body from abrasion
   b. protect the body from dehydration
   c. detect environmental stimuli
   d. bring about body movements
   e. serve as a blood reservoir for the rest of the body

6. Which of the following serve as shock pads and flex points in the human backbone?
   a. vertebrae
   b. cervical bones
   c. lumbar bones
   d. intervertebral disks

7. Which of the following is *not* a major type of muscle?
   a. smooth muscle
   b. integumentary muscle
   c. skeletal muscle
   d. cardiac muscle

8. The smallest unit of contraction in skeletal muscle is the _____.
   a. myofibril   c. muscle fiber
   b. sarcomere   d. myosin filament

9. Muscle contraction will not occur _____.
   a. in the absence of calcium ions
   b. in the absence of ATP
   c. both a and b
   d. neither a nor b

10. Match the terms on muscle structure and function.
   ___d___ myofibrils   a. contains many myofibrils
   ___ ___ sarcoplasmic   b. the contractile unit of muscle
          reticulum   c. drives the power stroke to slide
   ___b___ sarcomere   actin filaments past myosin
   ___a___ muscle cell   filaments
   ___ ___ ATP   d. composed of actin and myosin
                                  filaments
                               e. calcium ion storage site in a
                                  muscle

Selected Key Terms

actin *354*
appendicular skeleton *350*
axial skeleton *350*
cartilaginous joint *352*
dermis *346*
epidermis *346*
fibrous joint *352*
Haversian canal *350*
integumentary system *346*
intervertebral disk *350*

keratinization *347*
myofibril *354*
myosin *354*
osteocyte *349*
red marrow *349*
sarcomere *354*
sarcoplasmic reticulum *355*
sliding-filament model *354*
synovial joint *352*
yellow marrow *350*

Readings

Alexander, R.M. July–August 1984. "Walking and Running." *American Scientist* 72(4):348–354. The biomechanics of traveling on foot.

Eckert, R., D. Randall, and G. Augustine. 1988. *Animal Physiology: Mechanisms and Adaptations.* Third edition. New York: Freeman.

Hoyle, G. 1983. *Muscles and Their Neural Control.* New York: Wiley.

Huxley, H.E. December 1965. "The Mechanism of Muscular Contraction." *Scientific American* 213(6):18–27. Old article, great illustrations.

Luttgens, K., and K. Wells. 1982. *Kinesiology: Scientific Basis of Human Motion.* Seventh edition. Philadelphia: Saunders.

25 DIGESTION AND HUMAN NUTRITION

KEY CONCEPTS

1. Interactions among the digestive, circulatory, respiratory, and urinary systems supply the body's cells with raw materials, dispose of wastes, and maintain the volume and composition of extracellular fluid.

2. Most digestive systems have specialized regions for food transport, processing, and storage. Different regions are concerned with mechanical and chemical breakdown of food, absorption of the breakdown products, and elimination of unabsorbed residues.

3. To maintain an acceptable body weight and overall health, energy intake must balance energy output (by way of metabolic activity, physical exertion, and so on). Complex carbohydrates are the main energy source.

4. Nutrition requires the intake of vitamins, minerals, and certain amino acids and fatty acids that the body cannot produce itself.

It is just amazing. An Eskimo might eat only fish in a given day, a Nepalese might eat only rice, and an American might eat pizza, chocolate, kiwi, couscous, or snake meat. Yet through its metabolic magic, the body converts these and a dizzying variety of other substances into usable energy and tissues of its own.

With this chapter we start our tour of nutrition, which will take us through processes by which food is ingested, digested, absorbed, and converted to the body's carbohydrates, lipids, and proteins. The tour begins with the **digestive system**, which reduces foodstuffs to molecules small enough to move into the internal environment.

As Figure 25.1 shows, the digestive system does not act alone to meet the body's metabolic needs. A circulatory system picks up absorbed nutrients and transports them to cells throughout the body. A respiratory system nelps cells use the nutrients by supplying them with oxygen (for aerobic respiration) and relieving them of

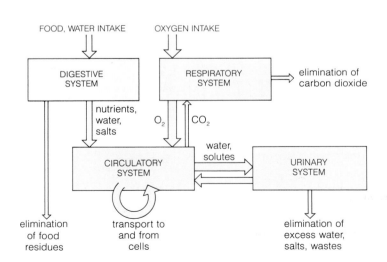

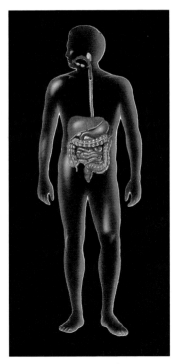

Figure 25.1 Links between the digestive, respiratory, circulatory, and urinary systems. These organ systems work together to supply the body's cells with raw materials and eliminate wastes. This chapter focuses on the digestive system; subsequent chapters will address the other systems shown here.

carbon dioxide wastes. Even when food and water intake varies, a urinary system helps maintain the composition and volume of extracellular fluid. Keep these vital interactions in mind as we proceed with our nutritional tour in this chapter and ones to follow.

DIGESTIVE SYSTEM

The digestive system of most animals is an internal tube or cavity divided into specialized regions for food transport, processing, and storage (Figure 25.2). In general, a digestive system serves these functions:

1. *Movement:* mechanical breakdown, mixing, and passage of ingested nutrients through different regions of the system, then elimination of undigested and unabsorbed residues.

2. *Secretion:* release of enzymes, hormones, and other substances required for the functions of the digestive tract.

3. *Digestion:* chemical breakdown of nutrients into particles, then into molecules small enough to cross the tube's lining and enter the internal environment.

4. *Absorption:* passage of digested nutrients, fluid, and ions across the tube wall and into the blood or lymph, which will distribute them through the body.

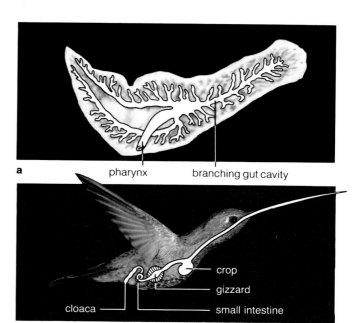

a   pharynx   branching gut cavity

crop
gizzard
cloaca
small intestine

b

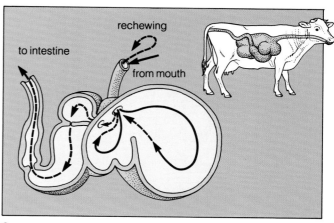

c

Figure 25.2 Examples of digestive systems.

(**a**) An "incomplete" digestive system has only one opening. In a flatworm, this means there is two-way traffic of food and undigested residues through the pharynx. The highly branched gut cavity serves both digestive and circulatory functions.

(**b**) A "complete" digestive system is a tube or cavity with regional specializations and an opening at each end. In birds, one such region is a crop (a food storage organ) and a gizzard (a muscular organ in which food is ground into smaller bits).

(**c**) Deer, cattle, and other *ruminants* are grazing animals with a complete digestive system that can handle the cellulose in plants. They swallow partially chewed plant material, which moves into two stomachlike chambers. Then they regurgitate the material, chew it, and swallow it again. The double chewing breaks apart the plant tissues and better exposes the cellulose fibers. The cellulose is broken down by digestive enzymes that are produced by symbiotic bacteria living in the gut. The double chewing gives the enzymes more time and more surface area upon which to act. Altogether, food is processed in four stomachlike chambers before being sent on to the small intestine, where it is absorbed.

Human Digestive System: An Overview

Figure 25.3 shows the human digestive system. Stretched out, this "tube" would be 6.5 to 9 meters (21 to 30 feet) long in adults. Its specialized regions are the mouth (oral cavity), pharynx, esophagus, stomach, small intestine, large intestine (colon), rectum, and anus. (The regions from the stomach down are often called the *gut.*) Enzymes and other substances from the salivary glands, liver, gallbladder, and pancreas are secreted into different parts of the tube. The secretions are necessary for digestion and absorption.

The tube wall has two layers of smooth muscle between an inner lining and an outer sheath of connective tissue. Contractions in the muscle layers mix food with assorted secretions and move it forward. During **peristalsis,** for example, rings of circular muscle contract behind food and relax in front of it. The food expands the tube wall, peristaltic movement forces the food onward and expands the next wall region, and so on.

Also built into the tube wall are **sphincters.** These rings of muscle mark the beginning and end of the stomach and other specialized regions. Sphincters help control the forward movement of food and prevent backflow.

The nervous and endocrine systems, together with local controls, govern the digestive system. They respond to changes in the volume and composition of food material in the gut. After a meal, for example, they call for muscle contractions or secretion of enzymes and other substances. The hypothalamus and other parts of the brain monitor such activities and coordinate them with other events.

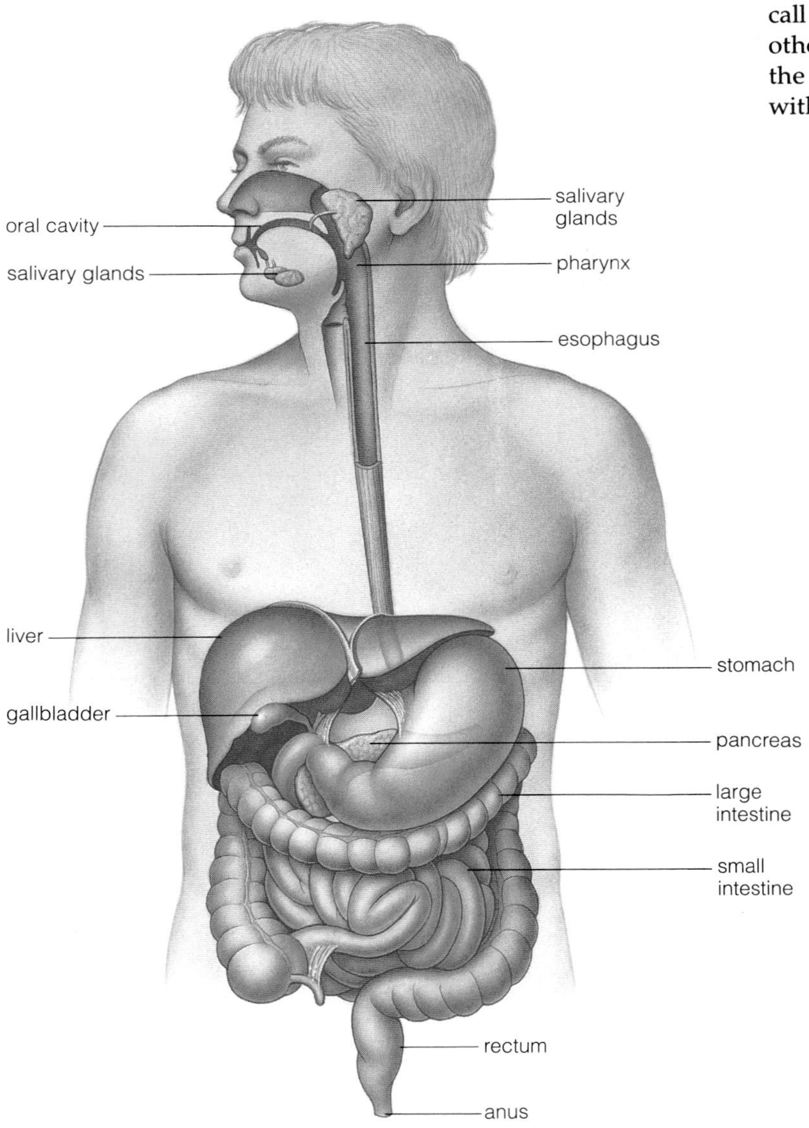

Figure 25.3 Simplified picture of the human digestive system.

Table 25.1	Components of the Human Digestive System
Organ	**Main Functions**
Mouth	Mechanically break down food, mix it with saliva
Salivary glands	Moisten food; start polysaccharide breakdown; buffer acidic foods in mouth
Stomach	Store, mix, dissolve food; kill many microorganisms; start protein breakdown; empty in a controlled way
Small intestine	Digest and absorb most nutrients
Pancreas	Enzymatically break down all major food molecules; buffer hydrochloric acid from stomach
Liver	Secrete bile for fat absorption; secrete bicarbonate, which buffers hydrochloric acid from stomach
Gallbladder	Store, concentrate bile from liver
Large intestine	Store, concentrate undigested matter by absorbing water and salts
Rectum	Control over elimination of undigested and unabsorbed residues

Into the Mouth, Down the Tube

Table 25.1 lists the organs of the digestive system and their functions. In the **mouth** (oral cavity), food starts getting pummeled and polysaccharide digestion begins. Only humans and other mammals *chew* food in the mouth. Adult humans normally have thirty-two teeth to do this. Each tooth is an engineering marvel, able to withstand many years of chemical insults and mechanical stress. It has an enamel coat (hardened calcium deposits), dentine (a thick bonelike layer), and an inner pulp (with nerves and blood vessels). The teeth called incisors bite off chunks of food, the cuspids tear it, and the molars grind it (page 214).

Food in the mouth becomes mixed with saliva, a fluid secreted from **salivary glands.** Saliva contains a starch-degrading enzyme (salivary amylase), bicarbonate (HCO_3^-), and mucins. The buffering action of HCO_3^- keeps the pH of your mouth between 6.5 and 7.5 even when you eat tomatoes and other acidic foods. Mucins (modified proteins) bind bits of food into a softened, lubricated ball that can be swallowed easily.

The Stomach

The **stomach,** a muscular, stretchable sac, has three main functions. First, it stores and mixes food. Second, its secretions help dissolve and degrade food. Third, it helps control passage of food into the small intestine.

Stomach Acidity. Each day, cells in the stomach lining secrete about 2 liters of substances, including hydrochloric acid (HCl), pepsinogens, and mucus. The substances make up the "gastric fluid" in the stomach. The HCl separates into H^+ and Cl^-, and the increase in acidity helps dissolve bits of food to form a solution called chyme. It also kills most of the microorganisms hitching rides into the body in food.

Stomach secretions begin when your brain responds to the sight, aroma, and taste of food (even to hungry thoughts about it) and fires off signals to secretory cells in the stomach lining. But most of the secretions occur in response to food in the stomach. When food stretches the stomach, it activates receptors in the stomach wall and so gives rise to neural signals that call for stepped-up secretions. Also, secretory cells are stimulated directly by certain substances, including partially dismantled proteins as well as caffeine, which is present in coffee, tea, chocolate, and cola drinks.

Muscle contractions of the tongue force the softened ball of food into the **pharynx.** This muscular tube connects with the **esophagus,** which leads to the stomach. Sensory receptors in the wall of the pharynx are stimulated, and they trigger contractions that initiate swallowing. The pharynx and esophagus do not have roles in digestion; contractions in their walls simply propel food into the stomach.

The pharynx also connects with the trachea, which leads to the lungs. Swallowing opens a sphincter at the start of the esophagus. You normally don't choke on food because a flaplike valve (epiglottis) closes off the trachea and keeps you from breathing while food is moving into the esophagus.

Protein digestion begins in the stomach. High stomach acidity changes the structure of proteins and exposes their peptide bonds. It also converts pepsinogens to active forms, called pepsins, that break down proteins. Protein fragments directly stimulate the secretion of gastrin, a hormone that acts on HCl-secreting cells. The more protein you eat, the more gastrin and HCl are released.

What protects the stomach lining itself from HCl and pepsin? Control mechanisms ensure that enough mucus and buffering molecules (bicarbonate ions especially) are secreted to protect the lining from their destructive effects. Sometimes, however, normal controls are blocked. When the surface of the stomach breaks down, H^+ diffuses into the lining and triggers the release of a chemical called histamine from tissue cells. Histamine acts on local blood vessels—and it stimulates more HCl secretion. A positive feedback loop is set up, tissues become damaged, and there may be bleeding into the stomach and abdomen. The outcome is a *peptic ulcer.*

Stomach Emptying. In the stomach, peristaltic waves mix the chyme and build up force as they approach the sphincter between the stomach and small intestine (Figure 25.4). The arrival of a strong contraction closes the

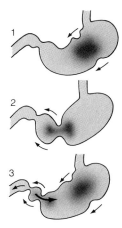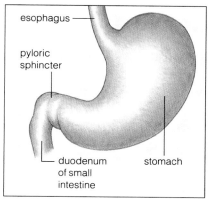

Figure 25.4 Peristaltic wave down the stomach, produced by alternating contraction and relaxation of muscles in the stomach wall.

sphincter, so most of the chyme is squeezed back. But a small amount moves into the small intestine.

The volume and composition of chyme affect how fast the stomach empties. For example, large meals activate more receptors in the stomach wall, these call for increases in the force of contraction, and the stomach empties faster. As another example, receptors in the small intestine sense increases in acidity, fat content, and so on, and they call for the release of hormones that slow down stomach emptying. Through such slow-downs, food is not moved along faster than it can be processed. Fear, depression, and other emotional upsets also can slow down stomach emptying.

The Small Intestine

Digestion is completed and most nutrients are absorbed in the **small intestine** (Figure 25.3). Secretions from the **pancreas** and **liver** enter a common duct that empties into the small intestine. Each day, about 9 liters of fluid travel from the stomach, liver, and pancreas into this part of the gut, and all but five percent of it is absorbed across the intestinal lining.

Digestion Processes. Enzymes secreted from the pancreas act on carbohydrates, fats, proteins, and nucleic acids. The pancreas also secretes bicarbonate, which helps buffer the HCl arriving from the stomach. Two hormones secreted from the pancreas (insulin and glucagon) do not function in digestion, but they have vital roles in nutrition, as described on page 456.

Bile is a secretion from the liver. It contains bile salts, bile pigments, cholesterol, and lecithin (a phospholipid). Bile salts assist in fat breakdown and absorption. Most fats in our diet are triglycerides, clumped into large fat globules. The globules are mechanically broken apart into droplets in the small intestine. Bile salts break down the droplets and keep them from clumping back together into globules. Through the emulsifying effect of bile salts, fat-degrading enzymes have greater access to more triglycerides, so fat digestion is enhanced. Between meals, bile is stored and concentrated in the **gallbladder** (Figure 25.3).

Absorption Processes. By the time proteins, lipids, and carbohydrates are halfway through the small intestine, mechanical action and enzymes have broken down most of them to smaller molecules. The breakdown

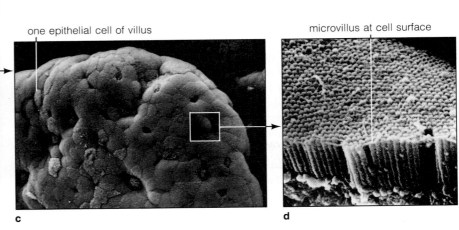

portion of the small intestine

folds

a

one epithelial cell of villus

microvillus at cell surface

b intestinal villi

c

d

Figure 25.5 Location of villi in the mammalian small intestine. (**a**) Surface view of the deep, permanent folds of the inner layer of the intestinal tube. (**b**) Some of the fingerlike projections (villi) that cover the inner layer. The villi are so dense and numerous, they give the surface a velvety appearance. (**c**) Tip of a single villus; individual epithelial cells are visible. (**d**) The dense crown of microvilli at the surface of a single cell.

products include glucose and other monosaccharides, amino acids, fatty acids, and monoglycerides, all of which can cross the intestinal lining. The lining itself is densely folded, and the folds are covered with **villi** (singular, villus). The villi are absorptive structures; they greatly increase the surface area available for interactions with chyme. Epithelial cells at the surface of these structures have a crown of **microvilli.** Each microvillus is a threadlike projection of the plasma membrane, and it further increases the surface area available for absorption (Figure 25.5).

At each villus, glucose and most amino acids cross the gut lining. (Active transport mechanisms carry them across the plasma membranes of the epithelial cells.) Then they diffuse through extracellular fluid and enter small blood vessels inside the villus (Figure 25.6). The free fatty acids and monoglycerides diffuse across the lipid bilayer of the membranes. Inside the epithelial cells, they recombine into triglycerides. The triglycerides and other fats cluster together as small droplets. Then the droplets leave the cells (by exocytosis) and enter lymph vessels, which drain into the general circulation (page 387). Water and mineral ions (such as Na^+, K^+, and Cl^-) also are absorbed at the intestinal villi.

Table 25.2 summarizes the regions where carbohydrates, fats, proteins, and nucleic acids are digested and absorbed. As you can see, *most nutrients are absorbed in the small intestine.*

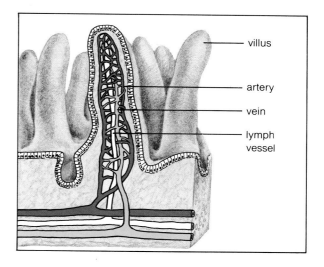

Figure 25.6 Location of blood vessels and lymph vessels in intestinal villi. Monosaccharides and most amino acids moving across the intestinal lining enter the blood vessels; fats enter the lymph vessels, which drain into the general circulation.

Table 25.2 Major Enzymes of Digestion				
Enzyme	Source	Where Active	Substrate	Main Breakdown Products*
Carbohydrate digestion:				
Salivary amylase	salivary glands	mouth	polysaccharides	disaccharides
Pancreatic amylase	pancreas	small intestine	polysaccharides	disaccharides
Disaccharidases	small intestine	small intestine	disaccharides	monosaccharides (e.g., glucose)
Protein digestion:				
Pepsins	stomach mucosa	stomach	proteins	peptide fragments
Trypsin and chymotrypsin	pancreas	small intestine	proteins and polypeptides	peptide fragments
Carboxypeptidase	pancreas	small intestine	peptide fragments	amino acids
Aminopeptidase	intestinal mucosa	small intestine	peptide fragments	amino acids
Fat digestion:				
Lipase	pancreas	small intestine	triglycerides	free fatty acids, monoglycerides
Nucleic acid digestion:				
Pancreatic nucleases	pancreas	small intestine	DNA, RNA	nucleotides
Intestinal nucleases	intestinal mucosa	small intestine	nucleotides	nucleotide bases, monosaccharides

*Yellow parts of table identify breakdown products that can be absorbed into the internal environment.

The Large Intestine

Material not absorbed in the small intestine moves into the **large intestine,** or **colon.** The colon concentrates and stores *feces,* a mixture of undigested and unabsorbed material, water, and bacteria. The concentrating mechanism involves the active transport of sodium ions across the lining of the colon; water follows passively as a result.

The colon is about 1.2 meters long and starts out as a cup-shaped pouch (Figure 25.7). The *appendix,* a narrow projection from the pouch, has no known digestive functions. (It may have roles in defense against infectious agents.) The colon ascends on the right side of the abdominal cavity, cuts across to the other side, then descends and connects with a short tube, the **rectum** (Figure 25.3). Distension of the rectal wall triggers the urge to expel fecal matter from the body. Expulsion is controlled by the nervous system, which can stimulate or inhibit contractions of a muscle sphincter at the *anus,* the terminal opening of the gut.

The average American diet does not include enough bulk. ("Bulk" is the volume of fiber and other undigested food material that cannot be decreased by absorption in the colon.) Without adequate bulk, it takes longer for feces to move through the colon, with irritating and perhaps carcinogenic effects.

Most people of rural Africa and India cannot afford to eat much more than whole grains—which are high in fiber content—and they rarely suffer colon cancer or appendicitis. (In *appendicitis,* the appendix becomes infected and may rupture; bacteria normally living in the colon may spread into the abdominal cavity and cause serious infection.) When those people move to wealthier nations and leave behind their fiber-rich diet, they are more likely to suffer colon cancer and appendicitis. And this example brings us to a survey of what constitutes good and bad nutrition.

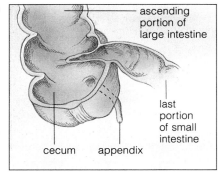

Figure 25.7 Location of the appendix, a narrow projection from the cecum (a cup-shaped pouch at the start of the large intestine).

ascending portion of large intestine

last portion of small intestine

cecum   appendix

HUMAN NUTRITIONAL REQUIREMENTS

The earliest human ancestors, it seems, dined on fresh fruits and other fibrous plant material. From this nutritional beginning, humans in many parts of the world have moved to diets rich in saturated fats, cholesterol, refined sugars, and salts—and low in fiber. Colon cancer is only one outcome of this long-term shift in diet. As you will see in later chapters, the dietary changes may also contribute to kidney stones, breast cancer, and circulatory disorders.

Energy Needs and Body Weight

The body grows and maintains itself when kept supplied with energy and materials from foods of certain types, in certain amounts. Nutritionists measure energy in units called "calories," which unfortunately are supposed to mean the same thing that other scientists call "kilocalories." A kilocalorie is 1,000 calories of energy, and that is the term we will use here.

To maintain an acceptable weight and keep the body functioning normally, caloric intake must be balanced with energy output. The output varies from one person to the next because of differences in physical activity, basic rate of metabolism, age, sex, hormone activity, and emotional state. Some of these factors are influenced by a person's social environment. But others have a genetic basis. For example, long-term studies were made of many identical twins (with identical genes) who were separated at birth and raised apart, in different households. At adulthood, the body weights of the separated twins were remarkably similar.

In most adults, energy input balances the output, so body weight remains much the same over long periods. As any dieter knows, the body behaves as if it has a set point for what that weight is going to be and works to counteract deviations from its set point.

How many kilocalories should you take in each day to maintain what you consider to be an acceptable body weight? First, multiply the desired weight (in pounds) by 10 if you are not very active physically, by 15 if you are moderately active, and by 20 if you are quite active. Then, depending on your age, subtract the following amount from the value obtained from the first step:

Age:	Subtract:
25–34	0
35–44	100
45–54	200
55–64	300
Over 65	400

For example, if you want to weigh 120 pounds and are highly active, $120 \times 20 = 2,400$ kilocalories. If you are

Man's Height	Size of Frame		
	Small	Medium	Large
5′ 2″	128–134	131–141	138–150
5′ 3″	130–136	133–143	140–153
5′ 4″	132–138	135–145	142–156
5′ 5″	134–140	137–148	144–160
5′ 6″	136–142	139–151	146–164
5′ 7″	138–145	142–154	149–168
5′ 8″	140–148	145–157	152–172
5′ 9″	142–151	148–160	155–176
5′10″	144–154	151–163	158–180
5′11″	146–157	154–166	161–184
6′ 0″	149–160	157–170	164–188
6′ 1″	152–164	160–174	168–192
6′ 2″	155–168	164–178	172–197
6′ 3″	158–172	167–182	176–202
6′ 4″	162–176	171–187	181–207

a

Woman's Height	Size of Frame		
	Small	Medium	Large
4′10″	102–111	109–121	118–131
4′11″	103–113	111–123	120–134
5′ 0″	104–115	113–126	122–137
5′ 1″	106–118	115–129	125–140
5′ 2″	108–121	118–132	128–143
5′ 3″	111–124	121–135	131–147
5′ 4″	114–127	124–138	134–151
5′ 5″	117–130	127–141	137–155
5′ 6″	120–133	130–144	140–159
5′ 7″	123–136	133–147	143–163
5′ 8″	126–139	136–150	146–167
5′ 9″	129–142	139–153	149–170
5′10″	132–145	142–158	152–173
5′11″	135–148	145–159	155–176
6′ 0″	138–151	148–162	158–179

b

Figure 25.8 (**a**) The "ideal" weights for adult men and women, at least according to one insurance company in 1983. The values shown are for persons 25 to 59 years old, wearing shoes with 1-inch heels and 5 pounds of clothing (men) or 3 pounds of clothing (women).

(**b**) Extremely obese individuals put severe strain on their circulatory systems. The body produces many additional capillaries to service the increased tissue masses, so blood volume is lowered. The heart is more stressed because it must pump harder to keep blood circulating.

thirty-five years old, then you should take in a total of (2,400 − 100), or 2,300 kilocalories a day.

By definition, **obesity** is an excess of fat in the body's adipose tissues, caused by imbalances between caloric intake and energy output. Yet what is too fat or too thin? What is a person's "ideal weight"? Many charts have been developed (Figure 25.8), mostly by insurance companies that want to identify overweight people who are considered to be insurance risks. People who are twenty-five percent heavier than the "ideal" are viewed as obese.

Some researchers who study causes of death suspect that the "ideal" actually may be 10 to 15 pounds heavier than the charts indicate. Some nutritionists are convinced the chart values should be less. Whatever the ideal range may be, serious disorders do arise with extremes at either end of that range (see *Commentary*).

To maintain an acceptable body weight, energy input (caloric intake) must be balanced with energy output (as through metabolic activity and exercise).

Carbohydrates

Complex carbohydrates are the body's main sources of energy. As we have seen, they can be readily broken down into glucose units (Chapter 6). Glucose is the primary energy source for your brain, muscles, and other body tissues. According to many nutritionists, the fleshy fruits, cereal grains, legumes (including beans and peas), and other fibrous carbohydrates should make up at least fifty-eight to sixty percent of the daily caloric intake.

Each year, the average American eats as much as 128 pounds of refined sugar, or sucrose. You may think this a far-fetched statement, but take a look at the ingredients listed on packages of cereal, frozen dinners, soft drinks, and other prepared foods. Many contain more sugar than you might realize. Sucrose is a simple sugar; it adds calories to the diet but does so without the fiber of complex carbohydrates.

Lipids

Fats and other lipids have important roles in the body. For example, phospholipids (such as lecithin) and cholesterol are components of animal cell membranes. Besides being used as energy reserves, fat deposits serve as cushions for many organs, including the eyes and kidneys, and they provide insulation beneath the skin. Fats from the diet also help the body store fat-soluble vitamins.

Lipids make up forty percent of the average diet in the United States. Most of the medical community agrees it should be less than thirty percent. The body can manufacture most of its own fats, including cholesterol,

Extreme Eating Disorders—Anorexia Nervosa and Bulimia

Millions of Americans are dieting. Unfortunately, in a growing number of cases, obsessive dieting leads to a potentially fatal eating disorder called *anorexia nervosa*. The disorder occurs primarily in women in their teens and early twenties.

Individuals with anorexia nervosa have a skewed perception of their body weight. They have an overwhelming fear of being fat and being hungry. They embark on a course of self-induced starvation and, frequently, overexercising. Emotional factors contribute to the disorder. Some individuals fear growing up in

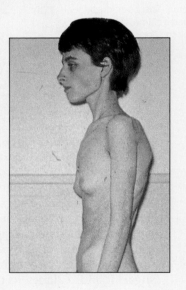

general and maturing sexually in particular; others have irrational expectations of what they can accomplish. Severe cases require psychiatric treatment.

Another eating disorder on the rise is *bulimia* ("an oxlike appetite"). At least twenty percent of college-age women are now suffering to varying degrees from this disorder. Those afflicted may look outwardly healthy, but their food intake is out of control. During an hour-long eating binge, a bulimic may take in more than 50,000 kilocalories. This is followed by vomiting or purging the body with laxatives, sometimes in doses of 200 tablets or more. The binge-purge routine may occur once a month; it may occur several times a day.

Some women start doing this because it seems like a simple way to lose weight. Others have emotional problems. Often they are well-educated, accomplished individuals, but they strive for perfection and may have problems with control exerted by other family members. According to one view, eating may actually be an unpleasant event for them, but the purging (which they themselves control) relieves them of anger and frustration.

Repeated purgings, however, can damage the gastrointestinal tract. Repeated vomiting, which brings stomach acids into the mouth, can erode teeth to stubs. At its most extreme, bulimia can lead to death through heart failure, stomach rupturing, or kidney failure. Psychiatric treatment and hospitalization may be required in severe cases. Unfortunately, bulimics tend to wait for years before seeking help.

from protein and carbohydrates. (That is exactly what it does when you eat too much protein and carbohydrates.) You only need to take in *one tablespoon a day* of a polyunsaturated fat, such as corn oil or olive oil. These oils contain linoleic acid, a fatty acid component of lecithin. It is one of the **essential fatty acids.** This means the body cannot produce it; the diet must provide it.

Butter and other animal fats are saturated fats, which tend to raise the level of cholesterol in the blood. Too much cholesterol may contribute to disorders of the circulatory system (page 381).

Proteins

Proteins should make up about twelve percent of the total diet. When proteins are digested and absorbed, their amino acids become available for the body's own protein-building programs. Of the twenty common amino acids, eight are **essential amino acids.** Our cells cannot build them; they must be obtained from food. The eight are cysteine (or methionine), isoleucine, leucine, lysine, phenylalanine (or tyrosine), threonine, tryptophan, and valine.

Figure 25.9 (**a**) Essential amino acids, which represent a small portion of the total protein intake. These eight amino acids must be available at the same time, in certain amounts, if cells are to build their own proteins.

Animal proteins (milk and eggs especially) are complete (they contain high amounts of all eight amino acids in the required proportions). Nearly all plant proteins are incomplete, so vegetarians should construct their meals carefully to avoid protein deficiency. They can do this, for example, by combining *different* foods from any two of the columns shown in (**b**). In addition, vegetarians who exclude dairy products and eggs from their diet should take vitamin B_{12} and B_2 (riboflavin) supplements.

Animal protein is a luxury in most traditional societies, yet workable combinations of plant proteins are present in their cuisines—including rice/beans, chili/cornbread, tofu/rice, lentils/wheat bread, and macaroni/cheese.

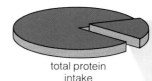

isoleucine
leucine
lysine
methionine
phenylalanine
threonine
tryptophan
valine

total protein intake

a

No Limiting Amino Acid	Low in Lysine	Low in Methionine, Other Sulfur-Containing Amino Acids	Low in Tryptophan
legumes: soybean tofu soy milk cereal grains: wheat germ nuts: black walnuts milk cheeses (except cream cheese) yogurt eggs meats	legumes: peanuts cereal grains: barley buckwheat corn meal oats rice rye wheat nuts, seeds: almonds cashews coconut English walnuts hazelnuts pecans pumpkin seeds sunflower seeds	legumes: beans (dried) black-eyed peas garbanzos lentils lima beans mung beans peanuts nuts: hazelnuts fresh vegetables: asparagus broccoli brussels sprouts green beans green peas mushrooms parsley potatoes soybeans Swiss chard	legumes: beans (dried) garbanzos lima beans mung beans peanuts cereal grains: corn meal nuts: almonds English walnuts fresh vegetables: corn green peas mushrooms Swiss chard

b

Most animal proteins are "complete" (they contain high amounts of all essential amino acids). Plant proteins are "incomplete." To get enough protein, vegetarians must eat certain combinations of different plants (Figure 25.9). Nutritionists use a measure called *net protein utilization*, or NPU, to compare proteins from different sources (Table 25.3). NPU values range from 100 (all essential amino acids present in ideal proportions) to 0 (one or more absent; the protein is useless when eaten alone).

You know that enzymes and other proteins are vital for the body's structure and function, so it should be readily apparent that protein-deficient diets are no joking matter. Protein deficiency is most damaging among the young, for rapid brain growth and development occur early in life. Unless enough protein is taken in just before and just after birth, irreversible mental retardation occurs. Even mild protein starvation can retard growth and affect mental and physical performance.

Table 25.3 Efficiency of Some Single Protein Sources in Meeting Minimum Daily Requirements

Source	Protein Content (%)	Net Protein Utilization (NPU)	Amount Needed to Satisfy Minimum Daily Requirement	
			(grams)	(ounces)
Eggs	11	97	403	14.1
Milk	4	82	1,311***	45.9***
Fish*	22	80	244	8.5
Cheese*	27	70	227	7.2
Meat*	25	68	253	8.8
Soybeans	34	60	210**	7.3**
Kidney beans	23	40	468**	16.4**
Corn	10	50	860**	30.0**

* Average values.
** Dry weight values.
*** Equivalent of 6 cups. The figure is somewhat misleading, for most of the volume of milk is water. Milk is actually a rich source of high-quality protein.

Table 25.4 Vitamins Required for Normal Cell Functioning

Vitamin	RDA* (milligrams)	Common Sources	Some Known Functions
Water-Soluble Vitamins:			
B_1 (Thiamine)	1.5	Lean meats, liver, eggs, whole grains, green leafy vegetables, legumes	Connective tissue formation; iron, folic acid utilization
B_2 (Riboflavin)	1.7	Milk, egg white, yeast, whole grains, poultry, fish, meat	Coenzyme action (FAD, FMN)
Niacin	20	Meat, poultry, fish; also peanuts, potatoes, green leafy vegetables, liver	Coenzyme action (NAD^+, $NADP^+$)
B_6	2	Meat, poultry, fish; also potatoes, tomatoes, spinach	Coenzyme role in amino acid metabolism
Pantothenic acid	10	In many foods, especially meat, yeast, egg yolk	Coenzyme role in glucose metabolism; fatty acid and steroid synthesis
Folic acid	0.4	Dark green vegetables, eggs, liver, yeast, lean meat, whole grains; produced by bacteria in gut	Coenzyme role in nucleic acid and amino acid metabolism
B_{12}	0.003	Meat, poultry, fish, eggs, dairy foods (not butter)	Coenzyme role in nucleic acid metabolism
Biotin	**	Legumes, nuts, liver, egg yolk; some produced by bacteria in gut	Coenzyme action in fat and glycogen formation; amino acid metabolism
Choline	**	Whole grains, legumes, egg yolk, liver	Component of phospholipids, acetylcholine
C (Ascorbic acid)	60	Citrus, papaya, cantaloupe, berries, tomatoes, potatoes, green leafy vegetables	Structural role in bone, cartilage, teeth; roles in collagen formation, carbohydrate metabolism
Fat-Soluble Vitamins:			
A (Retinol)	1	Formed from carotene in deep-yellow, deep-green leafy vegetables; already present in fish liver oil, liver, egg yolk, fortified milk	Role in synthesis of visual pigments; required for bone, tooth development; maintains epithelial tissues
D	0.01	Vitamin D_3 formed in skin cells (also in fish liver oils, egg yolk, fortified milk); converted to active form in other body regions	Promotes bone growth, mineralization; increases calcium absorption
E	15	Vegetable oils, margarine, whole grains, dark-green vegetables	Prevents breakdown of vitamins A, C in gut; helps maintain cell membranes
K	0.003	Most formed by bacteria in colon; also in green leafy vegetables, cauliflower, cabbage, pork liver	Role in clot formation; electron transport role in ATP formation

* Recommended daily allowance.
** Not established.

Vitamins and Minerals

Normal metabolic activity depends on small amounts of more than a dozen organic substances called **vitamins.** Most plant cells synthesize all of these substances. In general, animal cells have lost the ability to do so, so animals must obtain vitamins from food. Human cells need at least thirteen different vitamins, each with specific metabolic roles. Many reactions depend on several vitamins, and the absence of one can affect the functions of others (Table 25.4).

Metabolic activity also depends on inorganic substances called **minerals** (Table 25.5). For example, most cells use calcium and magnesium in many different reactions. All cells use potassium during protein synthesis, muscle activity, and nerve function. All cells require iron for cytochrome molecules, these being components of electron transport chains. Red blood cells require iron for hemoglobin, the oxygen-carrying pigment in blood.

The sensible way to supply cells with essential vitamins and minerals is to eat a well-balanced selection of carbohydrates, lipids, and proteins. About 250–500 grams of carbohydrates, 66–83 grams of lipids, and 32–42 grams of protein should do the trick. Some people claim the body will benefit from massive doses of certain vitamins and minerals. To date, there is no clear evidence that vitamin intake above recommended daily amounts

Mineral	RDA* (milligrams)	Common Sources	Some Known Functions
Calcium	800	Dairy products, dark-green vegetables, dried legumes	Bone, tooth formation; clotting, neural signals
Chlorine	?	Table salt; usually too much in diet	HCl formation by stomach; helps maintain body pH
Copper	2	Meats, legumes, drinking water	Used in synthesis of hemoglobin, melanin, transport chain components
Fluorine	2	Fluoridated water, seafood	Bone, tooth maintenance
Iodine	100–130	Marine fish, shellfish; dairy products	Thyroid hormone formation
Iron	10 (males) 18 (females)	Liver, lean meats, yolk, shellfish, nuts, molasses, legumes, dried fruit	Hemoglobin, cytochrome formation
Magnesium	300–350	Dairy products, nuts, whole grains, legumes	Coenzyme role in ATP–ADP cycle; role in muscle, nerve function
Phosphorus	800	Dairy products, red meat, poultry, whole grains	Component of bone, teeth, nucleic acids, proteins, ATP, phospholipids
Potassium	?	Diet provides ample amounts	Muscle, nerve function; role in protein synthesis; acid-base balance
Sodium	500?	Table salt; diet provides adequate to excess amounts	Key salt in solute-water balance, muscle and nerve function
Sulfur	?	Dietary proteins	Component of body proteins
Zinc	15	Seafood, meat, cereals, legumes, nuts, yeast	Component of digestive enzymes; roles in normal growth, wound healing, taste and smell, sperm formation

Table 25.5   Minerals Required for Normal Cell Functioning

* Recommended daily allowance.

leads to better health. To the contrary, excessive vitamin doses are often wasted and sometimes harmful.

For example, the body simply will not hold more vitamin C than it needs for normal functioning. Vitamin C is not fat soluble and the excess is eliminated in urine. In fact, any amount above the recommended daily allowance ends up in the urine almost immediately after it is absorbed from the gut! Abnormal intake of at least two other vitamins (A and D) can cause serious disorders. Like all fat-soluble vitamins, vitamins A and D can accumulate in tissues and interfere with normal metabolic function.

Similarly, sodium is present in plant and animal tissues, and it is a component of table salt. Sodium has roles in the body's salt-water balance, muscle activity, and nerve function. Yet prolonged, excessive intake of sodium can lead to high blood pressure as described on page 381.

Severe shortage or massive excess of vitamins and minerals can disturb the delicate balances in body function that promote health.

NUTRITION AND METABOLISM

Figure 25.10 summarizes the main routes by which nutrient molecules are shuffled and reshuffled once they have been absorbed into the body. With few exceptions (such as DNA), most carbohydrates, lipids, and proteins are broken down continually, with their component parts picked up and used again in new molecules. At the molecular level, your body undergoes massive and sometimes rapid turnover.

When you eat, the body builds up its pools of organic molecules. Excess carbohydrates and other dietary molecules are transformed mostly into fats, which are stored in adipose tissue. Some also is converted to glycogen in the liver and in muscle tissue. Most cells use glucose as their main energy source at this time; there is no net breakdown of protein in muscle or other tissues.

Between meals, there is a notable shift in the type of food molecules used to support cell activities. A key factor in this shift is the need to provide brain cells with glucose, the major nutrient they use for energy.

When glucose is being absorbed, its blood levels are readily maintained. How does the body maintain blood

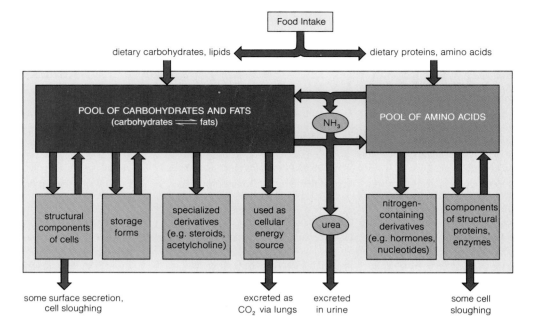

Figure 25.10 Summary of major pathways of organic metabolism. Urea formation occurs primarily in the liver. Carbohydrates, fats, and proteins are continually being broken down and resynthesized.

glucose concentrations between meals? *First,* glycogen stores, mainly in the liver, are rapidly broken down to glucose, which is released into blood. *Second,* body proteins are broken down to amino acids, which are sent to the liver for conversion to glucose that is released into the blood. The nervous system and endocrine system interact to control these and other aspects of organic metabolism (page 456).

Most cells use fats as the main energy source between meals. Fats stored in adipose tissue are broken down into glycerol and fatty acids, which are released into blood. The glycerol can be converted to glucose in the liver; the circulating fatty acids can be used in ATP production.

During a meal, glucose moves into cells, where it can be used for energy and where the excess can be stored.

Between meals, many cells use fat as the main energy source. Stored fats are mobilized, and brain cells are kept supplied with glucose (their major energy source).

As you can see, the liver is central to the storage and interconversion of absorbed carbohydrates, lipids, and proteins. Keep in mind that it has other vital roles (Table 25.6). For example, it helps maintain the concentrations of blood's organic substances and removes many toxic substances from it. The liver inactivates most hormone molecules and sends them on to the kidneys for excretion (in urine). Also, ammonia (NH_3) is produced when cells break down amino acids, and it can be toxic. The liver converts ammonia to urea. This less-toxic waste product leaves the body in urine.

SUMMARY

1. Nutrition includes all the processes by which the body takes in, digests, absorbs, and uses food.

2. A digestive system breaks down food molecules by mechanical and enzymatic means. It also absorbs the breakdown products into the internal environment, and it eliminates the unabsorbed residues.

3. The human digestive system includes the mouth, pharynx, esophagus, stomach, small intestine, large intestine (colon), rectum, and anus. Glands associated with digestion are the salivary glands, liver, gallbladder, and pancreas.

4. Controls over the digestive system operate in response to the volume and composition of food passing through. The response can be a change in muscle activity, the secretion rate of hormones and enzymes, or both.

5. Starch digestion begins in the mouth; protein digestion begins in the stomach. Digestion is completed and most nutrients are absorbed in the small intestine. Following absorption, monosaccharides (including glucose) and most amino acids are sent directly to the liver. Fatty acids and monoglycerides are combined as triglycerides and enter lymph vessels.

6. To maintain acceptable weight and overall health, caloric intake must balance energy output. Complex carbohydrates are the body's main energy source. The body produces fats as storage forms of carbohydrates and proteins. Eight essential amino acids, a few essential fatty acids, vitamins, and minerals must be provided by the diet.

Table 25.6	Some Activities That Depend on Liver Functioning

1. Carbohydrate metabolism
2. Control over some aspects of plasma protein synthesis
3. Assembly and disassembly of certain proteins
4. Urea formation from nitrogen-containing wastes
5. Assembly and storage of some fats
6. Fat digestion (bile is formed by the liver)
7. Inactivation of many chemicals (such as hormones and some drugs)
8. Detoxification of many poisons
9. Degradation of worn-out red blood cells
10. Immune response (removal of some foreign particles)
11. Red blood cell formation (liver absorbs, stores factors needed for red blood cell maturation)

Review Questions

1. Study Figure 25.1. Then, on your own, diagram the connections between metabolism and the digestive, circulatory, and respiratory systems. *360*

2. What are the main functions of the stomach? The small intestine? The large intestine? *362*

3. Name four kinds of breakdown products that are actually small enough to be absorbed across the intestinal lining and into the internal environment. *364–365*

4. A glass of milk contains lactose, protein, butterfat, vitamins, and minerals. Explain what happens to each component when it passes through your digestive tract. *365*

5. Describe some of the reasons each of the following is nutritionally important: carbohydrates, fats, proteins, vitamins, and minerals. *367–371*

Self-Quiz *(Answers in Appendix IV)*

1. The _____, _____, and _____ interact in supplying body cells with raw materials, disposing of wastes, and maintaining the volume and composition of extracellular fluid.

2. Different specialized regions of the digestive system function in _____ and _____ food, and in _____ unabsorbed food residues.

3. Maintaining good health and normal body weight requires that _____ intake be balanced by _____ output.

4. The main energy sources for the body are complex _____.

5. The human body cannot produce its own vitamins or minerals, and it also cannot produce certain _____ and _____.

6. Which glands are *not* associated with digestion?
   a. salivary glands
   b. thymus gland
   c. liver
   d. gallbladder
   e. pancreas

7. Digestion is completed and breakdown products are absorbed in the _____.
   a. mouth
   b. stomach
   c. small intestine
   d. large intestine

8. After absorption, fatty acids and monoglycerides move into the _____.
   a. bloodstream
   b. intestinal cells
   c. liver
   d. lymph vessels

9. _____ are storage forms of excess carbohydrates and proteins.
   a. Amino acids
   b. Starches
   c. Fats
   d. Monosaccharides

10. Match each digestive system component with its description.
    ___ liver
    ___ small intestine
    ___ human digestive system
    ___ nutrition
    ___ digestive system controls
    a. begins at mouth, ends at anus
    b. operate in response to food volume and composition
    c. functions are digestion, absorption, use of food
    d. where most digestion is completed
    e. receives monosaccharides and amino acids

Selected Key Terms

absorption *361*
digestion *361*
digestive system *360*
esophagus *363*
essential amino acid *368*
essential fatty acid *368*
gastric fluid *363*
kilocalorie *366*
large intestine (colon) *366*
liver *364*
microvillus *365*

mineral *370*
net protein utilization (NPU) *369*
obesity *367*
pancreas *364*
peristalsis *362*
salivary gland *363*
small intestine *364*
sphincter *362*
stomach *363*
villus *365*
vitamin *370*

Readings

Campbell-Platt, G. May 1988. "The Food We Eat." *New Scientist,* 19:1–4.

Cohen, L. 1987. "Diet and Cancer." *Scientific American* 257(5):42–68.

Hamilton, W. 1985. *Nutrition: Concepts and Controversies.* Third edition. Menlo Park, California: West.

Krause, M., and L. Mahan. 1984. *Food, Nutrition, and Diet Therapy.* Seventh edition. Philadelphia: Saunders.

26 CIRCULATION

1. Cells survive by exchanging substances with their surroundings. In complex animals, a closed circulatory system allows rapid movement of substances to and from all living cells. Most systems consist of a heart, blood, and blood vessels, and they are supplemented by a lymph vascular system.

2. The human body has two circuits for blood. In the pulmonary circuit, the heart pumps blood from body regions that have used up oxygen to the lungs (where it picks up oxygen), then the oxygen-enriched blood flows back to the heart. In the systemic circuit, the heart pumps oxygen-enriched blood to all body regions, then the oxygen-depleted blood flows back to the heart.

3. Arteries and veins are large-diameter transport tubes. Capillaries and venules, fine-diameter tubes, function in diffusion. Arterioles, with adjustable diameters, serve as control points for the distribution of different volumes of blood to different regions. (Metabolically active regions get more of the total volume at a given time.)

CIRCULATORY SYSTEM: AN OVERVIEW

Imagine what would happen if an earthquake or flood closed off the highways around your neighborhood. Grocery trucks couldn't enter and waste-disposal trucks couldn't leave—so food supplies would dwindle and garbage would pile up. Every living cell in your body would face similar predicaments if your body's highways were disrupted. These highways are part of the **circulatory system,** which functions in the rapid internal transport of substances to and from cells. Together with the other organ systems shown in Figure 26.1, the circulatory system helps maintain favorable neighborhood conditions, so to speak.

In most invertebrates and all vertebrates, the circulatory system has these components:

blood   *a fluid connective tissue composed of water, solutes, and formed elements (for example, blood cells and platelets)*

heart   *a muscular pump that generates the pressure required to keep blood flowing through the body*

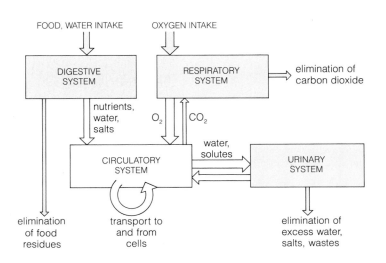

FOOD, WATER INTAKE   OXYGEN INTAKE

DIGESTIVE SYSTEM

RESPIRATORY SYSTEM → elimination of carbon dioxide

nutrients, water, salts

O_2 CO_2

CIRCULATORY SYSTEM

water, solutes

URINARY SYSTEM

elimination of food residues

transport to and from cells

elimination of excess water, salts, wastes

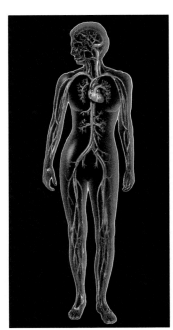

Figure 26.1 The central role of the circulatory system in transporting substances to and from the body's living cells. Together with the other systems shown, it helps maintain favorable operating conditions in the internal environment.

| blood vessels | *tubes of different diameters through which blood is transported (for example, large arteries and fine capillaries)* |

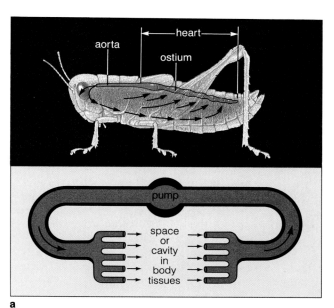

Circulatory systems can be open or closed. Arthropods and most mollusks have an open system, with fluid being pumped from the heart into tubes that dump it into a space or cavity in body tissues (Figure 26.2a). There, blood mingles with tissue fluids. The fluid has nowhere to go except through open-ended tubes leading back to the heart, which pumps it out again.

Most animals have a closed system, with the walls of the heart and blood vessels continuously connected. Before considering the components of this system, think about its overall "design." A heart pumps constantly, so the *volume* of flow through each part of the system is also constant. Yet the *rate* of flow to individual organs must be adjusted along the route. Blood flows rapidly through the large-diameter vessels, and it must be slowed somewhere in the system to allow enough time for substances to diffuse to and from cells. As you will see, the flow is rather leisurely at *capillary beds*, where the volume of blood is funneled through vast numbers of small-diameter tubes. As a result, the flow rate drops in each tube—and there is time enough for diffusion.

Finally, keep in mind that even a closed circulatory system is not completely sealed off. A small amount of fluid continuously filters out of capillaries, and materials continuously pass between capillaries and surrounding tissues. A supplementary network of tubes, the **lymph vascular system,** picks up excess fluid, reclaims proteins from tissues, and returns both to the circulatory system.

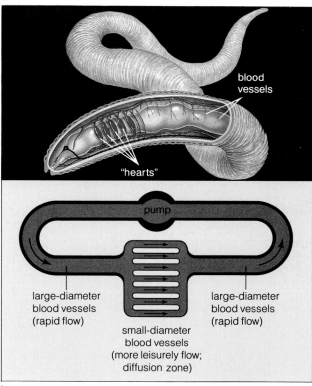

Figure 26.2 (**a**) Fluid flow through an open circulatory system. A grasshopper, for example, has a "heart" that pumps blood through a vessel (aorta), which dumps the blood into body tissues. Blood diffuses through the tissues, then back into the heart through openings (ostia).

(**b**) Fluid flow through a closed circulatory system. An example is the earthworm, with blood vessels leading away from and back to several muscular "hearts" near its head end. The walls of the hearts and blood vessels interconnect.

8 micrometer average diameter

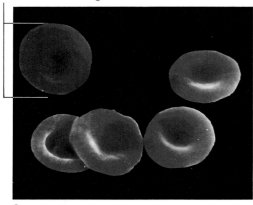

a

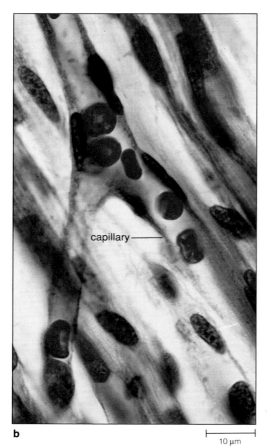

capillary———

b

|— 10 µm —|

Figure 26.3 (**a**) Scanning electron micrograph showing that red blood cells are concave on both sides. (**b**) Photomicrograph of red blood cells in capillaries.

Table 26.1 Components of Blood

Components	Relative Amounts	Functions
Plasma Portion (50%–60% of total volume):		
1. Water	91%–92% of plasma volume	Solvent
2. Plasma proteins (albumin, globulins, fibrinogen, etc.)	7%–8%	Defense, clotting, lipid transport, roles in extracellular fluid volume, etc.
3. Ions, sugars, lipids, amino acids, hormones, vitamins, dissolved gases	1%–2%	Roles in extracellular fluid volume, pH, etc.
Cellular Portion (40%–50% of total volume):		
1. Red blood cells	4,500,000–5,500,000 per microliter	O_2, CO_2 transport
2. White blood cells:		
Neutrophils	3,000–6,750	Phagocytosis
Lymphocytes	1,000–2,700	Immunity
Monocytes (macrophages)	150–720	Phagocytosis
Eosinophils	100–360	Roles in inflammatory response, immunity
Basophils	25–90	Roles in inflammatory response, anticlotting
3. Platelets	250,000–300,000	Roles in clotting

CHARACTERISTICS OF BLOOD

Functions of Blood

Blood itself has multiple functions. It carries oxygen as well as nutrients to cells, and it carries secretions (including hormones) and metabolic wastes away from them. Phagocytic cells travel the blood highways as mobile scavengers and infection fighters. Blood helps stabilize internal pH. In birds and mammals, blood helps equalize body temperature by carrying excess heat from regions of high metabolic activity (such as skeletal muscles) to the skin, where heat can be dissipated.

In most animals, blood is a transport fluid that carries raw materials to cells, carries products and wastes from them, and helps maintain an internal environment that is favorable for cell activities.

Blood Volume and Composition

The volume of blood in a person depends on body size and on the concentrations of water and solutes. Blood volume generally is about 5 liters, or a little more than 5 quarts, in an adult male of average weight.

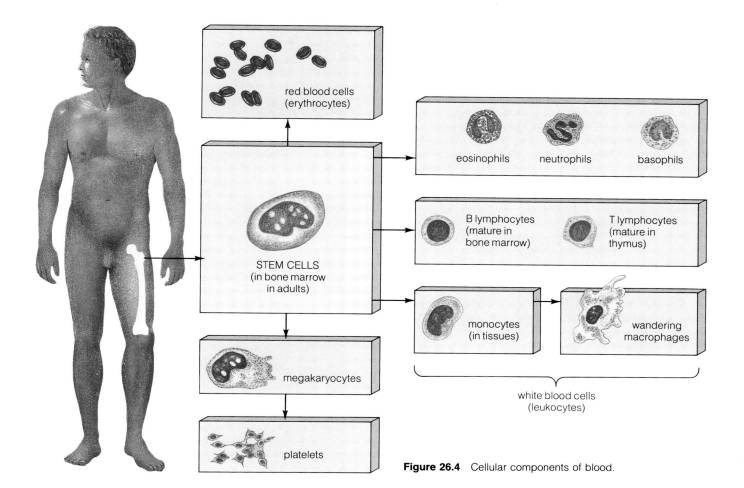

Figure 26.4 Cellular components of blood.

Human blood consists of red blood cells, white blood cells, platelets, and plasma (Table 26.1). Normally, plasma accounts for 50 to 60 percent of the total blood volume.

Plasma. The portion of blood called **plasma** is mostly water, which functions as a solvent. Dispersed through this straw-colored liquid are hundreds of different *plasma proteins*. Some proteins transport lipids and fat-soluble vitamins; others function in immune responses and blood clotting. Plasma also contains ions, glucose and other simple sugars, lipids, amino acids, vitamins, hormones, and dissolved gases (mostly oxygen, carbon dioxide, and nitrogen).

Red Blood Cells. Erythrocytes, or **red blood cells,** transport oxygen to cells. Your own red blood cells are shaped like doughnuts without the hole (Figure 26.3). Their red color comes from hemoglobin, an iron-containing protein (page 33). When oxygen from the respiratory system diffuses into the bloodstream, it binds with the hemoglobin. Oxygenated blood is bright red. Blood somewhat depleted of oxygen is darker red but

appears blue when observed through blood vessel walls (hence the "blue" veins visible at your wrists). Hemoglobin also transports some carbon dioxide wastes of aerobic metabolism.

Red blood cells form in bone marrow. Although they no longer have a nucleus at maturity, they remain functional for about 120 days. Phagocytic cells continually remove the oldest red blood cells from the bloodstream, but ongoing replacements keep the total number of red blood cells fairly stable in healthy adults.

White Blood Cells. Leukocytes, or **white blood cells,** function in day-to-day housekeeping and defense. Some scavenge dead or worn-out cells; others respond to tissue damage and invasion by bacteria, viruses, and other foreign agents. All white blood cells arise from immature cells (stem cells) in bone marrow. They travel the circulation highways, but they perform most housekeeping and defense functions after they squeeze out of blood capillaries and enter tissues.

There are five types of white blood cells, based on differences in size, nuclear shape, and staining traits.

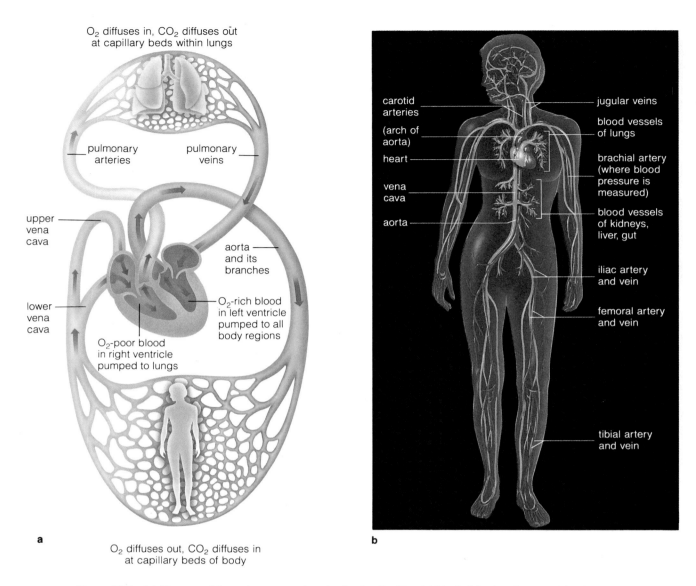

O₂ diffuses in, CO₂ diffuses out
at capillary beds within lungs

pulmonary
arteries

pulmonary
veins

upper
vena
cava

aorta
and its
branches

lower
vena
cava

O₂-rich blood
in left ventricle
pumped to all
body regions

O₂-poor blood
in right ventricle
pumped to lungs

O₂ diffuses out, CO₂ diffuses in
at capillary beds of body

a

carotid
arteries

(arch of
aorta)

heart

vena
cava

aorta

jugular veins

blood vessels
of lungs

brachial artery
(where blood
pressure is
measured)

blood vessels
of kidneys,
liver, gut

iliac artery
and vein

femoral artery
and vein

tibial artery
and vein

b

Figure 26.5 (**a**) Diagram of the pulmonary and systemic circuits. The right half of the heart pumps blood through the pulmonary circuit; the left half, through the systemic circuit. (**b**) Human circulatory system, showing the locations of the heart and some major blood vessels.

They are lymphocytes, neutrophils, monocytes, eosinophils, and basophils (Table 26.1 and Figure 26.4). Two major classes of lymphocytes, the "B cells" and "T cells," are central to immune responses, which are described in the next chapter.

Platelets. In bone marrow, some stem cells develop into "giant" cells (megakaryocytes), which then rupture. The cell fragments are **platelets.** Platelets help prevent blood loss from damaged blood vessels by releasing substances that have roles in clot formation.

HUMAN CIRCULATORY SYSTEM

Blood Circulation Routes

The human heart is divided into right and left halves, and this division is the basis of two cardiovascular circuits through the body. ("Cardiovascular" comes from the Greek *kardia*, meaning heart; and the Latin *vasculum*, meaning vessel.) In the **pulmonary circuit,** blood from the right half of the heart is pumped to the lungs, where it picks up oxygen and gives up carbon dioxide; then it flows to the left half of the heart. In the **systemic circuit,**

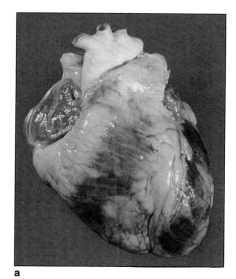

a

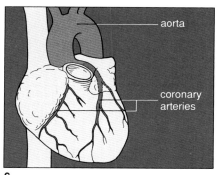

aorta

coronary
arteries

c

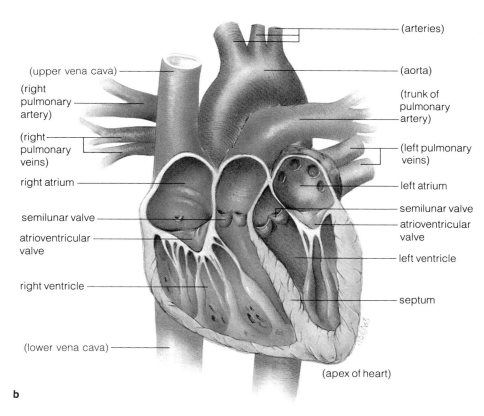

(arteries)

(aorta)

(upper vena cava)

(right pulmonary artery)

(trunk of pulmonary artery)

(right pulmonary veins)

(left pulmonary veins)

right atrium

left atrium

semilunar valve

semilunar valve

atrioventricular valve

atrioventricular valve

left ventricle

right ventricle

septum

(lower vena cava)

(apex of heart)

b

Figure 26.6 The human heart, external view (**a**) and partial view of the interior (**b**). Location of the coronary arteries (**c**).

the oxygen-enriched blood is pumped through the rest of the body (where oxygen is used and carbon dioxide is produced), then it flows to the right half of the heart. Both circuits begin with *arteries,* then continue with *arterioles, capillaries, venules,* and finally *veins.*

As Figure 26.5 suggests, a given volume of blood making either circuit generally passes through only one capillary bed. An exception is the blood passing through capillary beds in the digestive tract, where it picks up glucose and other substances absorbed from food. That blood moves on through another capillary bed, in the liver—an organ with a key role in nutrition. The decreased flow rate through this second bed gives the liver more time to monitor blood concentrations of absorbed substances.

The Human Heart

Heart Structure. During a seventy-year life span, the human heart beats some $2\frac{1}{2}$ billion times, and it rests only briefly between heartbeats. Its structure reflects its

role as a durable pump. The heart is mostly cardiac muscle tissue (Figure 26.6). Its inner chambers are lined with connective tissue and endothelium. (Endothelium is a layer of epithelial cells found only in the heart and blood vessels.)

Each half of the heart has two chambers called an *atrium* (plural, atria) and a *ventricle*. Between the atrium and ventricle is a membrane flap, or AV valve (short for "atrioventricular"). Another flap, the semilunar valve, is located between the ventricle and an artery leading away from the heart. Both types of valves help keep blood moving in one direction and so prevent backflow into the chambers.

Heart muscle cells are not serviced by the blood moving inside it. The heart has its own "coronary circulation," with two arteries leading into a large capillary bed (Figure 26.6c). Coronary arteries are the first to branch off the aorta, the major artery carrying oxygen-enriched blood away from the heart. The small diameter of coronary arteries can become clogged during some cardiovascular disorders, as described in the *Commentary* on page 381.

Cardiac Cycle. Each time the heart beats, its four chambers go through phases of contraction (systole) and relaxation (diastole). This sequence of muscle contraction and relaxation is a *cardiac cycle* (Figure 26.7).

When relaxed atria are filling, fluid pressure rises inside and forces the AV valves to open—so the ventricles fill completely when the atria contract. The AV valves snap shut when the filled ventricles contract, so pressure inside the ventricles rises sharply above that in blood vessels leading out from the heart. The semilunar valves open under the pressure, and blood flows out of the heart. After blood has been ejected, the cycle starts over.

During a cardiac cycle, atrial contraction simply helps fill the ventricles. *Ventricular* contraction is the driving force for blood circulation.

Mechanisms of Contraction. In skeletal muscle tissue, muscle cells are attached to bones. In cardiac muscle tissue, the cells branch and then connect with one another at their endings (Figure 26.8). Communication junctions occur where the plasma membranes of cardiac muscle cells are fused together. With each heartbeat, signals calling for contraction spread so rapidly across the junctions that the cells contract together as if they were a single unit.

Cardiac and skeletal muscle cells differ in another way. Signals from the nervous system bring about skeletal muscle contraction. But the nervous system can only *adjust* the rate and strength of cardiac muscle contraction. Even if all nerves leading to the heart are severed, the heart will keep on beating! How? Some cardiac muscle cells are self-excitatory; they produce and conduct signals for contraction (Figure 26.9).

Excitation begins in the SA node (short for "sinoatrial") of the right atrium. One wave of excitation follows another, producing the normal heart rate of 70 or 80 times a minute. Each wave spreads over both atria, causes them to contract, then reaches the AV node (for "atrioventricular"). The wave spreads more slowly here. The delay gives the atria time to finish contracting before the signal spreads over the ventricles.

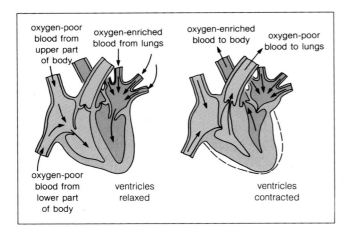

Figure 26.7 Blood flow through the heart during part of a cardiac cycle. The blood and heart movements generate vibrations, producing a "lub-dup" sound that can be heard at the chest wall. At each "lub," AV valves are closing as the ventricles contract. At each "dup," semilunar valves are closing as the ventricles relax.

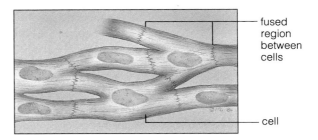

Figure 26.8 End-to-end regions between cardiac muscle cells. Communication junctions at these regions permit rapid signaling between cells that causes them to contract in unison.

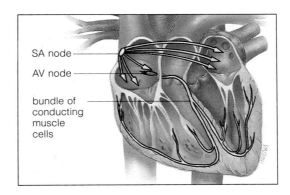

Figure 26.9 Location of cardiac muscle cells that conduct signals for contraction through the heart.

On Cardiovascular Disorders

More than 40 million Americans have cardiovascular disorders, which claim about a million lives every year. The most common cardiovascular disorders are *hypertension* (sustained high blood pressure) and *atherosclerosis* (a progressive narrowing of the arterial lumen). They are the major causes of most *heart attacks*—that is, the damage or death of heart muscle due to an interruption of its blood supply. (They also can cause *stroke*, or damage to the brain due to an interruption of blood circulation to it.)

Most heart attacks bring a "crushing" pain behind the breastbone that lasts a half hour or more. Frequently, the pain radiates into the left arm, shoulder, or neck. The pain can be mild but usually is excruciating. Often it is accompanied by sweating, nausea, vomiting, and dizziness or loss of consciousness.

Risk Factors in Cardiovascular Disorders

Cardiovascular disorders are the leading cause of death in the United States. Curiously, many factors associated with those disorders have been identified *and are controllable.* These are the known risk factors:

1. High level of cholesterol in the blood
2. High blood pressure
3. Obesity (page 367)
4. Lack of regular exercise
5. Smoking (page 414)
6. Diabetes mellitus (page 456)
7. Genetic predisposition to heart failure
8. Age (the older you get, the greater the risk)
9. Gender (until age fifty, males are at much greater risk than are females)

The last four factors obviously cannot be avoided; but the first five can be. The risk associated with all five can be minimized simply by watching your diet, exercising, and not smoking.

For example, the fatter you become, the more your body develops additional blood capillaries to service the increased number of cells, and the harder the heart has to work to pump blood through the increasingly divided vascular circuit. As another example, the nicotine in tobacco stimulates the adrenal glands to secrete epinephrine, which constricts blood vessels and so triggers an accelerated heartbeat and a rise in blood pressure. The carbon monoxide present in cigarette smoke has a greater affinity for binding sites on hemoglobin than does carbon dioxide—and its action means that the heart has to pump harder to rid the body of carbon dioxide wastes. In short, smoking can destroy not only your lungs, but also your heart.

The following descriptions will provide insight into the tissue destruction that can result from cardiovascular disorders.

Hypertension

Hypertension arises through a gradual increase in resistance to blood flow through the small arteries; eventually, blood pressure is sustained at elevated levels even when the person is at rest. Heredity may be a factor here (the disorder tends to run in families). Diet also is a factor; for example, high salt intake can raise the blood pressure in persons predisposed to the disorder. High blood pressure increases the workload of the heart, which in time can become enlarged and fail to pump blood effectively. High blood pressure also can cause arterial walls to "harden" and so influence the delivery of oxygen to the brain, heart, and other vital organs.

Hypertension has been called the silent killer because affected persons may show no outward symptoms; they often believe they are in the best of health. Even when their high blood pressure has been detected, some hypertensive persons tend to resist medication, corrective changes in diet, and regular exercise. Of 23 million Americans who are hypertensive, most are not undergoing treatment. About 180,000 will die each year.

Atherosclerosis

"Arteriosclerosis" refers to a condition in which arteries thicken and lose their elasticity. In atherosclerosis, conditions worsen because lipid deposits also build up in the arterial walls and shrink the diameter of the arterial lumen. How does this occur?

Recall that lipids such as fats and cholesterol are insoluble in water (page 29). Lipids absorbed from the digestive tract are picked up by lymph vessels that empty into the bloodstream. There, the lipids become bound to protein carriers that keep them suspended in the blood plasma. In atherosclerosis, abnormal smooth muscle cells have multiplied and connective tissue components have increased in arterial walls. Lipids have been deposited within cells and extracellular spaces of

the wall's endothelial lining. Calcium salts have been deposited on top of the lipids, and a fibrous net has formed over the whole mass. This *atherosclerotic plaque* sticks out into the lumen of the artery (Figures a,b).

Sometimes platelets become caught on the rough edges of plaques and are stimulated into secreting some of their chemicals. When they do, they initiate clot formation. As the clot and plaque grow, the artery can become narrowed or blocked. Blood flow to the tissue that the artery supplies diminishes or may be blocked entirely. A clot that stays in place is called a *thrombus;* if it becomes dislodged and travels the bloodstream, it is called an *embolus.*

With their narrow diameter, the coronary arteries and their branches (Figure c) are extremely susceptible to clogging through plaque formation or occlusion by a clot. When such an artery becomes narrowed to one-quarter of its former diameter, the resulting symptoms can range from mild chest pains (angina pectoris) to a full-scale heart attack.

Atherosclerosis can be diagnosed on the basis of several procedures. These include stress electrocardiograms, or EKGs (recording the electrical activity of the cardiac cycle while a person is exercising on a treadmill) and *angiography* (injecting a dye that will stain plaques and then taking x-rays of the arteries; see Figure a). Treatments of serious blockages include *coronary bypass surgery.* During this operation, a section of an artery from the chest is stitched to the aorta and to the coronary

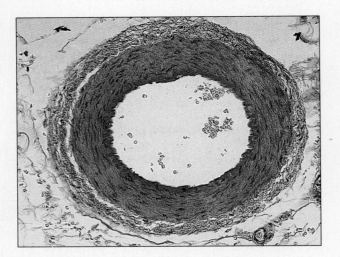

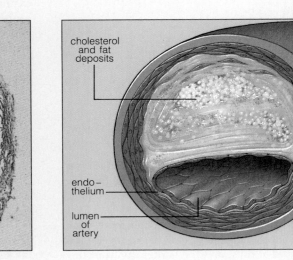

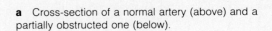

a Cross-section of a normal artery (above) and a partially obstructed one (below).

b Diagram of an atherosclerotic plaque.

artery below the narrowed or blocked region (Figure c). In another technique, called *laser angioplasty*, highly focused laser beams are used to vaporize the atherosclerotic plaques. *Balloon angioplasty* is more common; here, a small balloon is inflated within a blocked artery to break up plaques. All such procedures do not cure the underlying cardiovascular problem; they only buy time for the individual.

Plaque formation is related to cholesterol intake, but other factors are also at work here. For example, when cholesterol is transported through the bloodstream, it is bound to one of two kinds of protein carrier molecules: high-density lipoproteins (HDL) and low-density lipoproteins (LDL). High levels of LDL are related to a tendency toward heart trouble. LDLs, with their cholesterol cargo, have a penchant for infiltrating arterial walls. In contrast, the HDLs seem to attract cholesterol out of the walls and transport it to the liver, where it can be metabolized. (Atherosclerosis is uncommon in rats; rats have mostly HDLs. It is common in humans, who have mostly LDLs.) In addition, it appears that unsaturated fats, including olive oil and fish oil, can reduce the level of LDLs in the blood.

Arrhythmia

Arrhythmias are irregular or abnormal heart rhythms. They can be detected by an EKG (Figure d). Some arrhythmias are normal. For example, the resting cardiac rate of many athletes who are trained for endurance is lower than average, a condition called *bradycardia*. Inhibition of their cardiac pacemaker by the nervous system has increased as an adaptive response to ongoing strenuous exercise. A cardiac rate above 100 beats per minute (*tachycardia*) also occurs normally during exercise or stressful situations.

Serious tachycardia can be triggered by drugs (including caffeine, nicotine, and alcohol), hyperthyroidism, and other factors. Coronary artery disease also can cause arrhythmias.

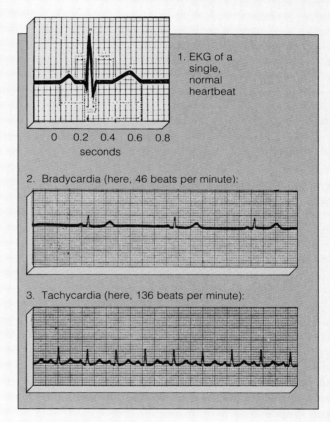

1. EKG of a single, normal heartbeat

2. Bradycardia (here, 46 beats per minute):

3. Tachycardia (here, 136 beats per minute):

d Examples of EKG readings.

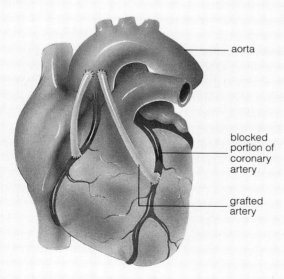

c Two coronary bypasses (green).

aorta

blocked portion of coronary artery

grafted artery

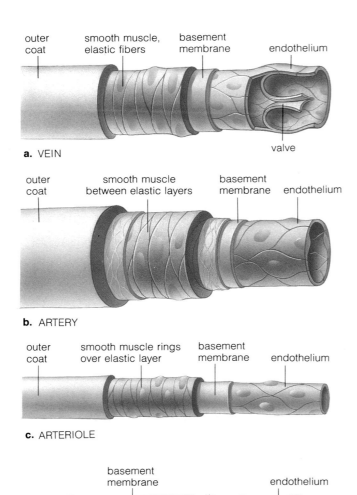

a. VEIN

b. ARTERY

c. ARTERIOLE

d. CAPILLARY

Figure 26.10 Structure of blood vessels. The basement membrane is collagen-containing connective tissue.

Blood Pressure in the Vascular System

Blood pressure, the fluid pressure generated by heart contractions, is not the same throughout the circulatory system. Pressures normally are high to begin with, then drop along the circuits away from and back to the heart. The pressure drops result from the loss of energy that is used to overcome resistance to flow as blood moves through blood vessels of the sort shown in Figure 26.10.

Arterial Blood Pressure. As we have seen, **arteries** conduct oxygen-poor blood into the lungs and oxygen-enriched blood to all body tissues. Arteries are pressure reservoirs that can "smooth out" the blood pressure changes of the cardiac cycle. Their thick, muscular wall

bulges somewhat under the pressure surges caused by ventricular contraction, then it recoils and forces blood onward. With their large diameters, arteries present little resistance to flow, so pressure does not drop much inside them. Figure 26.11 shows how blood pressure is measured at large arteries of the upper arms.

Resistance at Arterioles. Arteries branch into smaller diameter **arterioles.** The wall of an arteriole has rings of smooth muscle cells that respond to signals from the nervous system, endocrine system, or even changes in local chemical conditions. Different signals cause the arteriole diameter to increase or decrease. Thus, *arterioles serve as control points where adjustments can be made in the volume of blood to be delivered to different diffusion zones.* When you exercise, for example, arteriole diameter is enlarged in skeletal muscle tissues but decreased elsewhere. The adjustments allow more oxygen-enriched blood to be delivered to muscle cells that are rapidly using up oxygen for aerobic respiration. As Figure 26.12 shows, the greatest pressure drop in the blood circuit occurs at arterioles.

Capillary Function. Capillary beds are the diffusion zones for exchanges between blood and interstitial fluid; all other blood vessels are mainly transport tubes. A **capillary** has the thinnest wall, consisting mainly of a single layer of flat endothelial cells. Its diameter resists flow; it is so small that red blood cells squeeze through it single file (Figure 26.3). In a capillary bed, however, their combined diameters are greater than the diameters of arterioles leading into them. Thus capillaries present less total resistance to flow, and the drop in blood pressure is not great here.

Venous Pressure. Capillaries merge into "little veins," or **venules.** Some diffusion also occurs across the venule wall, which is only a little thicker than that of a capillary. Venules merge into large-diameter **veins,** the transport tubes leading back to the heart. Blood movement is assisted by valves inside the veins. When blood starts moving backward because of gravity, it pushes the valves closed and so prevents backflow (Figure 26.13).

With their large diameters, veins are low-resistance transport tubes. They also are blood volume reservoirs. Veins contain fifty to sixty percent of the total blood volume. The wall of a vein is thin and contains smooth muscle—and it can bulge more than an arterial wall. When body activities increase, blood must circulate faster. The smooth muscle cells in vein walls contract, the walls stiffen and don't bulge as much, so pressure in the veins rises and drives more blood to the heart. The filling pressure becomes greater in the heart—and the increased blood volume returned to the heart is now ejected.

Hemostasis

Don't even think about what would happen if the body could not repair ruptures or cuts even in its small blood vessels. Stopping the blood loss from damaged blood vessels, or **hemostasis,** requires blood vessel spasm, platelet plug formation, blood coagulation, and other mechanisms. First, smooth muscle in the wall of a damaged blood vessel contracts in a response called a spasm. The blood vessel constricts and the flow of blood is temporarily slowed or stopped. Second, platelets clump together, temporarily plugging the rupture. They also release substances that help prolong the spasm and attract more platelets. Third, blood coagulates—it converts to a gel—and forms a clot (Figure 26.14). Finally, the clot retracts into a compact mass, drawing the ruptured walls of the vessel together.

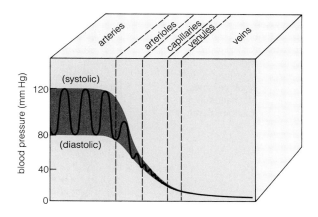

Figure 26.12 Drops in blood pressure in the systemic circulation.

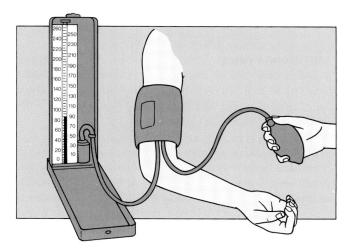

Figure 26.11 Measuring blood pressure with a sphygmomanometer. A hollow cuff, attached to a pressure gauge, is wrapped around the upper arm and inflated with air to a pressure above the highest pressure of the cardiac cycle (at systole, when the ventricles contract). Above the systolic pressure, no sounds can be heard through a stethoscope positioned above the artery.

Air in the cuff is slowly released, allowing some blood to flow into the artery. The turbulent flow causes soft tapping sounds, and when this first occurs, the value on the gauge is the systolic pressure—about 120mm Hg in young adults at rest. (This means the measured pressure would make a column of mercury rise a distance of 120 millimeters.)

More air is released until the sounds become dull and muffled. Just after this occurrence, blood flow is continuous; the turbulence and tapping sounds stop. The silence corresponds to the diastolic pressure (at the end of a cardiac cycle, just before the heart pumps out blood again). Generally the reading is about 80mm Hg. In this example, the *pulse pressure* (the difference between the highest and lowest pressure readings) is 120 − 80, or 40mm Hg.

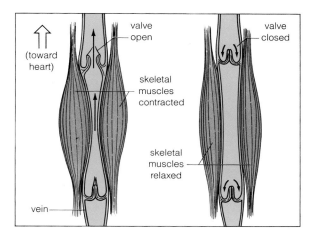

Figure 26.13 Role of skeletal muscle contractions and venous valves in returning blood to the heart.

Figure 26.14 Red blood cell caught in a fibrin net. Action by different enzymes leads to the formation of fibrin threads, which form a net at cuts or ruptured blood vessels. The net entangles blood cells and platelets and so forms a clot.

Figure 26.15 (**a**) Agglutination responses in blood types O, A, B, and AB when mixed with blood samples of the same and different types. (**b**) Micrographs showing the absence of agglutination in a mixture of two different but compatible types (above) and agglutination in a mixture of incompatible blood types (below).

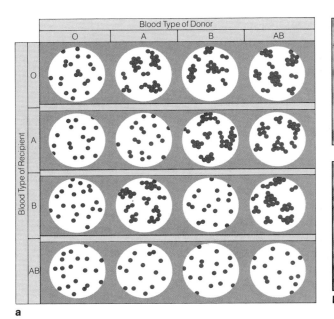

a

Figure 26.16 Development of antibodies in response to Rh⁺ blood.

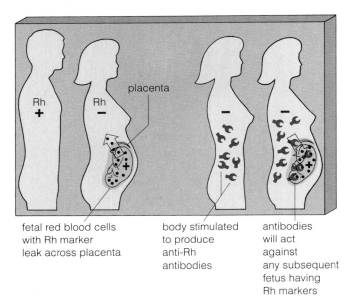

fetal red blood cells with Rh marker leak across placenta

body stimulated to produce anti-Rh antibodies

antibodies will act against any subsequent fetus having Rh markers

Blood Typing

All of your cells carry membrane proteins at their surface that serve as "self" markers; they identify the cells as being part of your own body. Your body also has proteins called **antibodies,** which can recognize markers on *foreign* cells (page 394). When the blood of two people mixes during transfusions, antibodies will act against any cells bearing the "wrong" marker. They will do the same thing during pregnancy, when antibodies diffuse from the mother's circulation system to that of her unborn child.

ABO Blood Typing. As we have seen, *ABO blood typing* is based on some of the surface markers on red blood cells (page 123). Type A blood has A markers on those cells, type B blood has B markers, type AB has both, and type O has neither one.

If you are type A, you do not carry antibodies against A markers—but you have antibodies against B markers. If you are type B, you have antibodies against A but not B markers. If you are type AB, you have no antibodies against A or B, so your body will tolerate donations of types A, B, AB, or O. If you are type O, however, you have antibodies against A and B markers—and those antibodies will act against cells bearing one or both types.

Figure 26.15 shows what happens when blood from different types of donors and recipients is mixed together. In a response called **agglutination,** antibodies act against the "foreign" cells and cause them to clump. Such clumps can clog small blood vessels; they may lead to kidney damage and death.

Rh Blood Typing. Other surface markers on red blood cells also can cause agglutination responses. For example, *Rh blood typing* is based on the presence or absence of an Rh marker (so named because it was first identified in the blood of *Rhesus* monkeys). Rh⁺ individuals have blood cells with this marker; Rh⁻ individuals do not. Ordinarily, people do not have antibodies that act against Rh markers. However, if someone has been given a transfusion of Rh⁺ blood, antibodies will be produced against it and will continue circulating in the bloodstream.

If an Rh⁻ female becomes pregnant by an Rh⁺ male, there is a chance the fetus will be Rh⁺. During pregnancy, some red blood cells of the fetus may leak into her bloodstream (page 477). If they do, they will stimulate her body into producing antibodies against the Rh markers (Figure 26.16). If the woman becomes pregnant *again*, Rh antibodies will enter the fetal bloodstream. If this second fetus happens to have Rh⁺ blood, the antibodies will cause red blood cells to swell and then rupture, releasing hemoglobin into the bloodstream.

In extreme cases of this disorder, called *erythroblastosis fetalis*, too many cells are destroyed and the fetus dies before birth. If it is born alive, all the blood of the newborn can be slowly replaced with blood free of the Rh antibodies. Currently, known Rh⁻ females can be treated right after their first pregnancy. A drug will protect her next fetus during pregnancy by inactivating any Rh antibodies circulating in her bloodstream.

LYMPHATIC SYSTEM

We conclude this chapter with a brief section on the **lymphatic system,** which supplements the circulatory system by returning excess tissue fluid to the bloodstream. This section is a bridge to the next chapter, on immunity, for the lymphatic system is also vital to the body's defenses against injury and attack.

The lymphatic system consists of transport vessels and lymphoid organs. The tubes make up the lymph vascular system, which supplements the pulmonary and systemic circuits. When tissue fluid has moved into these tubes, it is called **lymph.** The lymphoid organs, which take part in defense responses, are connected with both the blood and lymph vascular systems (Figure 26.17).

Lymph Vascular System

The **lymph vascular system** includes lymph capillaries, lymph vessels, and ducts that drain fluid back into the circulatory system. It serves these functions:

1. Return of excess filtered fluid to the blood.

2. Return of small amounts of proteins that leave the capillaries.

3. Transport of fats absorbed from the digestive tract.

4. Transport of foreign particles and cellular debris to disposal centers (lymph nodes).

At one end of the lymph vascular system are *lymph capillaries*, no larger in diameter than blood capillaries.

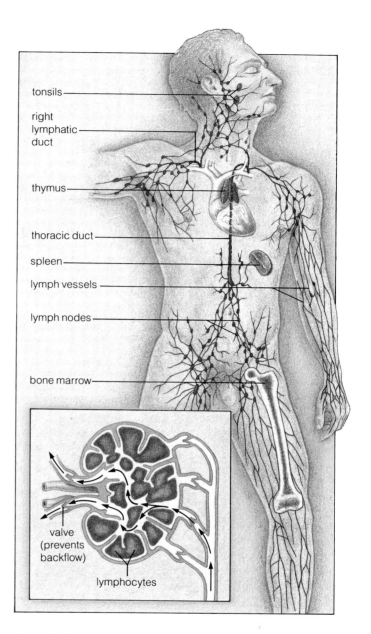

Figure 26.17 Lymphatic system, which includes the lymph vascular network and the lymphoid organs and tissues. Purple dots show some of the major lymph nodes. The inset illustrates the internal structure of a lymph node. Patches of lymphoid tissue in the small intestine and appendix also are part of the system.

They occur in the tissues of almost all organs and serve as "blind-end" tubes. They have no entrance at the end located in tissues; their only "opening" merges with larger lymph vessels (Figure 26.18).

Like veins, *lymph vessels* have smooth muscle in their walls and flaplike valves that prevent backflow. When you breathe, movements of the rib cage and skeletal muscle adjacent to the lymph vessels help move fluid through lymph vessels, just as they do for veins. Lymph vessels converge into collecting ducts, which drain into veins in the lower neck.

Lymphoid Organs

The **lymphoid organs** include the lymph nodes, spleen, thymus, tonsils, adenoids, and patches of tissue in the small intestine and appendix. These organs and tissue patches are production centers for infection-fighting cells, including lymphocytes; they also are sites for some defense responses.

Like all white blood cells, lymphocytes are derived from stem cells in bone marrow. The derivative cells enter the blood and take up residence in lymphoid organs. With proper stimulation, they divide by mitosis. (In fact, most new lymphocytes are produced by divisions in the blood and lymphoid organs, not in bone marrow.)

Lymph nodes are located at intervals along lymph vessels (Figure 26.17). All lymph trickles through at least one node before being delivered to the bloodstream. Each node has several inner chambers, and lymphocytes and plasma cells pack each chamber. Macrophages in the node help clear the lymph of bacteria, cellular debris, and other substances.

The largest lymphoid organ, the *spleen*, is a filtering station for blood and a holding station for lymphocytes.

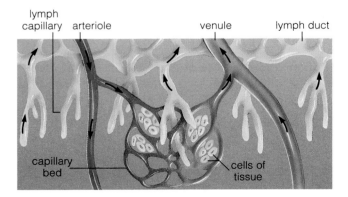

Figure 26.18 Lymph vessels near a capillary bed.

The spleen also has inner chambers, but these are filled with red and white "pulp." The red pulp contains large stores of red blood cells and macrophages. Red blood cells are produced here in developing human embryos.

The *thymus* secretes hormones concerned with the activity of lymphocytes. It also is a major organ where lymphocytes multiply, differentiate, and mature. The thymus is central to immunity, the focus of the chapter to follow.

SUMMARY

1. Multicelled animals have a circulatory system consisting of a muscular pump (heart or heartlike structure), blood, and blood vessels. In closed systems, blood flows inside the walls of these components; it exchanges substances with interstitial fluid only in diffusion zones.

2. Blood, a transport fluid, carries oxygen and other substances to cells; it also carries products and wastes (including carbon dioxide) from them. It helps maintain an internal environment favorable for cell activities. It consists of red and white blood cells, platelets, and plasma (water, plasma proteins, and diverse solutes, including dissolved gases).

3. The human heart is divided into two halves, each with two chambers (an atrium and a ventricle). The division is the basis of two cardiovascular circuits.
    a. In the pulmonary circuit, the right half of the heart pumps oxygen-poor blood to capillary beds inside the lungs, then oxygen-enriched blood flows back to the heart.
    b. In the systemic circuit, the left half of the heart pumps oxygen-enriched blood to all body regions, where it nourishes all tissues and organs, then oxygen-poor blood flows back to the heart.

4. Heart contractions (specifically, the contracting ventricles) are the driving force for blood circulation. Fluid pressure is high at the start of a circuit, then drops in arteries, arterioles, capillaries, then veins. It is lowest in the relaxed atria.

5. Arteries are pressure reservoirs that smooth out fluid pressure changes caused by heart contraction and relaxation. Arterioles are control points for the distribution of different volumes of blood to different body regions. Capillary beds are diffusion zones between the blood, interstitial fluid, and cells. Venules overlap capillaries and veins somewhat in function. Veins are blood volume reservoirs and help adjust volume flow back to the heart.

6. The lymph vascular system supplements the circulatory system by returning excess fluid that seeps out of blood vessels back to the circulation. Some of its components have major roles in immune responses.

Review Questions

1. What are some of the functions of blood? *376*

2. Describe the cellular components of blood. Describe the plasma portion of blood. *376*

3. Define the functions of the following:
   heart, cardiovascular system, and lymph vascular system *379–387*

4. Distinguish between the following:
   a. open and closed circulation *375*
   b. systemic and pulmonary circuits *378–379*
   c. lymph vascular system and lymphoid organs *387–388*

5. Label the component parts of the human heart: *379*

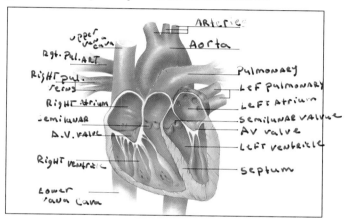

Self-Quiz   *(Answers in Appendix IV)*

1. In large, complex animals, a **Circulatory** system functions in the rapid exchange of substances to and from all living cells, and usually it is supplemented by a **lymph vascular** system.

2. **Arteries** and **veins** are large-diameter blood vessels for fluid transport; **Capillars** and **Venules** are fine-diameter, thin-walled blood vessels for diffusion; and **Arteios** serve as control points over the distribution of different blood volumes to different body regions.

3. Which of the following are *not* components of blood?
   a. red and white blood cells
   b. platelets and plasma
   c. assorted solutes and dissolved gases
   d. all of the above are components of blood

4. Red blood cells are produced in the _____ and function in transporting _____ and some _____.
   a. liver; oxygen; mineral ions
   b. liver; oxygen; carbon dioxide
   c. bone marrow; oxygen; hormones
   d. bone marrow; oxygen; carbon dioxide

5. White blood cells are produced in the _____ and function in both _____ and _____.
   a. liver; oxygen transport; defense
   b. lymph glands; oxygen transport; pH stabilization
   c. bone marrow; day-to-day housekeeping; defense
   d. bone marrow; pH stabilization; defense

6. In the pulmonary circuit, the _____ half of the heart pumps _____ blood to capillary beds inside the lungs, then _____ blood flows back to the heart.
   a. left; oxygen-poor; oxygen-enriched
   b. right; oxygen-poor; oxygen-enriched
   c. left; oxygen-enriched; oxygen-poor
   d. right; oxygen-enriched; oxygen-poor

7. In the systemic circuit, the _____ half of the heart pumps _____ blood to all body regions, then _____ blood flows back to the heart.
   a. left; oxygen-poor; oxygen-enriched
   b. right; oxygen-poor; oxygen-enriched
   c. left; oxygen-enriched; oxygen-poor
   d. right; oxygen-enriched; oxygen-poor

8. Fluid pressure in the circulatory system is _____ at the beginning of a circuit, then _____ in arteries, arterioles, capillaries, and then veins. It is _____ in the relaxed atria.
   a. low; raises; highest   c. low; drops; lowest
   b. high; drops; lowest   d. high; raises; highest

9. Match the type of blood vessel with its major function.
   __C__ arteries   a. diffusion
   __B__ arterioles   b. control of blood volume
   ____ capillaries   distribution
   __A__ venules   c. transport, blood volume reservoirs
   __C__ veins   d. overlap of capillary function
                               e. transport and pressure
                                  reservoirs

10. Match the circulation components with their descriptions.
   ____ capillary beds   a. two atria, two ventricles
   ____ lymph vascular   b. pressure reservoirs
        system   c. driving force for blood
   ____ heart chambers
   ____ veins   d. zones of diffusion
   ____ heart contractions   e. interstitial fluid
   ____ arteries   f. blood volume reservoirs

Selected Key Terms

aorta *379*	circulatory	plasma *377*
arteriole *384*	system *374*	platelet *378*
artery *384*	heart *374*	pulmonary circuit *378*
blood *374*	lymph *387*	red blood cell *377*
blood pressure *384*	lymph vascular	systemic circuit *378*
capillary *384*	system *387*	vein *384*
capillary bed *375*	lymphatic	venule *384*
cardiac cycle *380*	system *387*	white blood cell *377*

Readings

Brown, M. S., and J. L. Goldstein. November 1984. "How LDL Receptors Influence Cholesterol and Atherosclerosis." *Scientific American* 251(5):58–66.

Robinson, T. F., et al. June 1986. "The Heart as a Suction Pump." *Scientific American* 254(6)84–91.

Vander, A., J. Sherman, and D. Luciano. 1990. *Human Physiology.* Fifth edition. New York: McGraw-Hill.

27 IMMUNITY

KEY CONCEPTS

1. The vertebrate body defends itself against viruses, bacteria, and other foreign agents that enter the internal environment. Some defense responses are nonspecific, in that they occur when any kind of invasion is detected. Other responses are specific, with certain white blood cells being mobilized against a particular invader, not invaders in general.

2. White blood cells responsible for immune responses can distinguish between molecular markers on the body's own cells and antigens. An antigen is any large molecule that white blood cells perceive as foreign and that triggers an immune response. Different antigens occur at the surface of viruses, bacterial cells, and other foreign agents.

3. Antibody-mediated immune responses are made against antigens circulating in the body's tissues or attached to the surface of an invader. Cell-mediated immune responses are made against cells already infected, and possibly against mutant or cancerous cells. In most cases, both responses proceed simultaneously.

Until about a century ago, smallpox swept repeatedly through the world's cities. Some outbreaks were so intense that half or more of the people stricken with the contagious disease died. Those who survived had permanent facial scars. Intriguingly, the survivors were not vulnerable to subsequent attacks; they were "immune" to smallpox.

No one knew what caused smallpox. But in Asia, Africa, and then Europe, there were flurries of "inoculations," with healthy persons being intentionally infected with scrapings from sores of someone with a mild case of the disease. If all went well, inoculated persons did not get too sick and, from then on, they were immune to smallpox. But sometimes they became seriously ill. And sometimes they infected their family and friends—and triggered another epidemic.

While this immunological version of Russian roulette was going on, Edward Jenner was growing up in the English countryside. At the time it was known that cowpox, a rather mild disease, could be transmitted from cattle to humans. People who got cowpox never got smallpox. No one thought much about this until

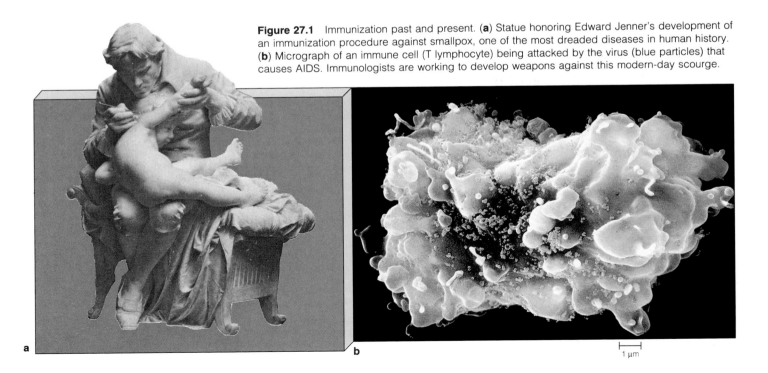

Figure 27.1 Immunization past and present. (**a**) Statue honoring Edward Jenner's development of an immunization procedure against smallpox, one of the most dreaded diseases in human history. (**b**) Micrograph of an immune cell (T lymphocyte) being attacked by the virus (blue particles) that causes AIDS. Immunologists are working to develop weapons against this modern-day scourge.

a

b

1 μm

1796, when Jenner, by then a physician, demonstrated the effectiveness of cowpox inoculations in providing immunity against smallpox (Figure 27.1). The French mocked Jenner's procedure, calling it "vaccination" (which translates as "encowment"). Much later, Louis Pasteur devised immunization procedures and also called them vaccinations; only then did the term become respectable.

Today we know the body defends itself against many *pathogens,* which are disease-causing agents such as viruses, bacteria, fungi, and protozoans. With his procedure, Jenner was actually mobilizing cells to make an immune response to a specific virus—one of the elegant defenses described in this chapter. Before we turn to the specific responses, however, let's start with the body's generalized defenses against attack.

NONSPECIFIC DEFENSE RESPONSES

Barriers to Invasion

The vertebrate body has impressive physical and chemical barriers against invasion, including the following:

1. Intact skin (very few bacteria can penetrate it).

2. Ciliated, mucous membranes in the respiratory tract (like sticky brooms, they sweep out airborne bacteria).

3. Secretions from exocrine glands in the skin, mouth, and elsewhere (one secretion, lysozyme, helps destroy the cell wall of many bacteria).

4. Gastric fluid (its acids destroy many pathogens that enter the stomach with food).

5. Normal bacterial inhabitants of the gut and, in females, the vagina (they outcompete many pathogens for resources and so help keep them in check).

Phagocytes

What happens when physical barriers to invasion are breached, as when skin is cut or scraped? Then, an invasion mobilizes phagocytic white blood cells. These cells engulf and destroy foreign agents. As Figure 26.4 shows, the phagocytic cells arise from stem cells in bone marrow. They include neutrophils, eosinophils, and monocytes that mature into macrophages, the "big eaters."

Phagocytes are strategically distributed cells. Some circulate within blood vessels, then enter damaged or invaded tissues by squeezing out through capillary walls. Some take up stations in lymph nodes and the spleen. (You may wish to review Figure 26.17, which shows the lymphatic system.) Other phagocytes are located in the liver, kidneys, lungs, and joints. The brain has its own phagocytic cells.

Complement System

When certain bacterial or fungal cells invade a tissue, about twenty plasma proteins interact as a system—the **complement system**—with roles in both nonspecific and specific defense responses. These circulating proteins are activated one after another in a cascade of reactions. (Each protein molecule switched on at the first reaction step switches on many more at the second step, each of which switches on many more at the third step, and so on.) The reactions have these results:

1. Chemical gradients (created by cascades of certain complement proteins) attract phagocytes to the scene.

2. Some complement proteins coat the surface of invading cells—and phagocytes zero in on the coat.

3. Other complement proteins help kill the pathogen by promoting lysis (they make the plasma membrane grossly "leaky").

Inflammation

Many cells and substances, including the complement proteins, take part in a series of events that destroy invaders and restore tissues and internal operating conditions to normal. These events, collectively called an **inflammatory response,** proceed during nonspecific *and* specific defense responses.

For example, while the complement system is gearing up, circulating basophils (and mast cells, their counterparts in tissues) release *histamine.* This potent substance

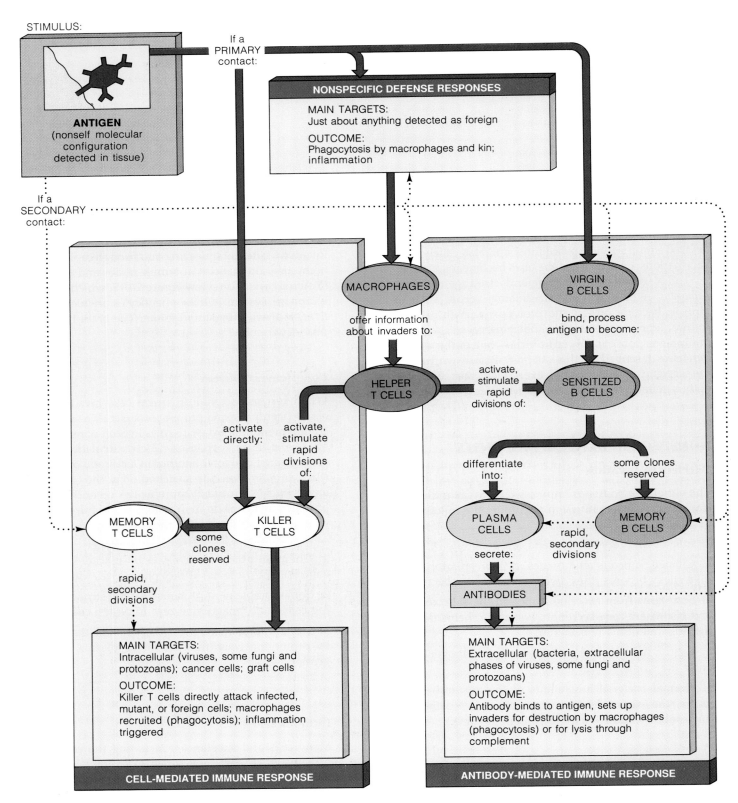

Figure 27.2 Overview of the cell-mediated and antibody-mediated branches of the vertebrate immune system. Brown arrows indicate a "primary" response, which follows a first-time encounter with a specific antigen. Dashed arrows indicate a "secondary" response to a subsequent encounter with the same kind of antigen. This illustration can be used as a road map as you make your way through the descriptions in the text. The details of the vertebrate immune system are astonishingly complex; even here, many events have been omitted so the main sequences can be seen clearly.

dilates capillaries and makes them "leaky," so fluid seeps out. The complement proteins and other substances used to fight an invasion are dissolved in this fluid. Also, clotting mechanisms (page 385) are working to keep blood vessels intact and to wall off infected or damaged tissues. In short, the inflammatory response involves these events:

1. Localized warmth and redness occur in damaged or invaded tissues when fine blood vessels dilate and become leaky.

2. Fluid seeping from blood vessels causes local swelling and also delivers infection-fighting proteins to tissues.

3. Phagocytes, following chemical gradients to affected tissues, engulf foreign invaders and debris.

4. Tissues are repaired, as by clotting mechanisms.

SPECIFIC DEFENSE RESPONSES: THE IMMUNE SYSTEM

Phagocytes engulf cellular debris and anything detected as foreign. Sometimes, however, their general attack response is not enough to stop the spread of an invader, and illness follows. When that happens, three types of white blood cells—the macrophages, T lymphocytes, and B lymphocytes—make precise counterattacks. Their interactions are the basis of the **immune system**. The hallmarks of this system are *specificity* (its cells zero in on specific invaders) and *memory* (they can mount a rapid attack if the same type of invader returns).

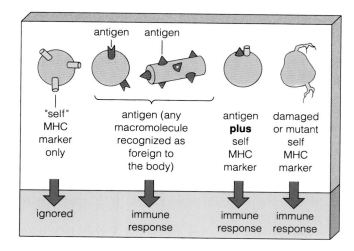

Figure 27.3 Molecular cues that stimulate lymphocytes to make immune responses.

Overview of the Defenders

Of every 100 cells in your body, one is a white blood cell. Here are the names and functions of the white blood cells responsible for immune responses:

1. **Macrophages**: phagocytes that alert helper T cells to the presence of specific foreign agents.

2. **Helper T cells**: master switches of the immune system; they stimulate the rapid division of B cells and killer T cells (which mount counterattacks).

3. **B cells**: lymphocytes responsible for producing molecular weapons, *antibodies*, that lock onto specific targets and tag them for destruction (by phagocytes or the complement system).

4. **Killer T cells** and **NK cells**: lymphocytes that directly destroy body cells already infected by certain viruses or parasitic fungi.

5. **Suppressor T cells**: controller lymphocytes (they slow down or prevent immune responses).

6. **Memory cells**: a portion of B-cell and T-cell populations produced during a first encounter with a specific invader but not used in battle; they circulate freely and respond rapidly to any subsequent attacks by the same type of invader.

The cells just listed belong to two fighting branches of the immune system. Both are called into action during most battles. T cells dominate one branch; they carry out a "cell-mediated" response. B cells dominate the other branch; they carry out an "antibody-mediated" response (Figure 27.2).

Recognition of Self and Nonself

Before we get into the immunological battles, think about an important question. How do the defenders distinguish *self* (the body's own cells) from *nonself* (any harmful foreign agent)? Such recognition is vital, for lymphocytes unleash extremely destructive immune reactions. We know this because on rare occasions the distinction is blurred and lymphocytes make an autoimmune response. (They turn on the body itself, causing disease and sometimes death.)

Among the surface receptors on your own cells are **MHC markers** (named after the genes coding for them). Your lymphocytes recognize these as "self" markers and normally ignore the cells bearing them (Figure 27.3). MHC markers are unique to each individual. Except in the case of identical twins, no one has the same kinds.

But viruses, bacteria, fungi, ragweed pollen, bee venom, cells of organ transplants, and just about any

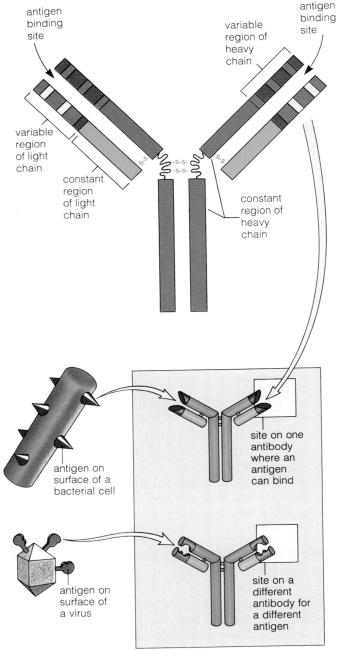

Figure 27.4 Structure of antibodies. An antibody molecule has four polypeptide chains joined into a Y-shaped structure. Some regions are always the same in all antibody molecules. But the molecular configuration varies in one region; this is the antigen-binding site. Specific gene segments on a parental chromosome code for the variable regions of antibody molecules. DNA recombinations at these segments give rise to millions of combinations of variable regions. This is why the body's immune system can respond to so many different antigens.

other foreign agent have antigens at their surface, which lymphocytes do not ignore. An **antigen** is any large molecule with a distinct configuration that triggers an immune response. Most antigens are protein or poly-saccharide molecules.

Lymphocytes will not attack cells bearing self-MHC markers only. They will mount an immune response when they encounter antigens. They will do so regardless of whether antigens are merely present in tissues or sticking out, flaglike, next to MHC markers on cell surfaces (Figure 27.3).

Primary Immune Response

A *first-time* encounter with an antigen elicits a **primary immune response** from macrophages, lymphocytes, and their products. Here we will first consider an antibody-mediated response to such an encounter, then a cell-mediated one.

Antibody-Mediated Immune Responses. The word **antibody** refers to a Y-shaped receptor molecule with binding sites for a specific antigen (Figure 27.4). Only B cells and some of their progeny, called **plasma cells,** make antibodies.

While each B cell is maturing in bone marrow, it makes many copies of just one kind of antibody, and these become positioned at its surface. (The tail of each "Y" is embedded in the plasma membrane, and the arms stick out above the surface.) The cell is released into the circulation as a "virgin" B cell, meaning it has membrane-bound antibodies and has not yet made contact with its specific antigen.

Suppose bacteria enter the body through a small cut. They might move undetected past any number of virgin B cells. But sooner or later they encounter the one with antibodies able to bind to the antigen on the bacterial cell surface. Once binding occurs, the B cell becomes sensitive to communication signals from macrophages and helper T cells.

During the general inflammatory response to invasion, macrophages had engulfed a few bacterial cells. Lysosomal enzymes digested the engulfed cells (page 51), but they did not completely destroy their antigens. Antigen fragments were transported to the macrophage surface, where they bound with MHC markers. *Each macrophage ended up displaying antigen-MHC complexes at its surface* (Figure 27.5).

Now surface receptors on helper T cells lock onto the antigen-MHC complexes. When this happens, macrophages secrete a substance, *interleukin-1,* and helper T cells secrete *lymphokines*. Both substances are com-

Labels in figure:

antigen binding site

variable region of heavy chain

antigen binding site

variable region of light chain

constant region of light chain

constant region of heavy chain

antigen on surface of a bacterial cell

site on one antibody where an antigen can bind

antigen on surface of a virus

site on a different antibody for a different antigen

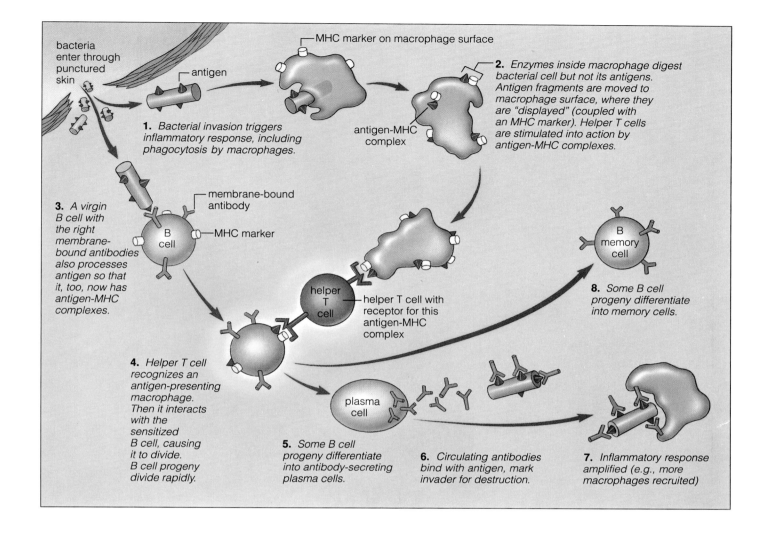

bacteria enter through punctured skin

antigen

MHC marker on macrophage surface

1. Bacterial invasion triggers inflammatory response, including phagocytosis by macrophages.

antigen-MHC complex

2. Enzymes inside macrophage digest bacterial cell but not its antigens. Antigen fragments are moved to macrophage surface, where they are "displayed" (coupled with an MHC marker). Helper T cells are stimulated into action by antigen-MHC complexes.

3. A virgin B cell with the right membrane-bound antibodies also processes antigen so that it, too, now has antigen-MHC complexes.

membrane-bound antibody

B cell

MHC marker

helper T cell

helper T cell with receptor for this antigen-MHC complex

B memory cell

8. Some B cell progeny differentiate into memory cells.

4. Helper T cell recognizes an antigen-presenting macrophage. Then it interacts with the sensitized B cell, causing it to divide. B cell progeny divide rapidly.

plasma cell

5. Some B cell progeny differentiate into antibody-secreting plasma cells.

6. Circulating antibodies bind with antigen, mark invader for destruction.

7. Inflammatory response amplified (e.g., more macrophages recruited)

munication signals; in their presence, any B cell that has become sensitized to the antigen will start dividing.

Rapid divisions among the B cell progeny give rise to a clone—a population of identical B cells, all making the same kind of antibody. Part of the population differentiates into plasma cells and they, too, make the same antibody. The plasma cells are weapons factories. For the next few days they secrete about 2,000 antibody molecules per second into their surroundings!

The circulating antibodies do not destroy invaders directly; they simply tag them for disposal by other means. As it happens, there are several different classes of antibodies (called the *immunoglobulins,* or Ig) that enlist the aid of different immune cells or chemical weapons. When bound to antigens, for example, IgM and IgG antibodies enlist the aid of macrophages and complement proteins, and IgE antibodies call histamine-secreting cells into action.

Figure 27.5 Amplification of the inflammatory response by specific immune reactions. This example is of an *antibody-mediated response* to a bacterial invasion. Plasma cells (the progeny of activated B cells) release antibodies, which circulate and mark invaders for destruction by other defense agents, including more macrophages recruited to the battle scene.

The main targets of an antibody-mediated response are bacteria and *extracellular* phases of viruses, some fungal parasites, and some protozoans. In other words, antibodies can't lock onto antigen if the invader has entered the cytoplasm of a host cell; antigen must be circulating in tissues or at the cell surface.

Cell-Mediated Immune Responses. The key fighters of this branch of the immune system are killer T cells and natural killer (NK) cells. They directly destroy body cells that are already infected, mainly by viruses. They also may attack mutant and cancerous cells (see the *Commentary*).

Killer T cells start developing in bone marrow. Then they move into the thymus gland and start making many copies of their own kind of antigen-binding receptor molecules. During an invasion, the ones with the right receptors are stimulated to divide when helper T cells raise the alarm. They ignore circulating antigens. But they directly destroy infected body cells, which display antigen-MHC complexes at their surface. They punch holes in these cells with perforin and other chemical secretions. In this way, infected cells are destroyed before the pathogen can reproduce inside them.

Killer cells might be the reason the body rejects foreign organ transplants. They recognize MHC markers of the donor as being foreign (unless the donor is an identical twin). Organ recipients take drugs to destroy killer cells, but this compromises their ability to mount immune responses to pathogens.

Control of Immune Responses. Antibody-mediated and cell-mediated responses are regulated events. When the tide of battle turns, antibody molecules are "saturating" the binding sites on pathogens that have not yet been disposed of. With fewer exposed antigens, fewer antibodies are secreted. Also, secretions from suppressor T cells call off the counterattack and keep the reactions from spiraling out of control.

Antibody Diversity and the Clonal Selection Theory

Your body can be invaded by an enormous variety of pathogens, each with many unique antigens. How do lymphocytes produce the millions of different receptors required to detect all the potential threats? The answer lies with DNA recombinations occurring in each B cell as it matures in bone marrow. Part of each arm of an antibody is a polypeptide chain, folded into a groove or cavity. All B cells have the same genes coding for the chain—but each shuffles the genes into one of millions

COMMENTARY

Cancer and the Immune System

Cancer cells might arise in your body at any time as a result of mutations induced by events such as viral attacks, chemical bombardment, or irradiation. "Cancer" refers to cells that have lost control over cell division. Through their berserk divisions, they destroy surrounding tissues (page 160).

Cancer cells can be destroyed by killer T cells and by natural killer cells. Sometimes, though, cancer cells simply are not detected. Maybe the mutation that triggered the cancerous transformation did not affect the cell surface markers; or maybe the markers became chemically disguised. Perhaps they were even released from the cell surface and began circulating through the bloodstream to lead the immune fighters down false trails. Sometimes, too, individuals are not genetically equipped to respond to a particular antigen. A person's age and overall state of health also seem to play a role in resistance to cancer.

At present, surgery, drug treatment (chemotherapy), and irradiation are the only weapons against cancer. Surgery works when a tumor is fully accessible and has not spread, but it offers little hope when cancer cells have begun wandering. When used by themselves, chemotherapy and irradiation destroy good cells as well as bad. *Immune therapy* is a promising prospect. The idea here is to mobilize the immune system by deliberately introducing agents that will set off the immune alarm.

Interferons, a group of small proteins, were early candidates for immune therapy. Most cells produce and release interferon following a viral attack. The interferon binds to the plasma membrane of other cells in the body and induces resistance to a wide variety of viruses. So far, however, interferon has been useful against only some rare forms of cancer.

Monoclonal antibodies hold promise for immune therapy. It is difficult to get normal, antibody-secreting B cells to grow indefinitely and thereby mass-produce pure antibody in useful amounts. But Cesar Milstein and Georges Kohler discovered a way to do this. They immunized a mouse with a specific antigen. (The point of doing this was to allow lymphatic tissues in the mouse—the spleen especially—to become enriched with

B cells specific for the immunizing antigen.) Later, B cells were extracted from the mouse spleen and were fused with a malignant B cell that showed indefinite growth. Some of the hybrid cells multiplied as rapidly as the malignant parent and produced quantities of the same type of antibodies as the parent B cells from the immunized mouse.

Clones of such hybrid cells can be maintained indefinitely and they continue to make the same antibody. Hence the name "monoclonal antibodies." All the antibody molecules are identical, and all are derived from the same parent cell.

Monoclonal antibodies are being studied for use in passive immunization against malaria, flu viruses, and hepatitis B. They also are candidates for cancer imaging. By using scanning machines along with radioactively labeled monoclonal antibodies that are specific for certain types of cancer, it is possible to pinpoint the exact location of cancer in the body. (See page 20, for example.) Such scans indicate whether cancer is present, where a tumor is located, and how big it is.

Monoclonal antibodies might also help overcome one of the major drawbacks to drug treatment of cancer. Such treatments have severe side effects because the drugs used are highly toxic and cannot discriminate between normal cells and cancerous ones. A current goal is to hook up drug molecules with a monoclonal antibody. As Milstein and Kohler speculated, "Once again the antibodies might be expected to home in on the cancer cells—only this time they would be dragging along with them a depth charge of monumental proportions." Such is the prospect of targeted drug therapy.

killer T cell

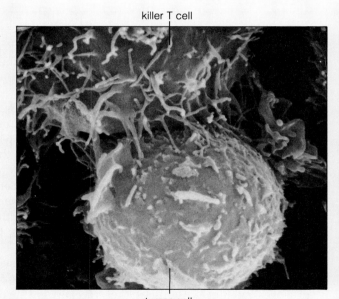

tumor cell

a A killer T cell recognizes and binds tightly to a tumor cell, then secretes pore-forming proteins that will destroy the integrity of the target cell membrane.

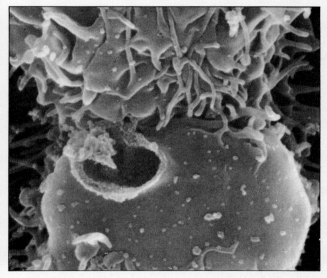

b The target cell has become grossly leaky and has ballooned under an influx of the surrounding fluid; soon there will be nothing left of it.

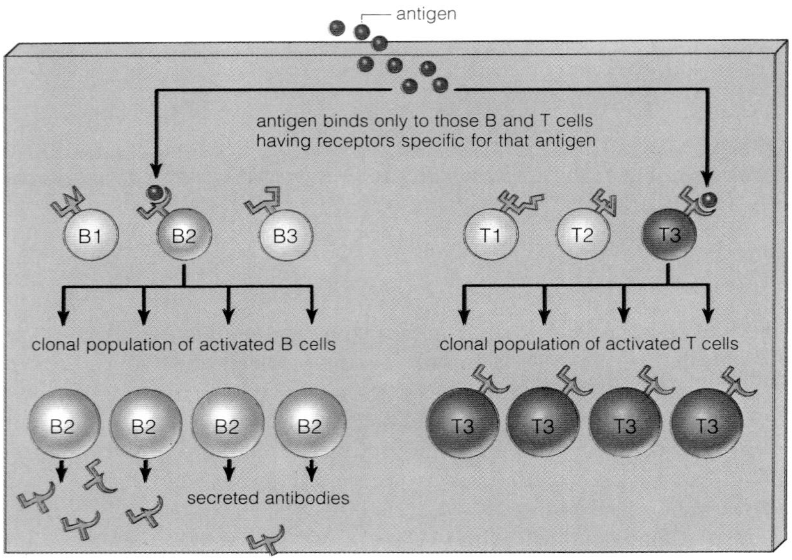

Figure 27.6 Clonal selection of lymphocytes having receptors for specific antigens. The proteins from which the receptors are constructed are produced through random shufflings of DNA segments while lymphocytes are maturing. Only antigen-specific lymphocytes will become activated and will give rise to a population of immunologically identical clones.

of possible combinations, so they can give rise to very large numbers of chain configurations (page 159).

Thus, it is not that you or any other individual inherited a limited genetic war chest from your ancestors, useful only against pathogens that were successfully fought off in the past. Even if you encounter an entirely new antigen (as might occur when an influenza virus has mutated), DNA recombinations in one of your maturing B cells may have produced the exact chain configuration that can lock onto the invader. By happy accident, you have the precise weapon needed.

According to the **clonal selection theory,** proposed by Macfarlane Burnet, an activated lymphocyte multiplies rapidly and all of its descendants will retain specificity against the antigen causing the activation. They constitute a *clone* of cells, immunologically identical for the antigen that "selected" them (Figure 27.6).

Secondary Immune Response

The clonal selection theory also explains how a person has "immunological memory," which is the basis of a more rapid response to a subsequent invasion by the same type of pathogen. A **secondary immune response** to a previously encountered antigen occurs more rapidly than a primary response, and it is greater and of longer duration (Figure 27.7).

Why is this so? During a primary immune response, some B and T cells of the clonal populations do not engage in battle. They continue to circulate for years, even decades in some cases, as patrolling battalions of **memory lymphocytes.** When a memory lymphocyte encounters the same type of antigen, it divides at once.

A large clone of active lymphocytes can be unleashed— and it can be unleashed in a matter of days or even hours.

IMMUNIZATION

Jenner didn't know why his cowpox vaccine provided immunity against smallpox. Today we know that the viruses causing the two diseases are related, and they bear similar antigens at their surface. Let's express in modern terms what is going on.

Immunization means deliberately introducing into the body an antigen that can provoke an immune response and the production of memory lymphocytes. A **vaccine** (a preparation designed to stimulate the appearance of memory lymphocytes) is injected into the body or taken orally. The first injection elicits a primary immune response. A second injection (the "booster" shot) elicits a secondary response, which provokes the production of more antibodies and memory cells to provide long-lasting protection against the disease.

Many vaccines are made from killed or weakened bacteria or viruses. (Sabin polio vaccine is a preparation of a weakened polio virus.) Other vaccines are made from the toxic but inactivated by-products of dangerous organisms, such as the bacterium responsible for tetanus.

Recently, selected antigen-encoding genes from pathogens were incorporated into the vaccinia virus. The virus was then used successfully to immunize laboratory animals against hepatitis B, influenza, rabies, and other serious diseases. A genetically engineered virus is not as potentially dangerous as a weakened but

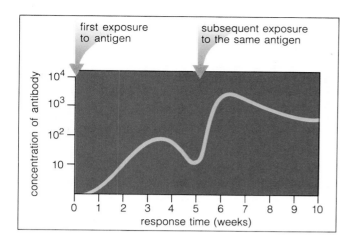

Figure 27.7 Differences in magnitude and duration between a primary and a secondary immune response to the same antigen. (The secondary response starts at week 5.)

still-intact pathogen (which might still revert to dangerous form).

For people already exposed to diphtheria, tetanus, botulism, and some other bacterial diseases, antibodies are injected directly to confer **passive immunity**. The effects are not lasting (the person's own lymphocytes are not producing the antibodies), but the antibodies may help counter the immediate attack.

ABNORMAL OR DEFICIENT IMMUNE RESPONSES

Allergies

Many of us suffer from **allergies,** in which a secondary immune response is made to a normally harmless substance. Exposure to dust, pollen, insect venom or secretions, drugs, certain foods, cosmetics, and other seemingly innocuous substances triggers the abnormal response. Some allergic reactions occur explosively within minutes; others are delayed. In either case, they can cause tissue damage.

Some individuals are genetically predisposed to allergies. But infections, emotional stress, even changes in air temperature can trigger or complicate reactions to dust and other substances that the body perceives as antigens. Immediately after the body is exposed to the antigen, IgE antibodies are produced. IgE initiates a local inflammatory response by provoking cells into secreting histamine, prostaglandins, and other potent substances. Besides promoting fluid seepage from capillaries, histamine also causes exocrine glands to secrete mucus. Prostaglandins constrict smooth muscle in dif-

ferent organs and contribute to platelet clumping. In *asthma* and *hay fever,* the resulting symptoms include congestion, sneezing, and a drippy nose, as well as labored breathing.

On rare occasions, the inflammatory response can be explosive and life-threatening. The few individuals who are hypersensitive to, say, wasp or bee venom can die within minutes following a single sting. Air passages leading to their lungs undergo massive constriction. Fluid escapes too rapidly from capillaries that have become grossly leaky when the clefts between their cells enlarge. Blood pressure plummets and this can lead to circulatory shock.

Allergy-producing substances often can be identified by tests, and in some cases the body can be stimulated to make a different type of antibody (IgG) that can block the inflammatory response. Over a long period of time, the allergy sufferer is given increasingly large injections of the antigen. Circulating IgG molecules bind with and mask molecules of the offending substance before they interact with IgE to produce the abnormal response.

Autoimmune Disorders

In an **autoimmune response,** lymphocytes are unleashed against the body's cells. An example is *rheumatoid arthritis,* in which movable joints especially are inflamed for long periods. Often, affected persons have high levels of an antibody (rheumatoid factor) that locks onto the body's IgG molecules as if they were antigens, then deposits them in membranes at the joints. The membranes become prime targets for abnormal events, including increased fluid seepage from capillaries. The accumulating fluid separates the membrane from underlying tissues, membrane cells divide repeatedly in response, and the joint thickens. These and other events continue in cycles of inflammation that do not end until the joint is totally destroyed.

Deficient Immune Responses

On rare occasions, cell-mediated immunity is weakened and the body becomes highly vulnerable to infections that might not otherwise be life-threatening. This is what happens in **AIDS** (acquired immune deficiency syndrome). AIDS is caused by the "human immune-deficiency virus," or HIV. The *Commentary* on page 400 describes some immunological aspects of HIV infection—how the virus replicates itself inside a human host and what the prospects are for treating or curing infected persons. Social implications of the worldwide pool of infection are described on page 485, in the *Commentary* on sexually transmitted diseases.

AIDS—The Immune System Compromised

AIDS is a constellation of disorders that follow infection by the human immune deficiency virus, or HIV. The virus cripples the immune system and leaves the body dangerously susceptible to opportunistic infections and some otherwise rare forms of cancer. Currently there is no vaccine against the known forms of this virus (HIV-1 and HIV-2) and no cure for those already infected.

From 1981 to late 1990, there were more than 150,000 cases and an estimated 1 million or more HIV-1 carriers in the United States alone. The World Health Organization (WHO) estimates there are already 5 million HIV carriers worldwide and as many as 600,000 already stricken with AIDS. No one can say how many people will be infected by HIV in the next decade; the number could be as high as 100 million.

HIV Replication Cycle

HIV compromises the immune system by attacking helper T cells (also called T4 lymphocytes) as well as macrophages. Sometimes the virus directly attacks the nervous system, causing mental impairment and loss of motor function.

HIV is a *retrovirus*; its genetic material is RNA rather than DNA. A protein coat surrounds the RNA and several copies of an enzyme (reverse transcriptase). The coat itself is wrapped in a lipid envelope derived from the plasma membrane of a host T4 cell. Once inside a host, the enzyme uses the viral RNA as a template for making DNA, which then is inserted into a host chromosome.

It may be up to fourteen months after infection has occurred before antibodies to several HIV proteins can be detected in the body. The antibodies do not eliminate infected cells or inactivate the circulating virus particles. Those cells can harbor the foreign DNA for months, even years. However, when the body is called upon to make a secondary immune response, the infected cell may be activated. It transcribes parts of its DNA—including the foreign insert. Transcription yields copies of viral RNA, which are translated into viral proteins. New viral particles are put together from the RNA and proteins. They bud from the plasma membrane of the host T4 cell or are released when the membrane ruptures (Figures a–d). With each new round of infection, more and more T4 cells are destroyed.

In time, the T4 cell population is depleted and the body loses its ability to mount immune responses. Initially there may be unexplained and persistent weight loss, flu-like symptoms of fatigue and malaise, fevers and bed-drenching night sweats, as well as enlarged lymph nodes. Opportunistic infections, such as a form of pneumonia caused by a protozoan (*Pneumocystis carinii*), may superimpose their own symptoms on the initial ones. Mainly in homosexual patients, blue-violet or brown spots may appear on the legs especially; these may be signs of Kaposi's sarcoma, a deadly form of skin cancer.

Modes of Transmission

Like any human virus, HIV requires a medium by which it can leave the body of its host, survive in the environment into which it is released, and enter a susceptible cell that can support its replication.

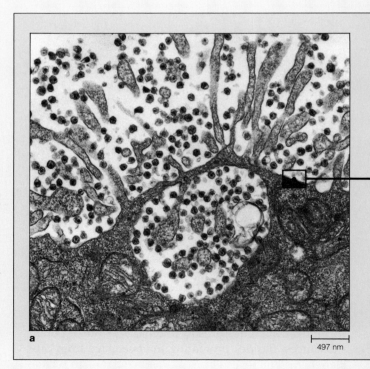

a

497 nm

HIV is transmitted when bodily fluids of an infected person enter another person's tissues. Initially in the United States, transmission occurred most often among male homosexuals and intravenous drug abusers who shared needles. The incidence is now increasing among heterosexuals in the general population. (Here you may wish to refer to the *Commentary* on sexually transmitted diseases, page 485.) HIV also has been transmitted from infected mothers to their infants during pregnancy, birth, and breast-feeding. Contaminated blood supplies accounted for some cases before screening for HIV was implemented in 1985. In several developing countries, HIV has spread through contaminated transfusions and through reuse of unsterile needles by health care providers.

HIV generally cannot survive for more than one or two hours outside the human body. Viral particles on needles and other objects are readily destroyed by disinfectants, including household bleach. At this time,

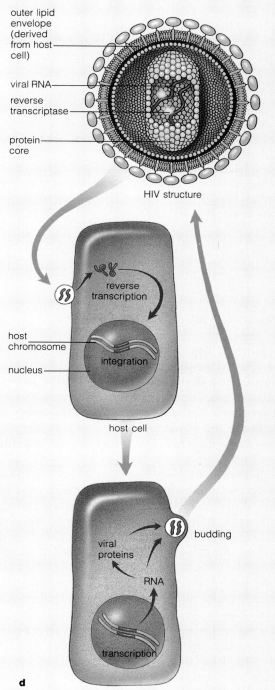

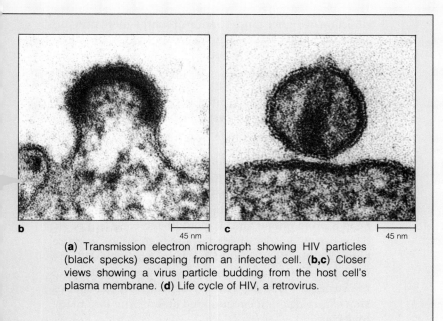

(a) Transmission electron micrograph showing HIV particles (black specks) escaping from an infected cell. (b,c) Closer views showing a virus particle budding from the host cell's plasma membrane. (d) Life cycle of HIV, a retrovirus.

there is no evidence that HIV can be effectively transmitted by way of food, air, water, or casual contact. The virus *has* been isolated from blood, semen, vaginal secretions, saliva, tears, breast milk, amniotic fluid, cerebrospinal fluid, and urine; it is likely to be found in other bodily fluids, secretions, and excretions. However, only infected blood, semen, vaginal secretions, and breast milk contain the virus in concentrations that seem to be high enough for successful transmission.

Prospects for Treatment

At present, the challenge of developing an effective vaccine is formidable. HIV mutates rapidly and it may be difficult to produce a vaccine that will work against all of its mutated forms. Even if a vaccine can be developed that could coax the body into producing antibodies to HIV, the antibodies may not protect against AIDS. There is evidence that antibodies do not neutralize the virus.

DNA cloning experiments have provided researchers with several growth factors that influence blood cell production. Also, researchers have isolated the stem cells that actually give rise to blood cells, including those of the immune system. Perhaps work of this sort will benefit AIDS patients by showing the way to increase the cell count of macrophages and other white blood cells. One problem is that the macrophages themselves can become infected with HIV and actually spread the virus.

The drug AZT (azidothymidine) is being used to prolong the life of AIDS patients, and, together with bone marrow transplants, it may turn out to be useful in developing a cure. At this writing, researchers are attempting to develop useful drugs for those who are already infected; until they do, checking the spread of HIV depends absolutely on implementing behavioral controls through education on a massive scale. We return to this topic on page 485.

Case Study: The Silent, Unseen Struggles

Let us conclude this chapter with a case study of how the immune system helps *you* survive attack. Suppose on a warm spring day you are walking barefoot to class. Abruptly you stop: a tack on the ground has punctured one of your toes. You pull out the tack at once, but the next morning the punctured area is red, tender, and swollen. A few days later, your foot is back to normal and you have forgotten the incident.

All that time your body had been struggling against an unseen enemy. During your walk, one of your feet had picked up some soil bacteria. And when the tack broke through your skin, it carried several thousand bacterial cells with it. Inside, the bacteria found conditions suitable for growth. They soon doubled in number and were on their way to doubling again. But their metabolic products were interfering with your own cell functions. If unchecked, the invasion would have threatened your life.

Yet even as your skin was punctured, your body's defenses were being mobilized. Blood from ruptured blood vessels began to pool and clot around the wound. Histamine and other secretions from basophils and mast cells caused capillaries to dilate and become more permeable to plasma proteins, including complement. Now

phagocytes crawled through clefts in capillary walls. Like bloodhounds on the trail, they moved in the direction of higher concentration of complement proteins and began engulfing bacteria, dirt, rust, bits of broken cells, and anything else not having the proper markers that mean, "I'm *self!*"

If there had been no bacteria on the tack or if the bacteria were unable to multiply rapidly in your tissues, then the inflammatory response would have cleaned things up. But bacterial cell divisions were outpacing the nonspecific defenses. The battle had to be turned over to lymphocytes.

If this had been your first exposure to the bacterial species, few B and T cells would have been around to respond to the call. The immune response would have been a primary one, and it would have been five or six days before B cells divided enough times to produce enough antibody. But when you were a child, your body did fight off this invader and it still carries vestiges of the struggle—memory cells. When the bacterial species showed up again, it encountered an immune trap ready to spring.

As inflammation progressed, B and T cells were also leaving the bloodstream. Most were specific for other antigens and did not take part in the battle. But some memory cells locked onto the antigens and became

Table 27.1 Summary of White Blood Cells and Their Roles in Defense

Cell Types	Take Part In	Main Characteristics
Lymphocytes:		
1. Killer T cell	Cell-mediated immune response	Each type equipped with membrane receptors specific for one type of antigen; each can directly destroy virus-infected cells (and possibly cancer cells) by punching holes in them
2. Helper T cell	Cell-mediated and antibody-mediated immune responses	Master switch of immune system; stimulates rapid proliferation of killer T cell and B cell populations
3. Suppressor T cell	Same as above	Modulates degree of immune response (slows down or prevents activity by other lymphocytes)
4. Virgin B cell	Antibody-mediated immune response	Not-yet-activated lymphocyte with *membrane-bound* antibodies (serving as antigen-specific receptors at plasma membrane)
5. Plasma cell	Same as above	*Antibody-secreting* descendant of an activated B cell
6. Memory cell	Cell-mediated or antibody-mediated immune responses	One of a clonal population of T cells or B cells set aside during a primary immune response and that can make a rapid, secondary immune response to another encounter with the same type of invader
Macrophages	Inflammatory, cell-mediated, and antibody-mediated immune responses	Phagocytic (engulfs foreign agents and infected, damaged, or aged cells); develop from circulating monocytes and take up stations in tissues; present antigens to immune cells; secretions trigger T cell and B cell proliferation
Neutrophil	Inflammatory response	Phagocytic; most abundant type of white blood cell; dominates early stage of inflammation
Eosinophil	Inflammatory response	Phagocytic (engulfs antigen-antibody complexes, certain parasites); combats effect of histamine in allergic reactions
Basophil and mast cell	Inflammatory response	Release histamine and other substances that contribute to vasodilation and a rapid inflammatory response
Natural killer cell (NK)	Cell-mediated immune response	Directly destroy tumor cells, some virus-infected cells; distinct from T and B cells

activated. They moved into lymph vessels with their cargo, tumbling along until they reached a lymph node and were filtered from the fluid. For the next few days, memory cells steadily accumulated, secreted communication signals, and divided rapidly in the node.

For the first two days the bacteria appeared to be winning; they were reproducing faster than phagocytes, antibody, and complement system were destroying them. By the third day, antibody production peaked and the tide of battle turned. For two weeks or more, antibody production will continue until the invaders are wiped out. After the response draws to a close, memory cells will go on circulating, prepared for some future struggle.

SUMMARY

1. The body makes nonspecific (general) and specific responses to foreign agents that enter the internal environment. It also has external lines of defense, including intact skin, exocrine gland secretions, gastric fluid, intestinal bacteria (which compete effectively against many invaders), and ciliated mucous membranes of the respiratory tract. Table 27.1 summarizes the cell types involved in specific responses.

2. During the inflammatory response, phagocytes, complement proteins, and other factors are mobilized to

destroy any agents detected as foreign, then to restore normal conditions in the tissue. They also are mobilized during immune responses.

3. White blood cells and their products are the basis of the immune system. Some of these cells show specificity (they attack only a particular pathogen, not invaders in general). They also show memory (they make rapid, secondary responses to the same pathogen whenever it is encountered again).

4. Cells of the immune system communicate with one another by chemical secretions (notably lymphokines), which stimulate rapid growth and division of certain lymphocytes (B cells, helper T cells, and killer T cells) into large armies against particular invaders.

5. Lymphocytes ignore the body's own cells, which have self markers (MHC markers). They mount immune responses when they encounter circulating antigens or cells bearing both MHC markers and antigens at their surface. An antigen is any large molecule that lymphocytes perceive as foreign and that triggers an immune response. Antigens occur at the surface of viruses, bacterial cells, fungal cells, and so on.

6. An *antibody-mediated immune response* is made against antigen circulating in the body's tissues or attached to the surface of an invading pathogen. First macrophages and virgin B cells become sensitized to the antigen (they display antigen-MHC complexes at their surface). When helper T cells encounter the complexes, they stimulate the virgin B cell to divide. Some of the B cell progeny develop into plasma cells, which secrete antibodies. Antibody molecules bind to specific antigens and mark the invaders for disposal (by macrophages or by complement proteins).

7. A *cell-mediated immune response* is made against cells already infected (and possibly against cancerous or mutant cells). Infected cells process the antigen of their invaders and display it at their surface. After macrophages engulf those cells, they in turn display the antigen in combination with MHC markers. Helper T cells recognize the antigen-MHC complexes on macrophages and secrete substances that trigger rapid divisions of killer T cells able to recognize that antigen. Killer T cells directly attack and destroy infected cells by punching holes in them.

8. Following a primary (first-time) immune response, portions of the B cell and T cell populations that were produced continue to circulate as memory lymphocytes. They are available for a rapid, amplified response to subsequent encounters with the same antigen (a secondary immune response).

Review Questions

1. The vertebrate body has physical and chemical barriers against invading pathogens. Can you name five such barriers? *391*

2. Phagocytes are cells that engulf and destroy foreign cells and substances by means of endocytosis. Where in the body are phagocytes located? Are they deployed during nonspecific defense responses or immune responses only? *391*

3. What are the four events that characterize an inflammatory response? What are some of the chemical factors associated with this response? *393*

4. The vertebrate immune system is characterized by *specificity* and *memory*. Can you describe what these two terms mean? *393*

5. Define the following types of white blood cells, which are central to immune responses: macrophages, helper T cells, B cells, killer T cells, suppressor T cells, and memory cells. *393*

6. Antibodies, lymphokines, and interleukins also are central to immune responses. Can you define them? *394*

7. What is an MHC marker? An antigen? An antibody? How do interactions among these three types of macromolecules allow the body to distinguish self from nonself? *393–394*

8. A secondary immune response to an antigen involves macrophages, lymphocytes, and their products. How does it differ from a primary immune response, which involves the same things? *398*

9. Two fighting branches of the immune system are deployed during most primary and secondary immune responses. One is the cell-mediated response, the other is the antibody-mediated response. Can you give an example of what goes on during each kind? *394–396*

10. What is immunization? What is a vaccine? *398*

11. What is the difference between an allergy and an autoimmune response? Is AIDS an autoimmune disease? *399*

Self-Quiz   *(Answers in Appendix IV)*

1. The body makes nonspecific, generalized counterattacks to invasion by foreign agents; it also makes specific, _____ responses against particular invaders.

2. _____ are able to distinguish between molecular markers (MHC markers) present on the body's own cells and _____.

3. _____ are any large molecules that white blood cells distinguish as foreign to the body and that elicit an _____ response.

4. Antibody-mediated immune responses are made against _____; cell-mediated immune responses, against _____.

5. External barriers to invasion include _____.
   a. unbroken skin
   b. exocrine gland secretions
   c. gastric fluid and intestinal bacteria
   d. ciliated mucous membranes
   e. all of the above

6. Inflammatory responses require _____, _____, and other factors to destroy foreign agents.
   a. red blood cells; complement proteins
   b. phagocytes; antigens
   c. red blood cells; antigens
   d. phagocytes; complement proteins

7. _____ and their products are the basis of the immune system.
   a. Red blood cells   c. White blood cells
   b. Blood platelets   d. Antigens

8. Some white blood cells show _____ in that they attack only particular pathogens, and they also exhibit _____ or the ability to recognize the same invader upon subsequent attacks.
   a. communication; perception
   b. specificity; memory
   c. general response; specific response
   d. flexibility; recognition

9. The body's own uninfected cells are ignored by its lymphocytes because they bear _____ at their surface.
   a. complement proteins   c. antigens
   b. self-MHC markers   d. antigen plus self-MHC marker

10. An antibody is _____.
   a. an activated plasma cell
   b. a receptor molecule with binding sites for virgin B cells
   c. a receptor molecule with binding sites for antigen
   d. an out-of-body experience

11. Which of the following is *not* a molecular cue that stimulates lymphocytes into making an immune response?
   a. self-MHC marker
   b. antigen
   c. antigen plus self-MHC marker
   d. damaged or mutant self-MHC marker
   e. all of the above serve as molecular cues

12. Match the immunity concepts.
   _____ killer T cells   a. stimulate virgin B cells to divide
   _____ helper T cells   b. destroy infected cells by punching holes in them
   _____ T cells   c. circulate as memory lymphocytes
   _____ some B cell progeny   d. origin of plasma cells
   _____ portions of B and   e. recognize antigen-MHC complexes on macrophages, trigger T cell divisions
          T cell populations

Selected Key Terms

AIDS *399*
allergy *399*
antibody *394*
antibody-mediated response *394*
antigen *394*
asthma *399*
autoimmune response *399*
B cell *393*
cell-mediated response *396*
clonal selection theory *398*
complement system *391*
helper T cell *393*
immune system *393*
immunization *398*
immunoglobulin *395*
inflammatory response *391*
killer T cell *393*
macrophage *393*
MHC marker *393*
memory lymphocyte *398*
natural killer (NK) cell *396*
passive immunity *399*
plasma cell *394*
primary immune response *394*
secondary immune response *398*
suppressor T cell *393*
vaccine *398*

Readings

Edelson, R., and J. Fink. June 1985. "The Immunologic Function of Skin." *Scientific American* 252(6):46–53.

Golub, E. 1987. *Immunology: A Synthesis.* Second edition. Sunderland, Massachusetts: Sinauer Associates.

Kimball, J. 1990. *Introduction to Immunology.* Third edition. New York: Macmillan.

Leder, P. May 1982. "The Genetics of Antibody Diversity." *Scientific American* 246(5):102–115. Describes how a few hundred DNA segments can be shuffled and recombined to make billions of different antibodies.

Roitt, I., J. Brostoff, and D. Male. 1989. *Immunology.* St. Louis: Mosby. Second edition. Lavishly illustrated.

Tizard, I. 1988. *Immunology: An Introduction.* Second edition. Philadelphia: Saunders.

Tonegawa, S. October 1985. "The Molecules of the Immune System." *Scientific American* 253(4):122–131.

28 RESPIRATION

KEY CONCEPTS

1. Animals are the most active organisms. The energy to drive their activities comes mainly from aerobic metabolism, which uses oxygen and produces carbon dioxide. In a process called respiration, animals move oxygen into their internal environment and give up carbon dioxide to the external environment.

2. All respiratory systems make use of the tendency of a gas to diffuse down its partial pressure gradient. Such a gradient exists between oxygen in the atmosphere (high pressure) and the metabolically active cells in body tissues (where oxygen is used rapidly; pressure is lowest here). Another gradient exists between carbon dioxide in body tissues (high pressure) and the atmosphere (with its lower amount of carbon dioxide).

3. All respiratory systems have a respiratory surface— a thin, moist layer of epithelium that gases can readily diffuse across. In most animals, oxygen is picked up by the general circulation and transported to body tissues, where carbon dioxide is picked up and transported back to the respiratory surface.

RESPIRATORY SYSTEMS

Specialized Respiratory Surfaces

You may not think you have too much in common with a flatworm, an aquatic animal, especially when you find yourself momentarily submerged in water. Like you, however, the flatworm uses muscle cell contractions for its movements. Even though the flatworm lives in water, its cells (like yours) use oxygen for aerobic metabolism— the only pathway that generates enough energy for swimming. You get oxygen from the air, the flatworm gets dissolved oxygen from water. Both of you also give up carbon dioxide from metabolism to the surroundings. In other words, you and the flatworm engage in **respiration**—the exchange of oxygen and carbon dioxide between the external world and the internal environment (Figure 28.1).

Like other substances, oxygen and carbon dioxide diffuse down concentration gradients—or, as we say for

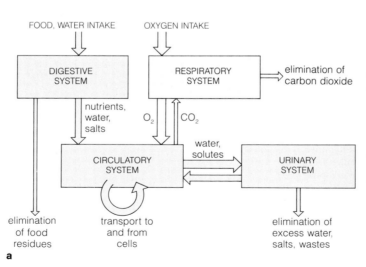

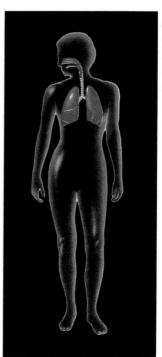

Figure 28.1 (**a**) Roles of the respiratory system in complex animals. (**b**) Unlike humans, flatworms are small enough that a circulatory system is not required; oxygen can reach individual cells simply by diffusing across the body surface. Unlike flatworms, humans would never survive on the low concentrations of oxygen dissolved in water.

gases, down pressure gradients. Those gases are not the only ones in water or air, so the pressure associated with each one is "partial" with respect to the combined pressure exerted by the total mix of gases. Any gas tends to diffuse from areas of high to low partial pressure. And respiratory systems take advantage of this tendency.

Exchange at Body Surface. Flatworms, earthworms, and many other animals do not have massive bodies or high metabolic rates, so their demands for respiration are not great. They rely on **integumentary exchange**, in which oxygen and carbon dioxide diffuse across a thin, vascularized layer of moist epidermis at the body surface. For the water dwellers, the surroundings keep the layer moist; for land dwellers, mucus and other secretions provide the moisture.

For other animals, the integument is too thick, too hardened, or too sparsely supplied with blood vessels to be a good respiratory surface. Also, animals larger than flatworms cannot depend on their integument alone to provide enough surface area for gas exchange. Without other adaptations in body plan, a massive animal would die; gases would not be able to diffuse across the integument fast enough to sustain the greater volume of interior cells. This is where gills, tracheas, lungs, and other specialized respiratory organs prove useful.

b

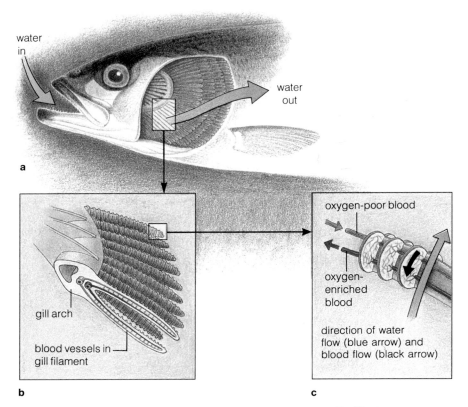

Figure 28.2 Respiratory system of many fishes. (**a**) Location of gills. The bony covering over them has been removed for this sketch. (**b,c**) Each gill has extensive capillary beds between two blood vessels. One vessel carries oxygen-poor blood into the gills, the other carries oxygen-rich blood back into the deeper body tissues. Blood flowing from one vessel to the other runs counter to the direction of water flowing over the gills. The arrangement favors the movement of oxygen (down its partial pressure gradient) into both vessels.

Gills. A typical **gill** has a moist, thin, vascularized layer of epidermis that functions in gas exchange. External gills project from the body of a few amphibians and some insects. The internal gills of fishes are rows of slits or pockets extending from the back of the mouth to the body surface. Water enters the mouth, moves down the pharynx, and flows out across the gills.

As Figure 28.2 shows, water moves *over* fish gills and blood circulates *through* them in opposite directions. Water first flows over capillaries at their "output" ends that lead back into the body. The blood inside has less oxygen than the surrounding water, so oxygen diffuses inward. Then the water continues on to flow over the "input" ends of capillaries that receive blood from the body. This blood has even less oxygen than the (by now) oxygen-poor water, so still more oxygen diffuses inward. With this opposing flow mechanism, fish get enough oxygen even though not much is dissolved in water, compared to air.

Tracheas. Insects and spiders are among the animals with air-conducting tubes called **tracheas.** Most insect tracheas are chitin-reinforced (Figure 28.3). They branch finely through the body and provide a rather self-contained system of gas conduction and exchange; assistance by a circulatory system is not required. Often a lid (spiracle) spans each opening at the body surface and helps keep the tubes moist by preventing evaporation.

Have you ever noticed how foraging bees stop every so often and pump the segments of their abdomen back and forth? The segments extend and retract like a telescope, forcing air into and out of the tracheal system. The stepped-up oxygen intake and carbon dioxide removal help support the high rate of metabolism required for insect flight.

Lungs. A **lung** is an internal respiratory surface in the shape of a cavity or sac. Simple lungs evolved more

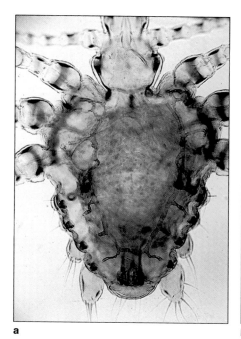

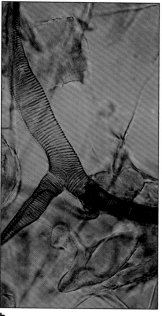

Figure 28.3 (**a**) Respiratory system of an insect (a louse). (**b**) A closer view of some chitin-reinforced tracheas.

than 450 million years ago and apparently assisted respiration in oxygen-poor habitats. In some lineages, the lungs developed into moist, thin-walled swim bladders. (Adjustments of gas volume in these organs help maintain the body's position in the water; some oxygen is also exchanged with blood and the surrounding tissues.) In other lineages, the lungs became complex respiratory organs.

In all animals with lungs, *airways* carry gas to and from one side of the respiratory surface of the lungs, and *blood vessels* carry gas to and from the other side:

1. Air moves by bulk flow into and out of the lungs, and new air is delivered to the respiratory surface.

2. Gases diffuse across the respiratory surface of the lungs.

3. Pulmonary circulation (the bulk flow of blood to and from the lung tissues) enhances the diffusion of dissolved gases into and out of lung capillaries.

4. In other tissues of the body, gases diffuse between blood and interstitial fluid, then between interstitial fluid and cells.

Let's focus now on the human respiratory system; its operating principles are the same for most vertebrates. The major exception is the respiratory system of birds, shown in Figure 28.4.

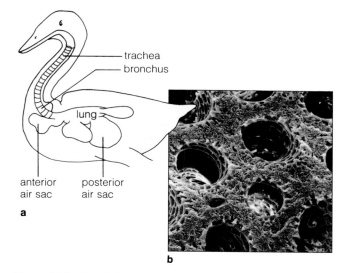

Figure 28.4 Respiratory system of birds. High metabolic rates and efficient gas exchange sustain flight and other activities. The rates are possible because of a unique ventilating system (**a**).

Typically, four air sacs are attached to each bird lung, which is somewhat small and inelastic. The sacs are not respiratory surfaces; they are more like bellows.

When the bird inhales, air is drawn into air sacs through small tubes (open at both ends) present in vascularized lung tissue. This is the respiratory surface (**b**). When the bird exhales, air is blown out of the sacs, through the small tubes, and out of the trachea.

Thus, air is not merely drawn into bird lungs: it is drawn *through* them. Air sacs and intricate lung airways make possible a *continuous* flow of air across the respiratory surface.

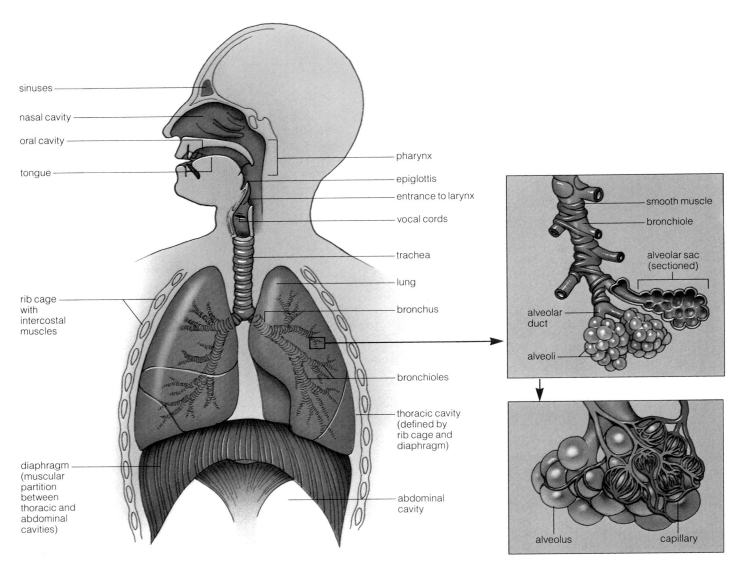

Figure 28.5 Human respiratory system.

Human Respiratory System

Air-Conducting Portion. The human respiratory system is shown in Figure 28.5. Air enters and leaves through the nose and, to a lesser extent, the mouth. Hairs and ciliated epithelium lining the two nasal cavities filter out dust and other large particles. Here also, incoming air becomes warmed and picks up moisture from mucus.

The filtered, warmed, and moistened air moves into the *pharynx*, or throat; this is the entrance to both the *larynx* (an airway) and the esophagus (a tube leading to the stomach). When you breathe, a flaplike structure attached to the larynx—the epiglottis—points up. When you swallow, the larynx moves upward and presses against the epiglottis, which partly covers the opening of the larynx. This helps prevent food from going down the wrong tube (Figure 28.6).

Vocal cords, two thickened folds of the wall of the larynx, contain the muscles used for speech. Air forced through the space between the vocal cords (the glottis) causes the vocal cords to vibrate and give rise to sound waves. The greater the air pressure on the vocal cords, the louder the sound. The greater the muscle tension on the cords, the higher the pitch of the sound.

From the larynx, air moves into the *trachea*, or windpipe, which branches into the two airways leading into the lungs. Each airway is a *bronchus* (plural, bronchi).

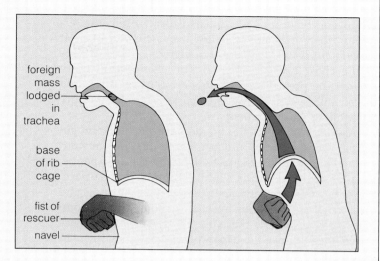

Figure 28.6 The Heimlich maneuver. Each year, several thousand people strangle to death when food enters the trachea instead of the esophagus (compare Figure 25.3). Strangulation can occur when the air flow is blocked for as little as four or five minutes. The Heimlich maneuver, an emergency procedure only, often can dislodge the misdirected chunks of food. The idea is to elevate the diaphragm forcibly, causing a sharp decrease in the chest cavity volume and a sudden increase in alveolar pressure. The increased pressure forces air up the trachea and may be enough to dislodge the obstruction.

To perform the Heimlich maneuver, stand behind the victim, make a fist with one hand, then press the fist, thumb-side in, against the victim's abdomen. The fist must be slightly above the navel and well below the rib cage. Next, press the fist into the abdomen with a sudden upward thrust. Repeat the thrust several times if needed. The maneuver can be performed on someone who is standing, sitting, or lying down.

Once the obstacle is dislodged, be sure the person is seen at once by a physician, for an inexperienced rescuer can inadvertently cause internal injuries or crack a rib. It could be argued that the risk is worth taking, given that the alternative is death.

Like the trachea, a bronchus has a cartilage-reinforced wall. Its epithelial lining contains cilia and mucus-secreting cells, both with housekeeping roles (Figure 28.7). Bacteria and airborne particles stick in the mucus. Then the cilia, beating upward, sweep the debris-laden mucus toward the mouth.

Gas Exchange Portion. Humans have two elastic, cone-shaped lungs, separated from each other by the heart. The lungs are located in the rib cage above the *diaphragm,* a muscular partition between the chest cavity and abdominal cavity. A thin membrane called the pleural sac surrounds each lung.

Inside the lungs, airways become shorter, more narrow, and more numerous. The first of the branchings that no longer contain cartilage-supported walls are the *bronchioles.* The terminal airways, the *respiratory bronchioles,* have cup-shaped outpouchings from their walls. Each outpouching is an **alveolus** (plural, alveoli). Most often, alveoli are clustered together, forming a larger pouch called an **alveolar sac** (Figure 28.5). These sacs are the major sites of gas exchange.

A dense mesh of blood capillaries surrounds the 150 million or so alveoli in each lung. Together, the alveoli provide a tremendous surface area for exchanging gases with the bloodstream. If they were stretched out as a single layer, they would cover the floor of a racquet ball court!

Figure 28.7 Color-enhanced scanning electron micrograph of cilia (gold) in the respiratory tract. Mucus-secreting cells (rust-colored) are interspersed among the ciliated cells. Foreign material sticks to the mucus-coated microvilli at the free surface of these cells, then the cilia sweep the mucus-laden debris back toward the mouth.

AIR PRESSURE CHANGES IN LUNGS

When you breathe, air is inhaled (drawn into the airways), then exhaled (expelled from them). The air movements result from rhythmic increases and decreases in the chest cavity's volume. The changing volumes reverse the pressure gradients between the lungs and the air outside the body—and gases in the respiratory tract follow those gradients.

As you start to inhale, the dome-shaped diaphragm contracts and flattens, and muscles lift the ribs upward and outward (Figure 28.8). As the chest cavity expands, the rib cage moves away slightly from the lung surface. Pressure in the narrow space between each lung and the pleural sac becomes even lower than it was, compared to atmospheric pressure. The pressure difference causes the lung itself to expand more, allowing fresh air to flow down the airways.

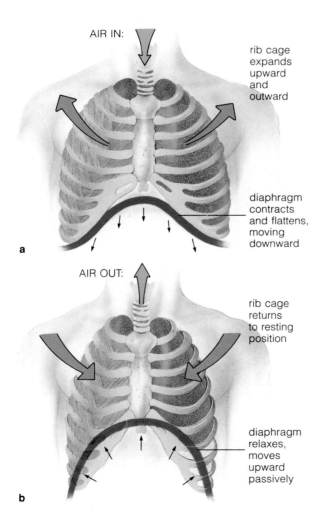

AIR IN:

rib cage expands upward and outward

diaphragm contracts and flattens, moving downward

a

AIR OUT:

rib cage returns to resting position

diaphragm relaxes, moves upward passively

b

Figure 28.8 Changes in the size of the chest cavity during breathing. Blue line indicates the position of the diaphragm during inhalation (**a**) and exhalation (**b**).

As you start to exhale, the elastic lung tissue recoils passively. The volume of the chest cavity decreases and compresses the air in alveolar sacs. The alveolar pressure becomes greater than the atmospheric pressure, so air follows the gradient and moves out from the lungs.

GAS EXCHANGE AND TRANSPORT

Gas Exchange in Alveoli

Each alveolus is only a single layer of epithelial cells, surrounded by a thin basement membrane. At most, a very thin film of interstitial fluid separates epithelial cells of the alveoli from the lung capillaries. Gases can diffuse rapidly across this narrow space (Figure 28.9).

Figure 28.10 shows the partial pressure gradients for oxygen and carbon dioxide through the human respiratory system. Passive diffusion alone is enough to move oxygen across the respiratory surface and into the bloodstream. And it is enough to move carbon dioxide in the reverse direction.

Driven by its partial pressure gradient, oxygen diffuses from alveolar air spaces, through interstitial fluid, and into the lung capillaries.

Carbon dioxide, driven by its partial pressure gradient, diffuses in the reverse direction.

Gas Transport Between Lungs and Tissues

Blood can carry only so much oxygen and carbon dioxide in dissolved form, so gas transport must be enhanced to meet the needs of the entire body. The hemoglobin of red blood cells increases oxygen transport by seventy times. Through some reversible reactions, carbon dioxide transport is increased by seventeen times by the red blood cells.

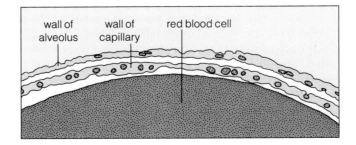

wall of alveolus   wall of capillary   red blood cell

Figure 28.9 Diagram of a section through an alveolus and an adjacent blood capillary. By comparison to the diameter of the red blood cell, the diffusion distance across the capillary wall, the interstitial fluid, and the alveolar wall is exceedingly small.

Oxygen Transport. There is plenty of oxygen and not much carbon dioxide in inhaled air that reaches the alveoli, but the opposite is true of blood in the lung capillaries. So oxygen diffuses into the blood plasma, then into red blood cells, where it rapidly binds with hemoglobin. A hemoglobin molecule tends to give up oxygen in tissues where the partial pressure of oxygen is lower than in the lungs. Hemoglobin's affinity for oxygen is also weaker in tissues where blood is warmer and shows a decrease in pH. These conditions occur to a greater extent in tissues with greater metabolic activity. That is why more oxygen is released in vigorously contracting muscle tissues, for example.

Carbon Dioxide Transport. The partial pressure of carbon dioxide in metabolically active tissues is greater than it is in blood flowing through the capillaries threading through them. Carbon dioxide diffuses into the capillaries, then it is transported to the lungs. Some remains dissolved in plasma or binds with hemoglobin. But most of the carbon dioxide is transported in the form of bicarbonate.

The bicarbonate forms when carbon dioxide combines with water in plasma to form carbonic acid (H_2CO_3), which separates into bicarbonate and hydrogen ions:

$$CO_2 + H_2O \rightleftharpoons H_2CO_3 \rightleftharpoons HCO_3^- + H^+$$

This reaction proceeds slowly in plasma. But much of the carbon dioxide diffuses into red blood cells, which contain an enzyme that increases the reaction rate by 250 times. The action of this enzyme helps maintain the gradient that keeps carbon dioxide diffusing from interstitial fluid into the bloodstream. The bicarbonate that forms tends to diffuse out of the red blood cells and into the plasma.

The reactions are reversed in the alveoli, where the partial pressure of carbon dioxide is lower than it is in the capillaries. Carbonic acid separates into water and carbon dioxide, which diffuses into the alveolar sacs. From there it is exhaled from the body.

CONTROLS OVER RESPIRATION

Matching Air Flow to Blood Flow

Gas exchange is most efficient when the rate of air flow is matched with the rate of blood flow. Both rates can be adjusted locally in the lungs and through the body as a whole.

The nervous system controls oxygen and carbon dioxide levels in arterial blood for the entire body. It does this by adjusting contractions of the diaphragm

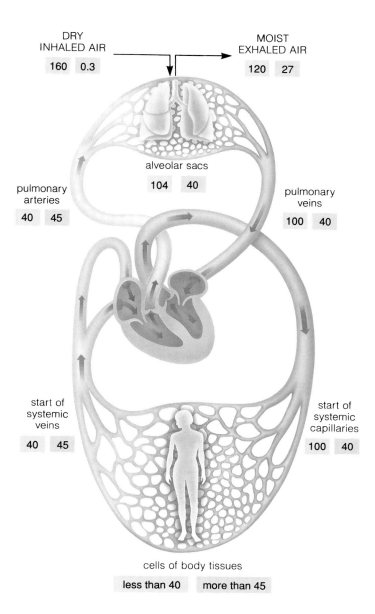

DRY INHALED AIR 160 0.3

MOIST EXHALED AIR 120 27

alveolar sacs 104 40

pulmonary arteries 40 45

pulmonary veins 100 40

start of systemic veins 40 45

start of systemic capillaries 100 40

cells of body tissues less than 40   more than 45

Figure 28.10 Partial pressure gradients for oxygen (blue boxes) and carbon dioxide (pink boxes) through the respiratory tract.

At sea level, the gases making up the atmosphere exert a total pressure of 760mm Hg. This simply means that the pressure is enough to hold the mercury (Hg) inside the narrow glass tube of a barometer at a height of 760 millimeters (mm). Oxygen makes up 21 percent of the atmosphere, so its partial pressure is (760 × 21/100), or about 160mm Hg. The partial pressure of carbon dioxide is about 0.3mm Hg.

The point to remember about the values shown is that *each gas moves from regions of higher to lower partial pressure.* That is why, for example, you become light-headed when you first visit places at high altitudes. The partial pressure of oxygen decreases with altitude, and your body does not function as well when the pressure gradient between the surrounding air and your lungs is lower than what you normally encounter. You simply cannot take in enough oxygen (page 412).

When the Lungs Break Down

In large cities, in certain occupations, even near a cigarette smoker, airborne particles and certain gases are present in abnormal amounts, and they put extra workloads on the lungs. Ciliated epithelium in the bronchioles is especially sensitive to cigarette smoke.

Bronchitis. A disorder called bronchitis can be brought on by smoking and other forms of air pollution that increase mucus secretions and interfere with ciliary action in the lungs. Mucus and the particles it traps—including bacteria—accumulate in the trachea and bronchi, and this triggers coughing. The coughing persists as long as the irritation does and it aggravates the bronchial walls, which become inflamed. Bacteria or chemical agents start destroying the wall tissue. Cilia are lost from the lining, and mucus-secreting cells multiply as the body works to fight against the accumulating debris. Fibrous scar tissue forms and can obstruct parts of the respiratory tract.

Emphysema. An acute attack of bronchitis can be treated easily if the person is otherwise in good health. When the irritation persists, however, fibrous scar tissue builds up and the bronchi become clogged with more and more mucus. Air becomes trapped in alveoli, and the alveolar walls break down. Inelastic fibrous tissue comes to surround the alveoli. The remaining alveoli enlarge and the balance between air flow and blood flow is abnormal. The outcome is emphysema, in which the lungs are so distended and inelastic that gases cannot be exchanged efficiently (compare Figures a and b). Running, walking, even exhaling can be difficult.

Poor diet, smoking, and chronic colds and other respiratory ailments sometimes make a person susceptible to emphysema later in life. And many who suffer from emphysema do not have a functional gene coding for antitrypsin. This substance inhibits tissue-destroying enzymes produced by bacteria.

Emphysema can develop slowly, over twenty or thirty years. By the time it is detected, the damage to lung tissue cannot be repaired. On average, 1.3 million people in the United States alone suffer from the disorder.

Effects of Cigarette Smoke. The table that follows lists some effects of cigarette smoke on the lungs and other organs. Cilia in the bronchioles can be kept from beating for several hours by noxious particles in smoke from one cigarette. The particles also stimulate mucus secretions, which in time can clog the airways. They can kill the infection-fighting phagocytes that normally patrol the respiratory epithelium. "Smoker's cough" is not the only outcome; the coughing can pave the way for bronchitis and emphysema. Marijuana smoke also can cause extensive lung damage.

(**a**) Normal appearance of human lung tissue. (**b**) A lung from someone afflicted with emphysema. (**c**) Cigarette smoke swirling down the human windpipe and into the two bronchial routes to the lungs.

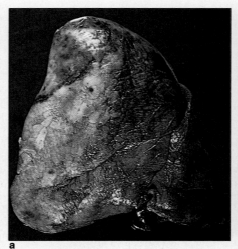

a

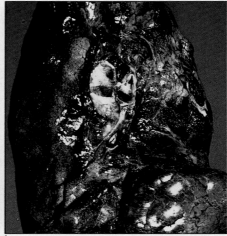

b

Cigarette smoke contributes to lung cancer. Inside the body, certain compounds in coal tar and cigarette smoke become converted to highly reactive intermediates. These are the real carcinogens; they provoke uncontrolled cell divisions in lung tissues. In its terminal stage, the pain associated with lung cancer is agonizing.

Susceptibility to lung cancer is related to how many cigarettes are smoked daily and to how many times and how deeply smoke is inhaled. Cigarette smoking is responsible for at least eighty percent of all lung cancer deaths. It is a disorder that only ten out of a hundred smokers will survive.

Risks Associated with Smoking	**Benefits of Quitting**
Shortened Life Expectancy: Nonsmokers live 8.3 years longer on average than those in midtwenties who smoke two packs daily	Cumulative risk reduction; after 10–15 years, life expectancy of ex-smokers approaches that of nonsmokers
Chronic Bronchitis, Emphysema: Smokers have 4–25 times more risk of dying from these diseases than do nonsmokers	Greater chance of improving lung function and slowing down rate of deterioration
Lung Cancer: Cigarette smoking the major cause of lung cancer	After 10–15 years, risk approaches that of nonsmokers
Cancer of Mouth: 3–10 times greater risk among smokers	After 10–15 years, risk is reduced to that of nonsmokers
Cancer of Larynx: 2.9–17.7 times more frequent among smokers	After 10 years, risk is reduced to that of nonsmokers
Cancer of Esophagus: 2–9 times greater risk of dying from this	Risk proportional to amount smoked; quitting should reduce it
Cancer of Pancreas: 2–5 times greater risk of dying from this	Risk proportional to amount smoked; quitting should reduce it
Cancer of Bladder: 7–10 times greater risk for smokers	Risk decreases gradually over 7 years to that of nonsmokers
Coronary Heart Disease: Cigarette smoking a major contributing factor	Risk drops sharply after a year; after 10 years, risk reduced to that of nonsmokers
Effects on Offspring: Women who smoke during pregnancy have more stillbirths, and weight of liveborns averages less (hence babies are more vulnerable to disease, death)	When smoking stops before fourth month of pregnancy, risk of stillbirth and lower birthweight eliminated
Impaired Immune System Function: Increase in allergic responses, destruction of macrophages in respiratory tract	Avoidable by not smoking

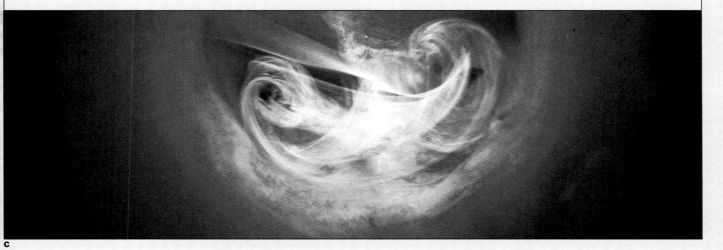

c

and muscles in the chest wall, and so adjusts the rate and depth of breathing. The brain receives input from sensory receptors that can detect rising carbon dioxide levels in the blood. It also receives input from receptors in the walls of certain arteries. These receptors can detect decreases in the partial pressure of oxygen dissolved in arterial blood. The brain responds by increasing the rate and depth of respiration, so more oxygen can be delivered to affected tissues and more carbon dioxide removed from them.

Local controls come into play in the lungs themselves when there are imbalances between air flow and blood flow. For example, when air flow in a lung region is hampered because of a diseased airway, oxygen partial pressure decreases in that region. The decrease affects smooth muscle in arterioles. The arterioles constrict, improving the match between the rates of air and blood flow.

Hypoxia

The partial pressure of oxygen decreases with increasing altitude. People who live at high altitudes produce more oxygen-carrying red blood cells, but visitors who have not had time to adapt to the "thinner air" can suffer *hypoxia*, or cellular oxygen deficiency. At 2,400 meters (about 8,000 feet) above sea level, they attempt to compensate for the oxygen deficiency by hyperventilating, or breathing much faster and more deeply than normal. At 3,650 meters (about 12,000 feet), oxygen deprivation causes headaches, nausea, and lethargy. At 7,000 meters (23,000 feet), hypoxia can lead to loss of consciousness and death.

Hypoxia also occurs when the oxygen content in arterial blood falls because of *carbon monoxide poisoning*. Carbon monoxide, a colorless, odorless gas, is present in exhaust fumes from cars and in smoke from tobacco, coal, or wood burning. It combines with hemoglobin at least 200 times faster than oxygen does. Even very small amounts can tie up half of the body's hemoglobin and so affect oxygen delivery to tissues.

SUMMARY

1. Aerobic metabolism requires oxygen and produces carbon dioxide. The process by which the body as a whole acquires oxygen and disposes of carbon dioxide is called respiration.

2. Air is a mixture of gases, each exerting a partial pressure. Each gas tends to move from areas of higher to lower partial pressure. Respiratory systems make use of this tendency.

3. Different animals use the body surface, gills, tracheas, or lungs as the basis of respiration. In all cases, oxygen and carbon dioxide diffuse across a moist, thin layer of epithelium (the respiratory surface). Inside the larger animals, blood vessels carry gas to and from the respiratory surface.

4. The interconnected airways of the human respiratory system are the nasal cavities, pharynx, larynx, trachea, bronchi, and bronchioles. Alveoli, located at the end of the airways, are the gas exchange portion of the system.

5. During inhalation, the chest cavity expands, the pressure in the lungs falls below atmospheric pressure, and air flows into the lungs. During normal exhalation, these processes are reversed.

6. Driven by its partial pressure gradient, oxygen brought into the lungs diffuses from alveolar air spaces into the pulmonary capillaries. Then it diffuses into red blood cells and binds with hemoglobin. When the oxygen-rich blood reaches body tissues, hemoglobin gives up the oxygen, which diffuses out of the capillaries, across interstitial fluid, and into cells. Hemoglobin combines with or releases oxygen in response to shifts in oxygen levels, pH, and temperature.

7. Driven by its partial pressure gradient, carbon dioxide diffuses from cells, through interstitial fluid, and into the bloodstream. Most reacts with water to form bicarbonate, but the reactions are reversed in the lungs, where carbon dioxide diffuses from the lungs into the air spaces of the alveoli.

Review Questions

1. What is the main requirement for gas exchange in animals? What types of systems are used for gas exchange in (a) water-dwelling animals and (b) land-dwelling animals? *406–409*

2. What governs the rate and depth of breathing? *413–416*

3. What drives oxygen from alveolar air spaces, through interstitial fluid, and across capillary epithelium? What drives carbon dioxide in the reverse direction? *412*

4. How does hemoglobin help maintain the oxygen partial pressure gradient during gas transport in the body? What reactions enhance the transport of carbon dioxide through the body? *413*

5. Label the component parts of the human respiratory system: *410*

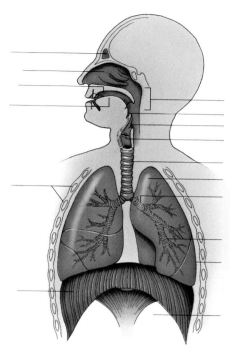

Self-Quiz   *(Answers in Appendix IV)*

1. _____ is used and _____ is produced in aerobic metabolism.

2. Operation of respiratory systems depends on the tendency of a _____ to diffuse down its _____.

3. Respiratory systems require thin, moist layers of _____ across which gases can easily _____.

4. Respiratory systems differ in their _____ for increasing gas exchange efficiency and the means for matching _____ to blood flow.

5. Which of the following is *not* related to the function of a respiratory system?
   a. air is a mixture of gases
   b. each gas in air exerts a partial pressure
   c. each gas in air tends to move from areas of higher to lower partial pressure
   d. all of the above are directly applicable

6. The basis of respiration in different animals might be _____.
   a. the body surface
   b. gills
   c. tracheas
   d. lungs
   e. all of the above

7. During inhalation, _____.
   a. the pressure in the thoracic cavity is greater than the pressure within the lungs
   b. the pressure in the thoracic cavity is less than the pressure within the lungs
   c. the diaphragm moves upward and becomes more curved
   d. the chest cavity volume decreases

8. Oxygen diffusing into pulmonary capillaries also diffuses into _____ and binds with _____.
   a. white blood cells; carbon dioxide
   b. red blood cells; carbon dioxide
   c. white blood cells; hemoglobin
   d. red blood cells; hemoglobin

9. Due to its partial pressure gradient, carbon dioxide diffuses from cells, into interstitial fluid, and into the _____; in the lungs, carbon dioxide diffuses into the _____.
   a. alveoli; bronchioles
   b. bloodstream; bronchioles
   c. alveoli; bloodstream
   d. bloodstream; alveoli

10. Match these respiratory components with their descriptions.
   _____ bronchi
   _____ alveoli
   _____ trachea
   _____ larynx
   _____ pharynx
   a. microscopically small air sacs where gases are exchanged
   b. contains vocal cords
   c. throat cavity behind mouth
   d. flexible windpipe reinforced with cartilage
   e. connects trachea to lungs

Selected Key Terms

alveolar sac *411*
alveolus *411*
bronchus *410*
diaphragm *411*
gill *408*
glottis *410*
integumentary exchange *407*
lung *408*
pleural sac *411*
respiration *406*
respiratory bronchiole *411*
trachea *408*
vocal cord *410*

Readings

American Cancer Society. 1980. *Dangers of Smoking; Benefits of Quitting and Relative Risks of Reduced Exposure.* Revised edition. New York: American Cancer Society.

Vander, A., J. Sherman, and D. Luciano. 1990. *Human Physiology: The Mechanisms of Body Function.* Fifth edition. New York: McGraw-Hill. Clear introduction to the respiratory system.

West, J. 1985. *Respiratory Physiology: The Essentials.* Third edition. Baltimore: Williams & Wilkins. Excellent, brief introduction to respiratory functions. Paperback.

29 SOLUTE-WATER BALANCE

KEY CONCEPTS

1. Vertebrate kidneys function to maintain the internal environment by balancing the body's intake and output of water and dissolved substances (solutes).

2. Urine forms in kidney nephrons. Its volume and composition depend on three processes, called filtration, reabsorption, and secretion.

3. First, water and solutes that are present in blood filter under pressure into nephrons. Second, most of the filtrate is reabsorbed by blood capillaries threading around the nephrons. Third, excess ions and a few foreign chemicals move from capillaries into tubular parts of the nephron for disposal by way of the urine.

4. Reabsorption occurs by way of concentration gradients that are maintained between nephrons and the interstitial fluid surrounding the nephrons. The amounts reabsorbed in a given interval are under neural control.

Judging from the fossil record, animals first evolved in shallow, ancient seas. From the beginning, then, their tissues and organs were geared to operating in a salty fluid. About 375 million years ago, some animals began invading the land. They were able to leave the seas behind partly because they carried salty fluid with them, as an "internal environment." But there were new challenges on land. Winds and radiant energy from the sun could dehydrate the animal body. Water was not always plentiful and most of it was fresh, not salty. Over time, the structures and functions of land-dwelling animals became modified in response to these threats, in ways that preserved the stability of the internal environment.

Today, as then, animals must respond to physical conditions over which they often have little or no control. In this chapter, we turn to the obligatory exchanges and adjustments that affect the body's water and solute balances. Our focus will be on the structure and function of the mammalian urinary system (Figure 29.1).

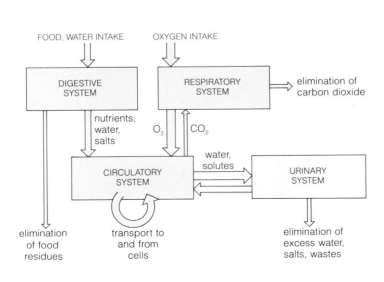

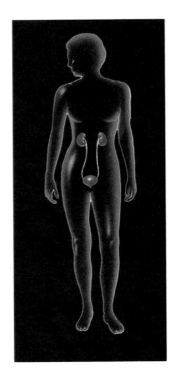

Figure 29.1 Links between the urinary system and other organ systems that maintain operating conditions in the internal environment.

CONTROL OF EXTRACELLULAR FLUID

Water Gains and Losses

Ordinarily, mammals take in just as much water as they lose on a daily basis (Table 29.1). Two processes account for the water gains:

1. Absorption of water from liquids and solid foods in the gastrointestinal tract.

2. Metabolism (specifically, the breakdown of carbohydrates and other organic molecules in reactions that yield water as a by-product).

Thirst behavior affects water intake. This behavior is under the control of the nervous system.

The body loses water by several processes, the most important of which are these:

1. Excretion by way of the urinary system.

2. Evaporation from the respiratory surface.

3. Evaporation through the skin.

4. Sweating.

5. Elimination by way of the gastrointestinal tract.

The process of greatest importance in controlling water loss is urinary **excretion,** the elimination of excess water and excess (or harmful) solutes from the internal environment by way of the kidneys. The next two processes listed above are called "insensible water losses" because the individual is not aware that they are taking place. Temperature control centers in the nervous system govern sweating (page 441). Normally, the large intestine reabsorbs nearly all water in the gastrointestinal tract, so very little water leaves the body in feces.

Solute Gains and Losses

Aside from oxygen (which is absorbed at the respiratory surface), solutes are added to the internal environment by three processes:

1. Absorption from the gastrointestinal tract. The absorbed substances include *nutrients* such as glucose (used as energy sources and in biosynthesis reactions), as well as drugs and food additives. They also include *mineral ions,* such as sodium and potassium ions.

2. Secretion of hormones and other substances.

3. Metabolism, including *waste products* of degradative reactions.

Carbon dioxide, the most abundant waste of metabolism, is eliminated at the respiratory surface. These are the other major metabolic wastes that must be eliminated:

1. *Ammonia,* formed in reactions whereby amino groups are stripped from amino acids. If allowed to accumulate in the body, ammonia can be highly toxic.

2. *Urea,* produced in the liver in reactions that link two ammonia molecules to carbon dioxide. Urea is the main nitrogen-containing waste product of protein breakdown and is relatively harmless.

3. *Uric acid,* formed in reactions that degrade nucleic acids. If allowed to accumulate, uric acid can crystallize and sometimes collect in the joints.

Table 29.1	Normal Balance Between Water Gain and Water Loss in Humans and in Kangaroo Rats				
Organism	Water Gain (milliliters)			Water Loss (milliliters)	
Adult human (measured on daily basis)	Ingested in solids:	850	Urine:	1,500
	Ingested as liquids:	1,400	Feces:	200
	Metabolically derived:	350	Evaporation:	900
		2,600		2,600
Kangaroo rat (measured over 4 weeks)	Ingested in solids:	6.0	Urine:	13.5
	Ingested as liquids:	0	Feces:	2.6
	Metabolically derived:	54.0	Evaporation:	43.9
		60.0		60.0

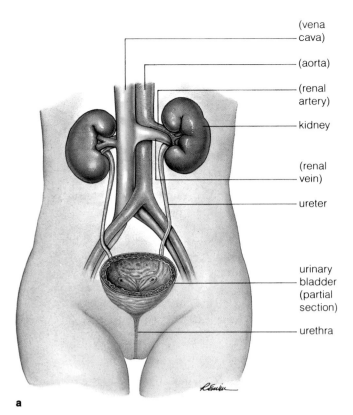

(vena cava)

(aorta)

(renal artery)

kidney

(renal vein)

ureter

urinary bladder (partial section)

urethra

a

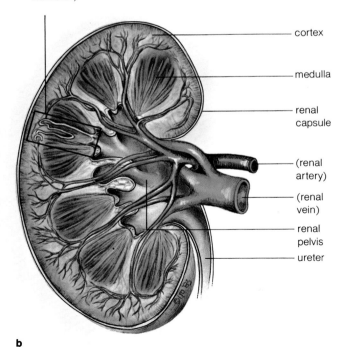

nephron (greatly exaggerated in diameter)

cortex

medulla

renal capsule

(renal artery)

(renal vein)

renal pelvis

ureter

b

Figure 29.2 (**a**) Components of the human urinary system. (**b**) Closer look at the kidney.

Urinary System of Mammals

A pair of organs called **kidneys** continuously filter water, mineral ions, organic wastes, and other substances from the blood. Only a tiny portion of the water and solutes going into the kidneys leaves as a fluid called **urine**; all but about one percent is returned to the blood. But the composition of the fluid that *is* returned has been adjusted in vital ways. *Through their action, kidneys regulate the volume and solute concentrations of extracellular fluid.*

Each kidney has two zones, a cortex and a medulla, enclosed within a tough coat (Figure 29.2). The human kidney is divided into several lobes, each containing many blood vessels and slender tubes called **nephrons**. Water and solutes filter out of the blood and enter the nephrons. Most of the filtrate is reabsorbed in nephrons, but some moves into the kidney's central cavity (renal pelvis); this fluid is the urine.

Urine flows from each kidney into a *ureter*, then into the *urinary bladder* (a storage organ). It leaves through a long tube, the *urethra*, which leads to the outside. The two kidneys, two ureters, urinary bladder, and urethra constitute the **urinary system** of mammals (Figure 29.2).

You may have heard about "kidney stones," these being deposits of uric acid, calcium salts, and other substances that settled out of urine and collected in the renal pelvis. At times the stones become lodged in the urethra, where they interfere with urine flow and intermittently cause pain. Kidney stones usually pass naturally from the body; if they do not, they can be eliminated by medical or surgical procedures.

Nephron Structure

Each fist-sized kidney has more than a million nephrons. A layer of epithelial cells makes up the nephron wall, but the cells and junctions between them are not all the same. Some wall regions are highly permeable to water and solutes, and others bar the passage of solutes except by way of specific transport systems built into the cell membranes (page 43).

Filtration starts at the **glomerulus**, where the nephron wall balloons around a cluster of blood capillaries (Figure 29.3). The nephron wall forms a cup (*Bowman's capsule*) for water and solutes being filtered from blood. The filtrate flows inside the nephron wall, first through a **proximal tubule**, then through a hairpin-shaped **loop of Henle** and a **distal tubule**, then through a collecting duct.

The capillaries that receive blood inside a Bowman's capsule do not send blood directly back to the general circulation. Instead, they converge to form an arteriole that branches into *another* set of capillaries. This set threads around the rest of the nephron and recaptures

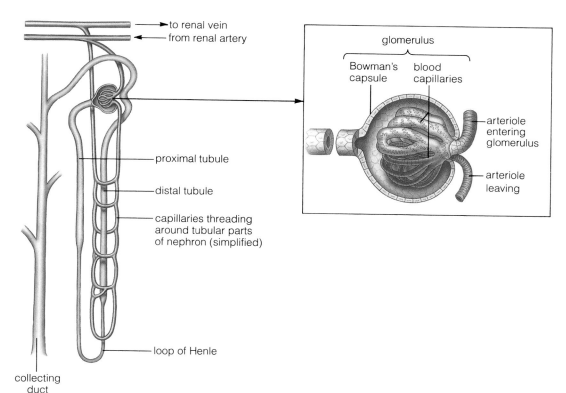

to renal vein
from renal artery

proximal tubule

distal tubule

capillaries threading
around tubular parts
of nephron (simplified)

loop of Henle

collecting
duct

glomerulus

Bowman's capsule   blood capillaries

arteriole entering glomerulus

arteriole leaving

Figure 29.3 Diagrams of a nephron, the functional unit of the kidney, and its association with blood capillaries.

water and essential solutes (Figure 29.3). Eventually these capillaries merge to form veins, which carry blood out of the kidney.

URINE FORMATION

Urine contains waste products as well as water and solutes in excess of the amounts necessary to maintain the extracellular fluid. As Figure 29.4 indicates, urine forms through three processes, called *filtration, reabsorption,* and *secretion.*

Filtration of Blood

Filtration occurs at the glomerulus. In this process, blood pressure generated by heart contractions forces water and solutes out of the capillaries and into Bowman's capsule. The blood is said to be filtered because blood cells, proteins, and other large solutes are left behind as smaller solutes (such as glucose, sodium, and urea) and water are forced out. The filtrate itself will flow into the proximal tubule.

More blood is forced through the human kidneys than through any other organ except the lungs. Kidneys filter about $1\frac{1}{2}$ quarts of blood every sixty seconds—

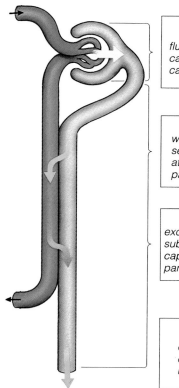

FILTRATION
fluid is filtered out of capillaries and into Bowman's capsule of a glomerulus

REABSORPTION
water, nutrients, salts are selectively returned to blood at capillaries around tubular parts of nephron

SECRETION
excess H^+, K^+, and some other substances are moved out of capillaries and into tubular parts of nephron

EXCRETION
excess water, solutes are eliminated by way of the urinary tract

Figure 29.4 Processes involved in the formation and excretion of urine.

about 45 gallons (180 liters) every day! Filtration is rapid for two reasons. First, blood entering a glomerulus is still under high pressure (arterioles leading into it have wider diameters—and less resistance to flow—than most arterioles). Second, capillaries in the glomerulus are highly "leaky," so they filter about 10 to 100 times more water and small solutes than other capillaries.

Reabsorption of Water and Solutes

Reabsorption occurs along tubular parts of the nephron, into the adjacent capillaries. In this process, water and solutes move *out* of the nephron (by diffusion or active transport), then into the capillaries. Most of the water and usable solutes are reclaimed and can be sent back to the general circulation. Some rather large quantities are involved (Table 29.2). If reabsorption were somehow stopped, all the water in the bloodstream would be urinated away in less than thirty minutes.

During reabsorption, urine volume and composition can be adjusted to compensate for shifts in the body's gains and losses of water and solutes. Suppose a person drinks an excessive amount of water. Then, the urine increases in volume and is less concentrated. Suppose

the person does not take in as much water as the body requires. Then, the urine decreases in volume and is more concentrated.

Water Reabsorption. Water reabsorption is controlled by the hypothalamus. This brain center triggers the secretion of **ADH** (antidiuretic hormone) when the solute concentration of extracellular fluid increases past a set point (page 455). When this happens, water is conserved because ADH makes both the end region of distal tubules and the collecting ducts more permeable to water. More water is reabsorbed, so the urine leaving the collecting ducts is more concentrated. When the body must dispose of excess water, ADH secretion is inhibited, so the urine is dilute.

The hypothalamus also controls thirst. When the hypothalamus detects a rise in solute levels (a drop in water volume), it issues a call for water-seeking behavior.

Sodium Reabsorption. Sodium is the major solute in extracellular fluid, and its reabsorption is controlled partly by way of a gland called the adrenal cortex (page 455). When the body loses more sodium than it takes in, the volume of all extracellular fluid drops. Sensory receptors in the walls of blood vessels and the heart detect the drop, and certain kidney cells are called into action (Figure 29.5). The cells secrete an enzyme (renin), which indirectly prods the adrenal cortex into secreting the hormone **aldosterone**. This hormone acts on the distal tubules and collecting ducts, causing them to reabsorb sodium.

Sodium retention is accompanied by water retention, and this leads to a rise in blood pressure. Abnormally high blood pressure, or *hypertension*, can adversely affect kidney function. (The "tension" part of the name refers to the muscle tone of walls of arteries and arterioles.) The wear and tear on the kidneys as well as the vascular

Table 29.2	Average Daily Reabsorption Values for a Few Substances		
	Filtered	Excreted	Proportion Reabsorbed
Water	180 liters	1.8 liters	99%
Glucose	180 grams	None, normally	100%
Sodium ions	630 grams	3.2 grams	99.5%
Urea	54 grams	30 grams	44%

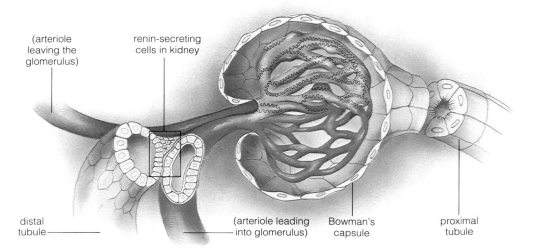

Figure 29.5 Location of renin-secreting cells that play a role in sodium reabsorption.

(arteriole leaving the glomerulus)   renin-secreting cells in kidney   distal tubule   (arteriole leading into glomerulus)   Bowman's capsule   proximal tubule

On Fish, Frogs, and Kangaroo Rats

Let's conclude this chapter with a look at how a few vertebrates maintain water and solute levels in entirely different settings—in the seas, in freshwater, and on land.

The tissues of herring, snapper, and other bony fishes have about three times less solutes than seawater does. These fishes continuously lose water (by osmosis) to their hypertonic environment, and continual drinking brings in replacements. (If the fishes are experimentally prevented from drinking, they die from dehydration in a few days.) Ingested solutes are excreted against concentration gradients. Although kidneys are present, they are too small to excrete much water or solutes. Most of the excess solutes are pumped out through membranes of fish gills, the cells of which actively transport sodium ions out of the blood (and potassium ions into it).

In fresh water, a hypotonic medium, lake trout and other bony fishes tend to gain water and lose solutes. The same is true of amphibians. These animals do not drink water; rather, water moves by osmosis into the body, through the thin gill membranes (or, in adult amphibians such as frogs, through the skin). Excess water leaves by way of well-developed kidneys, which excrete a large volume of dilute urine. Some solutes also are excreted, but the losses are balanced by solutes gained from food and by the active transport of sodium ions across the gills, into the body.

For desert-dwelling kangaroo rats, water is exceedingly scarce (see figure). The air is dry and temperatures can approach 45°C, so water losses could be devastating without behavioral and physiological adaptations for conserving water. Like many other desert animals, kangaroo rats spend the day in deep burrows, where the air temperature seldom exceeds 30°C. They look for food during the cooler hours of the night. Although some desert rodents eat moist plant parts, kangaroo rats eat primarily dry seeds, which contain very little water. They do not drink any water at all; most comes from the metabolic oxidation of carbohydrates, fats, and proteins in the seeds.

Most land-dwelling vertebrates lose water by way of the skin, respiratory tract, urine, and, to some extent, feces. Kangaroo rats and other desert rodents reduce all such losses. Their skin has no sweat glands, and it is thick and dry. Their nose is small, with narrow and twisted air passages. When the rodents inhale, dry air passing over the moist nasal tissues becomes warmed and saturated with water vapor. Evaporation from the nasal epithelium cools the tissues well below the body temperature. When the rodents exhale, the warm, humid air from the lungs is cooled as it passes over the cooled tissues. Water condenses on the nasal epithelium, like it does on the outside of a glass of ice water on a warm day. As a result, considerable respiratory water is recovered, not lost to the environment.

Kangaroo rats also do not give up much water in feces. Their urine can be twice as concentrated as that of laboratory rats and three times that of humans. This remarkable ability to conserve water loss is attributed to very long loops of Henle. The solute concentration in the surrounding interstitial fluid becomes very high. The osmotic gradient between that fluid and the urine is so steep that most of the water reaching the (equally long) collecting ducts is reabsorbed; only a small volume of concentrated urine leaves the kangaroo rat body.

A kangaroo rat, master of water conservation in deserts.

system and brain may proceed undetected for many years, with severe consequences (page 381). One of the ways to control blood pressure is to restrict salt intake.

Secretion

Secretion occurs at capillaries threading around the nephron. Excess amounts of hydrogen ions, potassium ions, and a few other substances move *out* of the capillaries and into the nephron. This highly regulated process also rids the body of foreign substances (such as penicillin), uric acid, the products of hemoglobin breakdown, and other wastes.

Acid-Base Balance

So far, we have considered the kidney's main function—how it reabsorbs water and sodium and so influences the total volume and distribution of body fluids. But the kidney has another function that has profound impact on health. *The kidneys help keep the extracellular environment from becoming too acidic or too basic.*

The overall acid-base balance is maintained by controls over the concentration of hydrogen ions (H^+) especially. Those controls are exerted through (1) buffer systems, (2) respiration, and (3) excretion by way of the kidneys.

Normally, the extracellular pH for the human body is between 7.35 and 7.45. At any time, different acidic and basic substances are entering the blood from the digestive tract and from normal cell metabolism. Many must be neutralized or eliminated, because acids lower the pH and bases raise it. Buffer systems can temporarily neutralize H^+ produced through metabolic reactions. Bicarbonate, for example, can combine with H^+ to form carbonic acid, which can separate into water and carbon dioxide:

$$H^+ + HCO_3^- \rightleftharpoons H_2CO_3 \rightleftharpoons H_2O + CO_2$$
$$\text{bicarbonate} \qquad \text{carbonic acid}$$

As we saw earlier, carbon dioxide can be disposed of in the lungs (page 413). Buffers can only neutralize the H^+ ions, not eliminate them entirely. *Only the urinary system can eliminate excess amounts of H^+.*

Kidney Failure

An estimated 13 million people in the United States alone suffer from kidney disorders, as when diabetes or autoimmune responses damage the glomeruli and interfere with urine formation. When the kidneys malfunction, ions and toxic by-products of protein breakdown can accumulate in the bloodstream. Nausea, fatigue, loss of memory and, in advanced cases, death may follow. A *kidney dialysis machine* can restore the proper solute balances. Like the kidney itself, the machine helps maintain extracellular fluid by selectively removing and adding solutes to the bloodstream.

"Dialysis" means the separation of substances across a membrane between solutions of differing concentrations. In *hemodialysis,* the machine is connected to an artery or a vein, then blood is pumped through tubes that are submerged in a warm-water bath. The precise mix of salts, glucose, and other substances in the bath sets up the correct gradients with the blood. In *peritoneal dialysis,* fluid of the proper composition is put into the abdominal cavity, then is drained out. Here, the lining of the cavity (the peritoneum) serves as the dialysis membrane.

Hemodialysis generally takes about four hours; blood must circulate repeatedly before solute concentrations in the body are improved. The procedure is usually performed three times a week. It is used as a bypass measure in acute, reversible kidney disorders. When kidney damage is severe and irreversible, the procedure must be used for the rest of the patient's life or until a functional kidney is transplanted. With treatment and controlled diets, many individuals are able to resume fairly normal activity.

SUMMARY

1. By balancing water and solute gains with water and solute losses, the body maintains its internal environment.

2. Water is gained by absorption from the gastrointestinal tract and by metabolism. A thirst mechanism controls water gain. Water is lost by evaporation from the lungs and skin, elimination from the gastrointestinal tract, and excretion of urine. Controls over water loss deal mainly with varying the composition and volume of urine.

3. In mammals, urine formation occurs in a pair of kidneys. Each human kidney contains about a million tubelike blood-filtering units called nephrons. Urine composition and volume depend on three processes. The first is filtration of blood at the glomerulus of a nephron, with blood pressure providing the force for filtration. In the second process, reabsorption, water and selected solutes move out of tubular parts of the nephron and back into adjacent blood capillaries. In the third process, secretion, excess ions and a few foreign substances move out of those capillaries and into the nephron, so that they are disposed of in urine.

4. Reabsorption occurs because of concentration gradients that are maintained between nephrons and the interstitial fluid surrounding them. Sodium ions and other ions follow gradients out of the nephron, and water follows passively, down its osmotic gradient.

5. The hormone ADH is secreted when the body must retain water; it acts on the nephron walls and makes them permeable to water. When the body must rid itself of excess water, ADH secretion is inhibited. The hormone aldosterone controls sodium reabsorption.

Review Questions

1. All animals have mechanisms for maintaining body fluid concentration and composition. In your own body, which organs cooperate in these tasks? *420*

2. Describe what happens during (a) filtration, (b) reabsorption, and (c) secretion in the kidney's nephron/capillary unit. What do these three processes influence? *421–424*

3. Which hormone is involved in the control of water reabsorption? *422* Which hormone plays a major role in sodium reabsorption? *422*

4. Which type of ion is especially important in maintaining the body's acid-base balance? *424*

Self-Quiz   *(Answers in Appendix IV)*

1. In vertebrates, maintaining the volume and composition of extracellular fluid depends on three kidney functions: _____, _____, and _____.

2. Urine formation occurs in _____.
   a. glomeruli
   b. loops of Henle
   c. nephrons
   d. ureters

3. The body gains water by _____.
   a. gastrointestinal absorption
   b. metabolism
   c. both a and b
   d. neither a nor b

4. The body loses water by _____.
   a. evaporation from lungs and skin
   b. elimination from gastrointestinal tract
   c. excretion of urine
   d. all of the above

5. Each human kidney contains about _____ nephrons.
   a. 1,000
   b. 10,000
   c. 100,000
   d. 1,000,000

6. The processes responsible for urine composition and volume occur in this order: _____.
   a. filtration, reabsorption, secretion
   b. secretion, reabsorption, filtration
   c. reabsorption, secretion, filtration
   d. secretion, filtration, reabsorption

7. Which of the following descriptions does *not* match the process named?
   a. *filtration:* blood enters Bowman's capsule of glomerulus
   b. *reabsorption:* water and solutes selectively returned to blood capillaries
   c. *secretion:* excess ions and some other substances move from blood capillaries into nephron
   d. all of the above match
   e. none of the above match

8. Concentration gradients that are maintained between nephrons and surrounding interstitial fluid are responsible for _____.
   a. blood filtration
   b. secretion
   c. reabsorption
   d. hypertension

9. The hormone ADH controls _____.
   a. nephron production
   b. sodium reabsorption
   c. secretion
   d. water retention

10. Match the solute-water balance concepts.
   _____ aldosterone
   _____ nephron
   _____ thirst mechanism
   _____ reabsorption
   _____ glomerulus

   a. blood filter of a nephron
   b. controls sodium reabsorption
   c. occurs at blood capillaries around the nephrons
   d. site of urine formation
   e. controls water gain

Selected Key Terms

ADH (antidiuretic hormone) *422*
aldosterone *422*
Bowman's capsule *420*
distal tubule *420*
excretion *419*
filtration *421*
glomerulus *420*
kidney *420*
loop of Henle *420*
nephron *420*
proximal tubule *420*
reabsorption *422*
secretion *424*
urinary system *420*
urine *420*

Readings

Schmidt-Nielsen, K. 1990. *Animal Physiology.* Fourth edition. New York: Cambridge. Chapters 8 and 9 provide an excellent introduction to water-solute balances in animals.

Smith, H. 1961. *From Fish to Philosopher.* New York: Doubleday. Available in paperback.

Valtin, H. 1983. *Renal Function: Mechanisms Preserving Fluid and Solute Balance in Health.* Second edition. Boston: Little, Brown. Paperback.

Vander, A., J. Sherman, and D. Luciano. 1990. "The Kidneys and Regulation of Water and Inorganic Ions" in *Human Physiology.* Fifth edition. New York: McGraw-Hill.

30 NEURAL CONTROL AND THE SENSES

KEY CONCEPTS

1. The vertebrate nervous system senses, interprets, and issues commands for response to specific aspects of the environment. Its communication lines are highly organized gridworks of neurons.

2. A polarity of charge exists across the plasma membrane of a neuron, and sudden, brief reversals in that polarity are the basis of messages sent through the nervous system. The reversals, called action potentials, occur when a neuron is adequately stimulated. They occur repeatedly, in sequence, from the point of stimulation to the neuron's junction with another cell.

3. At cell junctions called chemical synapses, action potentials trigger the release of a chemical substance that stimulates or inhibits the activities of the next cell in line.

4. As nervous systems evolved, so did sensory structures and motor structures. Their joint evolution was the foundation for more active, intricate life-styles.

WHAT NERVOUS SYSTEMS DO

Animals have marvelous ways of dealing with different aspects of the environment. They reach out or lunge after food; they pull back, crawl, swim, run, or fly when they are about to become food themselves. And think about what they have to do to find a mate and slow it down or hold its attention. (Think about all the things *you* have to do.) The more complex the environment and life-style, the more complex and rapid are the modes of sensory reception, integration, and response.

This chapter deals with the nervous system, its sensors, and its commands for response. The system is much more than a bunch of communication lines that are silent until signals are fed into them from the outside, much as telephone lines wait to carry calls from all over the country. The key point about a nervous system is not what it is made of but *what it does*—and it never stops doing things, even without proddings from the outside!

For example, nerve cells in brain regions concerned with breathing never do rest. They become active before birth, and nothing short of damage or death will stop them. Also before birth, nerve cells become organized in vast gridworks and begin a constant chatter among

Figure 30.1 A python of southern Asia, equipped with sensory receptors and a neural program for detecting, aiming, and striking at warm-blooded prey in the dark. Its heat-sensitive receptors are located inside the pits shown here, above and below its mouth.

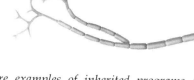
axon endings

themselves. *These are examples of inherited programs of activity in the nervous system.* The programs kick in early to help each animal respond to situations that members of its species are likely to encounter.

Consider the python shown in Figure 30.1. It eats small, night-foraging mammals that are a lot warmer than the night air. Its heat-detecting (infrared) receptors can pinpoint prey in the dark. The receptors notify the snake brain, which has a program for assessing signals about the location of objects. The program works; the snake's strike may be merely a few degrees off-center. Of course, the same snake might slither past a motionless, edible frog. Frog skin is cool and blends with background colors. The snake does not have receptors for detecting it or a neural program for responding to it.

Inherited neural programs help animals sense, interpret, and respond to specific aspects of the environment.

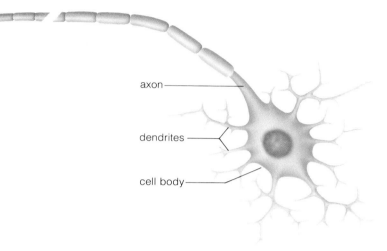
axon

dendrites

cell body

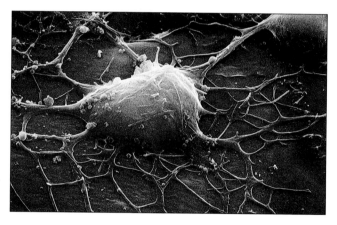

Figure 30.2 Component parts of a motor neuron. The scanning electron micrograph clearly shows the cell body and dendrites of this type of nerve cell.

NEURONS

Structure and Function of Neurons

The nerve cell, or **neuron**, is the basic unit of communication in vertebrate nervous systems. Neurons do not act alone. They collectively sense environmental change, integrate sensory inputs, then activate different body parts that can carry out responses. These tasks involve different classes of neurons, called sensory neurons, interneurons, and motor neurons.

We can define each class in terms of its role in a control scheme, described earlier, by which the nervous system monitors and responds to change. The scheme has receptor, integrator, and effector components:

INPUT: RECEPTORS INTEGRATORS EFFECTORS OUTPUT
stimulus sensory neurons inter-neurons activated by motor neurons *response*

Sensory neurons are *receptors* that can detect specific stimuli, such as light energy. They relay signals to the

brain and spinal cord—the *integrators* in our control scheme. In the brain and spinal cord are **interneurons,** which integrate information arriving on sensory lines and then influence other neurons in turn. **Motor neurons** relay information away from the integrators to muscle cells or gland cells, the body's *effectors,* which carry out responses.

Neurons differ greatly in structure. The ones described most often are motor neurons of the sort shown in Figure 30.2. Such neurons have many **dendrites** (short, slender extensions of the cell body) and an **axon** (a long, cylindrical extension). Their finely branched endings terminate next to muscle cells. Think of dendrites and

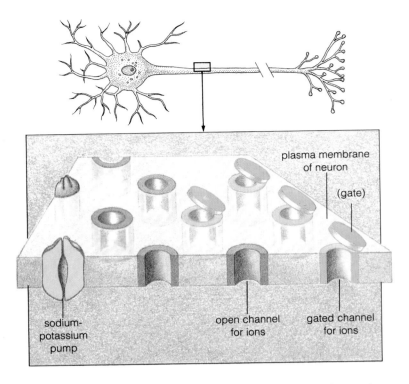

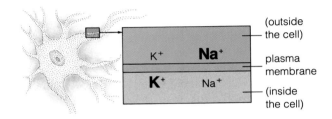

Figure 30.3 Pathways for ions across the plasma membrane of a neuron. These pathways are provided by proteins embedded in the lipid bilayer. Compare Figure 30.4 to this model of membrane structure.

the cell body as "input zones" where stimuli are received and the axon endings as "output zones," where messages are sent to other cells.

Neural Messages

Membrane Excitability. Like all cells, a neuron shows a polarity of charge across its plasma membrane (the inside is more negative than the outside). The polarity results from differences in the concentrations of potassium ions (K^+), sodium ions (Na^+), and other charged substances in the cytoplasm and extracellular fluid. The differences can be illustrated this way:

However, neurons show *excitability*. In response to stimulation, the polarity of charge across the plasma membrane can undergo a sudden reversal, called an **action potential.** For a fraction of a second, the inside of the neuron becomes more positive than the outside.

Excitability depends on three membrane properties. First, ions move into or out of the neuron through the

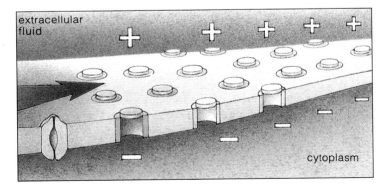

a Membrane at rest (inside negative with respect to the outside). An electrical disturbance (red arrow) spreads from an input zone to an adjacent trigger region of the membrane, which has many gated sodium channels (green).

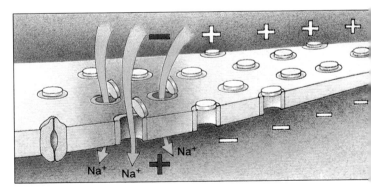

b A strong disturbance initiates an action potential. Sodium gates open, the inflow decreases the negativity inside; this causes more gates to open, and so on, until threshold is reached and the voltage difference across the membrane reverses.

Figure 30.4 Propagation of action potentials along the axon of a neuron. The plasma membrane is shown in yellow.

interior of proteins that serve as channels across the membrane; they cannot cross the lipid bilayer. Second, some channel proteins are always open, and others have "gates" where ion movements can be controlled (Figure 30.3). Third, transport proteins that span the membrane pump ions to maintain and restore the polarity of charge across the membrane in between action potentials. This is an example of an active transport mechanism, as described on page 43.

Neurons "At Rest." When nothing is bothering a neuron, most of the membrane channels for sodium are shut, and some of the channels for potassium are open. Potassium "leaks" out of the neuron, following its concentration gradient. This makes the interior more negative, and some of the potassium ions (which are positively charged) are attracted back inside. When the inward pull of opposite charge balances the outward force of diffusion, there is no more *net* movement of potassium across the membrane. Now there is a steady "voltage difference" across the membrane, this being an amount of energy inherent in the concentration and electric gradients between the two differently charged regions. This amount, the **resting membrane potential,** is about 70 millivolts for many neurons.

Action Potentials. Suppose our neuron "at rest" is a sensory neuron in your thumb, and suppose you disturb its input zone by pressing your thumb against a desktop.

The pressure affects ion movements across a small patch of membrane. The voltage difference changes slightly at this patch, producing a type of local signal. Local signals can be small or large, and they do not spread far. But if the disturbance is prolonged or intense, they can reach a nearby "trigger zone," a membrane patch loaded with ion channels. This is where action potentials are initiated.

Measurements of the voltage difference across a membrane before, during, and after an action potential reveal the following pattern:

1. The inside of a neuron at rest is more negative than the outside (its membrane is polarized).

2. During an action potential, the inside is more positive than the outside (the membrane is depolarized).

3. Following an action potential, resting conditions are restored (the membrane is repolarized).

An action potential is triggered when a disturbance causes the voltage difference to change by a certain minimum amount, a *threshold* level. The change occurs when gated channels for sodium ions open in an accelerating way (Figure 30.4). The inward flow of these positively charged ions makes the inside less negative. This causes more gates to open, more sodium to enter, and so on until the charge difference reverses. The accelerating flow of sodium is an example of positive feedback, whereby an event intensifies as a result of its own occurrence.

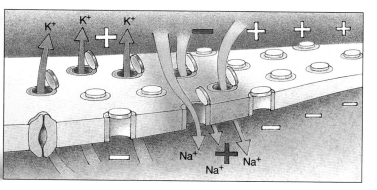

c The reversal causes sodium gates to shut and potassium gates (purple) to open at this site. Potassium follows its gradient (out of the neuron) and voltage is restored. Meanwhile, the electrical disturbance triggers another action potential at the adjacent membrane site, and so on, away from the point of stimulation.

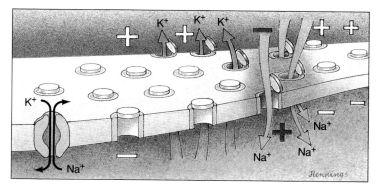

d The inside of the membrane becomes negative again following each action potential, but the sodium and potassium concentration gradients are not yet fully restored. Active transport at sodium-potassium pumps restores the gradients.

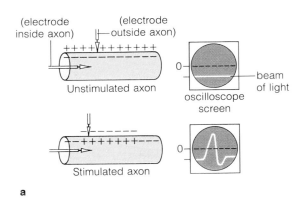

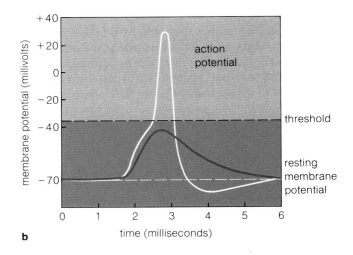

Figure 30.5 Action potentials. (**a**) Electrodes positioned inside and outside an axon can be used to detect voltage changes when the axon is stimulated. Changes register as deflections in a beam of light on an oscilloscope screen. (**b**) The white dashed line is the resting membrane potential for an unstimulated neuron. The solid white line is a recording of an action potential. The red line represents a local signal that did not reach threshold.

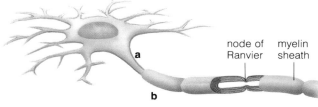

Figure 30.6 Propagation of an action potential along a motor neuron having a myelin sheath. (**a**) An action potential is initiated at a trigger zone in the axon membrane. (**b**) The sheath hinders ion movements across the membrane, so the disturbance spreads rapidly down the axon. (**c**) The small nodes of Ranvier are not sheathed, and they have very dense arrays of gated sodium channels. The voltage difference across the membrane reverses at these nodes. (**d**) The disturbance spreads rapidly to the next node in line, and so on down the axon (**e**).

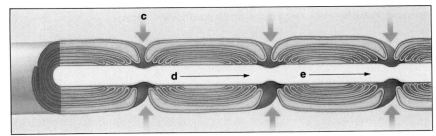

Once threshold is reached, the opening of more sodium gates no longer depends on the strength of the stimulus. It proceeds automatically because the positive-feedback cycle has started. That is why all action potentials in a given neuron "spike" to the same level above threshold as an *all-or-nothing event.* If threshold is reached, nothing can stop the full spiking. If it is not reached, the membrane disturbance will subside when the stimulus is removed (Figure 30.5).

An action potential ends abruptly after a few milliseconds. At the membrane region where it occurred, sodium gates shut, potassium gates open, and ion movements restore the original voltage difference across the membrane. Also, membrane pumps work to restore the original sodium and potassium gradients through active transport of those ions.

After an action potential occurs in a trigger zone, it repeats, or propagates, itself. The electrical disturbance spreads to adjacent membrane regions, where the opening of gated channels is repeated, and so on away from the stimulation site. Notice, in Figure 30.4, how action potentials travel *away* from the stimulation site. A *refractory period* following each one helps prevent backflow. During this period (when sodium gates are shut and potassium gates are open), the membrane is insensitive to stimulation.

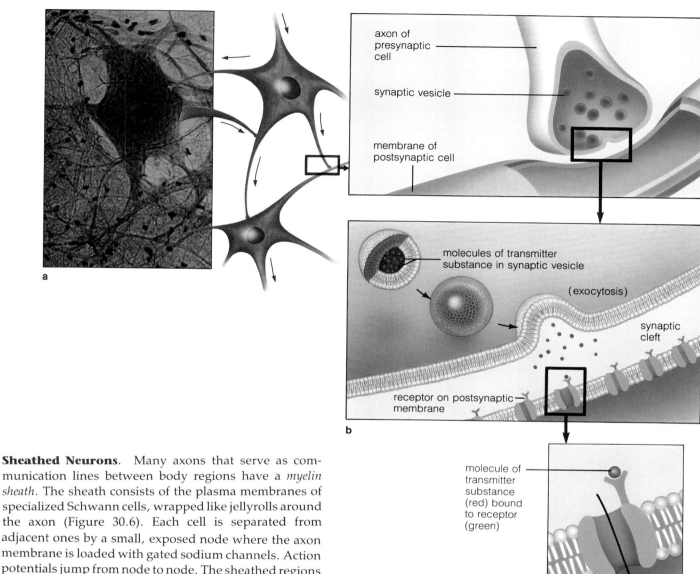

a

axon of
presynaptic
cell

synaptic vesicle

membrane of
postsynaptic cell

molecules of transmitter
substance in synaptic vesicle

(exocytosis)

synaptic
cleft

receptor on postsynaptic
membrane

b

molecule of
transmitter
substance
(red) bound
to receptor
(green)

Na⁺

c

Sheathed Neurons

Sheathed Neurons. Many axons that serve as communication lines between body regions have a *myelin sheath*. The sheath consists of the plasma membranes of specialized Schwann cells, wrapped like jellyrolls around the axon (Figure 30.6). Each cell is separated from adjacent ones by a small, exposed node where the axon membrane is loaded with gated sodium channels. Action potentials jump from node to node. The sheathed regions hinder the flow of ions across the membrane, and this forces the ions to flow along the length of the axon until they can exit at a node and generate a new action potential there. The node-to-node hopping takes less time to send a message over long distances.

Synapses

Action potentials can only be propagated along a plasma membrane. What happens when they reach axon endings? They trigger the release of a **transmitter substance,** a type of signaling molecule, into the junction between the neuron and an adjacent cell. These junctions are called **chemical synapses.** Some occur between two neurons, others between a neuron and a muscle cell or gland cell. Only a small space, the synaptic cleft, separates the two cells (Figure 30.7).

Figure 30.7 Chemical synapses. Typically, action potentials spread along axons, away from the neuron cell body (**a**). In (**b**), the axon terminates next to another neuron, this being an example of a chemical synapse. Information flows from the presynaptic cell to the postsynaptic cell by way of a transmitter substance (**c**).

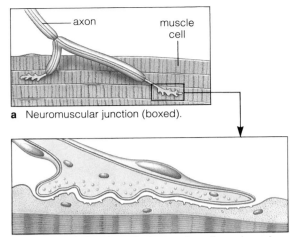

a Neuromuscular junction (boxed).

b Motor end plate (troughs in muscle cell membrane).

Figure 30.8 (**a**) One type of chemical synapse: the junction between a motor neuron and a muscle cell (boxed area). The myelin sheath of the axon stops at the junction, leaving the membranes of the two interacting cells exposed to each other. (**b**) Close-up of the troughs in the muscle cell membrane where the axon endings are positioned.

Consider the synapses between motor neurons and muscle cells. At this type of junction, the branched axon endings are positioned at the muscle cell membrane (Figure 30.8). An action potential traveling down the motor neuron spreads through all the endings and causes the release of acetylcholine (ACh), a transmitter substance, into the synaptic cleft. Molecules of ACh bind to receptors on the muscle cell membrane. The receptors are molecules that recognize the transmitter substance and cause ion channels to open. This initiates an action potential, which in turn leads to contraction (page 353).

At any chemical synapse, only one of the two cells usually releases transmitter molecules. Its action identifies it as the presynaptic cell, which relays signals to a postsynaptic cell. What happens next? The signal may have an excitatory or inhibitory effect on the postsynaptic cell. The outcome depends on which types of receptors are activated and on the nature of the signals being received by the membrane.

At each neuron, excitatory and inhibitory signals compete for control of the membrane. "Excitatory" signals bring the membrane closer to the threshold of an action potential. "Inhibitory" signals drive the membrane away from the threshold. In a process called **synaptic integration,** the competing signals are combined. Especially in the brain and spinal cord, hundreds of excitatory signals must be combined before a large postsynaptic cell will respond to stimulation with its own action potentials.

Synaptic integration is the moment-by-moment combining of excitatory and inhibitory signals acting on adjacent membrane regions of a neuron.

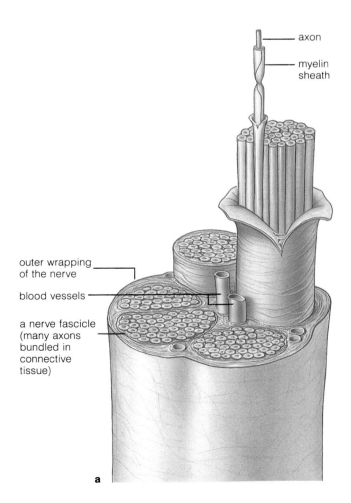

a

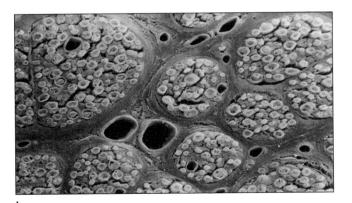

b

Figure 30.9 Structure of a nerve. The sketch (**a**) and the scanning electron micrograph (**b**) show bundles of axons in cylindrical wrappings of connective tissue inside the nerve.

When foreign substances interfere with synaptic integration, the consequences can be deadly. For example, on rare occasions the anaerobic bacterium *Clostridium tetani* enters the body through a puncture or cut. The bacterium can multiply if tissues around the wound become necrotic (die off). One of its metabolic products functions as a neurotoxin in this setting. It interferes with inhibitory synapses on motor neurons in the brain and spinal cord. The unbalanced excitation causes excessive contraction, as in lockjaw. Muscles cannot be released from contraction. The result is *tetanus,* a prolonged, spastic paralysis of muscles that can lead to death.

PATHS OF INFORMATION FLOW

Through synaptic integration, signals arriving at any given neuron in the body can be reinforced or dampened, sent on or suppressed. What determines the direction in which a given signal will travel? As the arrows in Figure 30.7 indicate, that depends on the organization of neurons into circuits or pathways.

The brain has many "local" circuits in whi chattering of neurons is confined to a single region. In contrast, signals between the brain or spinal cord and other body regions travel by cordlike communication lines called **nerves** (Figure 30.9). Axons of sensory neurons, motor neurons, or both are bundled together in a nerve. Within the brain and spinal cord, such bundles are called nerve tracts or pathways.

The sensory and motor neurons of many nerves take part in reflexes: simple, stereotyped movements made in response to sensory stimuli. In a **reflex arc,** sensory neurons directly synapse on motor neurons. The *stretch reflex* is an example; it works to contract a stretched muscle. Think about how you can hold out a large glass and keep it stationary when someone pours lemonade into it. As the lemonade adds weight to the glass and your hand starts to drop, a muscle in your arm (the biceps) is stretched. The stretching activates certain receptors in the muscle. These stretch-sensitive receptors are the input zone of sensory neurons that synapse with motor neurons in the spinal cord (Figure 30.10). Axons of the motor neurons lead right back to the stretched muscle, and action potentials that reach the axon endings

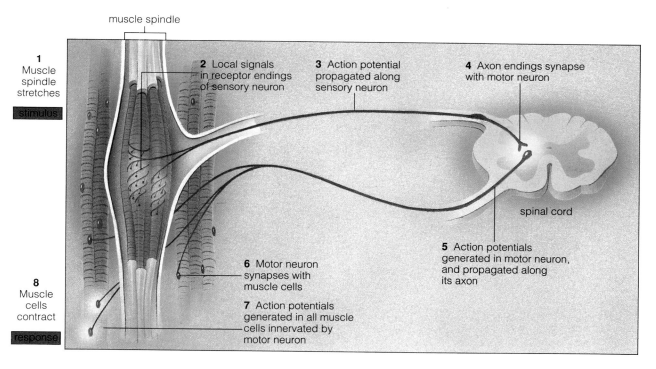

Figure 30.10 Simple reflex arc governing the stretch reflex. A sensory axon is shown in purple, a motor axon in red. Stretch-sensitive receptors of the sensory neuron are located in muscle spindles within a skeletal muscle. Stretching the muscle disturbs the receptors, and action potentials are generated in the sensory neuron. They travel to the axon endings that synapse with motor neurons—which have axons leading right back to the stretched muscle. Signals from the motor neuron can stimulate the muscle cell membrane and initiate contraction (page 353).

trigger the release of ACh, which initiates contraction. Continued receptor activity excites the motor neurons further, allowing them to maintain your hand's position.

In other reflexes, sensory neurons make connections with interneurons, which then activate or suppress the motor neurons necessary for a coordinated response. An example is the *withdrawal reflex,* a rapid pulling away from an unpleasant or harmful stimulus. If you have ever accidentally touched a hot stove, you know this reflex action can be completed even before you are conscious it has occurred.

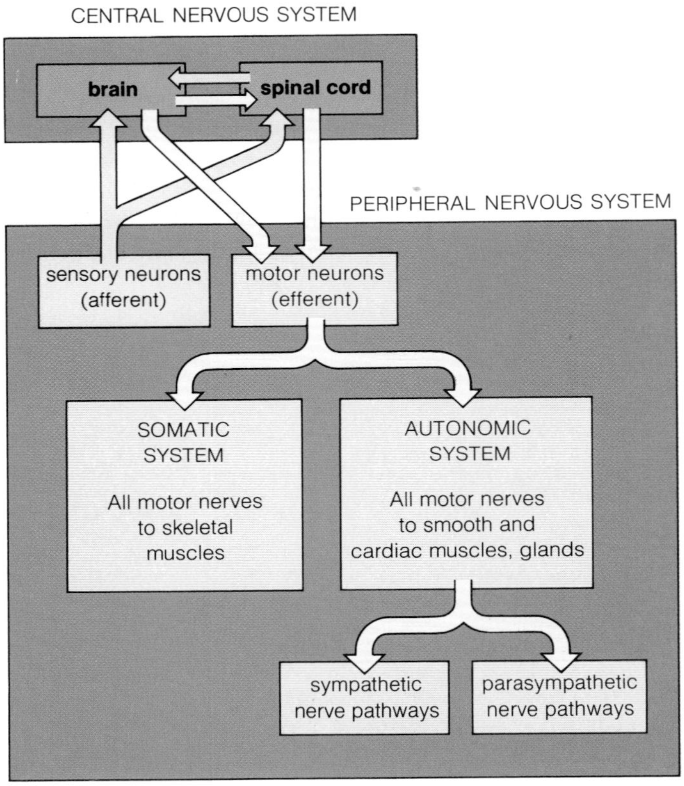

Figure 30.11 Divisions of the vertebrate nervous system.

VERTEBRATE NERVOUS SYSTEMS

During the evolution of complex vertebrates, additional nervous tissues became layered over the more ancient reflex pathways. The most recent layerings still deal with reflexes. But they also have complex, interrelated neural gridworks where information about experiences is stored, compared, and used to initiate *novel* responses. The gridworks are the basis of memory, learning, and reasoning.

For descriptive purposes, we can divide the nervous system into central and peripheral regions (Figure 30.11). The **central nervous system** includes the brain and spinal cord. The **peripheral nervous system** includes all the nerves carrying signals to and from the brain and spinal cord. Both divisions also have "neuroglial cells" that protect or assist neurons. The Schwann cell described earlier is an example. Neuroglial cells make up about half the volume of the nervous system.

Peripheral Nervous System

The peripheral nervous system has thirty-one pairs of spinal nerves, which connect with the spinal cord, and twelve pairs of cranial nerves, which connect directly with the brain. Some relay commands to skeletal muscles; they are called the *somatic system* of nerves. Others relay commands to smooth muscles and glands. They are called the *autonomic system* of nerves (Figure 30.12).

Nerves of the autonomic system play two roles in the body's overall functioning. Excitatory and inhibitory signals from its **parasympathetic nerves** tend to slow down the body overall and divert energy to basic "housekeeping" tasks, such as digestion. This nerve action dominates when the body is not receiving much outside stimulation. Signals from its **sympathetic nerves** tend to slow down housekeeping tasks and increase overall body activities during times of heightened awareness, excitement, or danger. Sympathetic nerves prepare the animal to fight, flee, or frolic; they are the basis of the "fight-flight" response.

Autonomic nerve action also brings about constant minor adjustments in internal organs. Even while low levels of sympathetic signals are causing your heart to beat a little faster, low levels of parasympathetic signals are opposing this effect. At any moment, your heart rate is the net outcome of opposing signals.

"Biofeedback" refers to conscious efforts to enhance or dampen autonomic and other physiological responses. For example, the state of contraction of neck muscles can induce tenseness. By consciously relaxing those muscles, you can make the feeling of tenseness go away.

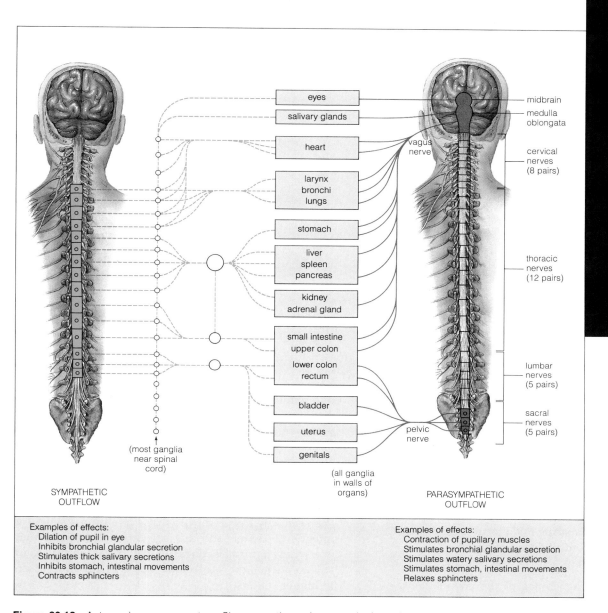

eyes
salivary glands
heart
larynx
bronchi
lungs
stomach
liver
spleen
pancreas
kidney
adrenal gland
small intestine
upper colon
lower colon
rectum
bladder
uterus
genitals

vagus nerve

midbrain
medulla oblongata

cervical nerves (8 pairs)

thoracic nerves (12 pairs)

lumbar nerves (5 pairs)

sacral nerves (5 pairs)

pelvic nerve

(most ganglia near spinal cord)

(all ganglia in walls of organs)

SYMPATHETIC OUTFLOW

PARASYMPATHETIC OUTFLOW

Examples of effects:
Dilation of pupil in eye
Inhibits bronchial glandular secretion
Stimulates thick salivary secretions
Inhibits stomach, intestinal movements
Contracts sphincters

Examples of effects:
Contraction of pupillary muscles
Stimulates bronchial glandular secretion
Stimulates watery salivary secretions
Stimulates stomach, intestinal movements
Relaxes sphincters

Figure 30.12 Autonomic nervous system. Shown are the main sympathetic and parasympathetic pathways leading out from the central nervous system to some major organs. Keep in mind that both systems have *paired* nerves leading out from the brain and spinal cord. Ganglia (singular, ganglion) are simply clusters of cell bodies of the neurons that are bundled together in nerves.

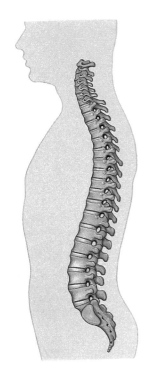

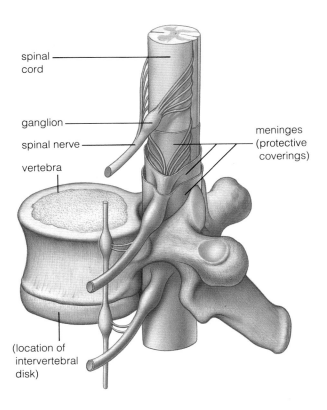

spinal cord

ganglion

spinal nerve

vertebra

meninges (protective coverings)

(location of intervertebral disk)

Figure 30.13 Organization of the spinal cord and its relation to the vertebral column.

Central Nervous System

The Spinal Cord. The spinal cord is protected by stacked bones of the vertebral column (Figure 30.13). Its gray matter (dendrites and cell bodies) deals mainly with the reflex connections necessary for limb movements (such as walking) and internal organ activity (such as bladder emptying). Its white matter includes major nerve tracts (bundles of sheathed axons) that ascend into and descend from specific brain centers, and these provide more refined control over activities.

Divisions of the Brain. Bones of the cranial cavity and membranes protect the brain, which begins as a continuation of the spinal cord. The brain has three major divisions: the hindbrain, midbrain, and forebrain (Table 30.1).

The **hindbrain** consists of the medulla oblongata, cerebellum, and pons. The *medulla oblongata* has reflex centers for respiration, blood circulation, and other vital tasks. Here also, motor responses and complex reflexes such as coughing are coordinated. Its centers influence other brain centers that help you sleep or wake up.

The *cerebellum* has reflex centers for maintaining posture and refining limb movements. It integrates signals from the eyes, muscle spindles, skin, and elsewhere. It keeps other parts of your brain informed about how your trunk and limbs are positioned, how much different muscles are contracted or relaxed, and in which direction the body or limbs happen to be moving. The *pons* means bridge. Nerve tracts pass through the pons on their way between brain centers.

The **midbrain** originally coordinated reflex responses to visual input. Its roof of gray matter, the tectum, still integrates signals from the eyes and ears. The tectum is important in fishes, amphibians, reptiles, and birds. (You can surgically remove a frog's cerebrum, its highest integrative center, and the frog can still do just about everything it normally does.) In mammals, sensory input still converges on the tectum, but it is rapidly sent on to higher centers.

The **forebrain** has the most recent layerings of nerve tissues. Originally, olfactory lobes dealing with the sense of smell dominated the forebrain. A brain center, the *cerebrum*, integrated input about odors and selected motor responses to it. Sensory signals were relayed and coordinated at the *thalamus*, a center below the cerebrum. (Some motor pathways also converged here.) The *hypothalamus* monitored internal organs and influenced forms of behavior related to their activities, such as thirst, hunger, and sex. In time, a thin layer of gray matter developed over each half of the cerebrum. This *cerebral cortex* expanded into information-encoding and information-processing centers in mammals. It has become most highly developed in the human brain.

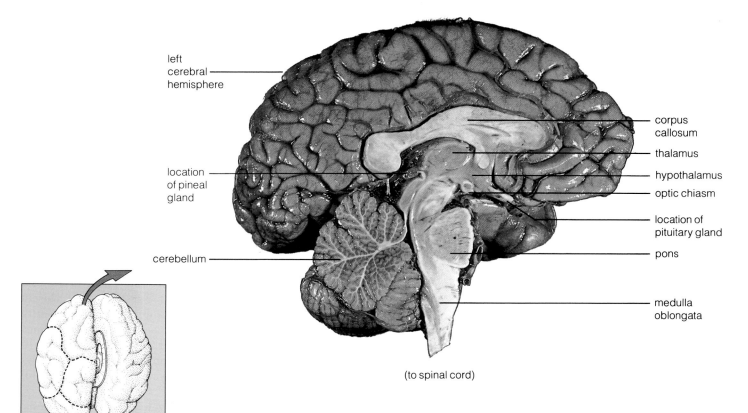

left cerebral hemisphere

location of pineal gland

cerebellum

corpus callosum

thalamus

hypothalamus

optic chiasm

location of pituitary gland

pons

medulla oblongata

(to spinal cord)

left cerebral hemisphere   right cerebral hemisphere

Figure 30.14 Inside view of the human brain. The corpus callosum is a major nerve tract between the two cerebral hemispheres. The boxed inset shows the two hemispheres pulled slightly apart; normally they are pressed together, with only a deep fissure between them.

THE HUMAN BRAIN

The Cerebral Hemispheres

Figure 30.14 shows how the human cerebrum is divided into two parts, the *cerebral hemispheres*. Much of its gray matter—the cerebral cortex—forms a thin surface layer on each hemisphere, which is further divided into lobes by folds and fissures. Its white matter consists of major nerve tracts that keep the hemispheres in communication with each other and with the rest of the body.

Different regions of the cortex have different functions (Figure 30.15). Its motor centers coordinate instructions for motor responses. (Experimentally stimulate different points on the motor cortex, and you will trigger contractions of different muscles.) Much of the motor cortex deals with thumb and tongue muscles, indicating the extent of control required for hand movements and verbal expression. Primary receiving centers of the cortex deal with sensory input, including that from the eyes, ears, and skin. Its association centers add information from memory stores to the primary sensory information.

Table 30.1	Regional Divisions of the Human Brain
Divisions	Main Components
Forebrain	Cerebrum
	Olfactory lobes
	Limbic system
	Thalamus
	Hypothalamus
	Pituitary gland
	Pineal gland
Midbrain	Tectum and other tissue regions
Hindbrain	Pons
	Cerebellum
	Medulla oblongata

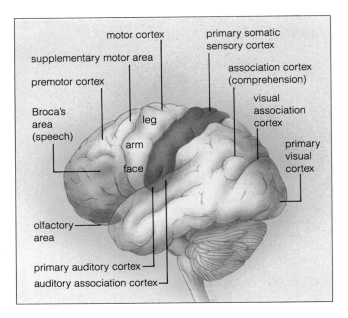

Figure 30.15 Primary receiving and association areas for the human cerebral cortex. Signals from receptors on the body's periphery enter primary cortical areas. Sensory input from different receptors is coordinated and processed in association areas. The text describes the main cortical regions. Also shown here are the premotor area, involved in intricate motor activity (as typified by a concert pianist performing); the supplementary motor area, which helps coordinate sequential voluntary movements; and Broca's area, which coordinates muscles required for speech.

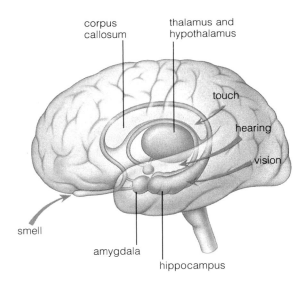

Figure 30.16 Some brain regions that play key roles in memory. Sensory input is processed by the cerebral cortex and sent into parts of the limbic system and parts of the forebrain. (The limbic system, our "emotional brain," includes the regions called the thalamus, hypothalamus, amygdala, and hippocampus.)

Memory

Memory is the storage and retrieval of information about previous experiences. It is part of your conscious experience and underlies your capacity for learning. Today we know that information becomes stored in stages, and "memory traces," or chemical and structural changes necessary for storage, occur in many different brain regions (Figure 30.16). Short-term storage lasts a few seconds to a few hours and seems to be limited to seven or eight bits of information—words of a sentence, numbers, and so on. Long-term storage lasts more or less permanently, and it seems to be limitless in its capacity.

Emotional States

Our emotions are governed by the cerebral cortex and by different brain regions collectively called the **limbic system** (Figure 30.16). The limbic system is still closely related to the sense of smell that figured so prominently in vertebrate evolution. That is why you sometimes "smell" a cologne all over again when you have a pleasant memory of the person who wore it, or why you smell a bad odor when you remember an awful confrontation with a skunk.

The hypothalamus is the gatekeeper of the limbic system; connections between the cerebral cortex and lower brain centers pass through it. Through these connections, the reasoning possible in the cerebral cortex can dampen rage, hatred, and other "gut reactions." The hypothalamus also monitors internal organs in addition to emotional states. This is what keeps your heart and stomach on fire when you are sick with passion (or indigestion).

States of Consciousness

States of consciousness include sleeping, dozing, daydreaming, and total alertness. The central nervous system governs these states, and psychoactive drugs can alter them (see *Commentary*). A brain region called the **reticular activating system** controls the changing levels of consciousness. It forms connections with the spinal cord, cerebellum, and cerebrum as well as back with itself. The flow of signals along these circuits—and the inhibitory or excitatory chemical changes accompanying them—affects whether you stay awake or fall asleep. Damage to parts of the circuits can lead to unconsciousness and coma.

Neurons of one of the "sleep centers" in the reticular formation release serotonin. This transmitter substance inhibits other neurons that arouse the brain and maintain wakefulness. High serotonin levels are linked to drowsiness and sleep. Substances released from another brain center counteract serotonin's effects.

Drug Action on Integration and Control

Classes of Psychoactive Drugs

Broadly speaking, a drug is any substance with no nutritive value that is introduced into the body to elicit some effect on physiological processes. A **psychoactive drug** affects parts of the central nervous system concerned with states of consciousness and behavior. For example, many affect a *pleasure center* in the hypothalamus. Every time we eat, drink, or engage in sexual activity, the pleasure center is activated by natural substances (including dopamine and norepinephrine), and the sense of pleasure we come to associate with those behaviors is reinforced. Artificial stimulation of the pleasure center also becomes reinforced; we call this addiction.

Four classes of psychoactive drugs are problems in our society. They are the stimulants, depressants and hypnotics, narcotic analgesics, and hallucinogens and psychedelics.

Stimulants. Stimulants include caffeine, nicotine, amphetamines, and cocaine. First they increase alertness and body activity, then they lead to depression.

Coffee, tea, chocolate, and many soft drinks contain caffeine, one of the most widely used stimulants. Low doses of caffeine stimulate the cerebral cortex first and cause increased alertness and restlessness. Higher doses act at the medulla oblongata to disrupt motor coordination and intellectual coherence.

Nicotine, a component of tobacco, has powerful effects on the central and peripheral nervous systems. It mimics acetylcholine and can directly stimulate a number of sensory receptors. Its short-term effects include water retention, irritability, increased heart rate and blood pressure, and gastric upsets. Its long-term effects can be devastating (pages 381 and 414).

Amphetamines (including "speed"), which resemble dopamine and norepinephrine, stimulate the pleasure center. In time, the brain produces less and less of its own signaling molecules and comes to depend on artificial stimulation.

Cocaine stimulates the pleasure center in a different way. It produces a rush of pleasure by blocking the reabsorption of dopamine, norepinephrine, serotonin, and other signaling molecules that are normally released at synapses. Receptor cells are incessantly stimulated over an extended period. Heart rate and blood pressure rise; sexual appetite increases. But then the effects change. The signaling molecules that have accumulated in synaptic clefts diffuse away—but the cells that produce them cannot make up for the extraordinary loss. The sense of pleasure evaporates as the receptor cells, which are now hypersensitive to stimulation, demand stimulation. The cocaine user becomes anxious and depressed. After prolonged, heavy use of cocaine, "pleasure" is impossible to experience. The addict loses weight and cannot sleep properly. The immune system becomes compromised, and heart abnormalities set in.

Granular cocaine, which is inhaled ("snorted"), has been around for some time. Crack cocaine, a cheaper but more potent form, is burned and the smoke inhaled. Crack is incredibly addictive; its highs are higher, but the crashes more devastating.

Depressants, Hypnotics. These drugs lower the activity in nerves and parts of the brain, so they reduce activity throughout the body. Some act at synapses in the reticular activating system and in the thalamus.

Depending on the dosage, most of these drugs can produce responses ranging from emotional relief, sedation, sleep (hypnosis), anesthesia and coma, to death.

Classes of Psychoactive Drugs	
Class	Examples
Depressants, hypnotics	Barbiturates (e.g., Nembutal, Quaalude) Antianxiety drugs (e.g., Valium, alcohol)
Stimulants	Caffeine Nicotine Amphetamines (e.g., Dexedrine) Cocaine
Narcotic analgesics	Codeine Opium Heroin
Psychedelics, hallucinogens	Lysergic acid diethylamide (LSD) *Cannabis sativa* (marijuana)

At low doses, inhibitory synapses are often suppressed slightly more than excitatory synapses, so the person feels excited or euphoric at first. Increased doses also suppress excitatory synapses, leading to depression. Depressants and hypnotics have additive effects; one amplifies another. For example, combining alcohol with barbiturates amplifies behavioral depression.

Alcohol (ethyl alcohol) differs from the drugs just described because it acts directly on the plasma membrane to alter cell function. Some persons mistakenly think of it as a harmless stimulant (it produces an initial "high"). But alcohol is one of the most powerful psychoactive drugs and a major cause of death. Small doses even over the short term can produce disorientation, uncoordinated motor functions, and diminished judgment. Long-term addiction destroys nerve cells and causes permanent brain damage; it can permanently damage the liver (cirrhosis).

Analgesics. When stress leads to physical or emotional pain, the brain produces its own analgesics, or natural pain relievers. Endorphins and enkephalins are examples. These substances seem to inhibit activity in many parts of the nervous system, including brain centers concerned with emotions and perception of pain.

The narcotic analgesics, including codeine and heroin, sedate the body and relieve pain. They are extremely addictive. The body develops a tolerance of these drugs; larger and larger doses are needed to produce the same effects. The body also becomes physically dependent on them. Deprivation following massive doses of heroin leads to fever, chills, hyperactivity and anxiety, violent vomiting, cramping, and diarrhea.

Psychedelics, Hallucinogens. These drugs, which alter sensory perception, have been described as "mind-expanding." Some skew acetylcholine or norepinephrine activity. Others, such as LSD (lysergic acid diethylamide), affect serotonin activity. Even in small doses, LSD dramatically warps perceptions. Marijuana is another hallucinogen. The name refers to the drug made from crushed leaves, flowers, and stems of the plant *Cannabis sativa*. In low doses marijuana is like a depressant. It slows down but does not impair motor activity, it relaxes the body, and it elicits mild euphoria. However, it can produce disorientation, increased anxiety bordering on panic, delusions (including paranoia), and hallucinations.

Like alcohol, marijuana can affect an individual's ability to perform complex tasks, such as driving a car. In one study, commercial pilots showed a marked deterioration in instrument-flying ability for more than two hours after smoking marijuana. Recent studies point to a link between marijuana smoking and suppression of the immune system.

Biological Perspective on Drug Abuse

A major problem in the modern world is drug abuse—the self-destructive use of drugs that alter emotional and behavioral states. The consequences show up in many ways—in babies born with crack addiction; in drivers whose perceptions, skewed by alcohol or amphetamines, cause tragic highway accidents; in victims of addicts who steal and sometimes kill to support a drug habit; in suicides who jump off buildings or bridges while under the LSD-induced delusion that they can fly.

Each of us possesses a body of great complexity. Its architecture, its functioning are legacies of millions of years of evolution. It is unique in the living world because of its highly developed nervous system—a system that is capable of processing far more than the experience of the individual. One of its most astonishing products is language, the encoding of *shared* experiences of groups of individuals in time and space. Through the evolution of our nervous system, the sense of history was born, and the sense of destiny. Through this system we can ask how we have come to be what we are, and where we are headed from here. Perhaps the sorriest consequence of drug abuse is its implicit denial of this legacy—the denial of self when we cease to ask, and cease to care.

Figure 30.17 (*Right*) Tactile receptors in human skin. Free nerve endings contribute to sensations of temperature, light pressure, and pain. Pacinian corpuscles contribute to sensations of deep, rapid pressure. Meissner corpuscles are stimulated at the onset and end of sustained pressure; the Ruffini endings react continually to ongoing stimuli.

SENSORY INPUT

During the evolution of nervous systems, sensory receptors and organs (such as eyes) as well as motor structures (such as legs and wings) evolved along with them. As we saw earlier, their joint evolution was the foundation for the more active, intricate life-styles of complex vertebrates. Let's now consider a few selected examples of sensory inputs into the neural gridworks.

Classes of Receptors

The body's receptors are finely branched endings of sensory neurons or specialized cells adjacent to them. Think of a **stimulus** as light, heat, mechanical pressure, or some other specific form of energy that the body can detect through its receptors. In sensory organs such as eyes, receptors and other tissues have become organized in ways that amplify or focus the stimulus energy. By using stimulus energy as a guide, we can define four major types of receptors:

1. **Chemoreceptors** detect chemical energy (ions or molecules dissolved in body fluids next to the receptor).

2. **Mechanoreceptors** detect mechanical energy associated with changes in pressure, position, or acceleration.

3. **Photoreceptors** detect photon energy of visible and ultraviolet light.

4. **Thermoreceptors** detect radiant energy (including infrared) associated with temperature.

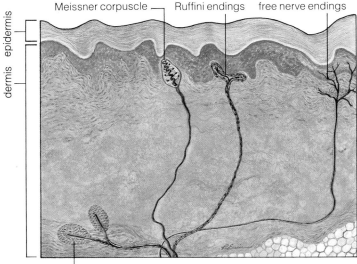

Meissner corpuscle — Ruffini endings   free nerve endings

epidermis

dermis

Pacinian corpuscle

Different animals do not have the same kinds or numbers of receptors, so they sample the environment in different ways and have different perceptions of it. You do not have photoreceptors for ultraviolet light, as bees do; and you do not "see" many flowers the way they do (page 321). Unlike the python in Figure 30.1, you do not have thermoreceptors that help you find food in the dark.

All sensory receptors convert stimulus energy to a form that can travel along the neural communication lines. Stimulation initiates local signals which, acting together, can trigger action potentials. But all action potentials are alike. How do they convey information that gives rise to *different* sensations, such as smell, taste, and color? There are three reasons.

First, inherited neural programs in the brain of each animal can only interpret incoming signals in certain ways. That is why you "see stars" when you accidentally poke your eye in the dark. Your brain always interprets action potentials arriving from your eyes, by way of a particular sensory pathway, as "light."

Second, strong stimulation of a receptor causes action potentials to fire more frequently. Thus, even though the same receptor detects energy associated with a throaty whisper or a wild screech, the brain senses the difference through frequency variations in the signals that the receptor sends to it.

Third, strong stimulation "recruits" more receptors in a larger area. You activate some receptors when you lightly touch skin on your arm, but you activate many more when you press hard on the same area. The increased disturbance sets off action potentials in many sensory axons at the same time, and the brain interprets the combined activity as an increase in stimulus intensity.

The brain uses signals arriving *from a particular nerve pathway* (not the action potentials themselves) to produce a particular sensation.

The brain uses differences in the *frequency* and *number* of action potentials coming from a given tissue to interpret variations in stimulus intensity.

Somatic Senses and Pain Perception

Mechanoreceptors and thermoreceptors have roles in the sensations of touch, pressure, temperature, and pain near the body surface. These are the somatic senses. Free nerve endings in skin contribute to sensations of light pressure, temperature, and pain; these and other receptors are shown in Figure 30.17.

Pain is the perception of injury to some body region. A live lobster certainly reacts to being dropped into a pot of boiling water, but no one knows whether it "feels" pain as we do. Like other invertebrates, it has no limbic system, which may mediate the perception of pain.

Most sensations of internal injury or malfunctions are based on misleading signals that cause "referred pain." The pain is felt in a tissue some distance from the real stimulation point. A heart attack, for example, might be felt as pain in the skin above the heart and along the left shoulder and arm.

Taste and Smell

Animals use chemoreceptors called *taste receptors* to detect the difference between nutritious and noxious substances. Depending on where the receptors are located, different animals "taste" with their mouth, antennae, legs, tentacles, or fins. Taste receptors on the tongue often are part of taste buds (Figure 30.18).

Olfactory receptors help animals detect odors. The human nose has about 10 million of these chemoreceptors; the German shepherd nose, about 220 million. Some receptors respond to molecules from food or predators. Others respond to pheromones: signaling molecules having targets outside the body, in other animals of the same species. Among other things, animals use pheromones to raise an alarm, attract a mate, and mark territorial boundaries. Receptors on a male silk moth can detect one molecule of a sex-attracting pheromone (bombykol) in 10^{15} molecules of air!

Hearing

Mechanoreceptors called **hair cells** give rise to action potentials when they are bent or tilted. Your sense of sound, or hearing, begins when specialized hair cells in

your ears bend under fluid pressure that is caused by vibrations. A vibration is a wavelike form of mechanical energy. For example, clapping produces waves of compressed air. Each time your hands clap together, molecules are forced outward and a low-pressure state is created in the region they vacated. The pressure variations can be depicted as a wave form, and its peaks correspond to loudness:

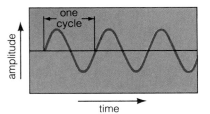

The frequency of a sound is the number of wave cycles per second. Each "cycle" extends from the start of one wave peak to the start of the next. The more cycles per second, the higher the frequency and the higher the perceived pitch of the sound.

When sound waves spread out through air, they rapidly become weaker with distance. Three regions of the mammalian ear receive, amplify, and sort out signals (Figure 30.19). The *outer ear* collects sound waves with its external flaps, and it channels them inward through a canal to an eardrum. The eardrum vibrates, and the vibrations are picked up and transferred inward by small bones of the *middle ear*. Pressure waves are sorted out in the coiled, three-part tube of the *inner ear*. Stimulation of hair cells in the tube gives rise to action potentials that the brain interprets as sound. Figure 30.20 shows what can happen to hair cells subjected to prolonged, intense sound.

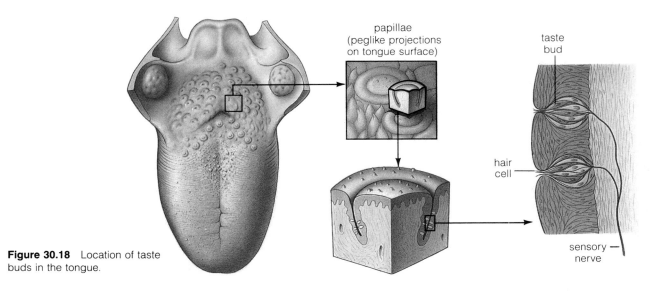

Figure 30.18 Location of taste buds in the tongue.

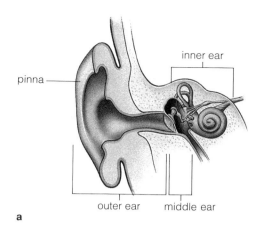

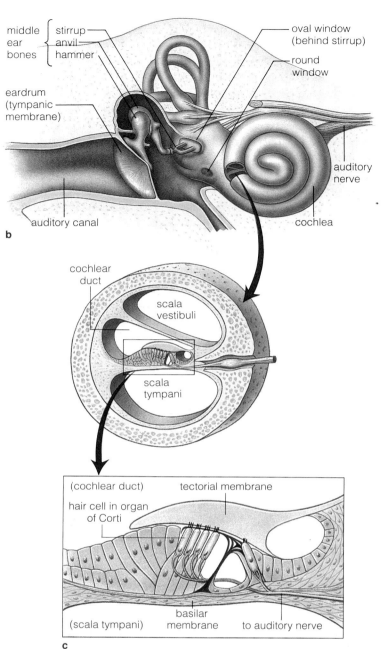

Figure 30.19 Sensory receptors in the human ear.

(**a,b**) Pressure waves funneled through the ear canal strike the eardrum (tympanic membrane), which bows in and out at the same frequency as the waves. This activates the middle earbones, a lever system in front of the oval window (an elastic membrane over the entrance to the coiled inner ear). Middle earbones amplify the stimulus by transmitting the force of pressure waves to the smaller surface of the oval window.

The oval window bows in and out, producing fluid pressure waves in two ducts in the inner ear (scala vestibuli and scala tympani). The waves reach a membrane (round window) that bulges under pressure. Without this bulging, fluid would not be able to move back and forth in the inner ear.

Pressure waves are sorted out at the third duct (cochlear duct) of the coiled inner ear. The duct's basement membrane (basilar membrane) starts out narrow and stiff, but it becomes broader and flexible deep in the coil. High-frequency waves set up membrane vibrations in the stiff region. So do low-frequency waves, but the vibrations are lower in amplitude and continue on, into the more elastic regions.

(**c**) Perched on the basilar membrane is the organ of Corti, which contains hair cells. Vibrations of different regions of the membrane push different patches of hair cells against an overhanging flap (the tectorial membrane). Signals from disturbed hair cells initiate action potentials that are carried to the brain by the auditory nerve.

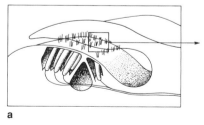

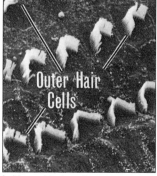

Figure 30.20 Effect of intense sound on the inner ear. Normal organ of Corti from a human (**a**) and from a guinea pig (**b**), showing two rows of outer hair cells. (**c**) Organ of Corti after twenty-four-hour exposure to noise levels approached by loud rock music (2,000 cycles per second at 120 decibels).

b Normal hair cells   **c** Scarred region

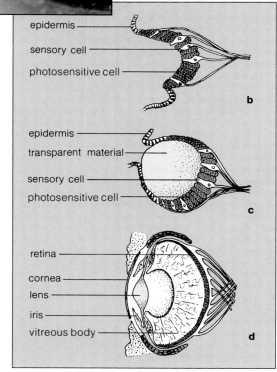

Figure 30.21 Photoreceptors in mollusks. (**a**) The well-developed eye of a shelled conch, here peering into the waters of the Great Australian Barrier Reef. Eyespot of a limpet (**b**) eyes of an abalone (**c**), and an octopus (**d**).

a

epidermis
sensory cell
photosensitive cell

b

epidermis
transparent material
sensory cell
photosensitive cell

c

retina
cornea
lens
iris
vitreous body

d

Vision

Photoreceptors have light-absorbing pigment molecules embedded in their cell membranes. All organisms, whether they see or not, are sensitive to light. (Even a single-celled amoeba will stop abruptly when you shine a light on it.) What we call "vision," however, requires a complex system of photoreceptors *and* a neural program in the brain to interpret signals from them. Those signals encode information about the position, shape, brightness, distance, and movement of a visual stimulus. Vision also requires an adjustable *lens,* a transparent cone or sphere that can be moved to focus incoming light precisely on a dense layer of photoreceptor cells behind it. **Eyes** are well-developed photoreceptor organs that contribute to some small or large degree to image formation.

Invertebrate Photoreceptors. Many invertebrates cannot see, but they do have *eyespots*—clusters of photosensitive cells arranged in a cuplike depression in the epidermis. Mollusks are the simplest animals with eyes (Figure 30.21). Complex mollusks, including squids and octopuses, have large, paired eyes capable of forming clear images.

The *compound eyes* of insects and crustaceans have closely packed photosensitive units. According to one theory, each unit samples a small part of the overall visual field. An image is built up according to signals about differences in light intensities across the field, with each unit contributing a small bit to a visual mosaic (Figure 30.22).

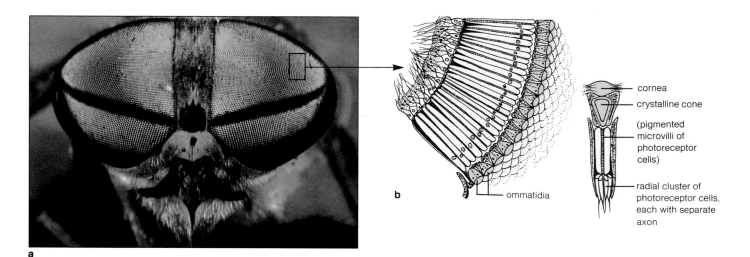

a

b

cornea
crystalline cone
(pigmented microvilli of photoreceptor cells)
radial cluster of photoreceptor cells, each with separate axon
ommatidia

Figure 30.22 (**a**) Compound eyes of a deerfly. The crystal-like cones of its photoreceptor units (ommatidia) act like a lens that focuses light on pigmented photoreceptor cells below (light blue in the sketch in **b**).

Table 30.2	Vertebrate Eye Components
Eye Region	Functions
Outer Layer:	
Sclera	Protect eyeball
Cornea	Focus light
Middle Layer:	
Choroid	
Pigmented tissue	Prevent light scattering
Iris	Control amount of light
Pupil	Entrance for light
Lens	Finely focus light on photoreceptors
Aqueous humor	Transmit light, maintain pressure
Vitreous body	Transmit light, support lens, eye
Inner Layer:	
Retina	Absorb, convert light
Fovea	Increase visual acuity
Start of optic nerve	Transmit signals to brain

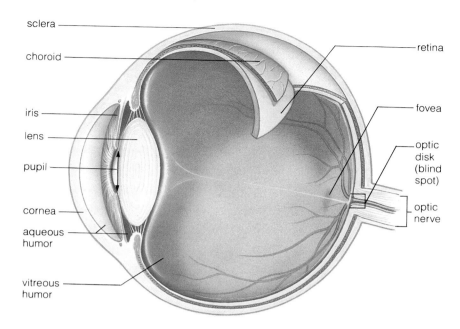

Figure 30.23 Structure of the human eye.

Figure 30.24 (*Right*) Visual accommodation in the human eye. (**a**) Close objects are brought into focus when eye muscles contract enough to slacken certain fibers interposed between them and the lens, and this causes the lens to thicken at its equator. (**b**) Distant objects are brought into focus when eye muscles relax, thereby putting tension on the fibers and stretching the lens into a flatter shape. (**c**) In the eyes of *farsighted* people, light from nearby objects is focused behind the retina. (**d**) In the eyes of *nearsighted* people, the retina is too far behind the lens; light from distant objects is focused in front of the retina.

Vertebrate Photoreceptors. As Table 30.2 and Figure 30.23 indicate, *vertebrate eyes* have a lens, sclera (outer coat), choroid (dark-pigmented tissue), and retina. A transparent cornea covers the front of the eyeball. Inward extensions of the choroid form an iris. The iris, with its abundant screening pigments and muscle fibers, is used to control the amount of incoming light. A clear fluid fills the space between the iris and cornea. The lens is positioned behind the iris, and a jellylike substance fills the chamber behind the lens. Lens adjustments focus light precisely onto the densely packed photoreceptor cells of the retina (Figure 30.24).

The photoreceptors are called **rod cells** and **cone cells** because of their shape (Figures 30.25 and 30.26). Rods are sensitive to very dim light. They contribute to coarse perception of movements by detecting changes in light intensity across the field of vision. Cones respond to intense light. They contribute to sharp daytime vision

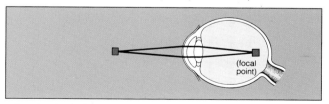

a Accommodation for near objects (lens bulges)

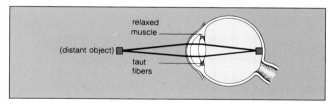

b Accommodation for distant objects (lens flattens)

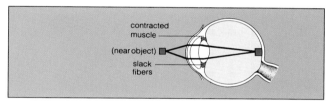

c Focal point in farsighted vision

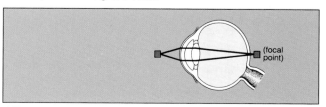

d Focal point in nearsighted vision

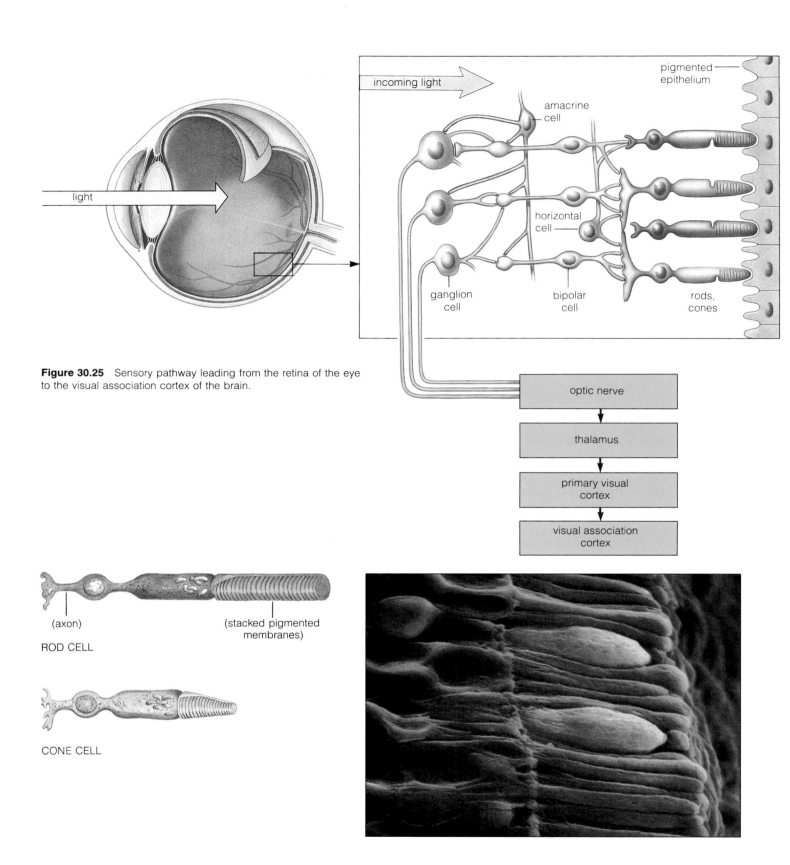

Figure 30.25 Sensory pathway leading from the retina of the eye to the visual association cortex of the brain.

incoming light

pigmented epithelium

amacrine cell

horizontal cell

ganglion cell

bipolar cell

rods, cones

light

optic nerve

thalamus

primary visual cortex

visual association cortex

(axon)

(stacked pigmented membranes)

ROD CELL

CONE CELL

Figure 30.26 Photoreceptors (rods and cones) of the vertebrate eye.

and color perception. Pigments in different cone cells are sensitive to wavelengths corresponding to red, green, and blue colors.

Cones of the human eye are densely packed in the fovea, a funnel-shaped depression near the center of the retina, where nerve tissue is thinner. Cones at the fovea contribute most to visual acuity (precise discrimination between adjacent points in space).

Rods contain molecules of rhodopsin in their membranes. Light absorption causes the breakdown of rhodopsin, and this leads to a voltage change across the membrane. The change signals the presence of light to neighboring neurons, which relay signals to cells that have long axons leading out from the eye. The axons come together to form the optic nerve. From there, signals travel to the thalamus, then on to visual processing centers in the cortex (Figure 30.25).

SUMMARY

1. Nervous systems detect, interpret, and respond directly to stimuli. Reflexes (simple, stereotyped movements made in response to sensory stimuli) are the basic operating machinery of nervous systems. In simple reflexes, sensory neurons directly signal motor neurons that act on muscle cells. In more complex reflexes, interneurons coordinate and refine the responses.

2. Sensory receptors respond to specific stimuli, such as light energy or chemical energy. Animals can respond to events in the outside world only if they have sensory receptors that are sensitive to the energy of the particular stimulus.

3. A neuron can receive and respond to stimuli because of its membrane properties. An unstimulated neuron shows a steady voltage difference across its plasma membrane (the inside is more negative by some value than the outside). At this time, the neuron maintains concentration gradients of potassium ions, sodium ions, and other charged substances across the membrane.

4. Under adequate stimulation, the voltage difference across the membrane changes dramatically, past a certain minimum amount (the threshold level). Then, gated sodium channels across the membrane open in an accelerating way and suddenly reverse the voltage difference, which registers as a spike (action potential) on recording devices.

5. Action potentials propagate themselves along the neural membrane until they reach an output zone, where axon endings form a junction (chemical synapse) with another neuron or a muscle or gland cell. The presynaptic cell releases a transmitter substance into the cleft, and

this has an excitatory or inhibitory effect on the post-synaptic cell. Integration is the moment-by-moment combining of all signals—excitatory and inhibitory—acting at all the different synapses on a neuron.

6. The direction of information flow through the body depends on the organization of neurons into circuits and pathways. Local circuits are sets of interacting neurons confined to a single region in the brain or spinal cord. Nerve pathways extend from neurons in one body region to neurons in different regions.

7. The central nervous system consists of the brain and spinal cord; the peripheral nervous system consists of nerves and ganglia in other body regions.

Review Questions

1. Define sensory neuron, interneuron, and motor neuron. *427*

2. Label the functional zones of a motor neuron: *427*

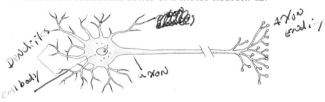

3. Distinguish between a local signal at the input zone of a neuron and an action potential. *429*

4. What is a synapse? Explain the difference between an excitatory and an inhibitory synapse. Define neural integration. *431–432*

5. What is a reflex? Describe the events in a stretch reflex. *433*

6. What constitutes the central nervous system? The peripheral nervous system? *434*

7. Distinguish among the following:
   a. neurons and nerves *433*
   b. somatic system and autonomic system *434*
   c. parasympathetic and sympathetic nerves *434*

8. What is a psychoactive drug? Describe the effects of one such drug on the central nervous system. *439*

9. What is a stimulus? Receptor cells detect specific kinds of stimuli. When they do, what happens to the stimulus energy? *441*

Self Quiz   *(Answers in Appendix IV)*

1. The vertebrate nervous system senses, interprets, and issues commands for responses to ENVIRONMENTAL STIMULI. Its communication lines are organized gridworks of nerve cells, or NEURONS

2. A neuron responds to adequate stimulation with ACTION POTENTIAL, a type of self-propagating signal.

3. When action potentials arrive at junctions between a neuron and another cell, they inhibit or stimulate the release of molecules of a TRANSMITTER SUBSTANCE that diffuse over to that cell.

4. The moment-by-moment combining of all signals acting on all the different synapses on a neuron is called _integration_

5. Interactions among neurons in your body (hence the nervous system itself) _____.
   a. have a genetic basis
   b. are mediated by memory
   c. are mediated by learning and reasoning
   d. all of the above are correct

6. The active, intricate life-styles of complex vertebrates emerged through the coevolution of _____ and _____.
   a. reptiles and mammals
   b. amphibians and mammals
   c. the nervous system; sensory and motor structures
   d. b and c are correct

7. In a simple reflex, _____ directly signal _____, which act on muscle cells.
   a. sensory neurons; interneurons
   b. interneurons; motor neurons
   c. sensory neurons; motor neurons
   d. motor neurons; sensory neurons

8. The accelerating flow of _____ ions through gated channels across the membrane is the actual trigger for an action potential.
   a. potassium
   b. sodium
   c. hydrogen
   d. a and b are correct

9. The peripheral nervous system consists of the _____; the central nervous system consists of the _____.
   a. nerves and ganglia; brain and spinal cord
   b. brain and spinal cord; nerves and ganglia
   c. spinal cord; brain
   d. nerves and interneurons; brain and spinal cord

10. _____ nerves slow down the body overall and divert energy to basic housekeeping tasks; _____ nerves slow down housekeeping tasks and increase overall activity during times of heightened awareness, excitement, or danger.
    a. autonomic; somatic
    b. sympathetic, parasympathetic
    c. parasympathetic; sympathetic
    d. peripheral; central

11. Match the central nervous system region with some of its functions.
    _e_ spinal cord
    _d_ medulla oblongata
    _b_ hypothalamus
    _c_ limbic system
    _a_ cerebral cortex

    a. receives sensory input, integrates it with stored information, coordinates motor responses
    b. monitors internal organs and related behavior (e.g., thirst, hunger, sex)
    c. governs emotions
    d. coordinates basic reflexes (e.g., for respiration, blood circulation)
    e. makes reflex connections for limb movements, internal organ activity

12. Match the sensory receptor with the appropriate stimulus.
    _____ olfactory receptor   a. dim light
    _____ rod cell   b. bending or tilting
    _____ free nerve ending   c. odors
    _____ cone cell   d. intense light
    _____ hair cell   e. pressure, temperature, pain

Selected Key Terms

action potential 428	nerve 433
axon 427	neuron 427
central nervous system 434	parasympathetic nerve 434
chemical synapse 431	peripheral nervous system 434
chemoreceptor 441	photoreceptor 441
cone cells 445	psychoactive drug 439
dendrite 427	reflex arc 433
forebrain 436	resting membrane potential 429
hair cell 442	rod cells 445
hindbrain 436	sensory neuron 427
interneuron 427	stimulus 441
limbic system 438	sympathetic nerve 434
mechanoreceptor 441	synaptic integration 432
midbrain 436	thermoreceptor 441
motor neuron 427	transmitter substance 431

Readings

Barlow, R., Jr. April 1990. "What the Brain Tells the Eye." *Scientific American* 262(4):90–95.

Churchland, P., and P. Churchland. January 1990. "Could a Machine Think?" *Scientific American* 262(1):32–37.

Julien, R. 1985. *A Primer of Drug Action.* Fourth edition. New York: Freeman. Effectively fills the gap between popularized (and often superficial or misleading) accounts of drug action and the upper-division books in pharmacology. Paperback.

Kalil, R. December 1989. "Synapse Formation in the Developing Brain." *Scientific American* 261(6):76–85.

Romer, A., and T. Parsons. 1986. *The Vertebrate Body.* Sixth edition. Philadelphia: Saunders. Insights into the evolution of vertebrate nervous systems.

Shepherd, G. 1988. *Neurobiology.* Second edition. New York: Oxford. Paperback.

KEY CONCEPTS

1. Hormones and other signaling molecules help integrate cell activities in ways that benefit the whole body. Some hormones help the body adjust to short-term changes in diet and levels of activity. Others have roles in long-term adjustments underlying growth, development, and reproduction.

2. Many endocrine gland secretions are integrated by interactions between the hypothalamus and pituitary gland, which serve as a neuroendocrine control center.

3. Neural signals, hormonal signals, local chemical changes, and environmental cues trigger hormone secretions. Only cells with receptors for a specific hormone are its targets. Steroid hormones trigger gene activation and protein synthesis in target cells; protein hormones alter activity of existing enzymes in target cells.

"THE ENDOCRINE SYSTEM"

Discovery of Hormones

The word "hormone" dates back to the early 1900s, when W. Bayliss and E. Starling were wondering what makes the pancreas secrete its juices exactly when food is traveling through the canine gut. Were signals from the nervous system or something else stimulating the pancreas into action? To find the answer, the researchers cut off the nerves leading to the small intestine but left the blood vessels intact. The pancreas still made the secretory response. More telling, extracts of cells taken from the lining of the small intestine also triggered the response. Those cells had to be the source of the signal, which came to be called secretin.

This work confirmed an idea that had been around for centuries: *Internal secretions released into the bloodstream influence the activities of tissues and organs.* Such internal secretions were named hormones (after the Greek *hormon*, meaning to set in motion). Later work led to the discovery of many more hormones and their sources.

As you can tell from Figure 31.1, many different organs and tissues, scattered throughout the body, are sources of hormones. Traditionally they have been viewed

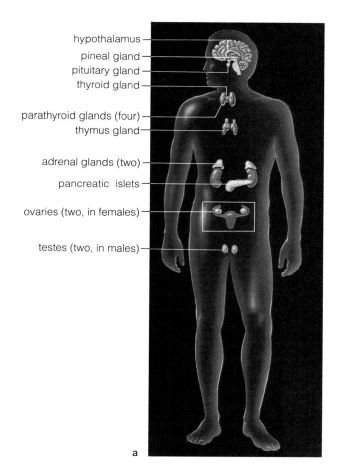

a

Figure 31.1 (**a**) Location of major endocrine glands and cells in the human body. (**b**) Closer view of the hypothalamus and pituitary, which together function as the neuroendocrine control center. Endocrine cells also occur in the liver, kidneys, heart, and intestinal epithelium.

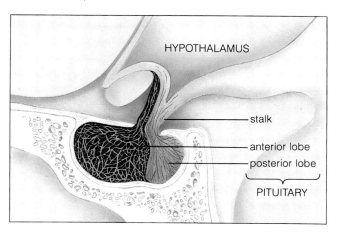

b

as "the endocrine system." The phrase suggests that they offer a separate means of control, apart from the nervous system. (The Greek *endon* means within; *krinein* means to separate.) Today we know the boundaries between the two systems are not so tidy. Glands thought to work independently are under neural control. Some neurons secrete "hormones." The pituitary, the so-called master gland of the endocrine system, is actually controlled by the hypothalamus.

We no longer say a complete division exists between the nervous and endocrine systems, given their overlapping. For this reason, we will begin with the secretions used as agents of control, regardless of their source.

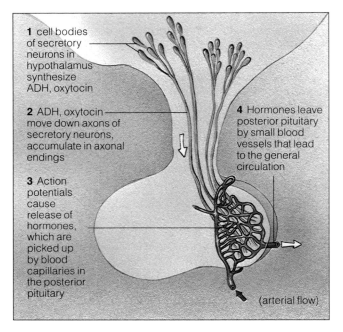

Figure 31.2 Functional links between the hypothalamus and the posterior lobe of the pituitary in humans.

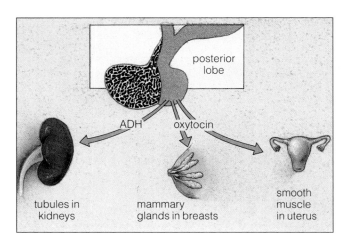

Types of Signaling Molecules

In complex animals, the nervous and endocrine systems work together to integrate the responses of thousands, millions, even many billions of cells in ways that benefit the whole body. Integration depends on *signaling molecules.* These are hormones and other secretions that alter the behavior of target cells. Any cell is a "target" if it has receptors to which specific signaling molecules can bind and elicit a response. A target may or may not be adjacent to the secreting cell. There are four main classes of signaling molecules:

---

1. Hormones are secreted from endocrine glands, endocrine cells, and some neurons, then are transported by the bloodstream to nonadjacent targets.

2. Transmitter substances are secreted from neurons, act on immediately adjacent target cells, then are rapidly degraded or recycled (page 431).

3. Local signaling molecules are secreted from cells in many different tissues. They alter chemical conditions in the immediate vicinity, then are rapidly degraded.

4. Pheromones, secreted by some exocrine glands, have targets outside the body. They diffuse through water or air and act on cells of other animals of the same species. They integrate social activities between animals.

THE HYPOTHALAMUS-PITUITARY CONNECTION

The hypothalamus and pituitary serve as a **neuroendocrine control center.** The hypothalamus, recall, monitors internal conditions and emotional states. The pituitary, which is about the size of a garden pea, is suspended from the base of the hypothalamus by a slender stalk (Figure 31.1).

The pituitary gland is lobed. Its *posterior lobe* stores and then secretes two hormones, which are actually produced in the hypothalamus. Its *anterior lobe* produces and secretes six hormones and controls the release of several more from other endocrine glands. (The pituitary of many vertebrates—not humans—also has an intermediate lobe. Often its secretions cause changes in skin or fur color.)

Figure 31.3 (*Left*) Secretions of the posterior lobe of the pituitary and some of their targets.

Posterior Lobe Secretions

Figure 31.2 shows the links between the hypothalamus and the posterior lobe. Notice the cell bodies of neurons in the hypothalamus and their axons, which extend down the pituitary stalk and into the posterior lobe. Antidiuretic hormone (ADH) and oxytocin are produced in the cell bodies and stored in the axon endings, next to a capillary bed. ADH or oxytocin released from the axon endings diffuses through the interstitial fluid and into the capillaries. They end up in different body regions and have effects on specific target cells (Figure 31.3).

Cells in the walls of kidney tubules are targets for ADH, which helps control the body's water and solute levels. Oxytocin has roles in reproduction. For example, it triggers muscle contractions during labor and milk release when the young are being nursed.

Anterior Lobe Secretions

The hypothalamus also produces and secretes **releasing hormones,** a type of signaling molecule with targets in the anterior lobe of the pituitary. Most releasing hormones stimulate secretions from a target cell, but some slow down its secretions (Table 31.1). As Figure 31.4 shows, the releasing hormones travel through two capillary beds before leaving the blood and binding to receptors on target cells in the anterior lobe.

Releasing hormones were identified by monumental research efforts that began in 1955, most notably by Roger Guillemin and Andrew Schally. Over one four-year period, Guillemin's team purchased 500 tons of sheep brains from meat processing plants and extracted 7 tons of hypothalamic tissue from it. They ended up with a single milligram of thyrotropin-releasing hormone.

In response to commands from the hypothalamus, different cells of the anterior lobe secrete the following hormones of their own:

Corticotropin-stimulating hormone (ACTH)
Thyrotropin-stimulating hormone (TSH)
Follicle-stimulating hormone (FSH)
Luteinizing hormone (LH)
Prolactin (PRL)
Somatotropin (STH); also called
    growth hormone (GH)

The first four hormones listed act on endocrine glands, which in turn produce other hormones. The last two

Table 31.1	Effect of Releasing Hormones on Anterior Pituitary Secretions	
Releasing Hormone	Influences Secretion of:	Effect*
Corticotropin-releasing hormone (CRH)	Corticotropin (ACTH)	+
Thyrotropin-releasing hormone (TRH)	Thyrotropin (TSH)	+
Gonadotropin-releasing hormone (GnRH)	Follicle-stimulating hormone (FSH)	+
	Luteinizing hormone (LH)	+
STH-releasing hormone (STHRH)	Somatotropin (STH); also called growth hormone (GH)	+
Somatostatin	Somatotropin, other hormones (PRL)	−
Dopamine	Prolactin (PRL)	−

* Stimulatory (+) or inhibitory (−).

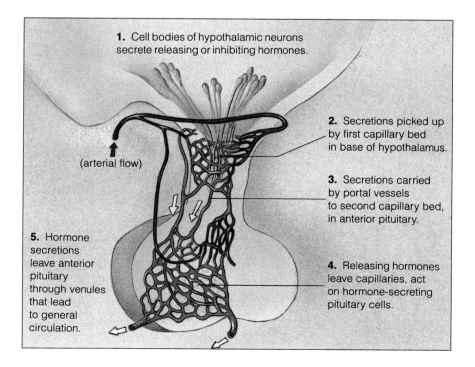

1. Cell bodies of hypothalamic neurons secrete releasing or inhibiting hormones.

(arterial flow)

2. Secretions picked up by first capillary bed in base of hypothalamus.

3. Secretions carried by portal vessels to second capillary bed, in anterior pituitary.

5. Hormone secretions leave anterior pituitary through venules that lead to general circulation.

4. Releasing hormones leave capillaries, act on hormone-secreting pituitary cells.

Figure 31.4 Functional links between the hypothalamus and the anterior lobe of the pituitary.

Table 31.2 Hormones Released from the Mammalian Pituitary Gland

Pituitary Lobe	Secretions	Abbreviation	Main Targets	Primary Actions
Posterior Nervous tissue (extension of hypothalamus)	Antidiuretic hormone	ADH	Kidneys	Induces water conservation required in control of extracellular fluid volume (and, indirectly, solute concentrations)
	Oxytocin		Mammary glands Uterus	Induces milk movement into secretory ducts Induces uterine contractions
Anterior Mostly glandular tissue	Corticotropin	ACTH	Adrenal cortex	Stimulates release of adrenal steroid hormones
	Thyrotropin	TSH	Thyroid gland	Stimulates release of thyroid hormones
	Gonadotropins:			
	Follicle-stimulating hormone	FSH	Ovaries, testes	In females, stimulates follicle growth, helps stimulate estrogen secretion, ovulation; in males, promotes spermatogenesis
	Luteinizing hormone	LH	Ovaries, testes	In females, stimulates ovulation, corpus luteum formation; in males, promotes testosterone secretion, sperm release
	Prolactin	PRL	Mammary glands	Stimulates and sustains milk production
	Somatotropin (also called growth hormone)	STH (GH)	Most cells	Has growth-promoting effects in young; induces protein synthesis, cell division; has role in glucose, protein metabolism in adults
Intermediate* Mostly glandular tissue	Melanocyte-stimulating hormone	MSH	Pigmented cells in skin, other surface coverings	Induces color changes in response to external stimuli; affects behavior

* Present in most vertebrates (not adult humans).

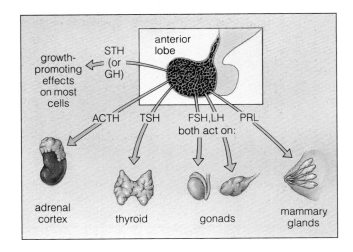

Figure 31.5 Secretions of the anterior lobe of the pituitary and some of their targets.

hormones, prolactin and somatotropin, have effects on body tissues in general (Figure 31.5 and Table 31.2).

For example, somatotropin (or growth hormone) stimulates protein synthesis and cell divisions in target cells. It profoundly influences overall growth, especially of cartilage and bone. Figure 31.6 shows what can happen with too little or too much somatotropin. *Pituitary dwarfism* results when not enough somatotropin was produced during childhood. The adult is similar in proportion to a normal person but much smaller. *Gigantism* results when excessive amounts of somatotropin were produced during childhood. The adult is similar in proportion to a normal person but much larger.

Acromegaly results from excessive secretion of somatotropin during adulthood, when long bones no longer can lengthen. Cartilage, bone, and other connective tissues of the hands, feet, and jaws thicken, as do epithelial tissues of the skin, nose, eyelids, lips, and tongue (Figure 31.7). Skin thickening is pronounced on the forehead and soles of the feet.

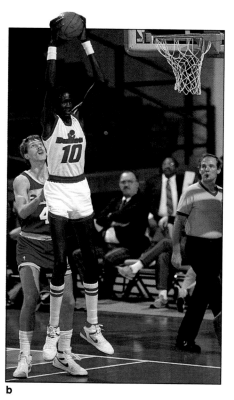

a

b

Figure 31.6 (**a**) Effect of somatotropin (STH) on overall body growth. The person at the center is affected by gigantism, which resulted from excessive STH production during childhood. The person at right displays pituitary dwarfism, which resulted from underproduction of STH during childhood. The person at the left is average in size.

(**b**) Manute Bol, an NBA center, is 7 feet 6¾ inches tall owing to excessive STH production during childhood.

| age nine | sixteen | thirty-three | fifty-two |

Figure 31.7 Acromegaly, which resulted from excessive production of somatotropin (STH) during adulthood. Before this female reached maturity, she was symptom-free.

Table 31.3 Hormone Sources Other Than the Mammalian Hypothalamus and Pituitary

Source	Its Secretion(s)	Main Targets	Primary Actions
Adrenal cortex	Glucocorticoids (including cortisol)	Most cells	Raises blood sugar level; helps reduce lipid, protein metabolism; mediates responses to stress
	Mineralocorticoids (including aldosterone)	Kidney	Promotes sodium reabsorption; controls salt, water balance
Adrenal medulla	Epinephrine (adrenalin)	Muscle, adipose tissue	Raises blood level of sugar, fatty acids; increases heart rate, force of contraction
	Norepinephrine	Smooth muscle of blood vessels	Promotes constriction or dilation of blood vessel diameter
Thyroid	Triiodothyronine, thyroxine	Most cells	Regulates metabolism; has roles in growth, development
	Calcitonin	Bone	Lowers calcium levels in blood
Parathyroids	Parathyroid hormone	Bone, kidney, gut	Elevates calcium levels in blood
Gonads:			
Testis (in males)	Androgens (including testosterone)	General	Required in sperm formation, development of genitals, maintenance of secondary sex traits; influences growth, development
Ovary (in females)	Estrogens	General	Required in egg formation; prepares uterine lining for pregnancy; other actions same as above
	Progesterone	Uterus, breasts	Prepares, maintains uterine lining for pregnancy; stimulates breast development
Pancreatic islets	Insulin	Most cells	Lowers blood sugar level
	Glucagon	Liver	Raises blood sugar level
	Somatostatin	Insulin-secreting cells of pancreas	Influences carbohydrate metabolism
Endocrine cells of stomach, gut	Gastrin, secretin, etc.	Stomach, pancreas, gallbladder	Stimulates activity of stomach, pancreas, liver, gallbladder
Liver	Somatomedins	Most cells	Stimulates overall growth, development
Kidney	Erythropoietin*	Bone marrow	Stimulates red blood cell production
	Angiotensin*	Adrenal cortex, arterioles	Helps control blood pressure, aldosterone secretion
	Vitamin D_3*	Bone, gut	Enhances calcium resorption and uptake
Heart	Atrial natriuretic hormone	Kidney, blood vessels	Increases sodium excretion; lowers blood pressure
Thymus	Thymosin, etc.	Lymphocytes	Has roles in immune responses
Pineal	Melatonin	Gonads (indirectly)	Influences daily biorhythms, sexual activity

* These hormones are not produced in the kidneys but are formed when *enzymes* produced in kidneys activate specific substances in the blood.

SELECTED EXAMPLES OF HORMONAL CONTROL

Table 31.3 lists hormones from endocrine glands other than the pituitary. From earlier chapters, we are already familiar with some of those glands. For example, we have seen how the thymus gland is a site where white blood cells grow and differentiate. But it also secretes thymosins, hormones that affect the functions of some lymphocytes. Here we will focus on a few examples of endocrine activity to show how hormonal controls work. These examples will lay the groundwork for understanding the interplays among hormones that govern human reproduction, a topic of the next chapter.

Adrenal Glands

Adrenal Cortex. Many endocrine glands are linked to the neuroendocrine control center by *homeostatic feedback loops.* In such loops, the hypothalamus or pituitary (or both) detects a change in the concentration of a hormone in some region and then responds by inhibiting or stimulating the gland that secretes the hormone.

Consider the **adrenal cortex,** the outer portion of a gland perched on top of each kidney (Figure 31.8). Glucocorticoids are among the hormones produced by the adrenal cortex. These hormones take part in metabolism and in the inflammatory response. Cortisol is an example; it helps maintain blood levels of glucose between meals. Cortisol blocks the uptake and use of glucose by muscle cells and makes more glucose in blood available to the brain. It also stimulates liver cells to form glucose from amino acids when glucose levels fall.

When blood levels of cortisol fall below a set point, the hypothalamus secretes a releasing hormone (CRH). The CRH prods the pituitary into secreting corticotropin (ACTH), a hormone that prods the adrenal cortex into secreting cortisol. When blood levels of cortisol are too high, CRH and ACTH secretions slow down the secretion of cortisol (Figure 31.8). This is an example of negative feedback.

Adrenal Medulla. Hormone-secreting neurons are located in the **adrenal medulla,** the inner region of the adrenal gland (Figure 31.8). Their secretions (epinephrine and norepinephrine) affect blood circulation and carbohydrate metabolism. The hypothalamus and other brain centers govern their secretion by issuing commands along sympathetic nerves.

The adrenal medulla helps mobilize the body during times of excitement or stress. For example, in response to epinephrine and norepinephrine, the heart beats faster and harder, blood flow increases to heart and muscle cells, airways in the lungs dilate, and more oxygen is delivered to cells throughout the body. These features of the "fight-flight" response complement those triggered by the sympathetic nerves (page 434).

Thyroid Gland

Earlier we saw how excessive or insufficient output from the pituitary affects body functioning. Abnormal secretions from other endocrine glands also can have profound effects on the body. Consider the human **thyroid gland,** positioned at the base of the neck (Figure 31.9). Its main hormones, thyroxine and triiodothyronine, affect overall metabolic rates, growth, and development.

Insufficient thyroid output leads to *hypothyroidism.* Hypothyroid adults are often overweight, sluggish, dry-

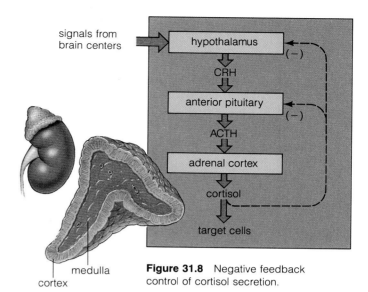

Figure 31.8 Negative feedback control of cortisol secretion.

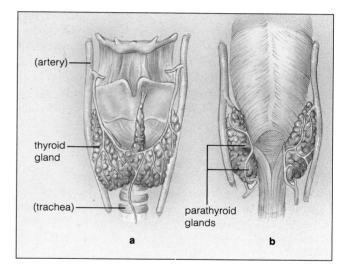

Figure 31.9 (**a**) Anterior view of the human thyroid gland. (**b**) Posterior view, showing the location of the four parathyroid glands embedded in it.

skinned, and intolerant of cold. If the disorder is present at birth and is not detected early in infancy, mental retardation and dwarfism may develop. Excessive thyroid output can lead to *hyperthyroidism.* Hyperthyroid adults suffer from increased heart rate, elevated blood pressure, weight loss despite normal caloric intake, intolerance of heat, and profuse sweating. Typically they show nervous, agitated behavior, and they have trouble sleeping.

Figure 31.10 A mild case of goiter, as displayed by Maria de Medici in 1625. During the late Renaissance, a rounded neck was considered a sign of beauty; it occurred regularly in parts of the world where iodine supplies were insufficient for normal thyroid function.

Thyroid hormones contain iodine and cannot be produced without it. In the absence of iodine, thyroid hormone levels in the blood decrease. The anterior pituitary responds by secreting thyroid-stimulating hormone (TSH). Excessive TSH overstimulates the thyroid gland and causes it to enlarge. The resulting tissue enlargement is a form of *goiter* (Figure 31.10). Goiter caused by iodine deficiency is no longer common in countries where iodized salt is widely used. Elsewhere, hundreds of thousands of people still suffer from the disorder.

Parathyroid Glands

Some glands are not stimulated directly by other hormones or nerves. Rather, they respond homeostatically to a chemical change in their surroundings. The **parathyroid glands,** embedded in tissues in the back of the thyroid gland, are like this (Figure 31.9). In response to a drop in extracellular levels of calcium ions, they secrete PTH (parathyroid hormone). By helping to restore blood calcium levels, the parathyroid glands influence the availability of calcium ions for gene activation, muscle contraction, and many other tasks.

PTH stimulates calcium and phosphate removal from bone and its movement into extracellular fluid. It increases the kidney's reabsorption of calcium. It also helps activate vitamin D. The activated form, which is considered to be a hormone, enhances calcium absorption from food moving through the gut. Vitamin D deficiency leads to *rickets,* a disorder arising from the lack of enough calcium for proper bone development.

When calcium levels rise in response to PTH-induced events, stimulation of the parathyroid glands decreases. At the same time, the thyroid gland secretes calcitonin, which promotes calcium deposition in bones.

Pancreatic Islets

Like endocrine glands, some *cells* have endocrine functions. For example, besides having exocrine tissues that secrete digestive enzymes, the pancreas also has about 2 million endocrine cell clusters, the **pancreatic islets.** The following are three types of hormone-secreting cells in the islets:

1. *Alpha cells* secrete **glucagon.** Between meals, cells use the glucose circulated to them by the bloodstream. The blood glucose level decreases, at which time glucagon secretions cause glycogen (a storage polysaccharide) and amino acids to be converted to glucose. In such ways, *glucagon raises the glucose level in the blood.*

2. *Beta cells* secrete **insulin.** After meals, when the blood glucose level is high, insulin stimulates uptake of glucose by liver, muscle, and adipose cells especially. It also promotes synthesis of proteins and fats, and inhibits protein conversion to glucose. Thus *insulin lowers the glucose level in the blood.*

3. *Delta cells* secrete **somatostatin,** which has a regulatory function. It can inhibit the secretion of insulin and glucagon.

Figure 31.11 shows how interplays among these hormones and others help keep blood glucose levels fairly constant, despite great variation in when—and how much—we eat. The importance of this function is clear if we consider what happens when the body cannot produce enough insulin or when insulin's target cells cannot respond to it. Disorders in fat, carbohydrate, and protein metabolism occur.

Insulin deficiency can lead to *diabetes mellitus.* Blood glucose levels rise; this promotes urination and so disrupts the body's water-solute balance. Glucose-starved cells start degrading proteins and fats for energy, and this leads to weight loss. Ketones (normal by-products of fat breakdown) accumulate. Extracellular pH decreases and the body becomes dehydrated. Brain function is disrupted; in extreme cases, coma and death may follow.

In "type 1 diabetes," insulin-secreting beta cells are destroyed. Genetic and environmental factors combine to produce the disorder, which is the less common but more serious of the two types of diabetes. The body mounts an autoimmune response against its own insulin-secreting cells after they have been damaged by a virus or toxin (page 399). Type 1 diabetics survive with insulin injections.

In "type 2 diabetes," insulin levels are close to or above normal—but target cells cannot respond to the hormone. They may have abnormal or too few insulin receptors. Type 2 diabetes usually occurs in middle age and is less severe than the other type. Affected persons can lead a normal life by controlling their diet.

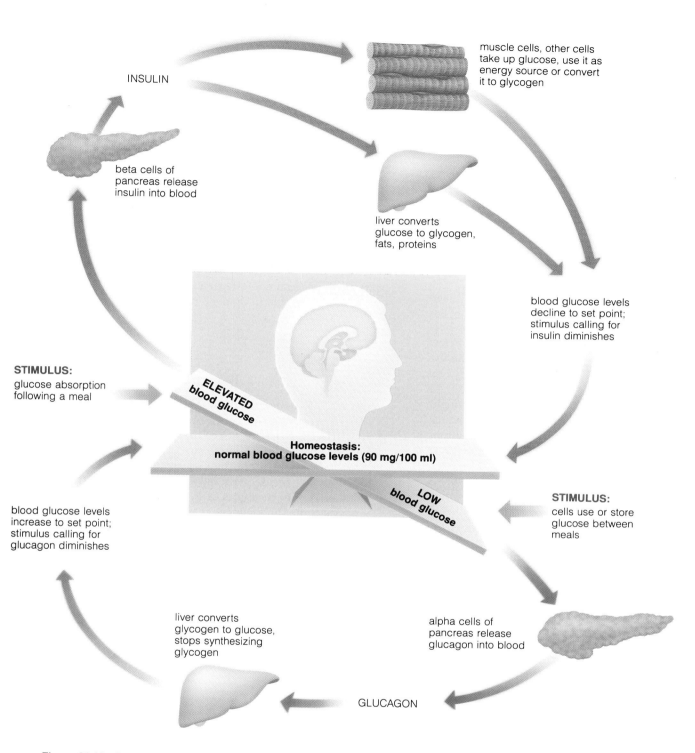

Figure 31.11 Some of the homeostatic controls over glucose metabolism. *Following* a meal, glucose enters the bloodstream faster than cells can use it. Blood glucose levels rise. Pancreatic beta cells are stimulated to secrete insulin. The hormonal targets (mainly liver, fat, and muscle cells) use glucose or store it as glycogen.

Between meals, blood glucose levels drop. Pancreatic alpha cells are stimulated to secrete glucagon, which enters the blood. Also, the hypothalamus prods the adrenal medulla into secreting other hormones that slow down the conversion of glucose to glycogen in liver, fat, and muscle cells.

Pineal Gland

So far, we have seen how endocrine glands and cells respond to other hormones, nerves, or chemical changes in their surroundings. Now we can start thinking about a larger picture, in which reproduction and development of the body are controlled by hormonal responses to environmental cues.

Until about 240 million years ago, vertebrates commonly had a third eye, on top of the head. Lampreys still have one, beneath the skin. A modified form of this photosensitive organ persists in mammals, birds, and most reptiles; we call it the **pineal gland** (Figure 31.1). The pineal gland secretes melatonin, a hormone influencing the development of reproductive organs (gonads) and reproductive cycles.

Melatonin is secreted in the absence of light. This means melatonin levels vary from day to night. It also means the levels change with the seasons, as when winter days are shorter than summer days. The hormonal effects are evident in hamster reproductive cycles. High melatonin levels in winter suppress sexual activity; in summer, when melatonin levels are low, sexual activity peaks.

In humans, decreased melatonin secretion might help trigger the onset of puberty, the age at which reproductive structures start to mature. Nearly a century ago, Otto Heubner performed an autopsy on a boy who had entered puberty prematurely (at age four) and discovered that a brain tumor had destroyed the boy's pineal gland. Normally, melatonin levels are high through age five. Then they steadily decrease until the end of puberty, when they stabilize at about one-fourth of the peak levels.

Table 31.4 Two Main Categories of Hormones

Type of Hormone	Examples
Steroid	Estrogens, testosterone, aldosterone, cortisol
Nonsteroid:	
Amines	Norepinephrine, epinephrine
Peptides	ADH, oxytocin, TRH
Proteins	Insulin, somatotropin, prolactin
Glycoproteins	FSH, LH, TSH

SIGNALING MECHANISMS

Hormones and other signaling molecules can trigger or alter the rate of protein synthesis or cause the functions of existing proteins to be modified. What dictates the nature of the target cell's response? That depends largely on two things. First, different signals activate different cellular mechanisms. Second, not all cells *can* respond to all types of signals. Most have receptors for insulin and some other hormones; that is why those hormones have such widespread effects. But only a few cell types have receptors for certain other hormones, which have highly directed effects.

Let's think about some responses to the following main categories of hormones (Table 31.4):

1. *Steroid hormones* or steroidlike compounds, which are assembled from cholesterol and are largely insoluble in water but soluble in lipids.

2. *Nonsteroid hormones,* which are assembled from amino acids and are water-soluble.

Steroid Hormone Action

Steroid hormones stimulate or inhibit protein synthesis by switching certain genes on or off. They don't alter the activity of already existing proteins, especially enzymes. Being lipid-soluble, steroid hormones diffuse easily across the plasma membrane of target cells (Figure 31.12). Once inside, they move into the nucleus, where they bind with receptors specific for them.

The three-dimensional shape of the hormone-receptor complex allows it to bind with chromosomal proteins. The complex activates certain gene regions. Transcription and then translation into specific proteins follow.

Testosterone is an example of a steroid hormone. It influences the development of male sexual traits. In *testicular feminization syndrome*, the receptor to which testosterone binds is defective. Genetically, affected individuals are males; they have functional testes that secret testosterone. But none of the target cells can respond to the hormone, so the secondary sexual traits that do develop are like those of females.

Nonsteroid Hormone Action

Protein hormones and other water-soluble signaling molecules cannot cross the plasma membrane of target cells. Some bind to receptors that trigger endocytosis at the patch of membrane bearing the receptors (page 44). Others bind to receptors that cause the opening of ion channels across the membrane. When the ions move inward, their concentration inside the cell changes. Cell activities change as a result.

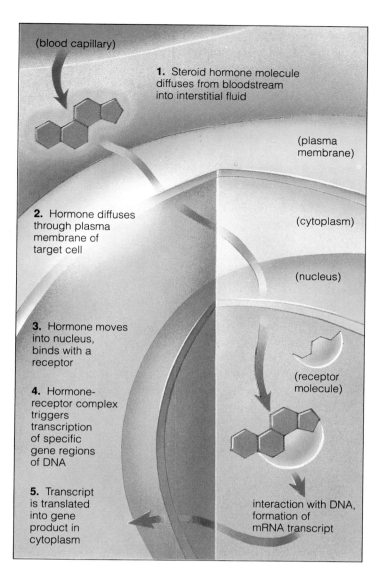

Figure 31.12 Proposed mechanism of steroid hormone action on a target cell. This same type of mechanism is also thought to occur for thyroid hormones, which have low solubility in water but more readily cross the lipid bilayer of the plasma membrane.

Within figure 31.12:

(blood capillary)

1. Steroid hormone molecule diffuses from bloodstream into interstitial fluid

(plasma membrane)

2. Hormone diffuses through plasma membrane of target cell

(cytoplasm)

(nucleus)

3. Hormone moves into nucleus, binds with a receptor

4. Hormone-receptor complex triggers transcription of specific gene regions of DNA

(receptor molecule)

5. Transcript is translated into gene product in cytoplasm

interaction with DNA, formation of mRNA transcript

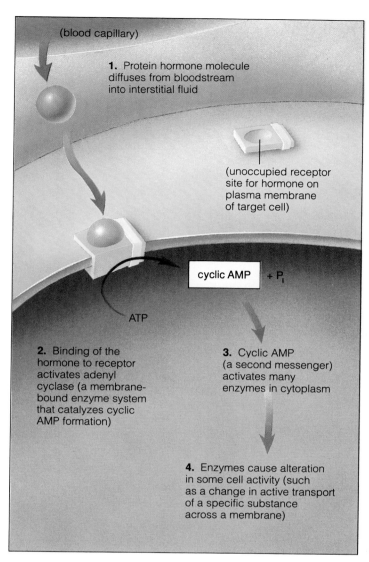

Figure 31.13 Proposed mechanism of protein hormone action on a target cell. The response is mediated by a second messenger inside the cell—in this case, cyclic AMP. Other chemical messengers may be involved, depending on the particular hormone and its particular target cell.

Within figure 31.13:

(blood capillary)

1. Protein hormone molecule diffuses from bloodstream into interstitial fluid

(unoccupied receptor site for hormone on plasma membrane of target cell)

cyclic AMP + P_i

ATP

2. Binding of the hormone to receptor activates adenyl cyclase (a membrane-bound enzyme system that catalyzes cyclic AMP formation)

3. Cyclic AMP (a second messenger) activates many enzymes in cytoplasm

4. Enzymes cause alteration in some cell activity (such as a change in active transport of a specific substance across a membrane)

Most of the protein hormones, including glucagon, activate **second messengers.** These are molecules inside the cell that mediate the response to a hormone. An example is *cyclic AMP* (cyclic adenosine monophosphate). First a hormone binds to a membrane receptor on a target cell. Binding alters the activity of a membrane-bound enzyme system (Figure 31.13). An enzyme (adenylate cyclase) is prodded into action; it speeds the conversion of ATP to cyclic AMP.

The hormone-receptor complex activates many molecules of the enzyme, not just one. Each enzyme molecule increases the rate at which many ATP molecules are converted to cyclic AMP. Each cyclic AMP molecule so formed then activates many enzymes. Each of the enzymes so activated can convert a very large number of substrate molecules into different enzymes, and so on.

Soon the number of molecules representing the final cellular response to the initial signal is very large. In this way, second messengers *amplify* a response to a signaling molecule.

SUMMARY

1. Cells constantly exchange substances with extracellular fluid. The nature and amount of those substances change with diet and levels of activity; they also change during development. All those cellular withdrawals and discharges must be integrated in ways that ensure survival for the whole body.

2. Integration is accomplished by signaling molecules. These are hormones and other secretions by one cell that adjust the behavior of another, target cell.

3. A neuroendocrine control center integrates many activities for the vertebrate body. It consists of the hypothalamus and pituitary gland.

4. Two hormones (ADH and oxytocin) produced by the hypothalamus are released from the posterior lobe of the pituitary. ADH influences kidney function. Oxytocin influences reproductive events, including muscle contractions during labor. Six other hormones produced by the hypothalamus (releasing hormones) control hormone secretions by cells of the anterior lobe of the pituitary.

5. The anterior lobe produces and secretes six hormones. Two (prolactin and somatotropin, or growth hormone) have general effects on body tissues. The remainder (ACTH, TSH, FSH, and LH) influence specific endocrine glands.

6. Hormones are secreted in response to neural signals, hormonal signals, local chemical changes, and environmental cues. Many endocrine gland secretions are controlled by negative feedback loops through the neuroendocrine control center.

7. Fast-acting hormones such as insulin generally work when the extracellular concentration of a substance must be homeostatically controlled. Slow-acting hormones such as somatotropin have more prolonged, gradual, and often irreversible effects, as during development.

8. Cells respond to specific hormones or other signaling molecules only if they have receptors for them. Steroid hormones have receptors in the nucleus of target cells. Protein hormones have receptors on the plasma membrane of target cells; responses to them are often mediated by a second messenger, such as cyclic AMP, inside the cell.

9. Steroid hormones trigger gene activation and protein synthesis; protein hormones alter the activity of enzymatic proteins already present in target cells. These cellular responses contribute to maintaining the internal environment or to the development or reproductive program.

1. Name the main endocrine glands and state where each is located in the human body. *449*

2. Define hormone. What functions do hormones serve? How do these functions differ from those of transmitter substances? *450*

3. There are two general classes of hormones: steroid and protein. How is each thought to act on a target cell? *458–459*

4. The hypothalamus and pituitary are considered to be a neuroendocrine control center. Can you describe some of the functional links between these two organs? *450*

5. How does the hypothalamus control secretions of the posterior lobe of the pituitary? The anterior lobe? *450*

6. Name three endocrine glands and a substance that each one secretes. What are the main consequences of their secretion? *454*

7. Name a hormone secreted by the anterior pituitary that has an effect on most body cells rather than on a specific cell type. What are the clinical consequences of too little or too much secretion of this hormone? *452*

8. Which secretions of the posterior and anterior lobes of the pituitary glands have the targets indicated? (Fill in the blanks; see pages 450 and 452.)

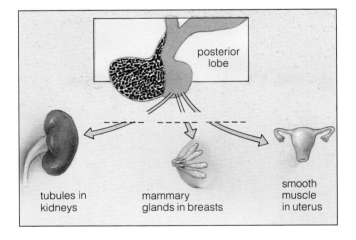

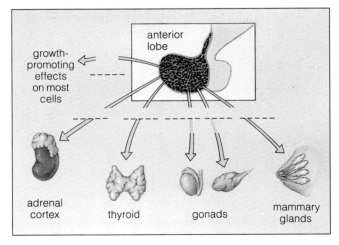

Self-Quiz   (Answers in Appendix IV)

1. _____, _____, _____, and _____ are types of signaling molecules that help integrate cell activities in ways that benefit the whole body.

2. The _____ and _____ gland interact as a neuroendocrine control center to integrate endocrine gland secretions.

3. Stimulation or inhibition of hormone secretions often involves _____ loops between the neuroendocrine control center and glands.

4. A target cell for a specific hormone or some other signaling molecule has _____ for that molecule.

5. The hypothalamus produces two hormones that are released from the anterior lobe of the pituitary gland. One hormone, _____, affects kidney function; the other, _____, affects some reproductive events.
   a. ADH; oxytocin
   b. prolactin; ADH
   c. oxytocin; ADH
   d. ADH; prolactin

6. The anterior lobe of the pituitary gland produces two hormones, _____ and _____, that have general effects on body tissues.
   a. ACTH; somatotropin
   b. prolactin; FSH
   c. ACTH; FSH
   d. prolactin; somatotropin

7. Which of the following does *not* stimulate hormone secretion?
   a. neural signals
   b. local chemical changes
   c. hormonal signals
   d. environmental cues
   e. all of the above can stimulate hormone secretion

8. Insulin is an example of a _____ hormone which must work to accomplish homeostatic control of an extracellular substance; somatotropin is an example of a _____ hormone which operates during body development.
   a. growth; metabolic
   b. fast-acting; slow-acting
   c. metabolic; growth
   d. slow-acting; fast-acting

9. Which of the following statements is true?
   a. steroid hormones have receptors on the plasma membranes of target cells
   b. protein hormones have receptors in target cell nuclei
   c. protein hormones alter the activity of genes
   d. steroid hormones activate genes and protein synthesis

10. Match the endocrine control concepts.
    _____ oxytocin   a. acts as neuroendocrine control
    _____ ADH   center
    _____ steroid   b. affects kidney function
            hormone   c. has general effects on growth
    _____ somatotropin   d. affects reproductive events
            (GH)   e. triggers protein synthesis
    _____ hypothalamus/
            pituitary

Selected Key Terms

adrenal cortex *455*
adrenal medulla *455*
anterior lobe of pituitary *450*
cyclic AMP *459*
endocrine system *449*
glucagon *456*
homeostatic feedback loop *455*
hormone *450*
insulin *456*
local signaling molecule *450*
neuroendocrine control
  center *450*

nonsteroid hormone *458*
pancreatic islet *456*
parathyroid gland *456*
pheromone *450*
pineal gland *458*
posterior lobe of pituitary *450*
releasing hormone *451*
second messenger *459*
steroid hormone *458*
thyroid gland *455*

Readings

Cantin, M., and J. Genest. February 1986. "The Heart As an Endocrine Gland." *Scientific American* 254(2):76–81.

Fellman, B. May 1985. "A Clockwork Gland." *Science 85* 6(4):76–81. Describes some of the known functions of the pineal gland.

Hadley, M. 1988. *Endocrinology.* Second edition. Englewood Cliffs, New Jersey: Prentice-Hall.

Sapolsky, R. January 1990. "Stress in the Wild." *Scientific American* 262(1):116–123. Study of hormonal effects on stress responses in baboons.

Snyder, S. October 1985. "The Molecular Basis of Communication Between Cells." *Scientific American* 253(4):132–141.

32 REPRODUCTION AND DEVELOPMENT

KEY CONCEPTS

1. Most animals reproduce sexually. Elaborate reproductive structures and forms of courtship and parental behavior have accompanied the separation into male and female sexes. The biological cost of those structures and behaviors is offset by the advantages of diversity among the offspring, at least some of which should prove to be adaptive in changing or new environments.

2. The reproductive capacity of human females is cyclic and intermittent, with eggs being released and the uterine lining being prepared for pregnancy on a monthly basis. The hormones estrogen, progesterone, FSH, and LH control this cyclic activity.

3. Human males produce sperm continuously from puberty onward. The hormones testosterone, LH, and FSH control male reproductive functions.

4. Development commonly proceeds through six stages: gamete formation, fertilization, cleavage, gastrulation, organ formation, and growth and tissue specialization.

5. The fate of embryonic cells is determined partly at cleavage (when daughter cells inherit qualitatively different regions of cytoplasm) and partly by cell interactions in the developing embryo. Cell differentiation and morphogenesis depend on those cleavages and cell interactions.

With a full-throated croak that only a female of its kind could find seductive, a male frog proclaims the onset of warm spring rains, of ponds, of sex in the night. By August the summer sun will have parched the earth, and his pond dominion will be gone. But tonight is the hour of the frog! Through the dark, a female moves toward the vocal male. They meet, they dally; he clamps his forelegs about her swollen abdomen and gives it a prolonged squeeze. Out streams a ribbon of hundreds of eggs. As the eggs are being released, the male expels a milky cloud of swimming sperm. Each egg joins with a sperm, and soon afterward, their nuclei fuse. With this fusion, fertilization is completed; a zygote has formed.

For the leopard frog, *Rana pipiens,* a drama now begins to unfold that has been reenacted each spring, with only minor variations, for many millions of years. Within a few hours after fertilization, the single-celled zygote begins dividing into two cells, then four, then many more to produce the embryo. Within twelve days the embryo becomes a larval form—a tadpole—that swims and feeds on its own (Figure 32.1). After several months, legs start to grow. The tail shortens, then disappears. The small mouth, once suitable for feeding on algae, develops jaws and now snaps shut on insects and worms. Eventually an adult frog leaves the water for life on land. With luck it will avoid predators, disease, and other threats in the months ahead. In time it may even find a pond formed by the new season's rains, and the cycle will begin again.

How does the single-celled zygote of a frog or any other complex animal become transformed into all the specialized cells and structures of the adult? With this question we turn to one of life's greatest mysteries—to the development of new individuals in the image of their parents. We are just starting to comprehend the underlying mechanisms.

fertilized egg embarking on its developmental journey

(cutaway view)

THE BEGINNING: REPRODUCTIVE MODES

We have already looked at the cellular basis of asexual and sexual reproduction. Let's turn now to some structural, behavioral, and ecological aspects of those reproductive modes.

Think about a new sponge produced asexually, by budding from the parent body. Or think about a flatworm dividing by fission into two parts that give rise to two new flatworms (Chapter 19). In both cases, the offspring are genetically identical to the parents. Having the same genes as parents is useful when parents are well adapted

Figure 32.1 Development of the leopard frog, *Rana pipiens*. (**a**) A male clasping a female in a behavior called amplexus. When the female releases her eggs into the water, the male releases his sperm over the eggs. (**b**) Frog embryos. (**c**) A larval form called a tadpole. (**d**) Transitional form between the tadpole and the young adult frog (**e**). The diagrams below only hint at the complex cell divisions, cell migrations, and changes in cell size, shape, and function that are necessary to transform the zygote into the adult.

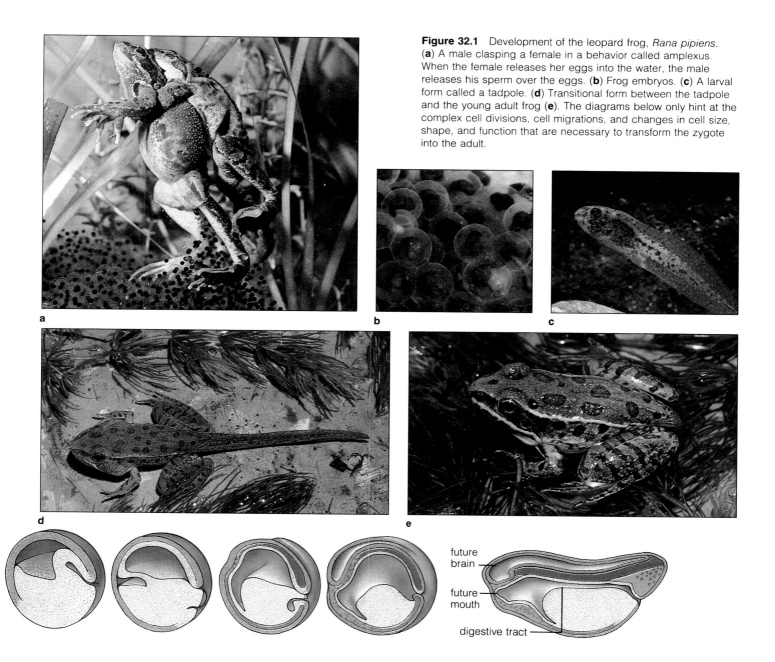

future brain

future mouth

digestive tract

to the surroundings—and when the surroundings remain stable.

But most animals live under changing, unpredictable conditions. They rely mainly on sexual reproduction, with sperm from a male fertilizing eggs from a female. Complete separation into male and female sexes is biologically costly. Getting sperm and eggs together depends on large energy investments in specialized reproductive structures and forms of behavior. Even so, the cost is offset by the variation among the resulting offspring, at least some of which are likely to survive and reproduce in a changing environment.

Consider the question of *reproductive timing*. How do mature sperm become available exactly when eggs mature in a separate individual? Timing depends on energy outlays for sensory structures and rather involved hormonal controls in both parents. Both must produce mature gametes in response to the same cues, such as changes in daylength that mark the onset of the best season for reproduction. Also, males and females of the same species must be able to recognize each other, so energy is invested in chemical signals, structural signals (such as feathers of certain colors and patterns), and often courtship routines.

Fertilization also comes at a cost with separate sexes. Most bony fishes simply release eggs and motile sperm into the water, and the chance of external fertilization would not be good if they produced only one sperm or one egg each season. They invest energy in producing very, very large numbers of gametes. Nearly all land-dwelling animals depend on internal fertilization. (This is understandable; for one thing, sperm released on dry land would not stand much of a chance of swimming over to an egg.) They invest energy in elaborate reproductive organs, such as a penis by which a male deposits its sperm inside the female.

Finally, energy is set aside for *nourishing some number of offspring*. Nearly all animal eggs contain yolk (nutritious material such as proteins and lipids for the embryo), but eggs of some species have more yolk than others. Sea urchin eggs are released in large numbers, and the biochemical investment in yolk for each one is limited. There is a premium on rapid development (sea stars feast on sea urchin offspring), and a self-feeding larval stage is reached in less than a day. Bird eggs are also released from the mother, but they have large yolk reserves that nourish the embryo through a longer period of development, inside an eggshell. Human eggs have almost no yolk. After fertilization, the egg attaches to the mother's body, then the embryo is nourished and supported by physical exchanges with her tissues during an extended pregnancy.

As these few examples suggest, animals show great diversity in their reproduction and development. How-

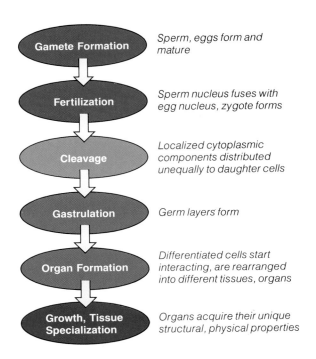

Figure 32.2 Overview of stages of animal development.

ever, some patterns are widespread in the animal kingdom, and they will serve as a framework for our study.

Separation into male and female sexes is an energetically costly mode of reproduction. Specialized structures and forms of behavior are required in getting sperm together with eggs and in lending nutritional support to offspring.

This reproductive mode is advantageous in unpredictable environments, for at least some of the diverse offspring may prove adaptive under new or changing conditions.

BASIC PATTERNS OF DEVELOPMENT

Stages in Development

Figure 32.2 lists the stages of animal development. In the first stage, **gamete formation**, sperm or eggs form and mature within the parents. An immature egg, or *oocyte*, is larger and more complex than a sperm. Its organelles and its stockpiles of proteins, RNA, and other cytoplasmic components have roles in early development. The very shaping and arrangement of the embryo's body parts will depend on where different components are positioned in the egg cytoplasm.

Fertilization, the second stage, starts when a sperm penetrates an egg. It ends when the sperm nucleus fuses with the egg nucleus and gives rise to the zygote, the first cell of the new individual. Next comes **cleavage**,

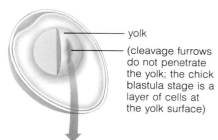
yolk

(cleavage furrows do not penetrate the yolk; the chick blastula stage is a layer of cells at the yolk surface)

Figure 32.3 Onset of organ formation in a chick embryo during the first 72 hours of development. The heart begins to beat at some time between 30 and 36 hours.

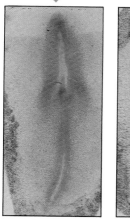
22 hours

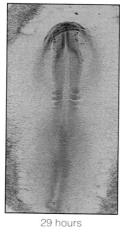
29 hours

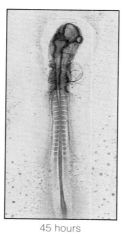
45 hours

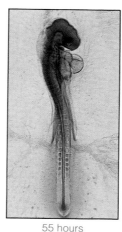
55 hours

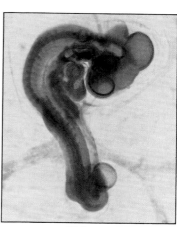
72 hours

when mitotic cell divisions convert the zygote to a ball of cells, the *blastula*. Figure 7.7 shows a frog zygote undergoing its first cleavage. Cells increase in number but not in size during this stage; the daughter cells collectively occupy the same volume as did the zygote.

As cleavage draws to a close, the pace of cell division slackens and gives way to **gastrulation,** a stage of major cell rearrangements. The organizational framework for the whole body is laid out as cells become arranged into two or three primary tissues, or *germ layers*. The human body arises from three such layers:

endoderm   *inner layer; gives rise to inner lining of gut and organs derived from it*

mesoderm   *intermediate layer; gives rise to muscle, the organs of circulation, reproduction, and excretion, most of the internal skeleton, and connective tissue layers of the gut and body covering*

ectoderm   *surface layer; gives rise to tissues of nervous system and outer layer of body covering*

Next, the germ layers split into subpopulations of cells, and this marks the onset of **organ formation**. Different lines of cells become unique in structure and function; they are the forerunners of distinct tissues and organs. The time line in Figure 32.3 hints at how rapidly this stage of development proceeds. Finally, during **growth and tissue specialization**, organs acquire specialized chemical and physical properties. This stage continues into adulthood.

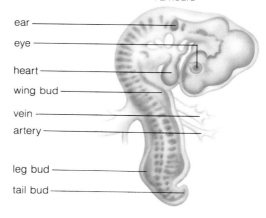
ear

eye

heart

wing bud

vein

artery

leg bud

tail bud

Mechanisms of Development

Two processes, cell differentiation and morphogenesis, are central to development. In **cell differentiation**, a single fertilized egg gives rise to diverse types of specialized cells. All cells produce a number of the same kinds of proteins. But differentiated cell types also produce some proteins that are *not* found in other cell types, and those proteins are the basis of distinctive cell structures and functions. Differentiated cells have the same number and kind of genes (they are all descended from the same zygote). Through controls over gene expression, however, restrictions are placed on *which* genes will be expressed in a given cell (page 157).

For example, while you were still an embryo and your eyes were developing, some cells started synthe-

Figure 32.4 Two experiments illustrating how qualitative differences in the cytoplasm of a fertilized egg help determine the fate of cells in a developing embryo. Frog eggs contain granules of dark pigment in their cortex (the plasma membrane and the cytoplasm just below it). The granules are concentrated near one pole of the egg; yolk is concentrated near the other pole. At fertilization, a portion of the granule-containing cortex shifts away from the yolk, and this exposes lighter colored cytoplasm in a crescent-shaped gray area:

pigmented cortex

yolk

(sperm penetrating frog egg)

gray crescent

Normally, the first cleavage divides the gray crescent between two daughter cells. (**a**) Even if the two daughter cells are separated from each other experimentally, each may still give rise to a complete tadpole. (**b**) But if a fertilized egg is manipulated so that the first cleavage plane misses the gray crescent entirely, only one of the two daughter cells gets the gray crescent and it alone will develop into a normal tadpole. The cell deprived of substances in the gray crescent will only give rise to a ball of undifferentiated cells.

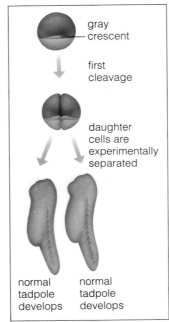

gray crescent

first cleavage

daughter cells are experimentally separated

normal tadpole develops

normal tadpole develops

a Experiment 1

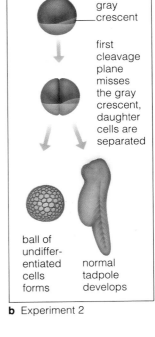

gray crescent

first cleavage plane misses the gray crescent, daughter cells are separated

ball of undifferentiated cells forms

normal tadpole develops

b Experiment 2

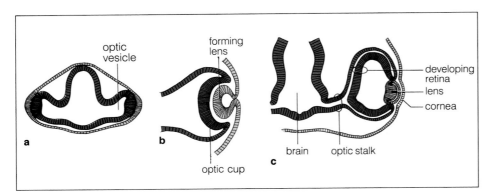

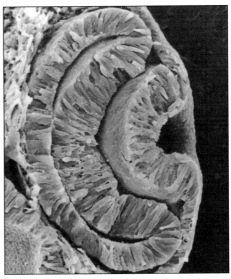

optic vesicle

forming lens

developing retina

lens

cornea

brain optic stalk

optic cup

a   **b**   **c**   **d**

Figure 32.5 Eye formation. The retina develops as an outgrowth of the brain; the lens, as an ingrowth of the ectoderm. (**a**) An optic vesicle grows out of the side of the brain. When it contacts the head ectoderm, it induces the elongation and inward folding of the ectodermal cells to form a lens vesicle (**b**). Meanwhile, the optic vesicle is induced to sink inward, forming the optic cup. (**c**) The cup's inner layer will form the retina. (**d**) Scanning electron micrograph of an optic cup and lens in a chick embryo.

sizing quantities of a protein (crystallin) that would be used in the construction of transparent fibers in the lens. No other cell type in your developing body could activate the genes necessary to do this. The transparent fibers caused the lens cells to elongate and flatten out and gave them their unique optical properties.

What determines the fate of an embryonic cell? That depends partly on cleavage, when different regions of the zygote cytoplasm are partitioned into daughter cells,

and partly on interactions between cells in the developing embryo. Figure 32.4 shows two experiments that illustrate the importance of cytoplasmic segregation at cleavage.

As the embryo continues its development, differentiated cells become organized into tissues and organs through processes called **morphogenesis**. The embryo's form and structure become more complex as those cells interact chemically and physically in two major ways.

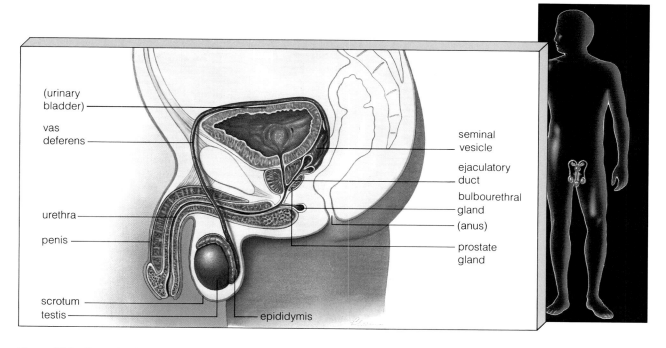

Figure 32.6 Reproductive system of the human male.

First, one cell type produces a hormone, growth factor, or some other substance that diffuses to other cell types and triggers changes in their behavior. Second, individual cells recognize specific proteins at the surface of other cells and either adhere to those cells or migrate over them. Such recognition is vital when body parts grow in certain directions, change shape, and become arranged according to predefined patterns.

Consider how signals from cells of one body region induce eye formation in an adjacent body region (Figure 32.5). Or consider how the forerunners of nerve cells migrated and established billions of precise connections with other cells to form your nervous system before you were even born. Those cells moved partly in response to recognition proteins and other cues on the surface of neighboring cells, and once the strongest adhesion was reached, their further migration was impeded.

Finally, consider how the development of hands and feet depends on controlled cell death, a mechanism by which cells and tissues required for only short periods in the embryo (or adult) are eliminated. Skin cells between the lobes of four "paddles" die on cue, leaving separate toes and fingers (Figure 7.1). This mechanism is genetically programmed. Between the time a death signal is sent and the actual time of death, protein synthesis declines dramatically in the doomed cells.

Development depends on cell differentiation and morphogenesis. Both are outcomes of cytoplasmic segregation during cleavage and interactions among embryonic cells.

HUMAN REPRODUCTIVE SYSTEM

Let's now focus on humans as an integrated example of reproduction and development. The male or female reproductive system consists of a pair of primary reproductive organs (gonads), accessory glands, and ducts. Male gonads are **testes** (singular, testis); female gonads are **ovaries**. Testes produce sperm; ovaries produce eggs. Both also secrete sex hormones, which influence reproductive functions and the development of secondary sexual traits. These traits are distinctly associated with maleness and femaleness but do not play a direct role in reproduction. Examples are the amount and distribution of body fat, hair, and skeletal muscle.

Gonads look the same in all early human embryos. But after seven weeks of development, activation of genes on the sex chromosomes and hormone secretions trigger their development into testes *or* ovaries. The gonads and accessory organs are already formed at birth, but they do not reach full size and become functional until twelve to sixteen years later.

Male Reproductive Organs

Sperm Formation. Figure 32.6 shows the male reproductive system and Table 32.1 lists its components. The testes start forming in the abdominal cavity, and before birth they descend into the scrotum, an outpouching of skin below the pelvic region. Sperm develop properly

Table 32.1	Organs and Accessory Glands of the Male Reproductive Tract
Organs:	
Testis (2)	Sperm production, sex hormone production
Epididymis (2)	Sperm maturation site, sperm storage
Vas deferens (2)	Rapid transport of sperm
Ejaculatory duct (2)	Conduction of sperm
Penis	Organ of sexual intercourse
Accessory Glands:	
Seminal vesicle (2)	Secretions make up large portion of semen
Prostate gland	Secretions make up a portion of semen
Bulbourethral gland (2)	Production of lubricating mucus

when the scrotum's interior is kept a few degrees cooler than the body's normal temperature. Through controlled contractions of muscles in the scrotum, the temperature stays at 95°F or so. When it is cold outside, contractions draw the pouch closer to the (warmer) body; when it is warm outside, the muscles relax and so lower the pouch.

Each testis is partitioned into as many as 300 wedge-shaped lobes. Each lobe contains two to three highly coiled tubes, the seminiferous tubules, where sperm develop (Figures 32.7 and 32.8). Although a testis is only about 5 centimeters long, 125 meters of tubes are packed into it!

From puberty onward, sperm are produced continuously, with many millions in different stages of development on any given day. A mature sperm has a tail, a midpiece, and a head with a DNA-packed nucleus. An enzyme-containing cap (acrosome) covers most of the head. Its enzymes help the sperm penetrate an egg at fertilization. Mitochondria in the midpiece supply energy for the tail's whiplike movements.

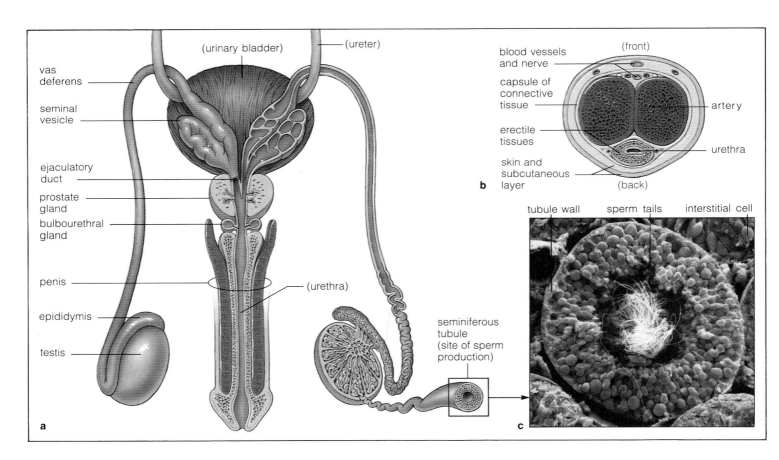

Figure 32.7 (**a**) Posterior view of the male reproductive system. (**b**) Cross-section of the penis. (**c**) Scanning electron micrograph showing the cells inside a seminiferous tubule.

Semen Formation. Sperm move from a testis into a long coiled duct, the epididymis. The sperm are not fully developed at this time, but secretions from the duct walls help them mature. Until sperm leave the body, they are stored in the last part of the epididymis. When they are to be released, they pass through a thick-walled tube (vas deferens), ejaculatory ducts, then through the urethra, the channel leading to the body's surface (Figure 32.7).

Secretions from glands along this route become mixed with sperm. This sperm-bearing mixture is semen. Secretions from the seminal vesicles contain fructose (which nourishes sperm) and prostaglandins (which may trigger contractions in the female tract and assist sperm movements). Secretions from the prostate gland probably help buffer acid conditions in the vagina. (Vaginal pH is about 3.5–4, but sperm motility and fertility improve when it is about 6.) Bulbourethral glands secrete some mucus-rich fluid into the urethra during sexual arousal. This fluid lubricates the penis, assisting its penetration into the vagina; it also assists sperm movements.

Hormonal Controls. Three hormones control male reproductive functions. One is **testosterone**, produced by endocrine cells in the testes. The others are **LH** and **FSH**, produced by the anterior pituitary (page 451).

Testosterone stimulates sperm production and controls the growth, form, and function of all parts of the male reproductive tract. This hormone has roles in normal sexual behavior and may tend to promote aggressive behavior. Beard growth, pubic hair growth, lowering of the voice, and other secondary sexual traits depend on testosterone secretions.

The hypothalamus governs testosterone secretion. When testosterone levels are low, the hypothalamus signals the anterior pituitary to secrete LH—which triggers testosterone secretion by the testes. Through negative feedback loops, high levels of testosterone inhibit the hypothalamic signals calling for LH secretion—so testosterone output slows down. The hypothalamus also governs FSH secretion, which in males stimulates the development of sperm during puberty.

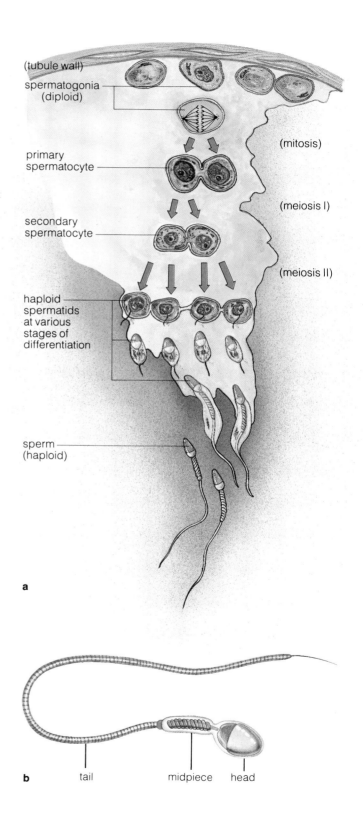

Figure 32.8 (**a**) Sperm formation in a seminiferous tubule. Undifferentiated diploid cells (spermatogonia) are closest to the tubule walls. They are forced away from it by ongoing mitotic cell divisions and are transformed into primary spermatocytes. Following meiosis I, they become secondary spermatocytes. Each chromosome in these haploid cells still consists of two sister chromatids (page 106). Sister chromatids separate from each other during meiosis II. The resulting spermatids gradually develop into mature sperm (**b**). The entire process takes about 9 to 10 weeks.

(tubule wall)
spermatogonia (diploid)
(mitosis)
primary spermatocyte
(meiosis I)
secondary spermatocyte
(meiosis II)
haploid spermatids at various stages of differentiation
sperm (haploid)
a
b   tail   midpiece   head

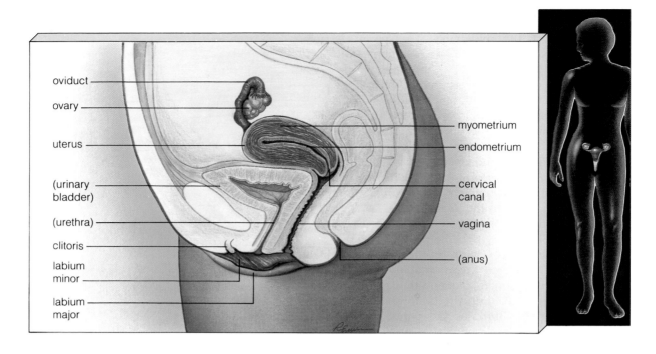

Figure 32.9 Reproductive system of the human female. The uterus has a thick layer of smooth muscle (myometrium) and an inner lining (endometrium).

Table 32.2	Female Reproductive Organs
Ovaries	Oocyte production, sex hormone production
Oviducts	Conduction of oocyte from ovary to uterus
Uterus	Chamber in which new individual develops
Cervix	Secretion of mucus that enhances sperm movement into uterus and (after fertilization) reduces the embryo's risk of bacterial infection
Vagina	Organ of sexual intercourse; birth canal

Female Reproductive Organs

Figure 32.9 shows the female reproductive system and Table 32.2 lists its components. The two ovaries are located in the abdominal cavity. They release eggs on a monthly basis, and they secrete the sex hormones **estrogen** and **progesterone**.

Even before a female is born, about 2 million eggs have started to form in her ovaries. These are immature eggs (oocytes). They have already entered meiosis I, but the division process has been arrested. By the time it resumes (starting at puberty), only about 30,000 or 40,000 oocytes are still around. And only about 400 of those will mature and be expelled from her body, one at a time on a monthly basis, over the next three decades or so. Even then, meiosis II will not be completed unless fertilization occurs.

An oocyte released from an ovary enters a nearby channel, an oviduct. Fingerlike projections from the oviduct extend over part of the ovary, and they sweep the oocyte into the channel. From there, the oocyte moves into a hollow, pear-shaped organ, the uterus. Following fertilization, the new individual grows and develops here. The uterus is mostly a thick layer of smooth muscle. Its interior lining, the **endometrium**, consists of connective tissue, glands, and blood vessels. The lower portion of the uterus (the narrow part of the "pear") is the cervix. A muscular tube, the vagina, extends from the cervix to the body surface. This tube receives sperm and functions as part of the birth canal.

At the body surface are the external genitalia (vulva), which include the clitoris and other organs for sexual stimulation.

Menstrual Cycle

Most mammalian females follow an "estrous" cycle. They can become pregnant only during estrus, when they are in heat (sexually receptive to males). This occurs only at certain times of year, when oocytes mature and

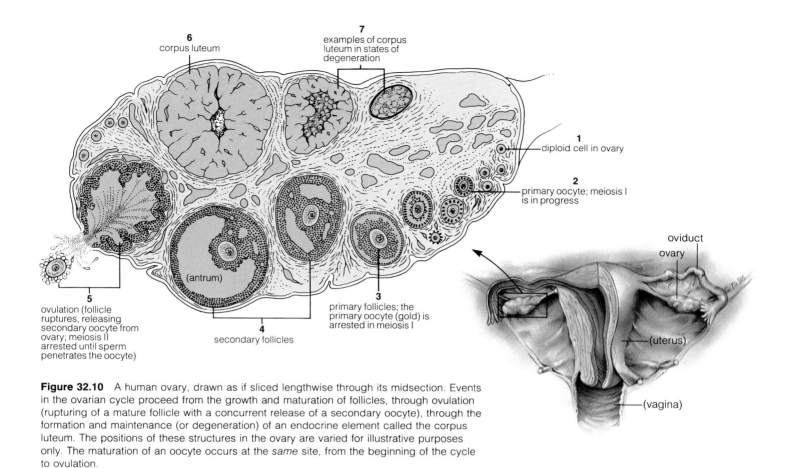

Figure 32.10 A human ovary, drawn as if sliced lengthwise through its midsection. Events in the ovarian cycle proceed from the growth and maturation of follicles, through ovulation (rupturing of a mature follicle with a concurrent release of a secondary oocyte), through the formation and maintenance (or degeneration) of an endocrine element called the corpus luteum. The positions of these structures in the ovary are varied for illustrative purposes only. The maturation of an oocyte occurs at the *same* site, from the beginning of the cycle to ovulation.

hormone action primes the endometrium to receive a fertilized egg.

The females of humans and other primates follow a **menstrual cycle**; the release of oocytes and priming of the endometrium is cyclic and intermittent. Unlike estrus, there is no correspondence between heat and the time of fertility. All female primates can be physically and behaviorally receptive to the male's overtures at any time. Human menstrual cycles begin at about age thirteen and continue until menopause (in the late forties or early fifties). On the average, it takes about twenty-eight days to complete one cycle (Table 32.3).

Ovarian Function. When an oocyte forms, it becomes surrounded by a layer of cells. This primary oocyte, together with the surrounding layer, is a **follicle**. Usually only one follicle reaches maturity during a menstrual cycle. Within that follicle, meiosis I resumes in the oocyte and two cells form. One is a secondary oocyte, which ends up with nearly all the cytoplasm. The other is a tiny polar body. The tiny cell functions only as a "dumping ground" for half the diploid number of chromosomes, so that both cells are haploid (Figures 8.8

Table 32.3 Events of the Menstrual Cycle

Phase	Events	Days of the Cycle*
Follicular phase	Menstruation; endometrium breaks down	1–5
	Follicle matures in ovary; endometrium rebuilds	6–13
Ovulation	Secondary oocyte released from ovary	14
Luteal phase	Corpus luteum forms; endometrium thickens and develops	15–28

* Assuming a 28-day cycle.

and 32.10). Neither cell will complete meiosis II until fertilization.

As the follicle develops, it secretes an estrogen-containing fluid. The fluid accumulates in the follicle and causes it to balloon outward from the ovary's surface, then rupture. The fluid escapes, carrying the secondary oocyte with it. The release of a secondary oocyte from

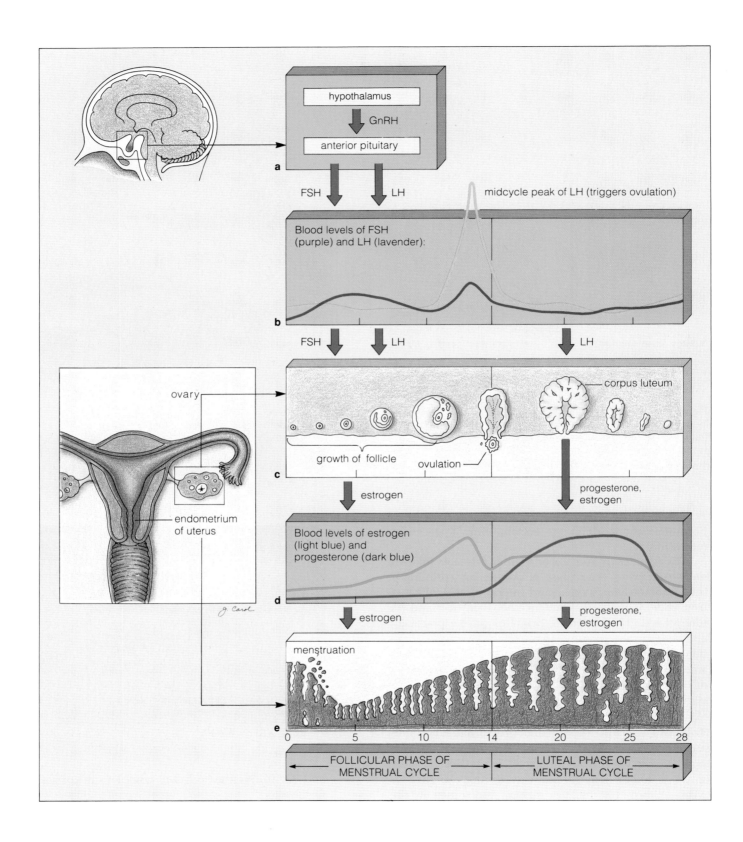

an ovary is called **ovulation**. The parts of the follicle left behind are transformed into a glandular structure, the **corpus luteum**, which secretes progesterone and some estrogen.

Feedback loops involving the hypothalamus, pituitary, and ovaries control events in the ovary. When the menstrual cycle begins, the hypothalamus signals the anterior pituitary to release LH and FSH, which in turn signal the ovary to secrete estrogen. About midway through the cycle, the increased estrogen level in the blood causes a brief outpouring of LH from the pituitary. *This midcycle surge of LH triggers ovulation* (Figure 32.11).

A corpus luteum can persist for about twelve days if fertilization does not follow ovulation. During that time, the hypothalamus signals the anterior pituitary to decrease its FSH secretions. This prevents other follicles from developing until the menstrual cycle is over.

The corpus luteum degenerates during the last days of the cycle if fertilization does not occur. Apparently it self-destructs by secreting prostaglandins, which interfere with its function. With the corpus luteum gone, progesterone and estrogen levels fall rapidly. Now FSH secretions can increase, another follicle can be stimulated to mature—and the cycle begins anew.

Uterine Function. The changing estrogen and progesterone levels just described cause profound changes that prepare the uterus for pregnancy. Estrogen stimulates the growth of endometrium and smooth muscle in the uterus. Progesterone causes blood vessels to grow rapidly in the thickened endometrium.

At ovulation, estrogen causes the cervix to secrete large amounts of a thin, clear mucus—an ideal medium through which sperm can travel. Right after ovulation, progesterone from the corpus luteum acts on the cervix. The mucus becomes thick and sticky, forming a barrier against vaginal bacteria that might enter the uterus through the cervix and endanger a new zygote.

When fertilization does not occur and the corpus luteum self-destructs, the endometrium starts to break down. Its blood vessels constrict and its tissues (deprived of oxygen and nutrients) die. Blood escapes from the ruptured walls of weakened capillaries. Menstrual flow consists of blood and sloughed endometrial tissues, and its appearance marks the first day of a new cycle. The menstrual sloughing continues for three to six days, until rising estrogen levels stimulate the repair and growth of the endometrium.

Each year, between four million and ten million American women are affected by *endometriosis*, the spread and growth of endometrial tissue outside the uterus. Estrogen acts on endometrial tissue wherever it occurs. This may lead to pain during menstruation, sexual relations, or urination. Endometrial scar tissue on the ovaries or oviduct can cause infertility. The disorder might arise when some menstrual flow backs up through the oviducts and spills into the pelvic cavity. Or perhaps some embryonic cells were positioned in the wrong place before birth and are stimulated to grow at puberty, when sex hormones become active.

Sexual Union

Suppose a secondary oocyte happens to be on its way down the oviduct when a female and male are engaged in sexual union (coitus). Within mere seconds of sexual arousal, the penis may undergo changes that help it penetrate into the vagina. The penis contains three cylinders of spongy tissue (Figure 32.7). The mushroom-shaped tip of one cylinder (the glans penis) is loaded with sensory receptors that are activated by friction. Between times of sexual arousal, blood vessels leading into the three cylinders are constricted and the penis is limp. Upon arousal, blood flows into the cylinders faster than it flows out and collects in the spongy tissue. The penis lengthens and stiffens as a result.

During coitus, pelvic thrusts stimulate the penis as well as the female's vaginal walls and clitoral region. The mechanical stimulation causes rhythmic, involuntary contractions in the male reproductive tract. The contractions force the contents of seminal vesicles and the prostate into the urethra, then ejaculation of semen into the vagina. (During ejaculation, a sphincter closes and prevents urine from being excreted from the bladder.)

Together, the muscular contractions, ejaculation, and associated sensations of release, warmth, and relaxation are called orgasm. Female orgasm involves similar events, including an intense vaginal awareness, involuntary uterine and vaginal contractions, and sensations of relaxation and warmth. Even if the female does not reach this state of excitation, she can still get pregnant.

Figure 32.11 (*Left*) Correlation between changes in the ovary and uterus with changing hormone levels during the menstrual cycle. Green arrows indicate which hormones dominate the follicular phase or the luteal phase of the cycle. A releasing hormone (GnRH) from the hypothalamus (**a**) controls the release of FSH and LH from the pituitary. The FSH and LH promote changes in ovarian structure and function (**b,c**), then estrogen and progesterone from the ovary promote changes in the endometrium (**d,e**).

Figure 32.12 Fertilization and implantation in the uterus. The left photograph shows a human sperm about to penetrate the membranous covering (zona pellucida) around a secondary oocyte. The right photograph shows the three polar bodies above a mature ovum; these products of meiosis will degenerate shortly.

a. A sperm penetrates the zona pellucida of a secondary oocyte and enters its cytoplasm. This triggers meiosis II in the first polar body and in the oocyte.

b. The sperm nucleus fuses with the egg nucleus at fertilization, producing the zygote. With the first cleavage, the fertilized egg enters the two-cell stage.

c. The second cleavage produces the four-cell stage.

d. Successive cleavages produce a solid ball of cells, the morula.

e. Fluid enters the ball and lifts some cells, forming a cavity. This produces the blastocyst, a ball of cells having a surface layer and an inner cell mass.

surface layer of cells

inner cell mass

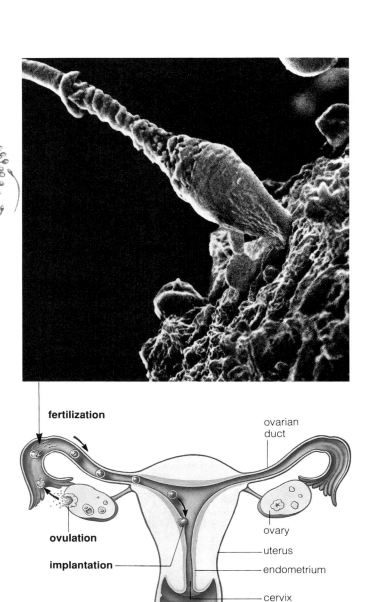

endometrium

blastocyst

(uterine cavity)

f. Implantation begins when the blastocyst attaches to and invades the endometrium. During the second week after fertilization, it slowly embeds itself in the endometrium.

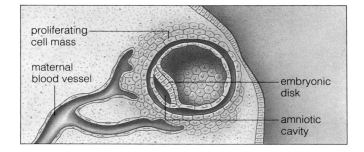

proliferating cell mass

maternal blood vessel

embryonic disk

amniotic cavity

g. During implantation, a slitlike cavity forms between the inner cell mass and the surface layer. The inner cell mass is transformed into a flattened, embryonic disk from which the embryo develops.

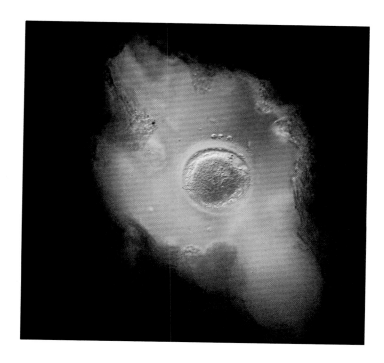

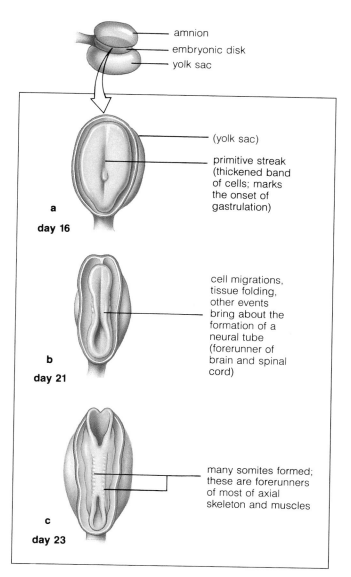

Figure 32.13 Transformation of the pancake-shaped embryonic disk into the early embryo. Shown are three posterior views (the embryo's back).

FROM FERTILIZATION TO BIRTH

Fertilization

Fertilization may occur if sperm enter the vagina anywhere from a few days before ovulation to a few days afterward. Within one-half hour after ejaculation, muscle contractions move sperm deeper into the female reproductive tract. As many as 150 million to 350 million sperm may be deposited in the vagina, but only a few hundred reach the upper region of the oviduct, where fertilization most commonly occurs.

When a sperm encounters a secondary oocyte, it releases digestive enzymes that clear a path through the membranous covering around the egg. Several sperm can reach the egg, but usually only one enters its cytoplasm. The arrival of that sperm stimulates the first polar body and secondary oocyte into completing meiosis II (Figures 8.8 and 32.12). There are now three polar bodies and a mature egg, or **ovum**. The sperm nucleus fuses with the nucleus of the mature ovum and their chromosomes intermingle, restoring the diploid number for the zygote.

Implantation

For the first three or four days after fertilization, the zygote travels down the oviduct, picking up nutrients from maternal secretions and undergoing the first cleav-

ages. By the time the cluster of dividing cells reaches the uterus, it is a solid ball (the morula). The ball becomes transformed into a blastocyst with a surface layer and an inner cell mass (Figure 32.12).

The blastocyst becomes implanted in the uterus before the end of the first week. It adheres to the uterine lining, and some of its cells send out projections that invade the maternal tissues. While the invasion is proceeding, the inner cell mass becomes transformed into an embryonic disk, which is shaped rather like an oval, flattened pancake (Figure 32.13). The disk will give rise to the embryo proper during the week following implantation, when all three germ layers will form.

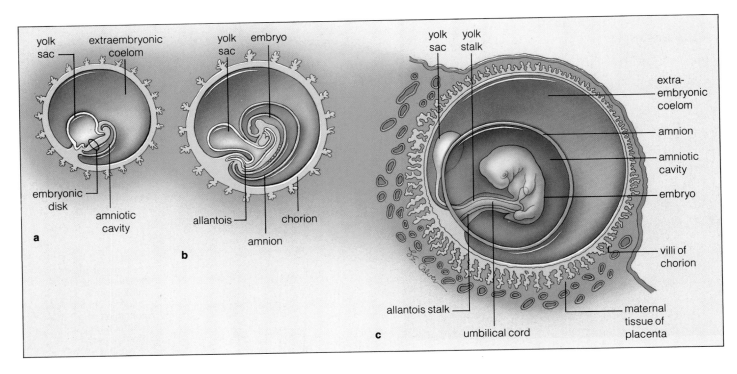

Figure 32.14 Stages in the development of extraembryonic membranes.

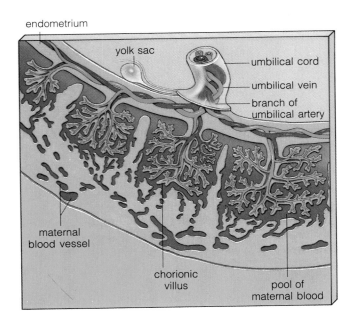

Figure 32.15 Relationship between fetal and maternal tissues in the placenta. The diagram shows how chorionic villi become progressively developed (from left to right across the illustration).

Membranes Around the Embryo

To understand what happens after implantation, think back on the shelled egg, which figured in the vertebrate invasion of land (page 285). Inside most shelled eggs are four membranes, the **yolk sac**, **allantois**, **amnion**, and **chorion**. These "extraembryonic" membranes protect the embryo and function in its nutrition, respiration, and excretion.

A human embryo is not housed in a shell or nourished by yolk, but it is still served by a yolk *sac* as part of its vertebrate heritage. The sac forms below the embryonic disk *as if* yolk were still there, and it helps give rise to a digestive tube (Figure 32.14).

In hard-shelled eggs, the allantois stores wastes from protein metabolism, and its blood vessels supply the embryo with oxygen. In humans, the allantois is not involved in waste storage but it still functions in oxygen transport.

The amnion of all land vertebrates is a fluid-filled sac that completely surrounds the embryo and keeps it from drying out. The fluid inside also absorbs shocks. The "water" flowing from the vagina just before childbirth is amniotic fluid, which is released when the amnion ruptures.

gill arches   somites

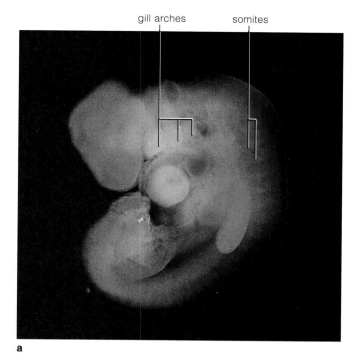
a

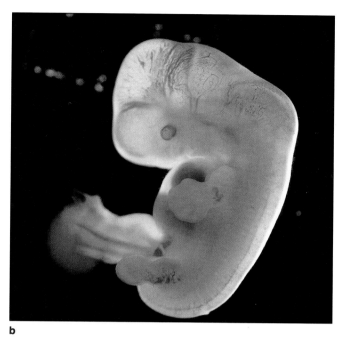
b

Figure 32.16 (**a**) Embryo at four weeks, about 7 millimeters (0.3 inch) long. Notice the tail and the gill arches, which vaguely resemble a double chin. These features emerge during embryonic development of all vertebrates. Arm and leg buds are also visible now. (**b**) Embryo at the end of five weeks, about 12 millimeters long. The head starts to enlarge and the trunk starts to straighten. Finger rays appear in the paddlelike forelimbs.

In time, the growing human embryo is connected to parts of the yolk sac, allantois, and amnion only by an **umbilical cord** that is well endowed with blood vessels (Figure 32.14). The chorion develops as a protective membrane around the embryo and other structures, and it becomes part of the placenta. A hormone secreted from the chorion maintains the corpus luteum, and progesterone secreted from the corpus luteum in turn maintains the uterine lining.

The Placenta

Three weeks after fertilization, almost one-fourth of the inner surface of the uterus has become a spongy tissue composed of endometrium *and* embryonic membranes, the chorion especially. By this tissue, the **placenta**, the embryo receives nutrients and oxygen from the mother and sends out wastes in return, which are disposed of through the mother's lungs and kidneys.

The tiny projections sent out from the blastocyst during implantation develop into many chorionic villi, each endowed with small blood vessels (Figure 32.15). After the embryo starts developing, its bloodstream will remain distinct from that of its mother. Substances simply will diffuse out of the mother's blood vessels, across the blood-filled spaces in the uterine lining, then into the embryo's blood vessels (and vice versa).

Embryonic and Fetal Development

Once the embryonic disk forms, development proceeds along the course described early in the chapter. Gastrulation (which starts during the second week) leads to the formation of three germ layers. Some surface cells migrate inward to form mesoderm. The remaining surface tissue (ectoderm) will give rise to the nervous system and other organs. Inside, endoderm will give rise to parts of the respiratory and digestive systems; mesoderm will develop into the heart, muscles, bone, and many other internal organs. After the third week, an early, tubelike heart is beating.

By the end of the fourth week, the embryo has grown 500 times its original size (Figure 32.16). Its growth spurt gives way to four weeks in which the main organs develop rather slowly. The nerve cord and the four heart chambers form; respiratory organs form but are not yet functional. Arms, legs, fingers, and toes are developing, along with the tail that emerges in all vertebrate embryos. The human tail gradually disappears after the eighth week.

umbilical cord   amniotic sac

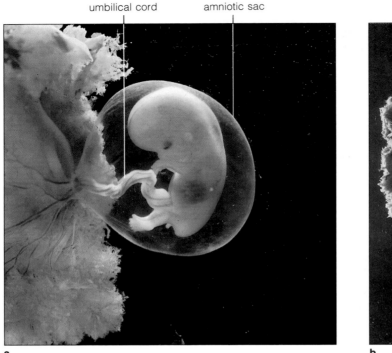

a

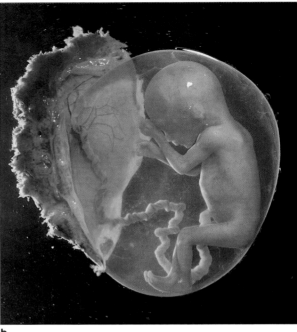

b

Figure 32.17 (**a**) The fetus at nine weeks, floating in fluid inside the amniotic sac. (Here, the chorion, which covers the amniotic sac, has been opened and pulled aside.) Notice the blood vessels in the umbilical cord. (**b**) The fetus at sixteen weeks.

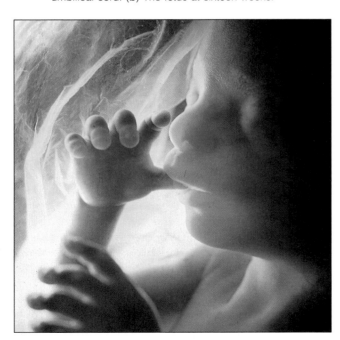

Figure 32.18 The fetus at eighteen weeks, about 18 centimeters (a little more than 7 inches) long. The sucking reflex begins during the earliest fetal stage, as soon as nerves establish functional connections with developing muscles. Legs kick, arms wave, fingers make grasping motions—all reflexes that will be vital skills in the world outside the uterus.

The "first trimester" is the developmental period extending from fertilization to the end of the third month of development. Figure 32.17a gives an idea of what the embryo looks like during this period. As indicated in the *Commentary* (page 480), the first trimester is a critical period of embryonic development. The "second trimester" extends from the start of the fourth month to the end of the sixth. All major organs have formed (Figure 32.17b), and the growing individual is now called a fetus, and the mother is quite aware of its movements. Figure 32.18 shows a fetus at eighteen weeks.

When the fetus is five months old, its heart can be heard through a stethoscope on the mother's abdomen. Soft, fuzzy hair (the lanugo) covers its body. Its skin is wrinkled, rather red, and protected from abrasion by a thick, cheesy coating. During the sixth month, eyelids and eyelashes form; during the seventh, the eyes open.

The "third trimester" extends from the seventh month until birth. Not until the middle of the third trimester will the fetus be able to survive on its own if born prematurely or removed surgically from the uterus. Although development appears to be relatively complete by the seventh month, few fetuses would be able to breathe normally or maintain a normal body temperature, even with the best medical care. By the ninth month, survival chances increase to about ninety-five percent.

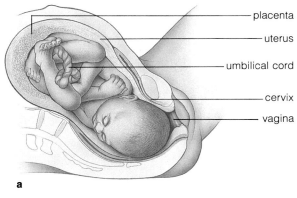

placenta

uterus

umbilical cord

cervix

vagina

a

Figure 32.19 Expulsion of the fetus during the birth process. The placenta, fluid, and blood are expelled shortly afterward (this is the "afterbirth").

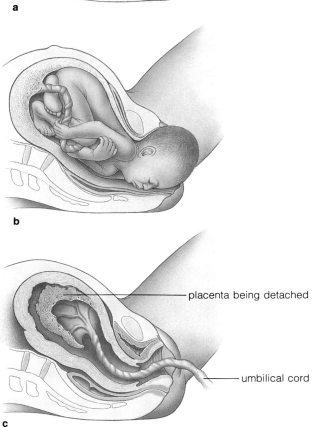

b

placenta being detached

umbilical cord

c

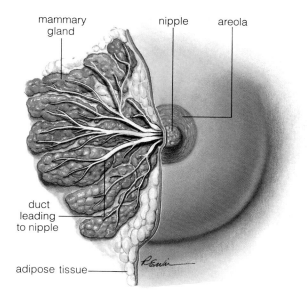

mammary gland

nipple

areola

duct leading to nipple

adipose tissue

Figure 32.20 Breast of a lactating female. This cutaway view shows the mammary glands and ducts.

Birth and Lactation

Birth takes place about thirty-nine weeks after fertilization, give or take a few weeks. The birth process begins when the uterus starts to contract. For the next two to eighteen hours, the contractions become stronger and more frequent. The cervical canal dilates fully and the amniotic sac usually ruptures. Birth typically occurs less than an hour after full dilation. Immediately afterward, uterine contractions force fluid, blood, and the placenta from the body (Figure 32.19). The umbilical cord—the lifeline to the mother—is now severed, and the newborn embarks on its nurtured existence in the outside world.

During pregnancy, estrogen and progesterone were stimulating the growth of mammary glands and ducts in the mother's breasts (Figure 32.20). For the first few days after birth, those glands produce a fluid rich in proteins and lactose. Then prolactin secreted by the pituitary stimulates milk production (page 451). When the newborn suckles, the pituitary also releases oxytocin, which causes breast tissues to contract and so force milk into the ducts. Oxytocin also triggers uterine muscle contractions that "shrink" the uterus back to its normal, prepregnancy size.

Mother as Protector, Provider, Potential Threat

Many safeguards are built into the female reproductive system. The placenta, for example, is a highly selective filter that prevents many noxious substances in the mother's bloodstream from gaining access to the embryo or fetus. Even so, from fertilization to birth, the developing individual is at the mercy of the mother's diet, health habits, and life-style.

Some Nutritional Considerations. During pregnancy, a balanced diet usually provides enough vitamins and minerals for normal development. The mother's vitamin needs are definitely increased, but the developing fetus is more resistant than she is to vitamin and mineral deficiencies. The placenta preferentially absorbs vitamins and minerals from her blood.

A few years ago, it was accepted medical practice for a pregnant woman to keep her total weight gain to 10 or 15 pounds. It is now clear that if the woman restricts her food intake too severely, especially during the last trimester, fetal development will be affected and the newborn will be underweight. Significantly underweight infants face more postdelivery complications than do infants of normal weight; in fact, they represent nearly half of all newborn deaths. They also will suffer a much higher incidence of mental retardation and other handicaps later in life. In most cases, a woman should gain somewhere between 20 and 25 pounds during pregnancy.

As birth approaches, the growing fetus demands more and more nutrients from the mother's body. During this last phase of pregnancy, the mother's diet profoundly influences the course of development. Poor nutrition damages most fetal organs—particularly the brain.

Risk of Infections. During pregnancy, antibodies transferred across the placenta protect the developing individual from all but the most severe bacterial infections. However, certain viral diseases can have damaging effects if they are contracted during the first six weeks after fertilization, the critical time of organ formation. For example, if the woman contracts German measles (rubella) during this period, there is a fifty percent chance that her embryo will become malformed. If she contracts the measles virus when the embryo's ears are forming, her newborn may be deaf. (German measles can be avoided by vaccination *before* pregnancy.) The likelihood of damage to the embryo diminishes after the first six weeks. The same disease, contracted during the fourth month or thereafter, has no discernible effect on the development of the fetus.

Effects of Prescription Drugs. During the first trimester, the embryo is highly sensitive to drugs. A shocking example of drug effects came during the first two years after *thalidomide* was introduced in Europe. Women using this prescription tranquilizer during the first trimester gave birth to infants with missing or severely deformed arms and legs. Once the deformities were traced to thalidomide, the drug was withdrawn from the market. However, there is evidence that other tranquilizers (and sedatives and barbiturates) might cause similar, although less severe, damage. Even the drug Accutane, used for treating acne, increases the risk of facial and cranial deformities. Tetracycline, a commonly prescribed antibiotic, causes yellowed teeth; streptomycin causes hearing problems and may affect the nervous system.

At no stage of development is the embryo impervious to drugs in the maternal bloodstream. Clearly, the woman should take no drugs at all during pregnancy unless prescribed by a knowledgeable physician.

Effects of Alcohol. As the fetus matures, its physiology becomes increasingly like that of the mother's. Alcohol passes freely across the placenta and has the same kind of effect on the fetus as on the woman who drinks it. *Fetal alcohol syndrome* (FAS) is a constellation of deformities that are thought to result from excessive use of alcohol by the mother during pregnancy. FAS is the third most common cause of mental retardation in the United States. It also is characterized by facial deformities, poor coordination, and, sometimes, heart defects. Between sixty and seventy percent of alcoholic women give birth to infants with FAS; some researchers now suspect that even two drinks a day during pregnancy may be dangerous for the fetus. Increasingly, physicians are urging total or near abstention during pregnancy.

Effects of Smoking.   Cigarette smoking has an adverse effect on fetal growth and development. Newborns of women who have smoked every day throughout pregnancy have a low birth weight. That is true even when the woman's weight, nutritional status, and all other relevant variables are identical with those of pregnant women who do not smoke. Smoking has other effects as well (see figure).

For example, for seven years in Great Britain, records were kept for all births during a particular week. The newborns of women who had smoked were not only smaller, they had a thirty percent greater incidence of death shortly after delivery and a fifty percent greater incidence of heart abnormalities. More startling, at age seven, their average ''reading age'' was nearly half a year behind that of children born to nonsmokers.

In this last study, the critical period was shown to be the last half of pregnancy. Newborns of women who had stopped smoking by the middle of the second trimester were indistinguishable from those born to women who had never smoked. Although the mechanisms by which smoking exerts its effects on the fetus are not known, its demonstrated effects are further evidence that the placenta—marvelous structure that it is—cannot prevent all the assaults on the fetus that the human mind can dream up.

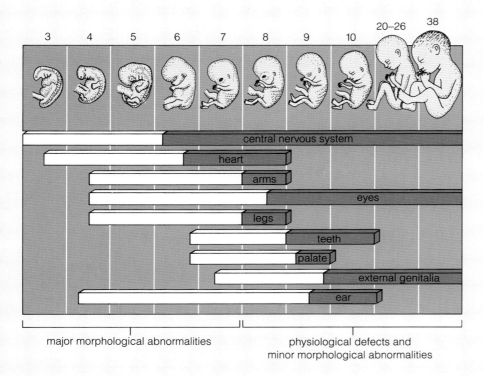

Critical periods of embryonic and fetal development. Red indicates periods in which organs are most sensitive to damage from cigarette smoke, alcohol, viral infection, and so on. Numbers signify the week of development.

POSTNATAL DEVELOPMENT, AGING, AND DEATH

Following birth, the new individual follows a prescribed course of further growth and development that leads to the adult, the sexually mature form of the species. Table 32.4 summarizes all the prenatal and postnatal stages. (Prenatal means before birth; postnatal means after birth.) Figure 32.21 shows how the human body changes in proportions as the course is followed.

Late in life, the body gradually deteriorates through processes called **aging**. Cell structure and function start to break down, and this is accompanied by structural changes and gradual loss of body functions. All organisms with extensively differentiated cells undergo aging.

Aging in humans involves loss of hair and teeth, increased skin wrinkling and fat deposition, and decreased muscle mass. Less obvious are gradual changes in physiology. Kidney cells falter, for example, so the body cannot respond as effectively as it once did to changes in fluid volume and composition. Or consider what happens to the collagen, which is present in nearly all tissues. Its structure changes with aging, and given that collagen may represent forty percent of the body's proteins, such changes are bound to have widespread physical effects.

No one knows what causes aging, but researchers give us interesting things to think about. In one study, cultured cells from a human embryo were allowed to divide. They all divided about fifty times, then they all died off. Other cells were frozen for years, thawed, and then allowed to divide. They divided about fifty times—then they all died on schedule. It seems mitosis is programmed to decline at a certain stage of life, but no one knows whether the decline is the cause or effect of aging.

It may also be that environmental insults to cells accumulate and gradually interfere with their DNA repair functions. Over time, mutations certainly could disturb the production of enzymes and other vital proteins. A mutation affecting collagen production alone would have repercussions throughout the body. A mutation that changes self-markers on the body's own cells might trigger autoimmune responses (page 399). Over time, such responses could lead to the increased stress and vulnerability to disease associated with aging.

Table 32.4 Stages of Human Development: A Summary	
Prenatal Period:	
1. Zygote	Single cell resulting from fusion of sperm nucleus and egg nucleus at fertilization
2. Morula	Solid ball of cells produced by cleavages
3. Blastocyst	Ball of cells with surface layer and inner cell mass
4. Embryo	All developmental stages from two weeks after fertilization until end of eighth week
5. Fetus	All developmental stages from the ninth week until birth (about thirty-nine weeks after fertilization)
Postnatal Period:	
6. Newborn	Individual during the first two weeks after birth
7. Infant	Individual from two weeks to about the first fifteen months after birth
8. Child	Individual from infancy to about twelve or thirteen years
9. Pubescent	Individual at puberty, when secondary sexual traits develop; girls between twelve and fifteen years, boys between thirteen and sixteen years
10. Adolescent	Individual from puberty until about three or four years later; physical, mental, emotional maturation
11. Adult	Early adulthood (between eighteen and twenty-five years), bone formation and growth completed. Changes proceed very slowly afterward.
12. Old age	Aging follows late in life

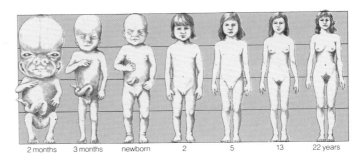

2 months   3 months   newborn   2   5   13   22 years

Figure 32.21 Diagram of changes in the proportions of the human body during prenatal and postnatal growth.

CONTROL OF HUMAN FERTILITY

Some Ethical Considerations

The transformation of a zygote into an intricately detailed adult raises profound questions. *When does development begin?* As we have seen, key developmental events occur even before fertilization. *When does life begin?* During her lifetime, a human female can produce as many as four hundred eggs, all of which are alive. During one ejaculation, a human male can release a quarter of a billion sperm, which also are alive. Even before sperm and egg merge by chance and establish the genetic makeup of a new individual, they are as much alive as any other form of life. It is scarcely tenable, then, to say "life begins" when they fuse. *Life began billions of years ago; and each gamete, each zygote, each mature individual is only a fleeting stage in the continuation of that beginning.*

This fact cannot diminish the meaning of conception, for it is no small thing to entrust a new individual with the gift of life, wrapped in the unique evolutionary threads of our species and handed down through an immense sweep of time.

Yet how can we reconcile the marvel of individual birth with the growing awareness of the astounding birth rate for our whole species? While this book is being written, an average of 2.6 infants are being born each second—155 each minute, 9,900 each hour. By the time you go to bed tonight, there will be 238,000 more people on earth than there were last night at that hour. Within a week, the number will reach 1,700,000—about as many people as there are now in the entire state of Massachusetts. *Within one week.* Worldwide population growth has outstripped resources, and each year millions face the horrors of starvation. Living as we do on one of the most productive continents on earth, few of us can know what it means to give birth to a child—to give it the gift of life—and have no food to keep it alive.

And how can we reconcile the marvel of birth with the confusion surrounding unwanted pregnancies? Even highly developed countries have inadequate educational programs concerning fertility control, and a good number of their members are not inclined to exercise control. Each year in the United States alone there are more than 100,000 "shotgun" marriages, about 200,000 unwed teenage mothers, and perhaps 1.5 million abortions. Many parents encourage early boy-girl relationships, at the same time ignoring the risk of premarital intercourse and unplanned pregnancy. Advice is often condensed to a terse, "Don't do it. But if you do it, be careful!"

The motivation to engage in sex has been evolving for more than 500 million years. A few centuries of moral and ecological reasoning that call for its suppres-sion have not prevented unwanted pregnancies. And complex social factors have contributed to a population growth rate that is out of control. How will we reconcile our biological past and the need for a stabilized cultural present? Whether and how fertility is to be controlled is one of the most volatile issues of our time. We will return to this issue in Chapter 33, in the context of principles governing the growth and stability of populations. Here, we can briefly consider some possible control options.

Birth Control Options. The most effective method of birth control is complete *abstention,* no sexual intercourse whatsoever. It is unrealistic to expect many people to practice it.

A modified form of abstention is the *rhythm method.* The idea is to avoid intercourse during the woman's fertile period, beginning a few days before ovulation and ending a few days after. Her fertile period is identified and tracked either by keeping records of the length of her menstrual cycles or by taking her temperature each morning when she wakes up. (It rises by one-half to one degree just before the fertile period.) But ovulation can be irregular, and miscalculations are frequent. Also, sperm deposited in the vaginal tract a few days before ovulation may survive until ovulation. The method *is* inexpensive (it costs nothing after you buy the thermometer) and does not require fittings and periodic checkups by a doctor. But its practitioners do run a large risk of getting pregnant (Figure 32.22).

Withdrawal, or removing the penis from the vagina before ejaculation, dates back at least to biblical times. But withdrawal requires very strong willpower and the method may fail anyway. Fluid released from the penis just before ejaculation may contain some sperm.

Douching, or rinsing out the vagina with a chemical right after intercourse, is next to useless. Sperm can move past the cervix and out of reach of the douche within ninety seconds after ejaculation.

Other methods involve physical or chemical barriers to prevent sperm from entering the uterus and moving to the ovarian ducts. *Spermicidal foam* or *spermicidal jelly* are toxic to sperm. They are packaged in an applicator and placed in the vagina just before intercourse. These products are not always reliable unless used with another device, such as a diaphragm or condom.

A *diaphragm* is a flexible, dome-shaped device, inserted into the vagina and positioned over the cervix before intercourse. A diaphragm is relatively effective when fitted by a doctor, used with foam or jelly before each sexual contact, and inserted correctly with each use.

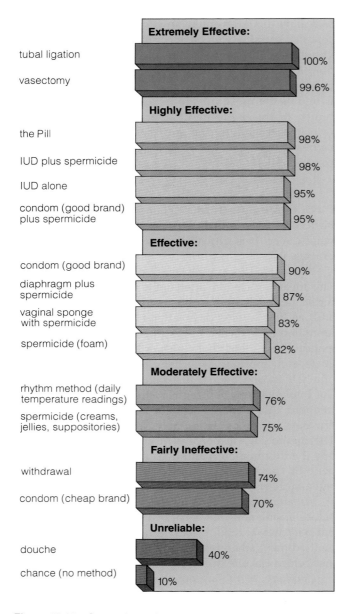

Extremely Effective:

tubal ligation — 100%

vasectomy — 99.6%

Highly Effective:

the Pill — 98%

IUD plus spermicide — 98%

IUD alone — 95%

condom (good brand) plus spermicide — 95%

Effective:

condom (good brand) — 90%

diaphragm plus spermicide — 87%

vaginal sponge with spermicide — 83%

spermicide (foam) — 82%

Moderately Effective:

rhythm method (daily temperature readings) — 76%

spermicide (creams, jellies, suppositories) — 75%

Fairly Ineffective:

withdrawal — 74%

condom (cheap brand) — 70%

Unreliable:

douche — 40%

chance (no method) — 10%

Figure 32.22 Comparison of the effectiveness of some contraceptive methods.

Condoms are thin, tight-fitting sheaths of rubber or animal skin, worn over the penis during intercourse. They are about eighty-five to ninety-three percent reliable, and they help prevent the spread of sexually transmitted diseases (see *Commentary,* page 485). However, condoms can tear and leak, and then are rendered useless.

The most widely used method of fertility control is *the Pill,* an oral contraceptive of synthetic estrogens and progesterones. It suppresses the normal release of these hormones from the pituitary and so stops eggs from maturing and being released at ovulation. The Pill is a prescription drug. Formulations vary and are selected to match each patient's needs. That is why it is not wise for a woman to borrow the Pill from someone else.

When a woman does not forget to take her daily dosage, the Pill is one of the most reliable methods of controlling fertility. It does not interrupt sexual intercourse, and the method is easy to follow. Often the Pill corrects erratic menstrual cycles and decreases cramping. However, the Pill has some side effects for a small number of users. In the first month or so of use, it may cause nausea, weight gain, tissue swelling, and minor headaches. Its continued use may lead to blood clotting in the veins of a few women (3 out of 10,000) predisposed to this disorder. Some cases of elevated blood pressure and some abnormalities in fat metabolism might be linked to a growing number of gallbladder disorders in Pill users.

In *vasectomy,* a tiny incision is made in the scrotum and each vas deferens is severed and tied off. The simple operation can be performed in twenty minutes in a doctor's office, with only a local anesthetic. After vasectomy, sperm cannot leave the testes and so will not be present in semen. So far there is no firm evidence that vasectomy disrupts the male hormone system, and there seems to be no noticeable difference in sexual activity. Vasectomies can be reversed, but half the men who have had the surgery develop antibodies against sperm and may not be able to regain fertility.

For females, surgical intervention includes *tubal ligation,* in which the oviducts are cauterized or cut and tied off. Tubal ligation is usually performed in a hospital. A small number of women who have had the operation suffer recurring bouts of pain and inflammation of tissues in the pelvic region where the surgery was performed. The operation can be reversed, although major surgery is required and success is not always assured.

Once conception and implantation have occurred, the only way to terminate a pregnancy is *abortion,* the dislodging and removal of the embryo from the uterus. *RU-486,* the "morning-after Pill," can induce the termination of pregnancy. It is administered under a doctor's supervision, at least in Europe. Those opposed to

Sexually Transmitted Diseases

(This Commentary is based on information from the Centers for Disease Control, Atlanta)

Sexually transmitted diseases (STDs) have reached epidemic proportions, even in countries with the highest medical standards. The disease agents are mostly bacteria and viruses, and they usually are transmitted from infected to uninfected persons during sexual intercourse. In 1989 in the United States alone, more than 12 million new cases of STDs were reported. No one can estimate the number of unreported cases.

The economics of this health problem are staggering. By conservative estimates, the cost of treatment is exceeding $2 billion a year—and this does not include the accelerating cost of treating AIDS patients. In Africa and many other developing countries, AIDS alone threatens to overwhelm health-care delivery systems and to unravel decades of economic progress.

The social consequences are sobering. Of every twenty babies born in the United States, one will start out in the world with a chlamydial infection. Of every 10,000 newborns, as many as three will contract systemic herpes. Half of those newborns may die early, and a fourth of the survivors will have serious neurological defects. Each year in the United States, 1 million adolescent and adult females are stricken with pelvic inflammatory disease, usually as a complication of gonorrhea and other STDs. Of those females, about 200,000 are hospitalized, more than 100,000 undergo pelvic surgery that results in permanent sterility, and 900 cannot recover, and die. The examples just given only hint at the alarming complications of many sexually transmitted diseases.

AIDS

Acquired immune deficiency syndrome (AIDS) is a set of chronic disorders that can follow infection by the human immune deficiency virus (HIV). The virus cripples the immune system, in the manner described in Chapter 27, and the body becomes highly vulnerable to infections, many of which would not otherwise be life-threatening. (Hence the description, "opportunistic" infections.)

AIDS is mainly a sexually transmitted disease, with most infections occurring through the transfer of bodily fluids during vaginal or anal intercourse. Such fluids include blood, semen, and vaginal secretions. The virus enters the body through cuts or abrasions in the penis, the vagina, or the rectum. Mucous membranes in the mouth may be another point of entry. Once inside the body, the virus locks onto cells that are capable of sustaining its replication (page 223). Helper T cells (the T4 lymphocytes), macrophages, brain cells, and epithelial cells of the cervix are known targets.

Unlike some other sexually transmitted diseases, AIDS cannot be effectively treated at this time and there are no vaccines against the causative agent. *There is no cure.* Infected persons may be symptom-free at first, but as many as half develop AIDS within five to ten years. Others develop symptoms that are milder than the ones characterizing AIDS. The milder symptoms are called the AIDS-related complex, or ARC. Will most or all of those infected eventually develop AIDS? We do not know enough about the natural history of the virus and the progression rates of the disease to discount that possibility.

It is the symptom-free carriers of HIV who unwittingly have been fueling a worldwide AIDS epidemic. *During the next decade, as many as 50 million to 100 million could be infected worldwide.*

HIV apparently has been present in localized regions of Central Africa for at least several decades. However, in the 1970s and early 1980s, it spread to different countries and has since reached epidemic proportions. (AIDS was not even identified until 1981.) Although in Africa the virus is transmitted primarily through heterosexual contact, the initial victims in the United States were male homosexuals. In the developed countries, HIV is still transmitted mainly through homosexual contact as well as through needle sharing by intravenous drug abusers. Unfortunately, the use of illicit drugs in conjunction with sexual practices is widespread. Today, in both rural and urban areas, more and more women are infected through needle sharing and relations with bisexual men.

Who, then, is *not* at risk? Only individuals who fall in these categories:

1. Individuals who are not drug abusers or who do not share unsterile needles or syringes

2. Individuals who abstain from sexual relations

3. Mutually monogamous couples who have had no other sexual partners since the 1970s (when AIDS began to spread dramatically)

4. Couples who are shown to be free of infection and who refrain from sexual relations with anyone else

Free or low-cost, confidential testing for AIDS is available through public health facilities and many physicians' offices. Keep in mind that there may be a time lag from a few weeks to six months or longer before detectable antibodies form in response to infection. The presence of antibodies indicates exposure to the virus, but this in itself does not mean that AIDS will develop. Even so, anyone who tests positive should be considered capable of spreading the virus.

Beyond this, there is confusion about what constitutes "safer" sex. Proper use of high-quality, latex condoms, together with a spermicide that contains nonoxynol-9, is assumed to be highly effective in stopping transmission—but there is still a small risk of irreversible infection. Open-mouthed, intimate kissing with a person who tests positive for the virus should be avoided. Caressing carries no risk—*if* there are no lesions or cuts through which the virus can enter the body. Such lesions commonly accompany other sexually transmitted diseases, and they apparently are correlated with increased susceptibility to HIV infection.

In sum, AIDS has reached epidemic proportions mainly for three reasons. First, we did not know that the virus is transmitted by semen, blood, and vaginal fluid and that *behavioral* controls can limit its spread. Second, we did not have tests that could be used to identify symptom-free carriers who could unwittingly infect others; we do now. Third, many thought AIDS was a threat associated only with homosexual behavior. The medical, social, and economic consequences of its rapid spread throughout the world make it everyone's problem.

Gonorrhea

Unlike AIDS, gonorrhea is a sexually transmitted disease that can be cured by prompt diagnosis and treatment. Gonorrhea ranks high among the reported communicable diseases in the United States, with 1 million new cases reported each year. (There may be anywhere from 3 million to 10 million unreported cases.)

Gonorrhea is caused by *Neisseria gonorrhoeae*. This bacterium can infect epithelial cells of the genital tract, eye membranes, and the throat. Since 1960, the incidence of gonorrhea in the population has been rising at an

alarming rate. The increase has coincided with the use of birth control pills and increased sexual permissivity.

Males have a greater chance than females do of detecting the disease in early stages. Within a week, yellow pus is discharged from the urethra. Urination becomes more frequent and painful because the urinary tract becomes inflamed. Females may or may not experience a burning sensation while urinating. They may or may not have a slight vaginal discharge; even if they do, the discharge may not be perceived as abnormal. Thus, in the absence of worrisome symptoms, gonorrhea often goes untreated. The bacteria may spread into the oviducts, the eventual outcome being violent cramps, fever, vomiting and, in many cases, sterility.

Complications arising from gonorrheal infection can be avoided with prompt antibiotic treatment. As a preventive measure, males who have multiple sexual partners can wear condoms to help prevent the spread of infection. Part of the problem is that the initial stages of the disease are so uneventful that the dangers are masked. Also, many infected persons wrongly believe that once cured of gonorrhea, they are safe from reinfection—which simply is not true. Multiple reinfections can and do occur.

Syphilis

Syphilis is caused by a motile bacterium, *Treponema pallidum* (a spirochete). As many as 300,000 humans may become infected in a given year in the United States, but only about 30,000 are reported. In the past 5 years, its incidence has nearly doubled among females between ages 15 and 24. The bacterium is transmitted by sexual contact. After it has penetrated exposed tissues, the body produces a chancre (that is, a localized ulcer) that teems with bacteria. The chancre, which is a symptom of the primary stage of syphilis, appears between one to eight weeks following infection.

During the second stage of infection, lesions can occur in mucous membranes, the eyes, bones, and central nervous system. Afterward, the infection enters a latent stage that has no outward symptoms. Syphilis can be detected only by serology tests during the latent stage, which can last many years. All the while, the immune system works against the bacterium. Sometimes the body does cure itself, but this is not the usual outcome.

If untreated, syphilis in its tertiary stage can produce lesions of the skin and internal organs, including the liver, bones, and aorta. Scars form; the walls of the aorta weaken. The brain and spinal cord are damaged in ways

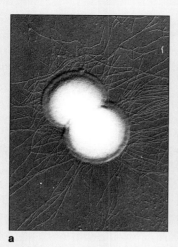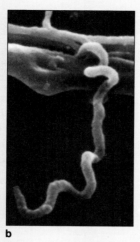

a   **b**

(**a**) *Neisseria gonorrhoeae*, a bacterium that typically is seen as paired cells, as shown here. The threadlike structures (pili) evident in this electron micrograph help the bacterium attach · to its host, upon which it bestows gonorrhea. (**b**) *Treponema pallidum*, a bacterium that causes syphilis.

that lead to various forms of insanity and paralysis. Women who have been infected typically have miscarriages, stillbirths, or sickly and syphilitic infants.

Chlamydial Infections

There are other infections of the genital tract besides gonorrhea and syphilis; they are called the "nonspecific" sexually transmitted diseases. The most prevalent are the chlamydial infections. Each year, anywhere from 3 million to 10 million Americans—college students particularly—are affected.

A bacterium, *Chlamydia trachomatis*, is the culprit behind a variety of diseases. Among other things, it infects cells of the genitals and urinary tract. Following infection, bacteria migrate to regional lymph nodes, which become enlarged and tender. The enlargement can impair lymph drainage and lead to pronounced tissue swelling. Chlamydial infections can be treated effectively with tetracycline and sulfonamides so that there will be no long-term complications. However, in some females the infection leads to pelvic inflammatory disease.

Pelvic Inflammatory Disease

A condition called pelvic inflammatory disease (PID) affects about 1.75 million women each year. It is one of the serious complications of gonorrhea, chlamydial infections, and other STDs. But it also can arise when normal vaginal microbes ascend into the pelvic region and when intrauterine devices (IUDs) malfunction and cause an inflammatory response in the uterine lining. Most often, the uterus, oviducts, and ovaries are affected. The pain may be so severe, infected women often think they are having an attack of acute appendicitis. The oviducts may become scarred, and this can lead to abnormal (ectopic) pregnancies as well as to sterility.

Genital Herpes

Genital herpes is an extremely contagious viral infection of the genitals. It is transmitted when any part of a person's body comes into direct contact with active *Herpes* viruses or sores that contain them. Mucous membranes (particularly of the mouth or genital area) are susceptible to invasion, as is broken or damaged skin. Transmission seems to require intimate sexual contact; the virus does not survive for long outside the human body.

There are an estimated 5 million to 20 million persons with genital herpes in the United States alone. From 1965 to 1979, the number of reported cases increased by 830 percent; 200,000 to 500,000 cases are still being reported annually. Newborns of infected mothers are among those cases. Contact with the mother's active lesions during normal vaginal delivery can lead to a form of herpes that is often fatal. Lesions arising in the infant's eyes can cause blindness. Chronic herpes infection of the cervix is now considered to be a contributing factor in the increased risk of cervical cancer.

There are many strains of *Herpes* viruses, which are classed as types I and II. The type I strains infect mainly the lips, tongue, mouth, and eyes. Type II strains cause most of the genital infections. Disease symptoms occur two to ten days after exposure to the virus, although sometimes symptoms are mild or absent. Among infected women, small, painful blisters appear on the vulva, cervix, urethra, or anal tissues. Among men, the blisters occur on the penis and anal tissues. Within three weeks, the sores crust over and heal without leaving scars.

After the first sores disappear, sporadic reactivation of the virus can produce new, painful sores at or near the original site of infection. Recurrent infections may be triggered by sexual intercourse, emotional stress, menstruation, or other infections. At present there is no cure for genital herpes. Acyclovir, an antiviral drug, decreases the healing time and often decreases the pain and viral shedding.

abortion are currently fighting its use in the United States.

At one time, abortions were generally forbidden by law in the United States unless the pregnancy endangered the mother's life. The Supreme Court has since ruled that the government does not have the right to forbid abortions during the early stages of pregnancy (typically up to five months). Before this ruling, there were dangerous, traumatic, and often fatal attempts to abort embryos, either by pregnant women themselves or by quacks.

Newer methods have made abortion relatively quick, painless, and free of complications when performed during the first trimester. Abortions in the second and third trimesters will probably remain extremely controversial unless the mother's life is clearly threatened. For both medical and humanitarian reasons, however, it is generally agreed in this country that the preferred route to birth control is not through abortion but through control of conception in the first place.

In Vitro Fertilization. Controls over fertility extend in the other direction—to childless couples who are desperate to conceive a child. In the United States, about fifteen percent of all couples cannot do so because of sterility or infertility. For example, hormonal imbalances may prevent ovulation in the female; or the sperm count in the male may be too low to assure fertilization.

With *in vitro fertilization*, external conception is possible, provided sperm and oocytes obtained from the couple are normal. A hormone is administered that prepares the ovaries for ovulation. Then a physician locates and removes the preovulatory oocyte with a suction device. Before the oocyte is removed, sperm from the male is placed in a solution that simulates the fluid in oviducts. When the suctioned oocyte is placed with the sperm, fertilization may occur a few hours later. About twelve hours later, the newly dividing zygote is transferred to a solution that will support further development, and about two to four days after that, it is transferred to the female's uterus. If all goes well, implantation may occur. It occurs in only about twenty percent of the attempts, and each attempt costs several thousand dollars.

SUMMARY

1. Most animals reproduce sexually. Separation into male and female sexes involves specialized reproductive structures and forms of behavior that help assure successful fertilization and that lend initial nutritional support to offspring.

2. Humans have a pair of primary reproductive organs (sperm-producing testes in males, egg-producing ovaries in females), accessory ducts, and glands. Testes and ovaries also produce hormones that influence reproductive functions and secondary sexual traits.

3. The hormones testosterone, LH, and FSH control sperm formation. They are part of feedback loops involving the hypothalamus, anterior pituitary, and testes. The hormones estrogen, progesterone, FSH, and LH control egg maturation and release, as well as changes in the lining of the uterus (endometrium). They are part of feedback loops involving the hypothalamus, anterior pituitary, and ovaries.

4. The menstrual cycle proceeds through a follicular phase, ovulation, and a luteal phase.
   a. Follicular phase: A follicle (an oocyte surrounded by a cell layer) matures and the endometrium starts to rebuild. (The endometrium breaks down at the end of each menstrual cycle when pregnancy does not occur.)
   b. Ovulation: A midcycle peak of LH triggers the release of a secondary oocyte from the ovary.
   c. Luteal phase: A corpus luteum forms from the remainder of the follicle; its secretions prime the endometrium for fertilization. When fertilization occurs, the corpus luteum is maintained and its secretions help maintain the endometrium.

5. Human development proceeds through gamete formation, fertilization, cleavage, gastrulation, organ formation, and growth and tissue specialization.

6. Development depends on cell differentiation and morphogenesis. Both processes are outcomes of the unequal distribution of localized cytoplasmic components during cleavage and of cell interactions in the developing embryo.
   a. In cell differentiation, initial populations of cells that are genetically the same give rise to subpopulations of cell lineages that differ in structure and function.
   b. In morphogenesis, differentiated cells become organized into tissues and organs according to predefined patterns. Local cell divisions, growth, cell migrations, changing cell shapes, and controlled cell death are required.

7. All tissues and organs arise from three germ layers—the endoderm, ectoderm, and mesoderm—of the early embryo. Embryonic development also requires four extraembryonic membranes (yolk sac, allantois, amnion, and chorion).

8. Control of human fertility raises many important ethical questions. These questions extend to the physical, chemical, surgical, or behavioral interventions used in the control of unwanted pregnancies.

Review Questions

1. Study Table 32.1. Then list the main organs of the human male reproductive tract and identify their functions. *468*

2. Which hormones influence male reproductive function? *469*

3. Label the component parts of the female reproductive tract: *471*

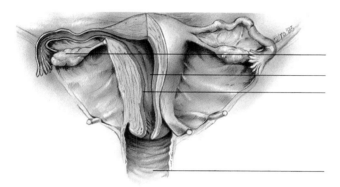

4. What is the menstrual cycle? Which four hormones influence this cycle? *471, 472*

5. List four events that are triggered by the surge of LH at the midpoint of the menstrual cycle. *472–473*

6. What changes occur in the endometrium during the menstrual cycle? *471*

7. Define and describe the main features of the following developmental stages: fertilization, cleavage, gastrulation, and organ formation and tissue specialization. *464*

8. Cell differentiation and morphogenesis are two processes that are critical for development. Define them. *465–466*

9. Which of your organs formed from endoderm? From mesoderm? From ectoderm? *465*

10. Describe these extraembryonic membranes: amnion, yolk sac, allantois, and chorion. *476*

Self-Quiz *(Answers in Appendix IV)*

1. The biological cost of elaborate sex-related structures and behaviors is offset by the advantages of _____ among offspring, at least some of which should prove to be adaptive in _____ or _____ environments.

2. Animal development commonly proceeds through six stages, called gamete formation, fertilization, _____, _____, and growth and tissue specialization.

3. The developmental fate of cells in the embryo is partly sealed at _____, when daughter cells inherit qualitatively different regions of cytoplasm. It also depends on _____ in the developing embryo.

4. In _____, initial populations of cells that are genetically the same give rise to subpopulations of cells that differ in structure and function.

5. In _____, differentiated cells become organized into tissues and organs according to predefined patterns.

6. Morphogenesis involves these events:
    a. cells undergoing divisions, growth, and changes in shape
    b. cells migrating about and dying in controlled ways
    c. wholesale remodeling of larval stages
    d. a and b are correct

7. The primary reproductive organs in humans are the _____ in males and the _____ in females
    a. penis; vagina
    b. testes; ovaries
    c. seminiferous tubule; uterus
    d. glans penis; clitoris

8. Sperm formation is controlled by _____.
    a. testosterone
    b. LH
    c. FSH
    d. all of the above affect sperm formation

9. Which is the correct order for the menstrual cycle?
    a. luteal phase, ovulation, follicular phase
    b. ovulation, luteal phase, follicular phase
    c. luteal phase, follicular phase, ovulation
    d. follicular phase, ovulation, luteal phase

10. Match the reproduction and development concepts.
    _____ embryonic germ layers
    _____ extraembryonic membranes
    _____ corpus luteum
    _____ cell differentiation
    _____ morphogenesis
    _____ follicle

    a. differentiated cells organize into tissues and organs
    b. oocyte and surrounding cell layer
    c. genetically identical cells give rise to cells differing in structure and function
    d. endoderm, ectoderm, mesoderm
    e. yolk sac, allantois, amnion, chorion
    f. its secretions help prepare endometrium for fertilization

Selected Key Terms

allantois 476
amnion 476
blastula 465
cell differentiation 465
chorion 476
cleavage 464
corpus luteum 473
endometrium 470
estrogen 470
estrus 470
fertilization 464
follicle 471
FSH 469
gastrulation 465
germ layer 465

LH 469
menstrual cycle 471
morphogenesis 466
ovary 467
oviduct 470
ovulation 473
ovum 475
placenta 477
progesterone 470
seminiferous tubule 468
testis 467
testosterone 469
umbilical cord 477
uterus 470
yolk sac 476

Readings

Alberts, B., et al. 1989. *Molecular Biology of the Cell*. Second edition. New York: Garland. Well-written introduction to molecular basis of development in Chapters 15 through 17.

Gilbert, S. 1988. *Developmental Biology*. Second edition. Sunderland, Massachusetts: Sinauer.

Nilsson, L., et al. 1986. *A Child Is Born*. New York: Delacorte Press/ Seymour Lawrence. Extraordinary photographs of embryonic development.

Spence, A. 1989. *Biology of Human Aging*. Englewood Cliffs, New Jersey: Prentice-Hall. Paperback.

Wassarman, P. December 1988. "Fertilization in Mammals." *Scientific American* 259(6):78–84.

Zack, B. 1981. "Abortion and the Limitations of Science." *Science* 213(4505).

FACING PAGE: *Dolphins in the waters off New Zealand, symbolic of the adaptations of organisms to one another and to their environment.*

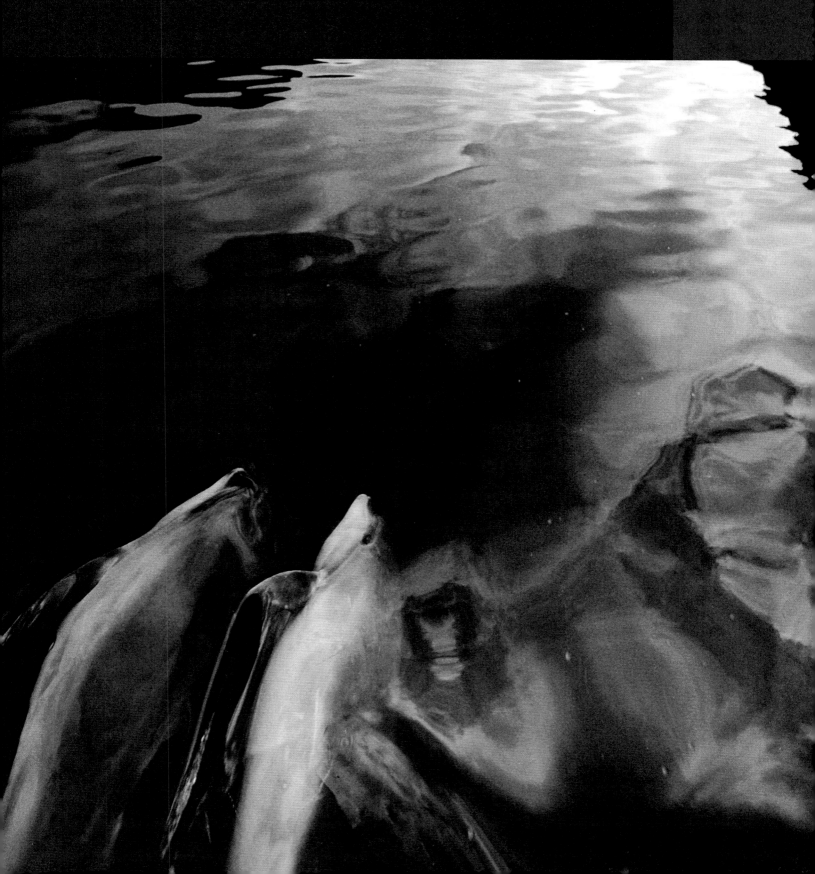

33

POPULATION ECOLOGY

KEY CONCEPTS

1. A population is a group of individuals of the same species occupying a given area at the same time. As we have seen, the population (not the individual or the species) is the unit of evolution. To gain insight into the changing nature of the human population and to predict its likely future, we can consider ecological principles that govern the growth and sustainability of all populations over time.

2. Population growth generally follows certain patterns. When the birth rate remains even slightly above the death rate and neither changes, a population shows exponential growth (it increases in size by ever larger amounts per unit of time). When the environment imposes limits on growth, the pattern may become one of logistic growth whereby a low-density population undergoes rapid increase, then levels off.

3. All populations face limits to growth, for no environment can indefinitely sustain a continucusly increasing number of individuals. Disease, predation, competition for resources, and other factors act as controls over population growth. The controls vary in their relative effects and they vary over time.

Suppose this year the federal government passes legislation to control population growth by limiting the size of each family to three children. Suppose they mandate that each father be sterilized after his third child is born, and that if he refuses, he will be sterilized without his consent. *It would never happen here,* you might be thinking. Such an invasion of privacy would never be tolerated in our society. Besides, family size is not much of an issue in North America, where standards of food production, hygiene, and medical care are among the world's highest. Most populations cannot take these things for granted; yet many are still growing at alarming rates (Figure 33.1).

For example, there already are more people in India than in North and South America combined. Most do not have adequate food or medical care, living conditions are often appalling, and unemployment is a nightmare. Each *week,* 100,000 people enter the job market. Each *day,* 100 acres of India's cropland are being removed from agriculture. Too many salts from irrigation water have accumulated in the soil, and much of India does not get enough rain to flush them out.

The government of India has supported population control programs for more than two decades, but these

Figure 33.1 A sampling of the more than 5.3 billion humans on earth. In this chapter we turn to the principles governing the growth and sustainability of populations, including our own.

492

have not worked well. Most people live in remote villages, so the programs are difficult to administer. Because so many people are illiterate, information must be conveyed by word of mouth. The very idea of limiting family size is met with resistance. Disease and starvation are so pervasive, many villagers believe that survival depends on having many children. Without large families, they ask, who will help a father tend fields? Who will go to cities and earn money to send back home? How can a father otherwise know he will be survived by a son, who must, by Hindu tradition, conduct the last rites so the soul of his dead father will rest in peace?

India's population may reach 1 billion in less than thirty years, and no one wants to think about how many then will face poverty, disease, and starvation. In 1976, out of desperation, the government passed a law calling for compulsory sterilization, although public outrage became so great the law eventually was rescinded.

Is there a way out of such dilemmas? Should other nations donate surplus food to the growing populations of less fortunate nations? Would donations help, or would they encourage greater increases in population size? Suppose the expanded populations came to depend on continuing support. What if the benefactor nations were hit by severe droughts year after year, and had trouble meeting even the demands of their own populations?

Whether we consider humans or any other organisms, *certain ecological principles govern the growth and sustainability of populations over time.* This chapter describes those principles, then shows how they apply to the past, present, and future growth of the human population.

ECOLOGY DEFINED

Ecology means the study of the interactions of organisms with one another and with the physical and chemical environment. In this unit of the book, we will consider ecological interactions at several levels of biological organization, which may be defined in the following way:

---

1. The **population:** a group of individuals of the same species occupying a given area at the same time. The place where a population (or individual) lives is its **habitat**.

2. The **community:** the populations of *all* species occupying a habitat. The term also is used for groups of organisms with similar life-styles in the habitat, such as the bird community and the plant community.

3. The **ecosystem:** a community and its environment. An ecosystem has a *biotic* component (all of its living members). It also has *abiotic* (nonliving) components, such as temperature, soil, and rainfall.

4. The **biosphere:** the entire realm in which organisms exist—the waters of the earth; the surface rocks, soils, and sediments of its crust; and the lower region of the atmosphere.

This chapter begins with relationships that influence the size, structure, and distribution of populations. We will return to the nature of human population growth at the chapter's end. Communities, ecosystems, and the biosphere will be our focus in chapters to follow.

POPULATION DYNAMICS

Populations as a whole display certain characteristics, including size, density, dispersion, and age structure. The *size* of a population is the number of individuals making up its gene pool. Its *density* is the number of individuals per unit area or volume, such as the number of rotifers in each liter of water in a pond. *Dispersion* refers to the general pattern in which the population's members are dispersed through its habitat. Most often, they live in clumps; less often, they are randomly or rather evenly spaced (Figure 33.2). Dispersion also varies with time, often in response to changing seasons.

Finally, the *age structure* of a population is the relative proportion of individuals of each age. These are often divided into pre-reproductive, reproductive, and post-reproductive age categories. The middle category represents the *reproductive base* for the population.

Population Size and Patterns of Growth

Over a given time span, the numerical size of a population depends on how many individuals enter (by birth or immigration) and how many leave (by death or emigration). Assume immigration and emigration are equivalent, so that we can put aside their effects on population size. Given this assumption, population size is stabilized when the birth rate is balanced over the long term by the death rate; there is *zero population growth*.

Exponential Growth. Populations can increase in size when the number of births exceeds the number of deaths. But how fast and how much can they increase in a given period? Think about a population with 1,000 members ($N = 1,000$). Suppose 50 members are born and 10 die each year. The actual rate of increase (r) for the population would be:

$$r = \frac{births - deaths}{N} = \frac{50 - 10}{1,000} = 0.04 \text{ or } 4\% \text{ per year}$$

As long as r remains positive, population size will increase by ever larger amounts. To see why this is so, start with one of the simplest "reproductive bases," a single bacterium, and put it in a culture flask with abundant nutrients. Within thirty minutes the bacterium divides in two; thirty minutes later the two divide into four. If no cells die between divisions, the number will double every thirty minutes. As Figure 33.3 shows, after $9\frac{1}{2}$ hours (nineteen doublings), population size exceeds 500,000; after 10 hours (twenty doublings), it soars past 1 million!

clumped   nearly uniform   random

a   b   c

Figure 33.2 Generalized dispersion patterns for individuals in a population. (**a**) Most commonly, members occur in clumps through their habitat. One reason is that resources, living space, hiding places, and other aspects of the habitat are usually patchy, not uniform. Another is that new generations (such as seeds and larvae) often do not spread far from their parents. Many animals, including the baboons shown here, clump into social groups.

(**b**) A population showing nearly uniform spacing, as you might see in an orchard, is rare in nature. Where it does occur, competition (for resources, social dominance, and so forth) seems to be the primary cause.

(**c**) Dispersion may be random when conditions are fairly uniform through a habitat and the members of a population are not attracting or repelling each other during a given time. Spiders randomly dispersed on a forest floor are an example.

When you plot the size increases, the result is a graph with a *J-shaped curve,* one that grows steeper with time (Figure 33.3). J-shaped curves are characteristic of all populations undergoing **exponential growth**, meaning they are increasing in size by ever larger amounts per unit of time.

When nothing stops its growth, a population will grow exponentially even when the birth rate only slightly exceeds the death rate. Start over with that single bacterium. If a fourth of the population dies between each doubling time, it will take almost two hours (instead of thirty minutes) for doubling to occur. *Only the time scale changes.* It now takes thirty hours (not ten) to get a million bacteria—but you still end up with a J-shaped curve (curve *b* in Figure 33.3).

As long as the birth rate remains even slightly above the death rate, a population will grow. If the rates remain constant, it will grow exponentially.

Biotic Potential. The **biotic potential** of a population is its *maximum* rate of increase, per individual, under ideal conditions. It is the rate that might be achieved when food and living space are abundant and when other organisms aren't interfering in access to those resources.

The biotic potential is not the same for every species. For many bacteria, it is 100 percent every half hour; for humans and other large mammals, it is between 2 and 5 percent per year. The differences arise through variations in (1) how soon the individual starts reproducing, (2) how often reproduction occurs, (3) how many offspring are born each time, and (4) how many of them tend to survive to reproductive age.

A population may grow exponentially even when it is not expressing its full biotic potential. Human females have not been bearing twenty children or more (as is biologically possible), and many have not reproduced at all. Yet the human population has been growing exponentially since the mid-eighteenth century. At any time, the *actual* rate of increase is influenced by environmental circumstances affecting human society.

Carrying Capacity. Most often, populations do not achieve their full biotic potential. Remember, a single sea star can release 2.5 million eggs each year, but the oceans never have been filled with sea stars. What restricts the biotic potential? This is not easy to determine in nature because of the complex interactions among populations. For that reason, start again with a lone bacterium in a culture flask, where you can control the variables.

Assume you enrich the culture medium with glucose and essential elements required for bacterial growth, then allow the bacteria to reproduce for many generations. At first growth is exponential, then growth tapers and population size remains rather stable. Then it starts

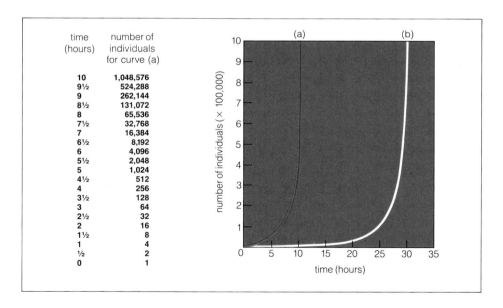

time (hours)	number of individuals for curve (a)
10	1,048,576
9½	524,288
9	262,144
8½	131,072
8	65,536
7½	32,768
7	16,384
6½	8,192
6	4,096
5½	2,048
5	1,024
4½	512
4	256
3½	128
3	64
2½	32
2	16
1½	8
1	4
½	2
0	1

Figure 33.3 (**a**) Exponential growth for a bacterial population that is dividing by fission every half hour. (**b**) Exponential growth of the population when division occurs every half hour, but when twenty-five percent die between divisions. Although deaths slow the rate of increase, in themselves they are not enough to stop exponential growth.

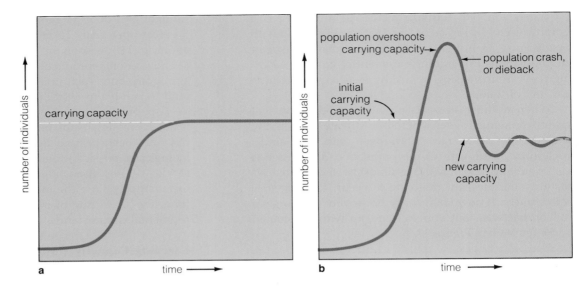

Figure 33.4 Idealized S-shaped curve characteristic of logistic growth (**a**). Following a rapid growth phase, growth slows and the curve flattens out as the carrying capacity is reached.

Sometimes a population grows rapidly and overshoots the carrying capacity (**b**). This happened to the reindeer population introduced on one of the Pribilof islands (Figure 33.5). At other times, the carrying capacity itself changes abruptly as environmental conditions change. This happened to the human population in Ireland before the turn of the century, when a disease (late blight, page 242) wiped out the potatoes that were the mainstay of the diet.

to decline rapidly and soon dies out. What happened? The culture dish held only so much glucose. As the population expanded faster and faster, it used more and more glucose. When food supplies began to dwindle, so did one of the foundations for growth.

When any essential resource is in short supply, it becomes a **limiting factor** on population growth. Predation, competition for living space, and pollution are other examples of factors that can limit population growth. The number of limiting factors can be enormous, many can be operating at the same time, and their relative effects can vary. Which factor has greatest influence also varies with time, with first one and then another setting the upper bound.

For example, glucose became a limiting factor in the culture flask. But suppose you attempted to sustain growth by keeping the bacterial population supplied with all necessary nutrients. It still would crash after its initial exponential growth. Like all organisms, bacteria produce metabolic wastes. The huge numbers of bacterial cells produced wastes in concentrations high enough to drastically alter chemical conditions in the flask; they polluted their own environment. They themselves put a stop to their growth.

All the limiting factors acting on a population collectively represent the environmental resistance to its growth. They define the **carrying capacity**, the number of indi-

viduals of a given species that can be sustained indefinitely in a given area.

The growth rate of a low-density population starts out slowly, goes through a rapid growth phase, and then levels off once the carrying capacity is reached. This pattern is called **logistic growth**. A plot of logistic growth gives us an *S-shaped curve* (Figure 33.4). This curve is only a simple approximation of what goes on in nature, however. Because environmental conditions vary, carrying capacity also can vary over time.

Checks on Population Growth

Density-Dependent Controls. When a population grows and its density increases, environmental conditions that dictate the carrying capacity are intensified, and together they lead to reduced birth rate and increased rates of death, dispersal, or both. Competition for resources, predation, parasitism, and disease are the main density-dependent factors. They generally have self-adjusting effects on population growth. Once density decreases, the pressures ease and population size may increase once more.

Later chapters describe the nature of competition within and between species. Here, we will simply consider what happened when 4 male and 22 female

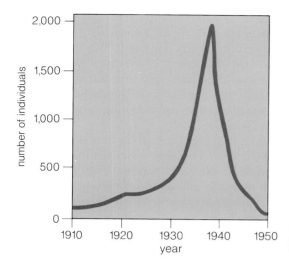

Figure 33.5 Rise and fall of a reindeer herd introduced on one of the Pribilof islands, Alaska. Rapid population growth led to an overshooting of the carrying capacity of the environment and its deterioration. Growth stopped abruptly, and the population size crashed to eight reindeer—eighteen fewer than were present in the starting population.

reindeer were introduced on one of the Pribilof islands of Alaska in 1910. The small herd increased to 2,000 within thirty years, and the reindeer had to compete for a dwindling supply of vegetation. Overgrazing almost wiped out the vegetation, and in 1950 the herd plummeted to 8 members. Figure 33.5 shows the growth pattern for this population, which "overshot" the carrying capacity and then rapidly crashed.

When prey or host populations become increasingly dense, their members face greater risk of being killed by predators, colonized by parasites, or infected by contagious disease. A classic example is the *bubonic plague* that killed 25 million Europeans during the fourteenth century. The disease agent (the bacterium *Yersinia pestis*) normally lives in wild rodents, and fleas transmit it to new hosts. The bacterium multiplies in the flea gut and blocks digestion, the fleas attempt to feed more and more often, and so the disease spreads. It spread like wildfire through the cities of fourteenth-century Europe because human habitats were crowded together, sanitary conditions were poor, and the rat populations were abundant.

Subsequent outbreaks of plague have not been as dramatic, but *Y. pestis* still lurks in many rodent populations, including ones from the Rocky Mountains to the west coast of the United States.

Density-Independent Controls. Some events tend to increase the death rate more or less independently of population density. An individual might be more competitive in staking out part of a dwindling food supply, but if it is caught out in the open in a severe storm, it is just as likely to be killed as any other member of the population. This happens every so often in the Colorado Rockies, when freak snowstorms in summer wipe out butterfly populations. Similarly, if members of a population bear ivory, attractive feathers, or some other attribute that humans value, they may be hunted down and killed regardless of how few and far between they become. Poachers, for example, have brought some populations of African elephants to extinction.

Survivorship Curves

Each species has a characteristic mode of reproduction. At one extreme, sea stars and other species produce huge numbers of small offspring that must feed and grow rapidly on their own, without nutritional support, protection, or guidance from the parents. Typically the death rate is high among these offspring, as when most of the sea star larvae become food for corals and other animals. At the other extreme, humans and other large mammals produce very few offspring, but they invest a great deal of energy and time to ensure that those few will reach reproductive age. For example, female elephants produce only four or five calves in a lifetime and devote several years of parental care to each one.

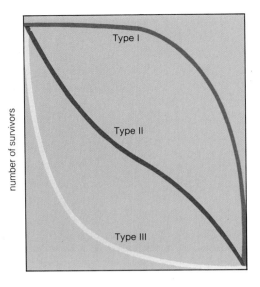

number of survivors

Type I

Type II

Type III

lifespan

Figure 33.6 Three generalized types of survivorship curves. For Type I populations, there is high survivorship until some age, then high mortality. Type II populations show a fairly constant death rate at all ages. For Type III populations, survivorship is low early in life.

Table 33.1	Life Table for the United States Population, 1982		
Age Interval	Number Alive at Start of Interval (per 100,000)	Number Dying During Interval	Average Number of Years of Life Remaining at Start of Interval
0–1	100,000	1,107	73.6
1–5	98,893	269	73.7
5–10	98,624	175	71.7
10–15	98,449	181	64.6
15–20	98,268	497	59.7
20–25	98,771	673	55.0
25–30	97,098	663	50.4
30–35	96,435	725	45.7
35–40	95,710	986	41.0
40–45	94,724	1,483	36.5
45–50	93,241	2,352	32.1
50–55	90,889	3,483	27.9
55–60	87,406	5,063	23.9
60–65	82,343	7,281	20.3
65–70	75,062	9,005	17.9
70–75	66,057	12,214	13.8
75–80	53,843	14,455	10.4
80–85	39,388	14,467	10.1
85+	24,921	24,921	9.2

We can construct **survivorship curves** that reflect the age-specific patterns of death for a particular population in a particular environment. Typically, this is done by following the fate of a group of newborn individuals until the last one dies, then plotting the number of survivors at each age. Similarly, birth schedules document the average number of offspring produced by individuals at each age.

Figure 33.6 shows three types of survivorship curves. A Type I curve is typical of large mammals that provide their offspring with extended parental care. It is characteristic of humans in communities with good health care services. Historically, and where health care is poor today, infant deaths cause a sharp drop at the start of the curve, which then levels off. A Type II curve is typical among organisms that are just as likely to be killed or die of disease at any age; it is the pattern for some songbirds, lizards, and small mammals. A Type III curve reflects a high death rate early in life. The pattern is typical of marine invertebrates, most insects, many fish, plants, and fungi—all of which produce large numbers of new individuals.

Survivorship and birth schedules may be put together in a *life table*, which shows the survival rate at each age group in a population (Table 33.1). Such tables were originally developed by life insurance and health insurance companies; they are used to help set the price of insurance for people of different ages.

HUMAN POPULATION GROWTH

In 1990, the human population reached 5.3 billion. In 1988 alone, almost 87 million more individuals were added to it. That amounted to an average of 1.7 million more per week, 238,000 per day, or 9,900 per hour. This staggering display of growth is occurring while at least one in six humans already on the planet is malnourished or starving, without clean drinking water, and without adequate shelter. It is occurring while health care delivery and sewage treatment facilities are nonexistent for a third of the population.

Suppose it were possible, by monumental efforts, to double food production to keep pace with growth. We would do little more than maintain marginal living conditions for most of the world, and death from starvation could still reach 10 million to 40 million a year. Even this would come at great cost, for we are introducing serious new pressures on resources into the environment that must sustain us. Salted-out cropland, desertification, deforestation, global pollution—these are some of the factors you will be reading about in Chapter 37, and they do not bode well for our future.

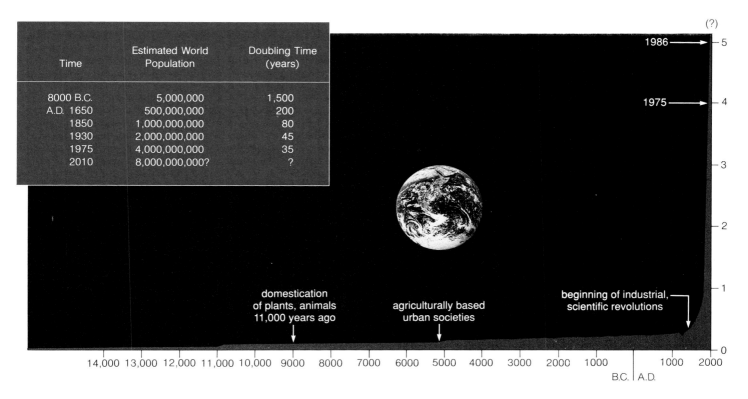

Time	Estimated World Population	Doubling Time (years)
8000 B.C.	5,000,000	1,500
A.D. 1650	500,000,000	200
1850	1,000,000,000	80
1930	2,000,000,000	45
1975	4,000,000,000	35
2010	8,000,000,000?	?

domestication of plants, animals 11,000 years ago

agriculturally based urban societies

beginning of industrial, scientific revolutions

14,000 13,000 12,000 11,000 10,000 9000 8000 7000 6000 5000 4000 3000 2000 1000 | 1000 2000
B.C. | A.D.

Figure 33.7 The curve of global human population growth. The vertical axis of the graph represents world population, in billions. (The slight dip between the years 1347 and 1351 shows the time when 25 million people died in Europe as a result of bubonic plague.) The growth pattern over the past two centuries has been exponential, sustained by revolutions in agriculture, industrialization, and improvements in health care.

For a while, it would be like the Red Queen's garden in Lewis Carroll's *Through the Looking Glass,* where one is forced to run as fast as one can to remain in the same place. But what happens when the human population doubles again? Can you brush this picture aside as being too far in the future to warrant your concern? It is no farther removed from you than your own sons and daughters.

How We Began Sidestepping Controls

How did we get into this predicament? Human population growth has been slow for most of human history, but in the past two centuries, there have been astounding increases in the rate of population growth (Figure 33.7). Why has our growth rate increased so dramatically? There are three possible reasons:

1. We steadily developed the capacity to expand into new habitats and new climate zones.

2. Carrying capacities increased in the environments we already occupied.

3. We removed several limiting factors.

Let's consider the first possibility. Early human populations apparently were restricted to savannas, and they were mainly vegetarians with scavenged bits of meat added opportunistically to the diet. By 150,000 years ago, small bands of hunters and gatherers had emerged; and within another 50,000 years, hunter-gatherers had radiated through much of the world (page 218).

For most animal species, such extensive radiations did not occur nearly as rapidly. Humans were able to do so with the application of learning and memory to problems such as how to build fires, assemble shelters, create clothing and tools, and plan a community hunt to exploit the abundance of wild game. Learned experiences were not confined to individuals but spread quickly from one band to another because of language— the ability for cultural communication. (It took less than seven decades from the time we first ventured into the air until we landed on the moon.) Thus, *the human population expanded into new environments, and it did so in an extremely short time span compared with radiations of other organisms.*

What about the second possibility? About 11,000 years ago, people began to shift from the hunting and gathering way of life to agriculture—from risky, demanding moves following the game herds to a settled, more dependable basis for existence in more favorable settings. A milestone was the domestication of wild grasses, including the species ancestral to modern bread wheat and rice. Seeds were harvested, stored, and planted in one place; animals were domesticated and kept close to home for food and for pulling plows. Water was diverted into hand-dug ditches to irrigate crops. These practices increased productivity, and with a larger, more dependable food supply, the rate of human population growth increased. Towns and cities emerged, and so did the social stratification that provided a labor base. Much later, food supplies increased again with the use of fertilizers and pesticides. Thus, *even in its simplest form, management of food supplies through agriculture increased the carrying capacity for the human population.*

What about the third possibility—removing a series of limiting factors? Consider what happened when medical practices and sanitary conditions improved. Until about 300 years ago, malnutrition, contagious diseases, and poor hygiene kept the death rate relatively high (especially among infants), and this more or less balanced the birth rate. Contagious diseases (density-dependent factors) spread rapidly through crowded settlements and cities. Without proper hygiene and sewage disposal methods, and plagued with such disease carriers as fleas and rats, population size increased only slowly at first. Then plumbing and sewage treatment methods developed. Bacteria and viruses were recognized as disease agents. Vaccines, antitoxins, and drugs such as antibiotics were developed.

And consider what happened when humans discovered how to harness the energy stored in fossil fuels, beginning with coal. This discovery occurred in the mid-eighteenth century, and within a few decades, large industrialized cities emerged in Western Europe and North America. After World War I, more efficient technologies developed. Cars, tractors, and other economically affordable goods were now mass-produced in factories; the use of machines reduced the number of farmers needed to produce food. Thus, *by bringing many disease agents under control and by tapping into concentrated, existing stores of energy, humans removed some of the factors that had previously checked their population growth.*

Present and Future Growth

What are the consequences of our farflung radiations and advances in agriculture, industrialization, and health care? It took *2 million years* for the human population to reach the first billion. It took only 130 years to reach the

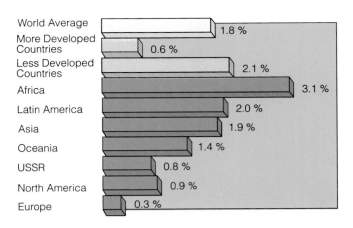

Figure 33.8 Average annual population growth rate in various groups of countries in 1989-1990.

second billion, 30 years to reach the third, 15 years to reach the fourth, *and only 12 years to reach the fifth!*

Figure 33.8 shows the annual growth rate for different parts of the world in 1990. If, as projected, the world average growth rate dips to 1.7 percent, we can expect the population to soar past 6 billion within the next decade. It may be very difficult to achieve similar increases in food production, drinkable water, energy reserves, and all the wood, steel, and other materials we use to make the planet habitable for us. There is evidence that harmful by-products of our existence—pollutants—are changing the land, seas, and the atmosphere in ominous ways (Chapter 37). From what we know of the principles governing population growth, we can realistically expect an imminent crash in our numbers; *although exponential growth continues, it is not sustainable.*

Controlling Population Growth

Today, there is widespread awareness of the links between overpopulation, resource depletion, and increased pollution. Many governments attempt to control their population size by restricting immigration from other countries; only a few (chiefly the United States, Canada, and Australia) allow large annual increases. Some attempt to reduce population pressures by encouraging emigration to other countries. But most efforts focus on decreasing the birth rate.

Two general approaches to decreasing birth rates are through economic development and family planning.

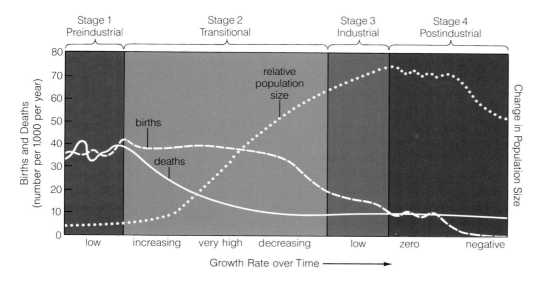

Stage 1
Preindustrial

Stage 2
Transitional

Stage 3
Industrial

Stage 4
Postindustrial

relative
population
size

births

deaths

Births and Deaths
(number per 1,000 per year)

Change in Population Size

low   increasing   very high   decreasing   low   zero   negative

Growth Rate over Time ⟶

Figure 33.9 The demographic transition model of changes in population size as correlated with changes in economic development.

The first involves providing more economic security and educational programs so that there will be less pressure on individuals to have large numbers of children to help them survive. Family planning involves educating individuals in ways to regulate when and how many children they will have.

Control Through Economics. In the **demographic transition model**, changes in population growth are linked with changes that unfold during four stages of economic development (Figure 33.9). In the *preindustrial stage,* living conditions are harsh and birth rates are high, but so are death rates; there is little population growth. In the *transitional stage,* industrialization begins, food production rises, and health care improves. Death rates drop, but birth rates remain high, so the population grows. Growth continues at high rates (2.5 to 3 percent, on the average) over a long period, but then starts to level off as living conditions improve. Growth slows in the *industrial stage,* when industrialization is in full swing, mostly because urban couples regulate family size. Many decide that raising children is expensive and having too many puts them at an economic disadvantage. In the *postindustrial stage,* zero population growth is reached. Then the birth rate falls below the death rate, and the population slowly decreases in size.

Today, the United States, Canada, Australia, Japan, the Soviet Union, New Zealand, and most countries of Western Europe are in the industrial stage, and their growth rate is slowly decreasing. Fourteen countries, including Sweden, the United Kingdom, and West Germany, are close to, at, or slightly below zero population growth.

Mexico and other less-developed countries are in the transition stage. At current growth rates, Mexico's population may reach 138 million by the year 2020. Like many countries in this stage, its huge population does not include enough skilled workers to compete effectively in today's technological markets. Fossil fuels and other resources that drive industrialization are being used up—they are not renewable—and fuel costs will become too high for countries at the bottom of the economic ladder. If population growth keeps outpacing economic growth, the death rate may increase. Thus the countries now stuck in the transitional stage won't stay there; they may well return to the harsh conditions of the preceding stage unless birth rates are controlled.

Control Through Family Planning. Family planning programs that are well conceived and carefully administered may bring about a faster decline in birth rates, at less cost, than economic development alone. Such programs vary from country to country, but all provide information on methods of birth control, as described on page 483.

Suppose family planning programs were successful beyond our wildest imagination, so that all couples decided to have only two children to replace them. (Actually, the average "replacement level for fertility" is slightly higher, for some female children die before reaching reproductive age. It is about 2.5 children per woman in less-developed countries, and 2.1 in more-

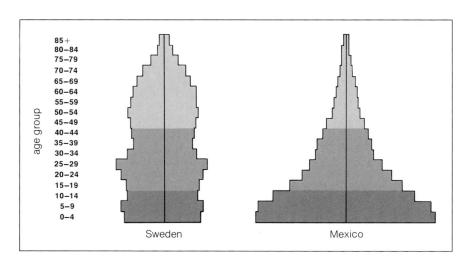

Figure 33.10 Age structure diagrams for two countries in 1977. Dark green indicates the pre-reproductive base. Purple indicates reproductive years; light blue, the post-reproductive years. The portion of the population to the left of the vertical axis in each diagram represents males; the portion to the right represents females. Mexico has a very rapid rate of increase. In 1980, Sweden showed zero population growth.

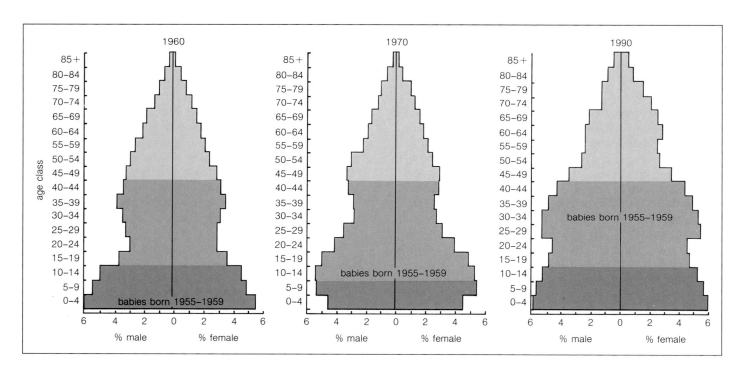

Figure 33.11 Age structure of the U.S. population in 1960, 1970, and 1990 (projected). The population bulge of babies between 1955 and 1959 will slowly move up.

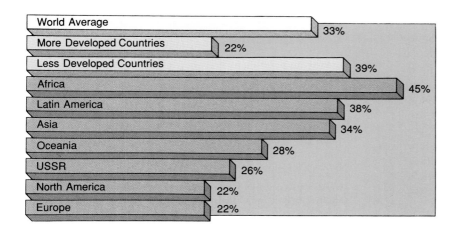

Figure 33.12 Percentage of individuals under age fifteen in various regions in 1987.

World Average 33%
More Developed Countries 22%
Less Developed Countries 39%
Africa 45%
Latin America 38%
Asia 34%
Oceania 28%
USSR 26%
North America 22%
Europe 22%

developed countries.) Even if the replacement level for fertility were achieved globally, the human population would keep on growing for at least another sixty years! Why? An immense number of already existing children will themselves be reproducing.

Take a look at Figures 33.10 and 33.11, which show age structure diagrams for three populations growing at different rates. (In these diagrams, ages 15 to 44 are used as the average range of childbearing years.) The one for Mexico, a rapidly growing population, has a broad base. It is filled not only with reproductive-age men and women but with a large number of children who will move into that category during the next fifteen years. As Figure 33.12 indicates, *more than one-third of the world population now has an age structure with a broad reproductive base.* This gives us an idea of the magnitude of the effort it will take to control population growth.

One way to slow down the birth rate is to encourage delayed reproduction—childbearing in the early thirties as opposed to the mid-teens or early twenties. This practice slows population growth by lengthening the generation time and by lowering the average number of children in each family.

China, for example, has established the most extensive family planning program in the world. Couples are strongly urged to postpone the age at which they marry. Married couples have ready access to free contraceptives, abortion, and sterilization; paramedics and mobile units ensure access even in remote rural areas. Couples who pledge not to have more than one child are given extra food, better housing, free medical care, and salary bonuses; their child will be granted free tuition and preferential treatment when he or she enters the job market. Those who break the pledge forgo benefits.

These may seem like draconian measures unless you know that family planning has been China's alternative to mass starvation. (Between 1958 and 1962 alone, an estimated 30 million Chinese died because of famine.) By 1987, the fertility rate had dropped from a previous high of 5.7 to 2.4 children per woman. Even so, the population time bomb has not stopped ticking. China's population now numbers 1.1 billion—and 340 million of its young women are about to move into the reproductive age category.

Questions About Zero Population Growth

For the human population, as for all others, the biological implications of exponential growth are staggering. Yet so are the social implications of achieving and maintaining zero population growth.

For instance, most members of an actively growing population fall in younger age brackets. Under conditions of constant growth, the age distribution means that there is a large work force. A large work force is capable of supporting older, nonproductive individuals with various programs, such as social security, low-cost housing, and health care. With zero population growth, far more people will fall in the older age brackets. How, then, can goods and services be provided for nonproductive members if productive ones are asked to carry a greater and greater share of the burden? These are not abstract questions. Put them to yourself. How much are you willing to bear for the sake of your parents, your grandparents? How much will your children be able to bear for you?

We have arrived at a major turning point, not only in our biological evolution but also in our cultural evolution. The decisions awaiting us are among the most difficult we will ever have to make, yet it is clear that they must be made, and soon.

All species face limits to growth. In one sense, we may think we are different from the rest, for our unique ability to undergo cultural evolution has allowed us to postpone the action of most of the factors that limit growth. But the key word here is *postpone*. No amount of cultural intervention can hold back the ultimate check of limited resources. We have repealed a number of the smaller laws of nature; in the process, we have become more vulnerable to those laws which cannot be repealed. Today there may be only two options available. Either we make a global effort to limit population growth in accordance with environmental carrying capacity, or we wait until the environment does it for us.

SUMMARY

1. The growth rate of a population depends on the birth rate, death rate, and the rates of immigration and emigration. If the birth rate per individual exceeds the death rate per individual by a constant amount, the population will grow exponentially (assuming immigration and emigration remain zero).

2. Carrying capacity is the number of individuals of a given species that can be sustained indefinitely in a given area. It is determined by resource availability, predation, competition, and other factors that limit population growth. Controls vary in their relative effects and over time, so carrying capacity also can change over time for a particular population.

3. Rapid growth of the human population during the past two centuries was possible largely because of our capacity to exploit new environments, and because of agricultural and technological developments that expanded the carrying capacity.

4. Some controls over population growth, such as competition for resources, disease, and predation, are density-dependent. Others tend to increase the death rate more or less independently of population density.

5. Currently, human population growth varies from zero in some of the more-developed countries to more than four percent per year in some of the less-developed countries. In 1989 the world average growth rate was 1.8 percent per year.

Review Questions

1. Why do populations that are not restricted in some way grow exponentially? *494*

2. If the birth rate equals the death rate, what happens to the growth rate of a population? If the birth rate remains slightly higher than the death rate, what happens? *494–495*

3. What defines the carrying capacity for a particular environment? Can you describe what happens when a low-density population shows a logistic growth pattern? *496*

4. At present growth rates, how many years will elapse before another billion individuals are added to the human population? *498–499*

5. How have human populations developed the means to expand steadily into new environments? How have humans increased the carrying capacity of their environments? How have they avoided some of the limiting factors on population growth? Or is the avoidance an illusion? *499–500*

6. If a third of the world population is now below age fifteen, what effect will this age distribution have on the growth rate of the human population? What sorts of humane recommendations would you make that would encourage this age group to limit the number of children they plan to have? *503*

7. Write a short essay about a hypothetical population that shows either one of the following age structures. Describe what might happen to younger and older age groups when members move into new categories. *502*

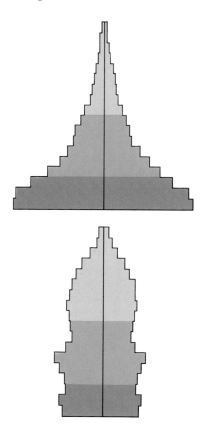

Self-Quiz (Answers in Appendix IV)

1. _ecology_ is the study of how organisms interact with one another as well as with their physical and chemical environment.

2. A _population_ is a group of individuals of the same species that occupy a certain area at the same time.

3. The rate at which a population grows or declines depends upon the _____.
   a. birth rate
   b. death rate
   c. immigration rate
   d. emigration rate
   e. all of the above

4. Populations grow exponentially when _____.
   a. birth rate remains above death rate and neither changes
   b. death rate remains above birth rate
   c. immigration and emigration rates are equal (a zero value)
   d. emigration rates exceed immigration rates
   e. both a and c combined are correct

5. The number of individuals of a species that can be sustained indefinitely in a given region is the _____.
   a. biotic potential
   b. carrying capacity
   c. environmental resistance
   d. density control

6. Which of the following factors does _not_ affect carrying capacity?
   a. predation
   b. competition
   c. available resources
   d. pollution
   e. each of the above can affect carrying capacity

7. Population growth controls such as resource competition, disease, and predation are said to be _____.
   a. density independent
   b. population sustaining
   c. population dynamics
   d. density dependent

8. At present, human population growth varies from about _____ percent in developed countries to more than _____ percent in some developing countries.
   a. four, zero
   b. zero; zero
   c. zero; four
   d. four; four

9. During the past two centuries, rapid growth of the human population has occurred largely because of _____.
   a. worldwide increased birth rate
   b. worldwide increased death rate
   c. carrying capacity reduction
   d. carrying capacity expansion

10. Match the population ecology terms.
    __d__ carrying capacity
    __e__ exponential growth
    __c__ population growth rate
    __a__ density-dependent controls
    __b__ population

    a. examples are disease and predation
    b. group of individuals of the same species occupying a given area at the same time
    c. depends on birth, death, immigration, emigration rates
    d. number of individuals of a given species that can be sustained indefinitely in a given area
    e. increases in population size by ever larger amounts per unit of time

Selected Key Terms

biosphere _493_
biotic potential _495_
carrying capacity _496_
community _493_
demographic transition model _501_
density-dependent controls _496_
density-independent controls _497_
ecology _493_
ecosystem _493_

exponential growth _495_
habitat _493_
J-shaped curve _495_
life table _498_
limiting factor _496_
logistic growth _496_
population _493_
S-shaped curve _496_
survivorship curve _498_
zero population growth _494_

Readings

Bush, M. May 1986. "The Cheetah in Peril." _Scientific American_ 254(5):84–92.

Krebs, C. 1985. _Ecology_. Third edition. New York: Harper & Row.

Miller, G. T. 1990. _Living in the Environment_. Sixth edition. Belmont, California: Wadsworth. This author consistently pulls together information on human population growth into a coherent picture.

Polgar, S. 1972. "Population History and Population Policies From an Anthropological Perspective." _Current Anthropology_ 13(2)203–241. Analyzes often-ignored cultural barriers to programs for population control.

Ricklefs, R. 1990. _Ecology_. Third edition. New York: Freeman. Chapters 15–19.

34

COMMUNITY INTERACTIONS

KEY CONCEPTS

1. Communities are associations of different populations that occupy the same habitat. We characterize a community by the kinds and diversity of species, as well as by the numbers and dispersion of their individuals through the habitat.

2. The properties of each community are influenced by rainfall, temperature, and other physical aspects of the habitat. They also are influenced by resource availability, adaptations of individuals that enable them to exploit resources, and species interactions.

3. Each species has its own niche, defined by the full range of its relations with other organisms and with its physical surroundings. Coexistence in the same habitat is often an uneasy balance, maintained partly through a partitioning of resources among species whose niches are similar.

4. The most stable state for a community in a particular habitat is reached by succession, which is the change in structure and composition leading toward the climax community.

CHARACTERISTICS OF COMMUNITIES

The Concepts of Niche and Habitat

Flying through the rain forests of New Guinea is an extraordinary pigeon with cobalt-blue feathers and plumes on its head (Figure 34.1). It is about as big as a turkey, and it flaps so slowly and noisily that its flight has been likened to the sound of an idling truck. Like eight species of smaller pigeons living in the same forest, it perches on branches to eat fruit. Why are there *nine* species of pigeons, all of which eat fruit? Wouldn't competition for food eventually leave one the winner?

In fact, each species has a different **niche** in the forest community, as defined by its relations with other

Figure 34.1 Tropical rain forest of New Guinea, habitat of many diverse species, including the turkey-size Victoria crowned pigeon and eight species of smaller pigeons. Within this habitat, each species has its own niche, as defined by the full range of its relations with other organisms and with its physical surroundings.

organisms and with its physical surroundings. The larger pigeons can perch on heavier branches when they feed, and they eat larger fruit. The smaller ones eat fruit hanging from branches too thin to support the weight of a turkey-size pigeon, and they have bills too small to open large fruit. The tree species vary in terms of fruit size and thickness of fruit-bearing branches, so the different kinds of pigeons end up foraging on different trees. Parts of the food supply used less by one kind of pigeon are used more by others in the same forest.

Leaf-eating, fruit-munching, and bud-nipping insects also have their own niches, as do nectar-drinking bats, birds, and insects that pollinate the trees. So do many decomposers living on the forest floor, where they extract energy from the remains and wastes of other

organisms and, by their metabolic activity, recycle nutrients to the trees.

This example reminds us that organisms interact as part of communities. A **community** is an association of populations, tied together directly or indirectly by way of competition for resources, predation, and other interactions. Each population in the community is adapted to living under the physical and biological conditions prevailing in a particular habitat at a particular point in time and space.

The **habitat** of an organism is the type of place where it normally lives. It is characterized by physical and chemical features, as well as by the general structure of the vegetation. Muskrats live in a streambank habitat, tree squirrels in an arboreal habitat, moles in an underground habitat. Humans live in disturbed habitats where the landscape has been deliberately altered for agriculture, urban development, and other purposes.

How many species occur in a habitat? How many individuals represent the species, and how are they dispersed through the habitat? That depends on the combined influence of these factors:

---

1. Geochemical and climatic processes that dictate rainfall, temperature, soil composition, and other physical characteristics of the habitat.

2. The kinds and amounts of food and other resources that are available throughout the year.

3. Adaptive traits that enable individuals of a species to exploit specific resources. These include evolutionary specializations in form, function, and behavior.

4. Interactions that occur among species in the habitat, including competition, predation, and mutualism.

Several community properties emerge as a result of these factors. One property is *diversity* (number of species). For example, the high humidity and temperature of tropical habitats allow the growth of many kinds of plants, which support many kinds of animals. In temperate and arctic habitats, diversity is much lower, for reasons still not fully understood. (There are more species of tall trees on a small island in a Panamanian lake than there are in all of Canada.)

Another community property is the *number of species at different "feeding levels,"* starting with producers and continuing up through different levels of consumers, such as herbivores, carnivores, and parasites. Two related properties are *relative abundances* (the number of individuals of each kind of organism) and *dispersion* (how individuals of each species are spaced in the habitat).

The next chapter describes the flow of energy through the community's feeding levels, and the chapter after that describes the geochemical properties affecting community structure. Here we will survey different types of species interactions, using the niche concept as a guide.

Table 34.1 Types of Interactions Between Two Species

Type of Interaction	Direct Effect of Interaction*	
	Species 1	Species 2
Neutral	0	0
Commensalism	+	0
Mutualism	+	+
Interspecific competition	−	−
Predation	+	−
Parasitism	+	−

* 0 indicates no direct effect on population growth,
  + indicates positive effect, − indicates negative effect.

Types of Species Interactions

The niche of each species includes the full range of physical and biological conditions under which its individuals live and reproduce. For a daytime predator, some of the conditions might be sunlight, hiding places, prey size and nutritional value, prey defenses, and so on. The conditions are not static; they shift in large and small ways over time.

When any two species in a community have some requirements or activities in common, they may interact to some degree. If they do not interact, the relationship is *neutral* (Table 34.1). Neither species directly affects the other even though they may be linked indirectly through interactions with other species. (Eagles and meadow grasses have neutral effects on each other. Eagles do eat grass-eating rabbits and so help the grasses, and the grasses help the eagles by fattening their prey, but these are both indirect interactions.)

Interactions may have good or bad consequences for one or both species. In *commensalism*, one species directly benefits from an interaction but the other is not helped or harmed much. Many birds use trees simply as roosting sites, but the tree gets nothing in return. In *mutualism*, both species directly benefit by interacting. In *interspecific competition*, the niches of two species overlap significantly and both species are harmed by the interaction. In

Figure 34.2 Mutualism in the high desert of Colorado. There are several species of yucca plants (**a**), but each is pollinated exclusively by only one kind of yucca moth species (**b**). The adult stage of the moth life cycle coincides with the blossoming of yucca flowers. Using mouthparts that have become modified for the task, the female moth gathers up the somewhat sticky pollen and rolls it into a ball. Then she flies to another flower and, after piercing the ovary wall, lays her eggs among the ovules. She crawls out the style and shoves the ball of pollen into the opening of the stigma. When larvae emerge (**c**), they eat a small portion of the yucca seeds. Then they gnaw their way out of the ovary to continue the life cycle. The seeds remaining are enough to give rise to a new yucca generation.

So refined is this mutual dependency that the moth and larva can obtain food from no other plant, and the flower can be pollinated by no other agent.

a

b

c

predation and *parasitism,* one species (the predator or parasite) benefits directly from the interaction, but the other (the prey or host) is directly harmed. Let's take a look at some examples of mutualism, competition between species, predation, and parasitism.

MUTUALLY BENEFICIAL INTERACTIONS

Chapter 22 described the evolutionary basis for the interactions between flowering plants and their pollinators. These interactions are classic examples of **mutualism,** in which positive benefits generally flow both ways between two species. Consider the yucca moth and yucca plant (Figure 34.2). The moth gets pollen only from the yucca plant; even its larval form eats only yucca seeds. The moth is the yucca plant's only pollinator. Thus, each species contributes to the other's reproductive success. The moth has a private energy source, available throughout its life. The plant has its pollen delivered to the doorstep, so to speak, instead of being spread rather haphazardly by pollinators that visit many different kinds of plants.

The interaction between the yucca plant and yucca moth is a form of **symbiosis.** The word implies that both species have become intimately and permanently dependent on each other for survival and reproduction. Pages 244 and 307 give other examples of symbiosis.

COMPETITIVE INTERACTIONS

Categories of Competition

As we saw in the preceding chapter, intraspecific competition (competition within a population of a species) is often fierce. Each individual can deplete or restrict access to some limited resource and so affect the ability of others to use it. This is not usually the case in **interspecific competition** (between species). Although requirements for different species might be similar, they differ more than they do for members of a single species, so competition is usually less intense.

There are two types of competitive interactions, regardless of whether they occur within or between species. In *exploitation competition,* all individuals have equal access to a required resource but they differ in how fast or efficiently they exploit it. In *interference competition,* certain individuals limit or prevent others from using the resource and so control access to it.

In pure exploitation competition, one species hampers the growth, survival, or reproduction of the other only indirectly, by reducing the common supply of resources. This interaction does not occur when shared resources are abundant. (If you and a friend were sharing a 10-gallon milkshake, each drinking from your own straw, you wouldn't care how fast your friend drank.) Interference in access to a resource complicates the picture. (Even if you shared a 10-gallon milkshake, you would object if your friend pinched your straw.)

Competitive Exclusion

When two species exploit the same resource, they may continue to coexist. Then again, one species may exclude the other from the habitat by interfering with its access to resources or by exploiting resources so effectively that the other starves to death. Any two species usually differ to some extent in their resource-securing adaptations, so one is likely to have the competitive edge. They are less likely to coexist in the same habitat when they are very similar in their use of scarce resources.

G. Gause tested this idea by growing two species of *Paramecium* separately, then together (Figure 34.3). Be-

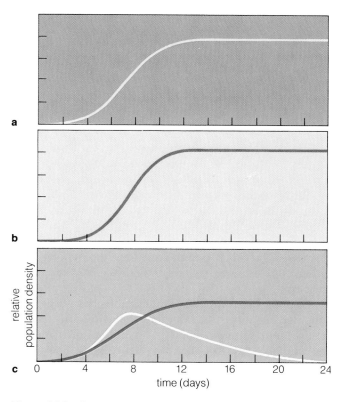

Figure 34.3 Competition between two species of *Paramecium.* When grown separately, *P. caudatum* (**a**) and *P. aurelia* (**b**) established stable populations. (**c**) When grown together, *P. aurelia* (red curve) drove the other species (gold curve) toward extinction.

Figure 34.4 (**a**) View of two adjacent habitats on the eastern slopes of the Sierra Nevada. Alpine tundra is at the highest elevation; below it, a lodgepole pine forest. (**b**) Different chipmunk species live in these and other habitats at lower elevations.

alpine chipmunk   lodgepole chipmunk   yellow pine chipmunk   least chipmunk

cause the two species exploited similar food, there was strong competition between them. The test results suggested that complete competitors cannot coexist indefinitely, a concept now called **competitive exclusion.**

What happened with two *Paramecium* species that did not overlap as much in their use of resources? (When grown together, one species fed on bacteria, the other on yeast cells in the culture tube.) The growth rate decreased for both populations—but not enough for either population to exclude the other. They continued to coexist.

Examples of competitive interactions have been documented in nature. Consider how four chipmunk species occupy four different habitats on the eastern slope of the Sierra Nevada in California (Figure 34.4). The alpine habitat is at the highest elevation; below it are the lodgepole pine habitat, the piñon pine and sagebrush habitat, and at the base of the mountains, the sagebrush habitat. The "least chipmunk" of the sagebrush habitat could actually live at higher elevations among the pines (it does so elsewhere). But the aggressively competitive behavior of the "yellow pine chipmunk" in the adjacent

habitat keeps it from doing so. (Dietary habits keep the yellow pine chipmunk out of the sagebrush.)

Species that compete for a limited resource cannot coexist indefinitely in the same habitat; differences in their adaptive traits will give one the competitive edge.

CONSUMER-VICTIM INTERACTIONS

"Predator" Versus "Parasite"

Of all community interactions, predation most rivets our attention—as well as the prey's, as Figure 34.5 suggests. A goat pulling up a thistle plant for breakfast, although less dramatic, is also a predator (its prey is a living organism, killed for food). But what about a horse grazing on but not killing plants? What about a mosquito taking blood from your arm before it flies off? What about ticks or fleas taking blood for long periods before they hop off or drop off their host and lay their eggs elsewhere? What about tapeworms (parasitic animals) and mistletoe (parasitic plants) that remain with the host?

For simplicity's sake, we will use only two broad definitions for all consumer-victim interactions. A **predator** gets food from other living organisms (its *prey*), which it may or may not kill, but it does not live on or in them. A **parasite** also gets food from other living organisms (its *hosts*), which it may or may not kill, but they do live on or in the host organism for much of their life.

Dynamics of Predator-Prey Interactions

Predator and prey populations interact in diverse ways. Some interactions lead to stable coexistence at steady population levels for both species, others to recurring cycles of abundance and population crashes, erratic population cycles, and predator or prey extinction. The outcome of the interaction is influenced by these factors:

1. Carrying capacity for the prey population, in the absence of predation.

2. Reproductive rates of the predator and prey.

3. The adaptive capacity of individual predators to respond to increases in prey density.

Stable coexistence tends to occur when predation keeps the prey population from overshooting its carrying capacity. This happens when predators can reproduce quickly and are capable of eating more when there are more prey organisms around. Fluctuations in population density tend to occur when predators do not reproduce as fast as their prey, when they can eat only so many prey organisms at a time no matter how many are around, and when the carrying capacity for the prey is high.

Figure 34.5 shows an idealized cycle of predator and prey abundance, caused by time lags in the predator's response to changes in prey abundance. There is indeed a correspondence between the rise and fall of predator and prey populations. But long-term studies of the Canadian lynx and snowshoe hare indicate that the cycling may be caused by more than predation alone.

Figure 34.5 Idealized cycling of predator and prey abundances. (The scale exaggerates predator density; predators usually are less common than their prey at all points in the cycle.)

The pattern arises through time lags in predator responses to changes in prey abundance. Starting at time *a*, prey population density is low, so predators are hungry and their population is declining. In response to the decline, prey start increasing, but the predator population does not start increasing until reproduction gets under way (time *b*).

Both populations grow until predation causes the prey population to decline (time *c*). Predators continue to increase and take out more prey animals, but the lower prey density leads to starvation among them and their growth rate slows (time *d*). At time *e*, a new cycle begins.

a

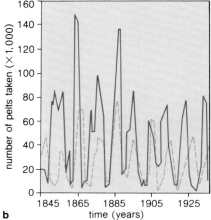

b

Figure 34.6 Predator-prey interactions between the Canadian lynx and snowshoe hare (**a**). The relative abundances of both populations, shown in (**b**), are based on counts of pelts that trappers sold to Hudson's Bay Company over a ninety-year period.

This figure is a good test of how willing you are to accept conclusions without questioning their scientific basis. (Remember the discussion of scientific methods in Chapter 1?) What other limiting factors could have influenced the relative abundances? Did climate vary greatly, with more rigorous winters imposing higher death rates? Do the prey counts take into consideration owls, martens, and foxes, which also prey on hares? Did trapping decrease during times of Indian uprisings? Were there major changes in the vegetation upon which the hares feed? (*See text* for some possible explanations.)

Take a look at the diagram in Figure 34.6. When the hare population density is at its peak, lynx usually are not abundant enough to cause an immediate drop in the prey population (compare the brown line for the hares with the blue line for the lynx). Lynx do not instantly reproduce and mature into efficient hunters, and it takes time for the mature hunters to migrate into areas where prey have become abundant.

If predation does not *trigger* the decline in prey abundance, what does? Apparently, recurring wildfires, floods, and insect outbreaks indirectly influence the hare-lynx cycle. These events destroy mature trees, creating favorable conditions for other shrubs and trees that are adapted to moving into exposed or disturbed habitats. Hares browse intensively on the newly estab-lished plants, which include alder, poplar, black spruce,

and birch species. In fact, the peaks in prey density correspond to the new plant growth. As it happens, the shoots of those new plants contain large amounts of toxins—which are density-dependent limiting factors for the hare population. The hares prefer to feed on harmless plant parts. But when their population density becomes high, the harmless parts are eaten up and the hares are forced to feed on toxic shoots. Hares ingesting large quantities of plant toxins rapidly lose weight, they become severely stressed, and they are more vulnerable to predation. The conditions may last long enough to trigger a decline in the hare population.

Thus, the lynx-hare cycle may be an outcome of hares preying on plants; perhaps the lynx merely amplify the cycle. With their built-in chemical defenses, the plants certainly help control increases in prey density.

Prey Defenses

Predators and prey exert continual selection pressure on each other. When some new, heritable means of defense spreads through a prey population, predators must be equipped to counter the defense or they won't eat. *When the prey evolves, the predator also evolves to some extent because the change affects selection pressures operating between the two.* This is an example of **coevolution,** the joint evolution of two (or more) species that are interacting closely in some ecological fashion. Let's take a look at some of the outcomes.

Warning Coloration and Mimicry. Predation pressure has favored the evolution of prey that are bad-tasting, toxic, or able to inflict pain (as by stingers) on their attackers. The conspicuous colors and bold patterns of many toxic prey species serve as warning signals to predators. Inexperienced predators might attack a prominently striped skunk, bright-orange monarch butterfly, or yellow-banded wasp—once. They quickly learn to associate the colors and patterning with pain or digestive upsets.

Weaponless prey species often have warning colors very similar to those of bad-tasting, toxic, or dangerous species. The resemblance of an edible species to a relatively inedible one is a form of **mimicry** (Figure 34.7).

Moment-of-Truth Defenses. When cornered, some prey animals defend themselves by startling or intimidating the predator with display behavior (Figure 34.5). Their behavior may create a moment of confusion, and a moment may be all it takes for a getaway. A bombardier beetle under attack sprays a noxious chemical. The adaptation works some of the time but not against grasshopper mice, which have learned to pick up the beetle, shove its tail end into the earth, and munch the head end.

a

b

c

d

Figure 34.7 Mimicry. Many animals avoid predation by having a bad taste, obnoxious secretion, or painful bite or sting. Each young predator learns about these traits the hard way, by unpleasant trials. If dangerous or unpalatable prey animals were not easy to recognize and remember, many would be lost while inexperienced predators were learning their lessons. Thus, they tend to have bright colors and bold markings, and often they do not even bother hiding.

Many less related—even totally unrelated—species avoid predation by mimicking the appearance and behavior of dangerous or unpalatable species. The aggressive yellowjacket shown in (**a**) is the probable model for similar-looking but edible flies (**b**), beetles (**c**), and wasps (**d**).

Figure 34.8 The fine art of camouflage. (**a**) Find the scorpion fish, a dangerous predator that lies motionless and camouflaged on the sea bottom, the better to surprise unsuspecting prey. (**b**) Find the plants (*Lithops*) that hide in the open from herbivores; they have the form, pattern, and coloring of stones. (**c**) A yellow crab spider lurked motionless against a yellow background and so escaped detection by this prey tidbit. (**d**) A caterpillar looks like an unappetizing bird dropping by virtue of its coloration and body positioning. (**e**) What bird??? With the approach of a predator, the least bittern stretches its neck (colored much like the surrounding withered reeds), thrusts its beak upward, and sways gently like reeds in the wind.

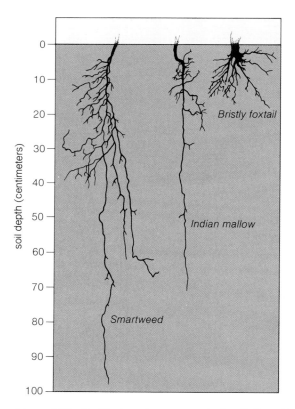

Figure 34.9 Partitioning of a resource (soil, with its nutrients and water) by three annual plant species that became established in a field plowed under the year before.

Chemicals produced by many plant and animal species serve as warning odors, repellents, and outright poisons. Earwigs, skunks, and skunk cabbages produce awful odors. Tannins in the foliage and seeds of certain plants taste bitter and make the plant tissues hard to digest. Nibble on a buttercup (*Ranunculus*) and you will badly irritate the lining of your mouth.

Camouflage. Predation pressure has also favored the evolution of prey that can hide in the open; they **camouflage** themselves. Adaptations in form, patterning, color, or behavior help the organism blend with its surroundings and escape detection. For example, one desert plant looks like a small rock (Figure 34.8). It flowers only during a brief rainy season, when other plants and water are available for the plant eaters. Camouflage, of course, is not the exclusive domain of prey. Stealthy predators also blend well with their backgrounds.

Parasitic Interactions

True Parasites. The blood flukes and tapeworms described in Chapter 19 are examples of true **parasites,** which live on or in a host organism and gain nourishment by tapping into its tissues. Sometimes parasites indirectly cause death, as when the host becomes so weakened by the attack that it succumbs to secondary infections.

e

over other insects in a community. In fact, many parasitoids have been successfully raised and released as an alternative to using chemicals to control pests. The alternative is promising, for many of the worst outbreaks of pests have *followed* applications of pesticides. The pest population soon recovers, but the applications also kill parasitoids—which help keep pest population density under control and which usually fail to recover as quickly.

COMMUNITY ORGANIZATION, DEVELOPMENT, AND DIVERSITY

Community stability is the result of forces that have come into balance—sometimes an uneasy balance. Resources are sustainable, as long as populations do not start dancing dangerously around their carrying capacity. Predators and prey coexist only as long as neither wins. Competitors have no sense of fair play (straw-pinching is common). Even mutualists are really antagonists; a flower gives as little nectar or pollen as necessary to attract a pollinator, and a pollinator takes as much nectar or pollen as it can for the least effort. Let's take a look at some community patterns arising from these conflicting forces.

Resource Partitioning

Think back on those nine species of fruit-eating pigeons in the same forest in New Guinea. In any community, similar species generally share the same kind of resource in different ways, in different areas, or at different times. This community pattern, called **resource partitioning,** arises in two ways. *First,* selective agents and other evolutionary forces act on established and competing populations, increasing the differences among them. *Second,* only species that are dissimilar from established ones can succeed in joining an existing community.

Consider how three species of annual plants partitioned resources (sunlight, water, and dissolved minerals) in a plowed, abandoned field. Each species exploited different parts of the habitat. Where soil moisture varied from day to day, foxtail grasses became established; their shallow, fibrous root systems absorb rainwater rapidly and help the plants recover from drought. Where deeper areas of soil were moist early in the growing season but drier later on, mallow plants took hold; their taproot system grows deeper in the soil. Where soil was continuously moist, smartweed prevailed; its taproot system branches in topsoil *and* in soil below the roots of the other species (Figure 34.9).

Generally, however, the attack causes death only when the parasite infects a novel host population, one with no coevolved defenses against it. A host that can live longer may spread about far more parasites than a vulnerable, rapidly dying host ever could. So parasite populations also tend to coevolve with their hosts, producing less-than-fatal effects.

Parasitoids. Unlike true parasites, **parasitoids** are insect larvae that kill their host by completely consuming its soft tissues while the larvae grow toward adulthood. Parasitoids sound horrendous; fortunately, their target hosts are not humans but the young stages of other insect species. They actually serve as natural controls

Hello Lake Victoria, Goodbye Cichlids

As the human population continues to grow exponentially, we keep looking for new ways to manage or control the populations of edible plants and animals that serve as our food production base. Although such efforts are well intentioned, they can have disastrous consequences when ecological principles are not taken into account.

Consider what happened several years ago, when someone thought it would be a great idea to introduce the Nile perch into Lake Victoria in East Africa. People had been fishing there for thousands of years using simple, traditional methods, but now they were overfishing the lake. In time there would not be enough fish to feed the local populations and no excess catches to sell for profit. But Lake Victoria is a very big lake, the Nile perch is a very big fish (more than 2 meters long), and this seemed an ideal combination for commercial fishermen from the outside, with their big, elaborate nets—right? Wrong.

The native fishermen had been harvesting a variety of native fishes called cichlids. Cichlids eat mostly detritus and aquatic plants, but the Nile perch eats other fish—including cichlids. Although the cichlids were native to the lake, the Nile perch was not. Having had no prior evolutionary experience with the perch, the cichlids simply had no defenses against it.

And so the Nile perch was able to eat its way through the cichlid populations and destroy the natural fishery. Dozens of cichlid species found nowhere else on earth became extinct. And the Nile perch, by wiping out its own food source, ended the basis for its own population growth and ceased to be a potentially large, exploitable food source for people living around the lake.

As if that weren't enough, the Nile perch is an oily fish. Unlike cichlids, which can be sun-dried, it required preservation by smoking—and smoking required firewood. And so the people started whacking away more rapidly at the trees in the local forests, which are not rapidly renewable resources. To add final insult to injury, the people living near the shores of Lake Victoria never liked to eat Nile perch anyway; they preferred the flavor and texture of cichlids.

What is the take-home lesson? A little knowledge and some simple experiments in a contained setting could have prevented the whole mess at Lake Victoria.

Effects of Predation on Competition

By reducing prey population densities, predation can reduce competition among prey species and promote their coexistence. Robert Paine kept sea stars out of experimental plots in a rocky intertidal zone for several years. (He left sea stars and their prey—fifteen invertebrate species—in control plots.) In the predator's absence, the number of prey species declined to eight. The strongest competitors were mussels—the main prey of sea stars.

Species Introductions

Many species have been introduced to different geographic regions, some intentionally and others by accident. In the 1880s, the water hyacinth from South America was put on display at the New Orleans Cotton Exposition. Flower fanciers from Florida and Louisiana carried home clippings of the blue-flowered plants and set them out for ornamental display in ponds and streams. Unchecked by their natural predators, the fast-growing hyacinths spread through the nutrient-rich

Table 34.2 Effects of Introducing a Few Species into the United States

Species Introduced	Origin	Mode of Introduction	Outcome
Water hyacinth	South America	Intentionally introduced (1884)	Clogged waterways; shading out of other vegetation
Dutch elm disease: The fungus *Ophiostoma ulmi* (the disease agent)	Europe	Accidentally imported on infected elm timber used for veneers (1930)	Destruction of millions of elms; great disruption of forest ecology
Bark beetle (the disease carrier)		Accidentally imported on unbarked elm timber (1909)	
Chestnut blight fungus	Asia	Accidentally imported on nursery plants (1900)	Destruction of nearly all eastern American chestnuts; disruption of forest ecology
Argentine fire ant	Argentina	In coffee shipments from Brazil? (1891)	Crop damage; destruction of native ant communities; mortality of ground-nesting birds
Camphor scale insect	Japan	Accidentally imported on nursery stock (1920s)	Damage to nearly 200 species of plants in Louisiana, Texas, and Alabama
Japanese beetle	Japan	Accidentally imported on irises or azaleas (1911)	Defoliation of more than 250 species of trees and other plants, including commercially important species such as citrus
Carp	Germany	Intentionally released (1887)	Displacement of native fish; uprooting of water plants with loss of waterfowl populations
Sea lamprey	North Atlantic Ocean	Through Erie Canal (1860s), then through Welland Canal (1921)	Destruction of lake trout and lake whitefish in Great Lakes
European starling	Europe	Released intentionally in New York City (1890)	Competition with native songbirds; crop damage; transmission of swine diseases; airport runway interference; noisy and messy in large flocks
House sparrow	England	Released intentionally (1853)	Crop damage; displacement of native songbirds; transmission of some diseases
European wild boar	Russia	Intentionally imported (1912); escaped captivity	Destruction of habitat by rooting; crop damage
Nutria (large rodent)	Argentina	Intentionally imported (1940); escaped captivity	Alteration of marsh ecology; damage to earth dams and levees; crop destruction

After David W. Ehrenfeld, *Biological Conservation*, 1970, Holt, Rinehart and Winston and *Conserving Life on Earth*, 1972, Oxford University Press.

waters and displaced many native species. In time they choked off ponds and streams, then rivers and canals. They are still thriving—now as far west as San Francisco—and they are still bringing river traffic in many areas to a halt. The *Commentary* describes another species introduction that had unforeseen and truly awful consequences.

Species successfully introduced into established communities do not always lead to such wholesale disasters, but few (if any) are without ecological consequences. As an example, "imported" honeybees have become part of existing communities in the United States, but they have displaced native bees in many areas. Aggressive African bees, accidentally released in South America, are about to invade Texas as this is being written, and their habit of group-stinging cattle and humans is causing some consternation. Table 34.2 lists other introduced species and their effects.

Succession

The potential repercussion of newly introduced species in a community raises an interesting question. How do communities come to exist in the first place? New communities may arise in habitats initially devoid of life, such as newly forming volcanic islands, or in disturbed but previously inhabited areas, such as abandoned pastures. Through a process called **succession,** the first species in the habitat thrive, then are replaced by other species, which are replaced by others in orderly progression until the composition of species becomes steady under prevailing conditions. This more or less stable array of species is the **climax community.**

Primary Succession. In *primary* succession, changes begin when pioneer species colonize a barren habitat, such as a recently deglaciated region (Figure 34.10). **Pioneer species** are adapted to growing in exposed, often windy areas with intense sunlight, wide swings in air temperature, and soil deficient in nitrogen and other nutrients. Pioneers typically are small plants with short life cycles. Each year they produce an abundance of small spores or seeds, which are quickly dispersed.

Once pioneers are established, they improve living conditions for other species—and commonly set the stage for their own replacement. For example, many are symbiotic with nitrogen-fixing soil microbes, and the gradual accumulation of plant litter adds nitrogen to the soil. Also, pioneers form low-growing mats that can shelter seeds of later species without shading out the seedlings.

Each new season at higher latitudes, many of the early successional species must start growing from scratch; they are annuals. This puts them at a disadvantage when perennial species, which dominate later successional stages, start to invade the habitat. Because perennials live for more than one growing season, they have a head start and can begin seasonal growth before many of the pioneers. They store more biomass and produce fewer seeds—but the seeds have nutrient reserves for the early growth period. In time, later successional species crowd out the pioneers, whose seeds travel as fugitives on the wind or water—destined, perhaps, for a new but equally temporary habitat.

Secondary Succession. In *secondary* succession, a community progresses back toward the climax state after parts of the habitat have been disturbed. This pattern of change occurs in abandoned fields and parts of established forests where falling trees or other disturbances have opened the canopy of leaves, letting sunlight reach the forest floor. Many plants arise from seeds or even seedlings already present when the process begins. (This is not the case in primary succession.)

a

b

c

d

Figure 34.10 Primary succession in the Glacier Bay region of Alaska (**a**), where changes in newly deglaciated regions have been carefully documented. A comparison of maps from 1794 onward shows that ice has been retreating at annual rates ranging from 3 meters (at the glacier's sides) to a phenomenal 600 meters at its tip over bays. (**b**) When a glacier retreats, the constant flow of meltwater tends to leach the newly exposed soil of minerals, including nitrogen. Less than ten years ago, the soil here was still buried below ice. (**c,d**) The first invaders of these nutrient-poor sites are the feathery seeds of mountain avens (*Dryas*), drifting over on the winds. Mountain avens is a pioneer species that benefits from the nitrogen-fixing activities of symbiotic microbes. It grows and spreads rapidly over glacial till.

(**e**) Within twenty years, young alders take hold. These deciduous shrubs also are symbiotic with nitrogen-fixing microbes. Young cottonwood and willows also become established (**f**). In time, alders form dense thickets (**g**). As the thickets mature, cottonwood and hemlock trees grow rapidly, as do a few evergreen spruce trees. (**h**) By eighty years, the spruce crowd out the mature alders. (**i**) In areas deglaciated for more than a century, dense forests of Sitka spruce and western hemlock dominate. By this time, nitrogen reserves are depleted, and much of the biomass is tied up in peat: excessively moist, compressed organic matter that resists decomposition and that forms a thick mat on the forest floor.

Figure 34.11 Surtsey, a volcanic island, at the time of its formation. Such islands are natural laboratories for population ecologists. The chart shows the number of species of mosses and vascular plants recorded on the new island from 1965 to 1973.

In secondary succession on land, both early and late species often are able to grow under prevailing conditions, but the later species simply are growing more slowly. In time, they will exclude the others through competition. In other cases, the early successional species might inhibit the growth of later ones, which only prevail if some disturbance removes the established competitors.

Cyclic Replacements. Even the most stable climax communities have successional patches brought about by major and minor disturbances. Winds, fires, insect infestations, and overgrazing all modify and shape the direction of succession by encouraging some species and eliminating others in different parts of the habitat. In fact, community stability over a broad area often requires episodes of *local instability* that permit the cyclic replacement of dominant species.

Giant sequoia trees grow in isolated groves in the Sierra Nevada. Some are more than 4,000 years old, so this type of climax community certainly has been stable over time. Stability is assured partly by brush fires that sweep through the forests every so often. Sequoia seeds germinate only in the absence of smaller, shade-tolerant plant species. Too much litter on the forest floor inhibits their germination. Modest fires eliminate trees and shrubs that compete with the sequoias but do not damage the sequoias themselves. Mature sequoias have very thick bark, which burns poorly and insulates them against moderate heat damage.

Fires once were prevented in many sequoia groves in national and state parks—not just accidental fires from campsites and discarded cigarettes, but also natural fires touched off by lightning. Litter builds up when small fires are stopped, fire-susceptible species take hold, and dense underbrush forms. Sequoia seeds cannot germinate in the underbrush—which also is fuel for hotter fires that can damage the giants. Controlled fires are now being set to eliminate underbrush and promote the conditions necessary for cyclic replacements in the climax community.

Episodes of instability, which permit the replacement of dominant species, often play a key role in maintaining a climax community over time.

Patterns of Species Diversity

Island Patterns. In 1965, a volcanic eruption formed a new island southwest of Iceland. Bacteria, fungi, seeds, flies, and some seabirds were established there within six months. The first vascular plant appeared after two years and the first moss two years after that. Soil conditions improved, and the number of plant species continued to increase. All species were colonists from Iceland; none evolved on the island, which was named Surtsey (Figure 34.11).

The number of species of plants and animals on Surtsey will not increase indefinitely. What will stop it? First, islands distant from source areas receive few colonizing species, and the ones that do arrive are adapted for long-distance dispersal. Second, larger islands tend to support more species than smaller islands at equivalent distances from source areas. The larger ones are physically more complex and often higher above sea level, so more kinds of habitats are available. Since many species live only in certain kinds of habitats, habitat diversity promotes species diversity. Also, larger islands probably "intercept" more colonists, being bigger targets.

Conversely, extinctions probably keep species diversity lower on small islands. There, populations are necessarily smaller—and more vulnerable to storms, volcanic eruptions, infectious diseases, and other stresses.

Mainland and Marine Patterns. The most striking patterns of species diversity on land and in the seas relate to distance from the equator. Most often, the number of coexisting species declines steeply as we move from the tropics to temperate and arctic regions. (Arctic regions get less rainfall and less direct sunlight through the year.)

Why is diversity more favored in the tropics? First, rainfall and direct sunlight are available throughout the year, so resource availability is more constant. For example, different kinds of trees can grow new leaves, flower, and fruit at different times during the year and so support different sets of herbivores, nectar foragers, and fruit consumers. In arctic regions, such specializations would be suicide.

Second, species diversity is self-reinforcing. When more plant species coexist, more species of herbivores emerge, partly because no one herbivore can overcome the chemical defenses of all plants. More kinds of predators and parasites evolve in response to the diversity of prey and hosts. The same is true of diversity on tropical reefs.

Finally, the overall rate of speciation in the tropics has long exceeded the rate of extinction from natural causes, whereas periodic mass extinctions at higher latitudes have helped to keep diversity low there. But keep in mind that millions of species in tropical forests may disappear in the next decade, for reasons that will be described in the chapter to follow.

SUMMARY

1. A community is an association of interacting populations of different species that live together. The populations are tied together directly or indirectly by way of mutualism, competition for resources, predation, parasitism, and other interactions.

2. Each species has its own niche, as defined by the full range of its relations with other organisms and with its physical surroundings. A habitat is the place where a particular species (or individual) lives.

3. When a resource is limited, two species that both require it are likely to compete, either by using up the resource as rapidly or efficiently as possible (exploitation) or by preventing the other from using it (interference). Competing species are more likely to coexist when their niches are not too similar.

4. The evolutionary races between predators and prey have produced a rich array of special adaptations for capture and escape. These include behavioral displays, chemical weapons, mimicry, and camouflage.

5. Coexistence in a community is often an uneasy balance. The balance is maintained in part because related species generally differ in how they use available resources; they partition what's available. Also, by reducing prey densities, predation reduces competition among different prey species and so promotes their coexistence.

6. Succession is the replacement of species in an orderly progression to the climax community, the most stable state for a particular habitat.

7. The number of species occupying an area is affected by the size of the area, the rate of colonization by new species, and the rate of extinction among existing species.

Review Questions

1. Define habitat. Why do you suppose it might be difficult to define "the human habitat"? *507*

2. What is the difference between the habitat and the niche of a species? *508*

3. Define mutualism and give a few examples of its occurrence in nature. *509*

4. Define interspecific competition. Explain how this form of behavior is incorporated into the concept of competitive exclusion. *509*

5. How might two species that compete for the same resource coexist? Can you think of some possible examples besides the ones used in the chapter? *509*

6. What effect does predation have on interspecific competition among prey species? *516*

7. Define coevolution. How might two species coevolve to the extent that they enter a mutualistic relationship? *513*

8. What is the difference between camouflage and mimicry? Can you give some examples of mimicry among insects? Camouflage among insects? *513–514*

9. Define primary and secondary succession. What is a climax community? *518*

Self-Quiz *(Answers in Appendix IV)*

1. _____ are associations of different populations that occupy the same habitat.

2. Each species has its own _____, defined by the full range of its relations with other organisms and with its physical surroundings.

3. An organism's _____, the type of place where it normally lives, is characterized by chemical features, physical features, and the general structure of the vegetation.

4. The properties of a community are influenced by _____.
   a. geochemical and climatic processes
   b. the amount and kinds of available resources
   c. adaptations of individuals for exploiting resources
   d. interspecific competition
   e. all of the above

5. When two species both require the same resource, _____ occurs.
   a. mutualism
   b. predation
   c. parasitism
   d. competition

6. Species competing for resources are most likely to coexist when their niches _____.
   a. are identical
   b. are very similar but not identical
   c. overlap almost completely
   d. are not too similar

7. Chemical weapons, mimicry, and camouflage are adaptations that _____.
   a. enhance mate selection
   b. predators and prey use for capture and escape
   c. are strictly for defensive behavior
   d. all of the above

8. Coexistence between organisms in a community is at best precarious; the balance is maintained mostly because _____.
   a. even closely related species differ in how they utilize available resources
   b. closely related species partition available resources
   c. prey densities can be reduced to reduce competition among different prey species
   d. all of the above are reasons coexistence is maintained

9. The most stable state for a particular habitat is the _____.
   a. pioneer community
   b. bare community
   c. climax community
   d. any early successional stage

10. Match the community interaction terms.
   _____ community
   _____ exploitation
   _____ succession
   _____ interference
   _____ climax

   a. orderly progression of community change
   b. most stable community configuration
   c. rapid use of resources by one species that is in competition with another
   d. prevention of resource use by one species that is in competition with another
   e. associations of different populations occupying the same habitat

Selected Key Terms

camouflage *514*
climax community *518*
coevolution *513*
commensalism *508*
community *507*
competitive exclusion *510*
habitat *507*
interspecific competition *509*
mimicry *513*
mutualism *508*
niche *506*

parasite *510*
parasitism *509*
pioneer species *518*
predation *509*
predator *510*
prey *510*
primary succession *518*
resource partitioning *515*
secondary succession *518*
succession *518*
symbiosis *509*

Readings

Barbour, M., J. Burk, and W. Pitts. 1987. *Terrestrial Plant Ecology.* Second edition. Menlo Park, California: Benjamin-Cummings.

Moore, P. 1987. ''What Makes a Forest Rich?'' *Nature* 329:292.

Ricklefs, R. 1990. *Ecology.* Third edition. New York: Freeman. Chapters 32–36 provide an excellent introduction to community ecology.

Sattaur, O. 1988. ''A New Crop of Pest Controls.'' *New Scientist* 14:48–54.

Smith, R. 1986. *Ecology.* Second edition. New York: Harper & Row.

35

1. An ecosystem is a complex of organisms and their physical environment, linked by the flow of energy and the cycling of materials. It is an open system, with inputs and outputs of both energy and nutrients.

2. There is a one-way flow of energy through an ecosystem, most often beginning when photosynthetic autotrophs harness sunlight energy and convert it to forms that they and other organisms of the ecosystem can use. These "primary producer" organisms directly or indirectly nourish some array of consumers, decomposers, and detritivores, which interact as part of grazing or detrital food webs.

3. Water and nutrients move from the physical environment, through organisms, then back to the environment. These movements, called biogeochemical cycles, are global in scale. Human activities are disrupting these movements and so are affecting ecosystem stability.

CHARACTERISTICS OF ECOSYSTEMS

The earth's surface is remarkably diverse. In climate, soils, vegetation, and animal life, its deserts differ from hardwood forests, which differ from tropical rain forests, prairies, and arctic tundra. Oceanic provinces, lakes, and rivers differ in their physical properties and arrays of organisms. *Yet despite the differences, each region functions as a system in much the same way as the others.*

With few exceptions, each system runs on energy from the sun. Plants and other photosynthetic autotrophs, the most common "self-feeders," capture sunlight energy and convert it to forms they can use to build organic compounds from simple inorganic substances. By securing energy from the physical environment, autotrophs serve as the **producers** for the entire system (Figure 35.1).

All other organisms in the system are heterotrophs, not self-feeders. They depend directly or indirectly on energy stored in the tissues of producers. Some heterotrophs are **consumers,** which feed on the tissues of other organisms. (The ones called *herbivores* eat plants,

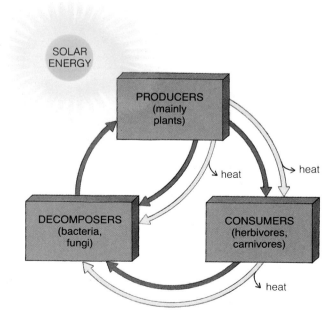

Figure 35.1 Generalized model of the one-way flow of energy and the cycling of materials through ecosystems.

a Summer in the arctic tundra

b Some of the producers of the tundra

c A ptarmigan (consumer)

d A lemming (another consumer)

Figure 35.2 Some of the producers, consumers, and decomposers of the arctic tundra—sedges, mosses, and other plants, along with the lemming (eater of plant parts); the snowy owl (eater of lemmings), and a fungal decomposer. As in all ecosystems, these organisms interact with one another and with the physical environment through a one-way flow of energy and a cycling of materials.

f Mushroom (a decomposer)

e Snowy owl (consumer)

carnivores eat animals, *parasites* reside in or on living hosts and extract energy from them, and *omnivores* partake of a variety of edibles.) Other heterotrophs, the **detritivores,** get energy from partly decomposed particles of organic matter. They include crabs, nematodes, and earthworms. Still other heterotrophs, the **decomposers,** include fungi and bacteria that extract energy from the remains or products of organisms (Figure 35.1).

Autotrophs secure nutrients as well as energy for the entire system. During growth, they take up water and carbon dioxide (as sources of oxygen, carbon, and hydrogen) along with dissolved minerals, including nitrogen and phosphorus. Such materials are building blocks for carbohydrates, lipids, proteins, and nucleic acids. When decomposers and detritivores get their turn at this organic matter, they can break it down completely to inorganic bits. If those bits are not washed away or otherwise removed from the system, they can be used again as nutrients by the autotrophs.

What we have just described in broad outline is an ecosystem. An **ecosystem** is a whole complex of organisms interacting with one another *and* with the physical environment through (1) a flow of energy and (2) a cycling of materials. Figure 35.2 shows some participants in one kind of ecosystem.

It is important to understand that ecosystems are *open* systems, and so are not self-sustaining. They require an *energy input* (as from the sun) and often *nutrient inputs* (as from minerals carried by erosion into a lake). Because

energy cannot be recycled, all ecosystems have *energy output,* usually as low-grade energy (heat). In time, most of the energy originally fixed by autotrophs is lost to the environment as metabolically generated heat. Although nutrients are typically recycled, some loss occurs from the system (as through soil leaching), so there also is a *nutrient output.* In this chapter, we consider the inputs, internal transfers, and outputs of ecosystems.

An ecosystem is a whole complex of producers, consumers, decomposers, and detritivores *and* the physical environment, interacting through energy flow and materials cycling.

STRUCTURE OF ECOSYSTEMS

Trophic Levels

The feeding relationships in all ecosystems are structured in much the same way. Each member of the system fits somewhere in a hierarchy of energy transfers called **trophic levels** (from *troph,* meaning nourishment). "Who eats whom?" we can ask. When organism B eats organism A, energy is transferred from A to B. All organisms at a given trophic level are the same number of transfer steps away from the energy input into the system.

Primary producers are the members closest to the initial energy source, so they make up the first trophic

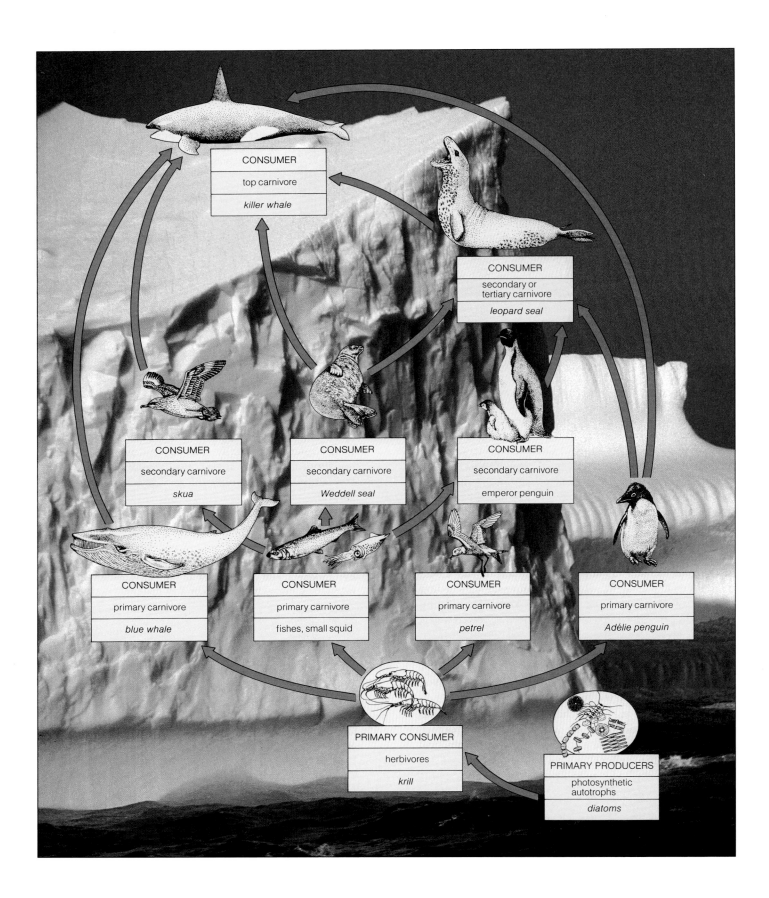

level. Photosynthetic autotrophs in a lake (including cyanobacteria and aquatic plants) are an example. Rotifers, snails, and other herbivores feeding directly on the producers are at the next trophic level. Birds and other primary carnivores that prey directly on the herbivores are at still another level.

Decomposers, humans, and other organisms can obtain energy from more than one source. They cannot be assigned to a particular trophic level; they are more like "trophic groups." Even so, the categories in Table 35.1 are a useful starting point for understanding feeding relationships.

Food Webs

The general sequence of who eats whom is sometimes called a **food chain,** but you will have a hard time finding such a simple, isolated sequence in ecosystems. Typically, the same food resource is part of more than one chain; this is especially true of resources at low trophic levels. The interconnected feeding relationships take the form of **food webs,** of the sort shown in Figure 35.3.

To understand the difference between a food chain and food web, think about a fisherman who nets some fish that were feeding on algae near the ocean's surface. Come lunchtime, he cooks some of the catch but later loses his footing and falls into the water, where other carnivores lurk. You might think this is just a simple food chain:

algae → fish → fisherman → shark

But the chain excludes alternative feeding relationships. Crustaceans also were grazing on the algae, small squids and mid-sized fishes were feeding on the crustaceans, and some larger fishes were feeding on the smaller ones. The sharks may have been moving in to feed on the large and mid-sized fishes. The fisherman, who cooked the fish in wine and herbs, shifted back and forth between herbivore and carnivore. He was even more omnivorous when he drank the alcoholic product of decomposers (yeasts whose fermentation activities yield wine from crushed grapes).

A food web is a network of crossing, interlinked food chains, encompassing primary producers and an array of consumers and decomposers.

Table 35.1   Examples of Trophic Levels

Members of Each Level	Energy Source	Representative Organisms
Primary producers:		
Photosynthetic autotrophs	Sunlight energy	Grasses, diatoms
Chemosynthetic autotrophs	Oxidation of inorganic substances	Nitrifying bacteria
Primary consumers:		
Herbivores	Primary producers	Grasshoppers, deer, krill
Secondary consumers:		
Primary carnivores	Herbivores	Spiders, foxes, small squids
Tertiary consumers:		
Secondary carnivores	Primary carnivores	Emperor penguins

Figure 35.3 (*Left*) Simplified picture of a food web in the Antarctic; there are many more participants, including an array of decomposer organisms.

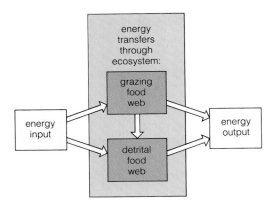

Figure 35.4 The one-way flow of energy through grazing food webs and detrital food webs in ecosystems.

Figure 35.5 Model of the energy input, transfers, and outputs for a grazing food web (**a**) and a detrital food web (**b**). In most ecosystems, the two types of food webs are interconnected.

ENERGY FLOW THROUGH ECOSYSTEMS

Primary Productivity

To get an idea of how energy flow is studied, consider just one type of ecosystem on land, for which multicelled plants are the primary producers. The plants capture light energy and convert it to the chemical energy of organic compounds. The ecosystem's **primary productivity** is the rate at which its producers capture and store a given amount of energy in a given length of time. How much actually gets stored depends on (1) the balance between photosynthesis and aerobic respiration in the plants, and (2) how many plants escape the attention of consumers and decomposers.

Gross **primary productivity is the total rate of photosynthesis for the ecosystem during a specified interval.**

Net **primary productivity is the rate of energy storage in plant tissues in excess of the rate of respiration by the plants themselves.**

Heterotrophic consumption affects the rate of energy storage.

Other factors influence the amount of net primary production, its seasonal patterns, distribution through the habitat, and the size and form of the primary producers themselves. These factors include the range of temperature and amount of rainfall during the growing season. The harsher the environment, the fewer shoots will be produced—and the lower the productivity.

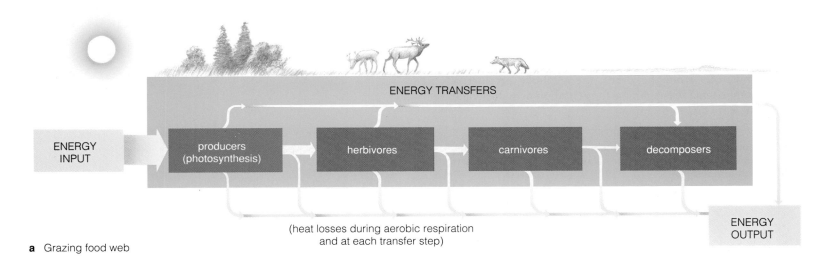

(heat losses during aerobic respiration and at each transfer step)

a Grazing food web

Major Pathways of Energy Flow

Figure 35.4 shows the direction of energy flow through ecosystems on land. Only a small part of the energy from sunlight becomes fixed in plants, and as a result of their own metabolic activities, the plants themselves use up to half of what they store. Other organisms tap into the fixed energy conserved in plant tissues, remains, or wastes, and they, too, lose heat to the environment.

The heat losses represent a one-way flow of energy out of the ecosystem. In **grazing food webs,** energy flows from plants to herbivores, then through some array of carnivores. In **detrital food webs,** it flows from plants through decomposers and detritivores (Figure 35.5). Usually, both kinds of food webs are interconnected in an ecosystem.

The amount of energy moving through the two kinds of food webs differs from one ecosystem to the next and often throughout the year. Usually, however, the largest portion of net primary production passes through detrital food webs. To be sure, about half of the net primary production passes through a grazing food web when cattle graze heavily on plants in a pasture—but the cattle don't *use* all the energy. Quantities of undigested residues become available in feces for decomposers and detritivores. In marshes, most of the stored energy is not used until plant parts die and become available for detrital food webs.

1. Energy flows into ecosystems from an outside source, which in most cases is the sun.

2. Energy flows through ecosystems by way of grazing food webs (based on the consumption of living tissues of photosynthesizers) and detrital food webs (based on the use of organic waste products and remains of photosynthesizers and consumers).

3. Energy leaves ecosystems through heat losses from each organism.

Ecological Pyramids

The trophic structure of an ecosystem is often diagramed as an "ecological pyramid," in which producers form a base for successive tiers of consumers above them. Some pyramids are based on a "head count" of all the ecosystem's members, with the lowest tier being the food production base for higher tiers. Others are based on "biomass," as determined by weighing all the members at each trophic level. Here is a pyramid of numbers for a bluegrass field:

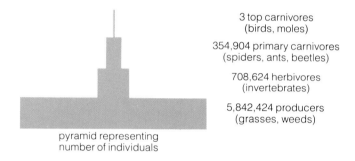

3 top carnivores
(birds, moles)

354,904 primary carnivores
(spiders, ants, beetles)

708,624 herbivores
(invertebrates)

5,842,424 producers
(grasses, weeds)

pyramid representing
number of individuals

Pyramids of numbers can be "upside down" when the reproductive rates, sizes, and weights of organisms differ

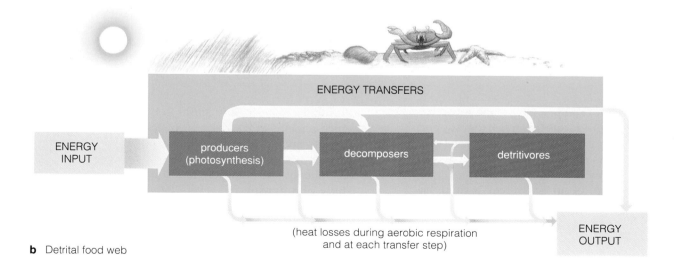

ENERGY TRANSFERS

ENERGY INPUT

producers (photosynthesis)

decomposers

detritivores

(heat losses during aerobic respiration and at each transfer step)

ENERGY OUTPUT

b Detrital food web

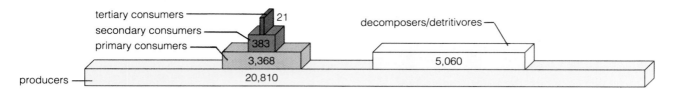

Figure 35.6 Pyramid of energy flow during one year at an aquatic ecosystem, Silver Springs in Florida.

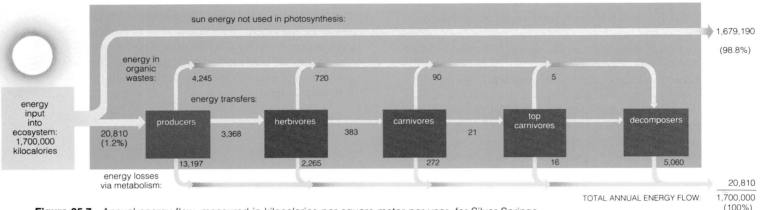

Figure 35.7 Annual energy flow, measured in kilocalories per square meter per year, for Silver Springs, Florida. The producers are mostly aquatic plants. The carnivores are insects and small fishes; top carnivores are larger fishes. The energy source (sunlight) is available all year long.

Photosynthetic autotrophs trap only 1.2 percent of incoming solar energy and use it to assemble new plant biomass. They themselves use more than 63 percent of what they trap. Herbivores get only 16 percent. Most of the stored energy in herbivores is used in metabolism or ends up with decomposers; only 11.4 percent enters carnivores, which burn up most of it. Only 5.5 percent reaches top carnivores. Decomposers recycle all the biomass from other trophic levels. In time, all of the 5,060 kilocalories transferred through the system will appear as metabolically generated heat.

This diagram is oversimplified, for no community is isolated from others. Organisms and materials constantly drop into the springs. Organisms and materials are slowly lost by way of a stream that leaves the springs.

at each trophic level. For example, a redwood forest has a small number of large producers (the trees) that support a large number of mostly small herbivores and carnivores (insects). If decomposers could be counted, they would outnumber all the rest by more than a billion.

The trophic structure of an ecosystem is more accurately represented by an **energy pyramid,** based on the energy flow at each transfer to a different trophic level. Energy pyramids are always "right-side up," with a large energy base at the bottom.

Energy pyramids are constructed by measuring how much energy each type of individual takes in, burns up

during metabolism, and stores in body tissues and how much remains in its waste products. Energy inputs and outputs are calculated so that energy flow can be expressed per unit of land or water *per unit of time.* Figure 35.6 shows one energy pyramid based on a long-term study of a grazing food web; Figure 35.7 shows some of the calculations used in its construction.

Many studies of this sort tell us that, given the metabolic demands of organisms and the amount of energy shunted into organic wastes, *only about 6 to 16 percent of the energy entering one trophic level becomes available to organisms at the next level.*

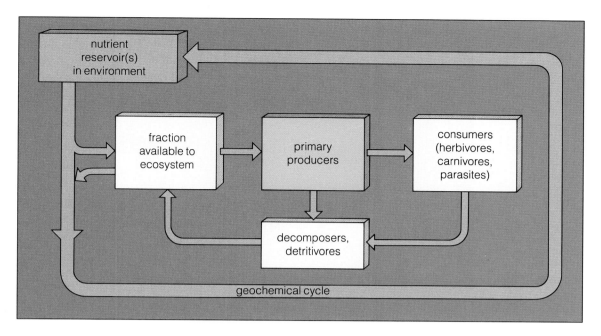

Figure 35.8 Generalized model of nutrient flow through a land ecosystem. The overall movement of nutrients from the physical environment, through organisms, and back to the environment constitutes a biogeochemical cycle.

BIOGEOCHEMICAL CYCLES

The trophic structure of ecosystems is profoundly influenced by the availability of nutrients as well as energy. Besides the carbon, oxygen, and hydrogen they get from water and air, plants require about thirteen mineral elements, including nitrogen and phosphorus (Table 12.1). Plant growth suffers with mineral deficiencies, which therefore affect primary productivity of the ecosystem at large.

The elements essential for life tend to move in **biogeochemical cycles,** meaning they are transferred from the environment, to organisms, then back to the environment. The physical environment serves as a large reservoir through which elements move rather slowly, compared to how rapidly they are exchanged between organisms and the environment. Figure 35.8 shows a model of the relationship between geochemical cycles and most ecosystems on land. The model is based on four factors:

1. The elements serving as nutrients are usually available to producers as mineral ions, such as ammonium (NH_4^+).

2. Nutrient reserves of an ecosystem are maintained by environmental inputs and recycling activities of decomposers and detritivores.

3. The amount of a nutrient being cycled within an ecosystem is greater than the amount entering or leaving in a given year.

4. Environmental inputs into an ecosystem occur by rainfall or snowfall, metabolism (including nitrogen fixation), and weathering of rocks. Outputs for land ecosystems include losses by runoff and evaporation.

There are three types of biogeochemical cycles. In the **hydrologic cycle,** oxygen and hydrogen move in the form of water molecules. In **atmospheric cycles,** a large portion of the element occurs in a gaseous phase, in the atmosphere. The cycling of carbon and nitrogen are examples of atmospheric cycles. In **sedimentary cycles,** the element does not have a gaseous phase; it moves from land to the seafloor and only "returns" to land through long-term geological uplifting (page 197). Phosphorus is such an element; it moves from land, to sediments in the seas, then back to the land. The earth's crust is the main storehouse for this and other minerals. Let's now look at some of these cycles.

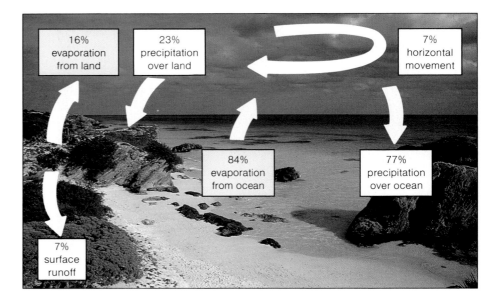

Figure 35.9 Global cycling of water. The percentages indicate the annual movements of water into and out of the atmosphere. Of the water entering, 84 percent is by way of evaporation from oceans. Of that, 7 percent is carried horizontally to land, which returns it to oceans by way of rivers and streams. The remaining 77 percent (that is, 84 − 7) leaves the atmosphere as precipitation over oceans.

16% evaporation from land

23% precipitation over land

7% horizontal movement

84% evaporation from ocean

77% precipitation over ocean

7% surface runoff

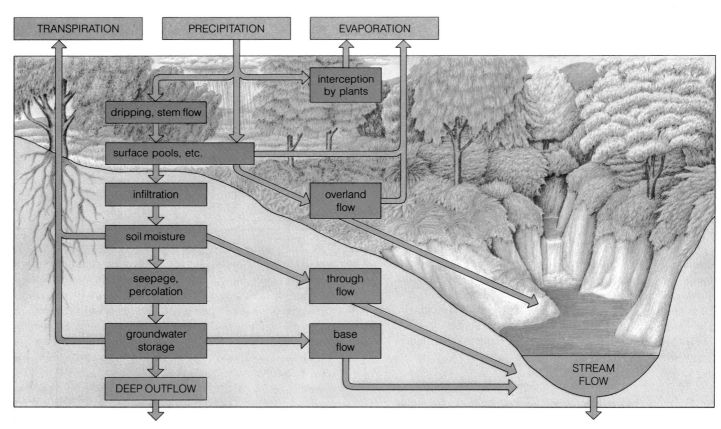

TRANSPIRATION   PRECIPITATION   EVAPORATION

interception by plants

dripping, stem flow

surface pools, etc.

infiltration

soil moisture

seepage, percolation

groundwater storage

DEEP OUTFLOW

overland flow

through flow

base flow

STREAM FLOW

Figure 35.10 Movement of water through a watershed.

a

b

c

Figure 35.11 Effects of different human activities on a forest ecosystem and its biogeochemistry. (**a**) These are experimental watersheds within the Hubbard Brook Valley of New Hampshire. Here, researchers have studied the effects of clear-cutting (foreground), progressive strip cutting (upper left), and deforestation (upper center). One of the gauging weirs used to collect all the water draining from an area under study is shown in (**b**). This watershed had been experimentally deforested, as shown in (**c**), then herbicides were applied to prevent regrowth during three years of studies. Two years after herbicide applications stopped, the vegetation recovered to the extent shown in (**b**).

Hydrologic Cycle

Cold and warm ocean currents, clouds, winds, and rainfall are all part of the global hydrologic cycle. Driven by solar energy, the waters of the earth move slowly and on a vast scale through the atmosphere, on or through the uppermost layers of land masses, to the oceans, and back again (Figure 35.9). Water moves into the atmosphere by evaporation, and there it remains aloft as vapor, clouds, and ice crystals. It falls back as precipitation—mostly rain or snow.

Water molecules do not stay aloft for more than ten days, on average, so the turnover rate is rapid for the airborne part of the cycle. Water released as precipitation remains on land for about 10 to 120 days, depending on the season and where it falls. Then some of it evaporates or is carried by rivers and streams to the seas where, with large-scale evaporation, the cycle begins again.

In itself, water is essential for life in any ecosystem. *But water is also an important medium by which nutrients move into and out of ecosystems.* This became clear through studies in **watersheds,** which are regions where all of the precipitation becomes funneled into a single stream or river. A watershed can be any size. The watershed of the Mississippi River extends across roughly one-third of the United States. Watersheds at the Hubbard Brook forest in New Hampshire average about 36 acres.

Water enters a watershed mainly as rain or snow, and most filters into the soil or runs off along the surface. It is absorbed by plant roots and is moved (by transpiration) to leaves, where evaporative water loss occurs. Water also seeps and percolates through soil to the water table (groundwater) or moves into a stream. Figure 35.10 shows these general movements of water. Figure 35.11 shows examples of studies of such movements at Hubbard Brook.

Plants greatly influence how fast nutrients move through the ecosystem phase of biogeochemical cycles. For example, you might think that water draining a watershed at Hubbard Brook would rapidly leach away calcium ions and other minerals. Yet in studies of watersheds with young, undisturbed forests, only about 8 kilograms of calcium were lost annually from each $2\frac{1}{2}$ acres. Tree roots were "mining" the soil, so that calcium became incorporated into growing plant parts and so became available for the food webs. Rainfall and the weathering of rocks brought calcium replacements into the watershed.

Nutrient outputs change when land is cleared. All the plants were removed from one watershed at Hubbard Brook, then herbicides were applied for three years to prevent regrowth. The soil itself was not disturbed; no organic material was removed. Yet the loss of calcium in the stream outflow was *six times* greater than in undisturbed watersheds. Given how slowly calcium and other nutrients move through geochemical cycles, *stripping the land of vegetation has long-term disruptive effects on nutrient availability for the entire ecosystem.*

Carbon Cycle

In the **carbon cycle,** carbon moves from reservoirs in the atmosphere and oceans, through organisms, then back to the reservoirs (Figure 35.12). Carbon enters the atmosphere by way of aerobic respiration, fossil fuel burning, and volcanic eruptions, which release carbon from rocks deep in the earth's crust. Carbon exists as a gas in the atmosphere (mostly as carbon dioxide, or

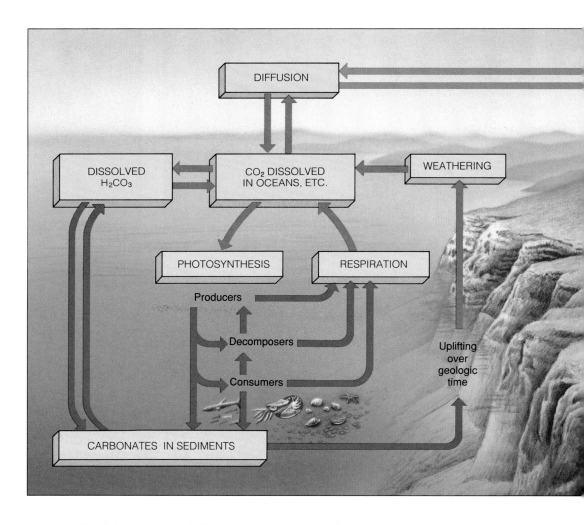

Figure 35.12 Global carbon cycle. To the left, the movement of carbon through marine ecosystems; to the right, its movement through terrestrial ecosystems.

CO_2). About half of all the carbon entering the atmosphere each year will move into two large "holding stations"—that is, into accumulated plant biomass and the oceans.

Each year, photosynthesizers capture airborne or dissolved carbon dioxide and incorporate billions of metric tons of its carbon atoms into organic compounds. However, the average length of time that a captured carbon atom remains in an ecosystem varies greatly. In tropical forests, decomposition and carbon uptake are rapid, so not much carbon is tied up in litter on the soil surface. In bogs, marshes, and other anaerobic settings, organic compounds are not broken down completely and carbon accumulates in forms such as peat (refer to Figure 34.10i).

In aquatic food webs, carbon becomes incorporated into shells and other hard parts. When the shelled organisms die, they sink and become buried in bottom sediments of different depths. Carbon in the deep oceans can remain buried for millions of years until geologic movements bring it to the surface. Still more carbon is slowly converted to long-standing reserves of gas, petroleum, and coal deep in the earth—reserves we tap for use as fossil fuels.

The worldwide burning of fossil fuels is putting more carbon into the atmosphere than can be returned to the global holding stations (oceans and plant biomass). This activity and others may be intensifying the greenhouse effect and may be triggering a global warming. The *Commentary* describes this effect and its consequences.

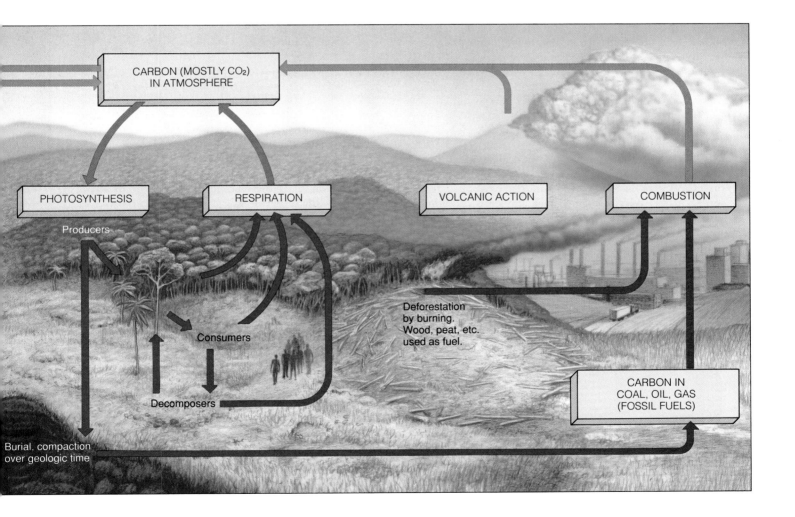

Greenhouse Gases and a Global Warming Trend

The atmospheric concentrations of carbon dioxide, water, ozone, methane, nitrous oxide, and chlorofluorocarbons profoundly influence the average temperature near the earth's surface, and that temperature influences global climates. Collectively, molecules of these gases act somewhat like a pane of glass in a greenhouse (hence their name, "greenhouse gases"). They allow wavelengths of visible light to reach the earth's surface, but they impede the escape of longer, infrared wavelengths—that is, heat—from the earth into space. They absorb infrared wavelengths, much of which gets reradiated back toward the earth (Figure a). In short, the greenhouse gases cause heat to build up in the lower atmosphere, a warming action called the **greenhouse effect.**

If there were no greenhouse gases, the earth would be a cold and lifeless planet. But there can be too much of a good thing. Largely as a result of human activities, the levels of greenhouse gases have been increasing (Figure b), and they may be contributing to an alarming increase in the global warming.

What is so alarming about a warmer planet? Suppose the temperature of the lower atmosphere were to rise by only 4°C (7°F). Sea levels would rise by about 2 feet, or 0.6 meter. Why? Ocean surface temperatures would increase—and water expands when heated. Also, global warming could cause partial melting of glaciers and Antarctic ice sheets. Low coastal regions would flood.

Imagine what a long-term rise in sea level, combined with high tides and storm waves, would do to the waterfronts of Vancouver, Boston, San Diego, Galveston, and other coastal cities. Huge tracts of Florida and Louisiana would face saltwater intrusions. Agricultural lowlands and deltas in India, China, and Bangladesh—where much of the world's rice is grown—would be submerged.

Global warming could affect world agriculture in other ways. Regional patterns of precipitation and temperature would change. Crop yields would decline in currently productive regions, including parts of Canada and the United States, and increase in others. There is speculation that warmer temperatures would promote insect breeding, with the increased population sizes of insect pests leading to more extensive crop losses.

In the late 1950s, a laboratory was set up on a mountaintop in the Hawaiian Islands to measure the concentrations of different greenhouse gases. The remote site was selected because it was free of local contamination and would represent average conditions for the Northern Hemisphere, and the monitoring activities are still going on. Consider what these studies tell us about carbon dioxide levels alone.

It turns out that the levels of atmospheric carbon dioxide follow the annual cycle of plant growth in the Northern Hemisphere. The levels are lower during

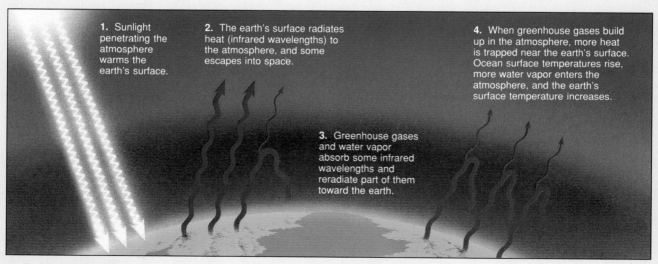

1. Sunlight penetrating the atmosphere warms the earth's surface.

2. The earth's surface radiates heat (infrared wavelengths) to the atmosphere, and some escapes into space.

3. Greenhouse gases and water vapor absorb some infrared wavelengths and reradiate part of them toward the earth.

4. When greenhouse gases build up in the atmosphere, more heat is trapped near the earth's surface. Ocean surface temperatures rise, more water vapor enters the atmosphere, and the earth's surface temperature increases.

a The greenhouse effect

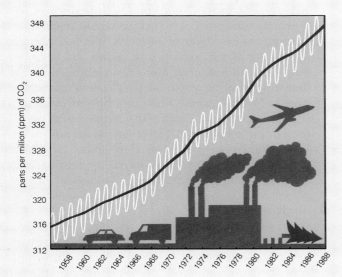

1. Carbon Dioxide (CO₂). By the year 2020, the relative contribution of the greenhouse gas CO_2 to the global warming trend is expected to be about fifty percent. Fossil-fuel burning, factory emissions, car exhaust, and deforestation are all contributing to the increased concentration.

2. Chlorofluorocarbons (CFCs). By 2020, this gas will probably be responsible for about twenty-five percent of the greenhouse effect. CFCs are used in plastic foams, air conditioners, refrigerators, and industrial solvents.

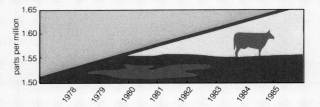

3. Methane (CH₄). By 2020, methane may be responsible for fifteen percent of the greenhouse effect. This gas is a natural by-product of bacteria (methanogens), as in feedlots.

4. Nitrous Oxide (N₂O). By 2020, this gas may be responsible for about ten percent of the greenhouse effect. It is a natural by-product of denitrifying bacteria; it is released from fertilizers and animal wastes, as in livestock feedlots.

b Relative contributions of different greenhouse gases to the global warming trend, projected to the year 2020.

summer, when plants are photosynthesizing most rapidly. They are higher in winter, when aerobic respiration continues even while photosynthetic activity declines. The lows and highs are represented by the peaks and troughs around the graph line in Figure b (part 1). *For the first time, scientists could see the integrated effects of the carbon balances of land and water ecosystems of a whole hemisphere.*

More disturbing, the peaks and troughs in the cycle showed a continuous increase. Here was evidence that a buildup of carbon dioxide in the atmosphere may intensify the greenhouse effect over the next century.

Increasing carbon dioxide levels are attributed mostly to the burning of fossil fuels, coal especially, throughout the world. Deforestation is another contributing factor. Today, vast tracts of tropical forests are being cleared and burned at a rapid rate (see, for example, Figure 37.7). Carbon is being released during the wood burning. And more importantly, the number of plants that absorb carbon dioxide during photosynthesis is plummeting.

Atmospheric concentrations of greenhouse gases are expected to continue increasing into the middle of the twenty-first century, and global warming by several degrees will probably follow. It is doubtful that we can sharply reduce fossil fuel burning and deforestation soon enough to slow down the rate of global warming. There is widespread agreement among scientists that we should begin preparing for the consequences. For example, research in genetic engineering could be intensified to develop drought-resistant and salt-resistant plants (page 168). Such plants may prove crucial in regions of saltwater intrusions and climatic change.

Nitrogen Cycle

Of all nutrients influencing the growth of land plants, nitrogen—a component of all proteins and nucleic acids—is often the one in shortest supply. Since the beginning of life, nitrogen has been present in the atmosphere and oceans but not in the earth's crust. Today, nearly all nitrogen in soils has been put there by nitrogen-fixing organisms.

The atmosphere is the largest nitrogen reservoir; about 80 percent of it is composed of gaseous nitrogen (N_2). Stable, triple covalent bonds hold the N_2 molecules together ($N{\equiv}N$), and few organisms can break them. Some bacteria and lightning can convert N_2 into forms that can be used in ecosystems. Nitrogen is lost from ecosystems through metabolic activities of bacteria that "unfix" the fixed nitrogen. Nitrogen is also lost by way of soil leaching on land, but it is gained by aquatic ecosystems with nutrient inputs from rivers and streams.

The Cycling Processes. Figure 35.13 shows the six processes of the nitrogen cycle. These are called nitrogen fixation, assimilation and biosynthesis, decomposition, ammonification, nitrification, and denitrification.

In **nitrogen fixation,** a few kinds of bacteria convert N_2 to ammonia (NH_3), which dissolves rapidly in water to produce ammonium (NH_4^+). The fixed nitrogen is used in the synthesis of amino acids, then of proteins and nucleic acids. *Anabaena, Nostoc,* and other cyanobacteria are the nitrogen fixers of aquatic ecosystems. *Rhizobium* (a symbiont with plants) and *Azotobacter* (which live in soil) are nitrogen fixers of many land ecosystems. These bacteria are small in size but mighty in numbers—collectively they fix about 200 million metric tons of nitrogen each year.

Fixed nitrogen becomes available to other organisms in the ecosystem. It moves into the tissues of plants that have entered mutually beneficial interactions with the free-living or symbiotic nitrogen fixers (page 307). These plants include peas, beans, and other legumes. Also, ammonium and other nitrogen-containing substances become available when the nitrogen fixers die and decompose. Such substances dissolve in soil water, from which they can be taken up by the roots of plants. Plants are the only nitrogen source for animals, which feed directly or indirectly on them.

Later, in **ammonification,** bacteria and fungi break down the nitrogen-containing wastes and remains of plants and animals. The decomposers use the released amino acids and proteins for growth and give up the excess as ammonia or ammonium, some of which is picked up by plants.

In **nitrification,** nitrifying bacteria strip ammonia or ammonium of electrons and so produce nitrite (NO_2^-). Then other nitrifying bacteria use nitrite in metabolism and produce nitrate (NO_3^-). Nitrification is an example of chemosynthesis (page 228).

Nitrogen Scarcity. The continual production of ammonia by nitrogen-fixing bacteria would seem to assure land plants of plenty of nitrogen. Yet soil nitrogen is scarce. Ammonia, nitrite, and nitrate are soluble and vulnerable to leaching. More fixed nitrogen is lost by **denitrification,** in which bacteria convert nitrate or nitrate to N_2 and a small amount of nitrous oxide (N_2O).

Most species of denitrifying bacteria ordinarily rely on aerobic respiration. However, when soil is waterlogged and poorly aerated, they switch to anaerobic pathways in which nitrate, nitrite, or nitrous oxide is used as the final electron acceptor instead of oxygen (Chapter 6). In doing so, the bacteria convert the fixed nitrogen to N_2.

In addition, nitrogen fixation comes at a high metabolic cost. Plants symbiotic with nitrogen-fixing bacteria don't get something for nothing; to gain nitrogen, they give up sugars and other photosynthetic products that can only be assembled with heavy investments of ATP and $NADPH_2$. Such plants are better off when soil nitrogen is scarce, but often they are displaced from nitrogen-rich soils by species that do not have to pay the metabolic price.

Nitrogen losses are great in agricultural regions. Some nitrogen leaves the fields with the plants at harvest time. Losses also occur through soil erosion and leaching. European and North American farmers traditionally have rotated crops, as when they alternate wheat with nitrogen-fixing legumes. Crop rotation has helped maintain soils in stable and productive condition, sometimes for thousands of years.

Modern agriculture depends on nitrogen-rich fertilizers. With plant breeding, fertilization, and pest control, the crop yields per acre have doubled and even quadrupled over the past forty years. With intelligent management, soil can sustain high yields indefinitely—as long as water and commercial nitrogen-containing fertilizers are available.

The catch is that we can't get something for nothing. Enormous amounts of energy go into fertilizer production—not energy from the unending stream of sunlight, but energy from fossil fuels. Fossil fuel supplies were once thought to be unending, so there was little concern about fertilizer costs. In fact, we still commonly pour more energy into the soil in the form of fertilizers than we are getting out of it in the form of food. However, as any hungry person might tell you, food calories are

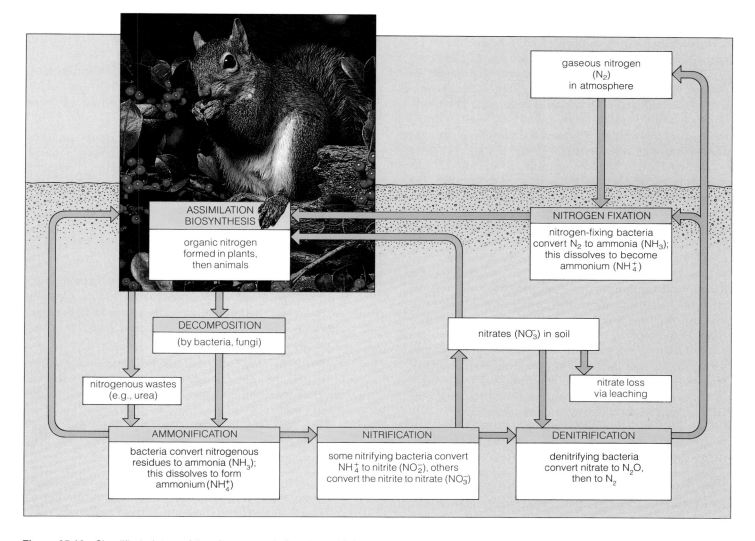

Figure 35.13 Simplified picture of the nitrogen cycle for a terrestrial ecosystem.

more basic to survival than are gasoline calories or perhaps, even, than a car. As long as the human population continues to grow exponentially, farmers will be engaged in a constant race to supply food to as many individuals as possible. Soil enrichment with nitrogen-containing fertilizers is essential in the race, as it is now being run.

Transfer of Harmful Compounds Through Ecosystems

Human activities have major and minor effects on the functioning of ecosystems. We will consider some of these effects in Chapter 37, but for now a simple example will underscore the point.

In 1955, the World Health Organization waged a major campaign to eliminate malaria-transmitting mosquitoes from the island of Borneo, now part of Indonesia. DDT is a chlorinated hydrocarbon compound that sends insects into convulsions, paralysis, and on to death. It has been instrumental in bringing mosquitoes and many other pests more or less under control.

DDT is a relatively stable compound; it is insoluble in water and breaks down very slowly. It is soluble in fat and tends to accumulate in fatty tissues. For this reason, DDT is a prime candidate for **biological magnification,** the increasing concentration of a nondegradable substance as it moves up through trophic levels. DDT that gets concentrated in tissues of herbivores such as insects becomes even more concentrated in tissues of carnivores that eat quantities of DDT-harboring herbivores. Concentration proceeds at each trophic level.

The decision to start a DDT-spraying program in Borneo was not made lightly. Nine out of ten people there were afflicted with malaria—an epidemic by anybody's standards. The program worked, insofar as the mosquitoes transmitting this terrible disease were brought almost entirely under control.

But DDT is a broad-spectrum insecticide; it kills nontarget as well as target species. Sure enough, the mosquitoes had company. Flies and cockroaches infesting the thatch-roofed houses on the island fell dead to the floor. At first there was much applause. Then the small lizards that also lived in the houses and preyed on flies and cockroaches found themselves presented with a veritable feast. Feast they did—and they died, too. So did the house cats that preyed on the lizards. With the house cats dead, the rat population of Borneo was rid of its main predator, and rats were soon overrunning the island.

The fleas on rats were carriers of still another disease, the sylvatic plague, which can be transmitted to humans. Fortunately, the threat of this new epidemic was averted in time. Someone got the inspired idea to parachute DDT-free cats into the remote parts of the island.

But on top of everything else, some home dwellers found themselves sitting under caved-in roofs. The thatch in their roofs was made of certain leaves that happen to be the food resource of a certain caterpillar. DDT did not kill the caterpillar, but it killed the wasps that were its natural predator. When the predator population collapsed, so did the roofs.

This brief example shows how disturbances to one part of an ecosystem can have unexpected effects on other, seemingly unrelated parts. A recent approach to predicting such unforeseen effects is through **ecosystem analysis.** This is a method of identifying crucial bits of information about the different components of a system and combining the information, through use of computer programs and models, in order to predict the outcome of the next disturbance. For example, an analysis of which species feed on which others in the food web shown in Figure 35.3 can be turned into a series of equations describing how much of each species is consumed. The equations can be used to predict the effect of, say, overharvesting whales.

As we attempt to deal with larger and more complex ecosystems, it becomes more difficult and expensive to run desired experiments in the field. The temptation is to run them instead on the computer. This is a valid exercise, in that the computer should hold all we know about a system. The danger is that the most important fact may be one that we do not yet know.

SUMMARY

1. An ecosystem is a whole complex of producers, consumers, detritivores, and decomposers *and* their physical environment, all interacting through a flow of energy and a cycling of materials.

2. Ecosystems are open systems, with inputs and outputs of energy as well as nutrients. With few exceptions, photosynthetic autotrophs are the primary producers; they secure energy from sunlight and take up the nutrients used by other members of the systems. Ecosystems generally are more open for inputs and outputs of water, carbon, and energy. Nutrients such as nitrogen and phosphorus are mostly recycled within the ecosystem.

3. Energy fixed by photosynthesis passes through grazing food webs and detrital food webs, which typically are interconnected in the same ecosystem. In both cases, energy is lost (as heat) through aerobic respiration and other metabolic activities.

4. Biogeochemical cycles include the movement of water, nutrients, and other elements and compounds from the physical environment, to organisms, then back to the environment.

5. Ecosystems on land have predictable rates of nutrient losses that generally increase when the land is cleared or otherwise disturbed.

6. Fossil fuel burning and conversion of natural ecosystems to cropland or grazing land are contributing to increased atmospheric concentrations of carbon dioxide, and this is contributing to a global warming trend.

7. Nitrogen availability is often a limiting factor for the total net primary productivity of land ecosystems. Gaseous nitrogen is abundant in the atmosphere, but it must be converted to ammonia and to nitrates that can be used by primary producers. A few species of bacteria, volcanic action, and lightning can cause the conversion.

Review Questions

1. Define ecosystem. Why do autotrophs play such a central role in an ecosystem? 525

2. Define trophic level. Can you name and give examples of some trophic levels in ecosystems? What is the energy source for each level? 525–527

3. Distinguish between a food chain and a food web. Can you imagine an extreme situation whereby you would be a participant in a food chain? 527

4. If you were growing a vegetable garden, what variables might affect its net primary production—that is, the amount of energy stored in the organic compounds of plant tissues? 528

Figure 36.16 African savanna, a region of warm grasslands punctuated by stands of shrubs and trees. More large ungulates (hooved, plant-eating mammals) live here than anywhere else. They include giraffes, which browse on leaves beyond the reach of other ungulates, and Cape buffalo, formidably horned animals that live in herds. Other ungulates are zebras and impalas, which are abundant and vulnerable to such predators as lions and cheetahs. Their remains (as well as the remains of lions and cheetahs) are picked over by hyenas, jackals, vultures, and other scavengers. The remains and wastes of all these organisms are broken down by detritivores and decomposers, which cycle nutrients back to the plant species.

soak up the brief, seasonal rainfall. The shortgrass prairie of the American Midwest was overgrazed and plowed under for wheat, which requires more moisture than the region can provide without irrigation. During the 1930s, strong prevailing winds, drought, and poor farming practices removed the tight vegetation cover and turned much of the prairie into a Dust Bowl (Figure 37.9).

Tallgrass prairie of the sort shown in Figure 36.15 once extended west from the temperate deciduous forests of North America. Legumes were abundant, and their symbiotic relations with nitrogen-fixing bacteria increased the net primary productivity for the region. Most tallgrass prairie has been converted to farmland.

Tropical grasslands include the broad belts of African *savanna*. In savanna regions of low rainfall, the main plant species are rapidly growing, tufted grasses of the sort shown in Figures 1.3 and 36.16. Acacia and other shrubs grow in scattered patches where there is slightly more rainfall; tall, coarse grasses, shrubs, and low trees grow where rainfall is high. Other tropical regions have comparable biomes called *monsoon grasslands*. "Monsoon" refers to a season of heavy rainfall that corresponds to a shift in prevailing winds over the Indian Ocean and parts of southern Asia; it alternates with a pronounced dry season.

Forests

The world's major forest biomes have tall trees growing close enough together to form a fairly continuous canopy over a broad region. There are three general types of forest trees, and which type dominates a region depends partly on distance from the equator. *Evergreen broadleafs* are dominant between 20° north and south latitudes;

bromeliad

a

Figure 36.17 (**a**) Tropical rain forest, with vines and aerial plants such as bromeliads in the canopy. Where light breaks through the canopy, luxuriant new growth occurs (Figure 34.1). Among the diverse occupants of different tropical rain forests are (**b**) jaguars and (**c**) *Rafflesia*, a leafless, foul-smelling plant with a fly-pollinated flower that can grow 3 meters across.

b

c

deciduous broadleafs are dominant at north and south midlatitudes. *Evergreen conifers* are the most common trees in the high latitudes of the Northern Hemisphere. Figures 36.17 and 36.18 show some forest biomes.

Evergreen Broadleaf Forests. These forest biomes occur in tropical parts of Africa, southwestern Asia, the East Indies, the Malay Archipelago, South America, and Central America. Annual rainfall can exceed 200 centimeters and is never less than 130 centimeters. Where regular, heavy rainfall coincides with high humidity (80 percent or more) and the annual mean temperature is about 25°C, you will find the highly productive biomes called *tropical rain forests* (Figure 36.17). A few of the trees native to these forests are sold in nurseries; they include the popular ornamental figs (notably *Ficus benjamina* and rubber plants), and tree ferns.

Some of the evergreen trees produce new leaves and shed old ones throughout the year. A few types near the fringes of the biome are periodically bare, but not for more than a few weeks. Because leaf production and leaf drop are generally continuous, tropical rain forests produce more litter than any other forest biome. However, decomposition and mineral cycling are rapid in the hot, humid climate, so the soil is not a significant reservoir of nutrients (see the *Commentary* on page 578).

The diversity in tropical rain forests is not confined to tree species. Different kinds of vines twist around tree trunks and grow toward the canopy, where more sunlight is available for photosynthesis. Orchids, mosses, lichens, and bromeliads (plants related to the pineapple) grow on tree branches, absorbing minerals dissolved in tiny pockets of water. (The minerals were released during the decay of bits of leaves, insects, and litter.) Entire

Spring

Summer

Autumn

Winter

communities of exotic insects, spiders, and amphibians live, breed, and die in the small pools of water that collect in the leaves of the aerial plants.

Compared to other biomes, tropical rain forests have the greatest variety of spectacularly plumaged birds, the heaviest insects (rhinoceros beetles), and the plants with the largest flowers (Figure 36.17c). There are not as many large herbivores and predators as in the open grasslands, but those in the forests are splendidly varied.

Figure 36.18 The changing character of a temperate deciduous forest in spring, summer, autumn, and winter. The one shown here is south of Nashville, Tennessee.

Deciduous Broadleaf Forests. As we move out from the tropical rain forests, we enter regions where temperatures remain mild but rainfall dwindles during part of the year. Semideciduous and deciduous trees are adapted to these conditions; they drop leaves during the pronounced dry season. The *monsoon forests* of India and southeastern Asia have such trees. Farther north, rainfall is even lower and temperatures become cold during the winter; here we find regions of *temperate deciduous forests* (Figure 36.18). Conditions are not as favorable for decomposition as they are in the humid tropics, so nutrients are conserved in the accumulated litter on the forest floor.

At one time, deciduous broadleaf forests stretched across northeastern North America, Europe, and eastern Asia. Ash, beech, birch, chestnut, elm, and deciduous oak trees dominated, but they largely disappeared as land was cleared for farming. In North America, species introductions wiped out nearly all chestnuts and many American elms (Table 34.2).

Evergreen Coniferous Forests. Evergreen conifers are cone-bearing trees, typically with needlelike leaves adapted to arid conditions. (Each needle has a thick cuticle and recessed stomata that help prevent water loss.) Conifers are primary producers in a variety of biomes, including boreal forests, montane coniferous forests, temperate rain forests, and pine barrens.

The sweeping expanses of coniferous trees in northern Europe, Asia, and North America are called *boreal forests*, or *taiga* (meaning "swamp forest"). Most boreal forests occur in glaciated regions punctuated by cold lakes and streams (Figure 36.19). It rains mostly in summer, and evaporation is low in the cool summer air. The cold, dry winters are more severe in eastern parts of the continents than in the west, where prevailing offshore winds moderate the climate.

Spruce and balsam fir dominate the boreal forests of North America. Deciduous birches and aspens are also abundant in burned-over or extensively logged areas. These forests become much less dense to the north, where nutrient-poor soils and bogs grade into tundra.

Montane coniferous forests extend southward through the great mountain ranges. Spruce and fir dominate in mountains paralleling the Pacific coast of Canada and the United States; fir and pine dominate in southern extensions of the Rockies and Cascades (Figure 36.20).

Some temperate lowlands also support coniferous forests. A *temperate rain forest* parallels the coast all the way from Alaska into northern California. It includes some of the world's tallest trees—Sitka spruce to the north and redwoods to the south (Figure 1.5d). *Pine barrens* grow in the sandy, nutrient-poor soil of New Jersey's coastal plain. This is a scrub forest, with grasses

Figure 36.19 Example of a boreal forest.

and low shrubs growing among open stands of pitch pine and oak trees. The pines recover quickly from frequent fires. Open stands of pine in the coastal plains from the Carolinas to Texas are adapted to frequent fires.

Tundra

Tundra is a word derived from the Finnish *tuntura* (treeless plain). *Arctic tundra* lies to the north, between the polar ice cap and huge belts of boreal forests in North America, Europe, and Asia. This biome is vast; it occupies twenty percent of the earth's land surface. Alaska's arctic tundra is largely flat, windswept, and desolate (Figure 35.2). Temperatures range from cool in summer to below freezing in winter. Little water vapor forms, so rainfall is sparse. Sunlight is nearly continuous for three summer months, short plants then grow profusely, flowers bloom, and seeds ripen quickly.

Although the tundra is not completely covered with snow all year long, summers are too short to warm

Figure 36.20 The montane coniferous forest of Yosemite Valley in the Sierra Nevada, beneath the first snows of winter.

much more than surface soil. Just beneath the surface is **permafrost,** a permanently frozen layer more than 500 meters thick in some regions. Permafrost forms an impenetrable basement beneath stretches of flat terrain, so water cannot drain. The low temperatures and anaerobic conditions of the waterlogged soil have major effects on nutrient cycling. Organic matter cannot completely decompose, and it accumulates in soggy masses (peat). All but about 5 percent of the carbon in the arctic is locked up in peat.

Alpine tundra occurs at high elevations in mountains throughout the world. The dominant plant species often form cushions and mats that can withstand buffeting from strong winds. The low temperatures and nutrient-poor soils do not favor much primary productivity. In contrast to arctic tundra, there is no permafrost layer below the soil surface.

THE WATER PROVINCES

As vast as the six biogeographic realms are, they cover less than 30 percent of the earth's surface. The water provinces are far more extensive and they, too, encompass diverse ecosystems. These include freshwater lakes, rivers, swamps, marshes, bogs and similar wetlands, as well as brackish seas. Along the coasts or in the oceans and seas are communities of organisms suspended in the water or tucked into the nooks and crannies of estuaries, rocky and sandy shores, coral reefs, and the ocean floor.

There is no such thing as a "typical" aquatic ecosystem. Lakes alone might be shallow enough to wade across; one is more than 1,700 meters deep. Although all aquatic ecosystems have gradients in light penetration, temperature, and dissolved gases, the gradients differ greatly. All we can do here is sample the diversity.

Lake Ecosystems

The topography, climate, and geologic history of a lake determine the kinds and numbers of organisms found there, how they are distributed, and how nutrients are cycled among them. All lakes form in land basins (Figure 36.21), and most are doomed to become filled with sediments over time.

Lake Zones. Deep lakes have littoral, limnetic, and profundal zones (Figure 36.22). The *littoral* is a shallow, usually well-lit zone extending all around the shore to the depth where rooted aquatic plants stop growing. Diversity is greatest here; the littoral is home to a variety of plants, decomposers, and snails, frogs, and other consumers. The *limnetic* zone includes the open, sunlit waters beyond the littoral, down to the depth where photosynthesis no longer is significant. Here we find **plankton,** communities of floating or weakly swimming organisms, mostly microscopic. Photosynthetic autotrophs (phytoplankton) include diatoms, green algae, and cyanobacteria. Heterotrophs (zooplankton) include rotifers and copepods.

The *profundal* is the deep, open-water zone where there is not enough light for photosynthesis. Detritus from the limnetic sinks through the profundal to the bottom ooze, which contains communities of bacterial decomposers. Through their activities, the decomposers release mineral ions into the water and enrich it with nutrients.

Seasonal Changes in Lakes. In temperate regions where warm summers alternate with cold winters, lakes undergo seasonal changes in density and temperature from the surface to the bottom. A layer of ice typically

Figure 36.21 Example of a lake basin carved by glacial action in the Canadian Rockies.

forms over a temperate lake in midwinter. The near-freezing water just beneath the ice is the least dense. Water has the greatest density at 4°C. Water at that temperature accumulates in the deeper layers of the lake, which are slightly warmer than the surface layer in midwinter.

In spring, daylength increases and the air becomes warmer. The ice melts and the surface water gradually warms to 4°C. Temperatures become uniform through the lake, and now winds acting on the lake surface cause a **spring overturn.** In such overturns, strong vertical movements carry dissolved oxygen from the surface layer of water to the depths, and nutrients released by decomposition are brought from the bottom sediments to the surface layer.

By midsummer, heat gain has been faster near the surface, compared to deeper layers. Now there is a *thermocline*, a middle layer where water temperature changes abruptly (Figure 36.23). Surface water actually floats on the thermocline, which prevents vertical mixing. During autumn, the upper layer begins to cool, increase in density, and sink, and the thermocline vanishes. The lake water mixes vertically during this **fall overturn,** when once again dissolved oxygen moves down and nutrients move up.

Cycles of primary productivity correspond to the seasonal changes. After the spring overturn, the longer daylight hours and the recycled nutrients support increases in primary productivity. As the growing season progresses, phosphorus, nitrogen, and other nutrients are used rapidly, but the thermocline cuts off vertical mixing. All the while, microscopic producers and consumers are dying, and nutrients tied up in their remains sink through deeper waters. As nutrient supplies dwin-

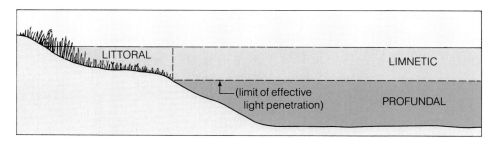

Figure 36.22 Lake zonation. The littoral includes all areas around the edge of the lake, from the shore to the depth where aquatic plants stop growing. The profundal includes areas below the depth of light penetration. Above the profundal are the open, sunlit waters of the limnetic zone.

dle in the upper waters, primary productivity declines until, by late summer, the shortages limit photosynthesis.

After the fall overturn, dissolved nutrients brought to the surface trigger another increase in primary productivity, but this does not last long. With autumn's shorter daylight hours and declining temperatures, the burst of activity is soon over, and primary productivity will not increase again until spring.

Trophic Nature of Lakes. Geologic processes give rise to lakes, as when the grinding action of advancing glaciers carves out basins in mountains. After the glaciers retreat, the exposed basins become filled with water. Other processes, such as erosion and sedimentation, change each lake's dimensions over time. Soils of the lake basin and the surrounding regions contribute to the type and amount of dissolved nutrients that will be available for the lake's primary producers.

Interactions among soils, basin shape, and climate produce conditions ranging from oligotrophy to eutrophy. *Oligotrophic lakes* are often deep, poor in nutrients, and low in primary productivity. *Eutrophic lakes* are often shallow, rich in nutrients, and high in primary productivity. When lake sediments accumulate, conditions may progress from oligotrophy to eutrophy, then on to a final successional stage with a completely filled-in basin.

Human activities can disrupt the balance between geologic, climatic, and biological forces that shape the trophic nature of lakes. This happens when sewage is dumped into lakes or when the surrounding land is logged over or cleared for farming. Field experiments dramatically demonstrate the effects of such disruption (Figure 36.24).

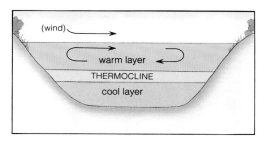

Figure 36.23 Thermal layering of water in a temperate lake in Connecticut during summer.

Figure 36.24 Field experiment demonstrating the effect of nutrient enrichment on a lake. A plastic curtain was stretched across a narrow channel between two basins of the same lake. Phosphorus, carbon, and nitrogen were added to the basin at the left; only carbon and nitrogen were added to the basin in the right foreground. Within two months, the phosphorus-enriched basin showed accelerated eutrophication, with a dense algal bloom.

Figure 36.25 New England salt marsh, with *Spartina* predominating. This salt marsh grass is the major producer, with its microbe-enriched litter providing the food for consumers in the creeks and sounds.

Marine Ecosystems

Like freshwater ecosystems, the oceans and seas vary in their physical and chemical properties, including light penetration and water temperature, depth, and salinity. Here we will focus on estuaries, then on ecosystems of the intertidal zone and open oceans.

Estuaries. An **estuary** is a partly enclosed coastal region where seawater mixes with fresh water from rivers or streams and runoff from the land. Generally, the fresh water flows out toward the sea over the saltier (and more dense) water beneath it. But tides, winds, and the earth's rotation influence the water movements.

Figure 36.25 shows an estuary along the New England coast, where coastal salt marshes are common. Its salt-tolerant plants, including *Spartina,* can withstand submergence at high tide.

Spartina is a major producer for the estuary, and it is largely cellulose—which few herbivores can digest. Thus it is the start of detrital (not grazing) food webs, in which bacterial and fungal decomposers are the first to feed. Its decomposed bits are food for nematodes, snails, crabs, and some fish. Clams, barnacles, and other filter-feeders live on the bits that become suspended in water by rising tides, and so on through food webs.

Estuaries are more than feeding grounds; they are nurseries for organisms belonging to food webs in the seas (Figure 36.26). These highly productive ecosystems are being rapidly altered by sewage, agricultural runoff, and diversions of fresh water for farm use—water that normally would drain into estuaries and keep salinity levels within tolerable ranges for the communities.

Life Along the Coasts. Along the rocky and sandy shores of coastlines are ecosystems of the **intertidal zone,** which is not exactly renowned for its creature comforts. The resident organisms are battered by waves, fiercely so during storms. They are alternately submerged, then exposed by the tides. The higher up they are on the shore, the more they may dry out, freeze in winter, or bake in summer, and the less food comes their way. The lower they are on the shore, the more they must compete for the limited space available. At low tides, birds, rats, and raccoons move in to feed on them; high tides bring the predatory fishes.

It is nearly impossible to generalize about life here, because bewildering arrays of habitats are constantly being resculpted by waves and tides. About the only feature common to all rocky and sandy shores is a vertical zonation—and even this varies in its details, with resident species often violating the artificial boundaries we impose on them.

Rocky shores often have three vertically arranged zones (Figure 36.26). The *upper littoral* is only submerged during the highest tide of the lunar cycle, and it is only sparsely populated. The *mid-littoral* is submerged during the highest regular tide and exposed during the lowest tide of each day. Red, brown, and green algae, sea anemones, snails, nudibranchs, hermit crabs, and small fishes typically live in the tide pools characteristic of this zone (Figure 36.27). The *lower littoral* is only exposed during the lowest tide of the lunar cycle; diversity is greatest here. Strong wave action prevents detritus from accumulating on rocky shores, so grazing food webs tend to predominate in all three zones.

Sandy and muddy shores are rather unstable stretches of loose sediments, continually rearranged by waves and currents. Few large plants grow in either place, so you won't find many grazing food webs. Bits of organic debris from offshore or nearby land ecosystems become trapped in the sediments and form the basis of detrital food webs. Vertical zonation occurs, but not much.

upper
littoral

mid-
littoral

lower
littoral

Figure 36.26 Vertical zonation in the intertidal zone of a rocky shore in the Pacific Northwest. The vertical difference between high and low tides varies from a few centimeters (in the Mediterranean Sea) to more than 15 meters (in the Bay of Fundy, next to Nova Scotia). It is about 1.2 meters for the area shown in this photograph.

In the sparsely populated upper littoral, the primary producers are cyanobacteria, green algae, and the algal part of lichens, all of which grow in mats or jellylike masses on the infrequently wet rocks. Small snails and limpets feed on the producers. Some large, aggressive shore crabs feed on the snails and limpets, and seabirds feed on snails, limpets, and smaller crabs.

The upper littoral may also be populated by barnacles that can get by with a few hours of filter feeding each month, when the highest tides carry plankton to them. They also grow profusely in the mid-littoral, along with mussels, sea stars, and red, brown, and green algae.

Diversity is greatest in the lower littoral, where abundant seaweeds form a leafy canopy for many organisms.

sea anemone

sea star

Figure 36.27 The kinds of residents of tide pools along the coast of Vancouver, British Columbia.

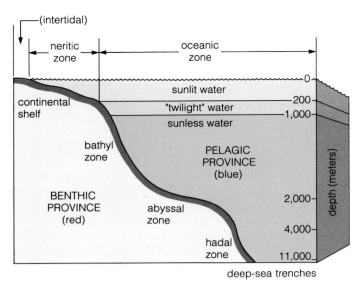

Figure 36.28 Zones of the ocean.

a

The Open Ocean. Beyond the intertidal are two vast provinces of the open oceans (Figures 36.28 and 36.29). The **pelagic province** includes the entire volume of ocean water. It is subdivided into the *neritic zone* (relatively shallow waters overlying the continental shelves) and the *oceanic zone* (water over the ocean basins). Tropical reefs occur in the neritic zone and around islands in the oceanic zone (Figure 36.30). The **benthic province** includes all sediments and rocky formations of the ocean bottom. It begins with the *continental shelf* and extends down through the zone of deep-sea trenches.

In the neritic and oceanic zones, photosynthetic activity is restricted to the upper surface waters, where phytoplankton often drift with the currents and form vast, suspended pastures for zooplankton. The zooplankton include copepods and shrimplike krill, all food for larger, strong-swimming carnivores such as squids and fishes. Organic remains and wastes sink down from waters near the ocean's surface to the benthic province, where they serve as nutrients for most communities in these zones.

Although we have yet to discover huge populations in the deep zones of the benthic province, remarkable communities do thrive at **hydrothermal vents,** where water is heated and spewed from fissures between two crustal plates (Figure 36.29). The primary producers of these unique communities do not use sunlight as the primary energy source. The producers, chemosynthetic bacteria, obtain energy through reactions involving hydrogen sulfide. They are consumed by clams, mussels, giant tube worms, and other organisms.

Upwelling. Primary productivity increases whenever currents stir ocean water and keep nutrients circulating back to the surface. This occurs with **upwelling,** an upward movement of deep, nutrient-rich water along the margins of continents. It occurs when winds force surface waters along the coast to move away from the shore. When the surface water moves out, deep water moves vertically to replace it.

For example, prevailing winds from the south and southeast force surface water away from the west coast of Peru. Cold, deeper water brought to the continental shelf by the Humboldt Current moves toward the surface and brings up tremendous amounts of nitrates and phosphates from below. Phytoplankton growing on the nutrients are the basis of one of the world's richest fisheries, with schools of small anchovies and other fishes reaching huge numbers.

Periodically, warm surface waters of the western equatorial Pacific move eastward. The massive displacement of warm water affects the prevailing winds, which accelerate the eastward movement. The movement is enough to displace the cooler waters of the Humboldt Current and prevent upwelling. This phenomenon, which local fishermen named El Niño, causes productivity to decline, with catastrophic effects on anchovy-eating birds as well as on the anchovy industry. The *Commentary* on page 566 gives a closer look at the El Niño phenomenon. With it, we come full circle to a concept presented at the start of this chapter—that interactions among the atmosphere, oceans, and land profoundly influence the world of life.

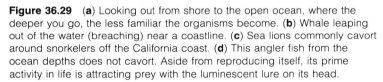

c

d

e

Figure 36.29 (**a**) Looking out from shore to the open ocean, where the deeper you go, the less familiar the organisms become. (**b**) Whale leaping out of the water (breaching) near a coastline. (**c**) Sea lions commonly cavort around snorkelers off the California coast. (**d**) This angler fish from the ocean depths does not cavort. Aside from reproducing itself, its prime activity in life is attracting prey with the luminescent lure on its head.

Hydrothermal vent ecosystems. In 1977, biologists discovered a distinct type of ecosystem deep in the Pacific Ocean, where sunlight never penetrates. John Corliss and his coworkers were exploring the Galápagos Rift, a volcanically active boundary between two of the earth's crustal plates. There, on the ocean floor, the near-freezing seawater seeps into fissures, becomes heated, and is spewed out through vents at high temperatures.

This hydrothermal outpouring results in deposits of zinc, iron, and copper sulfides as well as calcium and magnesium sulfates (all leached from rocks as pressure forces the heated water upward). In marked contrast to most of the deep ocean floor, these nutrient-rich, warm "oases" support diverse marine communities. Chemosynthetic bacteria and other microbes use the inorganic deposits as energy sources. They are the primary producers in a food web that includes tube worms, clams, sea anemones, crabs, and fishes (**e,f**).

So far, other hydrothermal vent ecosystems have been discovered near Easter Island (in the South Pacific Ocean); in the Gulf of California, about 150 miles south of the tip of Baja California, Mexico; near the Galápagos Islands; and in the Atlantic. In 1990, a team of United States and Soviet scientists discovered a unique hydrothermal vent ecosystem in Lake Baikal in Siberia which, at 1.7 kilometers, is the world's deepest lake. The tectonically formed lake basin seems to be splitting apart (hence the hydrothermal vents) and may mark the beginning of the formation of a new world ocean.

f

sea anemone

crown-of-thorns sea star

moray eel

pillar coral

green tube coral

coral reef   island   lagoon   open ocean

Figure 36.30 A sampling of the diversity characteristic of tropical reefs. Long ago, corals began to grow and reproduce in the warm, nearshore waters off the islands shown. The skeletons they left behind served as a foundation for more corals to grow upon. As skeletons and residues accumulated, the reef grew, and tides and currents carved ledges and caverns in it. Today, the reef's spine may be decked out with as many as 750 species of corals, such as the ones shown here.

Red algae typically encrust the coral foundation. In shallow waters behind the reef, red algae give way to blue-green forms. Many small, transparent animals feed on algae and other plants and in turn are food for predators, including the fishes, sea star, and sea anemone shown here. Fishes and other animals are food for the moray eel, which lurks in nooks and crannies in the reef.

El Niño and Oscillations in the World's Climates

Oceans cover more than seventy percent of the earth's surface, so it is not difficult to understand why variations in ocean surface temperatures affect climates around the world. But only within the past decade has numerical modeling begun to show how dramatic the effects may be.

In the winter of 1989–1990, for example, rainfall was below normal along the California coast, and by spring, strict water conservation and rationing programs were in the works. People were taking quick showers instead of baths, cutting back on flushing the toilet and running the dishwasher, and no longer deep-watering lawns and gardens because of the drought.

Drought is a relative term. If you live in Tucson, Arizona, or another desert city, "drought" is what you expect nearly year-round. If you live along the western coasts or in the interior of continents, where semiarid conditions prevail, prolonged drought conditions can be disastrous.

Yet "abnormally" dry seasons may be part of a recurring feedback relationship between sea surface temperature, drought-related conditions on land, and drought-sustaining atmospheric circulation patterns.

Consider the global climate system called the *El Niño Southern Oscillation* (ENSO). The "El Niño" part of this phenomenon is an irregular but episodic warming of surface waters in the eastern equatorial Pacific. The "Southern Oscillation" is a global-scale seesaw in atmospheric pressure at the earth's surface—specifically, at Indonesia, northern Australia, and the southeastern

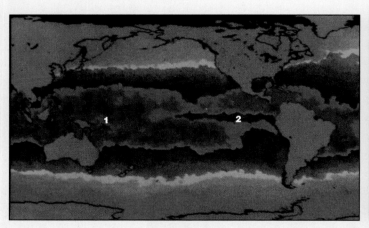

a Distribution of ocean surface temperatures in normal years, with the warmest waters found in the western equatorial Pacific (1), and a tongue of relatively cold water extending westward along the equator from South America (2).

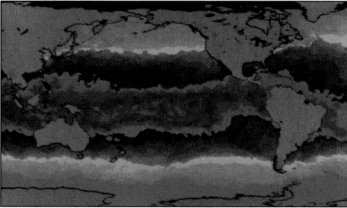

b Distribution of ocean surface temperatures that were associated with the 1982–1983 ENSO episode.

Pacific. This area of the Southern Hemisphere is the world's largest reservoir of warm water (Figure a), and more warm, moisture-laden air rises here than anywhere else. Rainfall is also heavy here, and it releases much of the heat energy that drives the world's air circulation system.

Every two to seven years, the warm reservoir and the associated heavy rainfall move eastward (Figure b). This causes prevailing surface winds in the western equatorial Pacific to pick up speed. The stronger winds have a more pronounced effect on "dragging" the ocean surface waters eastward. Upper ocean currents are affected to the extent that the westward transport of water slows down and the eastward transport increases. The outcome? *More* warm water in the vast reservoir moves east—and so on in a feedback loop between the ocean and the atmosphere.

The rainfall pattern in the Pacific and Indian oceans is massively dislocated when ENSO warm episodes occur. During the 1982–1983 episode, the vital monsoon rains hardly materialized over India—and record droughts occurred in Australia and nearby regions as well as in the Hawaiian Islands. Month after month, record rainfall drenched the arid and semiarid coasts of Ecuador and Peru. The ENSO also prolonged a devastating drought that already was under way in Africa, with its consequent and appalling human starvation.

Numerical models are now being used to study these and other episodes of climatic change a few seasons in advance. More reliable forecasting should follow when more and better observations are made of the interrelated systems of the ocean, land, and atmosphere.

SUMMARY

1. The biosphere is the narrow zone of water, the lower atmosphere, and the fraction of the earth's crust in which organisms live. It is composed of ecosystems, each influenced by the flow of energy and the movement of materials on a global scale.

2. The world distribution of species is a result of "accidents of history," climate, topography, and species interactions. Climate means prevailing weather conditions, including temperature, humidity, wind velocity, degree of cloud cover, and rainfall. It is an outcome of differences in the amount of solar radiation reaching equatorial and polar regions, the earth's daily rotation and its annual path around the sun, the distribution of continents and oceans, and the elevation of land masses.

3. Climatic factors interact to produce prevailing winds and ocean currents, which together influence global weather patterns. Weather in turn affects soil composition, sedimentation, and water availability, which influence the growth and distribution of primary producers. Through these interactions, climate is a major force governing ecosystems.

4. The earth's land masses can be classified as six major biogeographic realms, each with characteristic types of plants and animals and each more or less isolated by oceans, mountain ranges, or desert barriers.

5. Biomes are distinct vegetational subdivisions of the six major realms. They are created by regional variations in climate, landforms, and soil composition. Each is dominated by plant species adapted to a particular set of conditions. The major types are deserts, dry shrublands and woodlands, grasslands, forests, and tundra.

6. The water provinces, which cover more than 70 percent of the earth's surfaces, include standing fresh water (such as lakes), running fresh water (such as streams and rivers), and oceans and seas. Its marine ecosystems include estuaries, intertidal zones, rocky and sandy shores, tropical reefs, and scattered ecosystems of the open oceans. All freshwater and marine ecosystems have gradients in light penetration, temperature, salinity, and dissolved gases, features that vary daily and seasonally and that influence primary productivity.

7. In the open ocean, photosynthetic activity is greatest in shallow coastal waters and in regions of upwelling along the margins of continents. Upwelling is an upward movement of deep, cooler water that carries nutrients to the surface.

8. The interrelatedness of ocean surface temperatures, the atmosphere, and the land is especially clear through studies of the El Niño Southern Oscillation, a recurring phenomenon that is accompanied by abnormal weather conditions in many parts of the world.

Review Questions

1. Define these terms: biosphere, atmosphere, hydrosphere. *543*

2. Define climate. What interacting factors influence climate? What does climate in turn influence? *543*

3. How do prevailing air and ocean currents help dictate the distribution of different types of ecosystems? *544–546*

4. Describe a rain shadow. *546*

5. Distinguish between biogeographic realm and biome. In what type of biome would you say you live? *547*

6. How does the composition of regional soils affect ecosystem distribution? *549*

7. How do climatic conditions affect the character of the following biomes: desert, shrublands, tropical rain forests, temperate deciduous forests, evergreen coniferous forests, and tundra? *550–557*

8. What is the difference between the littoral and profundal zones of a lake ecosystem? What kinds of lakes are eutrophic? Oligotrophic? *558–559*

9. Define upwelling, and give an example of a region where upwelling has a profound effect on primary productivity. *562*

Self-Quiz *(Answers in Appendix IV)*

1. The primary energy source for ecosystems, and a determining factor in their distribution, is _____.

2. Heat energy derived from the sun warms the _____, and that energy drives the earth's _____.

3. Global air circulation patterns, ocean currents, and topography interact to create regional differences in _____.

4. The term "biosphere" includes _____.
   a. water zones
   b. the lower atmosphere
   c. the portion of the earth's crust inhabited by organisms
   d. all of the above are included in the term "biosphere"

5. "Accidents of history," climate, topography, and the interactions of species profoundly affect the world distribution of _____.
   a. water   d. weather
   b. oxygen   e. solar radiation
   c. species

6. Nearly all ecosystems are influenced by climate (prevailing weather conditions), which is shaped by _____.
   a. variations in amount of incoming solar radiation
   b. the earth's daily rotation and annual path around sun
   c. world distribution of continents and oceans
   d. elevation of land masses
   e. all of the above are factors that shape climate

7. The earth's major _____ have characteristic types of plants and animals and are, for the most part, isolated by oceans, desert barriers, and mountain ranges.
   a. biomes
   b. ecosystems
   c. biogeographic realms
   d. water provinces

8. Examples of _____ are deserts, shrublands, grasslands, forests, and tundra.
   a. biomes
   b. ecosystems
   c. biogeographic realms
   d. plant populations

9. Gradients in light penetration, temperature, salinity, and dissolved gases are all features of _____ ecosystems.
   a. marine
   b. forest
   c. desert
   d. freshwater
   e. both a and d are correct

10. Match the biosphere terms appropriately.
   _____ biomes
   _____ biogeographic realms
   _____ climate
   _____ biosphere
   _____ solar radiation
   _____ water

   a. earth's six major land regions
   b. covers 70 percent of the earth's surface
   c. temperature, humidity, wind velocity, degree of cloud cover, and rainfall
   d. zones of the earth's crust, waters, and atmosphere that support life
   e. primary energy source for ecosystems
   f. broad subdivisions of biogeographic realms

Selected Key Terms

atmosphere *543*
biogeographic realm *547*
biome *548*
biosphere *543*
climate *543*
El Niño Southern Oscillation (ENSO) *566*
fall overturn *558*
hydrosphere *543*
hydrothermal vent *562*
intertidal zone *560*
limnetic *558*
littoral *558*
oceanic zone *562*
permafrost *557*
plankton *558*
rain shadow *546*
soil *548*
spring overturn *558*
thermocline *558*
upwelling *562*

Readings

Barber, M., J. Burk, and W. Pitts. 1987. *Terrestrial Plant Ecology.* Second edition. Menlo Park, California: Benjamin-Cummings.

Colinvaux, P. May 1989. "The Past and Future Amazon." *Scientific American* 260(3):102–108.

Gibbons, B. September 1984. "Do We Treat Our Soil Like Dirt?" *National Geographic* 166(3):350–388.

Smith, R. 1989. *Ecology and Field Biology.* Fourth edition. New York: Harper & Row. Good descriptions of biomes.

Sumich, J. 1988. *Biology of Marine Life.* Fourth edition. Dubuque, Iowa: Brown.

HUMAN IMPACT ON THE BIOSPHERE

37

KEY CONCEPTS

1. The human population has undergone exponential growth since the mid-eighteenth century. Today we have the population size, the technology, and the cultural inclination to use energy and modify the environment at alarming rates.

2. The world of life ultimately depends on energy from the sun, which drives the complex interactions between the atmosphere, oceans, and land. The accumulation of human-generated pollutants in the atmosphere especially is disrupting those interactions, and those disruptions may have alarming consequences in the near future.

3. We as a species must come to terms with the principles of energy flow and resource utilization that govern all systems of life on earth.

ENVIRONMENTAL EFFECTS OF HUMAN POPULATION GROWTH

Of all the concepts introduced in the preceding chapter, the one that should be foremost in your mind is this: Complex interactions between the atmosphere, the oceans, and the land are the engines of the biosphere. Driven by energy from the sun, they produce the worldwide temperatures and circulation patterns on which life ultimately depends. With this chapter, we turn to a related concept of equal importance. Simply put, the human population has been straining the global engines without fully comprehending that engines can crack.

To gain perspective on what is happening, think about something we all take for granted—the air around us. The composition of the present atmosphere is the outcome of geologic and metabolic events, including photosynthesis, that began billions of years ago. At some time between 6 and 4 million years ago, the human species emerged. Like us, the first humans breathed oxygen from an atmosphere of ancient origins. Like us,

Figure 37.1 How do you view our world and the effects we have on it? Write your own caption for this photograph before reading this chapter, and write another one after the reading is done.

they were protected from harmful ultraviolet radiation by an ozone shield in the stratosphere. Their population sizes were not much to speak of, and their interactions with the biosphere were not significant. About 10,000 years ago, however, agriculture began in earnest, and it laid the foundation for rapid population growth. With agriculture, and with the medical and industrial revolutions that followed, human population growth became exponential in a mere blip of evolutionary time (Figure 33.7).

Today, our burgeoning population is placing unrealistic demands on the biosphere. As we take energy and resources from it, we give back wastes in monumental amounts. In the process, we are destroying the stability of ecosystems on land, contaminating the hydrosphere, and changing the composition of the atmosphere. Our carbon dioxide wastes alone are contributing to an amplified "greenhouse effect," described on page 536, and this may already be causing an unexpected warming of the entire planet.

In the developed countries of North America, Europe, and elsewhere, the rate of population growth has more or less stabilized, and resource use per individual is no longer increasing by much. But the resource utilization levels are already high. At the same time, population growth and demands for resources are increasing exponentially in the developing countries of Central America, South America, Asia, Africa, and elsewhere—even though many millions there are already starving to death and millions more suffer from serious malnutrition.

Many of the problems we will consider in this chapter are not going to go away soon. It will take decades, even centuries, to reverse some of the trends already in motion, and not everyone is ready to make the effort. An enlightened individual in Michigan or Alberta or New South Wales can make good attempts at conservation—but scattered attempts will not be enough. Individuals of every nation will make a concerted effort to reverse global trends only when they perceive that the dangers of *not* doing so outweigh the personal benefits of ignoring them.

Does this seem pessimistic? Think about the exhaust fumes being released into the atmosphere every time you drive a car. Think about the oil refineries, paper mills, and food-processing plants that supply you with goods but also release chemical wastes into our nation's waterways. Think about Mexico and other developing countries that produce cheap food by using an unskilled labor force and dangerous pesticides—which poison the people who work the land, and the land itself. Who changes behavior first? We have no answer to the question. We suggest, however, that a strained biosphere can rapidly impose the answer upon us.

Table 37.1 Major Classes of Air Pollutants	
Carbon oxides	Carbon monoxide (CO), carbon dioxide (CO_2)
Sulfur oxides	Sulfur dioxide (SO_2), sulfur trioxide (SO_3)
Nitrogen oxides	Nitric oxide (NO), nitrogen dioxide (NO_2), nitrous oxide (N_2O)
Volatile organic compounds	Methane (CH_4), benzene (C_6H_6), chlorofluorocarbons (CFCs)
Photochemical oxidants	Ozone (O_3), peroxyacyl nitrates (PANs), hydrogen peroxide (H_2O_2)
Suspended particles	Solid particles (dust, soot, asbestos, lead, etc.), liquid droplets (sulfuric acid, oils, dioxins, pesticides)

CHANGES IN THE ATMOSPHERE

If you were to compare the earth with an apple from the supermarket, the atmosphere would be no thicker than the layer of shiny wax applied to it. Yet this thin, finite wrapping of air around the planet receives more than 700,000 metric tons of pollutants each day in the United States alone. **Pollutants** are substances with which ecosystems have had no prior evolutionary experience, in terms of kinds or amounts, and so have no mechanisms for dealing with them. From the human perspective, pollutants are substances that adversely affect our health, activities, or survival.

Table 37.1 lists the major air pollutants. They include carbon oxides, sulfur oxides, nitrogen oxides, chlorofluorocarbons (CFCs), and photochemical oxidants.

Local Air Pollution

Whether air pollutants are dispersed throughout the atmosphere or concentrated at their source in a given time period depends on local climate and topography. Consider what happens during a **thermal inversion,** when a layer of dense, cool air becomes trapped beneath a layer of warm air (Figure 37.2). The pollutants cannot be dispersed by winds or rise higher in the atmosphere, so they accumulate to dangerous levels close to the ground. By intensifying a phenomenon known as smog,

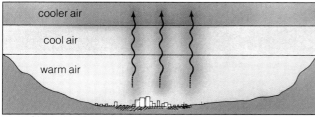

a Normal pattern

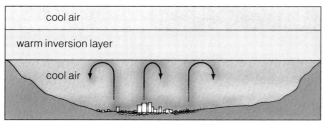

b Thermal inversion

Figure 37.2 Trapping of airborne pollutants by a thermal inversion layer. The photograph shows a Los Angeles freeway system under its self-generated blanket of smog at twilight.

thermal inversions have contributed to some of the worst local air pollution disasters.

There are two types of smog (gray air and brown air), both of which occur in major cities. **Industrial smog** is gray air that predominates in industrialized cities with cold, wet winters. London, New York, Pittsburgh, and Chicago are examples. These cities use fossil fuel for heating, manufacturing, and producing electric power. The burning fuel releases airborne pollutants, including dust, smoke, ashes, soot, asbestos, oil, bits of heavy metals, and sulfur oxides. The pollutants may reach lethal concentrations when winds and rain do not disperse them. Industrial smog was the cause of London's 1952 air pollution disaster; during that episode, 4,000 people died.

Photochemical smog is a brown and smelly trademark of large cities found in warm climates. When the surrounding land forms a natural basin, as it does around Los Angeles and Mexico City, photochemical smog can reach harmful concentrations. The main culprit is nitric oxide, which is produced chiefly by cars and other vehicles with internal combustion engines. Nitric oxide reacts with oxygen in the air to form nitrogen dioxide. When exposed to sunlight, nitrogen dioxide can react with hydrocarbons (spilled or partly burned gasoline, most often) to form photochemical oxidants. Other components of smog are ozone and PANs (short for *peroxyacyl nitrates*). PANs are similar to tear gas; even traces can sting the eyes and irritate the lungs.

Acid Deposition

Oxides of sulfur and nitrogen are among the most dangerous air pollutants. Coal-burning power plants, factories, and metal smelters are the main sources of sulfur oxides. Vehicles, power plants that burn fossil fuels, and nitrogen fertilizers are sources of nitrogen oxides.

Depending on climatic conditions, tiny particles of these substances may be airborne for a while and then fall to earth as **dry acid deposition.** Most sulfur and nitrogen oxides dissolve in atmospheric water to form weak solutions of sulfuric acid and nitric acid. Winds can distribute them over great distances before they fall to earth in rain and snow; this is called **wet acid deposition.** Acid rain can be four to forty times more acidic than normal rainwater, sometimes as much as lemon juice. The acids attack marble, metals, mortar, rubber, plastic, even nylon stockings. And they are disrupting ecosystems.

Because soils and vegetation are not identical in all watersheds, some regions are more sensitive to acid deposition than others. Highly alkaline soils neutralize some of the acids before runoff carries them into lakes, streams, and rivers. Water with high concentrations of carbonates also will help neutralize the acids. However, in watersheds throughout much of northern Europe, southeastern Canada, and scattered regions of the United States, thin soils overlie solid granite—

Figure 37.3 Average acidity of precipitation, and soil sensitivity to acid deposition, for regions of North America (1984).

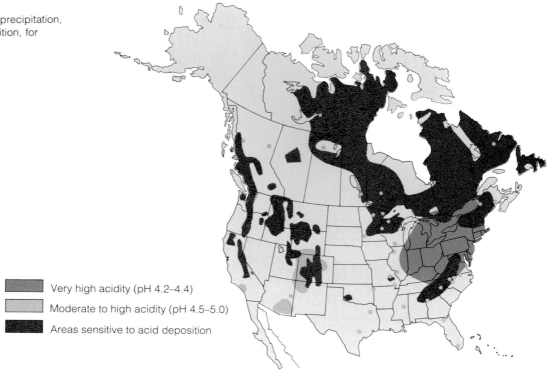

Very high acidity (pH 4.2–4.4)

Moderate to high acidity (pH 4.5–5.0)

Areas sensitive to acid deposition

Figure 37.4 Dead spruce trees in the Whiteface Mountains of New York (**a**) and in Germany (**b**). Acid deposition, perhaps in combination with prolonged exposure to other pollutants, seems to be contributing to the rapid destruction of trees in these and other forests. In some cases, prolonged exposure to multiple air pollutants directly damages the trees, especially conifers. In many cases, the pollutants weaken the trees and make them more susceptible to drought, disease, and insect attacks.

a

b

and such soils provide little buffer against the acids (Figure 37.3).

The precipitation in much of eastern North America is thirty to forty times more acidic than it was several decades ago, and croplands and forests are suffering (Figure 37.4). All fish populations have been wiped out in 300 lakes of the Adirondack Mountains of New York. Some Canadian biologists predict that within the next two decades, fish will disappear from 48,000 lakes in Ontario. Acidic pollutants originating in industrial regions of England and West Germany are acidifying lakes and streams, and damaging large tracts of forests in

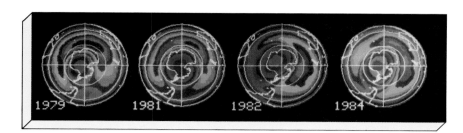

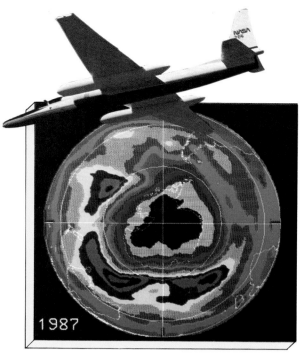

Figure 37.5 Expansion of the ozone hole over Antarctica from 1979 to October 1987, as recorded by special high-altitude planes. The lowest ozone values (the "hole") are indicated by pink colors in the plots for 1979 through 1984. In the plot for 1987, the lowest value ever recorded by that year is indicated by the black area at the center of the plot. The photograph shows the ice clouds over Antarctica that play a role in the formation of an ozone hole each spring.

northern Europe. They also are emerging as a serious problem in heavily industrialized parts of Asia, Latin America, and Africa.

Researchers confirmed years ago that power plants, factories, and vehicles are the main sources of acid depositions, and that the depositions are indeed damaging the environment. Not much has been done about it. Also, some of the responses to local air pollution standards have contributed to the problem, as when very tall smokestacks are added to power plants and smelting plants. The idea is to dump acid-laden smoke high in the atmosphere so winds can distribute it elsewhere—which winds readily do. The world's tallest smokestack, in Sudbury, Ontario, accounts for 1 percent of the annual worldwide emissions of sulfur dioxide.

But Canada cannot be singled out in this issue. Canada presently receives more acid depositions from industrialized regions of the northeastern United States than it sends across its southern border. Most of the acidic pollutants in Finland, Norway, Sweden, the Netherlands, Austria, and Switzerland are blown there from industrialized regions of western and eastern Europe.

Prevailing winds do not stop at national boundaries; the problem is of global concern.

Damage to the Ozone Layer

The ozone layer in the lower stratosphere absorbs most of the ultraviolet wavelengths from the sun—a form of radiation that is harmful to organisms (Figure 5.3). Yet this layer has been thinning since 1976. Each spring, an ozone "hole" appears over the Antarctic; it extends over an area about the size of the continental United States. Less pronounced thinning also occurs all the way into the midlatitudes.

Satellites and high-altitude planes have been monitoring the ozone hole since 1978. Some of the data are shown in Figure 37.5. By 1987, ozone levels above Antarctica had declined by 50 percent—this compared to the previous worst case of 40 percent in 1985.

The reduction in the ozone layer is allowing more ultraviolet radiation to reach the earth's surface, with potentially serious and wide-ranging consequences. Already there has been a dramatic increase in skin cancers, which almost certainly are related to increases in ultraviolet radiation. Cataracts may become more common, and it appears that ultraviolet radiation also can weaken the immune system, making individuals more vulnerable to some viral and parasitic infections. Reduction in the ozone layer also may adversely affect the world's populations of phytoplankton—the basis of food webs in freshwater and marine ecosystems and a factor in maintaining the composition of the atmosphere. (Collectively, these microbial photosynthesizers serve as a sink for carbon dioxide and a source for oxygen.)

The causes of ozone reduction are hotly debated in the scientific community. To be sure, large volcanic eruptions and cyclic changes in solar activity have some effects. But the prime suspects are chlorofluorocarbons (CFCs), which are compounds of chlorine, fluorine, and carbon. These odorless, invisible, and otherwise harmless compounds are widely used as propellants in aerosol spray cans, coolants in refrigerators and air conditioners, and industrial solvents; and they also are used in making plastic foams, including the Styrofoam cups and cartons used for packaging foods, drinks, and other consumer goods. CFCs enter the atmosphere slowly and resist breakdown. By some estimates, about 95 percent of the CFCs released between 1955 and 1987 are still making their way up to the stratosphere.

When a CFC molecule absorbs ultraviolet light, it gives up a chlorine atom. The chlorine can react with ozone to form an oxygen molecule and a chlorine monoxide molecule. When the chlorine monoxide reacts with a free oxygen atom, another chlorine atom is released that can attack another ozone molecule. Each chlorine atom released in the reactions can convert as many as 10,000 molecules of ozone to oxygen!

Recent studies show that chlorine monoxide levels above Antarctica are 100 to 500 times higher than at midlatitudes. Why? High-altitude clouds of ice form there during the frigid winters, and they are isolated from other latitudes by winds that rotate around the South Pole for most of the winter months (Figure 37.5). The ice provides a surface that facilitates the breakdown of chlorine compounds, so that chlorine is free to destroy ozone when the Antarctic air warms somewhat in the spring. (Hence the ozone hole.)

Since 1978, the United States, Canada, and most Scandinavian countries have banned the use of CFCs in aerosol spray cans. Aerosol uses have risen sharply in western Europe, however, as have nonaerosol uses of CFCs throughout the world. In late 1987, an international group assembled by the United Nations Environment Program agreed to a draft treaty to halve CFC emissions by the year 1999. Most nations seem certain to ratify its provisions. The treaty is a step in the right direction,

although some feel that it is too little and too late. CFCs already in the air will be there for over a century, before natural processes neutralize them. You, your children, and your grandchildren will be living with their destructive effects. Think about that, the next time you carry a Styrofoam container out of a fast-food restaurant.

CHANGES IN THE HYDROSPHERE

There is a tremendous amount of water in the world, yet two of every ten humans do not have enough water or, if they do, it is contaminated. Most water is saline (too salty) and cannot be used for human consumption or agriculture. For every 1 million liters of water, only about 6 liters are in a readily usable form.

Consequences of Large-Scale Irrigation

Expansion of agriculture is the basis of the exponential growth of the human population—and about half the food being produced today grows on irrigated land. Water is piped into agricultural fields from groundwater or from lakes and other sources of surface waters.

Irrigation can change the productivity of the land. If the irrigation water contains quantities of salts and the soil does not drain well, evaporation may cause salt buildup in the soil (salination), which can stunt growth, decrease yields, and eventually kill crop plants. Improperly drained irrigated lands can also become waterlogged. Water accumulating underground gradually raises the water table close to the soil surface, saturating the soil around plant roots with toxic saline water. Salinity and waterlogging can be corrected with proper management of the water-soil system, but the economic cost is high.

Groundwater is used for many purposes, but irrigation is often paramount. Consider what is happening to the Ogallala aquifer in the United States (Figure 37.6). Farmers withdraw so much water from the aquifer that the annual overdraft (the amount of water not replenished) is nearly equal to the annual flow of the Colorado River! As a result, the already low water tables in much of the region are falling rapidly, and stream and underground spring flows are dwindling. Where will the water come from when the aquifer is depleted?

Maintaining Water Quality

Not having enough water is serious enough, but the problem is being compounded by increased pollution of the water that is available. Water becomes unfit to drink

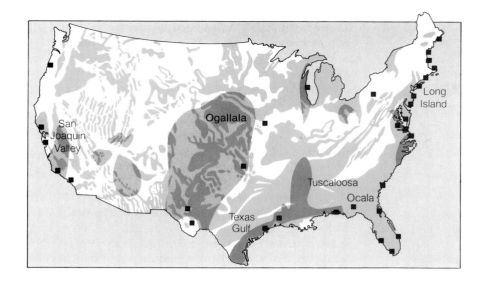

Figure 37.6 Major underground aquifers containing 95 percent of all fresh water in the United States. These aquifers are being depleted in many areas and contaminated elsewhere through pollution and saltwater intrusion. Blue areas indicate major aquifers; gold, the areas of groundwater depletion; and the black boxes indicate areas of saltwater intrusion.

(even to swim in) once it contains human sewage and animal wastes, which harbor pathogenic microbes. Agricultural runoff pollutes water with sediments, insecticides, herbicides, and plant nutrients. Power plants and factories pollute water with chemicals, radioactive materials, and excess heat (thermal pollution).

Pollutants accumulate in lakes, rivers, and bays before reaching the oceans. The polluted muck is dredged from the bottom of rivers and harbors to maintain shipping channels and then is often barged out to sea. Sewage sludge may be barged out, also. For many decades, dredge spoils, industrial wastes, and sewage sludge were dumped into the relatively shallow waters about 9 kilometers (nearly 12 miles) off the coasts of New York and New Jersey, near the mouth of the Hudson River. There, the ocean floor became coated with a black sludge, teeming with bacteria and viruses and highly contaminated with toxic metals and hospital wastes. The sludge makes its presence known when storms wash it ashore on Long Island and New Jersey beaches.

The United States has about 15,500 facilities for treating the liquid wastes from about 98 percent of the urban population and from 87,000 industries. Wastes from the remaining population (mostly suburban and rural) are treated in lagoons or septic tanks or discharged—untreated—directly into waterways.

There are three levels of wastewater treatment. In *primary treatment*, screens and settling tanks remove the sludge (coarse suspended solids), which is then burned, dumped in landfills, or treated further. Chlorine often is used to kill pathogens in the water, but it does not kill them all. Also, chlorine may react with certain industrial chemicals to produce chlorinated organic compounds, some of which may cause cancer.

In *secondary treatment*, microbial populations are employed to degrade the organic matter. After primary treatment (but before chlorination), the wastewater is either (1) sprayed and trickled through large beds of exposed gravel in which the microbes live or (2) aerated in tanks and seeded with microbes. The microbial "employees" are sometimes poisoned by toxic substances dissolved in the water. Then the treatment facilities are shut down until populations are reestablished.

Secondary treatment does not remove all of the oxygen-demanding wastes, suspended solids, nitrates, phosphates, viruses, and toxic substances, including heavy metals, pesticides, and industrial chemicals. Sometimes the water gets chlorinated before being released into the waterways, and sometimes not.

Tertiary treatment involves expensive and largely experimental methods of precipitating suspended solids and phosphate compounds, adsorption of dissolved organic compounds, reverse osmosis, stripping nitrogen from ammonia, and disinfecting the water through chlorination or ultrasonic energy vibrations. It may adequately reduce pollution levels, but it is used on only about 5 percent of the nation's wastewater.

What all this means is that most wastewater is not being properly treated. A typical pattern is repeated thousands of times along our waterways. Water for drinking is removed *upstream* from a city, and wastes from industry and sewage treatment are discharged *downstream*. It takes no great leap of the imagination to see that pollution intensifies as rivers flow toward the oceans. In Louisiana, where waters drained from the central states flow toward the Gulf of Mexico, pollution levels are high enough to be a threat to public health. Water destined for drinking does get treated to remove pathogens—but the treatment does not remove all the toxic wastes from numerous factories upstream. *You may find it illuminating to investigate where your own city's supply of water comes from and where it has been.*

CHANGES ON LAND

Solid Wastes

Resources are scarce in the developing countries, and very few materials are discarded. In the more affluent countries, the United States especially, a "throwaway" mentality prevails. Consumers use something once, discard it, and buy another.

Billions of metric tons of solid wastes are dumped, burned, and buried annually in the United States alone. About 60 billion beverage containers are part of it—50 billion of which are nonreturnable cans and bottles. Paper products represent half the total volume of solid wastes. If only half of those paper wastes were recycled, energy as well as trees could be conserved. The energy it takes to produce an equivalent amount of new paper

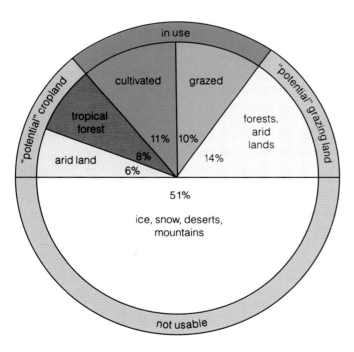

Figure 37.7 Classification of the earth's land. Theoretically, the world's cropland could be doubled in size by clearing tropical forests and irrigating arid lands. But converting this marginal land to cropland would destroy valuable forest resources, cause serious environmental problems, and possibly cost more than it is worth.

could be diverted to provide electricity to about 10 million homes each year. Also, it takes 150 acres of forest to produce the paper in each Sunday Edition of the *New York Times* alone.

Associated with the throwaway mentality is a problem that is rather unique in the world of life—what to do with the solid wastes. Instead of recycling materials, as is done in natural ecosystems, we bulldoze them down the sides of canyons or dump them in wetlands or landfills. There are only so many canyons; what happens when they are filled?

A transition from a throwaway mentality to one based on conservation and reuse is economically and technologically feasible. Consumers can put pressure on manufacturers by refusing to buy goods that are lavishly wrapped, excessively boxed, and designed for one-time use. Individuals can ask the local post office to turn off their daily flow of junk mail, which wastes an astounding amount of paper, time, and energy. Individuals can urge local governments to develop well-designed, large-scale resource recovery centers. Such a center has been operating in Saugus, Massachusetts. With such systems, existing dumps and landfills would be urban "mines" from which some materials might be recovered.

Conversion of Marginal Lands for Agriculture

The dumping of solid wastes is peculiar to affluent societies. There is another type of assault on the land that is occurring throughout the world, and that is the expansion of cultivation and grazing into areas that are only marginally suitable for agriculture (Figure 37.7). Almost 21 percent of available land is now being used for agriculture. Another 28 percent is said to be potentially suitable for cropland or grazing land, but its potential productivity is so low that conversion may not be worth the cost.

Asia and some other heavily populated regions suffer severe food shortages, yet more than 80 percent of the productive lands are already being intensively cultivated there. Those countries rely on subsistence agriculture (with energy inputs from sunlight and from human labor) and animal-assisted agriculture (with energy inputs from human labor and the work of draft animals, such as oxen). Valiant efforts have been made to improve crop production on existing land. Under the banner of the so-called **green revolution,** research has been directed toward (1) improving the varieties of crop plants for higher yields and (2) exporting modern agricultural practices and equipment to developing countries.

Modern agriculture is based on massive inputs of fertilizers and pesticides, and ample irrigation to sustain high-yield crops. It is based also on fossil fuel energy to drive the farming machines. Although crop yields are

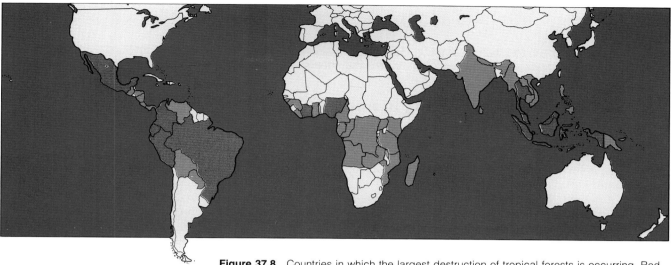

Figure 37.8 Countries in which the largest destruction of tropical forests is occurring. Red shading signifies regions where 2,000 to 14,800 square kilometers are deforested annually; orange signifies regions of more moderate deforestation (100 to 1,900 square kilometers).

four times as high, the modern practices use up a hundred times more energy and mineral resources.

Besides, the plain truth is that developing countries depend on subsistence farmers who cannot afford to take widespread advantage of the new crop strains. The ones who can afford to make the investment come to depend on industrialized producers of fertilizers and machinery. Of necessity, the costs of fertilizers and machinery are reflected in market food prices—which are too high for much of the country's own population.

Pressures to open up new land for farming are most intense in areas of Central and South America, Asia, the Middle East, and Africa. There, the human population is rapidly expanding into marginal lands, and the repercussions will extend beyond national boundaries, as the following discussion will make clear.

Deforestation

The world's great forests play major roles in the biosphere. As we saw in Chapter 36, forested watersheds are like giant sponges that absorb, hold, and release water gradually. By influencing the downstream flow of water, forests help control soil erosion, flooding, and sediment buildup in rivers, lakes, and reservoirs. Deforestation, especially on steep slopes, leads to loss of the fragile soil layer and disrupts the watershed. In the tropics, soil loss means long-term fertility loss as nutrients are quickly washed out of the system, leaving nutrient-poor soil behind (see the *Commentary* on page 578).

More than this, deforestation also can change regional patterns of rainfall as a result of altered rates of evaporation, transpiration, and runoff. For example, between fifty and eighty percent of the water vapor above tropical forests is released from the trees themselves. Without trees, annual precipitation declines and the region gets hotter and drier. Rain that does fall rapidly runs off the bare soil. As the local climate gets hotter and drier, soil fertility and moisture levels decline even more. Eventually, sparse grassland or even desertlike conditions might prevail where there had once been a rich tropical forest.

Clearing large tracts of tropical forests also may have global repercussions. Consider that these forests absorb much of the solar radiation reaching the equatorial regions of the earth's surface. When they are cleared, the land becomes shinier, so to speak, and reflects more incoming energy back into space. Consider also that the trees of these vast forested regions help maintain the global cycling of carbon and oxygen through their photosynthetic activities. When they are harvested or burned, the carbon stored in their biomass is released to the atmosphere in the form of carbon dioxide—and this may play a role in the amplified greenhouse effect.

Almost half of the world's expanses of tropical forests has already been cleared for cropland, grazing land, timber, and fuel. Deforestation is greatest in Brazil, Indonesia, Colombia, and Mexico (Figure 37.8). At present rates of clearing and degradation, only Brazil and Zaire will have large tracts of tropical forests in the year 2010; by 2035, those forests also will be gone.

Tropical Forests—Disappearing Biomes?

Tropical forests contain an incredible variety of organisms. Despite the diversity, they are one of the worst places to grow crops. Because of rapid decomposition in the hot, humid climates where such forests occur, there is practically no litter on the forest floor and very little nutrient storage in the subsoil. Minerals released during decomposition are rapidly picked up by roots and mycorrhizae in the top soil layers, and most become tied up in the standing biomass.

With *slash-and-burn agriculture*, forest biomass can be reduced to nutrient-rich ashes, then the ashes can be tilled into the soil. Even then, heavy rains wash away most nutrients from the exposed clay soils. After a few years, cleared plots become infertile and usually are abandoned. Because nutrients are so depleted, successional replacement is extremely slow.

Developing countries in Central America, Africa, and Southeast Asia have been clearing their tropical forests on a massive scale. By some estimates, 90 percent of the total human population growth is occurring in those countries. The land is being cleared for fuel and farmland (see Figures a and b). At present rates of clearing, most tropical rain forests will certainly disappear by the year 2035.

Clearing tropical forests means extinction for thousands of species. It will be our loss as well as theirs. A

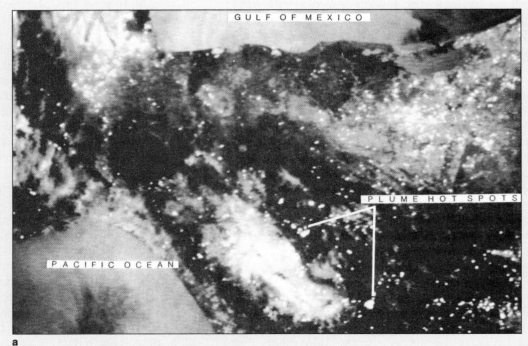

a

(**a**) Satellite photograph of southern Mexico and northern Guatemala taken on April 18, 1984. The well-defined white spots are fires, most of which are associated with land clearing for agriculture. The white area near the center of the photograph (the Grijalva Basin) is a major agricultural region; most of the burning here is for clearing previously cultivated land. The areas in the upper left (around Veracruz) and in the lower right (around the Guatemala border) are primarily virgin tropical forest, being cleared at great ecological cost for agriculture. (Satellite sensory devices penetrated the smoke to reveal the underlying fire activity.) (**b**) Tropical rain forest in Brazil being burned to clear the land for cattle grazing.

very small number of crop and livestock species represents most of the world's food supply. However, our food base can be broadened and made less vulnerable if we can develop new or hybrid crop plants. Through genetic engineering, diverse organisms of tropical rain forests can be tapped as genetic resources for developing improved varieties of crop plants, new antibiotics, and new vaccines.

Many alkaloids from tropical plants already are used in drug treatments of cardiovascular disorders, cancer, and other illnesses. Aspirin, probably the most widely used drug in the world, was formulated by using a chemical "blueprint" of a compound extracted from the leaves of tropical willow trees. Coffee, bananas, cocoa, cinnamon and other spices, sweeteners, and many other foods we take for granted originated in the tropics. So did latex, gums, resins, dyes, waxes, many oils, and other substances used in such diverse products as ice cream, toothpaste, shampoo, condoms, cosmetics, perfumes, records, tires, and shoes.

Several international groups have been working together on a comprehensive plan to preserve tropical forests, and the plan is already being implemented. A few developing countries have been reevaluating their agricultural policies. Brazil has designated about 100,000 square kilometers of tropical rain forest as unsuitable for agriculture. The region has been set aside for ecological research and for recreation.

b

Figure 37.9 Dust storm approaching Prowers County, Colorado, in 1934. The Great Plains of the American Midwest are normally dry, windy, and subject to severe recurring droughts. Beginning in the 1870s, the land was converted to agriculture. Overgrazing destroyed large regions of natural grassland, leaving the ground bare. In May 1934, the entire eastern portion of the United States was blanketed with a massive dust cloud of topsoil blown off the land, giving the Great Plains a new name—the Dust Bowl. About 9 million acres of cropland were destroyed and 80 million more severely damaged. Today, without massive irrigation and intensive conservation farming, desertlike conditions could prevail in this region.

Desertification

Desertification refers to the conversion of grasslands, rain-fed cropland, or irrigated cropland to desertlike conditions, with a drop in agricultural productivity of 10 percent or more. About 9 million square kilometers have turned into deserts over the past fifty years; at least 200,000 square kilometers are still being transformed each year. Prolonged drought may accelerate desertification, as it did in the American Midwest many decades ago (Figure 37.9). Today, however, large-scale desertification is occurring mainly as a result of overgrazing on marginal lands.

In Africa, for example, there are too many cattle in the wrong places. Cattle require more water than the wild herbivores that are native to the region, and so they move back and forth between grazing areas and watering holes. When doing this, they trample grasses and compact the soil surface (Figure 37.10). In contrast, gazelles, elands, and other native herbivores obtain most (if not all) of the water they require from the vegetation they eat. They also are better at conserving water; little is lost in feces, compared to cattle.

In 1978, the biologist David Hopcraft started a ranch composed of antelopes, zebras, giraffes, ostriches, and other native herbivores. He is raising cattle as "control groups" in order to compare costs and meat yields on the same land. Results are exceeding expectations. Native herds are increasing steadily and yielding meat. And range conditions are improving, not deteriorating. There are still problems to overcome. African tribes have their own idea of what constitutes "good" meat, and some tribes view cattle as symbols of wealth.

A QUESTION OF ENERGY INPUTS

Paralleling the J-shaped curve of human population growth is a steep rise in energy consumption. The rise is due not only to increased numbers of energy users, but also to extravagant consumption and waste.

For example, in one of the most temperate of all climates, a major university constructed seven- and eight-story buildings with narrow, sealed windows. The windows cannot be opened to catch the prevailing ocean breezes; the windows and the buildings themselves were not designed or aligned to take advantage of the abundant sunlight for passive solar heating and breezes for passive solar cooling. Massive energy-demanding cooling and heating systems are used instead.

Inefficiencies of this sort may be curtailed sooner than might be expected, for current energy supplies are limited. When you hear talk of abundant energy supplies, keep in mind that there is an enormous difference between the total supply and the net amount available. **Net energy** is the energy left over after subtracting the energy used to locate, extract, transport, store, and deliver energy to consumers. In addition, some sources of energy, such as direct solar energy, are renewable; others, such as coal and petroleum, are not. Currently, eighty-two percent of the energy stores being tapped fall in the second category (Figure 37.11).

Figure 37.10 Desertification in the Sahel, a region of West Africa that forms a belt between the dry Sahara Desert and tropical forests. This is savanna country that is rapidly undergoing desertification as a result of overgrazing and overfarming.

Fossil Fuels

Fossil fuels are carbon-containing remains of plants that lived hundreds of millions of years ago. The plants were buried and compressed in sediments and gradually transformed into coal, petroleum (oil), and natural gas. Often you hear about annual "production" rates for fossil fuels. How much oil, coal or natural gas do we really produce each year? None. We simply *extract* it from the earth.

Even with stringent conservation, known petroleum and natural gas reserves may be depleted during the next century. As petroleum and natural gas deposits become depleted in easy-to-reach areas, we begin to seek new sources, often in wilderness areas such as Alaska and in other fragile environments such as the continental shelves. The *net* energy decreases as costs of extraction and transportation increase; the environmental costs of extraction and transportation escalate. At present, the long-term impact of the 100-million-gallon spill from the tanker *Valdez* in Alaska's coastal waters is still not understood.

Colorado, Utah, and Wyoming probably have more potential oil than the entire Middle East. These states have vast deposits of oil shale, a type of buried rock containing the hydrocarbon kerogen. However, the extraction process may disfigure the land, increase water and air pollution, and tax existing water supplies in regions already facing water shortages.

What about coal? World reserves might meet the energy needs of the human population for at least several

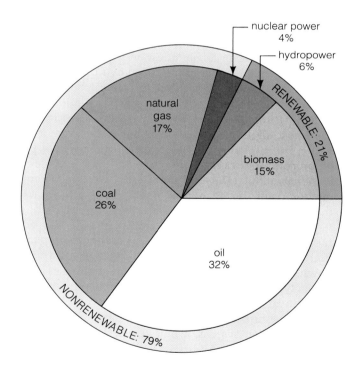

nuclear power 4%

hydropower 6%

RENEWABLE: 21%

natural gas 17%

coal 26%

biomass 15%

oil 32%

NONRENEWABLE: 79%

Figure 37.11 World consumption of nonrenewable and renewable energy sources in 1986.

centuries. But coal burning has been the largest single source of air pollution; most coal reserves contain low-quality, high-sulfur material. Unless the sulfur is removed before or after burning, sulfur oxides are released into the air and add to the global problem of acid deposition. Fossil fuel burning also releases carbon dioxide and amplifies the greenhouse effect.

Pressure is on to permit widespread strip-mining of coal reserves close to the earth's surface. Strip-mining limits the usefulness of the land for agriculture, grazing, and wildlife. Restoration is difficult and expensive in arid and semiarid lands, where most strip-mining occurs.

Nuclear Energy

As Hiroshima burned in 1945, the world recoiled in horror from the destructive potential of nuclear energy. Optimism replaced horror as nuclear energy became publicized during the 1950s as an instrument of progress. Today, nuclear power plants dot the landscape. Industrialized nations that are poor in energy resources, including France, depend heavily on nuclear power. Yet in most countries, plans to extend reliance on nuclear energy have been delayed or canceled. The cost, efficiency, environmental impact, and safety of nuclear energy are being seriously questioned.

The overall net energy produced by nuclear reactors is relatively low, and the cost of constructing nuclear power plants is high—much higher than initially expected. Using nuclear energy to generate electricity may soon cost slightly more than using coal—even if the coal-burning plants are equipped with expensive pollution control devices.

What about safety? Radioactivity escaping from a nuclear plant during normal operation is actually less than the amount released from a coal-burning plant of the same capacity. Also, nuclear plants do not add carbon dioxide to the atmosphere, as coal-burning plants do. However, there is the potential danger of a **meltdown.** As nuclear fuel breaks down, it releases heat, which typically is absorbed by water circulating over the fuel. Steam from the heated water drives electricity-generating turbines. Should a leak develop in the circulating water system, water levels around the fuel might plummet. The nuclear fuel would heat rapidly, past its melting point. The melting fuel would pour onto the floor of the generator, where it would come into contact with the remaining water and instantly convert it to steam. Formation of enough steam, along with other chemical reactions, could blow the system apart, releasing radioactive material. Also, the overheated reactor core could melt through its thick concrete containment slab and into the earth, causing groundwater contamination.

COMMENTARY

Biological Principles and the Human Imperative

Molecules, cells, tissues, organs, organ systems, multicelled organisms, populations, communities, ecosystems, the biosphere. These are the architectural systems of life, assembled in increasingly complex ways over the past $3\frac{1}{2}$ billion years. We are latecomers to this immense biological building program. Yet, during the relatively short span of 10,000 years, we have been restructuring the stuff of life at all levels—from recombining DNA of different species to changing the nature of the land, the oceans, and the atmosphere.

It would be presumptuous to think we are the only organisms that have ever changed the nature of living systems. Even during the Proterozoic Era, photosynthetic organisms were irrevocably changing the course of biological evolution by gradually enriching the atmosphere with oxygen. In the present as well as the past, competitive adaptations have assured the rise of some groups, whose dominance has assured the decline of others. Thus, change is nothing new to this biological building program. What *is* new is the accelerated, potentially cataclysmic change being brought on by the human population. We now have the population size, the technology, and the cultural inclination to use energy and modify the environment at frightening rates.

Where will rampant, accelerated change lead us? Will feedback controls begin to operate as they do, for example, when population growth exceeds the carrying capacity of the environment? In other words, will negative feedback controls come into play and keep things from getting too far out of hand?

Feedback control will not be enough, for it operates only when deviation already exists. Our explosive population growth and patterns of resource consumption are founded on an illusion of unlimited resources and a forgiving environment. A prolonged, global shortage of food or the passing of a critical threshold for the global engines can come too fast to be corrected. At some point, such deviations may have too great an impact to be reversed.

What about feedforward mechanisms? Many organisms have early warning systems. For example, skin receptors sense a drop in outside air temperature. Each sends messages to the nervous system, which responds by triggering mechanisms that raise the core temperature before the body itself becomes dangerously chilled. With feedforward control, corrective measures can begin before change in the external environment significantly alters the system.

Even feedforward controls are not enough for us, for they go into operation only when change is under way. Consider, by analogy, the DEW line—the Distant Early Warning system. This system is like a sensory receptor, one that detects intercontinental ballistic missiles that may be launched against North America. By the time the system detects what it is designed to detect, it may be far too late, not only for North America but for the entire biosphere.

It would be naive to assume we can ever reverse who we are at this point in evolutionary time, to de-evolve ourselves culturally and biologically into becoming less complex in the hope of averting disaster. However, there is reason to believe that we can avert disaster by using a third kind of control mechanism, one that is uniquely our own. We have the capacity to anticipate events *before* they happen. We are not locked into responding only after irreversible change has begun. We have the capacity to anticipate the future—it is the essence of our visions of utopia or of nightmarish hell. Thus we all have the capacity for adapting to a future that we can partly shape. We can, for example, learn to live with less. Far from being a return to primitive simplicity, it would be one of the most complex and intelligent behaviors of which we are capable.

Having that capacity and using it are not the same thing. We have already put the world of life on dangerous ground because we have not yet mobilized ourselves as a species to work toward self-control. Our survival depends on predicting possible futures. It depends on designing and constructing ecosystems that are in harmony not only with what we define as basic human values but also with the biological models available to us. Human values can change; our expectations can and must be adapted to biological reality. *For the principles of energy flow and resource utilization, which govern the survival of all systems of life, do not change.* It is our biological and cultural imperative that we come to terms at last with these principles, and ask ourselves what our long-term contribution will be to the world of life.

In 1986, the potential dangers of nuclear power were brought into sharp focus by a meltdown at the Chernobyl power station in the Soviet Union. Radiation was released into the atmosphere as the plant's containment structures were breached. A number of people died immediately and others died later of radiation sickness. Throughout Europe, people are still concerned about the long-term consequences. How long will the environment be contaminated? What are the risks of cancer from such exposure? We are in the midst of our first major, real-world experiment to find out.

The Chernobyl incident underscores the consequences of nuclear accidents. What about routine nuclear wastes? Nuclear fuel cannot be burned to harmless ashes, like coal. After about three years, the fuel elements of a reactor are spent. They still contain uranium fuel, but they also contain hundreds of new radioactive isotopes. Altogether, the wastes are an enormously radioactive, extremely dangerous collection of materials. As they undergo radioactive decay, they produce tremendous heat. They are immediately plunged into water-filled pools and stored for several months at the power plant. The water cools the wastes and keeps radioactive material from escaping. At the end of the holding period, the remaining isotopes are still lethal, and some must be isolated for 10,000 years. If a certain isotope of plutonium (^{239}Pu) is not removed, the wastes must be isolated for a quarter of a million years!

There are plans to seal radioactive wastes in ceramic material, place them in steel cylinders, then bury them deep underground in supposedly stable rock formations that are free from exposure to salt water and earthquakes for at least 10,000 years. No radioactive wastes have yet been put into permanent underground storage in the United States. Such facilities are not expected to be available until after the turn of the century.

Breeder reactors might eventually be developed; these would consume a rare isotope of uranium, and also convert a much greater amount of a common form of uranium (^{238}U) to an isotope of plutonium that can be used as nuclear fuel. Even though a conventional reactor cannot explode like an atomic bomb, a breeder reactor could undergo a very small nuclear explosion.

Also on the drawing boards is fusion power, in which hydrogen atoms are fused to form helium atoms with release of considerable energy. Such a power source would be something like reactions that produce heat energy in the sun. Without a major breakthrough, fusion power is not expected to be available to produce electricity on a commercial basis until the last half of the next century, if ever.

Quite probably, nuclear energy has given us the means of causing a mass extinction equal to those of past geologic eras (page 201). According to one scenario, a nuclear exchange involving about one-third of the existing American and Soviet arsenals would probably kill outright between 40 and 65 percent of the human population, along with a good portion of most other forms of life. Those escaping immediate death would have to remain in shelters for a week to three months or more to avoid exposure to dangerous radiation levels. The nuclear detonations would inject a huge, dark cloud of soot and smoke over most of the earth, especially the Northern Hemisphere, and would block out the sun. Much of the planet might be thrown into darkness, and temperatures would fall below freezing for months—an effect called **nuclear winter.** If the freezing lasted for a shorter time, we might have a nuclear "autumn"—but even then, the cold temperatures and darkness might be well beyond the tolerance limits of many plant and animal species.

SUMMARY

1. For more than a century, the human population has been growing exponentially, with concurrent increases in energy demands and pollution of the atmosphere, the hydrosphere, and the land.

2. Pollutants are substances with which ecosystems have had no prior evolutionary experience (in terms of kinds and amounts) and so have no mechanisms for dealing with them. By a narrower definition, pollutants are substances that adversely affect the health, activities, or survival of human populations.

3. Industrial and photochemical smogs are examples of regional air pollution. Acid deposition and depletion of ozone in the lower stratosphere are examples of air pollution with global effects.

4. The exponential growth of the human population has its foundations in the expansion of agriculture, which requires large-scale irrigation. Global supplies of fresh water are limited, and supplies are being polluted by agricultural runoff (which includes sediments, insecticides, herbicides, and fertilizers), industrial wastes, and human sewage.

5. Human populations are adversely affecting the land surface by the tremendous accumulation of solid wastes and by the conversion of marginal lands for agriculture. The wholesale destruction of tropical rain forest biomes is affecting regional soils and patterns of rainfall. The deforestation also may be amplifying the greenhouse effect. Millions of acres are undergoing desertification each year.

6. Energy supplies in the form of fossil fuels are non-renewable, they are dwindling, and their extraction and

use come at high environmental cost. Nuclear energy normally does not pollute the environment as much as fossil fuels do, but the risks associated with fuel containment and with storing radioactive wastes are enormous.

Review Questions

1. Make a list of the advantages you personally enjoy as a member of an affluent, industrialized society. Then list some of the drawbacks. Do you believe that the benefits outweigh the costs? (This is not a trick question.)

2. Describe some of the potential global consequences of acid deposition, reduction of the ozone layer, and amplification of the greenhouse effect. *571–574*

3. What are some of the consequences of deforestation? Of desertification? *577–580*

4. It is not likely that humans will turn their back on their technological advances and economic growth as a way of putting the biosphere on a stable footing. Is it possible that massive amounts of money and technology will be put to use to find answers as well as solutions? What prospects and problems do you foresee in making a global effort in this regard?

Self-Quiz *(Answers in Appendix IV)*

1. Since the mid-eighteenth century, the pattern of human population growth has been _____.

2. Human-generated _____ are disrupting the complex interactions of the sun, atmosphere, oceans, and land that are the engines of the biosphere.

3. Humans have not yet come to terms with the principles of _____ and _____ utilization, which govern all forms of life on earth.

4. For more than 100 years, the human population has been _____.
   a. growing exponentially
   b. increasing its energy demands
   c. polluting the atmosphere
   d. polluting the oceans and the land
   e. all of the above

5. Pollutants disrupt ecosystems because _____.
   a. they are composed of elements that differ from those of natural molecules
   b. of all the kinds of living things, only humans have uses for them
   c. ecosystems have not encountered them before and so do not have any evolved mechanisms that can handle them
   d. their only effect is on ecosystems but not humans

6. Industrial and photochemical smogs are examples of _____ air pollution; acid deposition and depletion of ozone in the lower stratosphere are examples of _____ air pollution.
   a. local; local   c. global; global
   b. local; global   d. global; local

7. Rapid deforestation, especially of rain forests, may be amplifying the _____.
   a. pollution of water
   b. already serious loss of energy
   c. greenhouse effect
   d. dwindling loss of fossil fuels

8. Energy from fossil fuels is _____; their extraction and use come at _____ cost to the environment.
   a. renewable; low
   b. nonrenewable; low
   c. renewable; high
   d. nonrenewable; high

9. Nuclear energy normally pollutes the environment _____ than fossil fuels; the problems associated with it are _____ than with fossil fuels.
   a. less; lesser
   b. more; greater
   c. more; lesser
   d. less; greater

10. Match the human ecological concepts.
   _____ atmospheric pollutants
   _____ agricultural runoff
   _____ exponential
   _____ deforestation
   _____ factors that modify environment

   a. includes sediments, insecticides, herbicides, fertilizers
   b. amplifies greenhouse effect
   c. current pattern of human population growth
   d. population size, cultural habits, technology, energy use
   e. disrupts interactions between sun, atmosphere, oceans, land

Selected Key Terms

acid rain *571*
chlorofluorocarbon (CFC) *570*
deforestation *577*
desertification *580*
fossil fuel *581*
green revolution *576*
industrial smog *571*
meltdown *582*

nuclear winter *584*
ozone hole *573*
PAN (peroxyacyl nitrate) *571*
photochemical smog *571*
pollutant *570*
slash-and-burn agriculture *578*
thermal inversion *570*
wastewater treatment *575*

Readings

Anderson, S. 1985. *Managing Our Wildlife Resources.* Columbus, Ohio: Merril.

Gribbin, J. 1988. *The Hole in the Sky.* New York: Bantam.

Gruber, D. 1989. "Biological Monitoring and Our Water Resources." *Endeavour* 13(3):135–140.

Miller, G.T. 1990. *Living in the Environment.* Sixth edition. Belmont, California: Wadsworth.

Mohnen, V. August 1988. "The Challenge of Acid Rain." *Scientific American* 259(2):30–38.

Repetto, R. April 1990. "Deforestation in the Tropics." *Scientific American* 262(4):36–42.

Wilson, E. 1988. *Biodiversity.* Washington, D.C.: National Academy of Sciences.

38 ANIMAL BEHAVIOR

KEY CONCEPTS

1. "Behavior" refers to all those observable, coordinated responses an animal makes to external and internal stimuli. It has a heritable basis, in that genes contain instructions for the development of the organ systems that are required to sense, interpret, and respond to stimuli. Thus, like other traits having a genetic basis, forms of behavior are subject to evolution by natural selection.

2. Behavior is not shaped by genes alone. It is shaped also by each animal's physical and chemical environment, its interactions with other species, and by interactions with others of its own species.

3. In the evolutionary view, the adaptive forms of behavior are the ones that promote reproductive success of individuals. The concept of individual selection can be applied even to self-sacrificing behavior that helps others of a social group but decreases an individual's chance to produce offspring of its own.

With the arrival of spring in a Canadian forest, a male white-throated sparrow whistles a song that sounds rather like "Sam Peabody, Peabody, Peabody." He repeats the song thousands of times and with such clarity and consistency, we might wonder how he does it, and what good it does him. We might also wonder why male swamp sparrows and white-crowned sparrows living in the same forest have distinctive songs of their own. Do all those birds automatically "know" what they are supposed to sing the first time they do it, or do they learn something from their surroundings that influences the way they sing? Questions of this sort lead us into the world of animal behavior studies.

Animal behavior refers to the observable, coordinated responses an animal makes to external and internal stimuli. Interactions among nervous, endocrine, and skeletal-muscular systems typically bring about the responses. *Heredity* provides the animal with its basic response mechanisms. During early development, the physical and chemical environment of the animal helps mold the glands, muscles, and other body parts involved in behavioral responses (page 465). *Learning* also influences behavior, with the animal modifying its responses as it interacts with the environment (Figure 38.1).

Figure 38.1 Two songbirds, a white-throated sparrow (**a**) and zebra finch (**b**), belting out their distinctive territorial songs. As with other forms of behavior, singing has genetic and learned components. The birds' capacity to sing is genetically based, when and if they sing is hormone-dependent, and while they are still juveniles, learning experiences smooth out the melodies.

a

GENES, HORMONES, AND BEHAVIOR

Genetic Basis of Behavior

The behavioral responses that any animal can make depend on the physical layout of its nervous system and the patterns of activity among the component neurons. Because genes dictate how the nervous system itself develops, they contribute in a major way to behavior.

To illustrate this point, suppose a researcher puts eggs from white-throated sparrows and white-crowned sparrows in an incubator, then hand-rears all the hatchlings under the exact same conditions. As they grow up, the young birds are allowed to listen to tape-recorded songs of adult white-throats and white-crowns. Before a year passes, the males are singing. The white-throats sing "Sam Peabody, Peabody, Peabody," and the white-crowns sing a different song.

Because the white-throated and white-crowned birds were raised under identical environmental conditions, the *difference* in their singing behavior must be the result of genetic differences between them. The number of genes underlying the difference is not important; even a one-gene difference can produce behavioral variations.

b

By analogy, imagine substituting garlic powder for sugar when following a recipe for chocolate cake. Even if all the other ingredients and all other cake-making steps are identical, that one substitution will have a dramatic effect on the cake.

Hormonal Effects on Behavior

Behavior is profoundly influenced by hormones, the signaling molecules of the endocrine system. Consider how melatonin affects reproductive behavior in white-throats and other songbirds. Melatonin, a hormonal secretion of the pineal gland, suppresses the growth and function of gonads (testes and ovaries). Its secretion is inhibited by sunlight, which acts on photoreceptors in the pineal. During spring, when daylight hours increase, melatonin levels decline, so the gonads are free to increase in size and start secreting hormones of their own. The hormones include estrogen and testosterone, and they trigger the physiological changes required for singing, mating, and other forms of behavior.

In very young male (not female) songbirds, estrogen influences the development of a "song system," which includes certain brain regions and nerve pathways leading to vocal organs. At the start of the breeding season, the enlarged male gonads also secrete more testosterone. It is the triggering mechanism that allows the bird to sing when it has staked out a territory, repel other males, and attract a female. Thus, estrogen *organizes* the development of the song system, then testosterone *activates* the song system and prepares the bird to sing when properly stimulated.

INSTINCT AND LEARNING

Instinctive Behavior

Animals can complete fairly complex, stereotyped responses to a number of environmental cues, even without having had prior experience with those cues. This capacity is called **instinct**. Instinctive behavior is automatically set in motion by rather simple cues, as the following examples suggest.

Cuckoo eggs are laid in the nests of other birds. A newly hatched cuckoo instinctively responds to the shape of any eggs it touches by pushing them out of

b

Figure 38.2 (**a**) A newly hatched cuckoo making a complex, innate behavioral response to an environmental cue (spherical objects in its foster parents' nest). The European cuckoo lays eggs in the nests of other species. Even before the newly hatched cuckoo opens its eyes, it responds to the shape of the host's eggs by shoving them out of the nest. (**b**) The foster parents keep feeding the usurper, even when it has grown larger than they are.

the nest—and so displaces its rightful occupants (Figure 38.2). Male stickleback fish guard eggs that they alone fertilize, they chase away rival males, and when opportunity strikes they eat the eggs that other males are guarding. The males have a red belly, and this is the cue for their aggressive behavior. In fact, sticklebacks chase almost any red object even if they have never before seen a red object of any sort. Humans, too, have instincts. Infants instinctively smile at a face-sized flat mask with two dark spots corresponding to where eyes would be on a face (one "eye" will not do the trick).

In the past, "instinctive" or "innate" behavior was said to be genetically determined, as if some forms of behavior could arise without environmental influence. However, no aspect of an individual—behavior included—can develop without both genetic and environmental contributions.

Learned Behavior

Many animals make connections between different experiences and the consequences of their actions. **Learning** is the adaptive modification of behavior in response to neural processing of information that has been gained from specific experiences. Even though a newly hatched peacock pecks the ground (an instinctive feeding response), it soon learns from personal experience that some areas of its habitat have more food to peck at, and that some times of day are better than others to go after it. The capacity to learn resides in the genetically prescribed wiring of the nervous system, where information is sorted and modified in ways that lead to altered behavior. And that wiring greatly influences what its owner *can* learn (page 434).

Learning takes place in a variety of ways. For example, animals can learn by association; they make a connection between a new stimulus and a familiar one. Dogs make a simple reflex response to a meat extract placed on their tongue; they salivate. If they regularly hear a bell just before getting the extract, however, they start salivating at the sound of the bell alone.

Associative learning often occurs by trial and error, when a chance behavior is followed by a stimulus that rewards or punishes the behavior. Put an earthworm in a simple T-maze (which has a base and two arms shaped like a "T"). When the worm turns down one arm, it gets an electric shock. When it turns down the other arm, it enters an earthworm-friendly chamber (moist and dark). After many trials and enough shocks, the "right" response becomes more frequent.

Learning also occurs through insight, whereby the brain integrates accumulated experiences and assesses responses that might work in new situations. Primates and many other animals seem capable of insight learning. Some chimpanzees were put in an enclosure with boxes scattered on the floor and a banana dangling from a high ceiling. They piled up the boxes under the banana and climbed up, having figured out (without trial-and-error practice) that the banana could be theirs.

Imprinting

Sometimes the capacity to learn specific information is pronounced during certain stages of development. This time-dependent form of learning is called **imprinting**.

a   b

Figure 38.3 (**a**) Human imprinting objects. No one can tell these goslings that Konrad Lorenz is not Mother Goose. (**b**) An imprinted rooster wading out to meet the objects of his affections. During a critical period of the rooster's life, he was exposed to a mallard duck. Although sexual behavior patterns were not yet developing during that period, the imprinting object became fixed in the rooster's mind for life. Then, with the maturation of sexual behavior, the rooster sought out ducks, forsaking birds of his own kind, and lending further support to the finding that imprinting may be one of the reasons why birds of a feather do flock together.

During a short sensitive period early in life, many kinds of birds are primed to form a learned attachment to a moving object—normally, their mother as she moves away from the nest. Konrad Lorenz separated newly hatched goslings from their mother and discovered that they formed an attachment to and followed a moving object, even a human (Figure 38.3). When the hatchlings were not offered any object to follow within a couple of days, they lost their readiness to imprint.

Imprinting has long-term consequences. When the male goslings matured, they directed their sexual behavior toward members of whatever species they followed as hatchlings. This normally would be a female member of their own species, but the sexually mature male goslings that had imprinted on Lorenz early in life preferred to court humans!

Imprinting also plays a role in migratory behavior, as when birds travel from a summer breeding area to a distant winter refuge. Like other migrating animals, the birds have a compass sense (they can travel in a constant direction) and a navigational sense (they have a sense of destination).

Consider the compass sense of one songbird, the indigo bunting, which breeds in the northeastern United States and spends winters in Mexico south to Panama. Young buntings imprint on the visual image of the night sky during a sensitive period. Then they leave on their own for the wintering grounds when they are just a few months old. They fly at night, using the position of the stars as an orientation guide. When young buntings are hand-reared and kept in a completely dark room for several months (or when they grow up in a planetarium under an artificial sky from which the North Star and the Big Dipper are missing), they cannot orient themselves in the proper migratory direction.

THE ADAPTIVE VALUE OF BEHAVIOR

From the biological perspective, many forms of behavior are adaptive traits that are subject to evolution by natural selection. The following definitions will help you understand the evolutionary mechanisms involved:

1. Reproductive success: survival and production of offspring.

2. Adaptive behavior: forms of behavior that promote reproductive success and that thereby tend to occur at increased frequency in successive generations.

3. Selfish behavior: forms of behavior by which an individual protects or increases its own chance of producing offspring, regardless of the consequences for the group to which it belongs.

4. Altruistic behavior: self-sacrificing behavior; the individual behaves in a way that helps others but, in so doing, decreases its own chance to produce offspring.

5. Natural selection: differential reproductive success among individual members of a group that vary in heritable traits—including many behavioral traits—that promote survival and reproduction.

a

b

c

Figure 38.4 (**a**) Banana slug of the Pacific coastal regions of California, food for garter snakes (**b**). (**c**) Newborn garter snake from a coastal population, tongue-flicking at a cube of banana slug.

"Selfish" or "altruistic" behavior does not mean the animal is consciously aware of what it is doing or knows its behavior is related ultimately to reproductive success. A lion will want to eat zebras without knowing that doing so is good for its survival and reproduction.

Adaptive Feeding Behavior

Stevan Arnold's studies of garter snakes are an elegant example of natural selection for feeding behavior. Garter snakes in moist habitats along the California coast eat banana slugs (Figure 38.4a). Garter snakes in inland habitats eat mostly tadpoles and small fish. When isolated, recently born garter snakes were offered a chunk of slug for their first meal, the coastal newborns almost always ate the slug, but the inland newborns rarely did. Also, unlike coastal snakes, the inland snakes tongue-flicked at cotton swabs drenched in essence of tadpole but didn't respond much to swabs drenched in essence of slug.

Male and female snakes from the two populations were induced to mate. By maintaining similar environments for the hybrid offspring, Arnold demonstrated that the populations differ in the genes controlling chemical responsiveness to slug odors.

Apparently, the genetic basis of slug acceptance by inland populations has been selected against over time. Slugs and leeches share some chemical similarity, and there are plenty of leeches inland. Leeches are hard to digest; some even remain alive in the gut and damage internal organs. Inland individuals with a genetic predisposition for slug/leech acceptance tended to leave fewer offspring than those endowed with different genes. Similarly, edible slugs (but not leeches) abound on the California coast. There, individuals attracted to slug odors probably gained access to a rich food supply, left more descendants, and helped spread the genetic basis for slug acceptance.

Anti-Predator Behavior

It does an animal no good to go out in the world and eat if it is attacked and killed in the process. By compromising their foraging efficiency to some extent, most animals improve the odds for escaping from predators or avoiding them even if detected.

The caterpillar of certain tropical moths eats frenziedly during the night but remains motionless on vines during the day. It is very large (nearly 15 centimeters when fully grown), so any movement certainly would attract day-foraging birds. Besides "hiding in plain sight," the caterpillar also makes a special behavioral response when poked sharply, as by a bird beak. It drops part way from the vine and puffs up the segments at its head end, so that it looks like a snake (Figure 38.5). The "snake" even strikes at whatever touches it, and small birds commonly hesitate before continuing an attack. A reluctance to deal with snakes generally prolongs the bird's life, a response that the caterpillar exploits to its own advantage.

Adaptive Reproductive Behavior

Sexual selection is the outcome of individuals of one sex competing for matings with individuals of the opposite sex. Males compete, often aggressively, for access to receptive females. If their reproductive success is measured in terms of offspring produced, the ones mating with the most females and fertilizing the most eggs are winners. In contrast, a female's reproductive success depends on how many eggs she produces or how many offspring she can care for. She usually goes for quality of the mate rather than quantity of matings. *Because the females produce a limited resource (eggs), they* *dictate the rules of male competition, with males attempting to fertilize as many eggs as possible.*

For example, a male white-throated sparrow must "convince" females that he rather than some other individual should fertilize her eggs. The female's reproductive success will depend on how well her nestlings are fed, and this means feeding them astonishing quantities of insects. And so the males compete for insect-rich patches of forest that will attract females. Competition involves vocally advertising their control over a resource-based territory. (A *territory* is an area in which an individual or a group establishes residency and fends off competitors.) The song of an established resident discourages rival males; it means they will have to fight for possession of the territory when there may be unoccupied areas elsewhere. And the song attracts females—which may check out the quality of the territory and the males for themselves.

By influencing the distribution of females, environmental factors affect male competition for access to mates. When resources sought by females are clumped rather than spread out through the habitat, we see **resource-defense behavior** in which the winners get mates and the losers are kept from reproducing (at least temporarily).

In contrast, males often show **female defense behavior** when females live together but do not become concentrated at patches of a useful resource. Male redwinged blackbirds, lions, elk, and bighorn sheep compete fiercely as a consequence of the female clustering behavior. The competition favors large males with formidable combat abilities, such as the readiness of bison and bighorn sheep to use their horns on each other. The return can be great; by winning possession of females, the males of these species have a ready-made harem (Figure 1.5f).

What does the female get out of all of this? With judicious mate selection, she gets genes, superior material benefits, or both. Consider the male sage grouse, which makes whistling-popping calls to females as he bobs up and down on a tiny patch of prairie. If he is selected by a female, he leaves her with his genes, period; the female goes off and raises the offspring by herself. How did the female "assess" his genetic quality? Perhaps the nature of his calls, the vigor of his bobbing, or his endurance were cues about his survival ability or foraging skills. To the extent that these behaviors are heritable, a male that puts on a superior show could endow his offspring with useful abilities; and females able to identify such a male would enjoy heightened reproductive success.

The hangingfly gives us another example of female benefits. Male hangingflies capture and kill a fly or a moth and hold it for a female, which they attract by releasing a sex pheromone (Figure 38.6). If females

Figure 38.5 A snake caterpillar showing predator avoidance behavior in response to being poked. The forward segments of the caterpillar let go of the vine and puff up like a snake head, which "strikes" at whatever touches it.

Figure 38.6 A male hangingfly (*Harpobittacus*) holding a prey insect that he will offer to a female if one comes to visit him. He will mate with the female if she accepts his food present.

Figure 38.7 Nasute termites, highly social insects with a soldier caste that defends the colony. A strand of glue, shot out from a soldier's pointy head, entangles and immobilizes ants and other invaders.

Figure 38.8 Example of the "yawns" by which male baboons threaten each other. Such facial displays often resolve a conflict without fighting.

choose mates partly for material benefits, they should prefer males that can feed them more. And female hangingflies do in fact choose males with the larger "nuptial gifts." When a female approaches a male, he offers her his prey; and if she begins to eat it, she allows mating to begin. However, she will not allow him to transfer *any sperm* until she has eaten for about five minutes. Thereafter, she accepts a steady flow of sperm into her reproductive tract, *but only as long as the food holds out.* If it does not, she will mate again—and dilute or replace her first partner's sperm.

SOCIAL BEHAVIOR

Social Communication

We turn now to **social behavior,** the tendency of individual animals to enter into cooperative, interdependent relationships with others of their kind. At the heart of social behavior is the ability of animals to *communicate* with one another, using a complex array of signals to exchange information.

Consider the termite. In one forest in southern Australia, on the trunk of a dead eucalyptus tree, nasute termites have glued together millions of pellets to form a continuous tunnel. The tunnel really is a covered highway extending from an underground nest to areas where worker termites gather dead wood. The workers deposit the wood in underground gardens, where they grow a fungus that is able to digest the cellulose fibers in wood as a food source. The termites harvest the mushrooms as food for themselves. The entire colony is defended by snout-shooters, soldier termites that shoot thin strands of glue out of their nose (Figure 38.7). The glue entangles and immobilizes ants that attempt to invade the tunnel.

Perhaps a million or more workers and soldiers in one colony cooperate to build tunnels, create and tend gardens, and defend the colony against predators. Not one worker or soldier will ever reproduce. In an underground bunker are one king and one queen termite— the original fertile founders of the colony, parents to all one million sterile progeny.

When a termite tunnel is breached, the workers bang their head against the ceiling and floor of the tunnel, creating vibrations that alert nearby soldiers. The soldiers point their nose in the direction of danger, as announced by the scent of ants or a sudden jostling of the tunnel wall. When a silky strand of glue shoots out of their nose, its odor attracts more soldiers. The glue serves as an alarm call; it is a type of communication signal.

A **communication signal** is a stimulus produced by one animal that changes the behavior of another indi-

Figure 38.9 Courtship behavior of various species of albatross. (**a**) The male spreads his wings as part of a courtship ritual that also includes pointing his head at the sky. These birds are at a future nest site in the male's territory (**b-d**). After pair formation is well advanced, the birds begin to touch bills. This contact display precedes copulation.

vidual of the same species. Communication signals are based on different sensory systems, of the sort described in Chapter 30. They include chemical, visual, tactile, and acoustical signals.

Chemical Signals. The alarm calls of termites, sex pheromones of male hangingflies, and trail markers of ants are examples of chemical signals. Often, receptive female insects release sex pheromones, which can bind with receptors on the large antennae of male insects. Males fly upwind, tracking the signal in a race to be first to reach the chemically beckoning female. Some primates also rely on chemical signaling. A female baboon announces her readiness to mate with chemical as well as visual signals.

Visual Signals. Visual signals abound among animals that are active during the day. Male baboons "yawn" at each other when they are competing for a receptive female (Figure 38.8). The yawn is a threat display; often it precedes an attack. When the receiver of the signal backs down, the signaler benefits by not being physically attacked and injured. The receiver may also benefit if, after mulling over the signal, he accurately judges that his rival can inflict a real beating.

Many animals use elaborate visual signals during courtship. Albatrosses and other birds assume exaggerated and often contorted postures with specific meaning, such as a readiness to mate (Figure 38.9). The posturing may minimize the possibility of confusion by providing the intended receiver with an unambiguous signal.

Animals that are active at night sometimes use *bioluminescent* signals, which are flashes of light energy released during special metabolic reactions. Fireflies have light-generating organs at the tip of the abdomen. After seeing a signal from a male, a receptive female waits a standard time and replies with a single flash. The two enter into a visual "dialogue" until the male precisely locates the female and mates with her. Females of a predatory species of firefly have broken this code. When they detect a signal from a male, they wait the standard interval, then give a "come-hither" answering flash. After luring the male to them, they eat it.

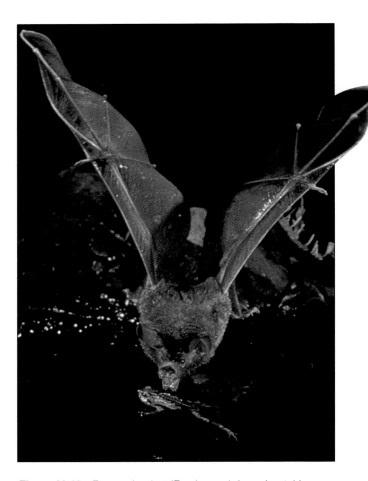

Figure 38.10 Frog-eating bat (*Trachops cirrhosus*) catching a singing tropical frog (*Physalaemus*). The bat can distinguish between poisonous and edible frogs, as well as locate them, by the frogs' calls. The frogs call to attract mates but face the dilemma of how not to attract the bat instead. Bat predation obviously has a major influence on frog courtship behavior.

Acoustical Signals. Acoustical signals are distinctive sounds with specific meaning. Worker termites bang their head to attract soldiers. Male birds sing as they stake out territories, attract females, and discourage rivals. Male frogs also use distinctive acoustical signals to communicate with rival males and receptive females.

Frog calling is not without problems. Many rival males may be densely clustered in the same area, creating intense competition for space on the "air waves." And some predators use the calls to track victims. One tropical frog has a two-part call, a whine followed by a "chuck." A solitary frog often drops the chuck, even though it gives better acoustical information about where he is located. Why? His call also attracts bats, which can snatch a frog from the water (Figure 38.10). Although a chuck-dropping male compromises his signal to females, he is more likely to live to call another day. He does give the complete call when part of a male chorus. Being only one of many likely to be grabbed that evening by a frog-eating bat, he may live *and* get to mate.

Tactile Signals. Tactile signals are distinctive patterns of touch. They become important when animals communicate over short distances. For example, when a foraging honeybee returns to the beehive after several trips to a rich nectar source, she may begin to dance. She moves in circles, jostling her way in the dark through the crowded mass of workers. Other bees keep in physical contact with the dancer and so acquire information about the distance and location of the food source.

When the food is close to the hive, bees perform a round dance; when it is far away, they perform a waggle dance (Figure 38.11). The more waggles and the faster the dance, the closer the food source. Both dances include a straight run up the surface of the comb in the beehive. The angle of the straight run provides information about the direction of the food relative to the sun and the hive. Thus, dancing bees transpose information about the position of the sun, hive, and food source into a code, and they convey information through this code by tactile signals to their hivemates.

Costs and Benefits of Social Life

Some animals live alone, others in small family units, and still others as loose associations of unrelated individuals. Social behavior (or its absence) can be explained in terms of how it affects *individual reproductive success* in different environments.

In the vast rookeries of egrets, gulls, or terns, hundreds of individuals form pairs and nest closely together, in apparent harmony. The harmony is illusory;

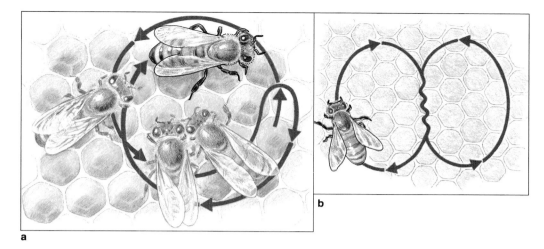

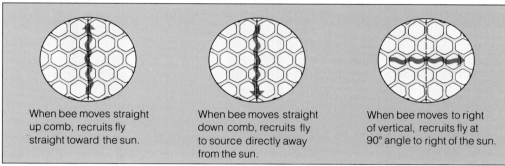

Figure 38.11 Information encoded in the tactile signaling of honeybee dancing, as discovered by Karl von Frisch. Marked foragers were trained to come to stations that were baited with sugar water. A series of stations with identical baits were set up at different distances or directions from the hive. Trained bees recruited others to the areas where food was abundant.

(**a**) Bees trained at stations close to the hive performed a round dance on the comb. Worker bees in physical contact with the dancing trained bees were stimulated to search for food near the hive, but not in any particular direction.

(**b**) Bees trained at stations more than 80 meters away from the hive performed a waggle dance. The faster the dance, and the more their abdomen waggled during the dance's straight runs, the closer the food source. (**c**) The angle of the straight run with respect to the vertical comb provided information about the direction of food relative to the sun and the hive.

When bee moves straight up comb, recruits fly straight toward the sun.

When bee moves straight down comb, recruits fly to source directly away from the sun.

When bee moves to right of vertical, recruits fly at 90° angle to right of the sun.

a nesting gull may eat any momentarily unguarded eggs or chicks of its neighbors. Competition is fierce for food resources. Given the crowded conditions in rookeries, the birds are also vulnerable to contagious diseases.

Predation and Sociality. Sociality clearly comes at a cost, but there are compensatory benefits. A pride of lions can, through cooperative hunting, bring down a giraffe or a Cape buffalo that a solitary lion would not dare attack.

Predation also is the main selection pressure that favors social behavior among prey species. Simply by being part of a large group, a prey animal "dilutes" the risk of being eaten. Fairy penguins of Australia hunt individually, then return to burrows on land. Surging waves deposit the penguins on the beach, where they stand about uncertainly. Penguins are not agile or swift on land, so they are vulnerable to large hawks and eagles patrolling the beaches. Each penguin improves its odds of survival by waddling inland along with the others. While an eagle is dispatching one victim, the others can waddle to safety.

Also, social animals collectively have many eyes to spot predators. When the first animal to detect danger

flees for cover, others can follow even if they themselves have detected nothing. As a flock member, a bird has more time to feed because less time is required for vigilance.

Often, prey animals can repel predators through coordinated group defense. Nasute termites work as a team to alert soldiers, which then combine forces against intruders. Musk oxen also employ group defense when threatened by attacking wolves (Figure 38.12).

The Selfish Herd. Some animals apparently live in groups simply to "use" others as living shields against predators. Such groups have been labeled "selfish herds." When hyenas or hunting dogs attack a wildebeest on an African savanna, the prey animals cluster into a tight mass and dash across the plain as if they were a single animal. Although their escape behavior seems to be a cooperative undertaking, it almost certainly results from the efforts of individuals not to be isolated from the group. Each apparently attempts to maneuver itself toward the center of the herd, where it will be safest. The predators separate a victim from the fringes of the herd rather than risk plunging into a mass of flying hooves.

Figure 38.12 Defensive formation of musk oxen. Predation pressure can favor the evolution of social life when members of the group are safer than are solitary animals.

Social Life and Self-Sacrifice

As the preceding discussion makes clear, many social animals do not make any personal sacrifice for others of their group. Yet others show self-sacrificing behavior. It is not that they deliberately make the sacrifice to promote the welfare of their group or species. Rather, *there is selection for individual behavior that promotes a particular genetic lineage,* as a few examples will show.

Parental Behavior. Parenting is a familiar form of self-sacrificing behavior. A pair of Caspian terns incubate their eggs, then shelter and feed the nestlings, defend them from predators, and accompany them when they start to fly and forage on their own. Such activities use up time and energy that might otherwise be spent on improving the parent's own chances of living to reproduce another time, but they also help at least some offspring survive to reproductive age. Thus, *even though the individual sacrifices some of its reproductive capacity, its genes still spread through the population.*

Cooperative Societies. Despite the existence of selfish herds and cannibalistic neighbors, nature also abounds with self-sacrificing individuals that direct friendly, helpful behavior to more than just offspring.

All the baboons in a troop recognize each other as individuals and actively seek out one another's company, even spending hours grooming a favored companion to remove parasites and burrs. At some risk to themselves, adult males join forces and confront an attacking leopard while the smaller females and juveniles flee to safety. Similarly, the members of a wolf pack have active greeting ceremonies after they have been separated. They coordinate hunting activities and share food from a kill, and the hunters often regurgitate food to pack members that stayed at the den to protect the pups.

Dominance Hierarchies. When a wolf brings back food to pack members that are *not* its offspring, it seems to be sacrificing time and food that could be used to further its own reproductive success. How can we account for such behavior in evolutionary terms? We must first recognize that the apparent harmony in even the most cooperative societies is, to some extent, superficial. Some "sacrifices" are made simply because some members are less dominant than others, and their competitiveness is held in check. There often is a **dominance hierarchy,** with some members subordinate to others, which in turn are dominated by others.

A low-ranking baboon that receives an aggressive signal from the top male of the troop absents itself quickly from a safe sleeping place, choice bits of food, or a receptive female. The second-ranking male can preempt all but the top male, the second-ranking can preempt the third, and so on. Subordinates seldom challenge a dominant male; often they must show appeasement gestures if they are to remain in his company (Figure 38.12).

The subordinates in the hierarchy almost never reproduce as much as dominant members do. What, then, is the advantage of subordinate behavior?

According to one view, subordinates within a group may move up to a higher position in the social hierarchy when dominant members die or become injured, old, or feeble. The advantage is that, with enough patience, a subordinate may ultimately reproduce. According to another view, accepting a low status in the group offers more advantages than other alternatives. For example, a young, small, or weak animal *could* compete for dominance, but challenging a stronger individual might mean injury or death—and no reproduction. An animal could strike out on its own, but usually it would end up being injured or killed before finding a mate and forming a new group. (A solitary baboon surely quickens the pulse of the first leopard to see it.)

Evolution of Altruism

Altruistic behavior reduces the individual's production of surviving offspring while increasing the reproductive success of the helped individual. Natural selection may favor altruistic behavior that is directed toward **kin**—close relatives.

Kin Selection. A nonreproducing, subordinate member of a group may help pass on copies of its genes, as long as the members benefiting from its altruistic behavior are relatives. Two siblings will be alike in about half their genes (they inherited some of the same genes from the same parents). An uncle and a nephew will be alike in about one-fourth of their genes. An individual can *indirectly* propagate its genes, including ones associated with helpful behavior, by helping to preserve and produce more relatives.

A flock of jays includes a breeding pair and as many as six nest helpers. The helpers—almost always older offspring of the pair—feed and protect younger brothers and sisters. They might never reproduce, yet they help perpetuate some "shared" genes when their siblings reproduce.

Suicide, Sterility, and Social Insects. Some animals are incapable of reproduction yet show altruistic behavior. Sterile workers occur among honeybees and other social insects. A queen bee is the reproducing female among tens of thousands of bees. Workers collect food and feed her, groom her, and distribute her socially binding pheromones through the colony. They also feed one another, build honeycomb, attend to the queen's eggs, feed larvae, remove dead or diseased pupae, and guard the hive. When a guard bee stings an intruder, the loss of its stinger and poison glands is fatal. Like ants, wasps, and soldier termites, these workers commit suicide when they protect their colony-mates!

Figure 38.13 Appeasement behavior among baboons. Notice the assured position of the dominant animal—and the abject stare and groveling posture of the subordinate one, who is intent on making little conciliatory smacking noises with its lips.

Suicidal behavior and sterility among these insects are adaptive traits, for close relatives are the beneficiaries. When workers sacrifice their own reproductive chances, they increase the number of siblings—which have a large number of genes in common with them. This is true of social insects generally.

Human Social Behavior

Can individual selection theory be applied to analysis of human behavior? Oddly, the question is controversial. Since Darwin's time, strong evidence has accumulated in support of the principle of biological evolution. Yet humans often are viewed as being so evolutionarily advanced that they are unique, even compared with their closest relatives in the animal kingdom.

Our cultural evolution has been extraordinary, and we are indeed unique in many ways. But beneath the elaborate, diverse layerings of culture, there still is a biological core for human behavior. Experiments strongly support the view that behavioral development has a genetic basis and therefore is subject to individual selection. Thus it can be argued that much of human behavior may also be ultimately explainable in these terms.

Think about some forms of behavior that occur in all societies. Smiling, for example, has an innate basis. It starts among newborns and apparently helps establish strong emotional ties between infants and the adults who help ensure their survival. Four or five weeks after birth, all infants make strong visual contact with their mother—even infants who are born blind. This behavior helps strengthen social bonds between the protector and the protected. Universally, humans express pleasure, anger, distress, surprise, and rage with the same kinds of facial movements. The meaning of these behavioral expressions is universally recognized; they are genetically based mechanisms that help promote survival and reproductive success in human social groups.

If you were to ask people why they behaved in a particular way, no one would ever say "My genes made me do it." Yet conscious awareness of a genetic contribution to some behavior is not a requirement for its expression. If it could talk, a baby egret battering its younger sibling to death (as baby egrets are prone to do) wouldn't say "My genes made me do it," either. It simply has a genetically based capacity to promote its individual survival and reproduction.

An evolutionary approach to human behavior may be worrisome to individuals who believe that its findings will be misused. They may think that to suggest a trait such as jealousy is adaptive (genetically advantageous) is to imply that it is desirable in moral or social terms.

Yet there is a clear difference between trying to explain something (by reference to its possible evolutionary history) and trying to justify it. "Adaptive" does not mean "moral." It simply means *more frequent transmission of an individual's genes.*

Research into the relationship between evolution, genes, and behavioral capacity is in its infancy. Yet it may help us resolve many critical issues related to our biology as a species, including overpopulation, exploitation of the environment, and aggression among modern nations. It remains to be seen how quickly our knowledge of human behavior will be gained and whether it will be used to alleviate negative aspects of the human condition.

SUMMARY

1. The observable, coordinated responses an animal makes to external and internal stimuli are collectively called animal behavior. Genes provide instructions for the development of the neural and endocrine systems, which influence an animal's behavior.

2. Instinctive and learned behavior are based on genetic information and environmental inputs. Instincts are not purely genetic, and learned responses are not purely environmental, as evidenced by imprinting behavior.

3. Forms of behavior are adaptive traits that are subject to evolution by natural selection of individuals.

4. Social species (those living in groups) depend on communication signals, by which a signaler and a receiver may benefit from the transfer of information between them. Chemical, visual, tactile, and acoustical signals are common.

5. Sociality carries with it higher risks of communicable diseases, exploitation by others, and increased competition for scarce resources. But if, say, predation pressure is severe, benefits arising from improved defense against predation can outweigh the costs of social life.

6. In most societies, individuals do not sacrifice their reproductive benefit to others in their group. In some species, parents gain by living with and helping their offspring. In other species, weaker subordinate animals gain by deferring reproduction to take advantage of the benefits that come from belonging to a group.

7. Altruistic behavior refers to helpful actions that reduce the individual's chances of reproductive success while increasing that of the helped individual. Such actions have selective advantage for the altruistic individual when they promote its genetic lineage.

Review Questions

1. Rephrase the statement "There is a gene for sexual imprinting by greylag geese." The reworded statement should avoid the implication that genes "make" behavioral traits in a one-to-one relationship. *588–589*

②. Why is song learning by white-throated sparrows a good illustration of the principle that genetic mechanisms contribute to the development of a learned response? *587*

3. You find an insect that looks and behaves like a piece of bark. How could you compare a number of insect species to test the hypothesis that the behavior of the animal is an adaptation to avoid being eaten by a predator? *590*

④. Develop a hypothesis for this observation: Male lions kill the offspring of females they acquire. They do so after they chase away the males that had been the mates of these females. *590*

5. A hyena places scent marks on vegetation in its territory by releasing specific chemicals from certain glands. What evidence would you need to demonstrate that this action is an evolved communication signal? *592*

6. Explain how a threat display can be considered an example of cooperation. *593*

7. Why don't the members of a "selfish herd" live apart if each member is "trying to take advantage" of the others? *595*

⑧. How can an animal be cooperative and still pass on more of its genes than a noncooperative individual? *596*

9. Is parental behavior always adaptive? *598*

⑩. What is true altruism? *597*

Self-Quiz   *(Answers in Appendix IV)*

1. **Animal behavior** _____ refers to the observable, coordinated responses an animal makes to external and internal stimuli.

2. Animal behavior is influenced by genetic instructions for the development of the **nervous** and **endocrine** systems.

3. Genetic information and environmental inputs provide the basis for **instinctive** behavior and **learned** behavior.

4. Studies of imprinting provide evidence that instincts are not purely **genetic**, and learned responses are not purely **environmental**.

5. Forms of behavior are **adaptive traits** _____ that are subject to evolution by natural selection.

6. Behavior is _____.
   a. an outcome of nerve impulses and muscular contractions
   b. an integral part of an animal's equipment for survival
   c. a species characteristic
   ⓓ all of the above

7. Social species use communication signals, by which the _____ and the _____ may both benefit from the transfer of information between them.
   a. predator species; prey species
   b. parasitic species; host species
   ⓒ signaler; receiver
   d. a and b are both correct

8. Which of the following is probably the *least* risk of sociality?
   a. communicable disease
   b. exploitation by others
   c. increased competition for scarce resources
   ⓓ attack by predators

9. In some societies, weaker subordinate animals may gain benefits by _____.
   a. consistently playing the signaler role
   b. defending the group against predators rather than reproducing
   ⓒ deferring reproduction but enjoying advantages of group living
   d. consistently being exploited by others in the group

10. When one individual in a society takes action that helps another while reducing its own chances of reproductive success, the behavior is _____.
    a. selfish   c. maladaptive
    ⓑ altruistic   d. reproductive

11. Match the animal behavior concepts.
    d communication signals   a. communicable disease, exploitation, and resource competition
    e altruism
    b basis of instinctive and learned behavior   b. genes and environmental influences
    c animal behavior   c. coordinated responses to external and internal stimuli
    a risks of sociality   d. information transfer in social species
                                               e. assisting another individual at one's own expense

Selected Key Terms

acoustical signal *594*
adaptive behavior *589*
altruistic behavior *589*
animal behavior *586*
bioluminescent signal *594*
chemical signal *593*
communication signal *592*
dominance hierarchy *596*
female defense behavior *591*
heredity *586*
imprinting *588*

instinct *587*
learning *588*
natural selection *589*
reproductive success *589*
resource-defense behavior *591*
selfish behavior *589*
sexual selection *590*
social behavior *592*
tactile signal *594*
territory *591*
visual signal *593*

Readings

Alcock, J. 1989. *Animal Behavior*. Second edition. Sunderland, Massachusetts: Sinauer.

Daly, M., and M. Wilson. 1983. *Sex, Evolution, and Behavior*. Second edition. Boston: Willard Grant Press. This book has two particularly good chapters on the evolutionary analysis of human behavior.

Frisch, K. von. 1961. *The Dancing Bees*. New York: Harcourt Brace Jovanovich. A classic on the natural history and behavior of honeybees.

Moffett, M. January 1990. "Dance of the Electronic Bee." *National Geographic* 177(1):134–140.

Trivers, R. 1985. *Social Evolution*. Menlo Park, California: Benjamin-Cummings. A complete and readable account of all the topics that make up an evolutionary approach to social behavior.

APPENDIX I
UNITS OF MEASURE

Metric-English Conversions

Length

English		Metric
inch	=	2.54 centimeters
foot	=	0.30 meter
yard	=	0.91 meter
mile (5,280 feet)	=	1.61 kilometer

To convert	multiply by	to obtain
inches	2.54	centimeters
feet	30.00	centimeters
centimeters	0.39	inches
millimeters	0.039	inches

Weight

English		Metric
grain	=	64.80 milligrams
ounce	=	28.35 grams
pound	=	453.60 grams
ton (short) (2,000 pounds)	=	0.91 metric ton

To convert	multiply by	to obtain
ounces	28.3	grams
pounds	453.6	grams
pounds	0.45	kilograms
grams	0.035	ounces
kilograms	2.2	pounds

Volume

English		Metric
cubic inch	=	16.39 cubic centimeters
cubic foot	=	0.03 cubic meter
cubic yard	=	0.765 cubic meters
ounce	=	0.03 liter
pint	=	0.47 liter
quart	=	0.95 liter
gallon	=	3.79 liters

To convert	multiply by	to obtain
fluid ounces	30.00	milliliters
quart	0.95	liters
milliliters	0.03	fluid ounces
liters	1.06	quarts

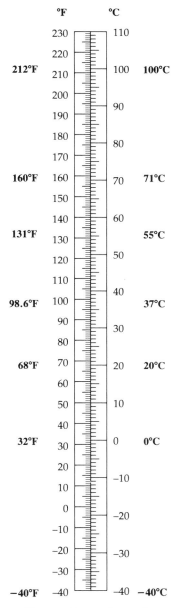

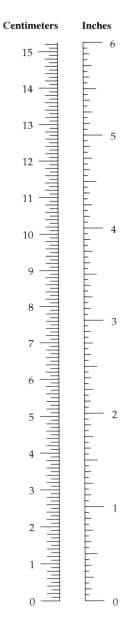

To convert temperature scales:

Fahrenheit to Celsius: $°C = 5/9 (°F - 32)$

Celsius to Fahrenheit: $°F = 9/5 (°C) + 32$

APPENDIX II
A CLASSIFICATION SYSTEM

This classification scheme is a composite of several used in microbiology, botany, and zoology. Although major groupings are more or less agreed upon, what to call them and (sometimes) where to place them in the overall hierarchy are not. As Chapter 15 indicated, there are several reasons for this. First, the fossil record varies in its quality and completeness, so certain evolutionary relationships are open to interpretation. Comparative studies at the molecular level are firming up the picture, but this work is still under way.

Second, since the time of Linnaeus, classification schemes have been based on perceived morphological similarities and differences among organisms. Although some original interpretations are now open to question, we are so used to thinking about organisms in certain ways that reclassification proceeds slowly. For example, birds and reptiles traditionally are considered separate classes (Reptilia and Aves)—even though there now are compelling arguments for grouping lizards and snakes as one class, and crocodilians, dinosaurs, and birds as another.

Finally, botanists as well as zoologists have inherited a wealth of literature based on schemes that are peculiar to their fields; and most see no good reason to give up established terminology and so disrupt access to the past. Thus botanists continue to use Division as a major taxon in the hierarchical schemes and zoologists use Phylum in theirs. Opinions are notably divergent with respect to an entire Kingdom (the Protista), certain members of which could just as easily be called single-celled forms of plants, fungi, or animals. Indeed, the term protozoan is a holdover from earlier schemes that ranked the amoebas and some other forms as simple animals.

Given the problems, why do we bother imposing hierarchical schemes on the natural history of life on earth? We do this for the same reason that a writer might decide to break up the history of civilization into several volumes, many chapters, and a multitude of paragraphs. Both efforts are attempts to impart structure to what might otherwise be an overwhelming body of information.

One more point to keep in mind: The classification scheme in this Appendix is primarily for reference purposes, and it is by no means complete (numerous phyla of existing and extinct organisms are not represented). Our strategy is to focus mainly on the organisms mentioned in the text, with numerals referring to some of the pages on which representatives are illustrated or described. A few examples of organisms are also listed under the entries.

SUPERKINGDOM PROKARYOTA. Prokaryotes (single-celled organisms with no nucleus or other membrane-bound organelles in the cytoplasm).

KINGDOM MONERA. Bacteria, either single cells or simple associations of cells; autotrophic and heterotrophic forms. *Bergey's Manual of Systematic Bacteriology*, the authoritative reference in the field, calls this "a time of taxonomic transition" and groups bacteria mainly on the basis of form, physiology, and behavior, not on phylogeny (Table 17.3 gives examples). The scheme presented here does reflect the growing evidence of evolutionary relationships for at least some bacterial groups.

SUBKINGDOM ARCHAEBACTERIA. Methanogens, halophiles, thermoacidophiles. Strict anaerobes, distinct from other bacteria in their cell wall, membrane lipids, ribosomes, and RNA sequences. 225, 230

SUBKINGDOM EUBACTERIA. Gram-negative and Gram-positive forms. Peptidoglycan in cell walls. Photosynthetic autotrophs, chemosynthetic autotrophs, and heterotophs. 225, 226

DIVISION GRACILICUTES.   Typical Gram-negative, thin wall. Autotrophs (photosynthetic and chemosynthetic) and heterotrophs. *Anabaena, Chlorobium, Escherichia, Shigella, Desulfovibrio, Agrobacterium, Pseudomonas, Neisseria.* 69, 226, 523, 525
DIVISION FIRMICUTES.   Typical Gram-positive, thick wall. Heterotrophs. *Staphylococcus, Streptococcus, Clostridium, Bacillus, Actinomyces.* 222, 225, 226, 228, 432
DIVISION TENERICUTES.   Gram-negative, wall absent. Heterotrophs (saprobes, parasites, pathogens). *Mycoplasma.* 226

SUPERKINGDOM EUKARYOTA. Eukaryotes (single-celled and multicelled organisms; cells typically have a nucleus and other organelles).

KINGDOM PROTISTA. Mostly single-celled eukaryotes. Some colonial.

PHYLUM GYMNOMYCOTA.   Heterotrophs.
   Class Acrasiomycota. Cellular slime molds.
      Dictyostelium. 69, 226, 232, 523
   Class Myxomycota. Plasmodial slime molds.
      Physarum. 232
PHYLUM EUGLENOPHYTA.   Euglenids. Mostly heterotrophic, some photosynthetic. Flagellated. *Euglena.* 229, 233
PHYLUM CHRYSOPHYTA.   Golden algae, yellow-green algae, diatoms. Photosynthetic. Some flagellated, others not. 229, 233
PHYLUM PYRROPHYTA.   Dinoflagellates. Mostly photosynthetic, some heterotrophs. *Gonyaulax.* 229, 234
PHYLUM SARCOMASTIGOPHORA.   Heterotrophs. Free-living, symbiotic, and parasitic forms.

Subphylum Mastigophora. Flagellated protozoans. *Trypanosoma.* 234
Subphylum Sarcodina. Amoeboid protozoans. Amoebas, foraminiferans, heliozoans, radiolarians. 234, 235
PHYLUM APICOMPLEXA. Sporozoans and some other parasitic protozoans, many intracellular. *Plasmodium.* 235
PHYLUM CILIOPHORA. Ciliated protozoans. *Paramecium, Didinium.* 236–237, 509, 510

KINGDOM FUNGI. Mostly multicelled eukaryotes. Heterotrophs (mostly saprobes, some parasites). All rely on extracellular digestion and absorption of nutrients.

DIVISION MASTIGOMYCOTA. All produce flagellated spores.
 Class Chytridiomycetes. Chytrids. 240–241
 Class Oomycetes. Water molds and related forms. *Plasmopora, Phytophthora.* 242
DIVISION AMASTIGOMYCOTA. All produce nonmotile spores.
 Class Zygomycetes. Bread molds and related forms. *Rhizopus, Pilobius.* 241, 242
 Class Ascomycetes. Sac fungi. Most yeasts and molds; morels, truffles. *Saccharomyces, Morchella.* 241
 Class Basidiomycetes. Club fungi. Mushrooms, shelf fungi, bird's nest fungi, stinkhorns. *Agaricus, Amanita.* 241–243
FORM-DIVISION DEUTEROMYCOTA. Imperfect fungi. All with undetermined affiliations because sexual stage unknown; if better known they would be grouped with sac fungi or club fungi. *Verticillium, Candida.* 242, 244

KINGDOM PLANTAE. Nearly all multicelled eukaryotes. Photosynthetic autotrophs, except for a few saprobes and parasites.

DIVISION RHODOPHYTA. Red algae. *Porphyra.* 202, 245–246, 561
DIVISION PHAEOPHYTA. Brown algae. *Fucus, Laminaria.* 239, 245–246, 247
DIVISION CHLOROPHYTA. Green algae. *Ulva, Spirogyra.* 39, 72, 202, 231
DIVISION CHAROPHYTA. Stoneworts.
DIVISION BRYOPHTYA. Liverworts, hornworts, mosses. *Marchantia, Sphagnum.* 248, 554
DIVISION RHYNIOPHYTA. Earliest known vascular plants; extinct. *Cooksonia, Rhynia.* 204
DIVISION PSILOPHYTA. Whisk ferns. 204
DIVISION LYCOPHYTA. Lycopods, club mosses. *Lycopodium, Selaginella.* 249
DIVISION SPHENOPHYTA. Horsetails. *Equisetum.* 106, 249–250, 251
DIVISION PTEROPHYTA. Ferns. 249, 250–251
DIVISION PROGYMNOSPERMOPHYTA. Progymnosperms. Ancestral to early seed-bearing plants; extinct. *Archaeopteris.* 250
DIVISION PTERIDOSPERMOPHYTA. Seed ferns (extinct fernlike gymnosperms).
DIVISION CYCADOPHYTA. Cycads. *Zamia.* 251, 252
DIVISION GINKGOPHYTA. Ginkgo. *Ginkgo.* 251, 252
DIVISION GNETOPHYTA. Gnetophytes. *Ephedra, Welwitschia, Gnetum.* 251, 252
DIVISION CONIFEROPHYTA. Conifers. 251–253
 Family Pinaceae. Pines, firs, spruces, hemlock, larches, Douglas firs, true cedars. *Pinus.* 253, 299, 304, 309
 Family Cupressaceae. Junipers, cypresses, false cedars.
 Family Taxodiaceae. Bald cypress, redwood, Sierra bigtree, dawn redwood. *Sequoia.* 520
 Family Taxaceae. Yews.
DIVISION ANTHOPHYTA. Flowering plants. 254, 294
 Class Dicotyledonae. Dicotyledons (dicots). Some families of several different orders are listed:
  Family Magnoliaceae. Magnolias, tulip trees.
  Family Ranunculaceae. Buttercups, delphinium.

Family Nymphaeaceae. Water lilies. 255
Family Papaveraceae. Poppies, including opium poppy.
Family Brassicaceae. Mustards, cabbage, radishes, turnips. 316
Family Malvaceae. Mallows, cotton, okra, hibiscus.
Family Solanaceae. Potatoes, eggplant, petunias. 242
Family Salicaceae. Willows, poplars. 301
Family Rosaceae. Roses, peaches, apples, almonds, strawberries.
Family Fabaceae. Peas, beans, lupines, mesquite, locust. 328
Family Cactaceae. Cacti. 316, 550
Family Euphorbiaceae. Spurges, poinsettia. 543
Family Cucurbitaceae. Gourds, melons, cucumbers, squashes. 312
Family Apiaceae. Parsleys, carrots, poison hemlock. 299
Family Aceraceae. Maples. 324
Family Asteraceae. Composites. Chrysanthemums, sunflowers, lettuces, dandelions. 293, 299, 329
 Class Monocotyledonae. Monocotyledons (monocots). Some families of several different orders are listed:
Family Liliaceae. Lilies, hyacinths, tulips, onions, garlic. 224, 255, 295
Family Iridaceae. Irises, gladioli, crocuses.
Family Orchidaceae. Orchids. 295, 322, 554
Family Arecaceae. Date palms, coconut palms. 295
Family Cyperaceae. Sedges.
Family Poaceae. Grasses, bamboos, corn, wheat, sugarcane. 254, 295, 326, 527
Family Bromeliaceae. Bromeliads, pineapple, Spanish moss. 325, 554

KINGDOM ANIMALIA. Multicelled eukaryotes. Heterotrophs (herbivores, carnivores, omnivores, parasites, decomposers, detritivores). 209, 259, 523

PHYLUM PLACOZOA. Small, organless marine animal. *Trichoplax.* 231, 261
PHYLUM MESOZOA. Ciliated, wormlike parasites, about the same level of complexity as *Trichoplax.*
PHYLUM PORIFERA. Sponges. 261–263
PHYLUM CNIDARIA
 Class Hydrozoa. Hydrozoans. *Hydra, Obelia, Physalia.* 258, 262, 264
 Class Scyphozoa. Jellyfishes. *Aurelia.* 190, 262
 Class Anthozoa. Sea anemones, corals. *Telesto.*
PHYLUM CTENOPHORA. Comb jellies. *Pleurobrachia.* 66
PHYLUM PLATYHELMINTHES. Flatworms. 260–261, 265–266
 Class Turbellaria. Triclads (planarians), polyclads. *Dugesia.* 265
 Class Trematoda. Flukes. *Schistosoma.* 265, 267
 Class Cestoda. Tapeworms. *Taenia.* 266–268
PHYLUM NEMERTEA. Ribbon worms.
PHYLUM NEMATODA. Roundworms. *Ascaris, Trichinella.* 259, 260, 266–268
PHYLUM ROTIFERA. Rotifers. 269, 558
PHYLUM MOLLUSCA. Mollusks. 269–271
 Class Polyplacophora. Chitons.
 Class Gastropoda. Snails (periwinkles, whelks, limpets, abalones, cowries, conches, nudibranchs, tree snails, garden snail), sea slugs, land slugs. 270
 Class Bivalvia. Clams, mussels, scallops, cockles, oysters, shipworms. 271
 Class Cephalopoda. Squids, octopuses, cuttlefish, nautiluses. *Loligo.* 271
PHYLUM BRYOZOA. Bryozoans (moss animals).
PHYLUM BRACHIOPODA. Lampshells.
PHYLUM ANNELIDA. Segmented worms.
 Class Polychaeta. Mostly marine worms. 203, 258
 Class Oligochaeta. Mostly freshwater and terrestrial worms, but many marine. *Lumbricus* (earthworm). 272–273, 375

Class Hirudinea. Leeches. 272, 590
PHYLUM TARDIGRADA. Water bears.
PHYLUM ONYCHOPHORA. Onychophorans. *Peripatus.*
PHYLUM ARTHROPODA. 260–261
  Subphylum Trilobita. Trilobites; extinct.
  Subphylum Chelicerata. Chelicerates. Horseshoe crabs,
    spiders, scorpions, ticks, mites. 273–274
  Subphylum Crustacea. Shrimps, crayfishes, lobsters, crabs,
    barnacles, copepods, isopods (sowbugs). 444–445
  Subphylum Uniramia. 273
    Superclass Myriapoda. Centipedes, millipedes. 276
    Superclass Insecta. 261
      Order Ephemeroptera. Mayflies.
      Order Odonata. Dragonflies, damselflies. 274
      Order Orthoptera. Grasshoppers, crickets,
        katydids. 375, 527
      Order Dermaptera. Earwigs.
      Order Blattodea. Cockroaches. 540
      Order Mantodea. Mantids.
      Order Isoptera. Termites. 592, 594–595
      Order Mallophaga. Biting lice.
      Order Anoplura. Sucking lice.
      Order Homoptera. Cicadas, aphids, leafhoppers,
        spittlebugs.
      Order Hemiptera. Bugs.
      Order Coleoptera. Beetles. 276, 321, 517
      Order Diptera. Flies. 321, 326
      Order Mecoptera. Scorpion flies. *Harpobittacus.* 591
      Order Siphonaptera. Fleas. 276, 540
      Order Lepidoptera. Butterflies, moths. 184, 276, 322
      Order Hymenoptera. Wasps, bees, ants. 321–322, 399, 594
PHYLUM ECHINODERMATA. Echinoderms. 261, 278, 279
  Class Asteroidea. Sea stars. 278, 279, 516
  Class Ophiuroidea. Brittle stars. 278
  Class Echinoidea. Sea urchins, heart urchins, sand
    dollars. 97, 278
  Class Holothuroidea. Sea cucumbers. 278
  Class Crinoidea. Feather stars, sea lilies.
  Class Concentricycloidea. Sea daisies.
PHYLUM HEMICHORDATA. Acorn worms.
PHYLUM CHORDATA. Chordates. 259, 261, 280, 281
  Subphylum Urochordata. Tunicates, related forms. 280, 281
  Subphylum Cephalochordata. Lancelets (amphioxus). 280, 281
  Subphylum Vertebrata. Vertebrates. 212–213, 281–283
    Class Agnatha. Jawless vertebrates (lampreys,
      hagfishes). 283–284
    Class Placodermi. Jawed, heavily armored fishes;
      extinct. 283
    Class Chondrichthyes. Cartilaginous fishes (sharks, rays,
      skates, chimaeras). 192, 284, 351
    Class Osteichthyes. Bony fishes. 423, 464
      Subclass Dipnoi. Lungfishes. 284
      Subclass Crossopterygii. Coelacanths, related
        forms. 283, 285
      Subclass Actinopterygii. Ray-finned fishes. 285
        Order Acipenseriformes. Sturgeons, paddlefish.
        Order Salmoniformes. Salmon, trout. 283, 285
        Order Atheriniformes. Killifishes, guppies.
        Order Gasterosteiformes. Sticklebacks,
          seahorses. 588
        Order Perciformes. Perches, wrasses, barracudas, tunas,
          freshwater bass, mackerels. 285
        Order Lophiiformes. Angler fishes. 563
    Class Amphibia. Mostly tetrapods; embryo enclosed in
      amnion. 200, 204, 282, 285
      Order Caudata. Salamanders. 285
      Order Anura. Frogs, toads. 285, 462–463
      Order Apoda. Apodans (caecilians). 285
    Class Reptilia. Skin with scales, embryo enclosed in
      amnion. 200, 204, 285–286

      Subclass Anapsida. Turtles, tortoises. 286
      Subclass Lepidosaura. *Sphenodon*, lizards, snakes. 174, 286
      Subclass Archosaura. Dinosaurs (extinct), crocodiles,
        alligators. 204–207, 286
    Class Aves. Birds. (In more recent schemes, dinosaurs,
      crocodilians, and birds are grouped in the same
      category.) 176, 192, 193, 201, 498
      Order Struthioniformes. Ostriches. 286
      Order Sphenisciformes. Penguins. 192, 527
      Order Procellariiformes. Albatrosses, petrels. 593–594
      Order Ciconiiformes. Herons, bitterns, storks,
        flamingoes. 192, 193
      Order Anseriformes. Swans, geese, ducks.
      Order Falconiformes. Eagles, hawks, vultures, falcons.
      Order Galliformes. Ptarmigan, turkeys, domestic fowl.
      Order Columbiformes. Pigeons, doves. 506–507
      Order Strigiformes. Owls.
      Order Apodiformes. Swifts, hummingbirds. 286
      Order Passeriformes. Sparrows, jays, finches, crows,
        robins, starlings, wrens. 586, 591
  Class Mammalia. Skin with hair; young nourished by milk-
    secreting glands of adult. 199–200, 204, 213, 282, 288
    Subclass Prototheria. Egg-laying mammals (duckbilled
      platypus, spiny anteaters). 288
    Subclass Metatheria. Pouched mammals or marsupials
      (opossums, kangaroos, wombats). 200, 288
    Subclass Eutheria. Placental mammals. 288
      Order Insectivora. Tree shrews, moles, hedgehogs. 193,
        215
      Order Scandentia. Insectivorous tree shrews. 215
      Order Chiroptera. Bats. 201, 594
      Order Primates.
        Suborder Strepsirhini (prosimians). Lemurs,
          lorises. 212, 213
        Suborder Haplorhini (tarsioids and anthropoids). 212,
          213
          Infraorder Tarsiiformes. Tarsiers. 212, 215
          Infraorder Platyrrhini (New World monkeys). 212
            Family Cebidae. Spider monkeys, howler
              monkeys, capuchin. 215
          Infraorder Catarrhini (Old World monkeys and
            hominoids). 212, 213
            Superfamily Cercopithecoidea. Baboons,
              macaques, langurs. 511, 592, 593, 597
            Superfamily Hominoidea. Apes and humans. 212,
              213
              Family Hylobatidae. Gibbons. 212, 215
              Family Pongidae. Chimpanzees, gorillas,
                orangutans. 212, 216
              Family Hominidae. Humans and most recent
                ancestors of humans. 216
      Order Carnivora. Carnivores. 525–527, 529
        Suborder Feloidea. Cats, civets, mongooses, hyenas.
        Suborder Canoidea. Dogs, weasels, skunks, otters,
          raccoons, pandas, bears. 213
      Order Proboscidea. Elephants; mammoths (extinct). 193,
        200
      Order Sirenia. Sea cows (manatees, dugongs).
      Order Perissodactyla. Odd-toed ungulates (horses,
        tapirs, rhinos). 193, 282
      Order Artiodactyla. Even-toed ungulates (camels, deer,
        bison, sheep, goats, antelopes, giraffes). 228, 361, 527
      Order Edentata. Anteaters, tree sloths,
        armadillos. 174
      Order Tubulidentata. African aardvark.
      Order Cetacea. Whales, porpoises. 193, 288, 526, 563
      Order Rodentia. Most gnawing animals (squirrels, rats,
        mice, guinea pigs, porcupines). 200, 423
      Order Lagomorpha. Rabbits, hares, pikas. 194, 200, 454

APPENDIX III
ANSWERS TO GENETICS PROBLEMS

Chapter Nine

1. a. *AB*
   b. *AB* and *aB*
   c. *Ab* and *ab*
   d. *AB*, *aB*, *Ab*, and *ab*

2. a. *AaBB* will occur in all the offspring.
   b. 25% *AABB*; 25% *AaBB*; 25% *AABb*; 25% *AaBb*.
   c. 25% *AaBb*; 25% *Aabb*; 25% *aaBb*; 25% *aabb*.

   d. $\frac{1}{16}$ *AABB* (6.25%)

   $\frac{1}{8}$ *AaBB* (12.5%)

   $\frac{1}{16}$ *aaBB* (6.25%)

   $\frac{1}{8}$ *AABb* (12.5%)

   $\frac{1}{4}$ *AaBb* (25 %)

   $\frac{1}{8}$ *aaBb* (12.5%)

   $\frac{1}{16}$ *AAbb* (6.25%)

   $\frac{1}{8}$ *Aabb* (12.5%)

   $\frac{1}{16}$ *aabb* (6.25%)

3. Yellow is recessive. Because the first-generation plants must be heterozygous and had a green phenotype, green must be dominant over the recessive yellow.

4. a. Mother must be heterozygous for both genes; father is homozygous recessive for both genes. The first child is also homozygous recessive for both genes.
   b. The probability that the second child will not be able to roll the tongue and will have free earlobes is $\frac{1}{4}$ (25%).

5. a. *ABC*
   b. *ABc* and *aBc*
   c. *ABC*, *aBC*, *ABc*, and *aBc*
   d. *ABC*, *aBC*, *AbC*, *abC*, *ABc*, *aBc*, *Abc*, and *abc*

6. The first-generation plants must all be double heterozygotes. When these plants are self-pollinated $\frac{1}{4}$ (25%) of the second-generation plans will be doubly heterozygous.

7. The most direct way to accomplish this would be to allow a true-breeding mouse having yellow fur to mate with a true-breeding mouse having brown fur. Such true-breeding strains could be obtained by repeated inbreeding (mating of related individuals; for example, a male and a female of the same litter) of yellow and brown strains. In this way, it should be possible to obtain homozygous yellow and homozygous brown mice.

   When true-breeding yellow and true-breeding brown mice are crossed, the progeny should all be heterozygous. If the progeny phenotype is either yellow or brown, then the dominance is simple or complete, and the phenotype reflects the dominant allele. If the phenotype is intermediate between yellow and brown, there is incomplete dominance. If the phenotype shows both yellow and brown, there is codominance.

8. a. The mother must be heterozygous ($I^A i$). The man having type B blood could have fathered the child if he were also heterozygous ($I^B i$).

   b. If the man is heterozygous, then he *could be* the father. However, because any other type B heterozygous male also could be the father, one cannot say that this particular man absolutely must be. Actually, any male who could contribute an O allele (*i*) could have fathered the child. This would include males with type O blood (*ii*) or type A blood who are heterozygous ($I^A i$).

9. a. F_1 genotypes and phenotypes: 100% *Bb Cc*, brown progeny. F_2 phenotypes: $\frac{9}{16}$ brown + $\frac{3}{16}$ tan + $\frac{4}{16}$ albino.

   F_2 genotypes: $\begin{cases} \frac{1}{16}\ BB\ CC\ +\ \frac{2}{16}\ BB\ Cc\ +\ \frac{2}{16}\ Bb\ CC\ +\ \frac{4}{16}\ Bb\ Cc; \\ (\frac{9}{16}\ \text{brown}) \\ \frac{1}{16}\ bb\ CC\ +\ \frac{2}{16}\ bb\ Cc;\ (\frac{3}{16}\ \text{tan}) \\ \frac{1}{16}\ BB\ cc\ +\ \frac{2}{16}\ Bb\ cc\ +\ \frac{1}{16}\ bb\ cc;\ (\frac{4}{16}\ \text{albino}) \end{cases}$

   b. Backcross phenotypes: $\frac{1}{4}$ brown + $\frac{1}{4}$ tan + $\frac{2}{4}$ albino.

   Backcross genotypes: $\begin{cases} \frac{1}{4}\ Bb\ Cc;\ (\frac{1}{4}\ \text{brown}) \\ \frac{1}{4}\ bb\ Cc;\ (\frac{1}{4}\ \text{tan}) \\ \frac{1}{4}\ Bb\ cc\ +\ \frac{1}{4}\ bb\ cc;\ (\frac{1}{2}\ \text{albino}) \end{cases}$

Chapter Ten

1. a. Males inherit their X chromosome from their mothers.
   b. A male can produce two types of gametes with respect to an X-linked gene. One type will lack this gene and possess a Y chromosome. The other will have an X chromosome and the linked gene.
   c. A female homozygous for an X-linked gene will produce just one type of gamete containing an X chromosome with the gene.
   d. A female heterozygous for an X-linked gene will produce two types of gametes. One will contain an X chromosome with the dominant allele, and the other type will contain an X chromosome with the recessive allele.

2. a. Because this gene is only carried on Y chromosomes, females would not be expected to have hairy pinnae because they normally do not have Y chromosomes.
   b. Because sons always inherit a Y chromosome from their fathers and because daughters never do, a man having hairy pinnae will always transmit this trait to his sons and never to his daughters.

3. A 0% crossover frequency means that 50% of the gametes will be *AB* and 50% will be *ab*.

4. The gene for hemophilia occurs on the X but not the Y chromosome. A male has only one X chromosome. Therefore, it would be impossible for a male simply to be a carrier; the allele associated with hemophilia would always be expressed.

5. Assuming the mother is heterozygous (most individuals with Huntington's disorder are), the woman has a $\frac{1}{2}$ (50%) chance of being heterozygous and therefore of later developing the disorder. Also, if this woman married a normal male, they would have a 50% chance of having a child with the disorder. Thus the *total* probability of their having a child with Huntington's disorder is (0.5)(0.5) = 0.25, or $\frac{1}{4}$ (25%).

6. The first child can only be color blind if it is a boy. Why? The probability of this happening is 25%. Similarly, their second child also has a 25% chance of being color blind. The probability that both will be color blind is $(0.25)(0.25) = 0.0625$, or 6.25%.

7. This indicates that genetic information other than that necessary for sex determination must reside on the X chromosome. Such information is necessary for survival regardless of whether one is male or female. Obviously, this is not true for the Y chromosome, in that individuals (females) survive quite nicely in its absence. A major function of the Y chromosome is to change what would have been a female individual into a male.

8. The only parent from whom this child could have received an X chromosome that bears a nonhemophilia allele is the mother. Therefore, nondisjunction must have occurred in the father.

9. A child with Klinefelter syndrome could be produced if a Y-bearing sperm fertilized an egg having two X chromosomes (as a result of nondisjunction during egg development). Such a child could also be produced if a normal egg (with one X chromosome) were fertilized by a sperm having an X and Y chromosome (as a result of nondisjunction during sperm development).

10. The Punnett square for this situation would be as follows:

	X-bearing sperm	Y-bearing sperm
XX-bearing egg	trisomic XXX	Klinefelter XXY
no X in egg	Turner XO	dies before birth; has only Y

11. a. An unaffected female selected at random has 1 chance in 50 (2%) of being heterozygous.

b. If you selected an unaffected male and unaffected female at random, the probability that both will be heterozygous is $(0.02)(0.02) = 0.0004$, or 0.04%. The probability that a pair of unaffected individuals selected at random could have a child affected by PKU is given by $(0.02)(0.02)(0.25) = 0.0001$, or 0.01%. This is the same thing as 1/10,000, which suggests that about one birth in every 10,000 will be a child with PKU, assuming that only heterozygous individuals have such children (an assumption which is not completely true).

APPENDIX IV
ANSWERS TO SELF-QUIZZES

Chapter 1

1. DNA
2. energy
3. Metabolism
4. Homeostasis
5. adaptations
6. mutations
7. reproductive capacity

8. b.
9. c.
10. c.
11. c.
12. c.

Chapter 2

1. electrons
2. electrons
3. electrons
4. Radioisotopes
5. a.
6. b.

7. d.
8. b.
9. c
   e
   b
   d
   a

Chapter 3

1. cells
2. plasma membrane
3. c.
4. d.
5. d.
6. b.
7. c.
8. a.
9. c.

10. a.
11. d.
12. b.
13. c
    h
    g
    f
    b
    d
    a
    e

Chapter 4

1. sun
2. thermodynamics
3. heat
4. b.
5. c.
6. e.
7. d.

8. c.
9. d.
10. c
    e
    d
    a
    b

Chapter 5

1. carbon atoms
2. carbon dioxide, sunlight
3. photosynthesis
4. Heterotrophs
5. d.
6. c.
7. b.

8. e.
9. c.
10. c.
11. c
    d
    e
    b
    a

Chapter 6

1. organic compounds
2. lactate fermentation, alcoholic fermentation
3. pyruvate, carbon dioxide, water
4. d.
5. c.
6. d.

7. c.
8. b.
9. b.
10. b
    c
    a
    d

Chapter 7

1. mitosis, meiosis
2. chromosomes, DNA
3. c.
4. d.
5. c.
6. a.
7. c.

8. d.
9. b.
10. d.
11. d
    b
    c
    a

Chapter 8

1. gamete, growth
2. sperm, eggs
3. diploid, two
4. c.
5. c.
6. c.
7. d.
8. c.

9. d.
10. c.
11. b.
12. d
    e
    a
    b
    c

Chapter 9

1. a.
2. c.
3. a.
4. c.
5. b.
6. c.

7. b.
8. d.
9. c.
10. c
    d
    e
    b
    a

Chapter 10

1. c.
2. e.
3. e.
4. c.
5. e.
6. c.
7. d.

8. d.
9. c.
10. c
    e
    d
    b
    a

Chapter 11

1. Hydrogen
2. e.
3. d.
4. d.
5. c.
6. a.
7. a.

8. d.
9. d.
10. d
    e
    b
    c
    a

Chapter 12

1. three
2. d.
3. d.
4. b.
5. d.
6. c.
7. a.

8. b.
9. a.
10. e
    c
    d
    b
    a
11. a.

Chapter 13

1. diversity
2. genetic engineering by recombinant DNA technology
3. Plasmids
4. cloning tools
5. c.
6. d.
7. a.
8. c.
9. e.
10. d
    e
    a
    c
    b

Chapter 14

1. population
2. differences
3. e.
4. a.
5. e.
6. d.
7. c.
8. c.
9. d.
10. d
    c
    e
    a
    b

Chapter 15

1. macroevolution
2. mass extinctions, adaptive radiations
3. gradual, bursts
4. e.
5. c.
6. e.
7. c.
8. d.
9. c.
10. d
    e
    a
    b
    c

Chapter 16

1. e.
2. b.
3. c.
4. c.
5. a.
6. b.
7. d.
8. d.
9. d.
10. d
    a
    e
    c
    b

Chapter 17

1. living host cells
2. e.
3. d.
4. d.
5. c.
6. c.
7. e.
8. b.
9. c.
10. c
    e
    d
    a
    b

Chapter 18

1. c.
2. c.
3. d.
4. b.
5. a.
6. c.
7. d.
8. d.
9. d.
10. c.
11. d
    b
    c
    e
    a
12. d
    e
    c
    f
    b
    a

Chapter 19

1. body symmetry, cephalization, type of gut, type of body cavity, segmentation
2. a.
3. b.
4. d.
5. a.
6. a.
7. b.
8. c.
9. b.
10. d
    a
    c
    e
    b

Chapter 20

1. dermal, ground, vascular
2. apical, lateral
3. b.
4. a.
5. a.
6. b.
7. c.
8. a.
9. d.
10. c
    d
    f
    a
    e
    b

Chapter 21

1. hydrogen bonds
2. stomata
3. c.
4. d.
5. e.
6. c.
7. b.
8. d.
9. d.
10. d.
11. e
    d
    c
    a
    b

Chapter 22

1. sporophyte
2. pollinators
3. d.
4. d.
5. c.
6. c.
7. c.
8. c.
9. b.
10. e
    d
    c
    a
    b
    f

Chapter 23

1. epithelial, connective, nervous, muscle
2. e.
3. d.
4. c.
5. b.
6. a.
7. d.
8. c.
9. a.
10. Receptors, integrator, effectors
11. d
    e
    c
    b
    a

Chapter 24

1. integumentary
2. Skeletal, muscular
3. protect, support, minerals, blood cells
4. smooth, cardiac, skeletal
5. d.
6. d.
7. b.
8. b.
9. c.
10. d
    e
    b
    a
    c

Chapter 25

1. digestive, circulatory, respiratory, urinary
2. breaking down, absorbing, eliminating
3. energy, energy
4. carbohydrates
5. amino acids, fatty acids
6. b.
7. c.
8. d.
9. c.
10. e
   d
   a
   c
   b

Chapter 26

1. circulatory, lymph vascular
2. Arteries, veins, capillaries, venules, arterioles
3. d.
4. d.
5. c.
6. b.
7. c.
8. b.
9. e.
10. d
   e
   a
   f
   c
   b

Chapter 27

1. immune
2. white blood cells, antigens
3. Antigens, immune
4. circulating antigen, cancer cells
5. e.
6. d.
7. c.
8. b.
9. b.
10. c.
11. a.
12. b
   e
   a
   b
   c

Chapter 28

1. Oxygen, carbon dioxide
2. gas, partial pressure gradient
3. epithelium, diffuse
4. adaptations, air flow
5. d.
6. e.
7. b.
8. a.
9. d.
10. d.
11. e
   a
   d
   b
   c

Chapter 29

1. filtration, reabsorption, secretion
2. c.
3. c.
4. e.
5. d.
6. a.
7. d.
8. c.
9. d.
10. b
   d
   e
   c
   a

Chapter 30

1. environmental stimuli, neurons
2. action potentials
3. transmitter substance
4. integration
5. d.
6. c.
7. c.
8. b.
9. a.
10. c.
11. e
   d
   b
   c
   a
12. c
   a
   e
   d
   b

Chapter 31

1. Hormones, transmitter substances, local signaling molecules, pheromones
2. hypothalamus, pituitary gland
3. negative feedback
4. receptors
5. a.
6. d.
7. e.
8. b.
9. d.
10. d
   b
   e
   c
   a

Chapter 32

1. diversity, new, changing
2. cleavage, gastrulation, organ formation
3. cleavage, cell differentiation
4. cell differentiation
5. morphogenesis
6. d.
7. b.
8. d.
9. d.
10. d
   e
   f
   c
   a
   b

Chapter 33

1. Ecology
2. population
3. e.
4. e.
5. b.
6. e.
7. d.
8. c.
9. d.
10. d
   e
   c
   a
   b

Chapter 34

1. Communities
2. niche
3. habitat
4. e.
5. d.
6. d.
7. b.
8. d.
9. c.
10. e
   c
   a
   d
   b

Chapter 35

1. ecosystem, energy, cycling
2. photosynthesis, one
3. energy, nutrients; energy, nutrients
4. biogeochemical cycles
5. d.
6. d.
7. e.
8. c.
9. e.
10. d
   c
   b
   a
11. c
   d
   e
   b
   a

Chapter 36

1. the sun
2. atmosphere, weather systems
3. rainfall
4. d.
5. c.
6. e.
7. a.
8. a.
9. e.
10. f
   a
   c
   d
   e
   b

Chapter 37

1. exponential
2. pollutants
3. energy flow, resource
4. e.
5. c.
6. b.

7. c.
8. c.
9. d.
10. e
    a
    c
    b
    d

Chapter 38

1. Animal behavior
2. nervous, endocrine
3. instinctive, learned
4. genetic, environmental
5. adaptive traits
6. d.
7. c.

8. d.
9. c.
10. b.
11. d
    e
    b
    c
    a

CREDITS AND ACKNOWLEDGMENTS

Page xi Stock Imagery / **Pages xii–xiii** James M. Bell/Photo Researchers / **Pages xiv–xv** S. Stammers/SPL/Photo Researchers / **Pages xvi–xvii** © 1990 Arthur M. Greene / **Page xviii** © Thomas D. Mangelsen 1985 / **Page xix** © Thomas D. Mangelsen 1987 / **Pages xx–xxi** Jim Doran

Page 1 NASA

Chapter 1

1.3 (left) Paul DeGreve/FPG; (right) Norman Meyers/Bruce Coleman, Inc. / **1.4** Jack deConingh / **1.5** (a) Tony Brain/SPL/Photo Researchers; (b) M. Abbey/Visuals Unlimited; (c) Edward S. Ross; (d) Dennis Brokaw; (e) Edward S. Ross; (f) Pat & Tom Leeson/Photo Researchers / **1.6** Levi Publishing Company

Page 15 James M. Bell/Photo Researchers

Chapter 2

2.1 Martin Rogers/FPG / **2.2** Jack Carey / **Page 19** (a) (left) Kingsley R. Stern; (right) Chip Clark / **Page 20** (b) Stanford Medical Center; (c) (left) Hank Morgan/Rainbow; (right) Dr. Harry T. Chugani, M.D., UCLA School of Medicine / **2.6** Art by L. Calver / **2.9** H. Eisenbeiss/Frank Lane Picture Agency / **2.10** Colin Monteath/Hedge Hog House New Zealand / **2.13** Michael Grecco/Picture Group / **2.24** Lewis L. Lainey / **2.29–2.30, 2.31** (b) A. Lesk/SPL/Photo Researchers

Chapter 3

3.1 Jan Hinsch/SPL/Photo Researchers / **3.2** (a) (left) National Library of Medicine; (right) Armed Forces Institute of Pathology, Neg. No. 66-1386-1; (b–e) Jeremy Pickett-Heaps, School of Botany, University of Melbourne / **3.4** Micrograph Keith R. Porter; art by K. Kasnot / **3.6** Micrograph M. Sheetz, R. Painter, and S. Singer, *Journal of Cell Biology*, 70:193, 1976, by copyright permission of The Rockefeller University Press / **3.7** Frank B. Salisbury / **3.12** (a) Micrograph G. Cohen-Bazire; (b) Gary Gaard and Arthur Kelman / **3.13–3.14** Art by Leonard Morgan / **3.15** Micrograph M. C. Ledbetter, Brookhaven National Laboratory; art by D. & V. Hennings / **3.16** Micrograph G. L. Decker; art by D. & V. Hennings / **3.17** Micrograph D. Fawcett, *The Cell*, Philadelphia: W. B. Saunders Co., 1966; art by D. & V. Hennings / **3.18** Art by K. Kasnot / **3.20** (b) W. Bloom and D. Fawcett, *A Textbook of Histology*, Philadelphia: W. B. Saunders Co., 1975; (c) Daniel S. Friend, M.D. / **3.21** Micrograph Gary W. Grimes / **3.22** Micrograph Keith R. Porter / **3.23** Micrograph L. K. Shumway / **3.24** J. Victor Small and Gottfried Rinnerthaler / **3.25** (a) Sidney L. Tamm; (b) art by D. & V. Hennings / **3.26** (a) After Alberts et al., *Molecular Biology of the Cell*, Garland Publishing Co., 1983; (b) Dianne T. Woodrum and Richard W. Linck / **3.27** Sketch by D. & V. Hennings after P. Raven et al., *Biology of Plants*, Third edition, Worth Publishers, 1981; micrograph P. A. Roelofsen

Chapter 4

4.1 (left) NASA; (right) Manfred Kage/Peter Arnold / **4.7** W. S. Bennett and T. A. Steitz / **4.11** Douglas Faulkner/Sally Faulkner Collection / **4.15** Art by L. Calver after B. Alberts et al., *Molecular Biology of the Cell*, Garland Publishing Co., 1983 / **4.16** Kathie Atkinson/Oxford Scientific Films

Chapter 5

5.1 Photograph Sam Zarember/Image Bank / **5.2** (a) Photograph Kjell Sandved; art by K. Kasnot; (b),(c) Harry T. Horner; (c),(d) art by Joel Ito / **5.3** (a) Barker-Blankenship/FPG; (b) art by Victor Royer / **5.4** Photograph E. R. Degginer / **5.5** Ron Thomas/FPG / **5.6** Art by Illustrious, Inc. / **5.10** (a) Art by L. Calver

Chapter 6

6.1 Janeart Ltd./Image Bank / **6.2** Photograph Tom McHugh / **6.3** (left) Art by L. Calver / **6.4** (a) Keith R. Porter; (b),(c) art by L. Calver / **6.7** Photograph Adrian Warren/Ardea, London / **6.10** David M. Phillips/Visuals Unlimited / **Page 91** R. Llewellyn/Superstock, Inc.

Page 93 © Lennart Nilsson

Chapter 7

7.1 (a–d) Lennart Nilsson from *A Child Is Born* © 1966, 1967 Dell Publishing Company, Inc. (e) Lennart Nilsson from *Behold Man*, © 1974 by Albert Bonniers Forlag and Little, Brown and Company, Boston / **7.2** (a) S. Brecher / **7.4** Andrew S. Bajer, University of Oregon / **7.5** Micrographs by Ed Reschke; art by K. Kasnot / **7.7** Micrograph H. Beams and R. G. Kessel, *American Scientist*, 64:279–290, 1976

Chapter 8

8.1 Courtesy of Kirk Douglas/The Bryna Company / **8.2** Art by L. Calver / **8.3** Art by K. Kasnot / **8.4** CNRI/SPL/Photo Researchers / **8.5** Art by K. Kasnot / **8.9** (b) © 1986 David M. Phillips/Visuals Unlimited / **8.11** Art by K. Kasnot

Chapter 9

9.1 Moravian Museum, Brno / **9.2** Jean M. Labat/Ardea, London / **9.11** David Hosking / **9.12** (b) F. Blakeslee, *Journal of Heredity*, 1914 / **Page 125** (b) David M. Phillips/Visuals Unlimited; (c) Bill Longcore/Photo Researchers

Chapter 10

10.3 Photographs Carolina Biological Supply Company / **10.5** (b) Photographs Keystone Press Agency / **10.9** After V. McKusick, *Human Genetics*, Second edition, copyright 1969. Reprinted by permission of Prentice-Hall, Inc. Englewood Cliffs, New Jersey / **10.10** Art by K. Kasnot / **10.11** (a) Used by permission of Carole Iafrate; (b–d) Courtesy of Peninsula Association for Retarded Children and Adults, San Mateo Special Olympics, Burlingame, CA / **10.12** (a) Cytogenetics Laboratory, University of California, San Francisco; (b) after Collman and Stoller, *American Journal of Public Health*, 52, 1962

Chapter 11

11.2 Art by D. & V. Hennings / **11.3** (b) A. K. Kleinschmidt / **11.6** Art by D. & V. Hennings; photograph A. Lesk/SPL/Photo Researchers / **11.7** Art by L. Calver / **11.8** (left) E. J. Dupraw; micrographs (left to right) J. G. Gall; B. Hamkalo; V. Foe

Chapter 12

12.1 (b) Lennart Nilsson © Boehringer Ingelheim International GmbH / **12.5–12.6** Art by L. Calver / **12.8** Peter Starlinger / **12.9** Art by Victor Royer / **12.10** Murray L. Barr / **Page 160** Lennart Nilsson © Boehringer Ingelheim International, GmbH

Chapter 13

13.1 Dr. Huntington Potter and Dr. David Dressler, Harvard Medical School / **13.6** Michael Maloney/San Francisco Chronicle / **13.7** W. Merrill / **13.8** R. L. Brinster and R. E. Hammer, School of Veterinary Medicine, University of Pennsylvania

Page 171 S. Stammers/SPL/Photo Researchers

Chapter 14

14.1 (a) By permission of the Darwin Museum, Down House, courtesy of George P. Darwin; (c) Christopher Ralling; (d) D. Barrett/Planet Earth Pictures; (f) C. P. Hickman, Jr.; (e), (g) Heather Angel / **14.3** (left) Field Museum of Natural History (CK21T) and the artist, Charles R. Knight; (right) Lee Kuhn/FPG / **14.4** (a) E. C. Williams/Visuals Unlimited; (b) David Cavagnaro; (c) Heather Angel; (d) Alan Root/Bruce Coleman Ltd.; (e) David Steinberg / **14.5** Photograph John H. Ostrom, Yale University / **14.6** Alan Solem / **14.8** After D. Futuyma, *Evolutionary Biology*, Sinauer, 1979 / **14.10** After M. Karns and L. Penrose, *Annals of Eugenics*, 15:206–233, 1951 / **14.11** (c),(d) Alex Kerstitch; (e) Thomas N. Taylor; (f) Edward S. Ross / **14.12** J. A. Bishop and L. M. Cook / **14.13** (a) Bruce Beehler; (b) Charles W. Fowler/National Marine Fisheries; (c) D. Avon/Ardea, London / **14.14** After F. Ayala and J. Valentine, *Evolving*, Benjamin-Cummings, 1979 / **14.15** After V. Grant, *Organismic Evolution*, W. H. Freeman and Co., 1977 / **14.16** After W. Jensen and F. B. Salisbury, *Botany: An Ecological Approach*, Wadsworth, 1972

Chapter 15

15.1 (a) Patricia G. Gensel; (b) Jonathan Blair/Woodfin Camp & Associates / **15.2** From T. Storer et al., *General Zoology*, Sixth edition, McGraw-Hill, 1979. Reproduced by permission of McGraw-Hill, Inc. / **15.3** Art by Victor Royer / **15.4** Art by Joel Ito / **15.5** (top) Douglas P. Wilson/Eric & David Hosking; (center) Superstock, Inc.; (bottom) E. R. Degginer / **15.6** After C. G. Sibley and J. E. Ahlquist, *Scientific American*, February 1986, copyright © by Scientific American, Inc., all rights reserved / **15.7** Chesley Bonestell / **15.9** (a) Sidney W. Fox; (b) W. Hargreaves and D. Deamer / **15.11** Art by Leonard Morgan / **15.13** After P. Dodson, *Evolution: Process and Product*, Third edition, Prindle, Weber & Schmidt / **15.14** Data from J. J. Sepkoski, Jr., *Paleobiology*, 7(1):36–53 and J. J. Sepkoski, Jr. and M. L. Hulver in

Valentine, ed., *Phanerozoic Diversity Patterns: Profiles in Macroevolution*, Princeton University Press, 1985 / **15.15** (a) Stanley W. Awramik; (b) M. R. Walter / **15.16** (a),(b) Neville Pledge/South Australian Museum; (c),(d) Chip Clark / **15.18** (a) H. P. Banks; (b) Patricia G. Gensel / **Page 205** (a) (left) NASA; (right) William K. Hartmann / **Pages 206–207** (b) © 1988 John Gurche / **15.19** After S. M. Stanley, *Macroevolution: Pattern and Process*, W. H. Freeman and Co., 1979

Chapter 16

16.1 Art by Illustrious, Inc. / **16.2** Art by D. & V. Hennings / **16.3** (a) Bruce Coleman Ltd.; (b) Tom McHugh/Photo Researchers; (c) Larry Burrows/Aspect Picture Library / **16.4** © Time Inc. 1965/Larry Burrows Collection / **16.6** (a) Dr. Donald Johanson, Institute of Human Origins; (b) Louise M. Robbins / **16.7** Art by D. & V. Hennings / **16.8** Photographs by John Reader copyright 1981

Page 221 © 1990 Arthur M. Greene

Chapter 17

17.1 Tony Brain/SPL/Photo Researchers / **17.2** Art by L. Calver / **17.3** C. McLaren and F. Siegel/Burroughs Wellcome Co. / **17.4** Breck's / **17.5** Art by L. Calver / **17.6** Micrograph J. J. Cardamone, Jr., University of Pittsburgh/BPS / **17.7** (a) John D. Cunningham/Visuals Unlimited; (b) Tony Brain/SPL/Photo Researchers; (c) P. W. Johnson and J. McN. Sieburth, University of Rhode Island/BPS / **17.8** Stanley W. Watson, *International Journal of Systematic Bacteriology*, 21:254–270, 1971 / **17.9** T. J. Beveridge, University of Guelph/BPS / **Page 230** (a) Art by Victor Royer / **Page 231** (b) H. Stolp; (c) Richard W. Greene; (d) Laszlo Meszoly in L. Margulis, *Early Life*, Jones and Bartlett, Publishers, Inc., Boston, © 1982 / **17.10** (a) Edward S. Ross; art by Joan Carol after Harold C. Bold and John W. LaClaire II, *The Plant Kingdom*, Fifth edition, © 1987, p. 243. Adapted by permission of Prentice-Hall Inc., Englewood Cliffs, New Jersey; (b) (first) John T. Bonner; (second) London Scientific Films; (lower three) Carolina Biological Supply Company / **17.11** (a) P. L. Walne and J. H. Arnott, *Planta*, 77:325–354, 1967; (b) T. E. Adams/Visuals Unlimited / **17.12** G. Shih and R. G. Kessel, *Living Images*, Jones and Bartlett Publishers, Inc., Boston © 1982 / **17.13** (a) Florida Department of Natural Resources, Bureau of Marine Research; (b) C. C. Lockwood / **17.14** (a) John D. Cunningham/Visuals Unlimited; (b) John Clegg/Ardea, London; (c) Biophoto Associates / **17.15** Art by Leonard Morgan; micrograph Steven L'Hernault / **17.16** (left two) Bruce Russell/BioMedia Associates; (right) Gary W. Grimes and Steven L'Hernault / **17.17** Gary W. Grimes and Steven W. L'Hernault

Chapter 18

18.1 Steven C. Wilson/Entheos / **18.4** (a) (above) David M. Phillips/Visuals Unlimited; (below) John D. Cunningham/Visuals Unlimited; (b) John E. Hodgin; (c) David M. Phillips/Visuals Unlimited; (d) Victor Duran / **Page 242** W. Merrill / **18.5** (a) Art by Leonard Morgan; (b) Jane Burton/Bruce Coleman Ltd.; (c) Roger K. Burnard; (d) Victor Duran / **18.6** (a) N. Allin and G. L. Barron; (b) Gary T. Cole, University of Texas/BPS; (c) G. L. Barron, University of Guelph / **18.7** (a) After Raven, Evert, and Eichhorn, *Biology of Plants*, Fourth edition, Worth Publishers, New York, 1986; (b) Ken Davis/Tom Stack & Associates / **18.9** (a) D. P. Wilson/Eric & David Hosking; (b) Douglas Faulkner/Sally Faulkner Collection; (c) Hervé Chaumeton/Agence Nature; (d) Dennis Brokaw / **18.10** Photograph D. J. Patterson/Seaphot Ltd.: Planet Earth Pictures; art by D. & V. Hennings / **18.11** Carolina Biological Supply Company / **18.12** Photograph Jane Burton/Bruce Coleman Ltd.; art by D. & V. Hennings / **18.13** John D. Cunningham/Visuals Unlimited / **18.14** (a) Field Museum of Natural History; (b) Edward S. Ross; (c) W. H. Hodge; (d) Jean Paul Ferrero/Ardea, London / **18.15** Art by D. & V. Hennings; photograph A. & E. Bomford/Ardea, London / **18.16** (a) Ed Reschke; (b) Kingsley R. Stern; (c) F. J. Odendaal, Duke University/BPS; (d) Edward S. Ross / **18.17** Photograph Edward S. Ross; art by D. & V. Hennings / **18.18** (a) Martin W. Grosnick/Ardea, London; (b) Hans Reinhard/Bruce Coleman Ltd.; (c) Dick Davis/Photo Researchers; (d) Heather Angel / **18.19** Art by D. & V. Hennings

Chapter 19

19.1 (left) John Kenfield/Bruce Coleman Ltd.; (right) Christopher Crowley / **19.3** Art by K. Kasnot / **19.6** David C. Haas/Tom Stack & Associates / **19.7** Art by Victor Royer / **19.8** (c) Frieder Sauer/Bruce Coleman Ltd.; (c) Kim Taylor/Bruce Coleman Ltd. / **19.9** (a) Walter Deas/Seaphot Limited: Planet Earth Pictures; (b) Bill Wood/Seaphot Limited: Planet Earth Pictures / **19.10** F. Stuart Westmorland/Tom Stack & Associates / **19.12** Photograph Andrew Mounter/Seaphot Limited: Planet Earth Pictures; art by Raychel Ciemma / **19.13** Art by Joan Carol after T. Storer et al., *General Zoology*, Sixth edition, © 1979 McGraw-Hill / **19.14** Art by K. Kasnot / **19.15** Cath Ellis, University of Hull/SPL/Photo Researchers / **Page 268** (b) Photograph Carolina Biological Supply Company; art by K. Kasnot; (c) Lorus J. and Margery Milne; (d) Dianora Niccolini / **19.17** Photograph J. Solliday/BPS; art by Raychel Ciemma / **19.18** Photograph Anthony and Elizabeth Bomford/Ardea, London / **19.19** (a), (b) Jeff Foott/Tom Stack & Associates; (c) Rick M. Harbo / **19.20** Art by Laszlo Meszoly and D. and V. Hennings / **19.21** J. Grossauer/ZEFA / **19.22** J. A. L. Cooke/Oxford Scientific Films / **19.23** Hervé Chaumeton/Agence Nature / **19.24** (a) Art by Ron Ervin; (b) After C. P. Hickman et al., *Integrated Principles of Zoology*, Sixth edition, St. Louis: C. V. Mosby Co., 1979 / **19.25** Art by D. & V. Hennings / **19.26** (a) John H. Gerard; (b) Ken Lucas/Seaphot Limited: Planet Earth Pictures; (c) P. J. Bryant, University of California, Irvine/BPS / **19.27** (a) Tom McHugh; (b) Agence Nature; (c) Hervé Chaumeton/Agence Nature / **19.28** (a) Leszczynski/Animals Animals; (b) Steve Martin/Tom Stack & Associates / **19.29** (a) David Maitland/Seaphot Limited: Planet Earth Pictures; (b–g), (i) Edward S. Ross; (h) C. P. Hickman, Jr. / **19.30** (a) John Mason/Ardea, London; (b), (c) Hervé Chaumeton/Agence Nature; (d) Ian Took/Biofotos / **19.31** Art by L. Calver / **19.33** (a–d) From *Living Invertebrates*, V. & J. Pearse and M. & R. Buchsbaum, The Boxwood Press, 1987. Used by permission. (e) Rick M. Harbo / **19.34** Photograph Hervé Chaumeton/Agence Nature; art by Laszlo Meszoly and D. & V. Hennings / **19.35** After C. P. Hickman, Jr. and L. S. Roberts, *Integrated Principles of Zoology*, Seventh edition, St. Louis: Times Mirror/Mosby College Publishing, 1984 / **19.36** Art by D. & V. Hennings after Romer and others / **19.38** (a) Peter Scoones/Seaphot Limited: Planet Earth Pictures; (b) art by Laszlo Meszoly and D. & V. Hennings / **19.39** Heather Angel / **19.40** (a) Allan Power/Bruce Coleman Ltd.; (b) Erwin Christian/ZEFA; (c) Bill Wood/Bruce Coleman Ltd.; (d) Robert and Linda Mitchell / **19.41** (a) © John Serraro/Visuals Unlimited; (b) Jerry W. Nagel / **19.43** (a) D. Kaleth/Image Bank; (b) W. A. Banaszewski/Visuals Unlimited; (c) C. B. & D. W. Frith/Bruce Coleman Ltd.; (d) Heather Angel; (e) Jane Burton/Bruce Coleman Ltd. / **19.44** (a) Hans Reinhard/ZEFA; (b) Wisniewski/ZEFA; (c) sketch after O. S. Pettinghill, *Ornithology in Laboratory and Field*, Burgess Publishing Company, 1970 / **19.45** (a) Warren Garst/Tom Stack & Associates; (b) Jack Dermid; (c) (above) Jean-Paul Ferrero/Ardea, London; (below) Alan Root/Bruce Coleman Ltd.; (d) Sandy Roessler/FPG / **Page 291** Bonnie Rauch/Photo Researchers

Chapter 20

20.2 (a–c) Biophoto Associates / **20.3** (a) G. Shih and R. G. Kessel, *Living Images*, Jones and Bartlett, Publishers, Inc., Boston, © 1982; (b) Chuck Brown / **20.4** Art by D. & V. Hennings / **20.6** (left) Art by D. & V. Hennings; (center) Carolina Biological Supply Company; (left) James W. Perry / **20.7** (left) Art by D. & V. Hennings; (center) Ray F. Evert; (right) James W. Perry / **20.8** (a) Robert and Linda Mitchell; (b),(c) Roland R. Dute / **20.9** (b–d) E. R. Degginger / **20.11–20.12** Art by D. & V. Hennings / **20.13** G. Shih and R. G. Kessel, *Living Images*, Jones and Bartlett, Publishers, Inc., Boston © 1982 / **20.14** John E. Hodgin / **20.15** Micrograph E. R. Degginger / **20.16** Sketch after T. Rost et al., *Botany: A Brief Introduction to Plant Biology*, Second edition, © 1984, John Wiley & Sons; micrographs Chuck Brown / **20.17** Ripon Microslides, Inc. / **20.19** (b) Jerry D. Davis / **20.21** (b) Biophoto Associates; (c) H. A. Core, W. A. Coté, and A. C. Day, *Wood Structure and Identification*, Second edition, Syracuse University Press, 1979

Chapter 21

21.1 Adrian P. Davies/Bruce Coleman Ltd. / **21.2** (a–c) Robert and Linda Mitchell / **21.3** Art by Leonard Morgan / **21.4** Micrograph Jean Paul Revel / **21.5** J. Mexal, C. Reid, and E. Burke, *Botanical Gazette*, 140: No. 3, University of Chicago Press, 1979 / **21.6** F. B. Reeves / **21.7** Micrographs H. A. Core, W. A. Coté, and A. C. Day, *Wood Structure and Identification*, Second edition, Syracuse University Press, 1979 / **21.8** Art by Leonard Morgan / **21.9** (a),(b) John Troughton and L. A. Donaldson / **21.12** Martin Zimmerman, *Science*, 133:73–79, © AAAS 1961

Chapter 22

22.3 F. D. Hess / **22.4** Art by D. & V. Hennings / **Page 320** (a) Harlo H. Hadow; (b) Ted Schwartz; (c) R. Taggart / **Page 321** (d) Thomas Eisner; (e) R. Taggart / **Page 322** (f) Edward S. Ross; (g) Ted Schwartz / **22.5** (a),(b) Patricia Schulz; (c),(d) Ray F. Evert; (e),(f) Ripon Microslides, Inc.; (far right) Kingsley R. Stern / **22.6** B. Bracegirdle and P. Miles, *An Atlas of Plant Structure*, Heinemann Educational Books, 1977 / **22.7** Janet Jones / **22.8** (a) B. J. Miller, Fairfax, VA/BPS; (b) R. Carr/Bruce Coleman Ltd.; (c) Richard H. Gross, Motlow State Community College / **22.9** (a) Kjell Sandved; (b) Edward S. Ross / **22.10** Photograph Carolina Biological Supply Company / **22.11** Photograph Hervé Chaumeton/Agence Nature / **22.13** Frank B. Salisbury / **22.14** John Digby and Richard Firn / **22.15** Frank B. Salisbury / **22.17** Jan Zeevaart / **Page 331** (above) Dennis Brokaw; (below) Edward S. Ross / **22.18** A. C. Leopold, et al., *Plant Physiology*, 34:570, 1958 / **22.19** R. J. Downs in T. T. Kozlowski, ed., *Tree Growth*, The Ronald Press, 1962

Page 335 © 1989 Kevin Shafer

Chapter 23

23.1 David Macdonald / **23.2** Manfred Kage/Bruce Coleman Ltd. / **23.3** Photographs (left) Lennart Nilsson from *Behold Man*, © 1974 by Albert Bonniers Forlag and Little, Brown and Company, Boston; (center) Manfred Kage/Bruce Coleman Ltd.; (right) Ed Reschke/Peter Arnold Inc. / **23.4–23.5** Photographs Ed Reschke / **23.6** Lennart Nilsson from *Behold Man*, © 1974 Albert Bonniers Forlag and Little, Brown and Company, Boston / **23.7** Art by L. Calver

Chapter 24

24.1 Art by L. Calver / 24.2 Art by Joel Ito / 24.3 (a) Chaumeton-Lanceau/Agence Nature; (b) CNRI/SPL/ Photo Researchers / 24.4 Michael Keller/FPG / 24.5 Art by Joel Ito; micrograph Ed Reschke / 24.6 Art by K. Kasnot / 24.7 Art by D. & V. Hennings / 24.9 Art by Ron Ervin / **Page 352** Photograph C. Yokochi and J. Rohen, *Photographic Anatomy of the Human Body*, Second edition, Igaku-Shoin Ltd., 1979 / 24.10 (b–d) Art by L. Calver / 24.11 (a) Ed Reschke; (b) D. Fawcett, *The Cell*, Philadelphia: W.B. Saunders Co., 1966 / 24.13 R.M. Jensen / **Page 357** Photograph Michael Neveux

Chapter 25

25.1 Art by L. Calver / 25.2 (a) Kim Taylor/Bruce Coleman Ltd.; (b) Wardene Weisser/Ardea, London / 25.3 Art by Richard Demarest / 25.4 After A. Vander, J. Sherman, and D. Luciano, *Human Physiology*, Fourth edition, McGraw-Hill, 1985. Used by permission. / 25.5 (a) Art by Victor Royer; (b), (c) From *Tissues and Organs: A Text-Atlas of Scanning Electron Microscopy* by R. G. Kessel and R. H. Kardon. Copyright © 1979 by W. H. Freeman and Company. Reprinted with permission; (d) J. D. Hoskins, W. G. Henk, and Y. Z. Abdelbaki, *American Journal of Veterinary Research*, 43:10, 1982 / 25.7 Art by L. Calver / 25.8 (b) Steven Jones/FPG / **Page 368** Photograph CNRI/Phototake / 25.9 (b) Photograph Ralph Pleasant/FPG / 25.10 Modified from A. Vander, J. Sherman, and D. Luciano, *Human Physiology*, Fourth edition, McGraw-Hill, 1985

Chapter 26

26.1 Art by L. Calver / 26.2 (b) (below) After M. Labarbera and S. Vogel, *American Scientist*, 70:54–60, 1982 / 26.3 (a) CNRI/SPL/Photo Researchers; (b) Lennart Nilsson from *Behold Man*, © 1974 by Albert Bonniers Forlag and Little, Brown and Company, Boston / 26.4 (left) Art by L. Calver and Victor Royer; (right) art by Victor Royer / 26.5 (a) Art by Leonard Morgan; (b) art by L. Calver / 26.6 (a) C. Yokochi and J. Rohen, *Photographic Anatomy of the Human Body*, Second edition, Igaku-Shoin Ltd., 1979; (b) art by Joel Ito / **Page 382** (a) (above) Ed Reschke; (below) F. Sloop and W. Ober/Visuals Unlimited / 26.10 Art by Robert Demarest based on A. Spence, *Basic Human Anatomy*, Benjamin-Cummings, 1982 / 26.13 After J. A. Gosling et al., *Atlas of Human Anatomy with Integrated Text*, copyright © 1985 by Gower Medical Publishing Ltd. / 26.14 Emil Bernstein, *Science*, cover 27 August 1971, copyright 1971 AAAS / 26.15 (a) After F. Ayala and J. Kiger, *Modern Genetics*, © 1980 Benjamin-Cummings; (b) Lester V. Bergman & Associates, Inc. / 26.16 After Gerard J. Tortora and Nicholas P. Anagnostakos, *Principles of Anatomy and Physiology*, Sixth edition, Copyright © 1990 by Biological Sciences Textbooks, Inc., A & P Textbooks, Inc. and Elia-Sparta, Inc. Reprinted by permission of Harper Collins Publishers / 26.17 Art by D. & V. Hennings

Chapter 27

27.1 (a) The Granger Collection; (b) Lennart Nilsson © Boehringer Ingelheim International GmbH / 27.4 Art by L. Calver and Victor Royer / 27.5 Art by L. Calver after S. Tonegawa, *Scientific American*, October 1985 / **Page 397** Photographs Dr. Gilla Kaplan / 27.6 Art by L. Calver after B. Alberts et al., *Molecular Biology of the Cell*, Garland Publishing Company, 1983 / **Pages 430–431** Micrographs Z. Salahuddin, National Institutes of Health; (d) Art by L. Calver

Chapter 28

28.1 (a) (right) Art by L. Calver; (b) (above) Peter Parks/Oxford Scientific Films/Animals Animals; (below) Steve Lissau/Rainbow / 28.2 Art by D. & V. Hennings after C. P. Hickman et al., *Integrated Principles of Zoology*, Sixth edition, St. Louis: C. V. Mosby Co., 1979 / 28.3 Ed Reschke / 28.4 Micrograph H. R. Duncker, Justus-Liebig University, Giessen, West Germany / 28.5 Art by L. Calver / 28.7 CNRI/ SPL/Photo Researchers / 28.8 Art by K. Kasnot / 28.10 Art by Leonard Morgan / **Page 414** Gerard D. McLane / **Page 415** Lennart Nilsson from *Behold Man*, © 1974 by Albert Bonniers Forlag and Little, Brown and Company, Boston / **Page 416** David Steinberg

Chapter 29

29.1 (right) Art by L. Calver / 29.2 (a) Art by Ron Ervin; (b) art by Joel Ito / 29.3 Art by Richard Demarest / 29.5 Art by Joel Ito / **Page 423** Tom McHugh/Photo Researchers

Chapter 30

30.1 Eric A. Newman / 30.2 Art by Leonard Morgan; micrograph Manfred Kage/Peter Arnold, Inc. / 30.3–30.4 Art by D. & V. Hennings / 30.6 Art by Leonard Morgan / 30.7 (a) Carolina Biological Supply Company; art by Leonard Morgan / 30.8 Art by Richard Demarest / 30.9 (a) Art by Richard Demarest; (b) From *Tissues and Organs: A Text-Atlas of Scanning Electron Microscopy* by R. G. Kessel and R. H. Kardon. Copyright © 1979 by W. H. Freeman and Company. Reprinted with permission / 30.10 Art by K. Kasnot / 30.12 (right) Art by L. Calver / 30.13 Art by Richard Demarest / 30.14 C. Yokochi and J. Rohen, *Photographic Anatomy of the Human Body*, Second edition, Igaku-Shoin Ltd., 1979 / 30.15 Art by Joel Ito / 30.16 Art by Richard Demarest / 30.17 Aart by Ron Ervin / 30.18–30.19 Art by Richard Demarest / 30.20 (b), (c) Robert E. Preston, courtesy Joseph E. Hawkins, Kresge Hearing Research Institute, University of Michigan Medical School / 30.21 (a) Keith Gillett/Tom Stack & Associates; (b–d) after M. Gardiner, *The Biology of Vertebrates* McGraw-Hill, 1972 / 30.22 E. R. Degginger / 30.23 Art by Richard Demarest / 30.25 Art by Richard Demarest / 30.26 Micrograph Lennart Nilsson © Boehringer Ingelheim International GmbH

Chapter 31

31.1 (a) Art by L. Calver; (b) art by Joel Ito / 31.2–31.3 Art by Joel Ito / 31.6 (a) Syndication International (1986) Ltd.; (b) Mitchell Layton / 31.7 Photographs courtesy of Dr. William H. Daughaday, Washington University School of Medicine. From A. I. Mendelhoff and D. E. Smith, eds., *American Journal of Medicine*, 20:133 (1956) / 31.9 Art by Joel Ito / 31.10 The Bettmann Archive / 31.11 Art by Leonard Morgan

Chapter 32

32.1 (a) Hans Pfletschinger; (b–e) John H. Gerard / 32.3 Photographs Carolina Biological Supply Company; sketch after M. B. Patten, *Early Embryology of the Chick*, Fifth edition, McGraw-Hill, 1971 / 32.5 (a–c) Adapted by permission of Macmillan Publishing Company from *Patterns and Principles of Animal Development* by John W. Saunders, Jr. Copyright © 1982 by John W. Saunders, Jr.; (d) S. R. Hilfer and J. W. Yang, *The Anatomical Record*, 197:423–433, 1980 / 32.6 (left) Art by Ron Ervin; (right) art by L. Calver / 32.7 Art by L. Calver; (c) From *Tissues and Organs: A Text-Atlas of Scanning Electron Microscopy* by R. G. Kessel and R. H. Kardon. Copyright © 1979 by W. H. Freeman and Company. Reprinted with permission / 32.8 Art by Ron Ervin / 32.9 (right) Art by Ron Ervin; (left) art by L. Calver / 32.11 Art by Joan Carol / 32.12 Art by Richard Demarest; (left) micrograph from Lennart Nilsson, *A Child Is Born*, © 1966, 1977 Dell Publishing Company, Inc.; (right) from Lennart Nilsson, *Behold Man*, © 1974 by Albert Bonniers Forlag and Little, Brown and Co., Boston / 32.13 Art by Richard Demarest / 32.14 Art by L. Calver; (c) after A. S. Romer and T. S. Parsons, *The Vertebrate Body*, Sixth edition, Saunders College Publishing, © CBS College Publishing / 32.15 Art by L. Calver after Bruce Carlson, *Patten's Foundations of Embryology*, Fourth edition, McGraw-Hill, 1981 / 32.16–32.18 From Lennart Nilsson, *A Child Is Born*, © 1966, 1977 Dell Publishing Company, Inc. / 32.19 Art by Richard Demarest / 32.20 Art by Ron Ervin / **Page 481** Modified from Keith L. Moore, *The Developing Human: Clinically Oriented Embryology*, Fourth edition, Philadelphia: W. B. Saunders Co., 1988 / 32.21 From L. B. Arey, *Developmental Anatomy*, Philadelphia: W. B. Saunders Co., 1965 / **Page 487** (a) Cheun-mo To and C. C. Brinton; (b) Joel B. Baseman

Page 491 David Doubilet

Chapter 33

33.1 Antoinette Jongen/FPG / 33.2 Photograph Fran Allan/Animals Animals / 33.5 Photograph E. Vetter/ ZEFA / 33.7 Photograph NASA / 33.10 After G. T. Miller, *Living in the Environment*, Sixth edition, Wadsworth, 1990 / 33.11 Data from Population Reference Bureau

Chapter 34

34.1 (left) Dona Hutchins; (right) Edward S. Ross / 34.2 (a),(c) Harlo H. Hadow; (b) Bob and Miriam Francis/Tom Stack & Associates / 34.3 After G. Gause, 1934 / 34.4 Photograph Clara Calhoun/Bruce Coleman Ltd. / 34.5 Photograph John Dominis, Life Magazine, © Time Inc. / 34.6 (a) (left) Ed Cesar/ Photo Researchers; (right) W. E. Ruth / 34.7 Edward S. Ross / 34.8 (a) Douglas Faulkner/Sally Faulkner Collection; (b) W. M. Laetsch; (c),(d) Edward S. Ross; (e) James H. Carmichael / **Page 516** R. Slavin/ FPG / 34.10 (a–f), (i) Roger K. Burnard; (g),(h) E. R. Degginger / 34.11 Photograph Dr. Harold Simon/ Tom Stack & Associates; (below) after S. Fridkriksson, *Evolution of Life on a Volcanic Island*, Butterworth: London, 1975

Chapter 35

35.2 (a) Lynn Eckmann, University of Washington/ BPS; (b–e) Roger K. Burnard / 35.3 Photograph Sharon R. Chester / 35.9 Photograph Steven D. Bach / 35.10 Art by Raychel Ciemma / 35.11 (a) Photograph Gene E. Likens from G. E. Likens and F. H. Bormann, *Proceedings First International Congress of Ecology*, pp 330–335, September 1974, Centre Agric. Publ. Doc. Wagenigen, The Hague, The Netherlands; (b),(c) photographs by Gene E. Likens from G. E. Likens et al., *Ecology Monograph*, 40(1):23–47, 1970 / 35.13 Photograph William J. Weber/ Visuals Unlimited

Chapter 36

36.1 Edward S. Ross / 36.3 (b) Art by L. Calver / 36.7 Art by Raychel Ciemma / 36.9 Art by D. & V. Hennings after G. T. Miller, *Environmental Science: An Introduction*, Wadsworth, 1986 / 36.10 After Whittaker; Bland; and Tilman / 36.11 Art by Joan Carol / 36.12 Harlo H. Hadow / 36.13 Dennis Brokaw / 36.14 Kenneth W. Fink/Ardea, London / 36.15 Ray Wagner/Save the Tall Grass Prairie, Inc. / 36.16 Jonathan Scott/Planet Earth Pictures / 36.17 (a) Thase Daniel; (b) Adolf Schmidecker/FPG; (c) E. R. Degginger / 36.18 Thomas E. Hemmerly / 36.19 Dennis Brokaw / 36.20 Ansel Adams / 36.21 D. W. MacManiman / 36.24 D. W. Schindler, *Science*, 184:897–899 / 36.25 E. R. Degginger / 36.26 Courtesy of J. L. Sumich, *Biology of Marine Life*, Fourth edition, William C. Brown, 1988 / 36.27 Phil Degginger /

36.29 (a) Dennis Brokaw; (b) McCutcheon/ZEFA; (c) Chuck Niklin; (d) William H. Amos; (e) Fred Grassle, Woods Hole Institution of Oceanography; (f) Robert Hessler / **36.30** (top right) Jim Doran; all other photographs Douglas Faulkner/Sally Faulkner Collection / **Page 566** Photographs R. Legeckis/ NOAA

Chapter 37

37.1 © 1983 Billy Grimes / **37.2** Photograph John Lawlor/FPG / **37.3** After G. T. Miller, *Environmental Science: An Introduction*, Wadsworth, 1986 and the Environmental Protection Agency / **37.4** (a) USDA Forest Service; (b) Heather Angel / **37.5** (Bottom: left) National Science Foundation; (Top: left; right) NASA / **37.6** From Water Resources Council / **37.7** Data from G. T. Miller / **37.8** After G. T. Miller, *Living in the Environment*, Sixth edition, Wadsworth, 1990 / **Page 578** National Oceanic and Atmospheric Administration/NESDIS / **Page 579** R. Bierregaard/ Photo Researchers / **37.9** USDA Soil Conservation Service/Thomas G. Meier / **37.10** Agency for International Development / **37.11** Data from G. T. Miller / **Page 583** J. McLoughlin/FPG

Chapter 38

38.1 (a) Photograph John S. Dunning/Ardea, London; sonogram J. Bruce Falls and Tom Dickinson, University of Toronto; (b) photograph Hans Reinhard/Bruce Coleman Ltd.; graph G. Pohl-Apel and R. Sussinka, *Journal for Ornithologie*, 123:211–214 / **38.2** (a) Eric & David Hosking; (b) Stephen Dalton/ Photo Researchers / **38.3** (a) Nina Leen in *Animal Behavior*, Life Nature Library; (b) F. Schultz / **38.4** (a) Eugene Kozloff; (b),(c) Stevan Arnold / **38.5** Lincoln P. Brower / **38.6–38.7** John Alcock / **38.8** Edward S. Ross / **38.9** (a) E. Mickleburgh/Ardea, London; (b–d) G. Ziesler/ZEFA / **38.10** Merlin D. Tuttle, Bat Conservation International / **38.11** Art by D. & V. Hennings / **38.12** Fred Bruemmer / **38.13** Timothy Ransom

GLOSSARY OF BIOLOGICAL TERMS

ABO blood type Blood type determined by particular recognition proteins at the surface of red blood cells; types A and B are genetically codominant. Type O results in homozygous recessives, who have neither A nor B proteins.

abortion Spontaneous or induced expulsion of the embryo or fetus from the uterus. Spontaneous abortions are also called miscarriages.

abscisic (ab-SISS-ik) **acid** Plant hormone that promotes seed and bud dormancy, and helps restrict water loss during short periods of drought.

abscission (ab-SIH-zhun) [L. *abscissus*, to cut off] Leaf (or fruit or flower) drop after hormonal action causes a corky cell layer to form where a leaf stalk joins a stem; nutrient and water flow is thereby shut off.

acid [L. *acidus*, sour] A substance that releases a hydrogen ion (H+) in solution.

acid deposition, dry The falling to earth of airborne particles of sulfur and nitrogen oxides.

acid deposition, wet The falling to earth of snow or rain that contains sulfur and nitrogen oxides.

acoelomate (ay-SEE-la-mate) Type of animal that has no fluid-filled cavity between the gut and body wall.

actin (AK-tin) A protein that functions in contraction; together with myosin, a component of the myofibrils of muscle cells.

action potential Nerve impulse; a sudden, dramatic reversal of the polarity of charge across the plasma membrane of neurons and some other cells.

activation energy The minimum amount of collision energy needed to boost reactant molecules to the point at which a reaction will proceed spontaneously.

active site A crevice on the surface of an enzyme molecule where a specific reaction is catalyzed.

active transport Movement of ions and molecules across a cell membrane, against a concentration gradient, by ATP expenditure. The ion or molecule is moved in a direction other than the one in which simple diffusion would take it.

adaptation [L., *adaptare*, to fit] The process of becoming adapted (or more adapted) to a given set of environmental conditions.

adaptive radiation A burst of evolutionary activity in geologic time, with lineages branching away from one another as they partition the existing environment or invade new ones.

adaptive trait An aspect of form, function, or behavior that helps an organism survive and reproduce under a given set of environmental conditions.

adaptive zone A way of life, such as "catching insects in the air at night." A lineage must have physical, ecological, and evolutionary access to an adaptive zone to become a successful occupant of it.

adenine (AH-de-neen) A purine; a nitrogen-containing base found in nucleotides.

adenosine diphosphate (ah-DEN-uh-seen die-FOSS-fate) ADP, a molecule involved in cellular energy transfers; typically formed by hydrolysis of ATP.

adenosine phosphate (ah-DEN-un-seen FOSS-fate) Any of several relatively small molecules, some of which function as chemical messengers within and between cells, and others as energy carriers.

adenosine triphosphate ATP, a molecule that is a major carrier of energy (by way of its phosphate groups) from one reaction site to another in all living cells.

ADH Antidiuretic hormone produced by the hypothalamus and released by the posterior pituitary; stimulates reabsorption in the kidneys and so reduces urine volume.

adrenal (ah-DREE-nul) **cortex** Outer portion of either of two adrenal glands; its hormones have roles in metabolism, inflammation, maintaining extracellular fluid volume, and other functions.

adrenal medulla Inner region of the adrenal gland; its hormones help control blood circulation and carbohydrate metabolism.

aerobic (air-OH-bik) [Gk. *aer*, air, + *bios*, life] Able to use free oxygen as a final electron acceptor in degradative reactions.

aerobic respiration Degradative pathway of ATP formation that requires oxygen for its operation. The pathway proceeds from glycolysis, through the Krebs cycle, and on through electron transport phosphorylation. Of all degradative pathways, aerobic respiration has the greatest energy yield, with 36 ATP typically formed for each glucose molecule degraded.

agglutination (Ah-glue-tin-AY-shun) Clumping of foreign cells, induced by the cross-linking of antigen-antibody complexes at their surface.

aging A range of processes, including the breakdown of cell structure and function, by which the body gradually deteriorates. Characteristic of all organisms showing extensive cell differentiation.

AIDS Acquired immune deficiency syndrome: A set of chronic disorders caused by the human immune-deficiency virus (HIV). The body's immune system becomes severely compromised and the individual becomes highly vulnerable to infections that might not otherwise be life-threatening.

alcoholic fermentation Anaerobic pathway of ATP formation in which pyruvate from glycolysis is broken down to acetaldehyde; the acetaldehyde then accepts electrons from NADH to become ethanol.

aldosterone (al-DOSS-tuh-rohn) Hormone secreted by the adrenal cortex that helps regulate sodium reabsorption.

alga, plural **algae** Any of the simple, mostly aquatic plants classified as red, brown, and green algae respectively.

allantois (ah-LAN-twahz) [Gk. *allas*, sausage] Vascularized extraembryonic membrane of reptiles, birds, and mammals that develops as a bladderlike pouch. Functions in excretion and respiration in reptiles and birds; functions in oxygen transport by way of the umbilical cord in placental mammals.

allele (uh-LEEL) One of two or more alternative forms of a gene at a given gene locus.

allele frequency The relative abundance in a population of each kind of allele that can occur at a given gene locus.

allergy An abnormal, secondary immune response to a normally harmless substance.

allosteric (AL-oh-STARE-ik) **control** Control of enzyme functioning via the binding of a specific substance at a control site on the enzyme molecule.

altruistic (al-true-ISS-tik) **behavior** Self-sacrificing behavior; the individual behaves in a way that helps others but, in so doing, decreases its own chance to produce offspring.

alveolar (al-VEE-uh-lar) **sac** Any of the pouch-like clusters of alveoli in lungs; the major sites of gas exchange.

alveolus (ahl-VEE-uh-lus), plural **alveoli** [L. *alveus*, small cavity] Any of the many cup-shaped, thin-walled outpouchings of the respiratory bronchioles (terminal airways) where gas exchange occurs between air in the lungs and the bloodstream.

amino (uh-MEE-no) **acid** A small organic molecule having an amino group, an acid group, and one or more atoms called its R group; the building block of proteins.

ammonification (uh-moan-ih-fih-KAY-shun) A decomposition process by which certain bacteria and fungi break down nitrogen-containing wastes and remains of other organisms.

amnion (AM-nee-on) In land vertebrates, an extraembryonic membrane in the form of a fluid-filled sac that surrounds the embryo, absorbing shocks and keeping it from drying out.

anaerobic (an-uh-ROW-bik) **cell** [Gk. *an*, without, + *aer*, air] A cell that either cannot use free oxygen as a final electron acceptor in degradative reactions or dies upon exposure to it.

anaphase (AN-uh-faze) The stage of mitosis when sister chromatids of each chromosome separate and move to opposite poles of the spindle.

anaphase I and II Stages of meiosis when each chromosome separates from its homologous partner (anaphase I) and, later, when the sister chromatids of each chromosome separate from each other (anaphase II).

androgen One of several sex hormones required in sperm formation, genital development, and maintenance of secondary sexual traits; also influences growth and development.

angiosperm (AN-gee-oh-spurm) [Gk. *angeion*, vessel, and *sperma*, seed] A flowering plant.

animal A heterotroph that eats or absorbs nutrients from other organisms; is multicelled, usually with tissues arranged in organs and organ systems; is usually motile during at least part of the life cycle; and goes through a period of embryonic development.

Animalia The kingdom of animals.

annual plant Vascular plant that completes its life cycle in one growing season.

anther [Gk. *anthos*, flower] In flowering plants, the pollen-bearing part of the male reproductive structure (stamen).

antibody [Gk. *anti*, against] Any of a variety of Y-shaped receptor molecules each with binding sites for a specific antigen (molecule that triggers an immune response); produced by B cells of the immune system.

anticodon In a tRNA molecule, a sequence of three nucleotide bases that can pair with an mRNA codon.

antigen (AN-tih-jen) [Gk. *anti*, against, + *genos*, race, kind] Any large molecule (usually a protein or polysaccharide) with a distinct configuration that triggers an immune response.

aorta (ay-OR-tah) [Gk. *airein*, to lift, heave] Main artery of systemic circulation; carries oxygenated blood away from the heart to all regions except the lungs.

apical dominance The influence exerted by a terminal bud in inhibiting the growth of lateral buds.

apical meristem (AY-pih-kul MARE-ih-stem) [L. *apex*, top, + Gk. *meristos*, divisible] In most plants, a mass of self-perpetuating cells at a root or shoot tip that is responsible for primary growth, or elongation, of plant parts.

appendicular (ap-en-DIK-you-lahr) **skeleton** In vertebrates, bones of the limbs, pelvic girdle (at the hips), and pectoral girdle (at the shoulders).

arteriole (ar-TEER-ee-ole) Any of the blood vessels that branch from arteries; arterioles serve as control points for adjusting the volume of blood to be delivered to different body regions.

artery Any of the large-diameter, thick-walled blood vessels that transport blood among the heart, lungs, and body tissues, and that branch into arterioles, which lead in turn to capillary beds.

asexual reproduction Mode of reproduction in which offspring arise from a single parent, and inherit the genes of that parent only.

atmosphere A region of gases, airborne particles, and water vapor enveloping the earth; eighty percent of its mass is distributed within 17 miles of the earth's surface.

atmospheric cycle A biogeochemical cycle in which a large portion of the element being cycled between the physical environment and ecosystems occurs in a gaseous phase in the atmosphere; examples include the cycling of carbon and nitrogen.

atom The smallest unit of matter that is unique to a particular kind of element.

atomic number The number of protons in the nucleus of each atom of an element; differs for each element.

ATP Adenosine triphosphate, an energy-carrying molecule composed of adenine, the sugar ribose, and three linked phosphate groups; directly or indirectly transfers energy to or from nearly all metabolic pathways.

ATP/ADP cycle Cycle in which ADP (adenosine diphosphate) is phosphorylated to ATP, usually by gaining a phosphate group; the ATP then donates a phosphate group elsewhere and becomes ADP.

australopith (OHSS-trah-low-pith) [L. *australis*, southern, + Gk. *pithekos*, ape] Any of the earliest known species of hominids; that is, the first species on the human evolutionary branch.

autoimmune response Abnormal immune response in which lymphocytes are unleashed against the body's own cells.

autonomic (auto-NOM-ik) **nervous system** Those nerves leading from the central nervous system to cardiac cells, muscle cells, smooth muscle cells (such as those of the stomach), and glands—that is, the visceral portion of the body.

autosomal dominant inheritance Condition in which a dominant allele on an autosome (not a sex chromosome) is always expressed to some extent.

autosomal recessive inheritance Condition in which a mutation produces a recessive allele on an autosome (not a sex chromosome); only recessive homozygotes show the resulting phenotype.

autosome Any of those chromosomes that are of the same number and kind in both males and females of the species.

autotroph (AH-toe-trofe) [Gk. *autos*, self, + *trophos*, feeder] An organism able to build all the organic molecules it requires using carbon dioxide (present in air and in water) and energy from the physical environment. Photosynthetic autotrophs use sunlight energy; chemosynthetic autotrophs extract energy from chemical reactions involving inorganic substances. Compare *heterotroph*.

auxin (AWK-sin) Any of a class of plant growth-regulating hormones; promotes stem elongation as one of its effects.

axial skeleton In vertebrates, the skull, backbone, ribs, and breastbone (sternum).

axon A long, cylindrical extension of a neuron with finely branched endings. Action potentials are propelled rapidly, without alteration, along an axon; their arrival at the axon endings can trigger the release of transmitter substances that may affect the activity of an adjacent cell.

bacterial fission Asexual reproduction by division of a bacterial cell into two equivalent parts.

bacterial flagellum Whiplike motile structure of many bacterial cells; unlike other flagella, it does not contain a core of microtubules.

bacterial fission Asexual reproduction by division of a bacterial cell into two equivalent parts.

bacteriophage (bak-TEER-ee-oh-fahj) [Gk. *baktērion*, small staff, rod, + *phagein*, to eat] Category of viruses that infect bacterial cells.

basal body Cell structure that organizes the growth of microtubules in flagella or cilia.

base A substance that, in solution, releases ions that can combine with hydrogen ions.

base pair Any of the pairs of nucleotide bases joined by hydrogen bonds in the DNA double helix. Two kinds of base pairings are possible: A—T (adenine with thymine) and G—C (guanine with cytosine).

behavior, animal Any coordinated response that an animal makes to external and internal stimuli. The responses are outcomes of the integration of sensory, neural, endocrine, and effector components, all of which have a genetic basis (hence are subject to natural selection); and they may be modified by learning processes.

benthic province All of the sediments and rocky formations of the ocean bottom; begins with the continental shelf and extends down through deep-sea trenches.

biennial (by-EN-ee-ull) Flowering plant that lives two growing seasons.

bilateral symmetry Body plan by which an animal has a head end and a tail end, and a left and right side.

biogeochemical cycle The transfer of elements necessary for life from the environment to organisms and then back to the environment; involves carbon, oxygen, hydrogen, and a range of mineral elements such as nitrogen and phosphorus.

biogeographic realm [Gk. *bios*, life, + *geographein*, to describe the surface of the earth] In one scheme, one of six major regions having a characteristic array of species that are generally isolated from the other realms by physical barriers that restrict dispersal.

biological clock Term applied to the apparent internal time-measuring mechanisms in many organisms that allows them to adjust to environmental change.

biological magnification The increasing concentration of a nondegradable substance in body tissues as it moves up through trophic levels.

biome A broad, vegetational subdivision of some biogeographic realm, shaped by climate, topography, ·and the composition of regional soils.

biosphere [Gk. *bios*, life, + *sphaira*, globe] Those regions of the earth's waters, crust, and atmosphere in which organisms live; the most inclusive level of biological organization.

biosynthetic pathway A metabolic pathway in which small molecules are assembled into lipids, proteins, and other large organic molecules.

biotic potential The maximum rate of increase in a population, per individual, under ideal conditions.

bipedalism Habitual use of two feet for standing and walking.

blastocyst (BLASS-tuh-sist) [Gk. *blastos*, sprout, + *kystis*, pouch] In mammalian development, a modified blastula stage consisting of a hollow ball of surface cells having inner cells massed at one end.

blastula (BLASS-chew-lah) In many animal species, an embryonic stage consisting of a hollow, fluid-filled ball of cells one layer thick.

blood pressure Fluid pressure, generated by heart contractions, that keeps blood circulating. Generally measured at large arteries of systemic circulation.

bronchus, plural **bronchi** (BRONG-cuss, BRONG-kee) [Gk. *bronchos*, windpipe] Tubelike branchings of the trachea (windpipe) that lead to the lungs.

bud An undeveloped shoot of mostly meristematic tissue; often protected by a covering of modified leaves.

buffer A substance that can combine with hydrogen ions, release them, or both, in response to changes in pH.

bulk flow In response to a pressure gradient, a movement of more than one kind of molecule in the same direction in the same medium (gas or liquid).

C4 plant A plant in which carbon dioxide fixation in some cells produces a four-carbon compound (oxaloacetate), which donates carbon to the Calvin-Benson cycle in other cells. C4 plants have a competitive edge on hot, dry days, when photorespiration prevails.

calorie (KAL-uh-ree) [L. *calor*, heat] The amount of heat needed to raise the temperature of one gram of water by 1°C. Nutritionists sometimes use "calorie" to mean kilocalorie (1,000 calories), which is a source of much confusion.

Calvin-Benson cycle Stage of the light-independent reactions of photosynthesis in which carbon-containing compounds are used to form carbohydrates (such as glucose) and to regenerate a sugar phosphate (RuBP) required in carbon dioxide fixation. The first product is a three-carbon compound (PGA).

cambium, plural **cambia** (KAM-bee-um) In vascular plants, one of two types of embryonic tissue masses that are responsible for secondary growth (increase in stem or root diameter). Vascular cambium gives rise to secondary xylem and phloem; cork cambium gives rise to periderm.

camouflage An outcome of form, patterning, color, or behavior that helps an organism blend with its surroundings and escape detection.

cancer Malignancy arising from cells having profound abnormalities in the plasma membrane and cytoplasm, abnormal growth and division, and weakened capacity for adhesion within the parent tissue.

capillary (L. *capillus*, hair) A small, extremely thin-walled blood vessel; component of capillary beds, the diffusion zones for exchanges of gases and materials between blood and interstitial fluid.

carbohydrate [L. *carbo*, charcoal, + *hydro*, water] A simple sugar or a large molecule composed of sugar units, and used universally by cells for energy and as structural materials.

carbon cycle Biogeochemical cycle in which carbon moves from reservoirs in land, atmosphere, and oceans, through organisms, then back to the reservoirs.

carbon dioxide fixation First stage of the light-independent reactions of photosynthesis, in which carbon dioxide is affixed to RuBP (ribulose bisphosphate), the starting point for subsequent reactions that lead to the formation of carbohydrates such as sucrose, starch, and cellulose.

carcinogen (kar-SIN-uh-jen) Ultraviolet radiation and many other agents that can act on proto-oncogenes and trigger cancer.

cardiac cycle [Gk. *kardia*, heart, + *kyklos*, circle] The sequence of muscle contractions and relaxation constituting one heartbeat.

cardiovascular system Of animals, an organ system composed of blood, one or more hearts, and blood vessels.

carnivore [L. *caro, carnis*, flesh, + *vovare*, to devour] An animal that eats other animals; a type of heterotroph.

carotenoid (kare-OTT-en-oyd) A pigment molecule that absorbs violet and blue wavelengths of light, but reflects yellow, orange, and red.

carpel (KAR-pul) One or more closed vessels that serve as the female reproductive parts of a flower. The chamber within a carpel is the *ovary* where eggs develop and are fertilized, and seeds mature.

carrying capacity The number of individuals of a given species that can be sustained indefinitely in a given area.

cartilaginous joints (car-tih-LADJ-in-us) Skeletal joints in which cartilage fills the space between bones and permits only slight movement.

cDNA Any DNA molecule copied from mRNA (as during DNA amplification).

cell [L. *cella*, small room]. The basic *living* unit. A cell has the capacity to maintain itself as an independent unit and to reproduce, given appropriate sources of energy and raw materials.

cell cycle A recurring sequence of events that extends from the time a cell forms until its own division is completed.

cell junction Any of various types of physical links between adjacent cells in multicelled organisms.

cell plate In plant cell division, a partition that forms at the equator of the mitotic spindle, between the two newly forming cells.

cell theory A theory in biology, the key points of which are that (1) all organisms are made of cells, (2) the cell is the basic living unit of organization, and (3) all cells arise from preexisting cells.

cell wall A rigid or semirigid supportive wall outside the plasma membrane; found among plants, fungi, protistans, and most bacteria.

central nervous system The brain and spinal cord of vertebrates.

central vacuole Fluid-filled organelle occupying fifty to ninety percent of the interior of a plant cell.

centriole (SEN-tree-ohl) A short cylinder of triplet microtubules near the nucleus in most animal cells; centrioles occur in pairs and may govern the plane of cell division.

centromere (SEN-troh-meer) [Gk. *kentron*, center, + *meros*, a part]. A special region of the chromosome serving as the attachment site for spindle microtubules during nuclear division.

cephalization (sef-ah-lah-ZAY-shun) [Gk. *kephalikos*, head] Differentiation of one end of the animal body into a head in which nervous tissue and sensory organs are especially concentrated.

cerebellum (ser-ah-BELL-um) [L. diminutive of *cerebrum*, brain] Hindbrain region that coordinates motor activity for refined limb movements, maintaining posture, and spatial orientation.

cerebrum (suh-REE-bruhm) In vertebrate forebrain, paired masses of gray matter; cerebral hemispheres overlying thalamus, hypothalamus, and pituitary. Includes primary receiving centers for receptors at body periphery, association centers for coordinating and processing sensory input, and motor centers for coordinating motor responses.

chemical bond A union between the electron structures of two or more atoms or ions.

chemical synapse (SIN-aps) [Gk. *synapsis*, union] A junction between two neurons, or between a neuron and a muscle cell or gland cell, that are separated by a small gap. The presynaptic neuron releases a transmitter substance into the gap, producing an excitatory or inhibitory effect on the postsynaptic cell.

chemiosmotic (kem-ee-oz-MOT-ik) **theory** Concept that an electrochemical gradient across a cell membrane drives ATP synthesis. Operation of electron transport systems builds up the hydrogen ion concentration on one side of the membrane. Then the electrical and chemical force of the H^+ flow down the gradient is linked to enzyme machinery that combines ADP with inorganic phosphate to form ATP.

chemoreceptor (KEE-moe-ree-sep-tur) Sensory cell that detects chemical energy (ions or molecules) dissolved in body fluids next to the cell.

chemosynthetic (KEE-moe-sin-THET-ik) **autotroph** One of a few kinds of bacteria able to build all the organic molecules it requires using carbon dioxide as the carbon source and certain inorganic substances (such as sulfur) as the energy source.

chlorophyll (KLOR-uh-fill) [Gk. *chloros*, green, + *phyllon*, leaf]. Light-trapping pigment molecule that acts as an electron donor in photosynthesis.

chloroplast (KLOR-uh-plast) The eukaryotic organelle that houses membranes, pigments, and enzymes of photosynthesis.

chorion (CORE-ee-on) In land vertebrates, the outermost membrane surrounding an embryo; secretes a hormone that maintains the corpus luteum (and thus the uterine lining).

chromatid Name given to each of the two parts of a duplicated chromosome as long as the two parts remain attached to each other at the centromere. Each chromatid consists of a DNA double helix and associated proteins, and each has the same gene sequence as its "sister" chromatid.

chromatid, sister Name given to each of the two parts of a duplicated chromosome as long as the two parts remain attached to each other at the centromere.

chromosome (CROW-moe-some) [Gk. *chroma*, color, + *soma*, body] In the unduplicated state, a DNA molecule and associated proteins. A duplicated chromosome consists of two DNA molecules and associated proteins; the two are called *sister chromatids*.

cilium, plural **cilia** (SILL-ee-um) [L. *cilium*, eyelid] Short, hairlike process extending from the plasma membrane and containing a regular array of microtubules. Some function as motile structures, others in creating currents of fluids; modified cilia are components of diverse sensory structures.

circulatory system An organ system consisting of a muscular pump, blood vessels, and blood itself; the means by which materials are transported to and from cells; in many animals, also helps stabilize body temperature and pH.

cleavage Stage of development when mitotic cell divisions convert the zygote to a ball of cells, the *blastula*.

cleavage furrow During cytokinesis, a depression that forms at the surface of a dividing animal cell; here, contractile microfilaments pull the plasma membrane inward and cut the cell in two.

climate Prevailing weather conditions in an ecosystem, including temperature, humidity, wind speed, cloud cover, and rainfall.

climax community For a given ecosystem, a more or less stable array of species that results from the process of succession.

clonal selection theory An explanation of immune system functioning in which activated lymphocytes rapidly multiply, giving rise to descendants (clones) that all retain the parent cell's specificity against the antigen causing the activation.

cloned DNA Multiple, identical copies of DNA fragments arising from a parent chromosome.

codominance Condition in which a pair of nonidentical alleles give rise to two different phenotypes; neither allele dominates expression of the other in heterozygotes.

codon One of a series of base triplets in an mRNA molecule.

coelom (SEE-lum) [Gk. *koilos*, hollow] Type of body cavity between the gut and the body wall that commonly occurs in bilateral animals; has a lining called the *peritoneum*.

coenzyme An organic molecule that serves as a carrier of electrons or atoms in metabolic reactions and that is necessary for proper functioning of many enzymes. NAD^+ is an example.

coevolution The joint evolution of two or more closely interacting species; when one species evolves, the change affects selection pressures operating between the two species and so the other also evolves.

cofactor A metal ion or coenzyme that either helps catalyze a reaction or serves briefly as a transfer agent.

cohesion Condition in which molecular bonds resist rupturing when under tension.

cohesion theory of water transport Explanation of how water moves up through tissues of vascular plants. Due to a state of molecular tension created as transpiration occurs, columns of water in xylem are pulled upward by the collective strength of hydrogen bonds between the water molecules confined in the xylem tubes.

colon (CO-lun) The large intestine.

commensalism [L. *com*, together, + *mensa*, table] Two-species interaction in which one species benefits significantly while the other is neither helped nor harmed to any great degree.

communication signal A stimulus produced by one animal that changes the behavior of another individual of the same species.

community The populations of all species occupying a habitat; in some contexts, the groups of organisms with similar life-styles in a habitat (such as the bird community). A community is characterized by the kinds of species present, their diversity, and the numbers and dispersion of their individuals through the habitat.

comparative morphology [Gk. *morph*, form] Detailed study of the differences and similarities in body form and structural patterns among major taxa.

competition, interspecific Two-species interaction in which both species can be harmed as a result of their overlapping niches (that is, they have some requirements or activity in common).

competition, intraspecific Interaction among individuals of the same species that are competing for the same resources.

competitive exclusion The concept that populations of two species competing for a limited resource cannot coexist indefinitely in the same habitat; the population better adapted to exploit the resource will enjoy a competitive (and hence reproductive) edge, and will eventually exclude the other species from the habitat.

complement system A group of about twenty proteins circulating in blood plasma that are activated during both nonspecific and specific immune responses to certain bacterial and fungal invaders; part of the *inflammatory response*.

concentration gradient For a given substance, a greater concentration of its molecules in one region than in another.

condensation Covalent linkage of small molecules in an enzyme-mediated reaction that can also involve formation of water.

cone cell One type of photoreceptor in vertebrate eyes; it responds to intense light and contributes to sharp daytime vision and color perception.

conjugation [L. *conjugatio*, a joining] In some bacteria, transfer of DNA between two different mating strains that have made cell-to-cell contact.

consumer [L. *consumere*, to take completely] An organism that is not a self-feeder but rather is heterotrophic, feeding on the tissues of other organisms; herbivores, carnivores, omnivores, and parasites are consumers.

continuous variation Small degrees of phenotypic variation in a genetic trait that occur over a more or less continuous range.

contractile vacuole (kun-TRAK-till VAK-you-ohl) [L. *contractus*, to draw together] In some single-celled organisms, a membranous chamber that takes up excess water in the cell body, then contracts, expelling the water through a pore to the outside.

convergence, morphological Outcome when dissimilar and only distantly related species adopt a similar way of life, and body parts that take on similar functions come to resemble one another.

cork cambium Type of lateral meristem that produces a tough, corky replacement for the epidermis as a plant ages.

corpus luteum (CORE-pus LOO-tee-um) A

grandular structure that develops from cells of a ruptured ovarian follicle and that secretes progesterone and some estrogen, both of which maintain the lining of the uterus (endometrium).

cortex [L. *cortex*, bark] In general, a rindlike layer; the kidney cortex is an example. In vascular plants, ground tissue that makes up most of the primary plant body, supports plant parts, and stores food.

cotyledon A so-called seed leaf that develops as part of a plant embryo; cotyledons provide food to the germinating seedling.

covalent (koe-VAY-lunt) **bond** [L. *con*, together, + *valere*, to be strong] A sharing of one or more electrons between atoms or groups of atoms. When electrons are shared equally, the bond is *nonpolar*. When they are shared unequally, the bond is *polar*––slightly positive at one end and slightly negative at the other.

crossing over In meiosis, event in which nonsister chromatids of a pair of homologous chromosomes break at one or more sites along their length and exchange corresponding segments at the breakage points. As a result, old combinations of alleles in a chromosome are broken up and new ones put together.

culture The sum total of behavior patterns of a social group, passed between generations by learning and by symbolic behavior, especially language.

cuticle (KEW-tih-kull) A body covering. In plants, a cuticle consisting of waxes and lipid-rich cutin is deposited on the outer surface of epidermal cell walls. Annelids have a thin, flexible cuticle. Arthropods have a thick, protein- and chitin-containing cuticle that is flexible, lightweight, and protective.

cyclic adenosine monophosphate (SIK-lik ah-DEN-uh-seen mon-oh-FOSS-fate) **cyclic AMP** A nucleotide that serves as an intracellular mediator of the cellular response to hormonal signals; a type of second messenger.

cyclic photophosphorylation (SIK-lik foe-toe-FOSS-for-ih-LAY-shun) Photosynthetic pathway in which electrons excited by sunlight energy move from a photosystem to a transport chain, then back to the photosystem. Energy released in the transport chain is coupled to ATP formation.

cytochrome (SIGH-toe-krome) [Gk. *kytos*, hollow vessel, + *chrōma*, color] Iron-containing protein molecule that occurs in electron transport systems used in photosynthesis and aerobic respiration.

cytokinesis (SIGH-toe-kih-NEE-sis) [Gk. *kinesis*, motion] The actual splitting of a parental cell into two daughter cells; also called *cytoplasmic division*.

cytokinin (SIGH-tow-KY-nun) Any of the class of plant hormones that stimulate cell division, promote leaf expansion, and retard leaf aging.

cytomembrane system [Gk. *kytos*, hollow vessel] The membranous system in the cytoplasm in which proteins and lipids take on their final form and are distributed. Components of the system include the endoplasmic reticulum, Golgi bodies, lysosomes, and a variety of vesicles.

cytoplasm (SIGH-toe-plaz-um) [Gk. *plassein*, to mold] All of the cell material enclosed by the plasma membrane, excepting the nucleus; it includes particles and filaments bathed in a semifluid substance, and has compartments in which specific metabolic reactions take place.

cytosine (SIGH-toe-seen) A pyrimidine; one of the nitrogen-containing bases in nucleotides.

cytoskeleton An interconnected system of filaments and tubules that provides cells with internal support and organization. Main components are microtubules, microfilaments, and intermediate filaments.

decomposer [L. *de-*, down, away, + *companere*, to put together] Generally, any of the heterotrophic bacteria or fungi that extract energy from the remains or products of other organisms.

deduction An "if-then" reasoning process that begins with a general statement and culminates by predicting consequences.

degradative pathway A metabolic pathway in which biological molecules are broken down in stepwise reactions that lead to products of lower energy.

deletion Loss of a chromosome segment; nearly always results in a genetic abnormality or disorder.

demographic transition model A model of human population growth in which changes in the growth pattern reflect the population's stage of economic development. The four stages range from a *preindustrial* stage, when both birth and death rates are high, through *transitional* and *industrial* stages, to the *postindustrial* stage, when the death rate exceeds the birth rate.

denaturation (deh-NAY-chur-AY-shun) Disruption of bonds holding a protein in its three-dimensional form, such that its polypeptide chain(s) unfolds partially or completely.

dendrite (DEN-drite) [Gk. *dendron*, tree] A short, slender extension from the cell body of a neuron; it has a branched ending where stimuli are received.

denitrification (DEE-nite-rih-fih-KAY-shun) The conversion of nitrate or nitrite to gaseous nitrogen (N_2) and a small amount of nitrous oxide (N_2O) by bacteria.

dentition (den-TIH-shun) The type, size, and number of an animal's teeth.

dermis The layer of skin underlying the epidermis, consisting mostly of dense connective tissue.

desertification (dez-urt-ih-fih-KAY-shun) The conversion of grasslands, rain-fed cropland, or irrigated cropland to desertlike conditions, with a drop of agricultural productivity of ten percent or more.

detrital food web A network of interlinked food chains in which energy flows from plants through decomposers and detritivores.

detritivore (dih-TRY-tih-vore) [L. *detritus*; after *deterere*, to wear down] An earthworm, crab, nematode, or other heterotrophic organism that obtains energy by feeding on partly decomposed particles of organic matter.

deuterostome (DUE-ter-oh-stome) [Gk. *deuteros*, second, + *stoma*, mouth] Any of the bilateral animals, including echinoderms and chordates, in which the first indentation in the early embryo develops into the anus.

diaphragm (DIE-uh-fram) [Gk. *diaphragma*, to partition] Muscular partition between the thoracic and abdominal cavities, the contraction and relaxation of which contribute to breathing. Also, a contraceptive device used temporarily to close off and thus prevent sperm from entering the uterus during sexual intercourse.

dicot (DIE-kot) [Gk. *di*, two, + *kotylēdōn*, cup-shaped vessel] Short for dicotyledon; class of flowering plants characterized primarily by seeds having embryos with two cotyledons (seed leaves), generally net-veined leaves, and floral parts generally arranged in fours, fives, or multiples of these.

differential reproduction The tendency of bearers of adaptive traits to reproduce successfully more than bearers of less adaptive traits. Because their offspring tend to make up an increasingly greater proportion of the reproductive base for each new generation, the adaptive traits increase in frequency also.

differentiation (cell) Process of development in which, through selective gene expression, a fertilized egg gives rise to diverse types of cells specialized in their composition, structure, and function.

diffusion Tendency of like molecules to move from their region of greater concentration to a region where they are less concentrated; occurs through random energetic movements of individual molecules, which tend to become dispersed uniformly in a given system.

digestive system In most animals, an internal tube or cavity divided into specialized regions for food transport, processing, and storage; its main function is to reduce food to molecules small enough to move from the gut cavity into the internal environment.

dihybrid cross In breeding experiments, a cross between two true-breeding organisms having contrasting forms of two traits.

diploid (DIP-loyd) Having two chromosomes of each type (that is, homologous chromosomes) in the somatic cells of sexually reproducing species. Except for sex chromosomes, the two homologues of a pair resemble their partner in length, shape, and which genes they carry. Compare *haploid*.

directional selection A shift in allele frequencies in a population so that forms at one end of the range of phenotypic variation become more common than the intermediate forms.

disaccharide (die-SAK-uh-ride) [Gk. *di*, two, + *sakcharon*, sugar] A carbohydrate; two monosaccharides covalently bonded.

disruptive selection Selection that favors forms at both ends of the phenotypic range in a population, and operates against intermediate forms.

distal tubule The tubular section of a nephron most distant from the glomerulus; a major site of water and sodium reabsorption.

divergence Accumulation of differences in allele frequencies between reproductively isolated populations of a species.

divergence, morphological Outcome when one or more homologous structures in different species have departed in appearance, function, or both from the ancestral form.

diversity, organismic Sum total of variations in form, functioning, and behavior that have accumulated in different lineages. Those variations generally are adaptive to prevailing conditions or were once adaptive to conditions that existed in the past.

DNA Deoxyribonucleic acid; usually a double-stranded molecule that twists helically about its own axis. DNA carries the chemical messages that are the instructions for assembling proteins, and, ultimately, new organisms.

DNA library A collection of DNA fragments produced by restriction enzymes and incorporated into plasmids.

DNA ligase (LYE-gase) Enzyme used in genetic engineering to permanently join base-paired chromosome fragments to cut plasmids.

DNA polymerase (poe-LIM-uh-race) Enzyme that governs nucleotide assembly on a parent strand of DNA during replication; also "proofreads" the growing strands for mismatched base pairs, which are replaced with correct bases.

dominance hierarchy Form of social organization in which some members of the group are subordinate to other members, which in turn are dominated by others.

dominant allele In a diploid cell, an allele whose expression masks the expression of its partner at the same gene locus on the homologous chromosome.

dormancy [L. *dormire*, to sleep] The temporary cessation of growth under conditions that could be quite suitable for growth.

double fertilization Events in flowering plants in which one sperm nucleus fuses with the egg nucleus to produce a diploid zygote, and a second sperm nucleus fuses with the two nuclei of the endosperm mother cell, which gives rise to triploid (3n) nutritive tissue.

duplication Type of chromosome rearrangement in which a gene sequence occurs in excess of its normal amount in a chromosome.

ecology [Gk. *oikos*, home, + *logos*, reason] Study of the interactions of organisms with one another and with the physical and chemical environment.

ecosystem [Gk. *oikos*, home] A community of organisms interacting with one another and with the physical environment through a flow of energy and a cycling of materials.

ecosystem analysis Identification of crucial pieces of information about the different components of an ecosystem, followed by computer processing and modeling which can then be used to predict the outcome of a subsequent disturbance.

ectoderm [Gk. *ecto*, outside, + *derma*, skin] In an animal embryo, an outermost cell layer that gives rise to the outer layer of skin and to tissues of the nervous system.

effector A muscle (or gland) that resends to nerve signals by producing movement (or chemical change) that helps adjust the body to changes in internal and/or external conditions.

egg A mature female gamete. •

electric charge A property of matter that enables ions, atoms, and molecules to attract or repel one another.

electron Negatively charged particle that orbits the nucleus of an atom.

electron transport phosphorylation (F OSS-for-ih-LAY-shun) Final stage of aerobic respiration, in which a transport system consisting of enzymes and other proteins takes up hydrogen ions and electrons derived from the Krebs cycle. The system drives the coupling of ADP and inorganic phosphate into ATP; the final electron acceptor in the system is oxygen.

electron transport system An organized array of membrane-bound enzymes and cofactors that accept and donate electrons in sequence. Operation of such systems leads to the flow of hydrogen ions (H^+) across a cell membrane, and this flow results in ATP formation and other reactions.

element Any substance that cannot be decomposed into substances with different properties.

embryo (EM-bree-oh) [Gk. *en*, in, + probably *bryein*, to swell] In animals generally, the early stages of development (including cleavage, gastrulation, organogenesis, and morphogenesis) after fertilization. In most plants, the young sporophyte, from the first cell divisions after fertilization until germination.

embryo sac In flowering plants, the female gametophyte.

endergonic (en-dur-GONE-ik) **reaction** Chemical reaction showing a net gain in energy.

endocrine (EN-doe-krin) **element** [Gk. *endon*, within, + *krinein*, to separate] Cell or gland that produces and/or secretes hormones.

endocrine system System of cells, tissues, and organs functionally linked to the nervous system and whose chemical secretions (hormones) help control body functioning.

endocytosis (EN-doe-sigh-TOE-sis) The process by which a region of the plasma membrane encloses substances (or cells, in the case of phagocytes) at or near the cell surface, then pinches off to form a vesicle that transports the substance into the cytoplasm.

endoderm [Gk. *endon*, within, + *derma*, skin] In an animal embryo, the innermost cell layer, which differentiates into the inner lining of the gut and organs derived from it.

endometrium (EN-doh-MEET-ree-um) [Gk. *metrios*, of the womb] Inner lining of the uterus, consisting of connective tissues, glands, and blood vessels.

endoplasmic reticulum or **ER** (EN-doe-PLAZ-mik reh-TIK-yoo-lum) System of membranous channels, tubes, and sacs in the cytoplasm that function in the processing of proteins destined for export, and in the manufacture of the protein and lipid components of most organelles. Rough ER has ribosomes on the surface facing the cytoplasm; smooth ER does not.

endoskeleton [GK. *endon*, within, + *sklēros*, hard, stiff] In chordates, the internal framework of bone, cartilage, or both. Together with skeletal muscle, supports and protects other body parts, helps maintain posture, and moves the body.

endosperm (EN-doe-sperm) Nutritive tissue that surrounds and serves as food for a plant embryo and, later, for the germinating seedling.

endospore A resistant body formed by some bacteria around their genetic material and a small amount of cytoplasm when environmental conditions are unfavorable; it germinates and gives rise to new bacterial cells when conditions become favorable.

energy The capacity to make things happen, to do work.

energy, net In human societies, the energy left over after subtracting the energy used to locate, extract, transport, store, and deliver energy to consumers.

energy pyramid A pyramid-shaped representation of the trophic (feeding) structure of an ecosystem, based on the decreasing energy flow at each upward transfer to a different trophic level; always depicted with the largest energy base at the bottom.

entropy (EN-trow-pee) A measure of the degree of disorder in a system—that is, a measure of how much energy in a system has become so dispersed (usually as low-quality heat) that it is no longer available to do work.

enzyme (EN-zime) A special class of proteins that speed up (catalyze) reactions between specific substances; they act only on specific substrates.

epidermis The outermost tissue layer of a multicelled plant or animal.

epistasis (uh-PISS-tih-sis) Masking of the expression of one gene pair by another gene pair, so that some expected phenotypes do not appear.

epithelium (EP-ih-THEE-lee-um) Sheet of cells, one or more layers thick, lining internal or external surfaces of the multicelled animal body.

equilibrium, dynamic [Gk. *aequus*, equal, + *libra*, balance] The point at which a chemical reaction runs forward as fast as it runs in reverse, so that there is no net change in the concentrations of products or reactants.

erythrocyte (eh-RITH-row-site) [Gk. *erythros*, red, + *kytos*, vessel] Red blood cell.

esophagus (ee-SOF-uh-gus) Digestive system structure that receives swallowed food and empties into the stomach.

essential amino acid Any of the eight amino acids that human cells cannot build and, hence, must obtain from food. These are cysteine (methionine), isoleucine, leucine, lysine, phenylalanine (tyrosine), threonine, tryptophan, and valine.

essential fatty acid Any of the fatty acids required by the body that must be obtained from food because cells cannot manufacture them.

estrogen (ESS-trow-jun) A major sex hormone required in egg formation, preparing the uterine lining for pregnancy, and maintaining secondary sexual traits; also influences growth and development.

estrus (ESS-truss) [Gk. *oistrus*, frenzy] For mammals generally, the cyclic period of a female's sexual receptivity to the male.

estuary (EST-you-ary) A partly enclosed coastal region where seawater mixes with fresh water from rivers or streams and runoff from the land.

ethylene (ETH-il-een) Plant hormone that stimulates fruit ripening and triggers abscission (the dropping of flowers, fruits, and leaves).

eukaryote (yoo-CARRY-oht) [Gk. *eu*, good, + *karyon*, kernel] A cell that has a "true nucleus" and other membranous organelles; comprises all cells except bacteria.

evaporation [L. *e-*, out, + *vapor*, steam] The changes by which a substance is converted from a liquid state into (and carried off in) vapor.

evolution [L. *evolutio*, act of unrolling] Changes within lines of descent over time; entails successive changes in allele frequencies in a population as brought about by occurrences such as mutation, genetic drift, gene flow, and selection pressure.

excretion Any of several processes by which excess water, excess or harmful solutes, or waste materials are passed out of the body. Compare *secretion*.

exergonic reaction (EX-ur-GONE-ik) A chemical reaction that shows a net loss in energy.

exocrine gland (EK-suh-krin) [Gk. *ex*, out of, + *krinein*, to separate] Secretory structure whose products travel through ducts that empty at a free epithelial surface.

exocytosis (EK-so-sigh-TOE-sis) The process by which substances are moved out of cells. A vesicle forms inside the cytoplasm, moves to the plasma membrane, and fuses with it in a way that dumps the vesicle contents outside.

exon Any of the portions of a newly formed mRNA transcript that are spliced together to form the mature mRNA molecule and are ultimately translated into protein.

exoskeleton [Gk. *exo*, out, + *skleros*, hard, stiff] An external skeleton, as in arthropods.

exponential (EX-poe-NEN-shul) **growth** Pattern of population growth in which the number of individuals increases in size by ever larger amounts per unit of time—that is, by successive doublings (2, 4, 8, 16, 32, . . .).

extracellular fluid In animals generally, all the fluid not inside cells; includes blood plasma and interstitial fluid (fluid occupying the spaces between cells and tissues).

extracellular matrix Internal meshwork of collagen and other fibrous proteins, along with a jellylike polysaccharide ground substance, that serves to hold animal cells and tissues together.

extinction, background The steady rate of species turnover that characterizes lineages through most of their histories.

extinction, mass An abrupt increase in the rate at which higher taxa disappear, with several higher taxa being affected simultaneously.

eyespot Sensory structure that contains light-absorbing pigments.

facilitated diffusion The movement of specific solutes across a cell membrane in the direction that diffusion would take them, but with the passive assistance of proteins that span the lipid bilayer of the membrane.

fall overturn The vertical mixing of a body of water in autumn as its upper layer cools, increases in density, and sinks; dissolved oxygen moves down and nutrients from bottom sediments are brought to the sur-face.

family pedigree A chart of the genetic relationships of the individuals in a family.

fat A lipid with one, two, or three fatty acid tails attached to a glycerol backbone.

fatty acid A compound having a long, un-branched carbon backbone (a hydrocarbon) with a —COOH group at the end.

feedback inhibition A mechanism of enzyme control in which the output of the reaction (such as a particular molecule) works in a way that inhibits further output.

female defense behavior Competition, typically involving physical combat, by which males of a species obtain and/or keep possession of a cluster of females.

fermentation [L. *fermentum*, yeast] Anaerobic pathway of ATP formation that begins with glycolysis and ends with the "spent" electrons being transferred back to one of the breakdown products or intermediates. Glycolysis provides the small ATP yield; the rest of the pathway serves to regenerate the NAD^+ necessary for glycolysis to proceed.

fertilization (L. *fertilis*, to carry, to bear) Fusion of sperm nucleus with egg nucleus. See also *double fertilization*.

fibrous joint Skeletal joint in which fibrous tissue unites the bones and no cavity is present.

fibrous root system Adventitious roots and their branchings.

filtration In urine formation, the process by which blood pressure forces water and solutes out of capillaries and into the cupped nephron wall (Bowman's capsule) of the glomerulus.

first law of thermodynamics [Gk. *therme*, heat, + *dynamikos*, powerful] Law that states that the total amount of energy in the universe remains constant. Energy cannot be created or destroyed, but can only be converted from one form to another.

flagellum (fluh-jell-um), plural **flagella** [L. *whip*] meaning Motile structure of many free-living eukaryotic cells; has a 9 + 2 microtubule array.

flower The often showy reproductive structure that distinguishes angiosperms from other seed plants.

fluid mosaic model Model of membrane structure, in which diverse proteins are embedded in a lipid bilayer or attached to one of its surfaces. The lipids give the membrane its basic structure and its relative impermeability to water-soluble molecules; packing variations and movements of lipids impart fluidity to the membrane. The proteins carry out most membrane functions, such as transport, enzyme action, and reception of chemical signals or substances.

follicle (FOLL-ih-kul) In a mammalian ovary, a primary oocyte (immature egg) together with the surrounding layer of cells.

food chain The linear sequence of who eats whom in an ecosystem.

food web A network of crossing, interlinked food chains, encompassing primary producers and an array of consumers and decomposers.

forebrain Brain region that includes the cerebrum and cerebral cortex, the olfactory lobes, and the hypothalamus.

fossil Recognizable evidence of an organism that lived in the distant past. Most fossils are skeletons, shells, leaves, seeds, and tracks that were buried in rock layers before they could be decomposed.

fossil fuel Coal, petroleum, or natural gas; formed in sediments by the compression of carbon-containing plant remains over hundreds of millions of years.

fruit [L. after *frui*, to enjoy] In flowering plants, the ripened ovary of one or more carpels, sometimes with accessory structures incorporated.

functional group An atom or group of atoms covalently bonded to the carbon backbone of an organic compound, contributing to its characteristic structure and properties.

Fungi The kingdom of fungi.

fungus A heterotroph that decomposes living or nonliving organic matter by way of extracellular digestion and absorption.

gall bladder Organ of the digestive system that stores bile secreted from the liver.

gamete (GAM-eet) [Gk. *gametēs*, husband, and *gametē*, wife] Mature haploid cell (sperm or egg) that functions in sexual reproduction.

gamete formation The first stage of animal development, in which sperm or eggs form and mature within the reproductive structures of the parents.

gametophyte (gam-EET-oh-fite) [Gk. *phyton*, plant] The haploid, multicelled, gamete-producing phase in the life cycle of most plants.

ganglion (GANG-lee-un) [Gk. *ganglion*, a swelling] A clustering of cell bodies of neurons into a distinct structure in body regions other than the brain or spinal cord. (Such clusterings in the brain or spinal cord are called nuclei.)

gastrulation (gas-tru-LAY-shun) Stage of embryonic development in which cells become arranged into two or three primary tissue layers (germ layers); in humans, the layers are an inner *endoderm*, an intermediate *mesoderm*, and a surface *ectoderm*.

gene [short for German *pangen*, after Gk. *pan*, all, + *genes*, to be born] Any of the molecular units of instruction for heritable traits. Each gene is a linear sequence of nucleotides in DNA that calls for the assembly of specific amino acids into a polypeptide chain.

gene flow Change in allele frequencies in a population due to immigration, emigration, or both.

gene frequency More precisely, allele frequency: the relative abundances of different alleles carried by the individuals of a population.

gene locus Particular location on a chromosome for a given gene.

gene mutation [L. *mutatus*, a change] Change in DNA due to the deletion, addition, or substitution of one to several bases in the nucleotide sequence.

gene pair In diploid cells, the two alleles at a given gene locus on homologous chromosomes.

gene pool Sum total of all genotypes in a population. More accurately, allele pool.

gene therapy Inserting one or more normal genes into existing cells of an organism as a way to correct some genetic defect.

genetic code [After L. *genesis*, to be born] The correspondence between nucleotide triplets in DNA (then in mRNA) and specific sequences of amino acids in the resulting polypeptide chains; the basic language of protein synthesis.

genetic drift Microevolutionary process in which allele frequencies in a population change randomly over time as a result of chance events.

genetic engineering Altering the information content of DNA through use of recombinant DNA technology.

genetic equilibrium Hypothetical state in a population in which allele frequencies for a trait remain stable through the generations; a reference point for measuring rates of evolutionary change.

genetic recombination Presence of a new combination of alleles in a DNA molecule compared to the parental genotype; the result of processes such as crossing over at meiosis, chromosome rearrangements, gene mutation, and recombinant DNA technology.

genome All the DNA in a haploid number of chromosomes of a species.

genotype (JEEN-oh-type) Genetic constitution of an individual. Can mean a single gene pair or the sum total of the individual's genes. Compare *phenotype*.

genus, plural **genera** (JEEN-US, JEN-er-ah) [L. *genus*, race, origin] A taxon (that is, a category of relationship based on phenotypic similarities, descent, or both) in which all species sharing (or exhibiting) certain characteristics are grouped.

germ cell Animal cell that may develop into gametes. Compare *somatic cell*.

germination (jur-min-AY-shun) The time at which an embryo sporophyte breaks through its seed coat and resumes growth.

gibberellin (JIB-er-ELL-un) Any of a class of plant hormones that promote stem elongation.

gill A respiratory organ, typically with a moist, thin, vascularized layer of epidermis that functions in gas exchange.

glomerulus (glow-MARE-you-luss) [L. *glomus*, ball] Cluster of capillaries in Bowman's capsule of a nephron, the functional unit of the kidney.

glucagon (GLUE-kuh-gone) Animal hormone secreted by alpha cells of the pancreas; it governs the conversion of glycogen (a storage polysaccharide) and amino acids to glucose.

glyceride (GLISS-er-eyed) A molecule having one, two, or three fatty acid tails attached to a backbone of glycerol. Glycerides—fats and oils—are the body's most abundant lipids and its richest source of energy.

glycerol (GLISS-er-oh) [Gk. *glykys*, sweet, + L. *oleum*, oil] Three-carbon molecule with three hydroxyl groups attached; combines with fatty acids to form fat or oil.

glycogen (GLY-kuh-jen) In animals, a starch that is a main food reserve; can be readily broken down into glucose subunits.

glycolysis (gly-CALL-ih-sis) [Gk. *glykys*, sweet, + *lysis*, loosening or breaking apart] The partial breakdown of glucose and other carbohydrate molecules into pyruvate, with a net yield of two ATP; the initial stage of both aerobic and anaerobic pathways of glucose breakdown.

Golgi (GOHL-gee) **body** Organelle where many newly forming proteins and lipids undergo final processing, then are sorted and packaged in vesicles.

gonad (GO-nad) Primary reproductive organ in which gametes are produced.

graded potential A neural signal that can vary in magnitude, depending on the stimulus. Many graded potentials acting at the same time can so change the voltage difference across the neural membrane that they initiate an *action potential*.

granum, plural **grana** A structure in chloroplasts; embedded in its membranes are chlorophyll and other light-trapping pigments, and reaction sites for ATP formation.

gravitropism (GRAV-ih-TROPE-izm) [L. *gravis*, heavy, + Gk. *trepein*, to turn] The tendency of a plant to grow directionally in response to the earth's gravitational force.

grazing food web A network of interlinked food chains in which energy flows from plants to herbivores, then through some array of carnivores.

greenhouse effect Warming of the lower atmosphere due to the buildup of so-called greenhouse gases—carbon dioxide, methane, nitrous oxide, ozone, water vapor, and chlorofluorocarbons.

green revolution In developing countries, the use of improved crop varieties, modern agricultural practices (including massive inputs of fertilizers and pesticides), and equipment to increase crop yields.

ground meristem (MARE-ih-stem) [Gk. *meristos*, divisible] A primary meristem that produces ground tissue, hence the bulk of the plant body.

guard cell Either of two adjacent cells having roles in the movement of gases and water vapor across the epidermis of leaves or stems. An opening (*stoma*) forms when both cells swell with water and move apart. When the cells lose water, they collapse against each other and the opening disappears.

gut A body region where food is digested and absorbed; in vertebrates, the tubular regions from the stomach onward.

gymnosperm (JIM-noe-sperm) [Gk. *gymnos*, naked, + *sperma*, seed] A seed plant that bears its seeds at the surface of reproductive structures. Unlike angiosperms (flowering plants), it does not have seeds enclosed in protective tissue layers. Fir and pine trees are examples.

habitat [L. *habitare*, to live in] The type of place where an organism normally lives, as characterized by physical and chemical features as well as by the general structure of the vegetation.

hair cell Type of hairlike mechanoreceptor that may give rise to action potentials when bent or tilted.

haploid (HAP-loyd) Having only one of each pair of homologous chromosomes that were present in the nucleus of a parent cell of sexually reproducing species; an outcome of meiosis. Compare *diploid*.

heart Muscular pump that keeps blood circulating through the animal body.

hemoglobin (HEEM-oh-glow-bin) [Gk. *haima*, blood, + L. *globus*, ball] Iron-containing protein that gives red blood cells their color; functions mainly in oxygen transport.

hemorrhage Bulk flow of blood from damaged vessels.

hemostasis (HEE-mow-STAY-sis) [Gk. *haima*, blood, + *stasis*, standing] The stopping of blood loss from a damaged blood vessel through coagulation, blood vessel spasm, formation of a platelet plug, and other mechanisms.

herbivore [L. *herba*, grass, + *vovare*, to devour] Plant-eating animal.

heterocyst (HET-er-oh-sist) A type of thick-walled, nitrogen-fixing cell that forms along the chains of cyanobacterial cells that show filamentous growth patterns.

heterotroph (HET-er-oh-trofe) [Gk. *heteros*, other, + *trophos*, feeder] Organism that must obtain carbon and energy from organic compounds already built by autotrophs (such as plants). Animals, fungi, many protistans, and most bacteria are heterotrophs.

heterozygote (HET-er-oh-ZYE-gote) [Gk. *zygoun*, join together] Individual having nonidentical alleles at a given gene locus.

hindbrain Brain region consisting of the medulla oblongata, cerebellum, and pons; includes reflex centers for respiration and blood circulation; also the site where motor responses and complex reflexes are coordinated.

histone Any of a class of structural proteins complexed with DNA in the eukaryotic chromosome.

homeostasis (HOE-me-oh-STAY-sis) [Gk. *homo*, same, + *stasis*, standing] In multicelled organisms, the capacity for maintaining the internal environment when conditions change.

hominid [L. *homo*, man] All species on the human evolutionary branch. *Homo sapiens* is the only living representative.

hominoid Apes, humans, and recent human ancestors.

homologous (huh-MOLL-uh-gus) **chromosome** [Gk. *homologia*, agreement; correspondence] In the nucleus of a diploid cell, one of a pair of chromosomes that resemble each other in size, shape, and the genes they carry, and which pair with each other during meiosis. X and Y chromosomes are also homologues.

homozygous (HOE-moe-ZYE-guss) **dominant** An individual having two dominant alleles at a given gene locus.

homozygous recessive An individual having two recessive alleles at a given gene locus.

hormone [Gk. *hormon*, to stir up, set in motion] Any of the signaling molecules secreted from endocrine cells or glands and some neurons and that travel the bloodstream to nonadjacent target cells.

hydrogen bond Type of chemical bond in which an atom of a molecule interacts weakly with a hydrogen atom already taking part in a polar covalent bond.

hydrogen ion A hydrogen atom that has lost its electron and so bears a positive charge (H^+); a "naked" proton.

hydrologic cycle A global biogeochemical cycle in which hydrogen and oxygen move, in the form of water molecules, through the atmosphere, on or through the uppermost layers of land masses, to the oceans, and back again; driven by solar energy.

hydrolysis (high-DRAWL-ih-sis) [L. *hydro*, water, + Gk. *lysis*, loosening or breaking apart] Reaction in which covalent bonds between parts of molecules are broken and an H^+ ion and an OH group derived from water become attached to the fragments.

hydrophilic [Gk. *philos*, loving] Having an attraction for water molecules; refers to a polar substance that readily dissolves in water.

hydrophobic [Gk. *phobos*, dreading] Repelled by water molecules; refers to a nonpolar substance such as oil that does not readily dissolve in water.

hydrosphere All the liquid or frozen water on or near the earth's surface.

hypothalamus [Gk. *hypo*, under, + *thalamos*, inner chamber or possibly *tholos*, rotunda] Region of vertebrate forebrain concerned with neural-endocrine control of visceral activities (e.g., salt-water balance, temperature control, reproduction).

hypha (HIGH-fuh), plural **hyphae** [Gk. *hyphe*, web] Filament of a mycelium; usually composed of elongated cells with chitin-reinforced walls.

hypothesis A general statement of a probable answer to a question or solution to a problem based on inductive reasoning ("educated guesses"). In science, hypotheses are tested through repeatable experiments.

immune system Three types of white blood cells (macrophages, T lymphocytes, and B lymphocytes) and their interactions and products; hallmarks of the system are *specific* responses to particular invaders, and *memory*—the ability to mount a rapid attack if the same type of invader returns.

immunization Deliberate introduction into the body of an antigen that can provoke an immune response and the production of memory lymphocytes; such *vaccines* often consist of killed or weakened bacteria or viruses.

imprinting Form of learning in which the capacity to learn specific information is pronounced during certain early stages of development.

incomplete dominance In a heterozygote, situation in which one allele of a pair only partially dominates expression of its partner.

independent assortment Mendelian principle that each gene pair tends to assort into gametes independently of other gene pairs located on nonhomologous chromosomes.

induced-fit model Model of enzyme-substrate interaction in which a bound substrate induces changes in the shape of the enzyme's active site, the result being a more precise molecular "fit" between the enzyme and its substrate.

induction A reasoning process that involves using logic to sort through observations and other information to arrive at a general statement.

infant dependency Reliance of young on adults for nourishment and protection and as models for behavior.

inflammatory response In specific and nonspecific immune system responses, the series of events that destroy foreign agents and restore tissues and internal operating conditions to normal; includes the *complement system* of circulating plasma proteins.

inheritance The transmission, from parents to offspring, of structural and functional patterns that have a genetic characteristic of each species.

inhibitor A substance that can bind with an enzyme and interfere with its functioning.

instinctive behavior The capacity of an animal to complete fairly complex, stereotyped responses to particular environmental cues without having had prior experience with those cues.

insulin Hormone, secreted by beta cells of the pancreas, that stimulates the uptake of glucose by cells; also promotes protein and fat synthesis and inhibits protein conversion to glucose. In general, lowers the glucose level in the blood.

integration, neural [L. *integrare*, to coordinate] Moment-by-moment summation of all excitatory and inhibitory synapses acting on a neuron; occurs at each level of synapsing in a nervous system.

integumentary (in-teg-you-MEN-tuh-ree) **exchange** Mode of respiration in some animals in which respiratory gases diffuse across a thin, vascularized layer of moist epidermis at the body surface.

integumentary system Animal skin and structures derived from it, such as hair, nails, oil glands, and sweat glands.

interneuron Any of the neurons in the vertebrate brain and spinal cord that integrate information arriving from sensory neurons and that influence other neurons in turn.

internode In vascular plants, the stem region between two successive nodes.

interphase Time interval (variable among species) in which a cell increases its mass, approximately doubles the number of its structures and organelles, and finally replicates its DNA, prior to nuclear division.

interspecific competition Competition between species for one or more limited resources.

interstitial fluid (IN-ter-STISH-ul) [L. *interstitus*, to stand in the middle of something] In vertebrates, that portion of the extracellular fluid occupying spaces between cells and tissues. (The remaining portion is blood plasma.)

intertidal zone Generally, the area on a rocky or sandy shoreline that is above the low water mark and below the high water mark; organisms inhabiting it are alternately submerged, then exposed, by the tides.

intervertebral disk One of a number of disk-shaped structures containing cartilage that serve as shock absorbers and flex points between bony segments of the vertebral column.

intron Noncoding portion of a newly formed mRNA transcript.

inversion Type of chromosome rearrangement in which a segment that has become separated from the chromosome is reinserted at the same place—but in reverse, so that the position and sequence of genes are altered.

invertebrate Animal without a backbone. The great majority of animals are invertebrate.

ion, negatively charged (EYE-on) An atom or a compound that has gained one or more electrons, hence has acquired an overall negative charge.

ion, positively charged An atom or a compound that has lost one or more electrons, hence has acquired an overall positive charge.

ionic bond An association between ions of opposite charge.

isotope (EYE-so-tope) An atom that contains the same number of protons as other atoms of the same element, but that has a different number of neutrons.

karyotype (CARRY-oh-type) Visual representation in which a photomicrograph of metaphase chromosomes in a cell has been cut apart and rejoined with the chromosomes arranged in order according to length, shape, banding patterns, and other features.

keratinization (care-AT-in-iz-AY-shun) Process in which cells in the mid-epidermal regions of skin die and become dead bags of the protein keratin. Keratinized cells at the skin surface form a barrier against dehydration, bacteria, and many toxic substances.

kidney In vertebrates, either of the paired organs that filter mineral ions, organic wastes, and other substances from the blood, and help regulate the volume and solute concentrations of extracellular fluid. The functional unit of the kidney is the *nephron*.

Krebs cycle Stage of aerobic respiration in which pyruvate is completely broken down to carbon dioxide and water. Resulting hydrogen ions and electrons are shunted to the next stage, which yields most of the ATP produced in aerobic respiration.

lactate fermentation Anaerobic pathway of ATP formation in which pyruvate from glycolysis is converted to the three-carbon compound lactate.

large intestine The colon; a region of the gut that receives unabsorbed food residues from the small intestine and concentrates and stores feces until they are expelled through the rectum.

larva, plural **larvae** A sexually immature, free-living animal that grows and develops into a sexually mature adult.

larynx (LARE-inks) Tube that leads to the lungs. In humans, contains vocal cords, the production site of sound waves used in speech.

lateral meristem Either vascular cambium or cork cambium, the meristems responsible for

secondary growth (increases in diameter) in plants.

leaf Structure having chlorophyll-containing tissue that is the major region of photosynthesis in most vascular plants.

learning The adaptive modification of behavior in response to neural processing of information that has been gained from specific experiences.

lichen (LY-kun) A composite organism comprising a fungus and a captive photosynthetic partner such as a green alga in permanent symbiotic relationship.

life cycle For any species, the genetically programmed sequence of events by which individuals are produced, grow, develop, and themselves reproduce.

light-dependent reactions The first stage of photosynthesis, in which sunlight energy is absorbed and converted to the chemical energy of ATP alone (by the cyclic pathway) or ATP and NADPH (by the noncyclic pathway).

light-independent reactions Second stage of photosynthesis, in which sugars and other compounds are assembled with the help of the ATP and NADPH produced during the first stage.

limbic system Brain regions that, along with the cerebral cortex, collectively govern emotions.

limiting factor Resource availability, predation, competition, or some other factor that can limit population growth. The number of such factors, the times at which they operate, and their relative effects can vary.

lineage (LIN-ee-age) A line of descent.

linkage Tendency of genes located on the same chromosome to stay together during meiosis and to end up together in the same gamete.

lipid A compound of mostly carbon and hydrogen that generally does not dissolve in water, but that does dissolve in nonpolar substances. Some lipids serve as energy reserves; others are components of membranes and other cell structures.

lipid bilayer A "back-to-back" arrangement of phospholipid molecules, in which the fatty acid tails are sandwiched between the hydrophilic heads; the structural basis of all cell membranes.

liver Glandular organ associated with the digestive system; has central roles in storing and interconverting absorbed carbohydrates, lipids, and proteins; maintaining blood; disposing of nitrogen-containing wastes; and other tasks.

locus (LOW-cuss) The particular location of a specific gene on a chromosome.

logistic (low-JIST-ik) **growth** Pattern of population growth in which the growth rate of a low-density population goes through a rapid growth phase and then levels off.

loop of Henle The hairpin-shaped, tubular region of a nephron; functions in reabsorption of water and solutes.

lung An internal respiratory surface in the shape of a cavity or sac.

lymph (LIMF) [L. *lympha,* water] Term applied to tissue fluid that has moved into the vessels of the lymphatic system.

lymph vascular system [L. *lympha,* water + *vasculum,* a small vessel] A network of vessels that supplements the blood circulation system, reclaiming proteins from tissues and fluid lost from capillaries and returning both to the circulatory system. It also transports fats absorbed from the digestive tract. Fluid in its vessels is called *lymph.*

lymphatic system System of lymphoid organs and vessels that supplements the circulatory system by returning excess tissue fluid to the bloodstream; also functions in defense responses.

lymphocyte Any of various white blood cells that take part in vertebrate immune responses.

lymphoid organs The lymph nodes, spleen, thymus, tonsils, adenoids, and patches of tissue in the small intestine and appendix.

lymphokine Any of a class of proteins by which the cells of the vertebrate immune system communicate with one another.

lysosome (LYE-so-sohm) Primary organelle of digestion inside the cell; lysosomes bud as vesicles from Golgi bodies and contain enzymes that can break down polysaccharides, proteins, nucleic acids, and some lipids.

macroevolution The large-scale patterns, trends, and rates of change among groups of species.

mass extinction An abrupt rise in extinction rates above the background level; a castastrophic, global event in which major groups of species are wiped out simultaneously.

mass number The total number of protons and neutrons in an atom's nucleus. (The relative masses of atoms are also called atomic weights.)

mechanoreceptor Sensory cell (or cell part) that detects mechanical energy associated with changes in pressure, position, or acceleration.

medusa (meh-DOO-sah) [Gk. *Medousa,* one of three sisters in Greek mythology having snake-entwined hair; this image probably evoked by the tentacles and oral arms extending from the medusa] Free-swimming, bell-shaped stage in cnidarian life cycles.

megaspore A type of spore that develops into a female gametophyte.

meiosis (my-OH-sis) [Gk. *meioun,* to diminish] Two-stage nuclear division process in which the parental number of chromosomes in each daughter nucleus becomes haploid— that is, with *one* of each type of chromosome that was present in the parent nucleus. Basis of gamete formation and of spore formation in plants. Compare *mitosis.*

meltdown In a nuclear reactor, a failure of the water-based cooling system that results in heating of the nuclear fuel past its melting point; could lead to melting of the overheated reactor core through its concrete containment slab.

memory The storage and retrieval of information about previous experiences; underlies the capacity for learning.

memory lymphocyte Any of the various B or T lymphocytes of the immune system that are formed in response to invasion by a foreign agent and that circulate for some period, available to mount a rapid attack if the same type of invader reappears.

menopause (MEN-uh-pozz) [L. *mensis*, month, + *pausa*, stop]. End of the period of a human female's reproductive potential.

menstrual cycle The cyclic release of oocytes and priming of the endometrium (lining of the uterus) to receive a fertilized egg; the complete cycle averages about 28 days in female humans.

menstruation Periodic sloughing of the blood-enriched lining of the uterus when pregnancy does not occur.

meristem (MARE-ih-stem) [Gk. *meristos*, divisible] In most plants, a mass of self-perpetuating cells not yet committed to developing into a specialized cell type.

mesoderm (MEH-so-derm) [Gk. *mesos*, middle, + *derm*, skin] In most animals embryos, a tissue layer between ectoderm and endoderm; gives rise to muscle, the organs of circulation, reproduction, and excretion, most of the internal skeleton (when present), and connective tissue layers of the gut and body covering.

metabolic pathway An orderly series of breakdown or synthesis reactions in cells, the steps of which are catalyzed (speeded up) by the action of specific enzymes.

metabolism (meh-TAB-oh-lizm) [Gk. *meta*, change] All those chemical reactions by which cells acquire and use energy as they synthesize, accumulate, break apart, and eliminate substances in ways that contribute to growth, maintenance, and reproduction.

metamorphosis (met-uh-MOR-foe-sis) [Gk. *meta*, change + *morphe*, form] Transformation of a larva into an adult form.

metaphase Stage of mitosis when spindle microtubules harness the duplicated chromosomes, orient the sister chromatids of each toward opposite spindle poles, and move them all to the spindle equator.

metaphase I and II Stages of meiosis when all duplicated chromosomes are moved to the spindle equator, with each chromosome and its homologous partner oriented at random toward one spindle pole or the other (metaphase I); and, later, when sister chromatids are oriented toward opposite poles and moved to the equator (metaphase II).

metastasis (muh-TAST-ih-sis) Invasion of other tissues by disease-causing cells, usually traveling via blood or lymph.

metazoan (MET-ah-ZOE-un). Multicelled animal.

MHC marker Any of the surface receptors that mark an individual's cells as "self";

except in the case of identical twins, they are unique to each individual.

microevolution Changes in allele frequencies brought about by mutation, genetic drift, gene flow, and natural selection.

microfilament [Gk. *mikros*, small, + L. *filum*, thread] Component of the cytoskeleton; involved in cell shape, motion, and growth.

microspore A type of spore that develops into a male gametophyte (pollen grain).

microtubular spindle An array of microtubules that helps establish the polarity necessary for chromosome movements during nuclear division.

microtubule Hollow cylinder of (mostly) tubulin subunits; involved in cell shape, motion, and growth; functional unit of cilia and flagella.

microvillus (MY-crow-VILL-us) [L. *villus*, shaggy hair] A slender, cylindrical extension of the animal cell surface that functions in absorption or secretion.

midbrain The brain region that includes the tectum (a roof of gray matter), where sensory input from the eyes and ears converges.

migration A cyclic movement between two distant regions at times of year corresponding to seasonal change.

mimicry (MIM-ik-ree) Situation in which one species (the mimic) bears deceptive resemblance in color, form, and/or behavior to another species (the model) that enjoys some survival advantage.

mineral An inorganic substance such as calcium or potassium that is required for the normal functioning of body cells.

mitochondrion, plural **mitochondria** (MY-toe-KON-dree-on) Organelle in which the second and third stages of aerobic respiration (that is, the Krebs cycle and preparatory conversions for it, as well as electron transport phosphorylation) occur.

mitosis (my-TOE-sis) [Gk. *mitos*, thread] Type of nuclear division that maintains the parental number of chromosomes for daughter cells. It is the basis of bodily growth and, in some cases, asexual reproduction of eukaryotes.

molecule A unit of two or more atoms of the same or different elements, bonded together.

molting The shedding of hair, feathers, horns, epidermis, or a shell or some other exoskeleton in a process of growth or periodic renewal.

Monera The kingdom of bacteria.

monocot (MON-oh-kot) Short for monocotyledon; a flowering plant in which seeds have only one cotyledon, whose floral parts generally occur in threes (or multiples of threes), and whose leaves typically are parallel-veined. Compare *dicot*.

monohybrid cross [Gk. *monos*, alone] A genetic cross between two parents that breed true for contrasting forms of a single trait; heterozygous offspring result.

monosaccharide (MON-oh-SAK-ah-ride) [Gk. *monos*, alone, single, + *sakharon*, sugar] A

one-unit sugar, the simplest carbohydrate. Glucose is an example.

monosomy Abnormal condition in which one chromosome of diploid cells has no homologue.

morphogenesis (MORE-foe-JEN-ih-sis) [Gk. *morphe*, form, + *genesis*, origin] Processes through which differentiated cells in an embryo become organized into tissues and organs, under genetic controls and environmental influences.

motor neuron Nerve cell that relays information away from the brain and spinal cord to the body's effectors (muscles or glands or both), which carry out responses.

mouth The oral cavity; in digestion, the site where polysaccharide breakdown begins.

multicelled organism An organism that has differentiated cells arranged into tissues, organs, and often organ systems.

multiple allele system More than two forms of alleles that can occur at a given gene locus.

muscle tissue Tissue having cells able to contract in response to stimulation, then passively lengthen and so return to their resting state.

mutagen (MEW-tuh-jen) Any of various environmental agents, such as viruses and ultraviolet radiation, that can permanently modify the structure of a DNA molecule.

mutation [L. *mutatus*, a change, + *-ion*, result of a process or an act] A heritable change in the kind, structure, sequence, or number of component parts of DNA.

mutualism [L. *mutuus*, reciprocal] A type of community interaction in which members of two species each receive benefits from the association; when the mutually beneficial interaction is intimate and involves a permanent dependency, it is called *symbiosis*.

mycelium (my-SEE-lee-um), plural **mycelia** [Gk. *mykes*, fungus, mushroom, + *helos*, callus] A mesh of tiny, branching filaments (hyphae) that is the food-absorbing part of a multicelled fungus.

mycorrhiza (MY-coe-RISE-uh) A symbiotic arrangement between fungal hyphae and the roots of forest trees and shrubs, in which the fungus obtains carbohydrates from the plant and in turn releases dissolved mineral ions to the plant roots.

myofibril (MY-oh-FY-brill) One of many threadlike structures inside a muscle cell; composed of actin and myosin protein molecules arranged as sarcomeres, the fundamental units of contraction.

myosin (MY-uh-sin) One of two types of protein filaments that make up sarcomeres, the contractile units of a muscle cell; the other is actin.

NAD$^+$ Nicotinamide adenine dinucleotide, a large organic molecule that serves as a cofactor in enzyme reactions. When carrying electrons and protons (H$^+$) from one reaction site to another, it is abbreviated NADH.

NADP$^+$ Nicotinamide adenine dinucleotide phosphate. When carrying electrons and protons (H$^+$) from one reaction site to another, it is abbreviated NADPH$_2$.

natural selection Differential survival and reproduction among the variant individuals of a population; one of the most important microevolutionary processes.

negative feedback mechanism Homeostatic mechanism by which detection of a change in some condition in the internal environment brings about a response that tends to return the condition to the original state.

nematocyst (NEM-add-uh-sist) [Gk. *nēma*, thread, + *kystis*, bladder, pouch] A stinging capsule that assists in capturing prey and that may serve in protection; a distinguishing feature of cnidarians such as jellyfishes.

nephridium (neh-FRID-ee-um) plural **nephridia** In invertebrates such as earthworms, a system for regulating water and solute levels.

nephron (NEFF-ron) [Gk. *nephros*, kidney] In the human kidney, any of the more than a million slender tubes in which water and solutes filtered from the blood are selectively reabsorbed, and urine is formed.

nerve Cordlike communication line of nervous systems, composed of axons of sensory or motor neurons (or both) packed tightly in bundles within connective tissue. In the brain and spinal cord, such bundles are called nerve pathways or tracts.

nerve impulse Action potential.

nerve net Cnidarian nervous system consisting of nerve cells and concerned primarily with feeding behavior.

nervous system Constellations of neurons oriented relative to one another in precise message-conducting and information-processing pathways.

neuroendocrine control center Those portions of the hypothalamus and pituitary that work together in hormonal control. Some hormones produced in the hypothalamus are stored in and secreted by the posterior lobe of the pituitary to act on distant body cells. Others (*releasing hormones*) act directly on the anterior lobe, triggering secretion of numerous pituitary gland hormones that, in turn, affect other endocrine glands or body tissues in general.

neuroglia (NUR-oh-GLEE-uh) Cells intimately associated with neurons and functioning in their structural and metabolic support, maintenance, and in some cases as axonal sheaths. In vertebrates they represent at least half of the volume of the nervous system.

neuromuscular junction The synapses between the splayed-out axon terminals of a motor neuron and a muscle cell. The terminals are positioned in troughs (in the muscle cell membrane) called the motor end plate.

neuron A nerve cell; the basic unit of communication in vertebrate nervous systems. Various classes of neurons collectively sense environmental change, integrate sensory inputs, then activate muscles or glands that can initiate or carry out responses.

neutral mutation Mutation in which the altered allele has no more measurable effect on survival and reproduction than do other alleles for the trait.

neutron Subatomic particle of about the same size and mass as a proton but having no electric charge.

niche (NITCH) [L. *nidas*, nest] The full range of physical and biological conditions under which a particular species can live and reproduce.

nitrification (nye-trih-fih-KAY-shun) Process by which certain soil bacteria strip electrons from ammonia or ammonium, thus releasing nitrite (NO$_2$); other soil bacteria then use nitrite for energy metabolism, yielding nitrate (NO$_3$).

nitrogen fixation Process in which a few kinds of bacteria convert gaseous nitrogen (N$_2$) to ammonia, which dissolves rapidly in water to produce ammonium. The fixed nitrogen is used by the bacteria themselves and also becomes available to other organisms.

node In vascular plants, the point on a stem where one or more leaves are attached.

noncyclic photophosphorylation (non-SIK-lik foe-toe-FOSS-for-ih-LAY-shun) [L. *non*, not, + Gk. *kylos*, circle] Photosynthetic pathway in which new electrons derived from water molecules flow through two photosystems and two transport chains, the result being formation of ATP and NADPH.

nondisjunction Failure of one or more chromosomes to separate during meiosis.

notochord (KNOW-toe-kord) In chordates, a long rod of stiffened tissue (not cartilage or bone) that serves as a supporting structure for the body.

nuclear envelope The double membrane (two lipid bilayers and associated proteins) that forms the outermost part of the cell nucleus.

nucleic (new-CLAY-ik) **acid** A large single- or double-stranded chain of nucleotide units; DNA and RNA are examples.

nucleolus (new-KLEE-oh-lus) [L. *nucleolus*, a little kernel] Within the nucleus of a non-dividing cell, a mass of proteins, RNA, and other material used in ribosome synthesis.

nucleosome (NEW-klee-oh-sohm) In a DNA molecule, an organizational unit consisting of a segment of DNA looped twice around a core of histone molecules.

nucleotide (NEW-klee-oh-tide) A small organic compound having a five-carbon sugar (deoxyribose), a nitrogen-containing base, and a phosphate group. Nucleotides are the structural units of adenosine phosphates, nucleotide coenzymes, and nucleic acids.

nucleotide coenzyme A protein that transports hydrogen atoms (protons) and electrons from one reaction site to another in cells.

nucleus (NEW-klee-us) [L. *nucleus*, a kernel] In atoms, the central core of one or more positively charged protons and (in all but hydrogen) electrically neutral neutrons. In

eukaryotic cells, the membranous organelle that houses the DNA.

obesity An excess of fat in the body's adipose tissues, caused by imbalances between caloric intake and energy output.

oligosaccharide A carbohydrate consisting of a small number of covalently linked sugar units. The subclass called *disaccharides* has two sugar units. Compare *monosaccharide* and *polysaccharide*.

omnivore [L. *omnis*, all, + *vovare*, to devour] An organism able to obtain energy from more than one source rather than being limited to one trophic level.

oncogene (ON-coe-jeen) Any gene having the potential to induce cancerous transformations in a cell.

oogenesis (oo-oh-JEN-uh-sis) Formation of a female gamete, from a germ cell to a mature haploid ovum (egg).

operator A short base sequence between a promoter and the start of a gene; interacts with regulatory proteins.

operon Any gene (or group of genes) together with its promoter and operator sequence.

organ A structure of definite form and function that is composed of more than one tissue.

organ formation Stage of development in which primary germ layers split into subpopulations of cells, and different lines of cells become unique in structure and function; during subsequent growth and tissue specialization, organs acquire specialized chemical and physical properties.

organ system Two or more organs that interact chemically, physically, or both in performing a common task.

organelle Any of various membranous sacs, envelopes, and other compartmented portions of cytoplasm that separate different, often incompatible metabolic reactions in the space of the cytoplasm and in time (through specific reaction sequences).

organic compound A compound of carbon-based molecules.

osmosis (oss-MOE-sis) [Gk. *osmos*, act of pushing] Passive movement of water across a differentially permeable membrane in response to solute concentration gradients, a pressure gradient, or both.

ovary (OH-vuh-ree) In female animals, the primary reproductive organ in which eggs form. In seed-bearing plants, the portion of the carpel where eggs develop, fertilization takes place, and seeds mature. A mature ovary (and sometimes other plant parts) is a fruit.

oviduct (OH-vih-dukt) Passageway through which eggs travel from the ovary to the uterus.

ovulation (AHV-you-LAY-shun) In a mammalian ovary, the release of a secondary oocyte (immature egg).

ovule (OHV-youl) [L. *ovum*, egg] Any of one or more structures that form on the inner wall of the ovary of seed-bearing plants and that, at maturity, are the seeds; contains the female gametophyte with its egg, surrounded by nutritive and protective tissues.

ovum (OH-vum) A mature female gamete (egg).

oxidation The loss of one or more electrons from an atom or molecule.

oxidation-reduction reaction An electron transfer from one atom or molecule to another. Often hydrogen is also transferred along with the electron or electrons.

pancreas (PAN-cree-us) Gland that secretes enzymes and bicarbonate into the small intestine during digestion, and also secretes the hormones insulin and glucagon.

pancreatic islets Any of the approximately 2 million clusters of endocrine cells in the pancreas; divided into three types of hormone-producing cells, respectively termed alpha cells, beta cells, and delta cells.

parasite [Gk. *para*, alongside, + *sitos*, food] An organism that obtains nutrients directly from the tissues of a living host, which it lives on or in and may or may not kill.

parasitoid An insect larva that kills its host by completely consuming the host's soft tissues while the larva grows and develops inside it.

parasympathetic nerve Any of the nerves of the autonomic nervous system which typically carry signals that tend to slow down overall body functioning and divert energy to basic processes.

parathyroid gland (PARE-uh-THY-royd) In vertebrates, either of two endocrine glands embedded in tissues at the back of the thyroid gland; they secrete parathyroid hormone, which helps restore blood calcium levels.

passive immunity Temporary immunity conferred by the deliberate introduction of antibodies into the body; such immunity can help counter an attack by some pathogens but does not produce lasting effects.

passive transport Movement of solutes across a cell membrane; solutes simply move down their concentration gradients, through the inside of channel proteins that span the membrane. No expenditure of ATP is required.

pathogen (PATH-oh-jen) [Gk. *pathos*, suffering, + *-genēs*, origin] Disease-causing organism.

pelagic province The entire volume of ocean water; subdivided into the *neritic zone* (relatively shallow waters overlying the continental shelves) and the *oceanic zone* (water over the ocean basins).

penis Component of the male reproductive system of many species; the copulatory organ by which sperm are deposited into a specialized duct of the female reproductive system.

perennial [L. *per-*, throughout, + *annus*, year] A plant that lives year after year.

pericycle (PARE-ih-sigh-kul) [Gk. *peri-*, around, + *kyklos*, circle]. One or more layers just inside the endodermis of the root vascular column, that gives rise to lateral roots and contributes to secondary growth.

periderm A protective covering that replaces epidermis during secondary growth of many plants.

peripheral (per-IF-ur-uhl) **nervous system** [Gk. *peripherein*, to carry around] In vertebrates, the nerves leading into and out from the spinal cord and brain and the ganglia along those communication lines.

peristalsis (pare-ih-STAL-sis) A rhythmic contraction of muscles that moves food forward through the animal gut.

permafrost A permanently frozen layer beneath the soil surface in arctic tundra; forms an impenetrable basement beneath stretches of flat terrain, so water cannot drain.

PGA Phosphoglycerate; a key intermediate of glucose metabolism.

PGAL Phosphoglyceraldehyde; a key intermediate of glucose metabolism.

pH Whole number referring to the number of hydrogen ions present in a liter of a given fluid.

pH scale A measure of the concentration of hydrogen ions in different solutions.

phagocytosis (FAG-uh-sigh-TOE-sis) [Gk. *phagein*, to eat, + *kytos*, hollow vessel] Engulfment of foreign cells or substances by amoebas and some white blood cells, by means of endocytosis.

pharynx (FARE-inks) A muscular tube by which food is taken into the gut; in humans, the gateway to the digestive tract and to the windpipe (trachea).

phenotype (FEE-no-type) [Gk. *phainein*, to show, + *typos*, image] Observable trait or traits of an individual; arises from interactions between genes, and between genes and the environment.

pheromone (FARE-oh-moan) [Gk. *phero*, to carry, + *-mone*, as in hormone] A chemical secreted by an exocrine gland that serves as a communication signal between individuals of the same species.

phloem (FLOW-um) The "food-conducting" tissue by which sugars and other solutes are transported through the body of a vascular plant; in flowering plants, the main components of phloem are sieve tube members and companion cells.

phospholipid A key component of cell membranes in plants and animals; a molecule with a glycerol backbone, two fatty acid tails, and a phosphate group to which an alcohol is attached.

phosphorylation (FOSS-for-ih-LAY-shun) Addition of one or more phosphate groups to a molecule.

photolysis (foe-TALL-ih-sis) [Gk. *photos*, light, + *-lysis*, breaking apart] First step in noncyclic photophosphorylation, when water is split into oxygen, hydrogen, and associated

electrons; photon energy indirectly drives the reaction.

photoreceptor Light-sensitive sensory cell.

photorespiration In photosynthetic cells, the attachment of oxygen instead of carbon dioxide to RuBP during the Calvin-Benson cycle, so that less PGA forms and plant growth suffers. Photorespiration occurs on hot, dry days when carbon dioxide is prevented from entering leaves and oxygen builds up inside leaves.

photosynthesis The trapping of sunlight energy and its conversion to chemical energy (ATP, NADPH, or both), which is used in manufacturing food molecules from carbon dioxide and water.

photosynthetic autotroph An organism able to build all of the organic molecules it requires using carbon dioxide as the carbon source and sunlight as the energy source. All plants, some protistans, and a few bacteria are photosynthetic autotrophs.

photosystem Functional light-trapping unit in photosynthetic membranes; contains pigment molecules and enzymes.

photosystem I A type of photosystem that operates during the cyclic pathway of photosynthesis.

photosystem II A type of photosystem that operates during both the cyclic *and* noncyclic pathways of photosynthesis.

phototropism [Gk. *photos*, light, + *trope*, turning, direction]. Adjustment in the direction and rate of plant growth in response to light.

phylogeny (fie-LAH-jun-ee) The evolutionary relationships among species. At its most encompassing, a phylogenetic scheme starts with the most ancestral species and includes all the branches leading to all of its descendants.

phytochrome Light-sensitive pigment molecule whose activation and inactivation trigger hormone activities governing leaf expansion, stem branching, stem length, and, in many plants, seed germination and flowering.

phytoplankton (FIE-toe-PLANK-tun) [Gk. *phyton*, plant, + *planktos*, wandering] Community of photosynthetic microorganisms in freshwater or saltwater environments.

pineal (py-NEEL) **gland** A photosensitive endocrine gland in vertebrates which secretes a hormone (melatonin) that influences the development of reproductive organs and cycles.

pioneer species Typically, any of various species of small plants with short life cycles that are adapted to growing in exposed, often windy areas with intense sunlight, wide swings in air temperature, and soils deficient in nitrogen and other nutrients; by improving living conditions in the areas they colonize, pioneers commonly set the stage for their own replacement by other species.

placenta (play-SEN-tuh) In the uterus, an organ made of extensions of extraembryonic membranes (the chorion especially) and the endometrium. Through this composite of embryonic and maternal tissues and vessels, nutrients reach the embryo and wastes are carried away.

plankton [Gk. *planktos*, wandering] Communities of floating or weakly swimming organisms, mostly microscopic; photosynthetic autotrophs (phytoplankton) include diatoms, green algae, and cyanobacteria. Heterotrophs (zooplankton) include rotifers and copepods.

plant Most often, multicelled autotroph able to build its own food molecules through photosynthesis.

Plantae The kingdom of plants.

plant spore A single haploid cell that develops into a male or female gametophyte.

plasma (PLAZ-muh) The liquid component of blood; consists of water, various proteins, ions, sugars, dissolved gases, and other substances.

plasma cell Any of the antibody-secreting progeny of B cells of the immune system.

plasma membrane The outermost membrane of a cell that separates internal metabolic events from the environment. The membrane has a lipid bilayer and a variety of proteins that carry out most membrane functions. Many substances move across the membrane, and the membrane also has receptors for external molecules that can alter cell activities.

plasmid A small, circular DNA molecule that carries a few genes and replicates independently of the bacterial chromosome.

plasmodesma (PLAZ-moe-DEZ-muh) In a multicelled plant, a junction between the linked walls of adjacent cells through which nutrients and other substances are transported.

plasticity In human evolution, the ability to be flexible and to adapt to a wide range of demands.

plate tectonics The arrangement of the earth's outer layer (lithosphere) in slablike plates, all in motion and floating on a hot, plastic layer of the underlying mantle.

platelet (PLAYT-let) Any of the cell fragments in blood which release substances that take part in clot formation.

pleiotropy (PLEE-oh-troh-pee) [Gk. *pleon*, more, + *trope*, direction] Form of gene expression in which a single gene exerts multiple effects on seemingly unrelated aspects of an individual's phenotype.

pollen grain [L. *pollen*, fine dust] In gymnosperms and flowering plants, the male gametophyte (gamete-producing body).

pollen sac In the anthers of a flower, any of the chambers in which pollen grains develop.

pollen tube A tube formed after a pollen grain germinates; carries the sperm into the ovule.

pollination The arrival of a pollen grain on female reproductive parts.

pollutant Any substance with which an ecosystem has had no prior evolutionary experience, in terms of kinds or amounts, and that can accumulate to disruptive or harmful levels. Can be naturally occurring or synthetic.

polymer (POH-lih-mur) [Gk. *polus*, many, + *meris*, part] A molecule composed of from three to millions of small subunits that may or may not be identical.

polymerase chain reaction Method of DNA amplification; DNA having the gene of interest is split into single strands, which enzymes (polymerase) copy; the enzymes then act on the copies to produce millions of copies of the gene.

polymorphism (poly-MORE-fizz-um) [Gk. *polus*, many, + *morphe*, form] In a population, the persistence of two or more forms of a trait, at a frequency that is greater than can be maintained by newly arising mutations alone; and that frequency, if changed, will return to its former value over several generations.

polyp (POH-lip) Vase-shaped, sedentary stage of cnidarian life cycles.

polypeptide Chain of amino acids linked by peptide bonds, which form through condensation reactions.

polypeptide chain Three or more linked amino acids.

polyploidy (POL-ee-PLOYD-ee) Condition in which offspring end up with three or more of each type of chromosome characteristic of the parental stock.

polyribosome During protein synthesis, a clustering of ribosomes engaged in translation of a messenger RNA molecule.

polysaccharide [Gk. *polus*, many, + *saccharon*, sugar] A straight or branched chain of hundreds of thousands of covalently linked sugar units, of the same or different kinds. The most common polysaccharides are glycogen, starch, and cellulose.

population A group of individuals occupying a given area and belonging to the same species.

positive feedback mechanism Homeostatic mechanism by which a chain of events is set in motion and intensifies the original condition.

predator [L. *prehendere*, to grasp, seize] An organism that obtains food from other living organisms (its *prey*), which it may or may not kill; but it does not live on or in its prey.

pressure flow theory Explanation of how organic compounds move through a vascular plant's phloem. The movement is said to be driven by differences in water pressure between source regions (such as photosynthetically active leaves) and sink regions (such as growing plant parts).

primary growth Plant growth orginating at root tips and shoot tips.

primary immune response The response from macrophages, lymphocytes, and their prod-

ucts elicited by a first-time encounter with an antigen; includes both antibody-mediated and cell-mediated response.

primary productivity The rate at which an ecosystem's producers capture and store a given amount of energy in a given time. *Gross* primary productivity is the total rate of photosynthesis for the ecosystem during a specified interval; *net* primary productivity is the rate of energy storage in plants minus the rate of aerobic respiration in those plants.

primates The group of mammals that includes prosimians (such as lemurs) and the anthropoids (monkeys, apes, and humans); most are tree-dwellers.

principle A major phenomenon of nature, such as gravity, increasing entropy, and evolution.

procambium (pro-KAM-bee-um) A primary meristem that gives rise to the primary vascular tissues of the plant.

producer An autotrophic organism; able to build its own complex organic molecules from simple inorganic substances in the environment. Plants, some protistans, and some bacteria are producers.

progesterone (pro-JESS-tuh-rown) A major sex hormone that maintains the uterine lining for pregnancy and stimulates breast development.

prokaryote (pro-CARRY-oht) [L. *pro*, before, + Gk. *karyon*, kernel] Single-celled organism that has no nucleus or other internal organelles; all bacteria are prokaryotes.

promoter A base sequence that signals the start of a gene; the site where RNA polymerase initially binds during transcription.

prophase The first stage of mitosis, when each duplicated chromosome becomes condensed into a thicker, rodlike form.

prophase I and II Stages of meiosis when each duplicated chromosome condenses and pairs with its homologous partner, followed by crossing over and genetic recombination among sister chromatids (prophase I); and a very brief stage after interkinesis during which each chromosome still consists of two chromatids (prophase II).

protein Molecule composed of one or more chains of amino acids (polypeptide chains).

Protista The kingdom of protistans.

protistan (pro-TISS-tun) [Gk. *prōtistos*, primal, very first] Single-celled eukaryote.

proton Positively charged unit of energy that is found in the atomic nucleus.

proto-oncogene A gene sequence similar to a potentially cancer-inducing gene (oncogene), but that codes for proteins necessary to normal cell functioning; only rarely triggers cancer, generally when specific mutations alter its structure or function.

protostome (PRO-toe-stome) [Gk. *proto*, first, + *stoma*, mouth] Any of the bilateral animals in which the first indentation in the early embryo develops into the mouth. Includes mollusks, annelids, and arthropods.

proximal tubule Tubular region of a nephron that receives water and solutes filtered from the blood.

psychoactive drug A substance that affects parts of the central nervous system concerned with states of consciousness and behavior.

pulmonary circulation Pathways of blood flow leading to and from the lungs.

Punnett-square method A diagramming technique for predicting the possible outcome of a mating or an experimental cross between individuals.

purine Nucleotide base having a double ring structure. Examples are adenine and guanine.

pyrimidine (pih-RIM-ih-deen) Nucleotide base having a single ring structure. Cytosine and thymine are examples.

pyruvate (PIE-roo-vate) Three-carbon compound produced by the initial breakdown of a glucose molecule during glycolysis.

radial symmetry Body plan in which the body can be divided into four or more roughly equivalent parts with respect to the structures they contain.

rain shadow A reduction in rainfall on the leeward side of high mountains, resulting in arid or semiarid conditions.

reabsorption Process in urine formation in which water and usable solutes move out of a nephron by diffusion or active transport and then into capillaries, where they return to the general circulation; regulated by the hormones ADH and aldosterone.

receptor Of cells, a molecule at the cell surface or within the cytoplasm that may be activated by hormones, viruses, or some other outside agent. Of nervous systems, a sensory cell or cell part that may be activated by a specific stimulus in the internal or external environment.

recessive allele [L. *recedere*, to recede] In heterozygotes, an allele whose expression is fully or partially masked by expression of its partner; recessive alleles can be fully expressed only in homozygotes.

recombinant DNA Whole molecules or fragments that incorporate parts of different parent DNA molecules, as formed by natural recombination mechanisms or by recombinant DNA technology.

rectum A short tube at the terminal end of the gut through which undigested food and other wastes are expelled to the outside.

red blood cells Erythrocytes; the oxygen-transporting cells in blood. Red blood cells form in bone marrow and get their color from the iron-containing protein hemoglobin.

red marrow In many bones, a substance that fills spaces in the spongy tissue and serves as a major site of blood cell formation.

reflex [L. *reflectere*, to bend back]. A simple, stereotyped, and repeatable movement that is elicited by a sensory stimulus.

reflex arc [L. *reflectere*, to bend back]. Type of neural pathway in which signals from sensory neurons are sent directly to motor neurons, without intervention by an interneuron.

regulatory gene A gene that controls the functioning of one or more other genes.

releasing hormone A type of signaling molecule with targets in the anterior lobe of the pituitary; most stimulate secretion from a target cell, although some slow down such secretion.

replication Process by which DNA is duplicated prior to cell division.

reproduction, asexual Production of new individuals by any process that does not involve gametes.

reproduction, sexual Process of reproduction that begins with meiosis, proceeds through gamete formation, and ends at fertilization.

reproductive isolating mechanism Any aspect of structure, functioning, or behavior that prevents successful interbreeding (hence gene flow) between populations or between local breeding units within a population.

reproductive success Survival and production of offspring.

resource defense behavior Behavior mechanism that manifests itself among competing males of a species; males that defeat their rivals for the resource get mates, while the losers are kept from reproducing (at least temporarily).

resource partitioning A community pattern in which similar species generally share the same kind of resource in different ways, in different areas, or at different times.

respiration [L. *respirare*, to breath] In most animals, the overall exchange of oxygen from the environment and carbon dioxide wastes from cells by way of circulating blood. Compare *aerobic respiration*.

resting membrane potential The steady voltage difference that exists across the plasma membrane of a neuron (or some other excitable cell) that is not being stimulated.

restriction enzymes Class of enzymes that function solely to cut apart foreign DNA molecules that may enter a cell; each makes its cut only at sites having a short, specific nucleotide sequence.

reticular activating system Brain region that controls the changing levels of consciousness; includes circuits that connect with the spinal cord, cerebellum, and cerebrum.

reverse transcription Mechanism by which the hereditary material of an RNA virus is duplicated in a host cell; a "DNA transcript" is assembled on the viral DNA, then inserted into a host chromosome. A viral enzyme (reverse transcriptase) is required. This enzyme also is used for DNA amplification in the laboratory.

ribonucleic (RYE-bow-new-CLAY-ik) **acid** RNA; a category of nucleotides used in translating the genetic message of DNA into actual protein structure.

ribosome In all cells, a structure having two subunits, each composed of RNA and protein molecules; the site of protein synthesis.

RNA Ribonucleic acid; a category of nucleotides used in translating the genetic message of DNA into protein.

rod cell A type of photoreceptor in the vertebrate eye; being sensitive to very dim light, rod cells contribute to coarse perception of movement.

root hair In a vascular plant, an extension of a specialized root epidermal cell; root hairs collectively enhance the surface area available for absorbing water and solutes.

RuBP Ribulose bisphosphate, a compound required for carbon fixation in the Calvin-Benson cycle of photosynthesis.

salivary gland Any of the glands in the mouth that secrete saliva, a fluid that initially mixes with food and begins the breakdown of starch.

salt An ionic compound formed when an acid reacts with a base.

saprobe Heterotroph that obtains its nutrients from nonliving organic matter. Most fungi are saprobes.

sarcomere (SAR-koe-meer) The basic unit of contraction in skeletal muscle, made up of repeating bands of actin and myosin that appear between Z lines.

sarcoplasmic reticulum (sar-koe-PLAZ-mik reh-TIK-you-lum) A membrane system surrounding the myofibrils of a skeletal muscle cell that stores calcium ions and releases them in response to signals from the nervous system. The calcium ions are required for contraction.

second law of thermodynamics When left to itself, any system along with its surroundings undergoes energy conversions, spontaneously, to less organized forms. When that happens, some energy gets randomly dispersed in a form (often evenly distributed, low-grade heat) that is not as readily available to do work.

second messenger Term applied to molecules inside cells that mediate and generally trigger amplified responses to hormones.

secondary growth In vascular plants, an increase in stem and root diameter, made possible by the activity of lateral meristems.

secondary immune response Response by cells of the immune system to a previously encountered antigen; occurs more rapidly than a primary response and is greater and of longer duration.

secretion Generally, the release of a stored product of cellular or glandular activity for use by the animal or plant producing it. (This differs from an *excretion*, which is expelled from the body as waste material.) In kidneys, secretion is a highly regulated stage in urine formation, in which ions and some other substances move from capillaries into nephrons.

sedimentary cycle A biogeochemical cycle in which an element does not have a gaseous phase; instead, it moves from land to the seafloor and returns to the land only through long-term geological uplifting.

seed In gymnosperms and flowering plants, a fully mature ovule (contains the plant embryo), with its integuments forming the seed coat.

segmentation In many animal species, a series of body units that may be externally similar to or quite different from one another.

segregation, allelic [L. *se-*, apart, + *grex*, herd] Mendelian principle that diploid organisms inherit a pair of genes for each trait (or a pair of homologous chromosomes) and that the two genes segregate from each other during meiosis and end up in separate gametes.

selfish behavior Forms of behavior by which an individual protects or increases its own chance of producing offspring, regardless of the consequences to the group to which it belongs.

semen (SEE-mun) [L. *serere*, to sow] Sperm-bearing fluid expelled from the penis during male orgasm.

semiconservative replication [Gk. *hēmi*, half, + L. *conservare*, to keep] Manner in which a DNA molecule is reproduced; formation of a complementary strand on each of the unzipping strands of a DNA double helix, the outcome being two "half-old, half-new" molecules.

senescence (sen-ESS-cents) [L. *senescere*, to grow old] The sum total of processes leading to the natural death of parts of an organism, or of the whole organism.

sensory neuron Any of the nerve cells that act as sensory receptors, detecting specific stimuli (such as light energy) and relaying signals to the brain and spinal cord.

sex chromosomes In most animals and some plants, chromosomes that differ in number or kind between males and females but that still function as homologues during meiosis. All other chromosomes are called autosomes.

sexual dimorphism Differences in appearance between males and females of a species.

sexual reproduction Production of offspring from the union of gametes from two parents, by way of meiosis, gamete formation, and fertilization; a new individual inherits two genes for every trait—one from each parent.

sexual selection Natural selection based on any trait that gives the individual a competitive edge in mating and thereby producing offspring.

shoot system The stems and leaves of plants; the system has internal channels that conduct water, minerals, and organic substances among roots, leaves, and other plant parts.

sink region In a multicelled plant, any part that is using organic compounds for growth and development or that is stockpiling them for later use.

sliding filament model Model of muscle contraction, in which actin filaments physically slide over myosin filaments within sarcomeres. The sliding movement depends on an energy boost from ATP and on the formation of cross-bridges between the actin and myosin; actin moves toward the center of the sarcomere during contraction and away from it during relaxation.

small intestine The region of the digestive system that receives food from the stomach; also the region where digestion is completed and most nutrients absorbed.

smog, industrial Gray-colored air pollution that predominates in industrialized cities with cold, wet winters; typically consists of dust, smoke, ashes, soot, asbestos, oil, bits of heavy metals, and sulfur oxides.

smog, photochemical Form of brown, smelly air pollution occurring in large cities in warm climates. Chief component is nitric oxide produced by internal combustion engines; in combination with oxygen forms nitrogen dioxide, which, when exposed to sunlight, can react with hydrocarbons to form photochemical oxidants.

social behavior The tendency of individual animals to enter into cooperative, interdependent relationships with others of their kind; based on the ability of animals to use various types of signals to communicate with one another.

social parasite An animal that depends on the social behavior of another species to gain food, care for young, or some other factor necessary to complete its life cycle.

sodium-potassium pump A transport protein spanning the lipid bilayer of the plasma membrane. When the protein receives an energy boost from ATP, its shape changes in such a way that it selectively transports sodium ions *out* of the cell and potassium ions *in*. This active transport process maintains the ion distributions (hence the voltage difference) across the membrane.

solute (SOL-yoot) [L. *solvere*, to loosen] Any substance dissolved in some solution. In water, this means its individual molecules are surrounded by spheres of hydration that keep their charged parts from interacting, so the molecules remain dispersed in the water.

solvent Fluid in which one or more substances is dissolved.

somatic (so-MAT-ik) **cell** [Gk. *sōma*, body] Any cell of the animal body that is not a germ cell (which develops by meiosis into sperm or eggs).

somatic nervous system Those nerves leading from the central nervous system to skeletal muscles.

sorus, plural **sori** A cluster of sporangia on the lower surface of a fern frond.

source region In a multicelled plant, generally any of the sites of photosynthesis.

speciation (spee-cee-AY-shun) The time at which a new species emerges, as by divergence or polyploidy.

species (SPEE-sheez) [L. *species*, a kind] For sexually reproducing organisms, one or more populations whose members interbreed under natural conditions and produce fertile offspring, and who are reproductively isolated from other such groups.

sperm [Gk. *sperma*, seed] Mature male gamete.

spermatogenesis (sperm-AT-oh-JEN-ih-sis) Formation of a mature sperm from a germ cell.

sphere of hydration Through positive or negative interactions, a clustering of water molecules around the individual molecules of a substance placed in water. Compare *solute*.

sphincter (SFINK-tur) Ring of muscle that serves as a gate between regions of a tubelike system (as between the stomach and small intestine).

spindle apparatus Structure composed of microtubules that is essential for mitosis and meiosis; it establishes two poles toward which the chromosomes are moved during nuclear division.

sporangium (spore-AN-gee-um), plural **sporangia** [Gk. *spora*, seed] The protective tissue layer that surrounds haploid spores in a sporophyte.

spore Generally, a reproductive cell that can undergo mitotic division and give rise to more haploid cells or to a multicelled haploid individual, which eventually produces gametes.

sporophyte [Gk. *phyton*, plant] Diploid, spore-producing stage of plant life cycles.

spring overturn Strong, vertical movements of water generated by winds in the spring; dissolved oxygen from the surface layer of water moves to the depths while nutrients from bottom sediments are brought to the surface.

stabilizing selection Mode of natural selection in which the most common phenotypes in a population are favored, and the underlying allele frequencies persist over time.

stamen (STAY-mun) In flowering plants, the male reproductive structure; commonly consists of pollen-bearing structures (anthers) positioned on single stalks (filaments).

steroid (STAIR-oid) A lipid with a backbone of four carbon rings. Steroids differ in the number and location of double bonds in the backbone and in their number, position, and type of functional groups.

stimulus [L. *stimulus*, goad] A specific form of energy, such as light, heat, and mechanical pressure, that the body can detect through its sensory receptors.

stoma (STOW-muh), plural **stomata** [Gk. *stoma*, mouth] A controllable gap between two guard cells in stems and leaves; any of the small passageways across the epidermis through which carbon dioxide moves into the plant and water vapor moves out.

stomach A muscular, stretchable sac of the digestive system in which considerable protein digestion occurs; usually located between the esophagus and intestine. Many invertebrates as well as vertebrates have some form of stomach.

stroma [Gk. *strōma*, bed] In chloroplasts, the semifluid matrix, surrounding the grana, where starch and other organic molecules are assembled.

substrate Molecule or molecules of a reactant on which an enzyme acts.

substrate-level phosphorylation Reaction in which a substrate molecule gives up a phosphate group to another molecule, as when an intermediate molecule that has formed during glycolysis donates phosphate to ADP, producing ATP.

succession (suk-SESH-un), **primary** [L. *succedere*, to follow after] The orderly changes in species composition from the pioneer species that inhabit an area previously devoid of life to the more or less constant array of species that constitutes the climax community.

succession, secondary Reestablishment of a climax community that has been disrupted in whole or in part.

surface-to-volume ratio In cells, a physical constraint on increased size: as the cell's linear dimensions grow, its surface area does not increase at the same rate as its volume (hence each unit of plasma membrane is called upon to serve increasing amounts of cytoplasm).

survivorship curve Mathematical curve that reflects the age-specific patterns of death for a particular population in a particular environment.

symbiosis (sim-by-OH-sis) [Gk. *sym*, together, + *bios*, life, mode of life] A mutually beneficial relationship involving continuous, intimate contact between interacting species.

sympathetic nerve Any of the nerves of the autonomic nervous system that typically are concerned with increasing overall body activities during times of heightened awareness, excitement, or danger.

synapse (SIN-aps), **chemical** [Gk. *synapsis*, union] A junction between two neurons, or between a neuron and a muscle or gland cell, that are separated by a small gap. At an *excitatory* synapse, a transmitter substance released from the first neuron produces changes in the receiving cell that bring its membrane closer to threshold. At an *inhibitory* synapse, a transmitter substance released from the first neuron produces changes in the receiving cell that drive membrane potential away from threshold.

synaptic integration The moment-by-moment combining of excitatory and inhibitory signals acting on adjacent membrane regions of a neuron.

synovial (sin-OH-vee-uhl) **joint** A freely movable area of contact (or near contact) between bones, in which a flexible capsule of dense connective tissue holds the bones near each other.

systemic circulation Blood circulation route in which oxygenated blood arriving from the lungs is pumped from the left side of the heart to the rest of the body (where it gives up oxygen to cells and takes on carbon dioxide); it then flows back to the right side of the heart.

taproot system A primary root and its lateral branchings.

telophase (TEE-low-faze) Final stage of mitosis, when the chromosomes decondense into threadlike structures and two daughter nuclei form.

telophase I and II Stages of meiosis when one of each type of duplicated chromosome has been moved to one or the other end of the spindle pole (telophase I); and when four daughter nuclei form, each with a haploid number of chromosomes (telophase II).

testcross Experimental cross in which hybrids of the first generation of offspring (F_1) are crossed with an individual known to be true-breeding for the same recessive trait as the recessive parent.

testis, plural **testes** Male gonad; primary reproductive organ in which male gametes and sex hormones are produced.

testosterone (tess-TOSS-tuh-rown) In male mammals, a major sex hormone that helps control male reproductive functions; an androgen.

theory In science, a related set of hypotheses that, taken together, form an explanation about some aspect of the natural world. In modern science, only explanations that can be relied upon with a very high degree of confidence are accorded the status of theory.

thermal inversion Situation in which a layer of dense, cool air becomes trapped beneath a layer of warm air; can cause air pollutants to accumulate to dangerous levels close to the ground.

thermoreceptor Sensory cell that can detect radiant energy associated with temperature.

thigmotropism (thig-MOTE-ruh-pizm) [Gk. *thigm*, touch] Growth oriented in response to physical contact with a solid object, as when a vine curls around a fencepost.

threshold value The minimum voltage change across the plasma membrane necessary to produce an action potential.

thylakoid membrane In chloroplasts, an internal membrane that is folded into flattened channels and disks called *grana*, and that contains the light-absorbing pigments of photosynthesis.

thymine Nitrogen-containing base found in some nucleotides.

thyroid gland Endocrine gland (located at the base of the neck in humans) which produces hormones that affect overall metabolic rates, growth, and development.

tissue A group of cells and intercellular substances that function together in one or more specialized tasks.

tonicity The relative concentrations of solutes in the fluid inside and outside the cell. When solute concentrations are *isotonic* (equal in both fluids), water shows no net osmotic movement in either direction. When one of the fluids is *hypotonic* (has less solutes than the other), the other is *hypertonic* (has more solutes) and is the direction in which water tends to move.

trachea (TRAY-kee-uh), plural **tracheae** In insects, spiders, and some other animals, an air-conducting tube that branches finely through the body and functions in respiration; in land vertebrates, the windpipe that carries air between the larynx and bronchi.

transcription [L. *trans*, across, + *scribere*, to write] The assembly of an RNA strand on one of the two strands of a DNA double helix; the resulting transcript has a nucleotide sequence that is complementary to the DNA region on which it is assembled.

translation The interaction of rRNA, tRNA, and mRNA in converting the DNA instructions encoded in the mRNA molecule into a particular sequence of amino acids to form a polypeptide chain.

translocation In cells, the transfer of part of one chromosome to a nonhomologous chromosome. In vascular plants, the conduction of organic compounds through the plant body by way of the phloem.

transmitter substance Any of the class of signaling molecules that are secreted from neurons, act on immediately adjacent cells, and are then rapidly degraded or recycled.

transpiration Evaporative water loss from stems and leaves.

transposition Genetic recombination in which transposable elements move from one site in the DNA to another. Transposable elements are also called jumping genes.

trisomy (TRY-so-mee) The presence of three of one type of chromosome in diploid cells.

trophic (TROW-fik) **level** [Gk. *trophos*, feeder] All organisms in an ecosystem that are the same number of transfer steps away from the energy input into the system.

true-breeding In sexually reproducing organisms, an indication that successive generations are exactly like the parents in one or more heritable traits.

tumor A tissue mass composed of cells that are dividing at an abnormally high rate.

turgor (TUR-gore) **pressure** [L. *turgere*, to swell] Internal pressure on a cell wall caused by osmotic movement of water into the cell body.

umbilical (um-BILL-uh-kul) **cord** In humans, a cordlike, vascularized tissue that connects a growing embryo to the extraembryonic membranes and that provides for its nutrition, respiration, and excretion.

upwelling An upward movement of deep, nutrient-rich water along the margins of continents.

uracil (YUR-uh-sill) Nitrogen-containing base found in RNA molecules; can base-pair with adenine.

urinary system An organ system concerned with regulating water and salt levels in the body; in mammals, it consists of the two kidneys, two ureters, urinary bladder, and urethra.

urine The fluid formed in the kidney through the processes of filtration, reabsorption, and secretion; consists of wastes and excess water and solutes.

uterus (YOU-tur-us) [L. *uterus*, womb] A chamber in which the developing em-bryo is contained and nurtured during pregnancy.

vaccine A preparation designed to stimulate the appearance of memory lymphocytes, and injected into the body or taken orally.

vacuole (VAK-you-ohle), **central** In plant cells, a membrane-bound, fluid-filled sac that may take up most of the cell interior; main function is to increase cell size and surface area, thereby enhancing absorption of relatively dilute concentrations of nutrients from the external environment.

vagina Part of the female reproductive system that receives sperm from the male penis; forms part of the birth canal, and acts as a channel to the exterior for menstrual flow.

vascular bundle One of several to many strandlike arrangements of primary xylem and phloem embedded in the ground tissue of roots, stems, and leaves.

vascular cambium In vascular plants, one of the lateral meristems that increase stem or root diameter.

vascular column In plant roots, the arrangement of the root's vascular tissues in a central cylinder.

vascular plant Plant having tissues that transport water and solutes through well-developed roots, stems, and leaves.

vein Within the circulatory system, any of the large-diameter vessels that lead back to the heart.

vertebra, plural **vertebrae** One of a series of hard bones that form the backbone in most chordates.

vertebrate Animal having a backbone made of bony segments called vertebrae.

vesicle (VESS-ih-kul) [L. *vesicula*, little bladder] In a cell, a small membranous sac that can transport or store substances within the cytoplasm.

villus (VIL-us), plural **villi** Any of several types of fingerlike projections from the membranous surface of some body part; those of the small intestine greatly increase the absorptive surface area.

viroid An infectious nucleic acid that has no protein coat; a tiny rod or circle of single-stranded RNA.

virus A noncellular infectious agent, consisting of DNA or RNA surrounded by a protective protein coat, that can replicate only after its genetic material enters a host cell and subverts the host's biosynthetic machinery.

vision Precise light focusing onto a layer of photoreceptive cells that is dense enough to sample details concerning a given light stimulus, followed by image formation in the brain.

vitamin In most animals, any of the organic substances that are required in small amounts for normal cell metabolism and that cannot be synthesized internally and so must be ingested in food; particular vitamins have specific metabolic roles.

water potential The sum of two opposing forces (osmosis and turgor pressure) that can cause the directional movement of water into or out of a walled cell.

watershed A region where all precipitation becomes funneled into a single stream or river.

wax A type of lipid; waxes have long-chain fatty acid tails and typically form protective, lubricating, or water-repellent coatings.

white blood cells Leukocytes; they arise from stem cells in the bone marrow and function in defense and as scavengers of dead or worn-out cells.

white matter Mainly axons of interneurons, so named because of the glistening myelin sheaths around them.

wild-type allele The normal or most common allele at a given gene locus.

wing In birds, a forelimb constructed of feathers, powerful muscles, and lightweight bones. In insects, the wing develops as a lateral fold of the exoskeleton.

X-linked gene Any gene on the X (female sex) chromosome.

X-linked recessive inheritance Recessive condition in which the responsible, mutated gene occurs on the X chromosome.

xylem (ZYE-lum) [Gk. *xylon*, wood] In vascular plants, a tissue that transports water and solutes through the plant body.

Y-linked gene Any gene on the Y (male sex) chromosome.

yolk sac In many vertebrates, an extraembryonic membrane that provides nourishment (from yolk) to the developing embryo; in humans, the sac does not include yolk but helps give rise to a digestive tube.

zygote (ZYE-gote) In most multicelled eukaryotes, the first diploid cell formed after fertilization (fusion of nuclei from a male and a female gamete).

INDEX

Italic numerals refer to illustrations.

A

Abalone, 444, *444*
Abiotic, defined, 195
Abiotic formation of organic
  compounds, 195, *195*
Abnormality, defined, 132
ABO blood group locus, 123
ABO blood typing, 123, 386
Abortion, 140, 484, 488
Abscisic acid, *327*, 328, 333
Abscission, 333
Absorption
  by digestive system, 361
  water-solute balance and, 419
Absorption spectrum, photosynthetic
  pigments, 71, *72*
Abstention, sexual, 483
Abyssal zone, *562*
Accutane, 480
Acetaldehyde, 88
Acetylcholine, 432
Acetyl-CoA (Coenzyme A), *84*, 85, *85*,
  89, 434
Achondroplasia, 134
Acid, defined, 25
Acid-base balance, 424
Acid deposition, 571–573
Acid group, 31, *31*
Acidity, stomach, 363
Acid rain, 26, *26*, 571–573
Acoelomate, 260–261
Acoustical signaling, 594
Acquired immune deficiency
  syndrome (*See* AIDS)
Acromegaly, *453*
Acrosome, 468, *469*
Actin, 52, *53*, 354, *354*, 355, *355*
Actinomycetes, *226*
Action potential (*See also* Membrane
  potential)
  all-or-nothing event, 430
  changes in membrane polarization
    during, 429ff.
  defined, 428
  frequency, 442
  propagation, *428–429*, 428–430,
    430
  refractory period, 430
  and stimulus intensity, 441
  threshold, 429
Activation energy, *62*, 62–63
Active site, 62
Active transport
  and ATP, 313
  defined, 43
  and intestinal absorption, 365–366
  in plant cells, 313–314, *314*
  by sodium-potassium pump, 43,
    428ff.
Acute pancreatitis, 63
Acyclovir, 487
Adaptive, defined, 2–3, 8
Adaptive behavior, 589ff.
Adaptive radiation
  amphibians, *200*
  apes, 216
  birds, *200*
  bony fish, *200*
  defined, 201

dinosaurs, 205
flowering plants, *199*
human, 499
insects, *199, 201, 204*
mammals, *199–200, 204*
plants, 204
protozoans, *200*
rabbits, *200*
reptiles, *200, 204*, 286
tribolites, *200*
whales, *200*
Adaptive zone, 201, 216
Addiction, 439–440
Adenine, 64, *64*, 145–147, *145–147*,
  151
Adenoid, 388
Adenosine
  diphosphate (*See* ADP)
  phosphate, 34, *34–35*
  triphosphate (*See* ATP)
Adenovirus, *224*
Adenylate cyclase, 459, *459*
Adhesion, cell, 160
Adipose tissue, 338, *338*, 367, 372
Adolescence (human), defined, *482*
ADP (adenosine diphosphate), 65,
  73, *74*, 77, *77*, 80, *86*
Adrenal gland
  cortex, 452, 455, *455*
  innervation, 435
  medulla, *454*, 455, *455*
Adrenalin (*See* Epinephrine)
Adrenal medulla, 455, *455*
Adrenocorticotropic hormone, *452*
Adult (developmental stage), *482*
Adventitious root, 299–300, 326
Aerobic respiration
  ATP formation by, 80–81
  defined, 4
  energy yield of, *81*
  links with photosynthesis, 69, *69*
  overview of, 80–81, *82*
  and oxygen, *72*, 76, 81, *82–84*, 85,
    85, 86, *86*, 90–91
  reactions of, *81*, 81–82, *82–86*
  stages of, *81–86*
  summary of, 81
Aerosol spray, 574
African sleeping sickness, *234*
Age, and cardiovascular disorders,
  381
Age structure, population, 494,
  502–503
Agglutination, blood, 386, *386*
Aging
  and bone turnover, 350
  and cancer resistance, 396
  defined, 482, *482*
  glandular secretions in, 348
  and human development, 482, *482*
    (*See* Aging)
  and mitosis, 482
  physical effects of, 482
  skin changes during, 348
  and tobacco smoke, 348
Agnathan, 283
Agriculture
  and crop rotation, 538
  and deforestation, 577, *577*, 578–579
  and desertification, 580
  and genetic engineering, 163, 168
  and green revolution, 576–577

and human population growth, 499–
  500
and irrigation, 168, 574
land available for, 168, 576–577
and nitrogen scarcity, 538–539
and pesticides, 16–17, 184
plant viruses, 225
and salination, 168
slash-and-burn, 578
Agrobacterium, 168, *226*
Agromegaly, 452
AIDS, *390*, 399 (*See also* Human
  immune-deficiency virus)
  carriers of, 402
  cases of, 402
  economics of, 485
  retrovirus causing, *224*
  spread of, 485
  testing for, 486
  transmission modes, 400–401, 485
  treatments for, 402, 485
AIDS-related complex, 485
Air circulation, 544, *545*
Air pollution, *570*, 570ff.
Air sac, bird, *409*
Alanine, *153*
Albatross, *593*, 593–594
Albinism, 125
Albino, *124*, 125
Albumin, 33, *286, 376*
Alcohol (ethyl alcohol)
  addiction, 440
  effects on central nervous system,
    440
  effects on heart, 383
  fetal alcohol syndrome, 480–481
  and pregnancy, 480
Alcoholic fermentation, 88, *88*
Aldosterone, 422, *454*, 458
Alfalfa stem, *296*
Alga (*See also* specific type)
  blue-green (cyanobacterium), 227–
    228, *228*
  brown, 239, 245–246, *247*, 251, *561*
  classification, 246
  evolution, *199*, 212
  golden, *229*, 233
  green, *39*, *72*, 202, 231, 244–245,
    246, 247, 251, 561
  red, 202, 245–246, *246*, 251, *561*,
    565
  species of, 245, *251*
  yellow-green, *229*, 233
Alkaloid, 579
Allantois, *286*, 476, *476*
Allele
  and crossing over, 107
  defined, 105, 118, *119*
  dominant, 123, 134
  frequencies in populations, *178*
  of homologous chromosomes, 107,
    119, *119*
  incompletely dominant, 123
  mutant, 166
  pool, 185
  recessive, 118–119, *130*, 132, *134*,
    134–135, *135*
  recombination at meiosis, 109
  segregation at meiosis, 119–121
  variable expression of, 123
  wild-type, defined, *130*
Allergy, 399

Alligator, 286, *286*
All-or-nothing event, 430
Allosteric enzymes, 63–64
Alpha cell, pancreas, 456, *457*
Alpine tundra, 557
Alternation of generations, *245*
Altruistic behavior, 580, 589, 597
Alveolar sac, *410*, 411, 413
Alveolus, *410*, 411–412, *412*
Amanita, 242
Amino acid
  abiotic formation of, 195, *195*
  blood, *376*
  defined, 31
  essential, 368–369, *369*
  intestinal absorption of, 365
  and polypeptide chain assembly,
    153–154
  R group, 31, *31*, 33
  right- vs. lefthanded, 196
  role in cell, 27
  and self-replicating systems, 196
  sequence for insulin, 32
  structure, 31, *31*
Amino group, 31, *31*,
Aminopeptidase, 365
Ammonia, 195, *195*, 228
  and amino acid degradation, 89,
    372, 419
  and nitrogen cycle, 538, *539*
Ammonification, 538, *539*
Amniocentesis, 140, *140*
Amnion, 140, *286*, 476, *476*
Amniotic egg, *286*, *286*
Amoeba, 3, *44*, 106, 229, 234, *235*
Amoebic dysentery, 235
Amphetamine, 439
Amphibian (Amphibia)
  adaptive radiation, *200*
  circulatory system, *282*
  comparative embryology, *190*
  endoskeleton, 285
  evolution, *199*, 204, 283, 285
  extinction, *200*
Amplexus, *463*
Amplification, DNA, 165
Amylose, *29*
Anabaena, 226, 228
Anabolic steroid, 357–358
Analgesic, narcotic, *439*, 440
Analogous structure, 193
Anaphase
  chromosome appearance, *97*
  meiosis, 106, 108–110, *114–115*
  mitosis, 97, *99*, 100, *114*
  nondisjunction at, *136*, 137
Androgen, *454*
Anemia, *166*
  sickle-cell (*See* Sickle-cell anemia)
Angina pectoris, 382
Angiography, 382, *382*
Angioplasty, 383
Angiosperm, *199*, 246, 250, *251*, 254,
  254–255 (*See also* Flowering
  plant)
Angiotensin), *454*
Angler fish, *563*
Anhidrotic ectodermal displasia, 159,
  159
Animal (Animalia)
  with armor plates, 202
  body cavities, *260*, 261

body symmetry, 259–260
cephalization, 259–260
characteristics, 259
classification, 209
as consumers, 4, 7
cytokinesis of, 101, 101
evolution, 199, 212ff., 258ff., 261, 418
genetic engineering, 168
germ, defined, 95
gut, 260–261
internal environment, 336–337
life cycle, generalized, 110, 110–111
organelles in, 46, 49
organ systems, 259ff., 340–341, 341
placental, 288
polyploidy in, 187
respiratory system, 408–409
segmentation, 261
with shells, 202
somatic, defined, 95
with spines, 202
types of, 337ff.
Animal development (See also
    Development; Human
    development)
blastula, 465
and cell death, 467
and cell differentiation, 465–466,
    466
cleavage, 464–465
deuterostome, 269
differentiation during, 158–159
and fertilization, 464
gamete formation, 464
gametogenesis, 110–112, 111
gastrulation, 465
germ layers, 258, 465
hormones and, 454ff.
mechanisms of, 465ff.
and morphogenesis, 466, 466–467
organ formation, 465, 465
overview of, 464
protostome, 269
tissue specialization in, 465
Animal reproduction (See also Life
    cycle, animal)
asexual, 259, 463–464
sexual, 95, 104–105, 259
strategic problems with, 464
Annelid (See also Earthworm)
Annelid (Annelida)
characteristics, 272–273
evolution, 260–261
major groups, 272–273
representative, 259
Anorexia nervosa, 368
Antagonistic muscle system, 356, 356
Antarctic food web, 526
Anterior lobe (See Pituitary gland)
Anther, stamen, 317–319
Antheridia, 248
Anthropoid, 212, 213
Antibiotic, sources of, 226, 229, 480
Antibody
circulating, 395
classes of, 395
defined, 386
formation, 394, 394–395, 403
functions of, 35
genes coding for, 394, 394–395
to HIV proteins, 400, 402
immunoglobulins, 395
monoclonal, 396–397
passive immunity, 399
polypeptide chain of, 396
during pregnancy, 480
production, 194
to Rh⁺ blood, 386
secretion, 396
structure, 394
Antibody-mediated immune response,
    392, 393–396, 395, 403

Anticodon, 153, 153
Antidiuretic hormone, 422, 450, 451,
    452 (See also Vasopressin)
Antigen
    and clonal selection theory, 396,
        398, 398
    defined, 394–395
    processing, 394, 395, 396, 398
Antigen-MHC complex, 394, 396
Anus, 362, 366, 513
Aorta, 379, 379, 420
Ape
    adaptive radiation, 216
    classification, 212, 213
    Miocene, 216
    skeletal organization, 214
Apex, of heart, 379
Apical dominance, 328
Apical meristem, 294, 294–295, 297,
        300, 323
Apodan, 285
Appendage, of arthropods, 274, 274
Appendicitis, 366
Appendicular skeleton, 350, 351
Appendix, 366, 366
Apple, 302, 324
Apple scab, 242
Aquatic ecosystem, 226, 558ff.
Aqueous humor, 445
Aquifer, 574, 575
ARC, (AIDS-related complex), 485
Archaebacterium, 227, 230
Archaeopteris, 190
Archaeopteryx, 176, 176, 286
Archean era, 191, 202
Archegonium, 248, 248
Arctic fox, 258
Arctic tundra, 524–525, 556–557
Argentine fire ant, 517
Arginine, 153
Armadillo, 174
Armillaria, 332
Armor plate, 202–203
Arnold, S., 590
Arrhythmia, 383
Arteriole, 379, 384–385
    control of blood volume in, 384
    defined, 384
    functions of, 384
    resistance at, 384
Artery
    atherosclerotic plaque, 382
    blood pressure in, 384
    coronary, 379, 379
    defined, 379
    functions of, 384
    hardening of, 381
    pulmonary, 379
Arthritis, 352, 399
Arthrobotrys, 244
Arthropod (See also Insect)
Arthropod (Arthropoda)
    adaptations, 273
    appendages, 274
    circulatory system, 375
    division of labor in, 273
    evolution, 260–261
    exoskeleton, 273
    groups of, 273
    metamorphosis, 273
    molting, 273
    representative, 259
    respiratory system, 274
    sensory structures, 274
Artificial selection, 9, 163
Asbestos, 570
Ascorbic acid, 370
Asexual reproduction (See also
        Reproduction)
    animal, 259
    Chlamydomonas, 247
    cnidarian, 264

defined, 104, 463
eukaryote, 95
flatworm, 463
flowering plant, 316–317
fungus, 240, 240, 244
plant, 316–317
prokaryote, 95
sponge, 463
sporophyte, 316–317
Asparagine, 153
Aspirin, 579
Association cortex, 438
Asteroid theory, of Cretaceous mass
    extinction, 206
Asthma, 399
Atherosclerosis, 31, 381–383
Atherosclerotic plaque, 382, 382
Athlete and anabolic steroids, 357
Athlete's foot, 242
Atmosphere
    and climate, 543–544, 544
    defined, 543
    early, 194–195
    greenhouse gases in, 536–537, 537
    and hydrologic cycle, 532, 533–534
Atmospheric cycle, 531
Atom
    bonds between (See Bond, chemi-
        cal)
    defined, 4, 17
    ions, 22
    isotopes, 18
    negatively charged, 22
    nucleus, 17–18
    positively charged, 22
    structure of, 17–18
Atomic number, 18, 18
ATP
    and active transport, 44
    formation, 46
    functions of, 34, 35
    in muscle contraction, 354–355
    and phosphorylation reactions, 65
    in plant function, 312, 313
    and second messengers, 459, 459
    structural formula for, 64
    turnover, 65
ATP/ADP cycle, 65
ATP formation
    by aerobic respiration, 65, 80–86,
        82–86
    by anaerobic electron transport, 87,
        87–88, 88
    by anerobic electron transport, 87,
        87–88, 88
    by bacteria, 226, 227–228, 228
    chemiosmotic theory of, 75, 86
    in chloroplasts, 71, 71, 73–74,
        74–75
    by photophosphorylation, 73–74, 76
    by photosynthesis, 5, 65, 70–71, 73–
        77
    role of electron transfers in, 66
    by substrate-level phosphorylation,
        82, 85
ATP synthase, 75, 84
Atrial natriuretic factor, 166, 454
Atrioventricular node, 380, 380
Atrioventricular valve, 379, 379, 380
Atrium, heart, 379, 379, 380
Auditory cortex, 438
Auditory nerve, 443
Australopithecus afarensis, 216, 217,
        218
A. africanus, 216, 217, 218
A. boisei, 216, 217, 218
A. robustus, 216, 217, 218
Autoimmune disorder, 399, 456
Autonomic nervous system, 434–435,
        434–435, 435
Autosomal dominant inheritance, 134,
        134

Autosomal recessive inheritance,
        132–134, 133
Autosome, 129
Autotroph
    chemosynthetic, 69, 226, 527
    defined, 69
    photosynthetic, 69, 523, 525, 527,
        558
Auxin, 327, 328–329
Avery, O., 145
Axial skeleton, 350, 351
Axillary bud, 292, 297
Axon, 427, 431, 432, 436
Azotobacter, 226, 229
AZT, 402

B

Baboon, 511, 592, 593, 597
Bacillus, 222, 225
Backbone, 174
Background extinction, 201
Bacteriophage, 144, 145, 223, 223
Bacterium (See also Prokaryote)
    aerobic, 229
    anaerobic, 88, 226, 229–230
    in animal gut (enteric), 226
    appearance in micrographs, 228
    archaebacterium, 226, 227, 230
    ATP formation, 227–228, 228
    autotrophic, 226
    beneficial, 87, 226, 227, 229
    binary fission, 227
    biotic potential, 495
    body plan of, 225
    capsule, 225
    cell wall, 55, 225–226
    characteristics, 225, 225–226
    chemosynthetic, 225–226, 226,
        228, 563
    chromosome, 163–164, 226
    classification, 226, 226–227
    commercial uses of, 88, 226, 229
    conjugation, 164, 164
    cylindrical, 225
    as decomposers, 4, 523, 525, 527
    disease-causing, 87, 226, 228–229
    DNA, 38, 163–164, 164
    endospore-forming, 226
    eubacteria, 226, 227–228, 230
    and eukaryotes, 230
    evolution, 199, 202
    fission, 226
    flagellum, 45
    genetically engineered, 167
    Gram-negative, 226, 226
    Gram-positive, 226, 226
    habitats, 226, 227–229
    halophile, 226, 230
    helical, 225
    heterocyst, 228
    heterotrophic, 226, 226, 229
    ice-minus, 167
    major groups of, 226, 226ff.
    methanogen, 226, 230
    and mitochondria, 230
    nitrifying, 226, 228, 228, 527, 538,
        539
    nitrogen-fixing, 229, 307, 307, 538,
        539
    pathogenic, 226
    photosynthetic, 202, 225, 226, 227–
        229
    plasmid, 163–164, 226
    protein synthesis, 225
    resting spore, 228
    shapes of, 225
    spherical, 225
    symbiotic relationship with plants,
        307
    thermoacidophile, 230, 226

Balloon angioplasty, 382–383
Banana slug, 590, *590*
Barbiturate, 480
Bark, 303, *303*
Barr body, *159*
Barrier, geographic, 186
Basal body, 54, *54*
Base vs. acid, 25, *25*
  nitrogen-containing, *145*, 145*ff.*
Basement membrane, 337, *384*
Base pairing, 146–147, *147*, 151
Basilar membrane, *443*
Basophil, *376*–377, 378, 391, 393, *403*
Bat
  adaptive radiation, *200*, 201
  frog-eating, 594, *594*
  pancreatic cell mitochondrion, *84*
  wing, 193, 201
Bathyl zone, *562*
B cell, *302*, 378, *392*, 393–395, *395*,
  396, *398*, 402, *402*
Beagle, H.M.S., *173*, 174
Beak, *175*
Bean, *328*
Becquerel, H., 19
Bee, *321–322*
  pollinator, 321–322
  tactile signaling, 594, *595*
  tracheal system, 408
  ultraviolet photoreceptors, 441
  venom, 399
Beet, sugar, 168
Beetle, *517*
  ladybird, *276*
  as pollinator, 321
  scarab, *276*
Behavior
  adaptive aspects of, 589*ff.*
  aggressive, 357
  altruistic, 580, 589, 597
  anti-predator, 590
  appeasement, *597*
  courtship, 464, *593*
  defined, 586
  disorders, anabolic steroid-related,
    357
  eating disorders, 368
  evolutionary approach to, 598
  feeding, *526*, 590
  female defense, 591
  fight-flight response, 434, 455
  genetic basis of, 587
  and heredity, 586
  hormonal effects on, *452*, 587
  and human population growth, 499–
    500
  imprinting, 588–589
  instinctive, 587–588
  learned, 588
  and learning, 586
  links with hypothalamus, 436
  melatonin effects on sexual activity,
    458
  and natural selection, 589
  parental, 596
  parental, of birds, *287*
  reproductive, 464
  resource-defense, 591
  safe sex, 486
  selfish, 580, 589
  sexual, *454*, 469, 483–487
  sexual selection, 590–591
  singing, 587
  smiling, 598
  social, 286, 592*ff.*
  testosterone effects, 469
Behavioral evolution, 215
Behavioral isolation, 186
Behavioral trait, 177
Benign tumor, 160
Benthic province, 562, *562*
Benzene, *570*

Beta cell, pancreas, 456, *457*
Bicarbonate, 363, 424
Biceps, *353*, 356, *356*
Bichir, 285
Bighorn sheep, *8–9*, 591
Bilateral symmetry, *259*, 260
Bile, 364
Bile salts, 364
Binary fission, 226, *227*
Biochemistry, comparative, 193
Biofeedback, 434
Biogeochemical cycle
  atmospheric, 531, *532*
  carbon, 534–535, *534–535*
  defined, 531
  hydrologic, 531–533
  nitrogen, 538–539
  sedimentary, 531
Biogeographic realm, 547, *547*
Biological clock, 328–329 (*See also*
  Photoperiodism)
Biological magnification, 539–540
Biological perspective
  on abortion, 483
  on drug abuse, 440
  on human gene therapy, 169
  on human life in the biosphere, 90–
    91, 582–583
  on human population growth, 498–
    504, 569–570
Biological rhythm, seasonal variation,
  544
Bioluminescence, *66*, 226, 594
Biomass, 529
Biome
  and biogeographic realms, 547
  defined, 547
  major types, 550*ff.*
  map, *548*
  and productivity gradients, *550*
  soil profiles of, 547, *549*
Biorhythm, *454*
Biosphere
  defined, 4, *4*, 493, 543
  energy flow through, 4–5, *5*
  human impact on, 569*ff.*
Biosynthetic pathway, defined, 61
Biotic potential, 495
Biotin, *370*
Bipedalism, 214, 216–217
Birch, 299
Bird (Aves) (*See also* specific type)
  adaptive radiation, *200*, 201
  beak, *175*
  brain size, *282*
  comparative embryology, *190*
  digestive system, *361*
  evolution, 176, *176*, 286
  extinction, 200
  fossils, 176, *176*
  imprinting, 588–589
  and morphological convergence,
    193
  and morphological divergence,
    192, 193
  pollinator, *320*, 321
  relationships among, *192*
  respiratory system, *409*
  skeleton, *287*
  song, *586*, 587, *587*
  survivorship patterns, 498, *498*
  tectum, 436
  tropical rain forest, 555
  wing, 193, 201, 288
Bird of paradise, *185*
Birth control, 483–484, 501, 503
Birth process, 479, *479*
Biston betularia, 184, *184*
Bivalve, 271, *271*
Blackbird, 591
Black widow spider, *275*
Bladder

cancer, *415*
  urinary, *435*, 436
Blade, *298*
Blastocyst, *482*
Blastula, 465
Blister, *348*
Blood (*See also* Extracellular fluid;
  Interstitial fluid)
  agglutination, 386, *386*
  cellular portion, *376–377*, 377–378
  clotting, 61, *130*, 339, 377–378, 384
  coagulation, 384
  components of, *376*
  composition, 376–377
  contaminated, 400
  filtration, *421*, 421–422
  functions, 339, 376, *376*
  and kidney failure, 424
  oxygenated, 377
  plasma, 342
  plasma portion, *376*, 376–377
  types of, 386
  typing, 123
  viscosity, 339
  volume, 376–377, 384
Blood cell (*See* Red blood cell; White
  blood cell)
Blood circulation (*See also*
  Cardiovascular system;
  Circulatory system)
  cardiac cycle, 380, *380*
  during coitus, 473
  control of, 455
  coronary, 379–380, *379–380*
  flow rate, 375
  invertebrate, 374–375
  pulmonary, *378*, 378–379
  systemic, *378*, 378–379, 384
  vertebrate, 374–375
Blood pressure
  arterial, 381, 384
  capillary role in, 384
  high, *166*, 371, 381
  in hypertension, 381
  and hyperthyroidism, 455
  and inflammatory response, 399
  measurement, *385*
  and obesity, 381
  and smoking, 381
  and sodium retention, 422
  variations in, 384
  venous, 384
Blood typing, 123, 386
Blood vessel (*See also* specific type)
  constriction, 384
  damage repair, 384
  in intestinal villi, 365
  plaque in, *30*
  ruptured, 384
  structure, 375, 384, *384*
Bloom, dinoflagellate, 234
Blue-green alga (*See*
  Cyanobacterium)
Blue lobster, *275*
Blue whale, *526*
Body building, 31, 357–358, *357*
Body weight, human, 366–367, *367*,
  480
Bol, M., 453
Bond, chemical
  covalent, 22–23
  defined, 21
  hydrogen, 23, *146*, 310, *311*
  ionic, 22, *22*
  peptide, *154*
Bone (osseous tissue) (*See also*
  Skeleton)
  calcium and, *454*
  cell, 349, *350*
  characteristics, 349*ff.*
  compact, 349, *349*
  development, 350, *350*

epiphyseal plate, *350*
  functions of, 339, 348–349
  Haversian system, 349, *349*
  hormonal effects on, 452, *454*, 456
  long, 349–350, *350*
  organization, 339
  osteocytes, 350
  red marrow, 349, 377
  remodeling, 350
  spongy, 349, *349*
  structure, *349*, 349–350
  tissue turnover, 350
Bony fish, 423, 464
Booster, vaccine, 398
Boreal forest, *548*, 556, *556*
Boron, *306*, 307
Borrelia, 228
Bottleneck, in genetic drift, 180
Botulism, 87, 399
Bowman's capsule, 420, *421–422*
Bradycardia, 383, *383*
Brain (*See also* specific region)
  connections with spinal cord, 433–
    434
  early *Homo,* 218
  evolution, 215, 282, *282*
  flatworm, *265*
  forebrain, 436
  hindbrain, 436
  hominid, 216
  Homo erectus, 218
  H. sapiens, 218
  human, 212–213, 215, 437*ff.*
  information processing, 342–343,
    433–434
  midbrain, *435*, 436
  sagittal section, *437*
  size of, 215, 282, *282*
  vertebrate, 282, *282*
Brain scan, *20*
Breastbone, *287*
Breathing (*See also* Respiration)
  mechanisms, 412
  neural control of, 413, 416
Brenner, S., 152
Brinster, R., 168
Brittle star, 278, *278*
Broca's area, *438*
Bromeliad, 554, *554*
Bronchiole, 411
Bronchitis, 17, 414, *415*
Bronchus, *409*, 410, *410*, 411
Brown alga (Phaeophyta), *239*,
  245–246, *247*, 251, *561*
Brown recluse spider, *275*
Brown rot, 242
Bryophyte, *246*, 248, 248–249, 251
Bubonic plague, 497, *499*
Bud, vascular plant, 297–298
  axillary, *292*, 297
  defined, 297
  lateral, 297
  terminal, *292*, 297, *297*
Buffer, 26, 424
Buffon, G. de, 174
Bulbourethral gland, *468*
Bulimia, 368
Bulk, dietary, 366
Burgess Shale, *203*
Burning, fossil fuel, 535
Bursa, 352
Butter, 368
Buttercup, 514
Butterfly, 322

C

C3 plant, 77
C4 plant, 77, *77*
Cacao, *325*
Cactus, 316, 550

Caffeine, effects of, 383, 439
Calcitonin, *454, 456*
Calcium
    atomic number, *18*
    in bone, 231, 350, *454*
    and calcitonin, *454*
    deficiency in plants, *306*
    in extracellular fluid, 342
    in human nutrition, 370, *371*
    mass number, *18*
    in muscle function, 355, *356*
    and parathyroid hormone, 456
    and plant function, *306*
    role in cell, 26
Caloric requirements, human, 366–367
Calorie, defined, 366
Caltha, 321
Calvin–Benson cycle, *76,* 76–77
Camarhynchus, 175
Cambium
    cork, *294*
    vascular, *294, 296*
Cambrian, *198–199,* 202, *203*
Camouflage, 514, *514*
Camphor scale insect, *517*
Cancer
    AIDS-related, 400
    bladder, *415*
    cells, *151*
    characteristics of, 160–161
    and cigarette smoking, 160
    colon, 366
    defined, 396
    drug treatment of, 396–397
    epidermal skin, *348*
    herpes virus-related, *224*
    imaging, 397, *397*
    and immune system, 396–397
    laryngeal, *415*
    liver, anabolic steroid-related, 357
    lung, 299, 415, *415*
    mouth, 299, *415*
    and nuclear accidents, 584
    pancreatic, *415*
    papovavirus-related, *224*
    plasma membrane changes in, 160–
        161
    radiation therapy, 20
    skin, *348,* 400
    and translocation, 136
    treatment, 20, *166,* 396–397
    and viruses, 160
Candida albicans, 242, 244
Canine, 213
Cannabis sativa, 439, 440
Capillary
    and blood circulation, 379
    defined, 384
    lung, 411
    micrograph of, *376*
    structure, 384
Capillary bed
    blood flow through, 375
    functions of, 384, *384*
    lymph vessels near, *388*
    structure, 384, *384*
Capsella, 323
Capsule, bacterial, *225*
Carapace, 274
Carbohydrate
    abiotic formation of, 195, *195*
    absorption, 364–365
    defined, *35*
    dietary, 367, 370
    digestion, 364, *365*
    disaccharides, 28
    functions of, 28, *35*
    metabolism, 88–89, *89, 365,* 367,
        371–372, *372*
    monosaccharide, 28, *28,* 30, *35*
    oligosaccharide, 28
    polysaccharide, 29

storage by plants, 313
transport by plants, 313
Carbohydrate breakdown
    degradative pathways, 80*ff.*
    disorders, 456
    and glycolysis, 81–82, *82–84*
    hormones and, *454,* 455
    and Krebs cycle, 85, 85*ff.*
Carbon
    and cell function, 69
    and cell structure, 68–69
    electron distribution, *21*
    numbering in sugar ring structures,
        145
    and plant function, 307
Carbon compound (*See also* Organic
        compound)
    assembly, *26*
    covalent bonding, 27
Carbon cycle, 534–535, *534–535,*
        536–537
Carbon dioxide
    acquisition by plants, 307
    atmospheric concentration, 534–
        535, 536
    diffusion across membranes, 42
    elimination from animal body, 419
    exchange in alveoli, 412
    fixation, 76, *76,* 77
    and global warming, 534–535, 536–
        537, *537*
    and hemoglobin, 416
    integumentary exchange, 407
    in Krebs cycle, 85
    partial pressures in body, 412, *413*
    and photosynthesis, 69, *71*
    in respiratory system, 406–407, 412
    stomatal control of, 299, 310, 332
    transport by blood, 412–413
    use by autotrophs, 525
Carbonic acid, 27, *27,* 413, 424
Carboniferous, *198–199,* 202, 204
[14]Carbon isotope, 19
Carbon monoxide, *570*
    in cigarette smoke, 381
    poisoning, 416
Carboxypeptidase, *365*
Carcinogen, defined, 160
Cardiac cycle, 380, *380*
Cardiac muscle tissue, *339,* 341, 353
Cardiovascular disorders, 381–383
        (*See also* Heart disease;
        specific disorder)
Cardiovascular system, 378*ff.* (*See
        also* Blood circulation;
        Circulatory system)
Carnivore, *527, 529*
    adaptive radiation, *200*
    defined, 525
    teeth of, 213
    and trophic levels, *526*
Carnivorous plant, *308*
Carotene, *73,* 348
Carotenoid, 71, 73, *73,* 247
Carp, *517*
Carpal, *351*
Carpel, *118,* 317, *317*
Carroll, L., 499
Carrot, 299
Carrying capacity, 495–496
Cartilage
    in growth and development, 350
    hormonal effects on, 452
    of intervertebral disk, 350
    at joint, 350, *350,* 352
    model, embryonic, *350*
    in vertebrate skeleton, *351*
Cartilaginous fish, 283–284, *284*
Cartilaginous joint, 352
Casparian strip, *308,* 309
Cataract, 574
Caterpillar, 540, 590, *591*

Catfish, 285
Cattail, 295
Cattle, *32, 361,* 580
Caudal fin, *284*
cDNA (*See* Complementary DNA)
Cecum, 366
Cell (*See also* specific cell type)
    cytoplasm (*See* Cytoplasm)
    cytoskeleton, 46, 52–54, 97, 160
    death, 467
    defined, 3, *4*
    differentiation, 158
    environmental pH, 26
    germ, 95, 105, *105*
    interior pH, 26
    junction, cardiac, 380, *380*
    membrane (*See* Membrane, cell)
    nucleus (*See* Nucleus, cell)
    origins of, 196, *196*
    reproduction, 99*ff.*
    size, 38
    somatic, 95
    water in functioning of, 24–25
Cell cycle, 97, *97*
Cell differentiation
    in animal development, 465–466
    in epidermal tissue, 347, *347*
    plant, 318–319, *319,* 323*ff.,* 331
Cell division
    and cancer, 160–161
    eukaryotic, 95, *96*
    hormonal stimulation of, 452
    and life cycles, 94–95, 110–112, 159
    mechanisms, 95*ff.*
    microtubule assembly during, 53
    in plants, 328, 331
    prokaryotic, 95
    rate of, 159
    rule for, 94
    signals for, 97
    in skin, 347
    spindle apparatus for, 97, *98–99,*
        100
Cell junction, 55
Cell-mediated immune response, *392,*
        393, 396, 399, *403*
Cell plate formation, 101
Cell theory, 38, *38–39*
Cellular slime mold, 232, *232*
Cellulose, 29
    in cell walls, 55
    digestion, *361*
    formation, *76–77*
    plates, 234
    structure, 29, *29*
Cell wall
    appearance in micrographs, *47*
    bacterial, 45, *45,* 55, 225, *225*
    functions of, *47,* 55
    fungus, 55
    plant, 55, *55,* 101
    protistan, 55
    structure, 55
    turgor pressure, 43
Cenozoic era, 191, *198–199,* 200, 201,
        204, 208
Centipede, 276, *276*
Central nervous system, 436*ff.*
    and autonomic system, *435*
    psychoactive drugs affecting, *439,*
        439–440
    and somatic system, *435*
Central vacuole
    appearance in micrographs, *47*
    defined, 47
    functions of, 52
Centriole, *98,* 100, *100*
    functions of, 54, *54*
    structure, 54, *54*
Centromere, 95
Cephalization, 259–260
Cephalopod, 271

Cerebellum, 436, *437*
Cerebral cortex, 436–438, *438*
Cerebral hemisphere, 437, *437–438*
Cerebrum, 436, *437* (*See also*
        Cerebral hemisphere)
Cervix, 470, *470,* 473, *474,* 479
Cestode, 266
CFC (*See* Chlorofluorocarbon)
Chain elongation (protein synthesis),
        154, *154–155*
Chameleon, *287*
Channel protein, 40, *40, 75*
Chaparral, 551
Chase, M., *144*
Cheek teeth, 213, 217
Cheetah, 180
Chelicerate (Chelicerata), 273–274
Chemical bond (*See* Bond, chemical)
Chemical equation, *18*
Chemical reaction (*See also*
        Metabolic reaction), 60, *60*
Chemical signaling, 593
Chemical synapse, *431,* 431–432
Chemiosmotic theory, 75, *86*
Chemoreceptor, 441–442, *442*
Chemosynthesis, 538
Chemosynthetic autotroph, 69,
        225–226, *226, 527*
Chemosynthetic bacterium, 227–228,
        563
Chemotherapy, 396
Chernobyl, 584
Cherry (*Prunus*), *317,* 324
Chestnut blight, 242
Chestnut blight fungus, *517*
Chicken, embryo, *466*
Chickenpox, *224*
Childbirth, 342, 352, 479
Chimpanzee, *106, 212,* 216
Chin fissure, *104*
Chipmunk, 510, *510*
Chitin, 240, 272, 408
Chlamydia, 226, 487
Chlamydomonas, 247
Chlorella, 247
Chlorine
    atomic number, *18*
    deficiency in plants, *306*
    electron distribution, *21*
    in human nutrition, *371*
    intestinal absorption, 365
    mass number, *18*
    and plant function, 307
Chlorine monoxide, 574
Chlorofluorocarbon (CFC), 536, *537,*
        570, 574
Chlorophyll, 71, *73,* 247
Chloroplast
    appearance in micrographs, *47*
    ATP formation, 71, *71, 73*–74, *74–75*
    color of, 50
    of *Euglena, 233*
    evolution, 230
    functions of, 46, *46,* 52, 70–71
    origin, 230
    photosynthesis in, 70*ff.*
    structure, 52, *53,* 70–71
Chocolate, 439, *439*
Cholecystokinin, *454*
Cholesterol
    in bile, 364
    and cardiovascular disorders, 381
    deposits, arterial, 381–382
    dietary, 368
    effects of anabolic steroids, 357
    functions of, 31, *35*
    plaque formation and, 382–383
    role in atherosclerosis, 31
Choline, 370
Chordate (Chordata)
    body plan, *280,* 281
    evolution, *261, 280*

invertebrate, 281
representative, *259*
Chorion, 476, *476*, 477
Chorionic villi sampling (CVS), 140
Choroid, 445, *445*
Chromatid
　defined, 95
　in meiosis, *106*, 106–110
　in mitosis, 95, *98–99*, 100
Chromatin, appearance in
　　micrographs, *49*
Chromosomal abnormality
　and human genetic disorders, 132,
　　132, 134–135, 135–136
　numerical, 137–139
　prenatal diagnosis, 140
　structural, 135–139
　types of, 132–139
Chromosome number
　chimpanzee, *106*
　corn, *106*
　diploid, 96
　disorders related to, 137–139
　earthworm, *106*
　Equisetum, 106
　examples of, 96
　fate during meiosis, 96, *106*, 122,
　　122
　and fertilization, 106
　fruit fly, *106*
　haploid, 96
　Homo sapiens, 106
　horsetail, *106*
　and nondisjunction, *136*, 137–138,
　　138
　Rana pipiens, 106
　Zea mays, 106
Chromosome(s)
　appearance in micrographs, *96–97*,
　　100, 108
　autosomes, 129
　bacterial, 163, 226
　crossing over, 107, 131, *132*
　defined, *49*, 95
　diploid number, 96
　DNA organization in, 148
　duplicated, 95–96, *96, 98*
　haploid number, 96
　homologous, 96, *98*, 105, 118–119
　human, *96, 108*
　karyotype, 129
　maternal, 109
　in meiosis, 96, 106*ff.*
　in mitosis, 96, 97*ff.*
　paternal, 109
　proteins of, 95
　shufflings at meiosis, 109, *122,*
　　122–123
　structural disorders, 135–136
　structure, *49*, 95–96, 129
　unduplicated, 95, *96*
　X, 96, 106–107, 129
　Y, 96, 106–107, 129
Chrysaora, 262
Chrysophyte (Chrysophyta), *229, 233*
Chyme, 363–364
Chymotrypsin, 365
Chytrid, *240*, 240–241, *241*
Cichlid, 516
Cigarette smoke, 381
　effects of, 414, *415*
　risks related to, *415*
Cigarette smoking (*See also* Smoking;
　　Tobacco smoking)
　and cancer, 160
　effects during pregnancy, 481
　and lung cancer, 415, *415*
　risks related to, 414–415, *415*
Ciliated protozoan, *229, 236, 236*
Ciliophora, *229*
Cilium
　bronchioles, 415

of ciliated protozoans, 236
epithelial, 411, *411*
functions of, 53
of gastropods, 270
lung, 53
organization, 54, *54*
of *Paramecium, 236*
structure, 53, *54*
Circadian rhythms, plant, 329, *329*
Circulatory system (*See also* Blood
　　circulation; Cardiovascular
　　system)
　amphibian, *282*
　arthropod, 375
　closed, 375, *375*
　components of, 374–375
　coronary, 379
　design of, 375
　disorders of, 368
　fish, *282*
　flow rate in, 375
　functions, 340
　human, *340*, 378*ff.*
　invertebrate, 374–375
　and lymph vascular system, 375
　mammal, *282*
　mollusk, 375
　open, 375, *375*
　pulmonary circuit, *378*, 378–379
　systemic circuit, *378*, 378–379
　vertebrate, 374–375
Citrate, *85*
Cladonia, 244
Cladophora, 72
Clam, 271, *271*
Classification
　of animals, *209*
　of bacteria, 7, *8*, 226
　and comparative biochemistry, 193
　and comparative morphology, 191–
　　193, *192*
　five-kingdom, 7, *209*
　of fungi, 7, *8–9, 209*, 240, *241*
　of monerans, *209*
　phylogenetic, 208–210, *209*
　of plants, 7, *8–9, 209*
　of protistans, 7, *8, 209, 229*
　of viruses, *224*
Class (taxon), *209*
Claviceps purpurea, 242
Clavicle, *351*
Cleavage
　defined, 464–465
　events of, 464–465
　human embryo, *474*
Cleavage furrow, 101, *101*
Cleft lip, 139
Climate
　and air circulation, 544, *545*
　atmospheric effects, 543–544, *544*
　global patterns of, 543*ff.*
　and ocean current, *546*
　oscillations in, 566–567
　seasonal variations in, 544
　shifts in, 186
　topography effects, 546
Climate zone, *545*
Climax community, 518
Clitoris, 470, *470*
Cloaca, *266, 361*
Clonal selection theory, 396, 398, *398*
Clone, 104, 160
Cloned DNA, 164–165, *165*, 165–166,
　402
Cloned genes, expression of, 166–167
Clostridium, 226, 228, 432
Clotting, blood, 339, 377–378, 384, 393
Club fungus, *241, 242–243, 243*
Club moss, 249

nematocysts, 264, *264*
phylogeny, *261*
representative, *259, 263–264*
reproduction, 264
Coagulation, blood, 384
Coal, 580–582
Coastal ecosystem, 560
Coat color, *63*, 123 (*See also* Fur; Hair)
Cocaine, effects of, 299, 439, *439*
Coccus, 225, *226*
Coccyx, *174, 351*
Cochlea, *443*
Cocklebur, 330, 333
Cockroach, 540
Coconut, 294
Codeine, 439, *439*
Codfish, *282*
Codominance, 123
Codon, 153, *153*
Coelacanth, *283*, 285
Coelom, *260*, 261
Coelomate, *260*, 261, *261*
Coenzyme, nucleotide
　and electron transfers, 82, *82, 84*,
　　85, *85–86*
　functions of, 34, *35*
　and self-replicating systems, 196
Coenzyme A (acetyl-CoA), *85*, 432,
　434
Coevolution
　defined, 321
　flowering plants and pollinators,
　　320–322
　reciprocal selection pressure in, 513
Cofactor, 61, 64
Coffee, 439, *439*
Cohesion, defined, 24
Cohesion theory of water transport,
　310, *311*
Coitus, 473
Cold (infection), 224, 414
Cold sore, *348*
Coleoptile, *323, 326, 328, 328*
Coleus, 297
Collagen, *35, 384*
　age effects, 482
　in connective tissue, 337–338, 348
　in extracellular matrix, 55
Collar cell, *262, 263*
Collenchyma, 293
Colon, 366 (*See also* Intestine, large)
Colon cancer, 366
Color
　chloroplast, 50
　corn, 157
　perception, 447
　and predation, 513
　skin, 348
　snail shell, *177*
Color blindness, 141
Coloration, warning, 513, *513*
Comb jelly, *66*
Color blindness, 141
Commensalism, 508, *508*
Communication
　acoustical, 594
　via bioluminescent signals, 594
　chemical, 593
　defined, 592
　signals, 593
　tactile, 594, *595*
　visual, 593–594
Communication junction, cardiac,
　380, *380*
Community
　climax, 518
　cyclic replacements, 520
　defined, 4, *4*, 493, 506
　feeding levels, 507–508
　habitat, 507
　patterns of diversity, 507, 520–521
　resource partitioning, *514*, 515
　and species introduction, 517

Companion cell, *313*
Comparative biochemistry, 193
Comparative morphology, 191–193
　analogous structures, 193
　homologous structures, 191–193
　stage of development, 191
Competition
　exploitative, 509
　interference, 509
　interspecific, 508, *508*, 509
　predation effects, 516
Competitive exclusion, 509–510
Complement, 391, 402
Complementary DNA, 165, *165*
Compound eye, 432, 444, *444*
Concentration gradient, 41
Condensation reaction, 27–28, *28*
Condom, 484, *484*
Cone photoreceptor, 445, *446*, 447
Conifer, 251, 252–253, *253*
Coniferous forest, *548–549*
Conjugation, 164, *164*
Connective tissue
　adipose, 338, *338*
　categories of, 337
　dense, 337, *338*, 348
　fatty, 337–338, *338*
　ground substance, 337
　of ligaments, 352
　loose, 337–338, *338*
Conservation of mass, *18*
Consumer
　defined, 523
　examples of, *526*
　and trophic levels, 526–527
Continental drift, 197, 201
Continental shelf, 562, *562*
Continuous variation, 125, *125*
Contraceptive methods, 483–484,
　501, 503 (*See also* Birth control)
Contractile vacuole, *236*
Contraction, muscle
　ATP and, 354–355
　and blood return, 385
　and calcium ion, 355
　and childbirth, 342, 479
　digestive system, 362
　heart, 380, *380*
　hormonal effects on, 342
　mechanism of, 353–355, *354–355*
　skeletal muscle, 353–355, *354–355*
　sliding-filament model, 354, *355*
　stomach, 363–364
　uterine, 479
Controlled variable, 11
Convergence, *192*, 193, *543*
Cooksonia, 204
Cooperative society, 596
Copepod, *275, 276*, 558
Copper, *306, 307*, 371
Coral, *263*, 565
Coral fungus, *243*
Cork, 303, *303*
Cork cambium, *294*, 303
Corliss, J., *563*
Corn (*See also Zea mays*)
　chromosome number, *106*
　developmental stages, 326
　fungal pathogens, 77
　grain of, *323*
　kernel color, 157
　leaf epidermis, *294*
　as protein source, 369
　root tip, *300*
　stem, *294, 296*
Cornea, 444, 445, *445, 466*
Corn oil, 368
Coronary artery, 357, 379, *379,*
　381–382, 383, *415*
Coronary bypass surgery, 382–383,
　383
Coronary circulation, 379

Corpus luteum, *471–472, 473,* 477
Cortex
  kidney, 420, *420*
  root, 300, *301*
  stem, *296,* 297
Corticotropin-releasing hormone, 455
Cortisol, *454,* 455, *455,* 458
Cottonwood, 550
Cotyledon, *292,* 295, *295, 323, 323,*
  327
Courtship behavior, 464, *593*
Covalent bond, 22, 23
Cowpox, *224,* 390–391, 398
Crab, *275*
Crabgrass, 77
Crab spider, *514*
Crack cocaine, 439
Cranium, *351*
Crayfish, *258–259*
Creosote, 550, *551*
Cretaceous, *198–199,* 204–205, 206
Crick, F., 145–146, 152
Cri-du-chat, 135
Crocodilian, 286, *286*
Crop bird, *361*
Crop, plants
  algae as, 247
  and dormancy, 333
  examples, *325,* 187, *187, 325*
  and fertilizers, 307
  fungal attack on, 242
  and genetic engineering, 163, 168
  and global warming, 536
  and green revolution, 576–577
  and insects, 16–17, 277
  and irrigation, 168, 574, 549
  salt-tolerant, 168, 549
  and soils, 168, 549
  viral diseases, 225
  viroid effects, 225
Cross-fertilization, 118, *118*
Crossing over
  and genetic recombination, 107, *132*
  importance of, 107
  and linkage, 131, *132*
  mechanisms of, 107, *107*
Crossopterygian, 284–285
Crown gall tumor, 168, *168*
Crustacean, 273–274, *275,* 276, 444,
  444, 445, *445*
Cryphonectria parasitica, 242
Crystallin, 466
Cuckoo, 587–588, *588*
Cucumber leaf, *312*
Culture, defined, 215
Cuspid, 363
Cuticle
  insect, 273
  plant, 30, 294
  roundworm, 266
Cutin, 30, *35, 55*
CVS (chorionic villi sampling), 140
Cyanobacterium, *226,* 227–228, *228,*
  229, *561*
Cycad (Cycadophyta), *251, 252, 252*
Cyclic AMP (cyclic adenosine
  monophosphate), 34, 459, *459*
Cyclic metabolic pathway, 60, *61*
Cyclic photophosphorylation, 74, *75*
Cyclic replacement, 520
Cyst, 235, *268*
Cysteine, *31, 153,* 368
Cytochrome, 64
Cytokinesis, 95, 101, 106
Cytokinin, *327,* 328, 333
Cytomembrane system, *50,* 50–52
Cytoplasm
  bacterial, 45, *45, 225*
  of cancer cell, 160
  and cytokinesis, 95, 101, 106
  defined, 38
  egg, localized differences, *466*

functions of, 38
Cytosine, 145–147, *145–147,* 151
Cytoskeleton
  of cancer cell, 160
  components of, 52
  functions of, 46, 53
  organization of, 54, 100
  permanent parts, 53–54
  structure, 46, 52, 53
  transient parts, 52–53

D

2,4-D, 328
Dandelion, *299*
Darwin, C., *598*
  and blending theory, 117
  development of evolutionary theory,
    8–9, 172*ff.,* 542
  portrait of, *172*
Daughter cell, 95
Daylength, and flowering, 330, *330,*
  332
Day lily, *318*
DDT, 539–540
Death, aging and, 482
Deciduous broadleaf forest, 548–549,
  554, 556
Decomposer
  bacteria, 4
  defined, 525
  fungi, *8,* 240
  role in biosphere, 240, 525
Deer, 228, *361, 527*
Deerfly, *432, 444*
Deforestation, 498, *533,* 534, 537,
  577, *577–578,* 578–579
Degradative pathway, 60–61
Dehydration synthesis (*See*
  Condensation reaction)
Deletion, chromosomal, 135
Deltoid muscle, *353*
Demographic transition model, 501,
  501
Denaturation, protein, 33
Dendrite, 427, *427,* 436
Denitrification, 538, *539*
Dense connective tissue, 337, *338,*
  348
Density, population, 494, 496–498
Dentition, 213–215, 217
Deoxyribose, 28, *145,* 146
Dependent variable, 11
Depressant, *439,* 439–440
Depression, 357, 439–440
Depth perception, 214
Dermis, animal
  component of skin, 346, *347*
  defined, 346
Desert
  biome, *548*
  plants, 550
  soil, *549*
Desertification, 498, 550, 580, *581*
Desulfovibrio, 226
Detrital food web, 527, 529
Detritivore, 525
Deuterostome, 269, 278
Development
  controls over, 157–159
  differentiation during, 158–159
  selective gene expression during,
    158–159

Diaphragm (contraceptive), 483
Diaphragm (human body), 410,
  411–412, *412, 484*
Diarrhea, 224
Diastole, 380
Diatom, 37, *37, 229,* 233, *233, 527*
Diatomaceous earth, 233
Dicot, *251*
  adaptive success of, 254
  floral structure, 254
  function, 295
  leaf, *295,* 298
  vs. monocot, *295*
  root system, 299
  seed of, 254
  species of, 254
  stem, *295–296*
  structure, 295
  vascular bundle, 295, *295–296,* 297
Dictyostelium discoideum, 232, *232*
Didinium, 237
Diet
  carbohydrates in, 367
  and diabetes, 456
  and dieting, 368
  and emphysema, 414
  lipids in, 367–368
  and mental retardation, 369
  and pregnancy, 480
  protein-deficient, 369
  proteins in, 367–368
  salt intake, 381
  vegetarian, 369, *369*
Dietary bulk, 366
Dieting, 368
Differentiation
  during animal development, 158–
    159
  and selective gene expression, 158–
    159
Diffusion
  across cell membrane, 41–42
  in capillary bed, 375
  concentration gradient and, 41
  defined, 41
  ethanol across membranes, 42
  facilitated, 43, *43*
  in gills, 408
  in human respiratory system, 412,
    412, 413
  and integumentary exchange, 407
  rate of, 413
  through ground substance, 55
Digestive system (*See also*
  Gastrointestinal tract)
  complete, *361*
  components of, 362*ff.*
  control of, 362
  earthworm, *272*
  and endocrine function, 364–365,
    365
  examples of, 361
  functions, *341,* 361
  human, *341,* 362*ff.*
  incomplete, *361*
  innervation, *435*
  insects, 277–278
  links with respiratory system, 360,
    360–361
  links with urinary system, 360, 361
  mollusks, 269
  rotifer, *269*
  ruminant, *361*
Digitalis, 299
Diglyceride
  functions of, 30
  structure, 30
Dihybrid cross, 122, *123*
Dinoflagellate, *229,* 234, *234*
Dinosaur, *199–200,* 204–207
Dioxin, *570*
Diphtheria, 399

Diploid state
  in animal life cycle, *110*
  defined, 96, 118
  reduction to haploid in meiosis, 96,
    106, *106*
Directional selection, *181,* 184, *184*
Disaccharidase, 365
Disaccharide, 28, *28*
Disease
  protozoan infection, 234–235
  sexually transmitted, 235, 400–401,
    485–487
  sporozoan infection, 235
Disorder, defined, 132
Dispersion, population, 494, *494,* 508
Disruptive selection, *181,* 184
Distal tubule, 420, *421–422*
Disulfide bridge, *32*
Divergence, 191, *192,* 193
  defined, 186
  and speciation, 185–186, *186*
Diversity of life
  classification scheme, 7
  defined, 9
  and tropical forests, 578–579
  evolutionary view of, 7–9
  molecular basis of, 147
  species diversity, 520–521
  of tropical reef, *565*
  unity and, 7, 147
Division of labor, in multicellularity, 231
Division (taxon), 7
DNA (deoxyribonucleic acid)
  and aging, 482
  amplification, *165,* 165–166
  bacterial, 38, 45, *45,* 163–164
  and biological organization, 3–4
  blueprints, 3, *3*
  cDNA (*See* Complementary DNA)
  and cell function, 94–95
  and chromosome structure, 95–96
  cloned, 164–165, 402
  coiled, *45*
  components of, *145,* 145–146
  condensed form, 49
  double helix, 95–96, 146–147, *146–*
    147, 147
  eukaryotic, 95–96
  eukaryotic cells, 49
  evolution, 196
  function, 94–95, 143–145
  functions of, 34, 45
  hybridization, 165, *165,* 193, *193*
  organization in chromosomes, 95–
    96, 148
  and organization of life, 91
  plasmid, 163–164
  prokaryotic, 94–95
  and protein synthesis, 150*ff.*
  proto-oncogenes of, 160
  restriction fragment length polymor-
    phisms, 166
  role in cell reproduction, 94–95
  and self-replicating systems, 195–
    196
  sperm, 468
  structure, 34, 95–96, 145–147, *146–*
    147
  threadlike, 49
  transposable elements, 157
  and unity and diversity of life, 147
  viral, *144,* 160, 223, *223*
DNA library, *165,* 165–166
DNA ligase, 165
DNA polymerase, 147–148
DNA replication
  assembly of nucleotide strands, 147
  bacterial, 226
  defined, 96
  enzymes of, 147–148
  prokaryotic, 95
  time of, 97, *97*

viral, 223, *223*
Dogwood, *297*
Dominance, apical, 328
Dominance, genetic
 codominance, 123
 defined, 123
 incomplete, 123
 in dihybrid crosses, 122–123, *123*
 in monohybrid crosses, 119–121, *121*
 vs. recessiveness, 118
 and variable expressivity, 123
Dominance hierarchy, 596–597
Dominant allele, 134
Double fertilization, 318–319, *319*
Double helix, DNA, 146–147, *146–147*, 147, *147*
Douching, 483, *484*
Douglas, K., *104*
Douglas fir, *332,* 333
Down syndrome, *137,* 137–138, 140
Dragonfly, 274, *276*
Drosophila melanogaster
 breeding experiments, *130–131,* 131, *132*
 chromosome number, *106*
 chromosomes, 130, *130, 130–131,* 131, *131*
 eye color, *130–131,* 131
 linked genes, *130–131*
 mutants, *130–131*
 photographs of, *130*
Drought, 68, 580
Drug(s)
 abuse, 299, 439–440
 from plants, 299, 579
 and pregnancy, 480
 psychoactive, 439–440
 use, and AIDS, 485
Drug therapy, 163, 396–397
Dry acid deposition, 571
Dryas, 518
Dryopith, 216
Duckbilled dinosaur, 205
Duck-billed platypus, *288*
Duck louse, *276*
Duodenum, *363*
Duplication, chromosomal, 135
Dust Bowl, *580*
Dutch elm disease, 242, *517*
Dwarfism, pituitary, *166,* 452, *453*
Dyhybrid cross, defined, 122
Dynamic equilibrium, 60, *61*

E

Ear, 442, *443*
Early *Homo,* 218, *219*
Early wood, 303
Earth
 early, 194–195, *194–195*
 evolution, 194*ff.,* *198–199,* 202*ff.*
 primordial, *194–195*
Earthworm, *106,* 272–273, *375,* 407 (*See also* Annelid)
Earwax, *35*
Earwig, *276*
Eating disorders, 368
Echinocereus, 542
Echinoderm (Echinodermata)
 body plan, 279, *279*
 characteristics, 279
 evolution, *261*
 representative, *259, 278*
Ecology, defined, 493
Ecosystem
 aquatic, 558*ff.*
 biogeochemical cycles, 531*ff.*
 and climate, 543*ff.*
 consumers (*See* Consumer)
 decomposer (*See* Decomposer)
 defined, 4, *4,* 493, 525

harmful compounds in, 539–540
 hydrothermal vent, 562, *563*
 inputs and outputs, 525
 intertidal, 560, *561*
 marine, 559*ff.*
 model of nutrient flow through, *531*
 as open system, 525
 primary productivity, 528
 producer (*See* Producer)
 and rainfall, 544
 structure, 525*ff.*
 trophic levels, 525, *526,* 527, *527*
 trophic structure, 525, *526,* 527, 529–530, *529–530*
Ecosystem analysis
 defined, 540
 at Hubbard Brook, 533, *533,* 534
 at Silver Springs, *530*
Ectoderm, 465
Eel, 285
Effector, in homeostatic control, 343, *343*
Egg (*See also* Oocyte; Ovum)
 amniotic, 286, *286*
 amphibian, 285
 animal, *111,* 112
 bird, 286
 blood fluke, 267, *267*
 cuckoo, 587–588
 flowering plant, 316, 318
 fluke, 267, *267*
 formation, 95, 110, *111,* 112
 frog, *463–464*
 human, 464, 475
 moth, *6*
 pinworm, 267
 plant, 316–317, *317*
 as protein source, *369*
 reptile, 286
 rotifer, *269*
 tapeworm, 266, *268*
 Wuchereria bancrofti, 268
 yolk, 464
Ejaculation, 473
Ejaculatory duct, *468,* 469
EKG, 382
Elastin, 337, 338, 348, *348*
Electrocardiogram, 382–383
Electron
 negative charge of, 18
 orbitals, 21
 shells, 21
 transfer, *65,* 65–66, *66,* 71, 73
Electron transport phosphorylation, 81, 86, *86,* 89
Electron transport system
 defined, 65
 and electron transport phosphorylation, 86, *86*
 functions, 65–66
 in photosynthesis, 71, *71,* 73–74, *74*
Element(s), chemical
 abundance in human body, 17, *17*
 atomic number of, 17, *17*
 essential, for human nutrition, 370–371, *371*
 essential, for plant nutrition, *306,* 307, *307*
 mass number for, 17, *17*
 symbol for, 17
Elephant, 193, *200*
Elephantiasis, 268, *268*
Elk, 591
Elm, 240, 316
El Niño, 562, 566–567
El Niño Southern Oscillation, 566–567
Elodea, 72
Embolus, 382
Embryo, animal
 amniotic egg, *286*
 development, *477,* 477–478, *478,* *481*

mammalian, 288
Embryo, plant, 318, *323,* 323–325, *325* (*See also* Plant development)
Embryology, comparative, *190*
Embryonic disk, *474–475,* 476, *476*
Embryo sac, 318, *319*
Emotional states, 438–440
Emperor penguin, *526*
Emphysema, 414, *414*
Endergonic reaction, 60, *60*
Endocrine gland, 337
Endocrine system (*See also* specific gland)
 control of digestive system, 362
 defined, 450
 functions, *340*
 hormones, summary of, *454*
 human, *340*
 links with nervous system, 450, *450*
Endocytosis
 in an amoeba, *44*
 receptor-mediated, 458
 vesicle formation during, 44, *44*
Endoderm, 300, 465
Endodermis, 300, *301*
Endometriosis, 473
Endometrium, *471–472,* 473, *474*
Endoplasmic reticulum
 appearance in micrographs, *47–48*
 functions of, 46, *47–48, 50*
 rough, 51, *51*
 smooth, 50, *51*
Endoskeleton, 285, 288 (*See also* Skeleton)
Endosperm, *255,* 318–319, *319, 323*
Endospore, *226,* 228
End product, defined, 61
Energy
 conversion, 59
 defined, 58
 hill, 62
 and laws of thermodynamics, 58–59
 light, in photosynthesis, 59, 69, *71*
 net, 580–581
 nuclear, 582, 584
 pyramid, 529–530
 quality of, 59
 sources of, 59
 staircase, 66, 74
 stores, 580
Energy carrier, 61
Energy flow (transfer)
 annual, *530*
 and organization of life, 4
 in photosynthesis, 5
 pyramid, 529–530, *530*
 through biosphere, 4–5, *5*
 through ecosystems, 528–529, *528–529*
Enerobius vermicularis, 267
Englemann, T., *72*
ENSO, 566–567
Entamoeba histolytica, 235
Entropy, defined, 59
Environment
 and behavior, 588–591
 cellular, 342
 effects of human population growth, 569*ff.*
 and evolution, 194*ff.*
 and gene expression, 125
 internal, of animals, 336–337, 342–343, 423
 internal, of fish, 423
 and phenotype, 125
 plant adaptations to, 307–309
 plant responses to, 328*ff.*
Enzyme (*See also* specific enzyme)
 and activation energy, *62,* 62–63
 active site of, 62
 activity, in cancer, 160

allosteric, 63–64
 bone cell secretion, 350
 control over activity, 63–64
 defined, 61
 digestive, 362, *365*
 of DNA replication, 147–148
 fat-degrading, 364–365
 feedback inhibition, 64, *64*
 functions of, 62–63
 heat sensitive, 179
 induced-fit model, 62, *62*
 inhibitors, 63–64
 lysosomal, 394
 nitrogen-fixing, 228
 and origins of life, 196
 pancreatic, 364
 pH effects on, 63
 for plant growth, 330
 restriction, 164–165, *165,* 166
 role in metabolic pathway, 61–64
 starch-degrading, 363
 stomach, 363
 structure of, 62–63
 substrates for, 61
 temperature effects on, 63
Enzyme-substrate complex, 62, *62*
Eocene, *198–199,* 215
Eosinophil, *377,* 378, 391, *403*
Epidermis, animal
 component of skin, 346, *347*
 defined, 294, 346
 melanin-producing cells, *348*
 rotifer, *269*
Epidermis, plant
 defined, 294
 examples of, *294, 296*
 formation, 294
 root, *292,* 300, *301*
Epidermophyton, 242
Epididymis, *468,* 469
Epiglottis, 410, *410*
Epinephrine, *454,* 455
Epiphyseal plate, *350*
Epistasis, 123, 125
Epithelium, 264, 337, *337–338*
Epstein-Barr virus, *224*
Equilibrium
 chemical, 60, *61*
 genetic, 179
Equisetum, 106, 183, 249, *249*
Ergotism, 242
Erythroblastosis, 386
Erythrocyte (*See* Red blood cell)
Erythropoietin, *166, 454*
Escherichia coli
 body plan, *45*
 chromosome, 163–164
 classification, *226*
 conjugation, 164, *164*
 gene activity in, 157–158
 lactose metabolism, 157–158
 lactose operon, 157–158, *158*
 micrographs of, *163*
 plasmids, 163–164
 toxins produced by, 229
 usefulness of, 229
 vitamins produced by, 229
Esophagus, 363
Essential amino acid, 368–369, *369*
Essential fatty acid, 368
Estrogen
 and ovarian function, 470, *472*
 and the pill, 484
 and protein synthesis, 458
 and song system of birds, 587
 source of, *454*
 targets of, *454*
 and uterine function, *472,* 473
Estrous cycle, 470–471
Estuary, 560, *560*
Ethanol, 42, 88
Ethiopian realm, *547*

Ethylene, *327*, 328–329
Ethyl group, *26*
Eubacterium, 227–229, *226*
Eucalyptus tree, *299*
Eugenic engineering, 169
Euglena (Euglenophyta), 229, 233, 233
Euglenid, *229, 233*
Eukaryote
  origin and evolution, 230–231
  vs. prokaryotes, 230
Eukaryotic cell (*See also* specific
    organelle)
  cell cycle, 97, *97*
  cell division, 95*ff.*
  chromosomes of, 95–96
  components of, 46
  defined, 45
  DNA, 49, 95–96
  origins of, 229–231
  photosynthetic, 46
Euphorbia, 543
European starling, *517*
European wild boar, *517*
Eutrophic lake, 559
Evaporation
  defined, 24
  energy required for, 24
  from watershed, *532*
  and water-solute balance, 423
Evergreen, 299
Evergreen broadleaf forest, *548,
    553–556*
Evergreen coniferous forest, *548*, 556
Evolution, biological (*See also*
    Macroevolution; Microevolution)
  of algae, *199*, 202
  of amphibians, 285
  of animals, 258*ff., 260–261*, 418,
    434*ff.*
  of bacteria, *199*
  of birds, 286
  and convergence, *192*, 193
  correlated with earth history, 194*ff.,
    198–199*
  correlated with Earth's history, 172
  and divergence, 191, *192*, 193
  DNA hybridization studies, 193
  effects of atmospheric oxygen, 202
  of fish, *199*, 283*ff.*
  of flatworms, 266
  of flowering plants, 204
  gradual model of, *208*
  Hardy–Weinberg principle, *178*, 179
  of hominids, 218–219
  and immunological comparisons,
    194
  of mammals, *199, 288*
  and mutation, 191
  parallel (*See* Convergence)
  photosynthetic pathways, 90–91
  of plants, *199*, 245–246
  of plasma membranes, 196
  and population size, 179–180
  of primates, 213–218
  principle of, 172*ff.*
  of prokaryotes, 230
  punctuational model of, *208*
  and regulatory gene mutations, 191
  of reptiles, 285–286
  and self-replicating systems, 195–196
  of species (*See* Speciation)
  symbiosis in, 230
  synthesis of biological molecules,
    195
  through gene flow, 179, *179*, 180
  through genetic drift, 179, *179*, 180,
    180
  through mutation, 179, *179*, 179–
    180, 193
  through natural selection, 7–9,
    105, 117, 175–176, 179, *179*,
    180–185

  time scale for, 174, 193, *198–199*
  and variation in populations, 105,
    117, 175–177
  of vertebrates, 281–283, 458
Evolution, chemical, 194–196, *199*
Evolution, human
  australopiths, 217, *217*
  behavior, 215
  brain, 215
  capacity for learning, 215
  cultural, 215, 219, 582–583
  dentition, 214–215
  early *Homo*, 218, *219*
  hands, 214
  hominids, 216–219
  Homo erectus, 218, *219*
  H. sapiens, 218, *219*
  infant dependency, 215
  life span, 215
  plasticity of, 217
  primate evolution and, 213–216
  skeletal changes, 214
  vision, 214
Exclusion, competitive, 509–510
Exercise, 87, 355, 366–367, 381
Exergonic reaction, 60, *60*
Exocrine gland, 337
Exocytosis, 44, *44*
Exon, 152, *152*, 166
Exoskeleton, arthropod, 273
Experiments, examples of
  artificial selection, 590, *590*
  auxin effects on cell elongation, *328*
  bacterial conjugation, 164, *164*
  bacterial transformation, 143–145,
    144
  bacteriophages, *144*
  behavioral selection, 590, *590*
  circadian rhythms, *328–329*
  circadian rhythms (plant), *328*, 330
  coleoptile growth, *328*
  competitive exclusion, 509, *509*
  cytoplasmic segregation at cleav-
    age, *466*
  DDT effects, 12
  delay of senescence, *332*
  dihybrid crosses, 122, *123*
  embryonic fate, 466, *466*
  flowering plants, 328, *328*
  genetic engineering, 163, 167–169
  global warming, 536–537
  gravitropism, *328*, 328–329
  Hubbard Brook Valley watersheds,
    533, *533*, 534
  ice-minus bacteria, 167
  intense sound effects on ear, *443*
  length of day/night effects on plant
    growth, *332*
  molecular basis of inheritance, *144,
    145*
  monohybrid crosses, 119–121, *121*
  nature's, 163
  numerical models of climate
    change, 567
  nutrient enrichment effects on lake,
    559, *559*
  origins of life, 195, *195*
  osmosis effects on plant cells, *43*
  photoperiodism, 329–330
  phototropism, 328, *328*
  plant breeding, 118*ff.*
  population growth, 494–495
  radioactive labeling, *144*
  reciprocal crosses, *130*
  stem elongation, *328*
  testcross, 121
  X-linked traits, *130–131*
  X-ray diffraction, 146
Experiments, scientific
  nature of, 10–11
  testing hypotheses with, 11
Exploitative competition, 509

Exponential growth, 494–495, *495,
    499,* 500
External fertilization, 464
External oblique, *353*
Extinction
  of amphibians, *200*
  background, 201
  of birds, *200*
  Carboniferous, 204
  chemical, *199*
  Cretaceous, *199*, 205
  Cretaceous–Tertiary boundary, 206
  Devonian, *199*, 204
  dinosaurs, 205–207
  of fish, *200*
  flowering plants, 201
  gymnosperms, 201
  of insects, *201*
  Jurassic, 204–205
  of mammals, *200*
  mass, *199*, 201, 205–207
  Ordovician, *199*
  Permian, 204
  Permkjain, *199*
  of protozoans, *200*
  of reptiles, *200*
  Triassic, 204
  of tribolites, *200*
Extracellular fluid (*See also* Blood;
    Interstitial fluid)
  and acid-base balance, 424
  and cell survival, 342
  composition, 342
  defined, 342
  of human body, 342
  in insulin deficiency, 456
  stability of, 342
  volume, 342, *452*
  water gains and losses, 419–420
Extracellular matrix
  components, 55
  functions of, 55
  structure, 55
Extraembryonic coelom, *476*
Eye
  abalone, 444, *444*
  color, *Drosophila, 130–131*
  compound, 432, 444, *444*
  evolution, 214
  farsightedness, *445*
  formation, *466*, 467
  human, 445–446, *445–446*
  innervation, *435*
  invertebrate, 444, *444*
  mollusk, 444, *444*
  nearsightedness, *445*
  octopus, 444, *444*
  snail, 444, *444*
  vertebrate, 445*ff.*
  visual accommodation in, *445*
Eyespot, 233, *233*, 444, *444*

F

F_1 and F_2 (genetic symbols), 119
Facial bones, *351*
Factor VIII, *166*
Factor IX, *166*
FAD (flavin adenine dinucleotide,
    oxidized), *84*, 85–86
$FADH_2$ (flavin adenine dinucleotide,
    reduced), *82, 84*, 85, *85–86*
Fall overturn, 558
Family pedigree, 132, *133*
Family planning, 501–502
Family (taxon), 7, *209*
Farsightedness, *445*
Fat (*See also* Lipid)
  as alternative energy sources, 88–
    89, *89*
  animal, 368

  in bloodstream, 381–382
  defined, 30
  deposits, arterial, 381–382
  digestion, 364, *365*
  functions of, 30, *35*
  metabolism, 89, *89*, 365
  metabolism disorders, 456
  polyunsaturated, 368
  saturated, 30
  storage by plants, 313
  transport by plants, 313
  unsaturated, 30, 383
Fat-soluble vitamin, *370*
Fatty acid, 372
  condensation into a triglyceride, *30*
  essential, 368
  intestinal absorption of, 365
  role in cell, 27
  structure, 29
Fatty connective tissue, 337–338, *338*
Feather, 177, 288, *347*
Feces, 235, 267, 366, 419
Feedback
  and body temperature, 342
  homeostatic, 455, *455*
  in human activities, 582
  negative, 342, *342*, 455, *455*, 469,
    582
  and ovarian function, 473
  positive, 342, 363
  in stretch reflex, 433–434
  in withdrawal reflex, 434
Feedback inhibition, 64, *64*
Feedforward control, 582–583
Feeding behavior
  adaptive, 590
  beak shape and, 175, *175*
  hominid, 216
Feeding relationships
  and Antarctic food web, *526*
  rocky shore, 560
  savanna, *553*
  and trophic levels, 525, *526*, 527
  tropical reef, *565*
Female defense behavior, 591
Femur, *351*, 352, *352*
Fermentation, 80, 202
Fermentation pathway
  alcoholic, *87*, 88, *88*
  ATP yield from, 88
  defined, 87
  lactate, *87*, 88
Fern, *249*, 250, *250–251*
Ferrous iron (Fe^{++}), 64
Fertility control, 483–484, 501, 503
Fertilization, animal
  and chromosome number, 106
  diploid chromosome number resto-
    ration at, 112, *112*
  external, 464
  gene shuffling at, 105, 109
  human, *474*, 475
  internal, 464
  in vitro, 488
Fertilization, plant
  cross-fertilization, 118
  defined, 95
  double, 318, *319*
  self-fertilization, 118
Fertilizer, nitrogen-rich, 307
Fetal alcohol syndrome, 480–481
Fetus (*See also* Human development)
  defined, 478, *482*
  development, *477*, 477–478, *478*
Fever, 63, *224*
Fever blister, *224*
Fiber
  dietary, 366
  sclerenchyma, 294
Fibrin, *385*
Fibrinogen, *376*
Fibrous joints, 352

Fibrous root system, *299, 300*
Fibula, *351*
Ficus benjamina, 554
Fight-flight response, 434, 455
Filament
  intermediate, 52
  stamen, *317–318*
Filtration, in kidney, 420–422, *421*
Fin, fish, 203–204, 283, *283–284*
Finch, *175,* 184, 201
Fingerprinting, genetic, 166
Fire ant, *517*
First law of thermodynamics, 59
Fish (*See also* specific type)
  adaptive radiation, *200*
  bony, 284–285, 423, 464
  cartilaginous, 283–284, *284*
  circulatory system, *282*
  comparative embryology, *190*
  evolution, *199,* 203, *281,* 283, 283*ff.*
  extinction, *200*
  fin, 203–204, 283, *283*
  gill, 408, *408,* 423
  internal environment, 423
  jawed, *281*
  jawless, *281,* 283–284
  lobe-finned, 203–204, 283, *283*
  as protein source, *369*
  respiratory system, 408, *408*
  skeleton, *283*
  skeletons, 284
  survivorship patterns, 498, *498*
  tectum, 436
  water-solute balance in, 423
Fish oil, 383
Fission
  bacterial, 226
  binary, 226, *227*
Five-toed limbs, 193
Flagella
  bacterial, *45,* 225, *225*
  functions of, 53
  organization, 54, *54*
  sponge, *262*
  structure, *54*
Flagellated protozoan, *229, 234,*
    234–235
Flame cell, 265
Flamingo, *193*
Flatworm (Platyhelminthes)
  anatomy, *265*
  characteristics, *260,* 261, 265–266
  digestive system, *361*
  evolution, *260–261,* 266
  integumentary exchange, 407
  major groups, 265–266
  oviduct, *265*
  penis, *265*
  representative, *259*
  reproduction, 104
  respiratory system, 406–407, *407*
Flavin adenine dinucleotide
  oxidized (*See* FAD)
  reduced (*See* FADH$_2$)
Flax, 293
Flea, *276,* 540
Flemming, W., 128
Floral structure, 317, *317,* 321
Florigen, *327, 328,* 330
Flower
  abscission, 333
  and biological clock, 328–329
  daylength and, 330, *330, 332*
  defined, 316
  of dicots, 254, *254, 295*
  gamete formation in, 317–318
  imperfect, 317
  of monocots, 254, *254, 295*
  perfect, 317
  pollination, 320–322
  soybean plant, *327*
  structure, 317

Flowering plant (*See also*
    Angiosperm; Plant)
  accessory, *324*
  aggregate, *324*
  buds, 298, *298*
  characteristics, *251*
  classes of, 254
  coevolution with pollinators, 320–
    322, *322*
  diversity of, 254
  double fertilization of, *255*
  evolution, *199,* 204
  flowering process, 330
  fruits from, *324*
  gametophytes, 316
  growth, secondary, *294*
  leaves, *292,* 297–299, *298*
  life cycle, *255*
  monocot vs. dicot, 295, *295*
  multiple, *324*
  pollination, 320–322
  primary, 294*ff.*
  root system, 293, 299–302, *302*
  shoot system, 293, 295–299, *295–*
    299
  simple, *324*
  sporophytes, 316
  stem growth, *302*
  woody, 303, *303,* 304, *304*
Fluid mosaic model, membrane, 40
Fluke, 265, *267*
Fluorescence microscopy, 53
Fluorine, in human nutrition, *371*
Fly, 321, *326* (*See also* specific types)
Fly agaric mushroom, *243*
Folic acid, *370*
Follicle, of oocyte, 471, *471*
Follicle-stimulating hormone (*See also*
    Gonadotropin; Luteinizing
    hormone)
  and male reproductive function, 469
  and ovulation, 473
  secretion, 469
  source of, *452*
  targets of, *452*
Fontanel, 352
Food chain, 527, *528*
Food poisoning, 242
Food production (*See* Agriculture;
    Crop, food)
Food web
  aquatic, 535
  defined, 527
  detrital, 529
  in estuaries, 560, *561*
  examples of, *528*
  grazing, 529
  in hydrothermal vents, *563*
  in intertidal zone, 560
  in lake zones, 558
  along tropical reef, *565*
Foraminiferan, *229, 234,* 235
Forebrain, 212
Forelimb
  modifications, 201
  morphological convergence in, 193
  morphological divergence in, *192,*
    193
Forest
  and acid deposition, 572, *572*
  biome, *548,* 553*ff.*
  soil, 549
  tropical rain, 506
Formula, chemical, *18*
Fossil
  Archaeopteryx, 176, *176*
  australopiths, 217
  bird, 193
  Burgess Shale, *203*
  Cooksonia, 204
  dating of, 190–191
  defined, 189

Devonian plants, 204
  early *Homo,* 218, *219*
  fuel, *534–535,* 535, *535,* 537, *581–*
    582
  fungus, 202
  green algae, 202
  handbone, 217
  Homo erectus, 218, *219*
  H. sapiens, 218–219
  living, *283, 287*
  marine worm, *203*
  plant, *204*
  plant spore, 202
  Psilophyton, 204
  radioactive dating, 19
  red algae, 202
  reptile, *192*
  Silurian plant, *204*
  sycamore leaf, *19*
  teeth, 213, 217
  tree fern, *19*
Fossil record
  bacteria, 227
  completeness of, 190
  of convergence, *192,* 193
  correlated with earth history, 194*ff.*
  of divergence, 191, *192, 193*
  early interpretations of, 174
  nature of, 189–190
  progymnosperms, 250
Founder effect, 180
Fovea, 445, *445*
Fox, *527*
Fox, S., 196
Franklin, R., 146
Free nerve ending, 441
Frog
  acoustical signaling, 594
  amplexus, *463*
  brain size, *282*
  characteristics, 285
  chromosome number, *106*
  embryonic development, 462–463,
    462–463
  evolution, 285
  protein, 194
  reproduction, 462–463, *462–463*
  singing tropical, *594*
  skeleton, 285
  skin, 285
  skin temperature, 427
  tadpole, *463*
  tectum, 436
  water-solute balance in, 423
Frog-eating bat, 594, *594*
Frond, 250
Frontal lobe, *437*
Fructose
  functions of, 28, *35*
  from seminal vesicle, 469
  structure, *28*
Fruit
  abscission, 333
  characteristics, 323, *324*
  coat, *323*
  defined, 323
  fleshy, 324
  from flowering plants, *324*
  formation, 323–324, *324*
  function, 324
  spin of, 324
Fruit fly (*See also* Drosophila), *106*
Fuel
  fossil, *534–535,* 535, *535,* 537, *581–*
    582
  nuclear, 584
Fumarate, *85*
Functional group, of organic
    compounds, 27, *27*
Fungal cell, 240
Fungus (Fungi) (*See also* specific
    type)

body plan, 240
  cell wall, 46, 55
  classification, 7, *209,* 240, *241*
  commercial uses of, 88, 241
  as decomposers, *8,* 239
  disease-causing, 241, *242,* 243
  evolution, *199*
  fossils, 202
  life cycle, *243*
  major groups of, 240–244
  medical uses of, 242, 243
  mode of nutrition, 240
  oak root, 332
  parasitic, 240
  reproduction, *240*
  saprobes, 240
  spores of, *240, 240*
  stinkhorn, *8–9*
  survivorship patterns, 498, *498*
Fungus-root, 309, *309*
Fur (*See also* Coat color; Hair)
Fur seal, *185*

G

G_1 and G_2 of cell cycle, *97*
Galactose, 133
Galactosemia, 133
Galileo, G., *38*
Gallbladder
  functions, *362,* 364
  human, *362,* 364
Galápagos Islands, *173,* 175
Galápagos Rift, *563*
Galápagos tortoise, *286*
Gamete (*See also* Egg; Oocyte;
    Sperm)
  chromosome number of, 96, 106
  formation, *96, 111,* 129
Gametogenesis
  animal, 110–112, *111*
  defined, 110, 112
Gametophyte
  compared to sporophyte, 316
  of flowering plant, 316
  in life cycle, 245–246, 318, *318*
Gamma radiation, 160
Ganglion (ganglia), vertebrate
    nervous system, *435*
Garter snake, 590, *590*
Gas exchange (*See* Respiratory
    system)
Gastric fluid, 25, 159, 363, 391
Gastrin, *454*
Gastrointestinal tract (*See also*
    Digestive system)
  endocrine function, *454*
  functions of, 361
  motility, 361, 363
  and nervous system, *435*
  organs of, 362*ff.*
  structure, 360–362, *362, 364*
Gastropod (Gastropoda), 270, *270*
Gastrulation, 465
Gated channel protein, 428
Gause, G., 509, *509*
Gender
  and cardiovascular disorders, 381
  determination, 129
Gene(s)
  allelic states of, 105, 118, *119*
  cloned, *166,* 166–167
  defined, 104, 118
  hemoglobin, 157
  interactions, 123, 125
  jumping, 157
  linkage, 131, *132*
  linked, 129, *130–131, 132*
  locus, defined, 118, *119*
  multiple effects of, 123, 125
  oncogenes, 160

regulator, 158
regulatory, 157–159, 191
X-linked, 129, *130, 130–131,* 131
Y-linked, 129, *130*
Gene amplification, 165–166
Gene expression
    allelic interactions in, 122–125
    and cancer, 160–161
    of cloned DNA, 166–167
    controls over, 157*ff.*
    and development, 158–159
    and differentiation, 158–159, 465–466
    and dominance relations, 122*ff.*
    in eukaryotes, 158*ff.*
    external environment effects, 125
    internal environment effects, 125
    in prokaryotes, 157–158, *158*
    selective, 158–159
Gene flow, 179, *179,* 180
Gene machine, 167
Gene mutation
    affecting embryonic stages, 191
    allele formation by, 105
    and cancer, 160
    defined, 155
    Drosophila, 130–131
    and evolution, 105
    and genetic disorders, 132–139
    mutagen-induced, 155
    and sexual reproduction, 104–105
Gene pair
    defined, 118, *119*
    interactions, 122–125, *124*
Gene regulation, eukaryotic
    and cancer, 160–161
    cell division, 159
    and differentiation, 158–159
    selective nature of, 158–159
    X chromosome inactivation, 159, *159*
Gene regulation, prokaryotic
    in lactose operon, 157–158, *158*
    mechanisms, 157–158
Gene sequencer machine, 167
Gene sequencing, 167
Gene shuffling, 105, 109, 112
Gene therapy, 168–169
Genetic code, 152–152
Genetic counseling, 139–140
Genetic disorder
    vs abnormalities, 132
    autosomal dominant, 134, *134*
    autosomal recessive, 132–134, *133*
    chromosome number-related, 137–139
    chromosome structure-related, 135–136
    counseling for, 139–140
    diagnosis of, 140
    diet modification of, 139
    and gene therapy, 168–169
    prenatal diagnosis of, 140
    preventive measures for, 139
    screening for, 139
    treatments for, 139
    X-linked recessive, *134,* 134–135, *135*
    XYY condition, 139
Genetic drift, 179, *179,* 180, *180*
Genetic engineering
    applications, 163, 167–169
    ethical implications of, 168–169
    potential risks and benefits, 163, 167–169
Genetic equilibrium, 179
Genetic fingerprinting, 166
Genetic recombination
    in antibody formation, *394*
    and crossing over (homologous recombination), *106,* 107, 131, *132*

by recombinant DNA technology, 163
Genetic screening, 139
Gene transfer, 168
Genitalia (*See* specific reproductive organ)
Genital infections, *224,* 487
Genome, 164, 167
Genotype, 118
Genus (taxon), 7, 208, *209*
Geographic isolation, 186
Geospiza conirostris, 175
    G. fuliginosa, 175
German measles, *224,* 480
German shepherd, 442
Germ cell, 95, 105, *105*
Germination, 326–327, *326–327*
Germ layer, 465
Giardia intestinalis, 235
Gibberellin, *327,* 328
Gibbon, *212,* 215
Gigantism, 452, *453*
Gill
    amphibian, 285
    bivalve, 271, *271*
    chordate, 281, *281*
    evolution, 282
    fish, 282, 408, *408,* 423
    gastropod, 270
    mollusk, 269
Ginkgo (Ginkgophyta), 251, 252, *252*
Gizzard, bird, *361*
Glacier, 203
Gland (*See also* specific type)
    endocrine, 337
    exocrine, 337
    innervation of, *435*
Global warming, 535–537
Globulin, *376*
Glomerulus, 420, *421,* 421–422, *422, 424*
Gloryblower, 321
Glucagon, *454,* 456
Glucocorticoid, *454,* 455
Glucose (*See also* Sucrose; Sugar)
    absorption, 419
    concentrations, 372
    diffusion across membranes, 42
    functions of, 28, *35*
    intestinal absorption of, 365
    metabolism, 82, *83–85,* 88–89, *89, 371–372, 372,* 455–456, *457*
    and photosynthesis, 70
    structure, *28,* 82
Glucose-1-phosphate, 133
Glutamate, *31, 153*
Glutamine, *153*
Glyceride, 30, *35*
Glycerol, 372
Glycine, *31, 153*
Glycogen
    function, 29
    and glucose metabolism, 87–88, 456
    and muscle cells, 355
    structure, 29
Glycolysis
    ATP yield from, 82, *83*
    defined, 81
    intermediates of, *82*
    links with photosynthesis and respiration, 69, *69*
    net energy yield, *83*
    reactions, 82, *83*
Glyptodont, *174*
Gnetophyte (Gnetophyta), 251, 252, *252*
Goiter, 456, *456*
Golden alga, *229,* 233
Golgi body, 46, *48, 50,* 51
Gonad
    development, 129, 458
    endocrine functions, *454*

as endocrine target, *452*
    melatonin effects, 587
Gonadotropin, *472* (*See also* Follicle-stimulating hormone; Luteinizing hormone)
Gondwana, 197, 203, *203*
Gonorrhea, 228, 486
Gorilla, *82*
    brain size, 215
    chromosomes, 96
    evolution, *212,* 216
    skeletal organization, *214*
Gradualism, *208*
Gram-negative bacterium, 226, *226*
Gram-positive bacterium, 226, *226*
Granum, *53,* 71, *71*
Grass, 295, *299,* 527
Grasshopper, *375, 527*
Grassland
    animal life, 552–553
    biome, *548*
    plants, 552–553, *552–553*
    soil, *549*
Gravitropism, 328–329, *329*
Gray matter, 436
Grazing food web, 529
Green alga (Chlorophyta), *39,* 72, 202, 231, 244–245, *246, 247,* 251
Greenhouse effect, 535–537, *536–537*
Greenhouse gases, 536–537, *537*
Green python, *287*
Green revolution, 576–577
Green tube coral, *564*
Griffith, F., 143, *144,* 145
Gross primary productivity, 528
Ground meristem, *294*
Ground tissue, plant, *293,* 293–294
Groundwater, 574, 582
Growth hormone, *also called* Somatotropin, 452, *452, 453*
Growth pattern
    exponential, 494–495, *495, 499,* 500
    logistic, 496, *496*
    plant, 326*ff.*
GTP, 85
Guanine, 145–147, *145–147,* 151
Guard cell, 312, *312*
Guillemin, R., 451
Gullet, 236
Gut, flatworm, *265*
Gymnomycota, *229*
Gymnosperm, *246,* 250, 251, 252–253
    adaptive radiations, *201*
    evolution, *199*
    extinctions, *201*

evolution, 214
    opposable movements, 214
    power grip, 214
    precision grip, 214
    prehensile movements, 214
Hangingfly, *591,* 591–592
Haploid state, 96, 106, 245
Hardy, G., *178,* 179
Hardy–Weinberg principle, *178*
Hare, 511, *512*
Harpobittacus, 591
Haversian system, *349*
Hawkmoth, 322
Hay fever, 399
Hearing, 442, *443*
Heart
    arteries, *379*
    beat, 20, 380, 383
    birds, 288
    circulation, 379
    communication junction, 380, *380*
    contraction, 380, *380*
    diastole, 380
    endocrine function of, *454*
    evolution, 282
    innervation, *435*
    mammalian, 288
    muscle cells, 379
    pacemaker for, 20
    rhythm, abnormal, 383, *383*
    signal conduction, 380, *380*
    in stress response, 455
    structure, *378,* 378–379, *379*
    systole, 380
    vertebrate, 282
Heart attack, *30, 166,* 381–382, 442
Heartwood, 303
Heimlich maneuver, *411*
Helianthus, 293
Heliozoan, *229*
Helium, electron distribution, *21*
Helper T cell, *392,* 393–394, *402*
Hemodialysis, 424
Hemoglobin, 370
    and carbon dioxide, 416
    carbon dioxide transport, 412–413
    functions of, *33, 35*
    genes for, 157
    osmosis effects, 42
    oxygen transport, 412–413
    polypeptide chains in, *33*
    quaternary structure, *33*
    sickle-cell, *124,* 125, *157*
    in skin, 348
    structure, 33, *33*
Hemophilia, 130, 135, *135,* 141, *166*
Hemorrhagic fever, 224
Hemostasis, 384
Hemp, 293
Henslow, J., 174
Hepatitis B, 397–398
Herbaceous plant, 303
Herbivore, 523, *526–527,* 529
Heredity, and human behavior, 586
Herniated disk, 350
Heroin, *439*
Herpes virus, 224, 225, 348, 487
Hershey, A., 144
Heterocyst, 228, *228*
Heterotroph, 69, 226, 523
Heterotrophic bacterium, 229
Heterozygote, defined, 118
High blood pressure, *166,* 371, 381
    (*See also* Hypertension)
High-density lipoprotein (LDL), 383
Hindbrain, 212
Hipbone, 217
Histamine, 363, 391, 393, 395, 399, 402
Histone, 148
HIV (*See* Human immune-deficiency virus)
H.M.S. *Beagle, 173,* 174

H

H⁺ (*See* Hydrogen ion)
Habitat
    defined, 506
    human, 493
Hagfish, 283–284
Hair, 177 (*See also* Coat color; Fur)
    epidermal, *347,* 348
    growth, hormonal effects, 348
    of leaves, 299
    protein deficiency and, 348
    root, 309, *309, 326*
    scalp, 348
    shaft, *347,* 348
    split ends, 348
    thinning, 348
Hair cell, 442, *443*
Half-life, defined, 19
Hallucinogenic drug, *439,* 440
Halobacterium, 226
Halophile, 227, 230
Hand
    australopith, 217
    development, *94,* 467

Homeostasis, defined, 5–6
Homeostatic mechanism
  components of, 342, *342–343*
  glucose metabolism, *457*
  and hormone secretion, 342
  and internal environment, 342
  negative feedback, 342, *342*, 455, *455*, 582
  and neuroendocrine control center, 455, *455*
  overview of, 342–343
  positive feedback, 342
Hominid, 213
  classification, *212*
  evolutions, 216*ff.*
  tree of descent for, *216*
Hominoid, *212*, 213
Homo erectus, 218, *219*
H. habilis, 218
Homologous chromosome
  alleles on, 118
  crossing over and genetic recombination, 107
  crossing over and recombination, 131, *132*
  defined, 96, 105
  examples of, 96
  genes on, 118–119, *119*
  human, 109
  maternal, 109
  in meiosis, 96, 106*ff.*
  at metaphase, *110*
  in mitosis, 96, *98*
  paternal, 109
  shuffling of, 109
Homologous structure, 191–193
H. sapiens
  chromosome number, *106*
  classification, *209*
  evolution, 218–219, *219*
Homozygote, 118
Homozygous dominant, 118
Homozygous recessive, 118
Honeydew, 313, *313*
Hooke, R., *38*
Hookworm, 266
Hopcraft, D., 580
Horizontal cell, *446*
Hormone, animal (*See also* specific type)
  and behavior, *452*, 587
  blood, *376*
  defined, 450
  in development, *454*
  effects on hair growth, 348
  and evolution, 458
  and male reproductive function, 469
  nonsteroid, 458–459, *459*
  in organic metabolism, *454*
  secretion, homeostatic control of, 342
  sources (summary), *454*
  steroid, 458, *459*
  targets (summary), *454*
Hormone, plant (*See also* specific type)
  defined, 327
  in growth and development, 327–328, *452*, *454*
  table of, *327*
Hornwort, 248
Horse, 193, *282*
Horsetail, *106*, 249, 249–250, *251*
House sparrow, *517*
Hubbard Brook Valley, 533, *533*, 534
Human
  blood typing, 123
  body plans, 172, 174
  body weight, 366–367
  brain, 212–213, 215, *437*, 437*ff.*
  brain size, 215
  circulatory system, *340*, 378*ff.*

classification, 213
digestive system, *341*, 362*ff.*
ear, *443*
emotional states, 438, 439
endocrine system, *340*
energy needs, 366–367
evolution, 212*ff.*
eye, *445–446*
genetic disorders, 132*ff.*
impact on biosphere, 569*ff.*
integumentary system, *340*
lymphatic system, *341*, 387, 387–388
muscular system, *340*, 353*ff.*
nervous system, *340*, 434*ff.*
nutritional requirements, 366*ff.*
organ systems, overview of, *340–341*, 341
population growth, 498*ff.*
reproductive system, *341*, 468
respiratory system, *341*, 410
sex determination, 129, *129*
sexual activity, 342, 439, 473
skeletal organization, *214*
skeletal system, *340*, 348*ff.*
skeleton, 212–213, 350, *351*
skin, 17, 125, 342, 346–348
social behavior, 177, 598
survivorship patterns, 498, *498*
teeth, 214–215
urinary system, *341*, 420*ff.*
vision, 445–446
Human chromosome
  abnormalities, 132–139
  karyotype, *128*
  number, 96, *106*
  sex, 129
Human development (*See also* Animal development; Human reproduction)
  aging (*See* Aging)
  alcohol effects on, 480–481
  beginnings of, 483–484
  birth, 479, *479*
  and bone formation, 350
  cleavage, *474*
  critical periods of (embryonic), *481*
  and death, 482
  embryonic, *474*, 476, 477, 477–478, *478*, 480, *481*
  extraembryonic membranes, 476, *476*
  fertilization, 475
  fetal, 477, 477–478, *478*, 480
  first trimester, 478
  gene activity in, 158–159
  gonadal, 129
  hormonal effects on, *452*, *454*
  implantation, *474*, 475
  lactation, 479, *479*
  and nutrition, 480
  postnatal, 482, *482*
  pregnancy, 475*ff.*
  prenatal, 475*ff.*
  prenatal diagnosis, 139–140
  protein deficiency and, 369
  second trimester, 478
  sex determination, 129, *129*
  smoking effects on, 481
  stages, summary of, *482*
  third trimester, 478
  and X chromosome inactivation, 159, *159*
Human gene therapy, 168–169
Human genome, mapping, 167
Human immune-deficiency virus (HIV), 399
  attack on T lymphocyte, *390*
  carriers of, 485
  fluids isolated from, 402
  mutation, 402
  replication cycle, 400, 485

spread of, 485
structure, 400, *401*
transmission modes, 400–401, 485
treatment of, 402
Human population
  control of, 500–503
  energy sources for, 580–582
  and food production, 168, 576*ff.*
  growth, 570
  impact on biosphere, 569*ff.*
  impetus for, 499–500
  magnitude of, 483, 492–493, 498–500
  zero, 503–504
Human reproduction (*See also* Animal development; Human development)
  control of, 483–484, *484*
  coitus, 469, 473
  and Down syndrome, *137*, 137–138
  female reproductive system, 471*ff.*
  fertilization, 474
  hormonal controls, 469
  male reproductive system, 467*ff.*
  menstrual cycle, 470–471
  oogenesis, 470–473, *472*
  orgasm, 473
  ovarian function, 471–473
  ovulation, 473
  semen formation, 469
  spermatogenesis, 467–469, *469*
  uterine function, *472*, 473
Humboldt Current, 562
Humerus, *287*, *351*
Humminbird, 288
Hummingbird, 286
Humus, *366*
Hunger, 436
Huntington's disorder, 134
Hybrid inviability, 186
Hybridization
  dihybrid cross, 122, *123*
  DNA, 165, *165*, 193, *193*
  monohybrid cross, 119–121, *120–121*
  and polyploidy, 187, *187*
Hydra (Hydrozoan), 258–259, 262
Hydration, spheres of, 25, *25*
Hydrochloric acid, 25, 363
Hydrogen
  in abiotic synthesis, 195, *195*
  atomic number, *18*
  electron distribution, 21, *21*
  mass number, *18*
  and plant function, 307
Hydrogen bond
  and cohesion theory, 310, *311*
  defined, 23
  in DNA double helix, 146, *146*
  in a polypeptide chain, *32*
Hydrogen ion (*See also* Proton)
  and acid-base balance, 424
  in aerobic respiration, 81, *84*, 85–86, *86*
  in ATP formation, 81, *84*, 85–86, *86*
  and carbon dioxide transport, 413
  in digestion, 363
  in extracellular fluid, 342
  and pH, 25, 25–27
  in photosynthesis, 74, *74–75*
  secretion, *421*, 424
Hydrogen peroxide, *570*
Hydrologic cycle, 531, *532*
Hydrolysis, 28, *28*
  into subunits, *28*
Hydrophilic interaction, 23
Hydrophobic interaction, 23, 40
Hydrosphere, 543, 574–576
Hydrothermal vent, 562, *563*
Hydroxide ion, 25
Hydroxyl group, 27, *27*
Hydrozoan, *264*
Hyoid bone, *351*

Hypodermis, 347
Hypertension (*See also* High blood pressure)
  factors affecting, 381
  and sodium retention, 422
Hyperthyroidism, 383, 455
Hypertonic solution, 42, *42*
Hyperventilation, 416
Hypha, *240*, 244
Hypnotic drug, *439*, 439–440
Hypocotyl, *327*
Hypothalamus
  functions of, 436, 438, 450, *472*
  and limbic system, 438, *438*
  links with pituitary, *450*, 450*ff.*, *451*
  location, 436, *437*
  pleasure center, 439
  releasing hormones, 455
  and testosterone secretion, 469
  and thirst, 422
  and water reabsorption, 422
Hypothesis, 11–12
Hypothyroidism, 455
Hypotonic solution, 42, *42*, 423
Hypoxia, 416

I

Ibis, *193*
Ice-minus bacterium, 167
Ichthyosaur, *190*
Ideal weight, 366–367, *367*
IgE, 395, 399
IgG, 395, 399
IgM, 395
Ileum, *366*
Immune response
  abnormal, 399*ff.*
    antibody-mediated, *392*, 393–396, *395*, 403
    cell-mediated, *392*, 393, 396, 399, *403*
  and complement system, 391
  control of, 396
  deficient, 399*ff.*
  inflammation during, 391, *392*, 393
  molecular clues to, *393*
  and pesticides, 17
  and plasma proteins, 377
  primary, *392*, 394–398, *399*
  secondary, *392*, 398, *399*
  targets, *392*, 396
Immune system
  antibodies of, *392*, 394
  cells of, *392*
  memory of, 393
  overview of, *392*
  self-nonself recognition by, *393*, 393–394
  smoking-related impairments, 415
  specificity of, 393
  ultraviolet radiation effects, 348
Immune therapy, 396
Immunity, passive, 399
Immunization, *390*, 398–399
Immunological comparison, 194
Imperfect fungus, *241*, 242–243, *244*
Implantation (human), *474*, 475
Imprinting, 588–589
Incisor, 213, 217, 363
Incomplete dominance, 123
Independent assortment, 122, *122*
Independent variable, 11
Indian corn (maize), *157*
Induced-fit model, 62, *62*
Industrial smog, 571
Industrial waste, and rain pH, 26
Infant dependency, mammalian, 213
Infant (human), defined, *482*
Infant mortality, 229
Infectious disease, *166*

Infectious mononucleosis, 224
Inflammatory response, 391, *392,
   393, 395,* 399, 402–403, *403*
Influenza, *224,* 225, 398
Inheritance
   autosomal dominant, 134, *134*
   autosomal recessive, 132–134, *133*
   basis of, 95, 104–105
   blending theory, 117
   defined, 7
   patterns of, 117*ff., 130–131*
   X-linked, 129, *130–131,* 131, *134,*
      134–135, *135*
Initiation, protein synthesis, 154,
   154–155
Innate behavior (*See* Instinctive
   behavior)
Inner cell mass, *474*
Inner ear, 442, *443*
Insect (*See also* Arthropod)
   adaptive radiations, 204
   body plan, 277
   classification, *261*
   coevolution with plants, 320–322
   compound eye, 445, *445*
   digestive system, 277–278
   evolution, *199, 201,* 204
   exoskeleton, 273
   extinctions, *201*
   jointed appendage, 274
   major orders of, *276*
   Malpighian tubule, 277
   metamorphosis, 273, 277
   molting, 273
   niches, 507
   pollinator, 277, 320–321, *321*
   reproductive capacity, 278
   respiratory system, 274, 408, *409*
   salivary gland, 234
   sensory structure, 274
   social, 597
   specialized appendage, *274,* 277
   species of, 277
   survivorship pattern, 498, *498*
   tropical rain forest, 555
   wing, 277
Insecticide, 17, 184, 299
Instinctive behavior, 587–588
Insulin
   amino acid sequence, 32
   cloned gene, *166*
   deficiency, 456
   and diabetes, 456
   functions of, *35*
   and glucose metabolism, 454, 456
   secretion, 456, *457*
Integration (synaptic)
   drug action on, 439–440
   and nervous system, 432*ff.*
   synaptic, 432–433
Integumentary exchange, 407
Integumentary system, *340,* 346–348
   (*See also* Skin)
Interference competition, 509
Interferon, *166,* 396
Interleukin, *166,* 394
Intermediate filament, 52
Intermediate lobe, pituitary, *452*
Internal environment (*See also* Blood;
   Extracellular fluid; Interstitial
   fluid)
   defined, 342
   homeostatic control of, 342–343
   seawater and, 418
   solute balance of, 423
   water balance of, 423
Internal fertilization, 464
Interneuron, 427
Internode, *292, 295,* 297, *297, 326*
Interphase
   activities during, 97
   chromosome appearance, *97*

defined, 97
   transition to mitosis, 97
Interspecific competition, 508, *508*
Interstitial cell, testis, *468*
Interstitial fluid (*See also* Blood;
   Extracellular fluid)
   composition (vertebrate), 342
   defined, 342
Intertidal zone, 560, *561–562*
Intervertebral disk, 350, *351*
Intestine, large (*See also* Colon), *362,*
   366, *435*
Intestine, small
   absorption processes, 364–365
   digestion processes, 364
   functions, 362
   human, *362*
   innervation of, *435*
   mammalian, 364
   microvilli, 365
   roundworm, *266*
   villi, *364,* 365
Intrauterine device, *484,* 487
Intron, 152, *152,* 166
Inversion, chromosomal, 136
Invertebrate (*See also* specific type)
   defined, 258
   photoreceptors, 444
   vs. vertebrates, 258
In vitro fertilization, 488
Iodine, *371,* 456
[123]Iodine, 20
Ionic bond, 22, *22*
Iris (eye), *295,* 445, *445*
Iron
   atomic number, *18*
   deficiency in plants, *306*
   in human nutrition, 370, *371*
   mass number, *18*
   and plant function, *307*
Irradiation, for cancer treatment, 396
Irrigation, 168, 574
Island patterns, species diversity,
   520–521
Isocitrate, *85*
Isolating mechanism, 186, *186*
Isoleucine, *153,* 368
Isotonic solution, 42, *42*
Isotope, 18–19, *144*
IUD, *484,* 487

J

Jaguar, *554*
Japanese beetle, *517*
Jaw
   Australopithecus robustus, 217
   Homo sapiens, 218
   rotifer, *269*
Jawed fish, 281, *281*
Jawless fish, 281, *281,* 283–284
Jellyfish (*See also* Cnidarian), 190, *262*
Jenner, E., 390, *390,* 391
Jet propulsion, cephalopod, 271
Joint, vertebrate, 352, *352*
Jointed appendage, of arthropods,
   274, *274*
J-shaped curve, 494–495, 580
Jumping gene, 157
Jurassic, *198–199,* 204–205

K

Kangaroo, *288*
Kangaroo rat, *419, 423,* 423
Kaposi's sarcoma, 400
Karyotype
   defined, 129
   Down syndrome, *137*
   preparation, 129

Kentucky bluegrass, 77
Keratin, *35,* 347–348
Keratinization, 347
Kerogen, 581
Ketone, 456
Kidney (*See also* Urinary system)
   and acid-base balance, 424
   capsule, *420*
   controls over, 435, 450, 452
   cortex, 420, *420*
   endocrine function, *454*
   functions of, 420
   human, *420,* 420–421
   hypertension effects, 422
   malfunction, 424
   medulla, 420, *420*
   renin-secreting cells, 422
   stones, 420
   structure, *420,* 420–421
   and water–solute balance, 419
Kidney bean, *369*
Kidney dialysis machine, 424
Killer T cell, 393, 396, *402*
Killer whale, *526*
Kilocalorie, defined, 366
Kingdom (taxon), 7, 208–210, *209*
Kin selection, 597
Klinefelter syndrome, 138, *138*
Knee cap (patella), 351–352, *352*
Kohler, G., 396
Komodo monitor lizard, 286
Koshland, D., 62
Krebs, H., 85
Krebs cycle, 81, *82,* 85
Krill, *526–527*
K-T boundary, 206

L

Labia majora, *470*
Labia minora, *470*
Lactate (lactic acid), 88
Lactate fermentation, 87, 88
Lactation, 479, *479*
Lactobacillus, 226, 229
Lactose, 28, 133, 157–158, *158*
Lactose metabolism, 157–158, *158*
Lactose operon, 157–158, *158*
Lacuna, *349*
Ladybird beetle, *276*
Lake
   and acid deposition, 572
   ecosystem, 558–559
   eutrophic, 559
   oligotrophic, 559
   pollution, 559, *559*
   primary productivity cycles, 558–
      559
   trophic nature of, 559
   zonation, 558, *559*
Lake Victoria, 516
Lamprey, 283, *283,* 284
Lancelet, *280,* 281, *281*
Land
   available for agriculture, 576–577
   classification of, 576, *576*
Land crab, *275*
Lanugo, 478
Large intestine (*See* Intestine, large)
Larva, *6,* 273, 464
Larynx, 125, 410, *410, 415*
Laser angioplasty, 382–383
Late blight, 242
Lateral meristem, 294, 295
Lateral root, *292, 299, 301,* 331
Late wood, 303
Latimeria, 283
Laws of thermodynamics, 59
Leaching of soil, 549
Leaf
   abscission, 333

arrangement of, 297
   compound, *298*
   dicot, *295, 298*
   formation, 297
   hair, 299
   internal structure, 299
   monocot, *295, 298*
   oak, 331
   primary, *327*
   scales, 299
   senescence, 333
   shape, *295,* 299
   of shoot system, *292*
   simple, *298*
   structure, *298,* 299
   vein, *295*
Leaflet, *298,* 299
Leakey, M., *217,* 218
Learned behavior, 213, 586–589
Lecithin
   in bile, 364
   role in body, 367
Leech, *272,* 590
Leeward, defined, 546
Lemming, *524*
Lemur, *212,* 213
Lens, eye, 445, *445, 466*
Leopard, *511*
Leopard seal, *526*
Leucine, *153,* 368
Leukemia, retrovirus causing, 224
Leukocyte (*See* White blood cell)
Lichen, 244, *244,* 554, *561*
Life
   diversity of (*See* Diversity of life)
   DNA blueprints for, 3, *3*
   molecular basis of, 147
   organization of, *3,* 3–7, 59
   origin and evolution, 194*ff.*
   perspective on, 582–583
Life cycle (*See also* Asexual
   reproduction; Human
   reproduction; Sexual
   reproduction)
   aging in (*See* Aging)
   animal, 110, *110,* 113, 259
   arthropod, 273
   bacteriophage, 223, *223*
   blood fluke, *267*
   bryophyte, *248*
   cell division during, 94–95, 110–112
   Chlamydomonas, 247
   conifer, 252–253, *253*
   Dictyostelium discoideum, 232
   fern, 250
   flowering plant, *255,* 316, *319*
   fungus, *240*
   herbaceous plant, 303
   human, 483
   human immune-deficiency virus,
      400–401, *401*
   hydrozoan, *264*
   Lilium, 255
   meristem, 294
   moss, *248*
   Obelia, 264
   Physarum, 232
   plant, 245
   Plasmodium, 235
   woody plant, 303
Life expectancy, *415*
Life table, 498, *498*
Ligament, 352, *352*
Light
   and dormancy, 333
   and flowering process, 330, *330*
   and melatonin secretion, 458
   and photoperiodism, *320,* 329–320
   and phototropism, 328, *328*
Light microscope, *39*
Lily (*Lilium*), *255, 295*
Limbic system, *437,* 438, *438,* 442

Limiting factor (population), 496, 500
Limnetic zone, 558, *559*
Lineage, defined, 208
Linear metabolic pathway, 60, *61*
Linkage, 131, *132*
Linnaeus, C., 208
Linoleic acid, 30, 368
Lion, 5, *5*, 591
Lipase, *365*
Lipid (*See also* Fat)
  absorption, 364–365
  bilayers, 40
  in blood, *376*
  defined, 29, *35*
  dietary, 367, 370
  with fatty acids, 29–30, *35*
  flow through cytomembrane system, *50*
  functions of, 29, *35*, 367
  metabolism, 367–368, 372, *372*
  with no fatty acids, 30, *35*
  and nutrition, 367–368
  in plaque formation, 381–382
Lipoprotein, 383
Liquid waste, 575–576
Listeria, 226
Lithops, 514
Littoral zone, 558, *559*
Liver
  bile secretion, 364
  cell cross-section, *48*
  damage, anabolic steroid-related, 357
  endocrine function, *454*
  functions, *362*, 372
  human, *362*
  innervation, *435*
  and somatomedins, *454*
  urea formation, 372, *372*
Liverwort, 248, *249*
Lizard, 286
Loam, *549*
Lobe-finned fish, 203–204, 283, *283*
Lobster, 275
Lockjaw, 432
Locus (gene), defined, 118, *119*
Logistic growth, 496, *496*
Loop of Henle, 420, *421*, 423
Loose connective tissue, 337–338, *338*
Louse, *409*
Low-density lipoprotein (LDL), 383
LSD, *439*, 440
Luciferase, *66*
Luciferin, *66*
Lucy (australopith), 217, *217*
Lumbar nerve, *435*
Lumbar vertebra, *351*
Lumbricus terrestris, 106
Luna moth, *276*
Lung
  air pressure changes in, 412
  cancer, 415, *415*
  cilia, 53
  defined, 408
  disorder, 414–415
  evolution, 282, 408–409
  gas exchange, 412
  gas transport, 409, 412
  innervation, *435*
  and oxygen transport, 412
  relation to pleural sac, *410*
  vertebrate, 282
Lungfish, 284
Luteinizing hormone (*See also* Follicle-stimulating hormone; Gonadotropin)
  and male reproductive function, 469
  and ovulation, 473
  secretion, 469
  source of, *452*
  targets of, *452*

Lycophyte, 249, *249*, 251
Lycopodium, 249, *249*
Lyell, C., 175
Lyme disease, 228
Lymph, 387
Lymphatic duct, *387*
Lymphatic system, 341, *387*, 387–388
Lymph capillary, *388*
Lymph node, *387*, 388
Lymphocyte
  of antibody-mediated immune response, *392*
  B, *392*
  of cell-mediated immune response, *392*
  clonal selection of, 398, *398*
  component of blood, 376–377
  derivation of, 388
  helper T, *392*, 393–394, *402*
  killer T, *392*, 393, 396, *402*
  in lymphatic system, *387*
  memory cell, 392, 393, 398, *403*, 403
  and MHC markers, 394
  plasma cell, *302*, *392*, 394–395, *395*, *403*
  roles in defense, *402*
  suppressor T, 393, 396, *402*
  T4, 400
  types of, 378, *402*
  virgin B, *392*, 394, *402*
Lymphoid organ, 388
Lymphokine, 394
Lymph vascular system, 375, 387–388
Lymph vessel, *365*, *387*, 388
Lynx, 511, *512*
Lyon, M., *159*
Lyonization, 159, *159*
Lysergic acid diethylamide, *439*, 440
Lysine, *153*, 368, *369*
Lysis, defined, 223
Lysogenic pathway, *223*
Lysosome
  appearance in micrographs, *48*
  defined, *44*
  functions of, 46, *50*, 51
Lystrosaurus, 205
Lytic pathway, *223*

M

Mace (spice), *325*
Macfarlane, B., 398
Macroevolution (*See also* Evolution, biological; Microevolution)
  and comparative biochemistry, 193
  and comparative morphology, 191–193
  evidence of, 189–190
  fossil record, 189–190
Macronutrient, *306*, 307
Macrophage, *376*, 391, 393–394, *395*, *403*
Madagascar hawkmoth, 322
Magnesium
  atomic number, *18*
  deficiency in plants, *306*
  electron distribution, *21*
  in human nutrition, 370, *371*
  and plant function, *306*
Magnification, biological, 539–540
Mainland patterns, species diversity, 520–521
Maize, *157*
Malaria, 235, *235*, 397
Malate, *85*
Malignant tumor, 160
Malpighian tubule, 277
Malthus, T., 175–176
Malus, *302*
Mammal (Mammalia)

adaptive radiation, *199–200*, 204
brain, *437*
circulatory system, *282*
coat color, 123–125
comparative embryology, *190*
dentition, 213
ear, *443*
egg-laying, 288
evolution, *199*, 288
extinction, *200*
infant dependency, 213
intestine, small, *364*
learning, 213
major groups of, 288
and mass extinction, *199*
and morphological convergence, *192*, 193
and morphological divergence, 191, *192*, 193
phenotype, 125
placental, 288
plant-eating, 204
pouched, 288
respiratory system, 288
skin, 125, 159, *159*
survivorship patterns, 498, *498*
Mammary duct, 479
Mammary gland, 288, *450*, *452*, 479, *479*
Manganese, *306*, 307
Manicougan crater, *205*
Mantle
  bivalve, 271, *271*
  mollusk, 269
Maple, 324, *325*
Mapping, human genome, 167
Marchantia, 249
Margulis, L., 230
Marijuana, *439*, 440
Marine ecosystem, 559*ff*.
Marine patterns, species diversity, 520–521
Marine worm, 203, 258
Marrow, 349–350, 377
Mars, 195
Marsh gas, 227
Marsh marigold, *321*
Marsupial, adaptive radiation, *200*
Mass extinction, 201, 578–579
Mass number, 18, *18*
Mast cell, 391, 393, *403*
McClintock, B., 157
Measles, *224*
Mechanical isolation, 186, *186*
Mechanoreceptor, 441, *441*, 441–442, *443*
Medicago sativa, 296
Medici, M. de, *456*
Mediterranean fruit fly, 276
Medulla, kidney, 420, *420*
Medulla oblongata, *435*, 436, *437*
Medusa, 263–264, *264*
Meerkat, 336, *336*, 342
Megakaryocyte, *377*, 378
Megaspore, 252, *253*, 318, *319*
Meiosis
  and chromosome number, 96, 106, *106*
  and crossing over, 107, *107*, 131, *132*, 177
  and cytokinesis, 95, 106
  defined, 95
  discovery of, 128
  function, 95, 113
  and gametogenesis, 110–112, *111*
  and genetic recombination, 177
  and independent assortment, 122, *122*, 177
  in life cycles, 111–112
  male gametophyte, *318*
  vs. mitosis, 96, 113
  nondisjunction during, 136, *136*, 187

and oogenesis, *111*, 112, 471, *471*, 473, 475
plant, *319*
and spermatogenesis, 110, *111*, 469, *469*
stages of, 106–110, *108–109*
summary of, *114–115*
Melanin
  and albinism, *124*, 125
  and coat color, *63*, 125
  in skin, 125, 348, *348*
Melanocyte-stimulating hormone, *452*
Melatonin, *454*, 458, 587
Meltdown, 582
Membrane, cell (*See also* Plasma membrane)
  components of, 40
  diffusion across, 41–42
  evolution, 196
  excitability, 428–429
  extraembryonic, 476, *476*
  fluid mosaic model, 40, *40*
  functions of, 40
  lipid bilayer of, 40
  origins of, 196
  permeability, 42
  proteins, 40, 43
  structure, 40
Membrane potential (*See also* Action potential)
  of neurons, 428*ff*.
  resting, 429
Membrane transport
  active, 43–44
  electric charge and, 42–43
  at endodermis, 300
  facilitated diffusion, 43, *43*
  functions of, 43
  passive, 43
  in plants, 309
  pressure effects, *43*
  solute concentration and, 42–43, *43*
Memory, 438
Memory cell, immunological, 392, 393, 398, 403, *403*
Memory trace, 438
Mendel, G., *117*, 128–129
Mendelian principle of independent assortment, 122
Mendelian principle of segregation, 121, *130*
Menstrual cycle, 470–471, *471–472*
Mental retardation, 60, 369
Meristem
  apical, *294*, 294–295, *297*, 300, *323*
  lateral, *294*, 295
  locations, *294*, 300
Mesoderm, 465
Mesoglea, 264
Mesozoic era, 191, 197, *198–199*, *200*, 201, 204
Mesquite, 550
Messenger RNA, 152–153, *153*, 154, *156*, 165
Metabolic pathway
  and ATP, 64–65
  biosynthetic, 61
  blocked, 133–134
  cyclic, 60, *61*
  defined, 60
  degradative, 60–61
  linear, 60, *61*
  mutation in, 179
  participants in, 61
Metabolic reaction (*See also* Chemical reaction)
  rate of, 60, *61*
  reversible, 60, *60*
Metabolism
  carbohydrate breakdown, 371–372, *372*, 455, 456
  defined, 4, 58

effects of thyroid disorders, 455–456
effects on extracellular fluid, 419
lipid, 371–372, *372*
organic, major pathways, *372*
photosynthesis (*See* Photosynthesis)
protein, 371–372, *372*
and water-solute balance, 419
Metacarpal, *351*
Metal ion, as cofactors, 64
Metamorphosis, 273, *280*
Metaphase
chromosome appearance, *97*
meiosis, 106–109, *114–115*
mitosis, 97, *99*, 100, *114*
Metastasis, 160
Metatarsal, *351*
Methane, *26*, 195, *195*, *537*, *570*
Methane gas, 227
Methanobacterium, 226
Methanogen, *226*, 227
Methionine, *31*, *153*, 368, *369*
Methyl group, *26*, 27, *27*
MHC (major histocompatiblity
complex) marker, *393*, 393–394
Microevolution (*See also* Evolution,
biological; Macroevolution)
defined, 179
major processes in, *179*
Microfilament, 52, 101, *101*
Micronutrient, 307, *307*
Micropyle, *319*
Microscopy, 38, *38–41*, *40*, 128
Microsphere, 196, *196*
Microspore, 252, *253*, 317, *318*
Microtubule, 52–53
assembly, 53
and cell division, *98–99*
and cleavage furrow, 101
of cytoskeleton, 100
of motile structures, *469*
organization of, 54
and spindle apparatus, 97, *98*, *98–99*, 100, 106, 109
structure, *54*
Microtubule organizing center, 54, 100
Microvillus
of collar cells, *262*
intestinal, 365
of photoreceptor, *444*
respiratory tract, *411*
Midbrain, 212, *435*
Middle ear, 442, *443*
Middle lamella, 101
Miescher, F., 143
Miller, S., 195, *195*
Millipede, *276*, *276*
Milstein, C., 396
Mimicry, 513, *513*
Mineral
absorption, 419
accumulation by plants, 312–313
and human nutrition, 370–371
intake during pregnancy, 480
recommended daily allowance, *371*
uptake by plants, 312–313
Mineralocorticoid, *454*
Minnow, 285
Miocene, *198–199*
Miscarriage, nondisjunction-related,
137
Mite, 274
Mitochondrion
appearance in micrographs, *47–48*, *52*, *84*, *233*
ATP formation in, *84*
and bacteria, 230
of *Euglena*, *233*
functional zones, *84*
functions of, 46, *47*, 52
membrane system, *84*, *86*
muscle cells, 355, *356*

origin of, 230
sperm, 468
structure, 52
Mitosis
and aging, 482
and cell cycle, 97, *97*
at cleavage, 464–465
defined, 95, *97*
duration of, 97
function, 97
vs. meiosis, *96*, 113
plant, *319*
stages of, 98–99
summary of, *114–115*
Molar, 213, 217, 363
Mole, 193
Molecular clock, 193
Molecule, defined, *4*, 17
Mollusk (Mollusca), *183*, 230
body plan, *270*
characteristics, 269–271
circulatory system, 375
diversity in, *270*
evolution, *260*
fossils of, 189–190
major groups, 269–271
photoreceptor, *444*
representative, *259*
Molting, arthropod, 273
Molybdenum
deficiency in plants, *306*
and plant function, *307*
Monilinia, 242
Monkey, *212*, 213ff.
Monoclonal antibody, *166*, 396
Monocot (Monocotyledonae)
adaptive success of, 254
characteristics, *251*
cotyledon, 295, *295*
vs. dicot, *295*
floral structure, 254
function, 295
leaf, *295*, 298
representative, *251*
root system, 299
seed of, 254
species of, 254
stem, *295–296*
structure, 295
vascular bundle, 295, *295–296*, 297
Monocyte, *376–377*, 378, 391
Monoglyceride, 30, 365
Monohybrid cross
defined, 119
garden pea, 119–121, *121*
Mendel's experiments with, *121*, *129*
Mononucleosis, 224
Monosaccharide, 28, *28*, *35*, 365
Monosomy, 137
Monsoon forest, 556
Montane coniferous forest, 548, 556,
557
Moray eel, *564*
Morgan, T. H., *130–131*
Morning-after pill, 484
Morphogenesis, 466–467
Morphological convergence, *192*, 193
Morphological divergence, 191, *192*,
193
Morphological trait, 177
Morphology, comparative, 191–193
Morula, 482
Mosaic tissue effect (Lyonization),
159, *159*
Mosquito, 235, 268
Moss, 248–249, 554
Moth, 6, *6–7*, 184, *184*, *276*, 590
Motor cortex, *438*

Motor end plate, *432*
Motor neuron, *340*, 341
defined, 427
functions of, 427
and nerves, 433
and neuromuscular junction, 432
Mount St. Helens, 292
Mouse, genetic engineering of, 168,
169
Mouth
cancer of, *415*
functions, *362*, 363
human, *362*, 363
Movement
by digestive system, 361
peristaltic, 362
mRNA (*See* Messenger RNA)
Mucin, 363
Mucus, 363, 399
Muddy shore, 560
Mule, 163
Multicellularity
defined, 231
division of labor in, 231
origin of, 231
Multicellular organism, 3–4
cell junctions, 55
defined, *4*
Mumps, 224
Muscle (*See also* specific type)
cardiac, *339*, 341, 353
fatigue, 87, 355
functions, 339
jaw, 217
rigor, 355
rigor mortis, 355
skeletal, *338*, 339, 353
smooth, *338*, 339, 353
sphincter, 362
Muscle cell (fiber)
chemical synapses, 431–432, *431–432*
contraction, 353–355, *354–355*,
432, *433*
and glycogen, 355
heart, 379
membrane system, *356*
mitochondria, 355
motor end plate of, *432*
structure, 353–355, *354–355*
Muscular system
functions, 339
human, *340*, 353ff.
Mushroom, 240, *243*, 525
Musk oxen, 596
Mussel, 271, 516, *561*
Mutagen, defined, 155
Mutation
base-pair substitution, 156
beneficial, 7
in corn, *157*
defined, 7, 179
and dominant alleles, 134, *134*
Drosophila, *130–131*
and environment, 179
and evolution, 179–180, 191
gene, 105, *130–131*, 132–139, 155ff.
harmful, 7, 179
human immune-deficiency virus,
402
and metabolic pathway, 134, 179
and molecular clock, 193
neutral, 193
and recessive alleles, 132–134,
133–134, 134–135, *135*
spontaneous, 155
Mutualism, 508, *508*, 509 (*See also*
Symbiosis)
Mycelium, 240
Mycoplasma, 226
Mycorrhiza, 244, 309, *309*, 332
Myelin sheath, 431, *432*

Myofibril, 354, *354, 356*
Myosin, 354, *354–355*
Myristica fragrans, 325
Myxococcus, 226
Myxococcus xanthus, 229

N

NADH (nicotinamide adenine
dinucleotide, reduced), 82, *82*,
84
NADH⁺ (nicotinamide adenine
dinucleotide phosphate,
oxidized), *35*, 64, 74, 77
NAD⁺ (nicotinamide adenine
dinucleotide, oxidized), 64, 82,
84, 88
functions of, 34, *35*
NADPH (nicotinamide adenine
dinucleotide phosphate,
reduced), 70–71, *72*, 74, 76–77
Narcotic analgesic, *439*, 440
Nasal cavity, *410*
Nasute termite, 592, *592*
Natural gas, 581
Natural killer cell, *393*, 396, *403*
Natural selection, 105 (*See also*
Selection)
Darwin's concept of, 8–9, 117, 176
defined, 9, 181
evidence of, 175–176, 181–185
evolution by, 7–9
modes of, 181–185
Wallace's concept of, 117, 176
Nautiloid, 203
Nautilus, 270
Nearsightedness, *445*
Nectar, *321*
Negative feedback, 342, *342*, 455,
455, 469, 582 (*See also*
Feedback)
Neisseria, 226, 487
Nematocyst, 264, *264*
Nematode (Nematoda) (*See also*
Roundworm)
characteristics, 266, *266*
representative, *259*
Nembutal, *439*
Neon, electron distribution, *21*
Nephridium, 273, *273*
Nephron, 420, *420*, 420–421, *421*,
422, 424
Neritic zone, 560, 562, *562*
Nerve(s)
autonomic, 434, *434*, 434–435, *434–435*, *435*
cervical, *435*
lumbar, *435*
and neurons, 433
parasympathetic, 434, *435*
pathways, 433
pelvic, *435*
sacral, *435*
sympathetic, 434, *435*
thoracic, *435*
tract, 433
vagus, *435*
Nerve cell (*See* Neuron)
Nerve cord, *265*, *280*, 280–281
Nerve impulse (*See* Action potential)
Nerve net, 264
Nervous system (*See also* specific
organism)
autonomic, 434, *435*
cephalopod, 271
control of digestive system, 362
control of explusion, 366
earthworm, 272
echinoderm, 279
functions of, *340*, 426–427
and heart activity, 380, *380*

human, *340*
  integration by, 433*ff.*
  links with endocrine system, 450, *450*
  peripheral, 434, *435*, 439–440
  somatic, 434, *434*
  and sweating, 419
  vertebrate, 282, 434*ff.*
  and water reabsorption, 422
Nervous tissue, 431, 436
Net primary productivity, 528
Net protein utilization (NPU), 369
Neuroendocrine control center, 450*ff.*
Neuroglial cell, 434
Neuromuscular junction, 432
Neuron
  and color perception, 447
  defined, 427
  functions of, 427–428
  hormone-secreting, 455
  interneuron, 427
  motor, *340*, 341, 427
  plasma membrane, 428*ff.*
  sheathed, 431
  sleep center, 438
  structure, *427*, 427–428
  trigger zone, 429–430
Neurotoxin
  Clostridium tetani, 433
  dinoflagellate, 234
  mechanism of, 433
Neurotransmitter (*See* Transmitter
  substance)
Neutral interaction, 508, *508*
Neutral mutation, 193
Neutron, 17–18
Neutrophil, *376–377*, 378, 391, *403*
Newborn (human), defined, *482*
New World monkey, *212*
Niacin, 370
Niche, defined, 506
Nicotinamide adenine dinucleotide
  oxidized (*See* NAD⁺)
  reduced (*See* NADH)
Nicotinamide adenine dinucleotide
  phosphate
  oxidized (*See* NADP⁺)
  reduced (*See* NADPH)
Nicotine effects
  on the central nervous system, 439
  on the heart, 383
  on the peripheral nervous system,
    439
Nitrate, 538, *539*
Nitric acid, 571
Nitric oxide, 570, *570*
Nitrification, 538, *539*
Nitrifying bacterium, *226*, 228, *228*
Nitrite, 228, 538, *539*
Nitrobacter, 228
Nitrogen
  in atmosphere, 538, *539*, 570
  atomic number, *18*
  deficiency in plants, *306*
  electron distribution, *21*
  mass number, *18*
  and plant function, *306*
Nitrogen cycle, 538–539, *539*
Nitrogen dioxide, 570, *570*
Nitrogen fixation, 538, *539*
Nitrogen-fixing bacterium, 229, 307, *307*
Nitrosoma, 226
Nitrous oxide, 536, 538, *570*
Node, stem, *292*, 297, *297*, *326–327*
Noncyclic photophosphorylation, 74, *74–75*
Nondisjunction, *136*, 137–139, 187
Nonpolar covalent bond, defined, 23
Nonsister chromatid, 107
Nonvascular plant, species of, *251*
Norepinephrine, *454*, 455

Northern fur seal, *185*
Nostoc, 226
Notochord, 280, *280*, 281, *281*
Nuclear autumn, 584
Nuclear energy, 582, 584
Nuclear envelope, 100
  appearance in micrographs, *47–49*
  defined, 49*f.*
  functions of, 49
  at metaphase, *97,* 100
  structure, 49
  at teleophase, 100
Nuclear medicine, 20
Nuclear winter, 584
Nucleic acid
  components, 145, *145*
  digestion, 364, *365*
  functions of, 34, *35*
  metabolism, *365*
  structure, 34, *34,* 145–147, *147*
Nucleolus, 49, *47–49, 98*
Nucleoplasm, 49
Nucleosome, *147,* 148
Nucleotide(s)
  adenosine phosphate, 34
  coenzymes, 34
  components of, 33
  defined, 33, *35*
  of DNA, *145*, 145–146
  functions of, *35*
  nucleic acids, 34
  of RNA, *151*
  role in cell, 27
  and self-replicating systems, 196
  strand assembly, 147
  triplet, genetic code, 153
  types of, 34
Nucleus, atomic, 17–18
Nucleus, cell
  animal, *48, 98*
  appearance in micrographs, *48*
  components of, 49, *49*
  defined, 38, 46
  Euglena, 233
  functions of, 38, 46
  micrographs of, *97*
  plant, *47*
Nudibranch, 270 (*See also* Sea slug)
Nutmeg, *325*
Nutria, *517*
Nutrition, human, 88–89, *89, 457*
  effects on scalp, 348
  human, 348, 366*ff.*
  and metabolism, 371–372, *372*
  and pregnancy, 480
  role of unsaturated fats, 30

O

Oak root fungus, 332
Oak tree, 317, 331–332
Oat, *328*
Obelia, 264
Obesity
  and cardiovascular disease, 381
  defined, 367
  and thyroid disorder, 455
Occipital lobe, *437*
Oceanic ridge, *197*
Oceanic zone, 562, *562*
Ocean (sea)
  current, 546, *546*
  ecosystem, 560*ff.*
  and hydrologic cycle, 533
  origin and evolution, 194
  zonation, 562, *562*
Ocotillo, 550, *551*
Octopus, 271, *271*, 444, *444*
Odor, receptors for, 442
Ogallala aquifer, 574, *575*
Oil

as fossil fuel, 581, *581*
  as organic compound, 30, *35*
Olduvai Gorge, 218, *219*
Old World monkey, *212*
Oleic acid, *29*
Olfactory lobe, 436, *437*
Olfactory receptor, 442
Oligocene, *198–199*, 216
Oligochaete, 272
Oligosaccharide, 28
Oligotrophic lake, 559
Olive oil, 368, 383
Ommatidium, *444*
Omnivore, 525
Oncogene, 160
Oocyte (*See also* Egg; Gamete;
  Ovum; Sperm)
  in animal life cycle, *111,* 113
  defined, 464
  follicle, 471, *471*
  human, *474*
  primary, 471
  secondary, *471,* 471–472, 475
Oogenesis, *111,* 112, *471*
Operator, 157–158, *158*
Operon, defined, 158
Ophiostoma, 242, *517*
Opium, *439*
Opossum, *288*
Optic chiasm, *437*
Optic cup, *466*
Optic nerve, *445–446*, 447
Oral cavity (*See* Mouth)
Orchid, 295, 322, 554
Order (taxon), 7, *209*
Ordovician, *198–199,* 202–203
Organ, defined, *4,* 337, 341
Organelle
  defined, *4,* 45
  evolution, 230–231
  functions of, 45–46
Organic compound (*See also* Carbon
  compound)
  condensation, 27–28
  families of, 27
  functional groups, 27, *27*
  hydrolysis, 28
  spontaneous assembly of, 195, *195*
Organic soup, 195
Organ of Corti, *443*
Organogenesis, 465, *465,* 469
Organ system
  defined, *4*
  human, overview of, *340–341,* 341
  and internal environment, 336–337, *342*
Orgasm, 473
Origin of life, 194–196
Osmosis, 42, *42, 43*
Osteoarthritis, 352
Osteoblast, 350
Osteocyte, *349,* 350
Osteoporosis, 350
Ostrich, 286
Outer ear, 442, *443*
Oval window, *443*
Ovary
  development, 129
  as endocrine gland, *454*
  flatworm, *265*
  human, 470*ff.,* *474*
  plant, 317, *317,* *319,* 323, *324–325,* 331
  rotifer, *269*
  structure, *471*
Oviduct, *265,* 470, *470–471*
Ovulation, 473, 483
Ovule, *319*
  conifer, 252
  defined, 318
  flowering plant, *255,* 320
  mature, 323–324

Ovum (*See also* Egg; Oocyte), 113, 475
Oxaloacetate, 77, 85, *85*
Oxidation-reduction reaction, 65
Oxidative phosphorylation (*See*
  Electron transport
  phosphorylation)
Oxygen
  and aerobic respiration, *72,* 76, 81,
    82–84, 85, *85,* 86, *86,* 90–91
  in atmosphere, 90
  atomic number, *18*
  deficiency, cellular, 416
  deprivation, 416
  diffusion across membranes, 42
  effect on evolution, 90, 202
  as electron acceptor, 86
  electron distribution, *21*
  exchange in alveoli, 412
  free, effect on biological molecules,
    194
  integumentary exchange, 407
  mass number, *18*
  partial pressures in body, 412, *413*
  and photolysis, *75*
  and photorespiration, 77
  and photosynthesis, 68–69, *69,* 74,
    76
  and plant function, 307
  transport by hemoglobin, 125, 413
Oxytocin, 342, *450,* 451, *452,* 479
Oyster, 271
Ozone, 536, 543, *544*
Ozone hole, 573, *573*
Ozone layer, 573–574

P

P680, 74, *74*
P700, 74, *74*
Pacemaker, heart, 20
Pacinian corpuscle, *440,* 441
Pain
  perception of, 441–442
  referred, 442
Paleocene, *198–199*
Paleozoic era, 191, 197, *198–199,*
    202–204
Palm, 295
Palmiter, R., 168
PAN (*See* Peroxyacyl nitrate)
Pancreas
  cancer of, *415*
  cell, of bat, *84*
  cell cross-section, 49
  disorders of, 63
  enzyme secretion, 364, *365*
  functions, 362
  human, *362*
Pancreatic islets
  alpha cells, 456, *457*
  beta cells, 456, *457*
  delta cells, 456
  and diabetes, 456
  endocrine function, *454*
Pancreatitis, acute, 63
Pangea, 197, 204
Pan, chromosome number, *106*
Papovavirus, 224
Paradisaea ragginana, 185
Paramecium, 54, 236, *236–237*, 509, *509,* 510
Parasite
  amoeboid protozoan, 235
  bacterial, *226*
  chytrid, *240,* 240–241
  defined, 510
  flagellated protozoan, 234–235
  flatworm, 267
  fluke, 265–267
  fungal, 240, *240*

habitat, 525
hookworm, 267
interactions, 514–515
pinworm, 267
vs. predator, 510
roundworm, 267–268
sporozoan, 235
tapeworm, 266, *266*, 267–268
true, 514
turbellarian, 265
water mold, 240–241
Parasitism, *508*, 509
Parasitoid, 515
Parasympathetic nerve, 434, *435*
Parathyroid gland, *454, 455*, 456
Parathyroid hormone, *454*, 456
Parazoa (*See* Sponge)
Parenchyma, 293, 295, 299, *302*, 331
Parental behavior, 596
Parietal lobe, *437*
Passive immunity, 399
Passive transport, 43
Pasteur, L., *222*, 391
Patella, *351–352*
Pathogen, 391
Pea, garden (*Pisum sativum*)
  chromosome number, *106*
  chromosome, 96
  dihybrid cross, *123*
  floral structure, 118, *118*
  Mendel's experiments with, 118*ff.*,
    128–129
  monohybrid cross, 119, *120–121*
  phototropism, *328*
  testcross, 121, *121*
Pear, 316–317
Peat, 535, *535*
Pectoralis major, *353*
Pedigree, 132, *133*
Pelagic province, 560, 562
Pelican, *193*
Pelvic girdle, 174, *174*, 351
Pelvic inflammatory disease (PID),
  485, 487
Penguin, *192, 526, 527*
Penicillin, 424
Penicillium, 243, *244*
Penis, *265, 468, 469*, 473
Peppered moth, 184, *184*
Pepsin, 63, 363, *365*
Pepsinogen, 363
Peptic ulcer, 363
Peptide bond, 154
Perch, 285
Percy, E., *352*
Pericycle, 300, *301*
Periderm, 294, 303, *303*
Peripheral nervous system, 434*ff.*,
  435, 439–440
Peristalsis, 362–363
Peritoneal dialysis, 424
Peritoneum, *260*
Permafrost, 557
Permian, *198–199*, 202, 204
Peroxyacyl nitrate (PAN), 570, *570*
Pesticide, 16–17, *16*, 184
PET scanner, 20, *20*
Petal, 317, *317, 324*
Petiole, 299
Petrel, *526*
Petroleum, 580–581
PGAL (*See* Phosphoglyceraldehyde)
Phagocyte, 402
  functions of, 376–377
  in inflammation, 393
  in nonspecific immune response,
    391
Phalange, *351*
Pharynx
  chordate, 280–281, *280–281*
  defined, 410
  earthworm, *272*

evolution, 282
flatworm, 265, *265, 361*
functions, 363
germ layers, 265
human, 363, 410, *410*
rotifer, *269*
roundworm, *266*
tunicate, *280*
Phenotype
  defined, 118
  and environment, 125
Phenylalanine, *31*, 60, 139, *153*, 368
Phenylketone, 60
Phenylketonuria (PKU), 60, 139
Phenylpyruvic acid, 139
Pheromone
  baboon, 593
  defined, 450
  functions of, 442, 450
  hangingfly, 593
  and reproductive behavior, 591–593
pH (H⁺ level)
  cell environment, 26
  cell interior, 26
  and enzyme activity, 63, *63*
  extracellular, 424
  of gastric juice, *25*
  internal, 376
  maintenance of, 43
  and oxygen transport, 413
  scale of, *25*, 25–26
  vaginal, 469
  of water, 26
Phloem
  companion cell, *313*
  function, 294
  nonvascular land plants lacking,
    248
  and plant evolution, 245
  pressure flow in, 313–314, *314*
  pressure flow theory, 314, *314*
  primary, 295, *296, 301*
  secondary, *302*
  structure, 294
  translocation, 313–314, *314*
  transport through, 313–314
  of vascular bundle, *296*
  woody stem, *303*
Phosphoglyceraldehyde, *76*, 76–77,
  77, 82, 89
Phosphoglycerate, *76*, 76–77, *77*
Phospholipid
  functions of, 30, *35*, 367
  structure, 30, *30, 35*, 40
Phosphorus
  atomic number, *18*
  deficiency in plants, *306*
  in DNA, *144*
  electron distribution, *21*
  in human nutrition, *371*
  mass number, *18*
  and plant function, *306*
Phosphorylation
  defined, 65
  substrate-level, 82, 85
Photochemical oxidant, 570, *570*
Photochemical smog, 571
Photolysis, 74, *75*
Photon energy, 71, 73–74
Photoperiodism, 329–330, *330* (*See
  also* Biological clock)
Photophosphorylation, 73, 74, *74, 75*
Photoreceptor
  compound eye, *432*
  cone, 445, *446*, 447
  dragonfly, 274
  function, 441
  invertebrate, 444
  mollusk, *444*
  rod, 445, *446*, 447
  vertebrate, 445*ff.*
Photorespiration, 77

Photosynthesis
  and aerobic respiration, 5, 68–69,
    69
  and *Cladophora* experiments, *72*
  defined, 4, 69
  electron transfers in, 71, 73–74, *74*
  by euglenids, 233
  in leaf, 297, *298*, 299
  light-dependent, 69, 71–76, *77*
  light-independent, 69–70, *71, 76*,
    76–77, *77*
  origin of, 202
  role in ecosystem, 535
  role in evolution, 90
  in soybean plant, *327*
  and stomatal action, 310, 312, *312*
  summary of, 70
Photosynthetic autotroph, 69, 225,
    523, 525, *526–527*, 558
Photosynthetic bacterium, 227–229
Photosynthetic pigment, 246–247
Photosystem, *71*, 73, 74, *74*
Phototropism, 328
Phylogeny, 208–210, *209*
Phylum (taxon), 7, *209*
Physalaemus, *594*
Physalia, *264*
Physarum, *232*
Phytochrome, 329–330, *330*, 332
Phytophthora infestans, *242*
Phytoplankton, 574
Pigeon, *10*, 506–507
Pigment, animal
  bile, 364
  carotene, 348
  and coat color, 123, 125
  hemoglobin, 124–125, 348
  melanin, *124*, 125, 348
  visual, 447
Pigment, plant
  anthocyanin, *73*
  carotene, *73*
  carotenoid, 71, 73, *73*, 247
  chlorophyll, 71, *73*, 247
  chrysophyte, 233
  photosynthetic, 234, 246–247
  xanthophyll, *73*, 246–247
Pill, the, 484, *484*
Pillar coral, *564*
Pilobolus, *241*
Pilus, 225
Pineal gland, *437, 449, 454*, 458
Pineapple, *325*
Pine barren, 556
Pine (*Pinus*), *253*, 299, *304, 309*
Pinna (pinnae), *443*
Pinworm, 267
Pioneer species, 518
Pisum sativum (*See* Pea, garden)
Pith, *296*, 297
Pituitary dwarfism, *166*, 452, *453*
Pituitary gland
  anterior lobe secretions, 450–452,
    452, 469, *472*
  intermediate lobe, *452*
  links with hypothalamus, 450,
    450*ff.*, *451*
  location, *437*, 450
  posterior lobe secretions, 450–451,
    452
  structure, 450
Placenta, 477
Placoderm, 283
Plakobranchus, 231
Planarian, 265, *265*
Plankton, 234, 558, *561*
Plant cell
  central vacuole in, 47, *47*
  cytokinesis, 101, *101*
  middle lamella, *101*
  organelles in, *46–47*
  osmosis effects, 43

wall characteristics, 46, *46*, 55, *55*,
  101
Plant development (*See also*
  Development)
  asexual reproduction, 316–317
  embryonic, 323–325, *323–325*
  endosperm formation, 318, *319*
  and enviromental rhythms, 327
  flowering process, 330
  fruit formation, 323–324
  gamete formation in, 317–318
  growth, factors affecting, 307
  growth patterns, 326*ff.*
  interrelated processes affecting, *312*
  life cycle, *245*
  and meristems, *294*, 294–295
  nutritional requirements, 307–309
  oak as case study, 331–332
  patterns of, 326*ff.*
  primary growth, 293*ff.*
  runner, 104, 316
  secondary growth, 303*ff.*
  seed dispersal, 323–324
  seed germination, *326*, 326–327
  sexual reproduction, 316*ff.*
  tropism, 328*ff.*
  and water absorption, 309
Plant (Plantae) (*See also* Flowering
  plant; specific division)
  abscission, 333
  adaptive radiations, 201, 204
  alternation of generations, *245*
  and biological clock, 328–329
  C3, 77
  C4, 77
  characteristics, 245
  classification, 7, *8–9, 209*
  crop, 168
  cuticle, 30
  day-neutral, 330
  dormancy, 333
  effect of mycorrhizal fungi, *309*
  and environment, 307–309
  evolution, *199*, 203–204, 245–246
  extinctions, *201*
  fertilization, 318, *319*
  form along productivity gradients,
    550
  fossils, *204*
  genetic engineering, 168, 579
  gene transfer in, 168
  herbaceous, 303
  mineral uptake and accumulation,
    312–313
  monocot vs. dicot, *295*
  nonvascular, 245
  nutritional requirements, 307–309
  photosynthesis in (*See* Photosynthe-
    sis)
  pollination, 318, *319*
  polyploidy in, 187
  reproductive structures, *105*
  salt-tolerant, 168
  seed-bearing, *204, 252*, 320
  seed-bearing species, 250
  senescence, 330, 333
  short-day, *330*
  storage of organic compounds, 313
  survivorship patterns, 498, *498*
  symbiotic relationship with bacteria,
    307
  tissues, 292*ff.*
  transpiration in, 310, *532*
  translocation in, 313–314
  trends in evolution, *246*
  tropical, destruction of, 578–579
  tropical forest, 555–556
  vascular, 245
  water absorption, 309
  woody, *303*, 303–304
Plant wilt, 242
Planula, 264, *264*

Plaque, atherosclerotic, 30, 382, *382*
Plasma, blood, 342, *376*, 377
Plasma cell, *302*, *392*, 394–395, *395*
Plasma membrane (*See also* Membrane, cell)
  appearance in micrographs, *41*, *47–48*
  bacterial, 45, *45*, *225*
  in cancer, 160–161
  at cell junctions, 431
  defined, 38
  electrical properties, 428*ff.*
  functions of, 38, *47–48*
  ion pathways across, *428*
  lysis, 223
  microvillus, 365
  of muscle cell, *356*
  of neuron, 428*ff.*
  origins of, 196
Plasma protein, *376*, 377
Plasmid
  bacterial, 226
  and bacterial conjugation, 164, *164*
  defined, 163
  and recombinant DNA technology, *165*
Plasmodesma, 300
Plasmodial slime mold, 232, *232*
Plasmodium, 235
Plasmolysis, *43*
Plasmopara viticola, 242
Platelet, *376–377*, 378
Plate tectonics, 197, 201, *203*
Platypus, *288*
Pleasure center, hypothalamic, 439
Pleiotropy, 125
Pleistocene, *198–199*
Pliocene, *198–199*
[238]Plutonium, 20, 584
Pneumonia, 143, 225, *226*, 400
Polar body, *111*, 112, *474*, 475
Polar covalent bond, defined, 23
Polarity, of water molecule, 23, *23*
Polio, *224*, 398
Pollen grain, *255*, 316–317, *318*, 320–322
Pollen sac, 317, *318*, 320
Pollen tube, *255*, 317, *319*, 331
Pollination, 252–253, *253*, *255*, 318, *319*, 331
Pollinator, 186, *186*, *254*, 277, 316–317, 320–322, *322*
Pollutant, defined, 570
Pollution
  air, 535, *536–537*, 570*ff.*
  solid waste, 576
  water, *559*, 575–576
Polychaete, *272*, 272–273
Polydactyly, *133*
Polymerase chain reaction, 166
Polymorphism, defined, 166
Polyp, 263–264, *264*
Polypeptide chain
  antibody, 396
  assembly, 153–154
  coiled, *32*
  hemoglobin, *33*
  hydrogen bonds in, *32*
  sheetlike, *32*
Polyploidy
  in animals, 187
  defined, 187
  and hybridization, 187, *187*
  in plants, 187
  speciation via, 187, *187*
Polysaccharide, defined, 29
Ponderosa pine, *253*
Pons, 436, *437*
Population
  and age structure, 494
  biotic potential, 495
  carrying capacity, 495–496

defined, 4, *4*, 177, 179, 493
density, 494
dispersion, 494, *494*
evolution, 175–176
geographic barriers, 186
in Hardy-Weinberg equilibrium, *178*, 179
life tables, 498, *498*
reproductive base for, 494
reproductive isolating mechanisms, 186
size, 494
survivorship curves, 498, *498*
variation (*See* Variation in populations)
Population growth
  control methods, 500*ff.*
  demographic transition model, 501, *501*
  density-dependent controls, 496–498
  and disease, 496–497
  disease control and, 500
  and economics, 501
  energy stores and, 500
  exponential, 494–495, *495*, 500
  and family planning, 501–502
  and food production, 497, 499–500
  future, 500
  global, 498–499, *499*, 500, *501*
  human, 498*ff.*
  lack of control over, 499–500
  limiting factors, 496, 500
  logistic, 496, *496*
  overshoot, 497, *497*
  patterns of, 494
  rate increase of, 499–500
  zero, 503–504
Porifera, *259* (*See also* Sponge)
Porpoise, *192*, 193
Portuguese man-of-war, *264*
Positive feedback, 342, 363 (*See also* Feedback)
Positron-emission tomography (PET scan), 20
Posterior lobe (*See* Pituitary gland)
Postsynaptic cell, *431*, 432
Potassium
  absorption, 365, 419
  active transport of, 428, *428*
  atomic number, *18*
  deficiency in plants, *306*
  in extracellular fluid, 342
  in human nutrition, *371*
  mass number, *18*
  and neural function, 428*ff.*
  and plant function, *306*
  role in cell, 26
  secretion, *421*, 424
  sodium-potassium pump, *43*
[40]Potassium, 19
Potato, 242
Power grip, hand, 214
Poxvirus, *224*
P (genetics symbol), 119
Prairie, *552*
Precipitation, *532*
Precision grip, 214
Predation
  defined, 509
  effects on competition, 516
  and population growth, 510*ff.*
  and sociality, 595
Predator
  anti-predator behavior, 590
  defined, 510
  vs. parasite, 510
Predator-prey interaction, 510–514
  leopard-baboon, *511*
  lynx-hare, *512*
  prey defenses, 513–514, *513–514*
Prediction, 10

Pregnancy
  and AIDS, 401
  alcohol effects, 480–481
  and antibodies, 386
  and birth, 479, *479*
  chorion, 476, *476*
  and connective tissue, 348
  control over, 483–484
  diagnosis of genetic disorders, 140
  and Down syndrome, 137–138
  ectopic, 487
  effects of prescription drugs, 480
  fertility control, 483–484, *484*
  fetal alcohol syndrome, 480–481
  fetal development, 477–478, *477–478*
  first week of, *474*
  and lactation, 479, *479*
  and nutrition, 480
  and orgasm, 473
  and Rh blood type, 386
  risks of infection during, 480
  role of placenta, 477
  smoking effects, *415*, 481
  stretch marks, 348
  trimesters of, 478
  unwanted, 483–484
Prehensile movement, 214
Premature birth, 478
Premolar, 213
Premotor cortex, *438*
Prenatal diagnosis, 140, *140*
Pressure, turgor, *43*
Pressure flow theory, 314, *314*
Presynaptic cell, *431*, 432
Prey defenses, 513–514 (*See also* Predator-prey interaction)
Pribilof Islands, 496–497, *497*
Prickly pear cactus, 550, *551*
Primary follicle, *471*
Primary immune response, *392*, 394–398, *399*
Primary meristematic tissue, *294*
Primary oocyte, *111*, 112, 471
Primary productivity, 528, 547–548, *550*, 558
Primary root, *292*, 299, *326–327*
Primary spermatocyte, *111*, 112, *469*
Primary structure, of proteins, 32
Primary succession, 518
Primary wall, 101
Primate (Primates)
  adaptive radiation, *200*
  adaptive zones, 216
  classification, *212*
  evolution, 215–216
  groups, *212*
  habitats, 213
Probability, defined, 120
Probe, radioactive, 166
Prochlorobacterium, *226*
Prochloron, *226*
Producer
  defined, 523
  in energy pyramid, *529*
  examples of, *524*
  and trophic levels, *527*
Profundal zone, 558, *559*
Progesterone
  and ovarian function, 470
  source of, *454*
  targets of, *454*
  and uterine function, *472*, 473
Proglottid, 266, *268*
Progymnosperm, 250
Prokaryote (*See also* Bacterium; Moneran)
  body plan, *45*
  cell wall, 45, *45*
  cytoplasm, 45, *45*
  defined, 45
  nucleoid, *45*

plasma membrane, 45, *45*
reproduction, 95
ribosomes, 45, *45*
Prolactin, 452, *452*
Proline, *153*
Promoter, 151, 157–158, *158*
Prophase
  chromosome appearance, *97*
  meiosis, 106–109, *114–115*
  mitosis, 97, *98*, 100, *114*
Prosimian, *212*, 213
Prostaglandin, 399, 469
Prostate gland, *468*, 469
Protein
  abiotic assembly of, 196
  absorption, 364–365
  animal, 369
  assembly, 31
  and cell structure, 94
  chromosomal, 95
  complement, 391, 395, 402
  complete, 369, *369*
  deficiency, 369
  defined, 31, *35*
  denaturation, 33
  dietary, 368–370
  digestion, 363–364, *365*
  and essential amino acids, 368–369, *369*
  fibrous, 33, *35*, 55
  flow through cytomembrane system, *50*
  formation, 32
  functions of, 31, *35*, 94–95
  gated, 428
  globular, 33, *35*
  hormone, 458–459, *459*
  human immune-deficiency virus, 400
  incomplete, 369, *369*
  membrane, 40, 43
  membrane, in cancer, 160
  metabolism, 89, *89*, *365*, 368–369, *372*, 372–372
  metabolism disorders, 456
  and nutrition, 368–369
  oncogene-encoded, 160
  peptide bonds, 32
  plant, 369
  plasma, *376*, 377
  primary structure, 32
  quaternary structure, 33
  recognition, 160
  regulatory, 157–158
  repressor, 157–158
  R group, 31, *31*, 33
  roles of, 31
  secondary structure, 32
  sources of, *369*
  storage by plants, 313
  tertiary structure, 33
  transport by plants, 313
Protein synthesis
  bacterial, 225
  and cancer, 160
  control of, 157*ff.*
  and DNA transcription, 150–151, *156*
  and DNA translation, 152–154, *156*
  genetic code for, 152–153
  hormonal stimulation of, 452
  mutation and, 155, 157
  overview of, 150–151
  role of cytomembrane system, 154
  role of RNA in, 150–152
  summary of, *156*
  templates for, 196
Proterozoic era, 191, *198–199*
Protistan (Protista)
  cell wall, 46, 55
  characteristics, *8*
  classification, 7, *209*, 229

defined, 229
evolution, 230
flagella, 53
Proton (*See also* Hydrogen ion)
in atomic nucleus, 17
mass number, 18
positive charge of, 17
Protonephridium, 265, *265,* 269
Proto-oncogene, 160, *348*
Protostome, *261,* 269
Protozoan
amoeboid, *229, 234*
ciliated, *229, 236, 236*
extinctions, *200*
flagellated, *229, 234, 234–235*
Proximal tubule, 420, *421–422*
Prunus (*See also* Cherry), *317*
Pseudomonas marginalis, 45
P. syringae, 167
Pseudopod, 235
Pseudocoelom, *260, 266*
Psilophyton, 204
Psychedelic drug, *439,* 440
Psychoactive drug, 439–440
Ptarmigan, *524*
Pterosaur, 193
Ptychodiscus, 234
Puberty, 6, 458, 468–469, *482*
Puccinia, 242
Pulmonary artery, *379*
Pulmonary circulation, *378,* 378–379
Pulmonary vein, *379*
Punctuation, *208*
Punnett-square method, 120, *120,*
121, 121
Pupa, moth, *6*
Pupil, *435, 445, 445*
Purine, 146
Purple sulfur bacteria, *226*
Pyloric sphincter, *363*
Pyramid, ecological, 529–530, *530*
Pyrimidine, 146
Pyrrophyta, 229
Pyruvate, 81, 85*ff., 89*
Python, *174, 426, 427,* 441

Q

Quaalude, *439*
Quadriceps, *353*
Quaternary Period, *198–199*
Quaternary structure, protein, 33
Quercus agrifolia, 331–332
Q. rubra, 304

R

Rabbit, 194, *200, 454*
Rabies, *224,* 398
Radial symmetry, *259,* 260
Radiation, harmful, 72, 299, 584
Radiation therapy, 20
Radioactive dating, 19, 191
Radioactive labeling, 166, 397
Radioactive waste, 584
Radioisotope, *144*
bacteriophage labeling with, 18–19,
144
half-life of, 19
in nuclear medicine, 20
scintillation counters, 20
as tracers, 20, *144*
used in dating, 19, 191
Radiolarian, *229*
Radish, 316
Radius, *351*
Radula, 269
Rafflesia, 554
Ragweed, *318*
Rainfall, 544
and deforestation, 577

global, 544
Rain forest, 303
Rain shadow, 546
Rana pipiens, 106, 462–463, *462–463*
Ranunculus, 301, 514
Rattlesnake, *124*
Ray, 284
Ray-finned fish, 285
Ray initial, *302*
Reabsorption, in kidney, 422, 424
Reactant, defined, 61
Receptacle, *317*
Receptor, sensory
chemoreceptors, 441–442, *442*
classes of, 441
in ear (human), 442–443, *443*
function, 427, 441
hair cell, *443*
hearing, 442, *443*
in homeostatic control, 342, *343*
mechanoreceptors, 441, *441,* 441–
442, *443*
pain, 441–442
pharynx, 363
photoreceptors, 441, 444, *444,* 445,
445–446
of python, *426,* 427
recruitment, 441
skin, 342
smell, 442
somatic, 441–442
stretch-sensitive, 433–434
taste, 442, *442*
thermoreceptors, 441, *441,* 441–442
withdrawal-sensitive, 434
Receptor protein, 40, *40*
Recessive allele, 118–119, *130,* 132,
134, 134–135, *135*
Reciprocal cross, *130*
Recognition protein, 40, *40,* 160
Recombinant DNA technology, 163,
169
applications, 163
defined, 163
genetic fingerprinting, 166
vs. natural selection mechanisms,
163
restriction fragment length polymorphisms, 166
Rectum, *362, 366, 435*
Red alga, 202, 245–246, *246, 251,*
561, 565
Red blood cell
carbon dioxide transport, 412–413
in fibrin net, *385*
formation, 377
function, 377
oxygen transport, 377, 412–413
shape of, *376,* 377
sickled, *124,* 125
surface markers, 386–387
Red marrow, 349, 377
Red oak, *304*
Red tide, 234, *234*
Redwood, *8–9*
Reef organism, 203
Referred pain, 442
Reflex arc
and central nervous system, 433–
434
and spinal cord, 433
and stretch reflex, 433
and withdrawal reflex, 434
Refractory period, 430
Regulatory gene, 158, 191
Regulatory protein, 157–158
Reindeer, 496–497, *497*
Releasing hormone, 451
Rembrandt tulip, *224*
Renal artery, 420
Renal pelvis, 420, *420*
Renal vein, *420*

Renin, 422, *422*
Repressor protein, 157–158
Reproduction (*See* Asexual
reproduction; Life cycle;
Sexual reproduction)
Reproductive behavior, 590–592
Reproductive isolating mechanism,
186
Reproductive rate, and population
growth, 499–503
Reproductive success, 589, 594
Reproductive system
female, 470–474, *470–474*
functions, *341*
human, *341,* 467*ff.*
male, 467–470, *467–470*
Reptile (Reptilia)
adaptive radiations, *200,* 204
comparative embryology, *190*
evolution, *199,* 204, 285–286
evolutionary links with birds, 176
extinction, *200*
tectum, 436
Research tools, examples of
bacteriophages, *144*
colchicine, *128*
computer simulations, 540
Drosophila, 130–131
electrodes, 430, *430*
gene sequencers, 167
Hardy-Weinberg principle, *178,* 179
oscilloscope, 430, *430*
polymerase chain reaction, 166
Punnett-square method, 120, *120–*
121
radioactive dating, 191
radioactive labeling, *144*
restriction enzymes, 164
reverse transcriptase, 165
rules of probability, 120–121
sheep brains, 451
viruses as, *144*
X-linked genes, *130–131*
Resource-defense behavior, 591
Resource partitioning, *514,* 515
Respiration (*See also* Breathing)
aerobic (*See* Aerobic respiration)
defined, 406
Respiratory system
and acid-base balance, 424
arthropods, 274
bivalves, 271
bony fish, 285
cephalopods, 271
and cigarette smoke, 414–415, *415*
controls over, 413, 416
evolution, 282
functions, *341*
gas exchange in, 406–409, 411–412
gas transport in, 412
gastropods, 270
human, *341, 410,* 410*ff.*
links with digestive system, *360,*
360–361
links with other organ systems, *406*
mammals, 288
mollusks, 269
partial pressures in, 412, *413*
reptiles, 286
roles of, *406*
vertebrate, 282
water loss by, 419, 423
Respiratory tract infections, *224*
Resting membrane potential, *428,*
429*ff., 430*
Restriction enzyme, 164–165, *165,* 166
Restriction fragment, 164, *164,*
165–166
Restriction fragment length
polymorphism (RFLP), 166
Reticular activating system, 438
Retina

formation, *466*
human, 445, *445–446,* 447
invertebrate, 445, *445*
vertebrate, 445, *445–446,* 447
Retrovirus, *224,* 400
Reverse transcriptase, 165, 400, *401*
Reverse transcription, RNA viruses,
223
RFPL (*See* Restriction fragment length
polymorphism)
R group, amino acid
defined, 31
interactions, 33
structure, 31
Rh blood typing, 386–387
Rheumatoid arthritis, 352, 399
Rhinovirus, 224
Rhizobium, 226, 229
Rhizome, 316
Rhizopus, 241, 242
Rhodopsin, 447
Rhythm method, of birth control, 483,
484
Rib cage, *351*
Riboflavin, *369*
Ribose, 28, 64, *64*
Ribosomal RNA, *156*
Ribosome
appearance in micrographs, *46*
bacterial, 45, *45*
and endoplasmic reticulum, 50, *51*
eukaryotic, 46
functions of, *47–48*
structure, 154, *154–155*
Ribulose bisphosphate, *76,* 76–77, *77*
Rice, 77
Rickets, 456
Rickettsia, 226
Ridge, oceanic, *197*
Rigor, 355
Rigor mortis, 355
Ringworm, 242
RNA
complementarity to DNA, 151
function, 34, 35
messenger (*See* Messenger RNA)
nucleotides of, *151*
ribosomal (*See* Ribosomal RNA)
and self-replicating systems, 195–
196
structure, 34
synthesis, 151, *152*
transcripts, 151–152, *152*
transfer (*See* Transfer RNA)
RNA polymerase, *152,* 157–158, *158*
RNA virus, 223, *223,* 400, *401*
Rockfish, 285
Rocky shore, 560, *561*
Rod (bacterial), *226*
Rodent, *200,* 423, *423*
Rod photoreceptor, 445, *446,* 447
Root
adventitious, 299–300, 326
cap, *292,* 300
cortex, 300, *301*
endodermis, 300, *301*
epidermis, *292,* 300, *301*
hairs, *292,* 300, 309, *309*
lateral, *292,* 299, *301,* 331
pericycle, 300, *301*
primary, *292,* 299, 326–327
senescence, 333
tip, *292,* 300, *323*
vascular column, 300, *301,* 308
vascular tissue, *292*
water movement in, *308*
Root system
defined, 299
diagram, *292*
fibrous systems, *299,* 300
functions, 245, 293, 299
structure, 293, 300, *301–302*

taproots, 299, *299*
of vascular plants, 245
Rosebud, 317
Rotifer, 266, *269*, 558
representative, *259*
Rough endoplasmic reticulum, 51, *51*
Round window, *443*
Roundworm (*See also* Nematode)
body plan, *266*
characteristics, 266
environmental pH for, 26
evolution, *260*
infections in humans, 266, 267–268
representative, *259, 268*
rRNA (*See* Ribosomal RNA)
RU-486 pill, 484
Rubella, 224
Ruminant, *361*
Runner, plant, 104, 316
Runner's knee, 352
Rye, 242, 299

S

Sabin polio vaccine, 398
Saccharomyces cerevisiae, 241
Sac fungus, 241, *241,* 242
Sacral nerve, 435
Sacrum, *351*
Safe sex, 486
Sage, *186*
Sage grouse, 591
Saguaro cactus, 550, *551*
Salamander, 285
Salination, 168, 574
Saliva, 363
Salivary amylase, 363, *365*
Salivary gland
functions, *362,* 363
human, *362,* 363
innervation of, *435*
insect, 234
Salix, 301
Salmon, 283, 285
Salmonella, 226
Salt (*See also* Solute)
defined, 26
dietary, 381
dissolved, 26
ionic bonding in, 22, *22*
role in cells, 26
Salt marsh, *560*
Salvia, 186
Sandy shore, 560
Saprobe, 240
Sapwood, *303*
Sarcoma, *224*
Sarcomastigophora, *229*
Sarcomere, 354, *354–355*
Sarcoplasmic reticulum, 355, *356*
Sarcoscypha, 241
Sardine, 283
Sartorius, *353*
Saturated fat, 30
Savanna, *553*
Scala tympani, *443*
Scala vestibuli, *443*
Scale
of fish, 283
of leaves, 299
Scallop, 271
Scalp
hair, 348
human, 348
nutrition effects on, 348
Scanning electron microscope, *39*
Scapula, *351*
Scarab beetle, *276*
Scarlet cup fungus, *241*
Schally, A., 451
Schistosoma japonicum, 267

Schistosomiasis, 267
Schleiden, M., *38*
Schlerophyllous woodland, *548*
Schwann, T., *38*
Schwann cell, 431, *432,* 434
Science, methods of, 143*ff.*
Scientific explanation, 13
Scientific method, 10–13
Sclera, 445, *445*
Sclerenchyma, 293, *296*
Scolex, 266, *266, 268*
Scorpion, 274
Scorpion fish, *514*
Scotch broom, *322*
Scrotum, 467
Scrubjay, 331–332
Sea anemone, *263,* 263–264, *564*
Sea cucumber, 278
Seafloor spreading, 197
Sea horse, *284*
Seal, *185, 526*
Sea lamprey, *283*
Sea lettuce, 247
Sea lion, *563*
Sea nettle, *262*
Sea palm, *239*
Sea slug, 270, *270* (*See also*
Nudibranch)
Seasonal variation
in climate, 544, *545*
in lakes, 558–559
Sea squirt (*See* Tunicate)
Sea star, 278, *278,* 279, 516, *561, 564*
Sea turtle, *286*
Sea urchin, 97, 278, *278*
Secondary immune response, *392,*
398, *399*
Secondary oocyte, *111,* 112, *471,*
471–472, 475
Secondary spermatocyte, *111,* 112, *469*
Secondary structure, protein, 32
Secondary succession, 518, 520
Second law of thermodynamics, 59
Second messenger, 459, *459*
Secretin, *454*
Secretion
digestive system, 361, 363–364
hydrochloric acid, 363
of signaling molecules, 450*ff.*
Sedimentary cycle, 531
Seed
angiosperms, 253
coat, *323,* 326, *327*
defined, 323
dispersal, 323–324
flowering plants, 253, 323–324
formation, 323, 331
germination and early growth, 326–
327, *326–327*
leaves, 323
oak, 331–332
pine, 253
sequoia, 520
Seedling
growth responses of, 326*ff.*
oak, 332
phototropism, 328
sunflower, *329*
Segmentation, in body plan, 261
Segregation, allelic, 119
Selection (*See also* Natural selection)
artificial, 9, 163
and convergence, *192,* 193
defined, 185
directional, *181,* 184, *184*
disruptive, *181,* 184
and divergence, 185–186, *186,* 191,
192, 193
and evolution's ratchet, 185
kin, 597
natural, 117, 175–176, 181–185, 589
nature of, 185–186

sexual, 184–185, *185,* 590–591
and sexual reproduction, 105
stabilizing, *181,* 182, *183*
and variation in populations, 105,
117, 175–176
Self-fertilization, 118
Selfish behavior, 580, 589
Selfish herd, 595
Self-nonself recognition, *393*
Self-replicating system, 195–196
Semen, 469
Semilunar valve, *379,* 380, *380*
Seminal vesicle, *468,* 469
Seminiferous tubule, 468, *468–469*
Senescence, 330, 333
Sensory neuron (*See also* Receptor,
sensory), 427, 433
Sepal, 317, *317, 324*
Septum, *379*
Sequoia, 520
Serine, *153*
Serotonin, 438, 440
Seta (setae), 272, *272,* 272–273
Sex cell (*See* Gamete)
Sex chromosome
abnormalities, 138–139
vs. autosome, 129
of *Drosophila melanogaster, 131*
homologous pairings of, 96, 106
human, *129*
Sex (gender) determination, 129, *129*
Sex hormone, 31
Sex-linked gene, 129 (*See also*
X-linked gene)
Sexual characteristics, 129, 458
Sexually transmitted disease (STD),
400–401, 485–487
Sexual reproduction (*See also* Life
cycle; Reproduction)
animal, 259, 464
bacterial conjugation, 164, *164*
characteristics of, 104–105, 113
Chlamydomonas, 247
cnidarian, 264
defined, 95, 104
flatworm, 265–266
flowering plant, 316*ff.*
fungus, 240, *240*
gamete formation, 95–96, 129
gene mutation and, 104–105
and genetic variation, 105, 129
human (*See* Human reproduction)
oxytocin role in, 451
planarian, 265
plant, 245, 316*ff.*
Spirogyra, 248
sponge, 263
watersilk, 248
Sexual selection, 184–185, *185,*
590–591
Shark, *192,* 284, *284,* 351
Sheep, bighorn, *8–9*
Shelf fungus, 243
Shell
amniotic egg, *286*
animals with, 202
coconut, 294
diatom, *233*
snail, 177, *177*
Shepherd's purse, *323*
Shigella, 226
Shingles, *224*
Shock, circulatory, 399
Shoebill, *193*
Shoot system
defined, 293
flowering plants, *292,* 295–299
functions, 293
vascular plants, 245
Shortgrass prairie, *552,* 552–553
Shrubland, *548, 551,* 551–552
Siamese cat, *63*

Sickle-cell anemia, *124,* 125
abnormal cell shape in, *124,* 125
amniocentesis for, 140
detection, 166
mutations causing, 157
Sickle-cell hemoglobin, *157*
Sieve plate, *313*
Sieve tube, *313*
Signal conduction, heart, 380, *380*
Signaling molecule
defined, 450
responses to, 458–459, *459*
types of, 450
Silica, 233, 250
Silk moth, 442
Silurian, *198–199,* 202
Sink region, plant, 314, *314*
Sinoatrial node, 380, *380*
Sister chromatid
defined, 96, 106
in meiosis, 106
in mitosis, 100
separation of, 110
Skate, 284
Skeletal muscle
contraction, 353–355, *354–355*
fine structure, *354,* 354–355
sarcoplasmic reticulum, 355, *356*
tissue, *338,* 339, 353
Skeletal-muscular system
interactions, 355–356, *356*
major skeletal muscles of, *352*
Skeletal system (*See also* Bone)
functions, *340,* 348*ff.*
human, *340,* 348*ff.*
Skeleton (*See also* Bone)
appendicular, 350, *351*
axial, 350, *351*
bird, *287,* 288
cartilaginous, 283–284
coral, 264
endoskeleton, 285, 288
evolution, 214, 231
fish, 284
in fossil record, 189–190
frog, 285
human, 212–213, 350*ff.*
internal, 231
intervertebral disks, 350, *351*
joints of, 352, *352*
Lucy (australopith), 217
mammal, generalized, *351*
reptile, 286
structure, 350*ff.*
vertebrate, 212–213
Skin (*See also* Integumentary system)
in aging, 348
amphibian, 285
anhidrotic ectodermal displasia,
159, *159*
as barrier, 391
cancer, *348*
color, 125, 348
frog, 427
functions, 337, 342
functions of, 346
hormonal effects on, 452
integumentary exchange, 407
pigmentation, 125, *452*
radiation damage, 299, 348, *348,*
573–574
receptors, 342
structure of, 346–348, *347*
tactile receptors in, *441,* 441–442
tissue, 337
ultraviolet radiation effects, 348
Skua, *526*
Skull
Australopithecus afarensis, 218
A. africanus, 218
A. boisei, 218
A. robustus, 218

australopiths, 217
bones, *437*
early *Homo,* 219
H. erectus, 218, *219*
H. sapiens, 218, *219*
human vs. chimp, 191, *191*
Slash-and-burn agriculture, 578
Sleep, 438, 455
Sliding-filament model, 354, *355*
Slime mold, 229, 232
Slug, 270, *270*
Small intestine (*See* Intestine, small)
Smallpox, 224
Smell, sense of, 436, 438
Smog, 570–571
Smoke
    carbon monoxide in, 381
    cigarette, 381, 414, *415*
    marijuana, 414
Smoker's cough, 415
Smoking (*See also* Cigarette smoking;
        Tobacco smoking)
    and bronchitis, 414
    and cardiovascular disorders, 381
    effects of, 439
    and emphysema, 414
    marijuana, 440
Smooth endoplasmic reticulum, 50, *51*
Smooth muscle
    hormonal effects on, *450,* 473
    tissue, *338,* 339, 353
Smut, 242
Snail
    alleles, *177,* 179
    characteristics and development, 270
    eye, 444, *444*
    variations in, 177, *177*
Snake
    adaptive radiation, 286
    albino, *124*
    characteristics, 286
    feeding behavior, *287, 590*
    pelvic girdle, 174, *174*
Snapdragon, 123, *123*
Snowy owl, *525*
Social behavior
    costs and benefits of, 594–595
    defined, 592
    human, 598
    and predation, 595
    and self-sacrifice, 596–598
Sodium
    absorption, 365, 419
    active transport of, 366, 428, *428*
    atomic number, *18*
    electron distribution, 21, *21*
    and high blood pressure, 371
    and human nutrition, 371, *371*
    intake, 371
    and neural function, 428*ff.*
    reabsorption, 422, *422*
    reabsorption in kidney, 422, 424
Sodium chloride (NaCl)
    dissociation of, 24–25
    formation, 26, *26*
    ionic bonding, 22, *22*
Sodium hydroxide, 25
Sodium-potassium pump, 43, 428*ff.*
Soft drinks, 439, *439*
Soft spots (newborn), 352
Soil
    and acid deposition, 571–572, *572*
    characteristics, 547, *549*
    and ecosystem distribution, *549*
    and irrigation, 168
Solar radiation, 543–544, *545*
Soldier fish, *284*
Solid waste, 576
Solute, 24 (*See also* Salt)
Solvent
    defined, 24
    water as, 24–25

Somatic cell, defined, 95
Somatic senses, 441–442
Somatic system, 434, *434*
Somatomedin, *454*
Somatostatin, *454*
Somatotropin, *also called* Growth
        hormone
    binding effects, 40
    cloned gene for, *166,* 168, *169*
    effects, *452*
    excessive, 452, *453*
    secretion of, *452*
    targets of, 452, *452*
Song, bird, *586,* 587, *587*
Sorus, 250
Source region, plant, 314, *314*
Soybean plant
    development, *327*
    as protein source, *369*
    senescence, *332*
Spanish influenza, 225
Sparrow, *586,* 591
Spartina, 560, *560*
Speciation
    and divergence, 185–186, *186*
    and geographic barriers, 186
    and latitude, 521
    by polyploidy, 187, *187*
    and reproductive isolating mecha-
        nisms, 186
    in the tropics, 521
Species
    classification of, 7, 208, *209*
    destruction in tropical forests, 578–
        579
    dispersion, 508
    diversity of, 7
    diversity patterns, 507–508, 520–521
    evolutionary view of, 175–176
    interactions, 508*ff.*
    introductions, 516–517, *517*
    local instability, 520
    pioneer, 518
    relative abundances, 508
    succession, 518, *518–519,* 520, *520*
Sperm (*See also* Gamete; Oocyte)
Sperm, animal
    defined, 95
    formation, 95, *111,* 112, 467–468
    human, *469*
    maturation, 469
    penetration of egg, *474*
    and reproductive strategies, 464
    structure, 468, *469*
Sperm, plant, 316, 318, *318*
Spermatid, *111,* 112, *469*
Spermatocyte, *111,* 112, *469*
Spermatogenesis, 110, *111,* 467–469,
        469
Spermatogonium, *469*
Spermicidal foam, 483, *484*
Spermicidal jelly, 483, *484*
Sphagnum moss, 26
Sphenophyte, *183*
Spheres of hydration, 25, *25*
Sphincter, 362, *363,* 366, *435*
Sphygmomanometer, *385*
Spicule, roundworm, *266*
Spider, 274, *275, 527*
Spider monkey, 215
Spinach, 330, *330*
Spinal cord
    animals with, 202
    functional connections in, 436
    and nerves, *435*
    organization, 436, *436*
Spindle, microtubular, 97, *98–99, 100,*
        106, *108,* 109
Spiracle, *281,* 408
Spirillum, 225
Spirochete, *226,* 228
Spirogyra, 248

Spleen, *387,* 388
Split ends, hair, 348
Sponge, *261,* 261–263, *262*
Spontaneous assembly of organic
        compounds, 195, *195*
Sporangium, 248, 250
Spore
    conifer, 252
    fungal, 240, *240*
    plant, 202, 245
Sporophyte
    compared to gametophyte, 316
    of flowering plants, 316, *319*
    of vascular plants, 245–246
Sporozoan, *229,* 235, *235*
Sporozoite, 235
Sports medicine, 352
Spring overturn, lake, 558
Spruce tree, *572*
Squid, 270, *527*
S-shaped (sigmoid) curve, 496, *496*
Stabilizing selection, *181,* 182, *183*
Staining, in microscopy, *128*
Stamen, 118, *118,* 317, *317,* 324
Staphylococcus, *226*
Starch
    formation, 45, 70, *76–77*
    function, 29
    storage by plants, 313
    structure, 29, *35*
    transport by plants, 313
Starvation, 168, 180, 369, 483, 498
Stearic acid, structure, *29*
Stem
    buds, *292,* 297, *297–298*
    dicot, *295–296*
    growth, 302
    internode, *292*
    leaves, *297,* 297–299, *298*
    monocot, *295–296*
    node, *292*
    oak, 331
    primary structure, *296*
    senescence, 333
    of shoot system, *292*
    vascular bundles, 295, *295–297*
Stem cell, *377,* 378
Stephanotis, *322*
Sterility, 597–598
Sternum, *351*
Steroid
    anabolic, 357
    functions of, 31, *35*
    structure, 31, *31, 35*
Steroid hormone, 357, 458, *459*
Stickleback, 588
Stigma, *317*
Stimulant (drug), 439, *439*
Stimulus
    defined, 441
    receptors for, 441
    types of, 441
Stinkbug, *276*
Stinkhorn fungus, *8–9*
Stoma
    defined, 299
    of flowering plants, 294, 299
    function, 299
    guard cells, 312, *312*
    location, 299
    location of, 312, *312*
    transpiration through, 310
    of vascular plants, 245
Stomach
    acidity, 363
    emptying, 363–364
    enzymes active in, 363
    functions, *362,* 363
    gastric fluid, 363
    human, *362–363,* 363–364
    hydrogen ion effects, 363
    lining, 363

protein digestion, 363
ruminant, *361*
Stomatolite, 202, *202*
Stone, kidney, 420
Stone tool, 218, *219*
Stork, *192*
Strawberry plant, 104, 167, *167,* 324,
        325
Streptococcus, 226
Streptococcus pneumoniae, 144
Streptomyces, *226*
Streptomycin, 480
Stress
    and allergies, 399
    and skin disorders, *348*
Stress electrocardiogram, 382
Stress response, 455
Stretch marks, 348
Stretch reflex, 433
Strip mining, 582
Stroke, *166,* 381
Stroma, *53,* 71, *71,* 76
Style, of carpel, 317
Styrofoam, 574
Subatomic particle, defined, *4*
Suberin, in cell walls, *55*
Suborder (taxon), *209*
Substrate
    defined, 61
    transition state of, 62
Substrate-level phosphorylation, 82, 85
Succession, of species, 518, *518–519,*
        520, *520*
Succinate, *85*
Succinyl-CoA, *85*
Sucrose (*See also* Glucose; Sugar)
    formation, 70, *76–77*
    functions, 28, *35*
    storage in plants, 313
    structure, 28
    transport in plants, 313
Suction pad, cephalopod, 271
Sugar, 27 (*See also* Glucose; Sucrose)
Sugarbeet, 168, 299
Sugarbird, *185*
Sugarcane, 18, 77, 254
Sugar phosphate, *71,* 76, 77
Suicide, 597–598
Sulfolobus, 226
Sulfur
    atomic number, *18*
    in bacteriophage proteins, *144*
    deficiency in plants, *306*
    electron distribution, *21*
    in human nutrition, *371*
    mass number, *18*
    and plant function, *306*
Sulfur dioxide, *26,* 570, *573*
Sulfuric acid, *570,* 571
Sulfur oxide, 571
Sulfur trioxide, *570*
Sun, 59
Sunburn, *348*
Sunflower, *293, 329*
Sunlight (*See* Energy, light)
Supercontinent, 197
Superfamily (taxon), *209*
Supernatural, defined, 13
Suppressor T cell, 393, 396, *402*
Surgery
    cancer treatment using, 396
    for skin cancers, *348*
Surtsey, *520,* 521
Survivorship curves, 498, *498*
Sweat gland, 159, *159*
Sweating, 419, *437,* 455
Sycamore leaf, *19*
Sylvatic plague, 540
Symbiosis (*See also* Mutualism)
    bacteria and plants, 307, *307*
    examples of, 244, 307, *307*
Symbiotic origin

of chloroplasts, 230
of mitochondria, 230
Sympathetic nerve, *435*
Synapse, chemical
  defined, 431
  excitatory, 432
  inhibitory, 432
  at neuromuscular junction, *431,* 431–432, *432*
  psychoactive drugs affecting, *439,* 439–440
  and transmitter substance, 431, 438
Synaptic cleft, 431, *431,* 432
Synaptic integration, 432–433
Synaptic vesicle, *431*
Syphilis, 228, 486–487
Systematic observation, 11
Systemic circulation, *378,* 378–379, 384
Systole, 380

T

Tachycardia, 383, *383*
Tactile signaling, 594, *595*
Taena saginata, 268
Taiga, 556
Tallgrass prairie, *552,* 553
Tangelo, 163
Tanning, skin, 348, *348*
Tapeworm, 266, *266,* 267, *268*
Taproot system, 299, *299*
Target cell, defined, 450
Tarsal, *351*
Tarsier, *212, 215*
Taste bud, 442, *442*
Taste receptor, 442, *442*
T cell
  classification, 378
  helper, 393–394, *402*
  and immune response, 402
  killer, 393, 396, *402*
  suppressor, 393, 396, *402*
TDF (testis determining factor), 129
Tea, 439, *439*
Tectorial membrane, *443*
Tectum, 436, *437*
Teeth
  Australopithecus afarensis, 217
  Australopithecus boisei, 217
  early *Homo,* 218
  evolution, 214–215
  Homo sapiens, 218
  human, 214–215, 363
  mammalian, 213
  monkey, 214
Teleost, 285
Telesto, 263
Telophase
  chromosome appearance, *97*
  meiosis, 106, 108–109, *115*
  mitosis, 97, *99,* 100, *115*
Temperate climate zone, *545*
Temperate deciduous forest, *548,* 555, 556
Temperate rain forest, 556
Temperature, animal body
  control of, 342
  and enzyme function, 63, *63*
  regulation, 376, 419
  and spermatogenesis, 468
Temperature, environmental
  and dormancy, 333
  and greenhouse effect, 536–537
  and rainfall, 544
  and seed germination, 326
Temperature-stabilizing effects, of water, 24
Temporal lobe, *437*
Terminal bud, *292*
Termite, 592, *592,* 594–595
Tertiary Period, *198–199*

Tertiary structure, of proteins, 33
Testability, of hypotheses, 11
Testcross, 121
Testicular feminization syndrome, 458
Testing
  for AIDS, 486
  organization of results, 11–12
  randomization, 11
  through experiments, 10
Testis
  anabolic steroid effects, 357
  development, 129
  endocrine function, *454*
  flatworm, *265*
  formation, 467–468
  functions of, *468*
  hormone production by, 469
  location, *468*
  roundworm, *266*
  structure of, *468*
  testosterone secretion, 469
Testis determining factor, 129
Testosterone
  and hair growth, 348
  and male reproductive function, 469
  megadoses of, 357
  primary actions, *454*
  and protein synthesis, 458
  pubertal production of, 125, 357
  secretion, 469
  and song system of birds, 587
Tests
  generalizing from results, 12
  significant differences, 12
  statistical, 12
Tetanus, 87, 399, 432
Tetracycline, 480
Thalamus, 436, *437*
Thalidomide, 480
Theobroma cacao, 325
Theory, defined, 12
Thermal inversion, 570–571
Thermoacidophile, *226,* 227
Thermocline, 558, *559*
Thermodynamics, laws of, 59
Thermoplasma, 226
Thermoreceptor, 441, *441,* 441–442
Thigmotropism, 329
Thirst, 422, 436
Thoracic duct, *387*
Thoracic nerves, *435*
Thoracic vertebrae, *351*
Thorium-232 isotope, 19
Thorn forest, *548*
Threonine, *153,* 368
Threshold, 429
Thrombin, 61
Thrombus, 382
Thylakoid membrane, 70, *70–71*
Thymine, 145–147, *145–147,* 151
Thymosin, 454, *454*
Thymus
  endocrine function, *454*
  and immunity, 388, *454*
  location, *387*
  and lymphocytes, 388
Thyroid gland
  and body temperature, 455
  disorders, 20, 455–456
  endocrine function, *454,* 455–456, *459*
  as endocrine target, *452*
  location, 455
  structure, *455*
Thyroid-stimulating hormone, *452*
Thyrotropin-releasing hormone, 451, *452,* 456
Thyroxine, *454,* 455
Tibia, *351,* 352, *352*
Tibialis anterior, *353*
Ticks, 274
Timothy grass, *47*

Tissue, animal (*See also* specific type)
  blood, 339
  bone, 339
  cartilage, 339
  and cell adhesion, 160
  connective, 337–339
  defined, *4,* 337
  epithelial, 337
  formation, 465
  germ layers, 465
  lung, 411–413
  muscle, 339, 341
  overcrowding with cancer, 160
Tissue, plant (*See also* specific type)
  defined, *4*
  dermal, *292*
  meristem, *294*
  organization, 294*ff.*
  of primary growth, 294*ff.*
  of secondary growth, 303*ff.*
  vascular, *292*
  woody, 303
Tissue plasminogen factor, *166*
T4 lymphocyte, 400
Toad, 285
Tobacco smoking (*See also* Cigarette smoking; Smoking)
  benefits of quitting, 415
  effects during pregnancy, 481
  effects of, 439
Togavirus, *224*
Tomato, 242
Tongue, taste buds, 442, *442*
Tonsils, *387,* 388
Tools
  early *Homo,* 218, *219*
  stone, 218, *219*
Topography, and climate, 546
Topsoil, *549*
Torsion, gastropod, 270
Tortoise, *286*
Toxin
  Clostridium botulinum, 228
  Escherichia coli, 229
Tracer, radioisotopes, 20
Trachea
  arthropod, 274
  bird, *409*
  human, 363, 410
  insect, 408, *409*
  spider, 408
Tracheid, *310*
Trachops cirrhosus, 594
Tradewinds, 545
Trait, adaptive, 8
Transcription
  controls over, 157, *158*
  mechanisms of, *156*
  rate of, 157
Transcript processing, 151–152, *152*
Transfer RNA, 152–154, *156*
Transformation
  cancerous, 160–161
  genetic, 144
Transformed cell, 144
Transfusion, blood, 386
Transition state, of substrates, 62
Translation
  defined, 151
  and genetic code, 152–153
  stages of, 154, *156*
Translocation, 136, 313, *313*
Transmission electron microscope, *39*
Transmitter substance
  and chemical synapses, 431–432
  defined, 431, 450
  release of, 431, *431*
  and sleep, 438
Transpiration, 310, *532*
Transplant, organ, 396
Transport protein, 40, *40*
Transport vesicle, *50*

Transposable element, 157, *157*
Tree fern, *19,* 554
Tree ring, 304, *304*
Tree shrew, 215
Treponema, 226
Treponema pallidum, 486, *487*
Triassic, *198–199,* 204
Trilobite, *200,* 202–203, *203,* 273
Triceps, *353,* 356, *356*
Triceratops, 206
Trichinella spiralis, 267–268, *268*
Trichomonad, 8, 235
Trichomonas vaginalis, 235
Trichonomad, *8*
Trichophyton, 242
Trichoplax, 231, *261*
Triglyceride
  condensation of a fatty acid into, *30*
  dietary, 364
  functions of, 30
  metabolism, 364
  structure, 30
Triiodothyronine, *454,* 455
Triphosphate, 64, *64*
Triploid nuclei, plant, 318–319, *319*
Trisomy, 137
Trisomy 21, *137,* 137–138
Triticum, 187
tRNA (*See* Transfer RNA)
Trophic level, 525, *526, 527, 527*
Trophic structure, 529–530, *529–530*
Trophoblast, *474*
Tropical climate zone, *545*
Tropical deciduous forest, *548*
Tropical forest, *506, 549, 554,* 554–555, *577–578, 578–579*
  species diversity, 554–556
Tropical reef, *565*
Tropical savanna, *548*
Tropical scrub forest, *548*
Tropism, plant, *328,* 328–329
Trout, 283–284
True-breeding, defined, 118
Trypanosome, 234, *234*
Trypsin, 63, *365*
Tryptophan, *31,* 64, *153,* 368, *369*
T tubule, *356*
Tuatara, *287*
Tubal ligation, 484, *484*
Tubastraea, 263
Tube feet, echinoderm, 279
Tubulin, 52, *53*
Tulip, *224*
Tumor
  animal virus-related, *224*
  benign, 160
  crown gall, 168, *168*
  defined, 160
  malignant, 160
  retrovirus-related, *224*
Tuna, 285
Tundra, 556–557
  biome, *548*
Tunicate, *280,* 281
Turbellarian, 265
Turgor pressure, *43*
Turner syndrome, 138, *138*
Turtle, 286
Tympanic membrane, *443*
Typing, blood, 386
Tyrannosaurus rex, 206
Tyrosine, *31,* 139, *153*

U

Ulcer, peptic, 363
Ulna, *351*
Ultrasaur, 205
Ultraviolet radiation
  as carcinogen, 160
  effects on skin, *348*

and ozone layer, 573–574
receptors for, 441
Ulva, 247
Umbilical cord, *476, 477, 479, 479*
Ungulate, *553*
Uniramian, 273
Unity of life
molecular basis of, 147
relation to diversity, 147
Unsaturated fat, 30, 383
Upper littoral, 560, *561*
Upwelling, 562
Uracil, *150*, 151, *151*
Uranium-238 isotope, 19, 584
Urea
formation, 372, *372*
as metabolic waste, 419
reabsorption, 422, *422*
role in organic metabolism, 372, *372*
Ureter, 420, *420*
Urethra, *420, 469, 470, 473*
Uric acid, 419–420, 424
Urinary bladder, 420, *420, 470*
Urinary system (*See also* Kidney)
functions, *341*
human, *341, 420, 420ff.*
links with digestive system, *360*, 361
links with other organ systems, *418*
mammalian, *420ff.*
Urine
components of, 421
defined, 420
formation, 421–423
role in organic metabolism, 372, *372*
secretion, 424
Uterus
at childbirth, *452, 479, 479*
function, *470, 472, 473*
hormonal effects on, *472, 473*
lining, 477
structure, *471*

V

Vaccination, 163, 480
Vaccine, 398
Vaccinia virus, 398
Vacular cambium, *303*
Vacuole
central, 47, *47*, 52
contractile, *236*
Vagina, 469–470, *470–471, 474*
Vagus nerve, 435
Valdez oil spill, 581
Valine, *31, 153, 368*
Valium, *439*
Variable, categories of, 11
Variable expressivity, 123, 125
Variation in populations
behavioral traits, 177
categories of, 177
continuous, 125, *125*
and crossing over, 107
Darwin's perception of, 117, 175–176
and evolution, 105, 117, 175–176
genetic basis of, 105, 177
morphological traits, 177
and natural selection, 175–176
phenotypic, 177
physiological traits, 177
and sexual reproduction, 104–105, 129
sources of, 177, 179
Vascular bundle
defined, 295
dicot, 295, *295–296*, 297
monocot, 295, *295–296*, 297
veins, 299, *299*
Vascular cambium, *294, 296, 302, 303, 304*
Vascular column, 300, *301–302, 308*

Vascular plant (*See also* specific type)
body plan, *292*, 293–294
characteristics, *251*
with flowers, *251*
function, *292*, 245
with naked seeds, *251*
seedless, *249*
species of, *251*
structure, 245, *292*
tissues, *292ff.*
Vascular tissue, plant, *292ff.*
Vas deferens, *468*, 469
Vasectomy, 484, *484*
Vasopressin, *452* (*See also* Antidiuretic hormone)
Vegetarian diet, *369*
Vein
in cardiovascular circuits, 379
defined, 384
eucalyptus tree leaf, *299*
in leaves, 299, *299*
pulmonary, *379*
Vena cava, *379, 420*
Venous pressure, 384
Venous valve, 384, *385*
Ventricle, heart, 379, *379*
Venturia inequalis, 242
Venule, 379
Venules, 384, *384–385*
Venus flytrap, *308*, 309
Vertebral column (backbone), 350, *351*
Vertebrate (Vertebrata)
classification, 212–213
comparative embryology, 191
defined, 212
evolution, 191*ff.*, 281–283
key characteristics, 212–213
nervous system, 212–213, 434*ff.*
organ system, 212–213
photoreceptors, 445*ff.*
Verticillium, 242
Vesicle
and cell plate formation, 101, *101*
defined, 44
endocytic, *50*
exocytic, *50*
formation during endocytosis, 44, *44*
fusion during exocytosis, 44, *44*
transport, 46
Villus, intestinal, *364*
Viral infection, *166*
Virchow, R., *39*
Virgin B cell, *392, 394, 402*
Viroid, 225
Virus (*See also* specific type)
animal, 225
bacteriophage, *144*
and cancer, 160
characteristics, *144*
classification, *224*
components, 223, *223*
defined, 223
DNA, *144, 223, 223*
genetically engineered, 399
lysogenic pathway, *223*
lytic pathway, *223*
oncogenes of, 160
plant, 225
replication, *144*, 400
RNA, 223, *223*
skin, ultraviolet radiation effects, *348*
structure, *144, 223*
Vision
color perception, 447
defined, 444
evolution, 214
invertebrate, 444
primate, 214
vertebrate, 445*ff.*
Visual cortex, *438*

Visual perception, 444*ff.*
Vitamin
in blood, *376*
excessive doses of, 370
fat-soluble, *370*
and human nutrition, 370
intake during pregnancy, 480
recommended daily amount, 370
synthesis, 370
water-soluble, *370*
Vitamin A, *370*, 371
Vitamin B$_1$, *370*
Vitamin B$_2$, 369–370
Vitamin B$_6$, *370*
Vitamin B$_{12}$, 369–370
Vitamin C, *370*, 371
Vitamin D, *370*, 371, 456
Vitamin D$_3$, and calcium levels, *454*
Vitamin E, *370*
Vitamin K, 229, *370*
Vitreous body (humor), *445*
Vocal cords, 410
Volkmann's canal, *349*
Voltage difference, 429, *430*
Volvox, 231
Vomiting, 224, 368
Vulture, 5, *5*, 192
Vulva, 470

W

Wallace, A., 117, 176
Walnut tree, *297*
Wart, 224
Wasp venom, 399
Waste, radioactive, 584
Wastewater treatment, 575–576
Water
absorption by roots, 309
atmospheric, 536
cohesive properties, 24
diffusion across membranes, 42
evaporation, 24
formation in aerobic respiration, *84, 85, 85*, 86, 90–91
hydrogen bonds of, 310, *311*
and hydrologic cycle, 533–534
loss, insensible, 419
loss by plants, 310–312, *311*
loss by urinary excretion, 419, 423
movement in roots, *308*
and origin of life, 194
parasite-contaminated, 235
pH of, 26
and pressure flow theory, 313–314, *314*
properties of, 91
quality, 574–576
reabsorption in kidney, 422, *422*, 424
and seed germination, 326
solvent properties, 24–25
temperature stabilizing effects, 24
and transpiration, 310
transport, cohesion theory of, 310, *311*
transport by plants, 310–312, *311*
Water and solute balance
gains and losses, 481–419
and urinary system, 419*ff.*
Water hyacinth, *517*
Water lily, *254*
Water mold, 240–241, *241*, 242
Water molecule
diffusion, 41
hydrogen bonding, 23, *23*
hydrophilic interactions, 23
hydrophobic interactions, 23
polarity of, 23, *23*
Watershed
and acid deposition, 571–572

defined, 533
evaporation from, *532*
Hubbard Brook Valley, 533, *533*, 534
Water-soluble vitamin, *370*
Water–solute balance
in diabetes, 456
gains and losses, 423, 481–419
and urinary system, 419*ff.*
Water-vascular system, 279, *279*
Watson, J., 145–146
Wavelength, light
absorbed by chloroplasts, 73, *73*
absorbed by plant pigments, 71, *72*
Wax, *35*, 55
Weather system, 544
Weddell seal, *526*
Weight
energy needs and, 366–367, *367*
ideal, 366–367, *367*
Weinberg, W., *178*, 179
Weismann, A., 128
Welwitschia, 252
Went, F., 328
Westerlies, *545*
Wet acid deposition, 571
Whale, *200, 288, 526, 563*
Wheat, 77, 187, *187*
Wheat rust, 242
White blood cell
formation, 377
functions of, 377
in immune response, 393*ff.*
types of, 377–378
Whitefish cell, mitosis, *98–99*
White matter, 436
White-throated sparrow, *586*, 591
Whittaker, R., 208, *209*
Wild-type allele, defined, *130*
Willow, *168*, *301*
Wilson's disorder, 139
Windward, 546
Wing
bird, 288
evolution, 193, 201
Withdrawal, as birth control method, 483, *484*
Withdrawal reflex, 434
Wolf spider, *275*
Woodland, *548, 551, 551–552*
Woody plant, *303*, 303–304
Woolly mammoth, 207
Wuchereria bancrofti, 268, *268*

X

Xanthophyll, *73*, 246–247
X chromosome
as homologue of Y, 96, 106, 129
inactivation, 159, *159*
and sex-linked traits, 129
Xenopus laevis, 285
X-linked gene, 129, *130, 130–131* (*See also* Sex-linked genes)
X-linked recessive inheritance, *134*, 134–135, *135*
X-ray diffraction, 146
X rays, as carcinogen, 160
XXY condition (Klinefelter syndrome), 138
Xylem
function, 294
nonvascular land plants lacking, 248
and plant evolution, 245
primary, 295, *296, 301*
secondary, *302*
structure, 294
of vascular bundle, *296*
woody plants, 303
XYY condition, 139

Y

Y chromosome
    as homologue of X, 96, 106, 129
Yeast, *88*, 241, *241*
Yellow fever, *224*
Yellow-green algae, *229*, 233
Yellow marrow, 350
Yersinia pestis, 497
Y-linked gene, 129, *130*
Yolk, 464
Yolk sac, *286*, 476, *476*
Yucca moth, *508*
Yucca plant, *508*

Z

Zea mays (*See also* Corn)
    chromosome number, *106*
    epidermis, *294*
    grain of, *323*
Zebra, 5, *5*
Zebra finch, *587*
Zero population growth, 503–504
Zinc
    deficiency in plants, *306*
    in human nutrition, *371*
    and plant function, *307*
Z line, *355–356*
Zona pellucida, *474*
Zooplankton, 558
Zygospore-forming fungus, 241, *241*,
    242
Zygote
    defined, 95, *482*
    diploid, 318
    oak, 331

APPLICATIONS INDEX

ABO blood typing, 123, 386
Abortion, 140, 483, 484, 488
Achondroplasia, 134
Acid rain, 4, 26, *26*, 571–573
Acromegaly, 452, *453*
Accutane, 480
Acute pancreatitis, 63
Acyclovir, 487
Aerosol sprays and ozone layer, 574
African sleeping sickness, *234*
Aging
　and cancer resistance, 396
　physical effects of, 482
　skin changes during, 348
　and tobacco smoke, 348
Agriculture
　and crop rotation, 538
　and deforestation, 577, *577*,
　　578–579
　and desertification, 580, *580*, *581*
　and genetic engineering, 163, 168
　and green revolution, 576–577
　and human population growth,
　　499–500
　and irrigation, 168, 574
　land available for, 168, 576–577, *576*
　and pesticides, 16–17, 184
　and nitrogen scarcity, 538–539
　slash-and-burn, 578
AIDS, 399, 400–402, 485–486
AIDS-related complex, 485
Air pollution, 26, 570–574, *570–573*
Albinism, 125
Alcohol (ethyl alcohol)
　addiction, 440
　effects on nervous system, 440
　effects on heart, 383
　fetal alcohol syndrome, 480–481
　and pregnancy, 480
Algae, uses of, 246–247
Allergy, 17, 399
Amniocentesis, 140, *140*
Amoebic dysentery, 235
Amphetamine, 439
Anabolic steroids, 357
Angina pectoris, 382
Angiography, 382, *382*
Angioplasty, 383
Anhidrotic ectodermal displasia,
　159, *159*
Anorexia nervosa, 368
Antibiotic, *226*, 229, 480
Antidepressant drug, 439–440
Appendicitis, 366
Apple scab, 242
Arrhythmia, 383
Arteriosclerosis, 381
Arthritis, 352, 399
Asthma, 17, 399
Atherosclerosis, 31, 381–383
Aspirin, 579
Athlete's foot, 242
Atmosphere, greenhouse gases in,
　536–537, *537*
Autoimmune response, 393, 399

Bacteria
　beneficial, 87, *226*, 227, 229
　commercial uses of, 88, *226*, 229
　disease-causing, 87, *226*, 228–229

Balloon angioplasty, 382–383
Barbiturate, 480
Benign tumor, 160
Biological perspective
　on abortion, 483
　on drug abuse, 440
　on human gene therapy, 169
　on human life in the biosphere,
　　90–91, 582–583
　on human population growth,
　　498–504, 569–570
Birth control, 483–484, 501, 503
Blood pressure
　high, *166*, 371, 381
　in hypertension, 381
　and hyperthyroidism, 455
　measurement, *385*
　and obesity, 381
　and smoking, 381
Blood typing, 123, 386
Body building, and steroids, 31,
　357–358, *357*
Body weight (human), 366–367, *367*,
　368, 480
Botulism, 87, 229
Bradycardia, 383, *383*
Brain scan, *20*
Breast cancer, 366
Bronchitis, 17, 414, *415*
Brown rot, 242
Bubonic plague, 497, *499*
Bulimia, 368

Caffeine, effects of, 383, 439
Calcium deficiency, 350
Cancer
　AIDS-related, 400
　bladder, *415*
　breast, 366
　and cigarette smoking, 160
　colon, 366
　imaging, 397, *397*
　and immune system, 396–397
　liver, anabolic steroid-related, 357
　lung, 299, 415, *415*
　mouth, 299, *415*
　and nuclear accidents, 584
　skin, 348, 400
　treatment, 20, *166*, 396–397
Candida infection, 242, *244*
Capillary (blood), and obesity, 379
Carbon cycle, and global warming,
　535, 536–537
Carbon monoxide, *570*
　in cigarette smoke, 381
　poisoning, 416
Cardiovascular disorders, 381–383
Cellular oxygen deficiency, 416
Cheetah, vulnerability to extinction, 180
Chernobyl, 584
Chestnut blight, 242
Childbirth, 342, 352, 479, *479*
Chin fissure, *104*
Chlamydial infection, 487
Chlorofluorocarbon (CFC), 536, *537*,
　570, 574
Chocolate, 439, *439*
Cholesterol
　and cardiovascular disorders, *30*, 31,
　　381–382
　dietary, 367–368
　effects of anabolic steroids, 357
　plaque formation and, 382–383

Chorionic villi sampling (CVS), 140
Chromosome abnormality, 132, *132*,
　134–135, 135–139, 140
Cigarette smoking
　and cancer, 160, 299
　effects during pregnancy, 481
　and lung cancer, 415, *415*
　risks related to, 414–415, *415*
Cleft lip, 139
Clotting, blood, 339, 377–378, 384
Cocaine, 299, 439, *439*
Codeine, *439*
Coffee, 439, *439*
Color blindness, 141
Condom, 484, *484*, 486
Conifer, commercial uses of, 253
Contraceptive methods, 483–484, 501,
　503
Coronary bypass surgery, 382–383,
　383
Cowpox, *224*, 390–391, 398
Crack cocaine, 439
Cri-du-chat, 135
Crop plants
　algae as, 247
　examples, 187, *187*, *325*
　and fertilizers, 307
　and genetic engineering, 163, 168
　and global warming, 536
　and insect attack, 16–17, 277
　and irrigation, 168, 549, 574
　salt-tolerant, 168, 549
　and soils, 168, 549
　viral diseases of, 225
　viroid attack on, 225
　and weeds, 16

DDT, 12, 17, 539–540
Death, aging and, 482
Deforestation, 498, *533*, 534, 537, 577,
　577–578, 578–579
Depressant, *439*, 439–440
Depression, 357, 439–440
Desertification, 498, 550, 580, *581*
Dexedrine, *439*
Diabetes, *166*
　mellitus, 381, 456
Dialysis, 424
Diaphragm (contraceptive), 483
Diet
　carbohydrates in, 367
　and diabetes, 456
　and dieting, 368, *368*
　and emphysema, 414
　lipids in, 367–368
　and mental retardation, 369
　and pregnancy, 480
　protein-deficient, 369
　proteins in, 367–368
　salt intake, 381
　vegetarian, 369, *369*
Dieting, 368
Digitalis, 299
Diptheria, 399
Down syndrome, *137*, 137–138, 140
Drought, 68, 580
Drug(s)
　abuse, 299, 439–440
　from plants, 299, 579
　and pregnancy, 480
　psychoactive, 439–440
　use, and AIDS, 485

Drug therapy, 163, 396–397
Dry acid deposition, 571
Dust Bowl, *580*
Dutch elm disease, 242, *517*
Dwarfism, pituitary, *166*, 452, *453*
Dysentery, 235

Electrocardiogram, 382–383
Elephantiasis, 268, *268*
El Niño, 562, 566–567
Embolus, 382
Emphysema, 414, *414*
Endometriosis, 473
Entamoeba infection, 235
Ergotism, 242
Erythroblastosis fetalis, 386
Essential amino acids, 368–369, *369*
Estuary, pollution of, 560
Eugenic engineering, 169
Exercise
　and cardiovascular disorders, 381
　and diet, 366–367
　and muscle fatigue, 87, 355

Farsightedness, *445*
Fetal alcohol syndrome, 480–481
Fever, 63, *224*
Fever blister, *224*
Fight-flight response, 434, 455
Fluke, parasitic, 265, *267*
Fossil fuel, 500, *534–535*, 535,
　537, 581–582
Fungus
　commercial use of, 88, 241
　disease-causing, 241, *242*, 243
　medical use of, 242, 243

Galactosemia, 133
Gene therapy, 168–169
Genetic counseling, 139–140
Genetic disorder (human)
　vs abnormality, 132
　autosomal dominant, 134, *134*
　autosomal recessive, 132–134, *133*
　chromosome number-related,
　　137–139
　chromosome structure-related,
　　135–136
　counseling for, 139–140
　diagnosis of, 140
　and diet modification, 139
　and gene therapy, 168–169
　prenatal diagnosis of, 140
　preventive measures for, 139
　screening for, 139
　treatments for, 139
　X-linked recessive, *134*, 134–135,
　　135
Genetic engineering
　applications, 163, 167–169, 579
　potential risks and benefits, 163,
　　167–169
Genetic fingerprinting, 166
Genital herpes, 487
German measles, *224*, 480
Giardia infection, 235
Gigantism, 452, *453*
Global warming, 536–537
Goiter, 456, *456*
Gonorrhea, 228, 486
Greenhouse effect, 4, 535–537,
　536–537, 577
Green revolution, 576–577
Groundwater contamination, 582